Handbook of Numerical Analysis

Volume 23

Numerical Control: Part A

Handbook of Numerical Analysis

Volume 23

Numerical Control: Part A

Edited by

Emmanuel Trélat

Sorbonne Université

Laboratoire Jacques-Louis Lions

Paris, France

Enrique Zuazua

Department of Data Science

Friedrich-Alexander-Universität Erlangen-Nürnberg (FAU)

Erlangen, Germany

Fundación Deusto

Bilbao, Basque Country, Spain

Universidad Autónoma de Madrid

Madrid, Spain

ELSEVIER

N·H

North-Holland

An imprint of Elsevier

North-Holland is an imprint of Elsevier
Radarweg 29, PO Box 211, 1000 AE Amsterdam, Netherlands
The Boulevard, Langford Lane, Kidlington, Oxford OX5 1GB, United Kingdom

Library of Congress Cataloging-in-Publication Data
A catalog record for this book is available from the Library of Congress

British Library Cataloguing-in-Publication Data
A catalogue record for this book is available from the British Library

ISBN: 978-0-323-85059-9
ISSN: 1570-8659

For information on all North-Holland publications
visit our website at https://www.elsevier.com/books-and-journals

Publisher: Zoe Kruze
Acquisitions Editor: Sam Mahfoudh
Editorial Project Manager: Naiza Ermin Mendoza
Production Project Manager: James Selvam
Designer: Matthew Limbert

Typeset by VTeX

In memoriam Roland Glowinski (1937–2022)

We are deeply saddened by the passing of our fellow series editor, Professor Roland Glowinski, who played a leading role in our work on the Handbook of Numerical Analysis series.

Roland was a giant of modern computational and applied mathematics, whose wisdom, breadth of knowledge and boundless energy have been an inexhaustible source of inspiration to generations of mathematicians.

His untimely departure is a tragic and irreplaceable loss.

Qiang Du, Michael Hintermüller, and Endre Süli

Contents

2. Modeling, control, and numerics of gas networks

Martin Gugat and Michael Herty

3. Optimal control, numerics, and applications of fractional PDEs

Harbir Antil, Thomas Brown, Ratna Khatri, Akwum Onwunta, Deepanshu Verma, and Mahamadi Warma

4. Optimal control of PDEs and FE-approximation

Eduardo Casas, Karl Kunisch, and Fredi Tröltzsch

5. Numerical solution of multi-objective optimal control and hierarchic controllability problems

Enrique Fernández-Cara

6. Numerics for stochastic distributed parameter control systems: a finite transposition method

Qi Lü, Penghui Wang, Yanqing Wang, and Xu Zhang

15. Data-driven modeling and control of large-scale dynamical systems in the Loewner framework

Ion Victor Gosea, Charles Poussot-Vassal, and Athanasios C. Antoulas

16. Machine learning and control theory

Alain Bensoussan, Yiqun Li, Dinh Phan Cao Nguyen, Minh-Binh Tran, Sheung Chi Phillip Yam, and Xiang Zhou

Contributors

Harbir Antil, Dept. of Mathematical Sciences, Center for Mathematics and Artificial Intelligence, George Mason University, Fairfax, VA, United States

Athanasios C. Antoulas, Electrical and Computer Engineering (ECE) Department, Rice University, Houston, TX, United States
Max Planck Institute, Magdeburg, Germany
Baylor College of Medicine, Houston, TX, United States

Francesco Ballarin, Department of Mathematics and Physics, Catholic University of the Sacred Heart, Brescia, Italy

Alain Bensoussan, International Center for Decision and Risk Analysis, Jindal School of Management, The University of Texas at Dallas, Richardson, TX, United States
School of Data Science, City University of Hong Kong, Kowloon Tong, Hong Kong

Umberto Biccari, Chair of Computational Mathematics, Deusto Foundation, Bilbao, Basque Country, Spain
University of Deusto, Bilbao, Basque Country, Spain

Thomas Brown, Dept. of Mathematical Sciences, Center for Mathematics and Artificial Intelligence, George Mason University, Fairfax, VA, United States

Leon Bungert, Hausdorff Center for Mathematics, University of Bonn, Bonn, Germany

Martin Burger, Department of Mathematics, Friedrich–Alexander University, Erlangen, Germany

Eduardo Casas, Department of Applied Mathematics and Computer Science, University of Cantabria, Santander, Spain

Timm Faulwasser, Institute for Energy Systems, Energy Efficiency and Energy Economics, Department of Electrical Engineering and Information Technology, TU Dortmund University, Dortmund, Germany

Enrique Fernández-Cara, EDAN and IMUS, Universidad de Sevilla, Sevilla, Spain

Ion Victor Gosea, Data-Driven System Reduction and Identification (DRI) Group, Max Planck Institute (MPI), Magdeburg, Germany

Lars Grüne, Chair of Applied Mathematics, Mathematical Institute, University of Bayreuth, Bayreuth, Germany

Martin Gugat, FAU Erlangen-Nürnberg, Lehrstuhl für Dynamics, Control and Numerics (Alexander von Humboldt-Professorship), Erlangen, Germany

Didier Henrion, CNRS-LAAS, Toulouse, France
Faculty of Electrical Engineering, Czech Technical University in Prague, Prague, Czech Republic

Michael Herty, RWTH Aachen University, Institut für Geometrie und Praktische Mathematik, Aachen, Germany

Zhuo Jin, Department of Actuarial Studies and Business Analytics, Macquarie University, North Ryde, NSW, Australia

Ratna Khatri, Optical Sciences Division, U.S. Naval Research Laboratory, Washington, DC, United States

Milan Korda, CNRS-LAAS, Toulouse, France
Faculty of Electrical Engineering, Czech Technical University in Prague, Prague, Czech Republic

Karl Kunisch, Institute for Mathematics and Scientific Computing, University of Graz, Graz, Austria

Jean Bernard Lasserre, CNRS-LAAS, Toulouse, France
Institute of Mathematics of Toulouse, Toulouse, France

Martin Lazar, Department of Electrical Engineering and Computing, University of Dubrovnik, Dubrovnik, Croatia

Yiqun Li, Department of Statistics, The Chinese University of Hong Kong, Shatin, Hong Kong

Jérôme Lohéac, Université de Lorraine, CNRS, CRAN, Nancy, France

Qi Lü, School of Mathematics, Sichuan University, Chengdu, China

Idriss Mazari, CEREMADE, UMR CNRS 7534, Université Paris-Dauphine, Université PSL, Place du Maréchal De Lattre De Tassigny, Paris cedex 16, France

Grégoire Nadin, CNRS, Sorbonne Université, Laboratoire Jacques-Louis Lions, Paris, France

Dinh Phan Cao Nguyen, Department of Mathematics, Southern Methodist University, Dallas, TX, United States
Faculty of Information Technology, Nha Trang University, Nha Trang, Viet Nam

Akwum Onwunta, Department of Industrial and Systems Engineering, Lehigh University, Bethlehem, PA, United States

Gilles Pagès, Sorbonne Université, Laboratoire de Probabilités, Statistique et Modélisation, Paris, France

Olivier Pironneau, Sorbonne Université, Laboratoire Jacques-Louis Lions, Paris, France

Charles Poussot-Vassal, Information Processing and Systems Department, ONERA, Toulouse, France

Yannick Privat, IRMA, Université de Strasbourg, CNRS UMR 7501, Strasbourg, France
Institut Universitaire de France (IUF), Paris cedex 05, France

Gianluigi Rozza, mathLab, Mathematics Area, SISSA, Trieste, Italy

Maria Strazzullo, mathLab, Mathematics Area, SISSA, Trieste, Italy

Ky Tran, Department of Applied Mathematics and Statistics, SUNY Korea, Incheon, Republic of Korea

Minh-Binh Tran, Department of Mathematics, Southern Methodist University, Dallas, TX, United States

Fredi Tröltzsch, Institute for Mathematics, Technical University of Berlin, Berlin, Germany

Deepanshu Verma, Department of Mathematics, Emory University, Atlanta, GA, United States

Penghui Wang, School of Mathematics, Shandong University, Jinan, China

Yanqing Wang, School of Mathematics and Statistics, Southwest University, Chongqing, China

Mahamadi Warma, Dept. of Mathematical Sciences, Center for Mathematics and Artificial Intelligence, George Mason University, Fairfax, VA, United States

Sheung Chi Phillip Yam, Department of Statistics, The Chinese University of Hong Kong, Shatin, Hong Kong

George Yin, Department of Mathematics, University of Connecticut, Storrs, CT, United States

Xu Zhang, School of Mathematics, Sichuan University, Chengdu, China

Xiang Zhou, School of Data Science, City University of Hong Kong, Kowloon Tong, Hong Kong
Department of Mathematics, City University of Hong Kong, Kowloon Tong, Hong Kong

Enrique Zuazua, Chair for Dynamics, Control and Numerics, Alexander von Humboldt-Professorship, Dept. of Data Science, Friedrich-Alexander-University Erlangen-Nürnberg, Erlangen, Germany
Chair of Computational Mathematics, Deusto Foundation, Bilbao, Basque Country, Spain
Dept. of Mathematics, Autonomous University of Madrid, Madrid, Spain

Preface

These volumes are dedicated to Roland Glowinski, who sadly, left this world in January 2022. He was a great man, and outstanding scientist, who had a great influence on the scientific career of many of the contributors of this volume. He is the author of some fundamental key contributions that started a new era in the theory of control of PDE and its numerical counterpart.

His contributions are numerous, and an in-depth description of them goes much beyond the limits of this preface. One can refer for instance, to his contributions in collaboration with his mentor Jacques-Louis Lions, another science giant, to the development of efficient numerical solvers for controllability problems for PDEs, including ill-posed equations, like heat processes in the backward sense of time, bi-grid algorithms for wave control and augmented Lagrangian and splitting methods.

He was also one of the leading editors of this Handbook series and at the origin of the project of gathering the contributions in this first volume and the one to follow on the recent advances on numerical methods for control and related topics. He is also the co-author of a chapter included in the second volume, that he had just submitted before passing away.

Roland, these two volumes are dedicated to you.

This is the first volume of the "Handbook of Numerical Analysis and Scientific Computing" on "Numerical Methods in Control Optimization, applications and related topics".

In the last 60 years, our society has witnessed the emergence of new mathematical tools for modeling, understanding, simulating and controlling complex models and systems like financial markets, fluid and solid mechanical systems, social networks, climate evolution, biological or ecological systems, computational chemistry models, quantum systems, just to name a few of the prominent or emerging societal and industrial challenges. Those models often rely on ordinary or partial differential equations, and are able to reflect complex phenomena on which one may wish to act by means of an adequate control. They could also be monitored by appropriate sensors with the aim of predicting the future evolution. Most physical or biological systems are inherently multiscale, because critical events often occur at very different spatial and temporal levels. This inherent complexity of the models raises new fascinating open problems to be investigated by mathematicians.

Among the mathematical approaches developed to tackle these challenges, control and optimization have revealed their powerful potential to provide decision-makers with reliable tools to predict and act on their environment for the good of humanity.

Numerical methods and techniques of scientific computing are the key to reliably simulate complex systems, and their importance in recent decades has grown significantly in all application fields.

Moreover, over the past few years, the emergence of big-data, machine and deep learning combined with the considerable evolution of IT resources has made possible the design of new data-driven control and optimization techniques, and new powerful algorithms able to solve challenging problems that were not tractable 10 years ago. We thus think it is the right time to take stock of the situation.

The present Handbook participates in this great scientific movement of innovation and progress of humankind, by presenting a non-exhaustive, albeit sufficiently broad account of some of the prominent and most promising research directions in control and optimization.

We have gathered chapters written by experts reporting on key aspects of control theory and their numerical issues. These chapters cover a number of aspects of ODE and PDE theory, from a numerical perspective, with connection and application in fields related to control and optimization. The boundary between those fields, and with others, such as the calculus of variations, inverse problems, imaging, is not well defined. Therefore, we have adopted a broad perspective, collecting the most relevant numerical methods and active fields nowadays. Some contributions underlie the increasing need of integrating data-driven methods and thus design new strategies for control and optimization.

Emmanuel Trélat
Paris

Enrique Zuazua
Erlangen

January 2022

Chapter 1

Control and numerical approximation of fractional diffusion equations

Umberto Biccari[a,b,*], Mahamadi Warma[c], and Enrique Zuazua[d,a,e]

[a]*Chair of Computational Mathematics, Deusto Foundation, Bilbao, Basque Country, Spain,* [b]*University of Deusto, Bilbao, Basque Country, Spain,* [c]*Dept. of Mathematical Sciences, Center for Mathematics and Artificial Intelligence, George Mason University, Fairfax, VA, United States,* [d]*Chair for Dynamics, Control and Numerics, Alexander von Humboldt-Professorship, Dept. of Data Science, Friedrich-Alexander-University Erlangen-Nürnberg, Erlangen, Germany,* [e]*Dept. of Mathematics, Autonomous University of Madrid, Madrid, Spain*

**Corresponding author: e-mail address: umberto.biccari@deusto.es*

Contents

Handbook of Numerical Analysis, Volume 23, ISSN 1570-8659, https://doi.org/10.1016/bs.hna.2021.12.001

Abstract

The aim of this chapter is to give a broad panorama of the control properties of fractional diffusive models from a numerical analysis and simulation perspective. We do this by surveying several research results we obtained in the last years, focusing in particular on the numerical computation of controls, though not forgetting to recall other relevant contributions which can be currently found in the literature of this prolific field. Our reference model will be a non-local diffusive dynamics driven by the fractional Laplacian on a bounded domain Ω. The starting point of our analysis will be a Finite Element approximation for the associated elliptic model in one and two space-dimensions, for which we also present error estimates and convergence rates in the L^2 and energy norm. Secondly, we will address two specific control scenarios: firstly, we consider the standard interior control problem, in which the control is acting from a small subset $\omega \subset \Omega$. Secondly, we move our attention to the exterior control problem, in which the control region $\mathcal{O} \subset \Omega^c$ is located outside Ω. This exterior control notion extends boundary control to the fractional framework, in which the non-local nature of the models does not allow for controls supported on $\partial\Omega$. We will conclude by discussing the interesting problem of simultaneous control, in which we consider families of parameter-dependent fractional heat equations and we aim at designing a unique control function capable of steering all the different realizations of the model to the same target configuration. In this framework, we will see how the employment of stochastic optimization techniques may help in alleviating the computational burden for the approximation of simultaneous controls. Our discussion is complemented by several open problems related with fractional models which are currently unsolved and may be of interest for future investigation.

Keywords

Fractional Laplacian, Fractional diffusion equation, Finite Element Method, Interior Control, Exterior Control, Simultaneous Control, Numerical approximation

MSC Codes

35K05, 35R11, 35S05, 49M29, 49M41, 93B05

1 Introduction

A non-local PDE is a particular type of differential equation involving integral or pseudo-differential terms. For this reason, these models are often referred as integro-differential equations.

In recent years, there has been a growing interest on non-local models because of their relevance in a wide spectrum of practical applications. Indeed, there is a plethora of situations in which a non-local equation gives a significantly better description than a local PDE of the problem one wants to analyze. A widely studied class of non-local models involves fractional order operators, which have nowadays emerged as a modeling alternative in various branches of science. They usually describe anomalous diffusion. Typical examples in which non-local or fractional equations appear are models in turbulence (Bakunin, 2008), population dynamics (De Roos and Persson, 2002), image processing

(Gilboa and Osher, 2008), laser design (Longhi, 2015), and porous media flow (Vázquez, 2012). Besides, a number of stochastic models associated with fractional operators have been introduced in the literature for explaining anomalous diffusion. Among them we quote the fractional Brownian motion, the continuous time random walk, the Lévy flights, the Schneider gray Brownian motion, and more generally, random walk models based on evolution equations of single and distributed fractional order in space (see e.g. Dubkov et al., 2008; Gorenflo et al., 2007; Mandelbrot and Van Ness, 1968; Schneider, 1990). In general, a fractional diffusion operator corresponds to a diverging jump length variance in the random walk. Finally, we can refer to Antil and Rautenberg (2019); Weiss et al. (2020) for the relevance of fractional operators in geophysics and imaging science.

This surge of interest towards non-local and fractional models has opened a very challenging field in the applied mathematical research, since most of the existing techniques in PDE analysis were not adapted to treat non-local effects.

In the wide spectrum of non-local and fractional models, of particular interest are the ones involving the fractional Laplacian. From a mathematical perspective, there is nowadays a well established and rich literature on this operator and its employment in PDE models. Among many other contributions, we remind here the works by Biccari and Hernández-Santamaría (2018); Biccari et al. (2017, 2018); Caffarelli and Silvestre (2007, 2009); Leonori et al. (2015); Ros-Oton and Serra (2014); Servadei and Valdinoci (2012, 2013); Warma (2015). Besides, it is well known that the fractional Laplacian is the generator of s-stable Lévy processes, and it is often used in stochastic models with applications, for instance, in mathematical finance (Levendorskii, 2004; Pham, 1997).

In addition to that, control problems in the non-local setting have been largely considered in recent years. An incomplete bibliography at this respect includes (Antil et al., 2020a; Biccari, 2021; Biccari and Hernández-Santamaría, 2019a,b; Biccari and Warma, 2020; Biccari et al., 2020b; Chaves-Silva et al., 2017; Fernández-Cara et al., 2016; Lü et al., 2017; Micu and Zuazua, 2006; Miller, 2006; Warma, 2019; Warma and Zamorano, 2020a,b).

The aim of this chapter is to give a broad panorama of the control properties of fractional diffusive models from a numerical analysis and simulation perspective. We will do this by surveying several research results we obtained in the last years, focusing in particular on aspects related with the numerical computation of the controls, though not without recalling other relevant contributions which can be currently found in the literature of this prolific field. In more details, we will consider the following specific situations.

1. **Elliptic problems**: we will start in Section 2 by considering the elliptic problem associated with the fractional Laplace operator, for which we will present a Finite Element (FE) approximation in one and two space dimensions. This FE scheme will be at the basis of all our numerical simulations in the sections afterward. Our discussion in this section is based on the con-

tributions (Acosta et al., 2017; Acosta and Borthagaray, 2017; Biccari and Hernández-Santamaría, 2019a).

2. **Interior control of fractional heat equations**: secondly, we will consider in Section 3 the interior control of fractional heat equations. We will start by recalling the theoretical controllability results we have obtained in Biccari and Hernández-Santamaría (2019a); Biccari et al. (2020b, 2021). In a second moment, we will deal with the numerical computation of controls in one and two space dimensions by means of the penalized Hilbert Uniqueness Method.
3. **Exterior control of fractional heat equations**: thirdly, we will deal in Section 4 with the exterior control properties of fractional heat equations. As we shall see, this is the equivalent of the boundary controllability in the framework of the fractional Laplacian. Also in this case, we will first remind the theoretical controllability results we have obtained in Antil et al. (2020a); Warma and Zamorano (2020b). After that, we will focus on the numerical computation of controls and discuss some relevant differences with respect to the interior control problem of Section 3.
4. **Simultaneous control of parameter-dependent fractional heat equations**: finally, we will address in Section 5 the simultaneous control of parameter-dependent fractional heat equations, in which we aim at designing a unique parameter-independent control capable of steering different realizations of the same fractional dynamics to zero in finite time. Our discussion here will focus on the associated optimal control problem and its computational aspects. In particular, we will propose the employment of stochastic optimization algorithms for the computation of the simultaneous controls, and discuss the advantages and disadvantages of these techniques with respect to some more classical ones (namely, Gradient Descent and Conjugate Gradient) typically applied in PDE control.

Finally, this chapter will be complemented with Appendices A and B, in which we gather some results on fractional Sobolev spaces and the fractional Laplacian which will be used in our analysis.

2 Finite Element approximation of the fractional Laplace operator

In this section, we give an abridged presentation of the fundamental aspects required to convey a complete FE analysis of the following elliptic problem

$$\begin{cases} (-\Delta)^s u = f & \text{in } \Omega, \\ u = 0 & \text{in } \Omega^c, \end{cases} \tag{2.1}$$

where $\Omega \subset \mathbb{R}^d$ $(d = 1, 2)$ is a bounded and $C^{1,1}$ domain, $\Omega^c := \mathbb{R}^d \setminus \Omega$, f is a smooth enough function whose regularity will be specified later, and where for all $s \in (0,1)$ we denote with $(-\Delta)^s$ the fractional Laplace operator defined as

$$(-\Delta)^s u(x) := C_{d,s} P.V. \int_{\mathbb{R}^d} \frac{u(x) - u(y)}{|x-y|^{d+2s}} \, dy, \qquad (2.2)$$

$C_{d,s}$ being an explicit normalization constant (see (A.1)).

Numerical approximation for the fractional Laplacian on bounded domains has been extensively addressed in the last years. In Huang and Oberman (2014), the authors proposed a method combining finite differences with numerical quadrature to obtain a discrete convolution operator which accurately approximates the singular integral defining the fractional Laplacian. Nevertheless, it is noteworthy that the convergence of the proposed algorithm is proved assuming that solutions to (2.1) are of class C^4, which is generally not the case.

These regularity requirements can be avoided if we discretize the fractional Laplacian (2.2) via a FE approach, which is based on the weak variational formulation associated with (2.1) (see A.1). This issue was firstly addressed in Acosta et al. (2017); Acosta and Borthagaray (2017); Borthagaray (2017), in the 2-D case, by combining techniques borrowed from the theory of the Boundary Element Method (Sauter and Schwab, 2010) together with an appropriate numerical treatment of the integrals involving the unbounded domain Ω^c. An alternative yet complementary approach, allowing to treat also 3-D problems, has been proposed in Bonito et al. (2019), where the authors based their FE approximation on a different (but equivalent) variational formulation of the Dirichlet problem. Moreover, in Biccari and Hernández-Santamaría (2019a), motivated by control purposes, we specifically focused on the one-dimensional fractional Laplacian. In this setting, we showed that the stiffness matrix discretizing the operator can be computed entirely offline, without need of any numerical integration, with entries depending only on their position, fractional exponent s and the mesh-size h.

For completeness, we shall also mention some works dealing with the so-called *spectral* Dirichlet fractional Laplacian, that is, the fractional s-power of the realization in $L^2(\Omega)$ of the Laplace operator $-\Delta$ with null Dirichlet boundary conditions. To discretize this operator, in Nochetto et al. (2015) the authors proposed a FE approach combined with the famous *Caffarelli-Silvestre extension* (see Caffarelli and Silvestre, 2007) which allows to work in a local framework. Nevertheless, we have to stress that the spectral fractional Laplacian is different from the integral one that we consider in this work (see e.g. Servadei and Valdinoci, 2014 and the references therein).

In what follows, we will summarize the main steps for the FE approximation of the fractional Laplacian in space dimension $d = 1, 2$, as presented in Acosta et al. (2017); Acosta and Borthagaray (2017); Biccari and Hernández-Santamaría (2019a); Borthagaray (2017). We will start by discussing in Subsection 2.1 the construction of the stiffness matrix. In the one-dimensional case, we will follow the procedure of Biccari and Hernández-Santamaría (2019a), while the presentation of the two-dimensional problem is based on Acosta et al. (2017); Acosta and Borthagaray (2017); Borthagaray (2017). Secondly, we will give in Subsection 2.2 an overview of the error analysis and the convergence rate of the FE scheme in H^s and L^2. Finally, Subsection 2.3 will be devoted to some numerical experiments.

2.1 Computation of the stiffness matrix

As in any FE scheme, the first step is to introduce the variational formulation associated with the problem we aim to solve. In our case, this variational formulation reads as follows (see Definition A.1): find $u \in H_0^s(\Omega)$ such that, for all $v \in H_0^s(\Omega)$,

$$a(u,v) := \frac{C_{d,s}}{2}\int_{\mathbb{R}^d}\int_{\mathbb{R}^d}\frac{(u(x)-u(y))(v(x)-v(y))}{|x-y|^{d+2s}}\,dxdy = \int_\Omega f v\,dx. \tag{2.3}$$

Let $\mathfrak{M} = \{T_i\}_{i=1}^N$ be a partition of the domain Ω (i.e. $\overline{\Omega} = \bigcup_{i=1}^N T_i$) composed by N elements as follows:

- In dimension $d = 1$, we consider a uniform partition of $\Omega = (-1, 1)$
$$-1 = x_0 < x_1 < \ldots < x_i < x_{i+1} < \ldots < x_{N+1} = 1,$$
with $x_{i+1} = x_i + h$, $i = 0, \ldots N$, and we denote $T_i := [x_i, x_{i+1}]$.
- In dimension $d = 2$, we consider triangular elements T_i, $i = 1, \ldots, n$. We indicate with h_i the diameter of the element T_i and with ρ_i its inner radius, i.e. the diameter of the largest ball contained in T_i. We define
$$h = \max_{i\in\{1,\ldots,N\}} h_i.$$
Moreover, we require that the triangulation satisfies the usual regularity and local uniformity conditions:
$$\begin{aligned}&\text{there exists } \sigma > 0 \text{ s.t. } h_i \le \sigma\rho_i \text{ for all } i = 1,\ldots,N,\\ &\text{there exists } \kappa > 0 \text{ s.t. } h_i \le \kappa h_j \text{ for all } i,j = 1,\ldots,N,\ T_i \cap T_j = \emptyset.\end{aligned} \tag{2.4}$$
Naturally the second condition is a consequence of the first one. In this way κ can be expressed in terms of σ.

Consider the discrete space

$$\mathbb{V} := \left\{ v \in H_0^s(\Omega) \,:\, v|_{T_i} \in \mathcal{P}^1 \right\},$$

$\mathcal{P}^1$ being the space of the continuous and piece-wise linear functions. We approximate (2.3) with the following discrete problem: find $u_h \in \mathbb{V}$ such that

$$\frac{C_{d,s}}{2}\int_{\mathbb{R}}\int_{\mathbb{R}}\frac{(u_h(x)-u_h(y))(v(x)-v(y))}{|x-y|^{1+2s}}\,dxdy = \int_\Omega f v\,dx, \tag{2.5}$$

for all $v \in \mathbb{V}$. Let $\{\phi_i\}_{i=1}^N$ be the standard nodal basis of $\mathbb{V}$ corresponding to the internal nodes $\{x_1, \ldots, x_N\}$, that is $\phi_i(x_j) = \delta_{i,j}$ (see Fig. 1). By decomposing

$$u(x) = \sum_{j=1}^N u_j \phi_j(x) \quad \text{and} \quad f(x) = \sum_{j=1}^N f_j \phi_j(x),$$

and taking $v = \phi_i$, (2.5) becomes the linear system $\mathcal{A}_h \mathbf{u} = \mathcal{M}_h \mathbf{f}$, where $\mathbf{u} = (u_1, \ldots, u_N) \in \mathbb{R}^N$, $\mathbf{f} = (f_1, \ldots, f_N) \in \mathbb{R}^N$, and the stiffness and mass matrices $\mathcal{A}_h, \mathcal{M}_h \in \mathbb{R}^{N \times N}$ have components

$$\begin{aligned} a_{i,j} &= \frac{C_{d,s}}{2} \int_{\mathbb{R}^d} \int_{\mathbb{R}^d} \frac{(\phi_i(x) - \phi_i(y))(\phi_j(x) - \phi_j(y))}{|x-y|^{1+2s}} \, dx dy \\ m_{i,j} &= \int_\Omega \phi_i(x) \phi_j(x) \, dx \qquad\qquad i, j = 1, \ldots, N. \end{aligned} \tag{2.6}$$

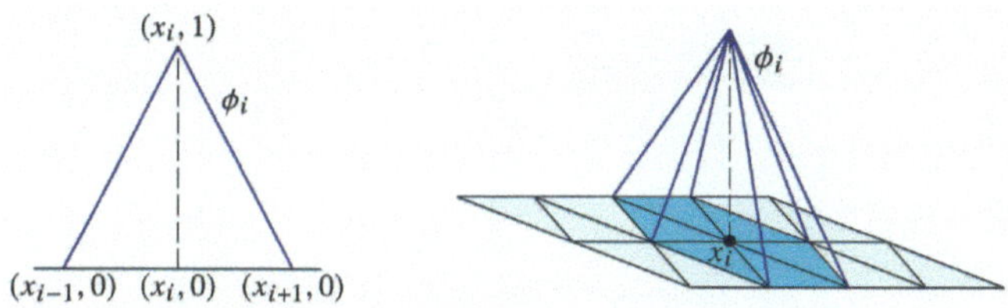

FIGURE 1 Basis functions $\phi_i(x)$ in space dimension $d = 1$ (left) and $d = 2$ (right).

Notice that $a(\phi_i, \phi_j)$ in (2.6) provides non-local interactions between the basis functions on the entire real line. Consequently, the stiffness matrix $\mathcal{A}_h$ will be full.

Remark 2.1. Although in our presentation we focused on uniform meshes, adapted grids could have also been employed. This would allow to treat problems with corner singularities or to cope with boundary layers arising for instance in convection-dominated problems (see Acosta and Borthagaray, 2017). Nevertheless, for the problems that we are considering here, non-uniform meshes will not be necessary. Besides, we stress that, in the 1-D case, the employment of a uniform mesh actually allows to compute explicitly the entries of the stiffness matrix $\mathcal{A}_h$, without need of using any quadrature formula (see Biccari and Hernández-Santamaría, 2019a).

We now give an abridged presentation of how to compute the elements of $\mathcal{A}_h$. To this end, we shall consider the one and two-dimensional cases separately.

2.1.1 1-D case

In space dimension $d = 1$, it is possible to identify three well defined regions in which we have different intersections among the support of the basis functions, thus generating different values of the coefficients $a_{i,j}$.

1. In the upper triangle in red in Fig. 2, corresponding to $j \geq i+2$, we have $supp(\phi_i) \cap supp(\phi_j) = \emptyset$ and we get

$$a_{i,j} = -C_{1,s} \int_{x_{j-1}}^{x_{j+1}} \int_{x_{i-1}}^{x_{i+1}} \frac{\phi_i(x)\phi_j(y)}{|x-y|^{1+2s}} dxdy. \tag{2.7}$$

2. In the upper diagonal in purple in Fig. 2, corresponding to $j = i+1$, we have

$$a_{i,i+1} = \frac{C_{1,s}}{2} \int_{\mathbb{R}} \int_{\mathbb{R}} \frac{(\phi_i(x) - \phi_i(y))(\phi_{i+1}(x) - \phi_{i+1}(y))}{|x-y|^{1+2s}} dxdy. \tag{2.8}$$

3. In the diagonal in green in Fig. 2, corresponding to $j = i$, we have

$$a_{i,i} = \frac{C_{1,s}}{2} \int_{\mathbb{R}} \int_{\mathbb{R}} \frac{(\phi_i(x) - \phi_i(y))^2}{|x-y|^{1+2s}} dxdy. \tag{2.9}$$

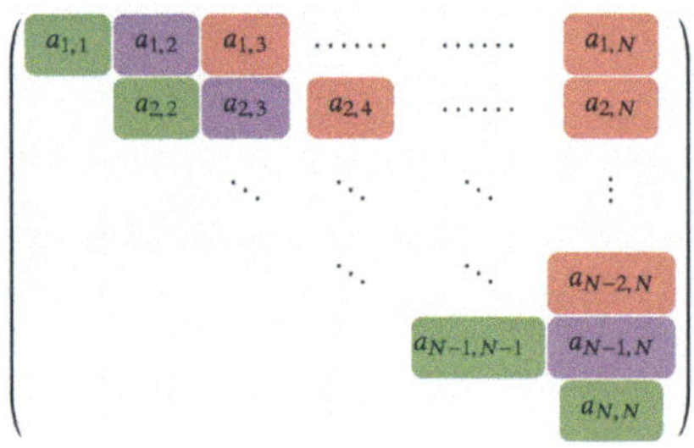

FIGURE 2 Structure of the stiffness matrix $\mathcal{A}_h$. The red, purple, and green elements are computed through the integrals (2.7), (2.8), and (2.9), respectively.

Let us remark that the integrals in (2.8) and (2.9) are defined on the entire real line and, therefore, they need to be treated with special attention. Those integrals, as well as (2.9), have been computed exactly in Biccari and Hernández-Santamaría (2019a) yielding to explicit values for the elements $a_{i,j}$, which only depend on i, j, s, and h. For the sake of brevity, the complete computations are omitted here.

Remark 2.2. It is worth noticing that for our implementation, we used $\mathcal{P}^1$ elements, meaning that we are considering piece-wise linear and continuous functions for the FE approximation. Actually, for $s < 1/2$, we could have used $\mathcal{P}^0$ (that is, piece-wise constant) elements too. Indeed, in this case, the spaces $H_0^s(-1,1)$ and $H^s(-1,1)$ coincide (see Grisvard, 2011, Chapter 1 and Lions and Magenes, 1968, Chapter 11), therefore it is not necessary that the discrete functions v_h vanish on the boundary in order to have a conforming method. On the other hand, a $\mathcal{P}^1$ basis is indeed necessary when $s \geq 1/2$. The case of $\mathcal{P}^0$ for $s < 1/2$ is not addressed here in order to present a unified approach for the whole range $0 < s < 1$.

2.1.2 2-D case

We now focus on the Dirichlet problem (2.1) in dimension $d = 2$, for which the FE procedure has been described in Acosta et al. (2017); Acosta and Borthagaray (2017); Borthagaray (2017). In this case, we have

$$a_{i,j} = \frac{C_{2,s}}{2} \int_{\mathbb{R}^2} \int_{\mathbb{R}^2} \frac{(\phi_i(x) - \phi_i(y))(\phi_j(x) - \phi_j(y))}{|x - y|^{2+2s}} \, dxdy.$$

Moreover, since $\phi_i = 0$ in Ω^c, the integral on $\mathbb{R}^2 \times \mathbb{R}^2$ is reduced to integrals on the set $(\Omega \times \Omega) \cup (\Omega \times \Omega^c) \cup (\Omega^c \times \Omega)$ and, taking into account that the interactions in $\Omega \times \Omega^c$ and $\Omega^c \times \Omega$ are symmetric with respect to x and y, we get

$$\begin{aligned} a_{i,j} = & \frac{C_{2,s}}{2} \int_{\Omega} \int_{\Omega} \frac{(\phi_i(x) - \phi_i(y))(\phi_j(x) - \phi_j(y))}{|x - y|^{2+2s}} \, dxdy \\ & + C_{2,s} \int_{\Omega} \int_{\Omega^c} \frac{\phi_i(x)\phi_j(x)}{|x - y|^{2+2s}} \, dxdy. \end{aligned} \tag{2.10}$$

Notice that the second integral in (2.10) has to be computed over the unbounded domain Ω^c. To do that, it is convenient to introduce a ball B containing Ω as in Fig. 3, since this allows to employ polar coordinates and exploit symmetry properties.

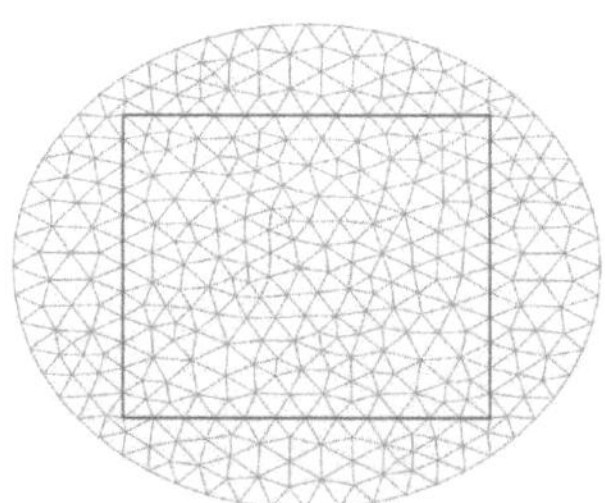

FIGURE 3 A square domain Ω (delimited by the blue lines) and an auxiliary ball containing it. Regular triangulations $\mathcal{T}$ and $\mathcal{T}_A$ for Ω and $B \setminus \Omega$ are shown.

We indicate with N_B the number of elements on the triangulation of B. Then, recalling (2.10), the coefficients $a_{i,j}$ are given by the expression

$$a_{i,j} = \frac{C_{2,s}}{2} \sum_{\ell=1}^{N_B} \left(\sum_{m=1}^{N_B} \mathcal{I}_{\ell,m}^{i,j} + 2\mathcal{J}_{\ell}^{i,j} \right),$$

where, for $1 \leq \ell, m \leq N_B$, we defined

$$\mathcal{I}_{\ell,m}^{i,j} := \int_{T_\ell} \int_{T_m} \frac{(\phi_i(x) - \phi_i(y))(\phi_j(x) - \phi_j(y))}{|x - y|^{2+2s}} \, dxdy, \tag{2.11a}$$

$$\mathcal{J}_\ell^{i,j} := \int_{T_\ell}\int_{B^c} \frac{\phi_i(x)\phi_j(x)}{|x-y|^{2+2s}}\,dxdy. \tag{2.11b}$$

The computations for each of the above integrals are challenging for different reasons: (2.11a) involves a singular integrand if $\overline{T}_\ell \cap \overline{T}_m \neq \emptyset$, while (2.11b) needs to be calculated on an unbounded domain. A complete discussion on the actual computation of (2.11a) and (2.11b), together with the corresponding algorithm, can be found in Acosta et al. (2017); Borthagaray (2017). For the sake of brevity, we omit this discussion here.

2.1.3 The fractional Laplacian with non-homogeneous exterior condition

We conclude this section with an abridged discussion on the FE approximation of the fractional Laplacian with non-homogeneous exterior condition. That is, for any $f \in L^2(\Omega)$ and $g \in H^s(\Omega^c)$, we consider the following elliptic problem

$$\begin{cases} (-\Delta)^s u = f & \text{in } \Omega \\ u = g & \text{in } \Omega^c. \end{cases} \tag{2.12}$$

There are several possibilities to discretize (2.12) in FE:

1. A first possibility is to start from the variational formulation associated with (2.12), as we did before for the problem with homogeneous exterior condition. Employing the integration by parts formula in Proposition B.2, this variational formulation is given by: find $u \in H^s(\mathbb{R}^d)$ such that $u = g$ in Ω^c and, for all $v \in H_0^s(\Omega)$,

$$\int_\Omega f v\,dx = a(u,v), \tag{2.13}$$

where $a(\cdot,\cdot)$ is the bilinear form given in (2.3). Moreover, notice that because of the non-homogeneous exterior datum in (2.12), this bilinear form reads as follows

$$\begin{aligned} a(u,v) = {} & \frac{C_{d,s}}{2}\int_\Omega\int_\Omega \frac{(u(x)-u(y))(v(x)-v(y))}{|x-y|^{d+2s}}\,dxdy \\ & + C_{d,s}\int_\Omega\int_{\Omega^c} \frac{(u(x)-g(y))(v(x)-v(y))}{|x-y|^{d+2s}}\,dxdy. \end{aligned} \tag{2.14}$$

One then needs to introduce some appropriate FE space, with the corresponding basis on which to decompose the functions u, f, and g, so that (2.14) is reduced to solve a linear system. We stress that this may be a delicate issue, due to the fact that these functions are supported on different domains. As a matter of fact, to our knowledge this approach has never been used in the literature.

2. A second possibility would be to convert the non-homogeneous exterior problem (2.12) into a homogeneous one. This can be done in the following way: let $G \in H^s(\mathbb{R}^d)$ such that $G|_{\Omega^c} = g$ and denote u_G the solution to the Dirichlet problem

$$\begin{cases} (-\Delta)^s u_G = f - (-\Delta)^s G & \text{in } \Omega \\ u_G = 0 & \text{in } \Omega^c. \end{cases}$$

Notice that, since $f - (-\Delta)^s G \in H^{-s}(\Omega)$, then we have from Proposition A.2 that $u_G \in H_0^s(\Omega)$. Moreover, u_G can be approximated using the FE scheme for the homogeneous exterior problem that we have introduced in the previous section. Then, finally, the solution of (2.12) would simply be given by $u = G + u_G$.
3. A third possibility is to approximate the exterior Dirichlet problem (2.12) with an exterior Robin one, as it has been done in our recent contributions (Antil et al., 2019, 2020b). This is the approach that we will use later for our simulations. A more detailed presentation will be given in Section 4.1.1.

2.2 Error analysis

We devote this section to a brief discussion on the approximation error of the FE discretization we just introduced. In particular, we will present the convergence rates of the FE scheme in the $H_0^s(\Omega)$ and in the $L^2(\Omega)$-norms. The former is the energy norm associated with the elliptic problem (2.1). In this case, we have the following result.

Theorem 2.3 (Acosta and Borthagaray, 2017, Theorem 4.6). *Let f satisfy the following regularity assumptions:*

$$\begin{aligned} &f \in C^{\frac{1}{2}-s}(\Omega), && \text{if } 0 < s < 1/2, \\ &f \in L^\infty(\Omega), && \text{if } s = 1/2, \\ &f \in C^\beta(\Omega) \text{ for some } \beta > 0, && \text{if } 1/2 < s < 1, \end{aligned}$$

where $C^{s-\frac{1}{2}}(\Omega)$ and $C^\beta(\Omega)$ denote the standard Hölder spaces of order $s - 1/2$ and β, respectively. Then, for the solution u of (2.3) and its FE approximation u_h on a uniform mesh with size h, we have the following a priori estimates

$$\begin{aligned} &\|u - u_h\|_{H_0^s(\Omega)} \le \frac{C(s,\sigma)}{\varepsilon} h^{\frac{1}{2}-\varepsilon} \|f\|_{C^{\frac{1}{2}-s}(\Omega)}, && \forall \varepsilon > 0, \text{ if } \; s < 1/2, \\ &\|u - u_h\|_{H_0^s(\Omega)} \le \frac{C(\sigma)}{\varepsilon} h^{\frac{1}{2}-\varepsilon} \|f\|_{L^\infty(\Omega)}, && \forall \varepsilon > 0, \text{ if } \; s = 1/2 \\ &\|u - u_h\|_{H_0^s(\Omega)} \le \frac{C(s,\beta,\sigma)}{\sqrt{\varepsilon}(2s-1)} h^{\frac{1}{2}-\varepsilon} \|f\|_{C^\beta(\Omega)}, && \forall \varepsilon > 0, \text{ if } \; s > 1/2, \; \beta > 0 \end{aligned}$$

where C is a positive constant not depending on h and σ is given by (2.4).

The L^2 convergence rate, instead, will come into play in Section 3 when addressing the implementation of the null controllability problem for the fractional heat equation.

Theorem 2.4 (Borthagaray, 2017, Proposition 3.3.2). *Let* $\alpha := \min\{s, 1/2 - \delta\}$, *with* $\delta > 0$ *arbitrary small. If* $f \in L^2(\Omega)$ *and* u *is the solution to* (2.1), *for its FE approximation on a uniform mesh with size* h *it holds that*

$$\|u - u_h\|_{L^2(\Omega)} \leq C(s, \alpha) h^{2\alpha} \|f\|_{L^2(\Omega)}.$$

2.3 Numerical experiments

We conclude this section with some numerical experiments. Here we focus on the one-dimensional case on the interval $\Omega = (-1, 1)$. Analogous simulations in dimension $d = 2$ can be found in Acosta and Borthagaray (2017). To test the efficiency of our FE scheme, we consider the Dirichlet problem (2.1) with $f = 1$, whose explicit solution is given by (see Getoor, 1961)

$$u(x) = \gamma_s \left(1 - x^2\right)^s \cdot \chi_{(-1,1)}, \qquad \gamma_s = \frac{2^{-2s}\sqrt{\pi}}{\Gamma\left(\frac{1+2s}{2}\right)\Gamma(1+s)}. \tag{2.15}$$

In Fig. 4, we show a comparison for different values of s between the exact solution (2.15) and the computed numerical approximation. One can notice that for large $s \geq 1/2$ the FE scheme provides a good approximation. On the other hand, when $s < 1/2$, the computed solution is to a certain extent different from the exact one, as there is a discrepancy approaching the boundary.

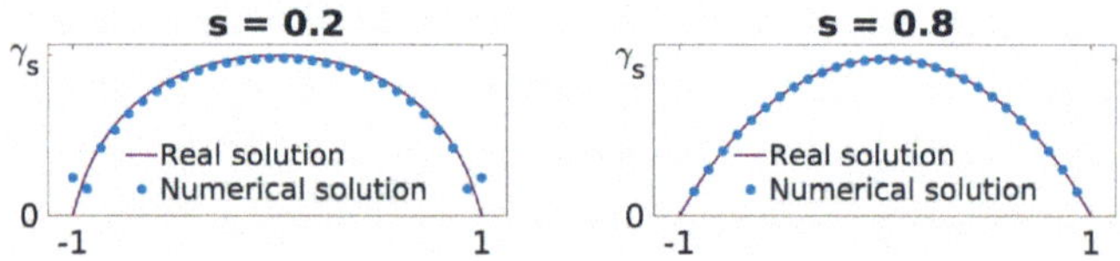

FIGURE 4 Exact and numerical solution of the Dirichlet problem (2.1) with $f = 1$ for different values of $s \in (0, 1)$.

Despite this fact, we can see in Fig. 5 that, for all $s \in (0, 1)$ the H^s error of our FE approximations decreases with h at a rate $\|u - u_h\|_{H_0^s(-1,1)} \sim \sqrt{h}$, which is the expected one according to Theorem 2.3. This is a numerical evidence of the accuracy of the proposed FE scheme when discretizing the elliptic problem (2.1).

3 Interior controllability properties of the fractional heat equation

In this section, we discuss the interior controllability properties of the fractional heat equation. That is, for any $y_0 \in L^2(\Omega)$, $\Omega \subset \mathbb{R}^d$ being a bounded and $C^{1,1}$

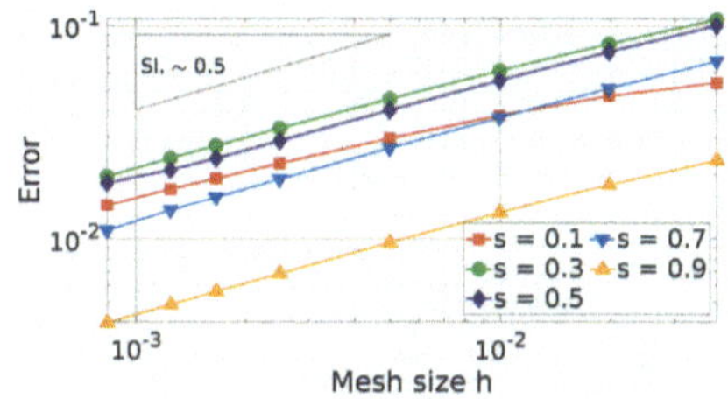

FIGURE 5 Convergence in $H^s(-1,1)$ of the approximation error for different values of $s \in (0,1)$. The rate $\|u-u_h\|_{H_0^s(-1,1)} \sim \sqrt{h}$ is in accordance with the theoretical result.

domain, and $\omega \subset \Omega$ a nonempty and open subset, we are going to consider the following control system

$$\begin{cases} y_t + (-\Delta)^s y = u\chi_\omega & \text{in } \Omega \times (0,T), \\ y \equiv 0 & \text{in } \Omega^c \times (0,T), \\ y(\cdot,0) = y_0 & \text{in } \Omega. \end{cases} \tag{3.1}$$

Here u is the control function belonging to some functional space to be specified later. We will start by recalling in Subsection 3.1 the theoretical controllability results which are presently available in the literature. Secondly, in Subsection 3.2, we will give an abridged presentation of the well-known *penalized Hilbert Uniqueness Method* that we shall employ for the computation of our numerical controls. Finally, Subsection 3.3 is devoted to the presentation and discussion of our numerical experiments.

3.1 Review of theoretical controllability results

We summarize the theoretical controllability results which can be currently found in the literature for the fractional heat equation (3.1).

Theorem 3.1. *For the fractional heat equation* (3.1), *the following results hold.*

1. ***Approximate controllability***. *Let $\omega \subset \Omega$ be any nonempty and open subset of Ω and $s \in (0,1)$. For any $T > 0$, $y_0, y_T \in L^2(\Omega)$ and $\varepsilon > 0$, there exists a control function $u \in L^2(\omega \times (0,T))$ such that the unique solution y of* (3.1) *satisfies $\|y(\cdot,T) - y_T\|_{L^2(\Omega)} \leq \varepsilon$.*
2. ***L^2 null controllability in 1-D***. *Let $\Omega = (-1,1)$ and $\omega \subset (-1,1)$ be any nonempty and open subset. For any $T > 0$ and $y_0 \in L^2(-1,1)$, there exists a control function $u \in L^2(\omega \times (0,T))$ such that the unique solution y of* (3.1) *satisfies $y(\cdot,T) = 0$ a.e. in $(-1,1)$, if and only if $s \in (1/2,1)$.*
3. ***L^∞ null controllability in 1-D***. *Let $\Omega = (-1,1)$ and $\omega \subset (-1,1)$ be any nonempty and open subset. For any $T > 0$ and $y_0 \in L^2(-1,1)$, there exists a control function $u \in L^\infty(\omega \times (0,T))$ such that the unique solution y of* (3.1) *satisfies $y(\cdot,T) = 0$ a.e. in $(-1,1)$, if and only if $s \in (1/2,1)$.*

4. ***L^2 null controllability in multi-D.*** *Let $\omega \subset \Omega$ be a neighborhood of $\partial\Omega$. For any $T > 0$ and $y_0 \in L^2(\Omega)$, there exists a control function $u \in L^2(\omega \times (0,T))$ such that the unique solution y of* (3.1) *satisfies $y(\cdot,T) = 0$ a.e. in Ω, if and only if $s \in (1/2, 1)$.*

The approximate controllability property has been established in Biccari and Hernández-Santamaría (2019a, Theorem 1.2). As for any analytic semi-group (as it is the one generated by $-(-\Delta)^s_D$, see Appendix A or Biccari et al., 2017, Lemma 2.5), this is a consequence of the unique continuation of the fractional Laplacian proved in Fall and Felli (2014), which yields to the unique continuation of the corresponding parabolic problem. The null controllability in space dimension $d = 1$ with $L^2(\omega \times (0,T))$ controls has been proved in Biccari and Hernández-Santamaría (2019a, Theorem 1.1), while the same result with $L^\infty(\omega \times (0,T))$ controls has been obtained in Biccari et al. (2020b, Theorem 2.2). Here we shall highlight the restriction $s \in (1/2, 1)$. At this regard, we remark that the exponent s determines the diffusion strength of the fractional heat semi-group. The limitation $s \in (1/2, 1)$ then means that, when dealing with heat-like processes involving the fractional Laplacian, a minimal amount of diffusivity is required to obtain positive null-controllability results. On a mathematical perspective, this arises from the application of moments techniques and parabolic Ingham inequalities in our proofs, and from the lack of an asymptotic gap between the eigenvalues of the one-dimensional Dirichlet fractional Laplacian when $s \leq 1/2$. Finally, the null controllability for $d \geq 2$ has been obtained in Biccari et al. (2021, Theorem 1.1) through a combination of partial observability results for the fractional wave equation involving the fractional Laplacian, transmutation techniques, and the Lebeau-Robbiano strategy. In this multi-dimensional case, apart from the limitation $s \in (1/2, 1)$, the employment of transmutation techniques imposes the further restriction that ω has to be a neighborhood of the boundary $\partial\Omega$.

In addition to the above results, the controllability property under state/control constraints has been investigated for the one-dimensional fractional heat equation in Biccari et al. (2020b). In particular, the following has been obtained.

Theorem 3.2 (1-D constrained controllability, Biccari et al., 2020b, Theorem 2.1). *Let $s \in (1/2, 1)$, $y_0 \in L^2(-1, 1)$ and let $\widehat{y}$ be a positive trajectory, i.e., a solution of* (3.1) *with initial datum $0 < \widehat{y}_0 \in L^2(-1, 1)$ and right-hand side $\widehat{u} \in L^\infty(\omega \times (0,T))$. Assume that there exists $\varpi > 0$ such that $\widehat{u} \geq \varpi$ a.e. in $\omega \times (0,T)$. Then, the following assertions hold.*

1. *There exists a minimal strictly positive controllability time $T_{\min} > 0$ such that, for all $T > T_{\min}$, we can find a non-negative control $u \in L^\infty(\omega \times (0,T))$ whose corresponding solution y of* (3.1) *satisfies $y(\cdot,T) = \widehat{y}(\cdot,T)$ a.e. in $(-1,1)$. Moreover, if $y_0 \geq 0$, then $y(x,t) \geq 0$ for every $(x,t) \in (-1,1) \times (0,T)$.*

2. *For* $T = T_{\min}$*, there exists a non-negative control* $u \in \mathcal{M}(\omega \times (0, T_{\min}))$*, the space of Radon measures on* $\omega \times (0, T_{\min})$*, such that the corresponding solution* y *of* (3.1) *satisfies* $y(\cdot, T) = \widehat{y}(\cdot, T)$ *a.e. in* $(-1, 1)$.

For completeness, we shall mention that the null controllability properties of *spectral fractional heat-like equations* have been also analyzed in Micu and Zuazua (2006); Miller (2006) in the case where the integral fractional Laplace operator $(-\Delta)^s$ is replaced by the spectral Dirichlet fractional Laplacian mentioned in Section 2. Recall that this operator is different from the integral one considered in this work. Nevertheless, when addressing the null control of fractional diffusion, it shows an analogous behavior as $(-\Delta)^s$. In particular, also for the spectral operator null controllability requires a minimum amount of diffusivity, $s = 1/2$ being the critical case, in which the property fails. Consequently, also the spectral fractional heat equation turns out to be null controllable if and only if $1/2 < s < 1$. This spectral fractional Laplacian is easier to handle and well understood since its eigenfunctions are the same as those of the Laplacian with the Dirichlet boundary condition and the associated eigenvalues are the s-power of the eigenvalues of the Laplacian with the Dirichlet boundary condition. Thus, for the spectral fractional Laplacian, the existing observability results for the classical heat equation and its eigenfunctions can be reused to show that null controllability is preserved in the range $1/2 < s < 1$. By the contrary, for the integral fractional Laplacian, the eigenfunctions depend genuinely on s and the same techniques inherited from the Laplacian do not apply in general. At this regard, let us stress that for the Laplace operator, many fundamental properties are based on Carleman inequalities that have not been properly developed for the fractional setting.

3.2 The penalized Hilbert Uniqueness Method

We devote this section to an abridged description of the *penalized Hilbert Uniqueness Method* (HUM) that we shall employ for computing the controls for the fractional heat equation. Here we will mostly refer to the works (Boyer, 2013; Glowinski and Lions, 1994; Glowinski et al., 2008).

Let $\mathcal{H}$ be a Hilbert space and $\mathcal{A} : \mathcal{D}(\mathcal{A}) \subset \mathcal{H} \to \mathcal{H}$ be an unbounded operator such that $-\mathcal{A}$ generates an analytic semi-group. Let $\mathcal{U}$ be another Hilbert space and $\mathcal{B} : \mathcal{U} \to \mathcal{D}(\mathcal{A})^\star$ be a bounded operator. Let $T > 0$ be given and, for any $y_0 \in \mathcal{H}$ and $u \in L^2(0, T; \mathcal{U})$, let us consider the Cauchy problem

$$y_t + \mathcal{A}y = \mathcal{B}u \ \text{ in } [0, T], \quad y(0) = y_0. \tag{3.2}$$

The penalized HUM approach consists in finding the control of minimal $L^2(0, T; \mathcal{U})$ norm for (3.2) by means of the following optimization problem:

$$u_\beta = \min_{u \in L^2(0,T;\mathcal{U})} F_\beta(u) \tag{3.3}$$

$$F_\beta(u) := \frac{1}{2}\int_0^T \|u(t)\|_{\mathcal{U}}^2\,dt + \frac{1}{2\beta}\|y(T)\|_{\mathcal{H}}^2.$$

Notice that, if (3.2) is controllable (either null, exactly, or approximately), then for any $\beta > 0$, the functional F_β is strictly convex, continuous, and coercive. Hence, it has a unique minimizer $u_\beta \in L^2(0,T;\mathcal{U})$. Moreover, we have the following result.

Theorem 3.3 (Boyer, 2013, Theorem 1.7). *The following controllability properties hold.*

1. *Problem* (3.2) *is approximately controllable at time* $T > 0$ *from* $y_0 \in \mathcal{H}$ *if and only if* $\|y_\beta(T)\|_{\mathcal{H}} \to 0$ *as* $\beta \to 0$*, where* y_β *denotes the solution corresponding to* u_β.
2. *Problem* (3.2) *is null-controllable at time* $T > 0$ *from* $y_0 \in \mathcal{H}$ *if and only if*

$$\mathcal{E}_{y_0} := 2\sup_{\beta>0}\left(\inf_{u\in L^2(0,T;\mathcal{U})} F_\beta(u)\right) < +\infty. \tag{3.4}$$

In this case, we have

$$\|u_\beta\|_{L^2(0,T;\mathcal{U})} \le \sqrt{\mathcal{E}_{y_0}} \tag{3.5a}$$

$$\|y_\beta(T)\|_{\mathcal{H}} \le \sqrt{\mathcal{E}_{y_0}\beta}. \tag{3.5b}$$

Moreover, as $\beta \to 0$, $u_\beta \to \bar{u}$ *strongly in* $L^2(0,T;\mathcal{U})$, $\bar{u}$ *being the optimal control obtained from the functional* (3.3) *without the second penalization term.*

According to Theorem 3.3, there is then an essential difference between approximate and null controllability in this penalization context. In both cases, the solution at time T of (3.2) corresponding to u_β converges to zero in $\mathcal{H}$ as $\beta \to 0$. Nevertheless, for null-controllability, this convergence has a precise rate $\sqrt{\beta}$ which, as we will see in Section 3.3 and had already been observed in Boyer (2013), is typically violated when only approximate controllability holds. Furthermore, when the problem is null controllable, we also have that the control cost $\|u_\beta\|_{L^2(0,T;\mathcal{U})}$ remains uniformly bounded which, together with (3.5b), yields the convergence of u_β to the solution $\bar{u}$ of the non-penalized problem.

Let us recall that $-(-\Delta)_D^s$ generates an analytic semi-group (see Biccari et al., 2017; Claus and Warma, 2020). Hence, Theorem 3.3 applies to (3.1) when selecting $\mathcal{H} = L^2(\Omega)$ and $\mathcal{U} = L^2(\omega)$.

Furthermore, to solve the optimization problem (3.3), we will apply duality and the Fenchel-Rockafellar theory (see e.g. Ekeland and Temam, 1999, Chapters VI to VII) to build an equivalent dual minimization problem defined on the space $L^2(\Omega)$. Although this duality argument is nowadays classical, for completeness we shall recall below its main steps.

Step 1. Since the problem (3.1) is linear, we can write its solution as $y = \xi + z$, with

$$\begin{cases} \xi_t + (-\Delta)^s \xi = 0 & \text{in } \Omega \times (0,T), \\ \xi \equiv 0 & \text{in } \Omega^c \times (0,T), \\ \xi(\cdot, 0) = y_0 & \text{in } \Omega \end{cases} \tag{3.6}$$

and

$$\begin{cases} z_t + (-\Delta)^s z = u\chi_\omega & \text{in } \Omega \times (0,T), \\ z \equiv 0 & \text{in } \Omega^c \times (0,T), \\ z(\cdot, 0) = 0 & \text{in } \Omega. \end{cases} \tag{3.7}$$

Moreover, in what follows, for the solution of (3.6) we will use the notation

$$\xi(\cdot, t) = e^{-(-\Delta)^s_D t} y_0,$$

where $(-\Delta)^s_D$ is the realization in $L^2(\Omega)$ of the fractional Laplacian with the zero Dirichlet exterior condition (see (A.3)).

Step 2. Let $\mathcal{L}_T : L^2(\omega \times (0,T)) \to L^2(\Omega)$ be the linear continuous operator defined as $\mathcal{L}_T(u) = z(\cdot, T)$, with z solution of (3.7). Then, the adjoint operator $\mathcal{L}_T^\star : L^2(\Omega) \to L^2(\omega \times (0,T))$ is given by $\mathcal{L}_T^\star(p_T) = p\chi_\omega$ where, for all $p_T \in L^2(\Omega)$, p solves

$$\begin{cases} -p_t + (-\Delta)^s p = 0 & \text{in } \Omega \times (0,T), \\ p \equiv 0 & \text{in } \Omega^c \times (0,T), \\ p(\cdot, T) = p_T & \text{in } \Omega. \end{cases} \tag{3.8}$$

Step 3. With this notation, we have $F_\beta(u) = \widehat{F}(u) + G_\beta(\mathcal{L}_T u)$, where

$$\widehat{F}(u) := \frac{1}{2}\int_0^T \|u(t)\|^2_{L^2(\omega)}\, dt \quad \text{and}$$

$$G_\beta(\mathcal{L}_T u) := \frac{1}{2\beta}\left\| \mathcal{L}_T u + e^{-(-\Delta)^s_D T} y_0 \right\|^2_{L^2(\Omega)}.$$

Since both $\widehat{F}$ and G_β are convex functionals, Fenchel-Rockafellar theory (see Boyer, 2013, Proposition 1.5) yields that

$$u_\beta = p_\beta \chi_\omega, \tag{3.9}$$

with p_β solution of (3.8) corresponding to the initial datum

$$p_{T,\beta} = \min_{p_T \in L^2(\Omega)} J(p_T),$$

and $J(p_T) := \widehat{F}^\star(\mathcal{L}_T^\star p_T) + G_\beta^\star(-p_T)$, $\widehat{F}^\star$ and $G_\beta^\star$ being the convex conjugates

$$\widehat{F}^\star(u) = \sup_{v\in L^2(\omega\times(0,T))} \left\{ \langle u, v\rangle_{L^2(\omega\times(0,T))} - \widehat{F}(v)\right\}, \qquad u \in L^2(\omega\times(0,T))$$

$$G_\beta^\star(-p_T) = \sup_{q_T\in L^2(\Omega)} \left\{ -\langle p_T, q_T\rangle_{L^2(\Omega)} - G_\beta(q_T)\right\}, \quad q_T \in L^2(\Omega). \tag{3.10}$$

Step 4. It can be readily checked using (3.10) that

$$\widehat{F}^\star(\mathcal{L}_T^\star p_T) = \frac{1}{2}\int_0^T \|p(t)\|^2_{L^2(\omega)}\,dt$$

$$G_\beta^\star(-p_T) = \langle p_T, e^{-(-\Delta)^s_D T} y_0\rangle_{L^2(\Omega)} + \frac{\beta}{2}\|p_T\|^2_{L^2(\Omega)}.$$

Collecting everything, we then obtain that $J_\beta(p_T)$ is given by

$$J_\beta(p_T) = \frac{1}{2}\int_0^T \|p(t)\|^2_{L^2(\omega)}\,dt + \frac{\beta}{2}\|p_T\|^2_{L^2(\Omega)} + \langle p_T, e^{-(-\Delta)^s_D T} y_0\rangle_{L^2(\Omega)}. \tag{3.11}$$

To compute the numerical control, let us introduce the fully-discrete version of (3.1). Given a uniform mesh $\mathfrak{M}$ of size h on Ω and any integer $M > 0$, we set $\delta t = T/M$ and we approximate (3.1) through an implicit Euler method:

$$\begin{cases} \mathcal{M}_h \dfrac{y_h^{m+1} - y_h^m}{\delta t} + \mathcal{A}_h y_h^{m+1} = \mathcal{B}_h u_h^{m+1}, & \text{for all } m \in \{1,\dots,M-1\} \\ y_h^1 = y_{0,h}, \end{cases} \tag{3.12}$$

where $y_{0,h} \in \mathbb{R}^N$ is the projection of the initial datum $y_0 \in L^2(\Omega)$ on the mesh $\mathfrak{M}$, $\mathcal{A}_h$ and $\mathcal{M}_h$ are the stiffness and mass matrices given in Section 2, while the matrix $\mathcal{B}_h$ has entries (see Boyer et al., 2011, Section 1.4.2)

$$b_{i,j} = \int_\omega \phi_i(x)\phi_j(x)\,dx, \quad i, j = 1, \dots, N.$$

In (3.12), $u_h = (u_h^m)_{m=1}^M \in \mathbb{R}^{N\times M}$ is a fully-discrete control function, whose cost is given by the discrete $L^2(\Omega\times(0,T))$-norm defined by

$$\|u_h\|_{L^2_{h,\delta t}} := \left(\sum_{m=1}^M \delta t |u_h^m|^2_{L^2_{h,\mathcal{M}_h}}\right)^{1/2}, \tag{3.13}$$

and where $|\cdot|_{L^2_{h,\mathcal{M}_h}}$ is the norm associated with the L^2-inner product on $\mathfrak{M}$ and the mass matrix $\mathcal{M}_h$:

$$\text{for all } v=(v_i)_{i=1}^N \in \mathbb{R}^N \quad \text{and} \quad w=(w_i)_{i=1}^N \in \mathbb{R}^N$$

$$\langle v, w\rangle_{L^2_{h,\mathcal{M}_h}} = \langle \mathcal{M}_h v, w\rangle_{L^2_h} = h\sum_{i=1}^N (\mathcal{M}_h v)_i w_i \longrightarrow |v|^2_{L^2_{h,\mathcal{M}_h}} = \langle v, v\rangle_{L^2_{h,\mathcal{M}_h}}.$$

With the above notation, given some penalization parameter $\beta > 0$ we can introduce the fully-discrete primal and dual functionals

$$\begin{aligned} F_{\beta,h}(u_h) &= \frac{1}{2}\|u_h\|^2_{L^2_{h,\delta t}} + \frac{1}{2\beta}|y_h^M|^2_{L^2_{h,\mathcal{M}_h}} \\ J_{\beta,h}(p_h^M) &= \frac{1}{2}\|\mathcal{B}_h p_h\|^2_{L^2_{h,\delta t}} + \frac{\beta}{2}|p_h^M|^2_{L^2_{h,\mathcal{M}_h}} + \left\langle p_h^M, e^{\mathcal{A}_h T} y_{0,h}\right\rangle_{L^2_{h,\mathcal{M}_h}}, \end{aligned} \tag{3.14}$$

with $p_h = (p_h^n)_{n=1}^M \mathbb{R}^{N\times M}$ solution to the adjoint system

$$\begin{cases} \mathcal{M}_h \dfrac{p_h^m - p_h^{m+1}}{\delta t} + \mathcal{A}_h p_h^m = 0, & \text{for all } m \in \{1,\dots,M-1\} \\ p_h^M = p_{T,h}, \end{cases} \tag{3.15}$$

where $p_{T,h} \in \mathbb{R}^N$ is the projection of $p_T \in L^2(\Omega)$ on the mesh $\mathfrak{M}$.

3.3 Numerical experiments

To address the actual computation of the fully-discrete controls for (3.12), we apply an optimization algorithm to the dual functional $J_{\beta,h}(p_h^M)$. This functional being quadratic and coercive, the conjugate gradient (CG) method is a natural choice. At this regard, we recall (see Glowinski and Lions, 1994; Glowinski et al., 2008) that the implementation of the CG algorithm requires the gradient of $J_{\beta,h}(p_h^M)$, which is given by $\nabla J_{\beta,h}(p_h^M) = y_h^M + \beta p_h^M$, where y_h^M is the solution at time T of (3.12) corresponding to the initial datum $y_{0,h}$ and the control $u_h = \mathcal{B}_h p_h$. Hence, to compute $\nabla J_{\beta,h}(p_h^M)$ requires to solve two parabolic equations, one forward and the other backward in time.

In what follows, we will consider the 1-D fractional heat equation on the interval $(-1,1)$ and the 2-D one on the unit ball $B_0(1)$. In both cases, we will discuss the control properties of (3.12) from the viewpoint of Theorem 3.3. In the context of the fully-discrete problem (3.12) this may be a delicate issue. As a matter of fact, it is well known that, in general, we cannot expect for a given bounded family of initial data that the fully-discrete controls are uniformly bounded when h, δt, and β tend to zero independently. Instead, we expect to obtain uniform bounds by taking $\beta = \phi(h)$ that tends to zero in connection with the mesh size not too fast (see Boyer, 2013) and a time step δt verifying some weak condition of the kind $\delta t \leq \zeta(h)$ where ζ tends to zero logarithmically when $h \to 0$ (see Boyer et al., 2011). In particular, it is crucial to choose properly the penalization parameter β. Following the discussion in Boyer (2013), a reasonable practical rule is to take $\beta = \phi(h) \sim h^{2p}$, where p is the order of accuracy in space of the numerical method employed for the discretization of the fractional Laplacian. We refer to Boyer (2013, Section 4.2) for some heuristic comments on the choice of this penalization parameter. Nevertheless, we have to stress that, as far as the authors know, the effectiveness of this choice has only been demonstrated numerically and is still not supported by rigorous mathematical results.

To select the correct value for p, let us recall that the solution y to (3.1) with $y_0 \in L^2(\Omega)$ and $u \in L^2(\omega \times (0,T))$ belongs to $L^2(0,T; H_0^s(\Omega)) \cap C([0,T]; L^2(\Omega))$. In particular, we have that $y(\cdot,T) \in L^2(\Omega)$. Therefore, we shall choose the value of p as the convergence rate in the L^2-norm for the approximation of the elliptic problem (2.1). By virtue of Theorem 2.4, the appropriate value of p that we shall employ is thus

$$p = \begin{cases} 2s, & \text{for } s < \frac{1}{2} \\ 1 - 2\delta, & \text{for } s \geq \frac{1}{2} \end{cases} \quad \longrightarrow \quad \beta = h^{2p} = \begin{cases} h^{4s}, & \text{for } s < \frac{1}{2} \\ h^{2-4\delta}, & \text{for } s \geq \frac{1}{2}, \end{cases} \tag{3.16}$$

with $\delta > 0$ arbitrary small. With this choice of β, by observing the behavior of the norm of the control, the optimal energy (3.4) and the norm of the solution at time T, we will obtain numerical evidences for the null and approximate controllability of (3.1), which are in accordance with the theoretical results we recalled in Section 3.1. Nevertheless, we shall mention that these numerical evidences are, at present time, not supported by a rigorous mathematical analysis, since the controllability of fully-discrete fractional heat equations is still an open question. We refer to Section 6 for more details.

3.3.1 1-D simulations

Let us present here our numerical simulations in one space dimension. To this end, we introduce a uniform N-points mesh discretizing the space domain $(-1,1)$. The time interval $(0,T)$ is discretized with a uniform partition as well, this time composed by M points, on which we will implement an implicit Euler method. We begin by considering $s = 0.2$, for which we know that (3.1) is only approximately controllable. We set $\omega = (-0.3, 0.8)$, $T = 0.3$, and $y_0(x) = \sin(\pi x)$. We then run the CG algorithm to compute the optimal control $\hat{u}_h$ on several uniform meshes with decreasing mesh-size $h \to 0$, and display in Fig. 6 the following quantities of interest in dependence of h:

- The cost of control $\|\hat{u}_h\|_{L^2_{h,\delta t}}$ (see (3.13)).
- The optimal energy $F_{\beta,h}(\hat{u}_h)$.
- The size $|\hat{y}^M|_{L^2_{h,\mathcal{M}_h}}$ of the corresponding solution at time T.

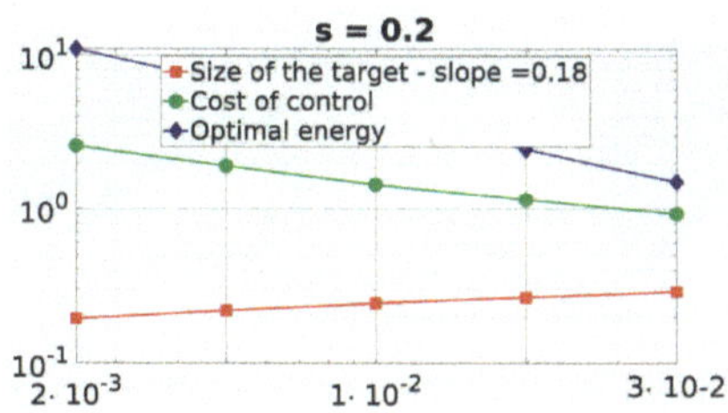

FIGURE 6 Behavior with respect to the mesh size h of the cost of control, optimal energy and size of the solution to (3.12) in space dimension $d = 1$ at time T when $s = 0.2$.

We observe that the discrete L^2 norm of $\hat{y}^M$ tends to zero as $h \to 0$, confirming computationally the approximate controllability of (3.1). However, we also see that the cost of the control and the optimal energy increase as $h \to 0$. Hence, numerical evidence indicates that the null controllability is not fulfilled, in agreement with Theorem 3.3.

Let us now take $s = 0.8$. In this case, as we can see in Fig. 7, the situation changes. Indeed, we can observe that this time the control cost and the optimal energy remain bounded as $h \to 0$. Furthermore, we also see that

$$|\hat{y}^M|_{L^2_{h,\mathcal{M}_h}} \sim h = \sqrt{\beta},$$

which is the expected convergence rate given in (3.16) for the discrete L^2 norm of $y(\cdot, T)$. According to Theorem 3.3, all these facts confirm that, for $s = 0.8$, (3.1) is indeed null controllable as we already proved in Theorem 3.1.

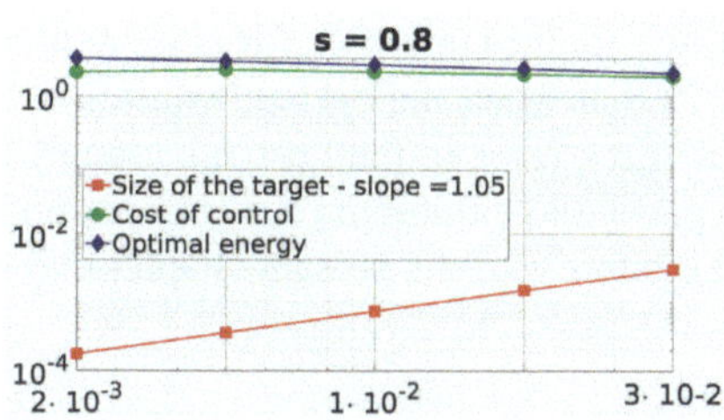

FIGURE 7 Behavior with respect to the mesh size h of the cost of control, optimal energy and size of the solution to (3.12) in space dimension $d = 1$ at time T when $s = 0.8$.

This positive controllability result is also appreciated in Fig. 8, where we illustrate the time evolution of the uncontrolled solution as well as the controlled one. We can clearly see that the uncontrolled solution diffuses under the action of the fractional heat semi-group, but does not reach zero at time T. On the other hand, the introduction of a control modifies the dynamical behavior of y in such a way that we achieve $y(\cdot, T) = 0$.

Finally, in Fig. 9, we see the evolution of the control function which, as we can observe, is almost inactive for a large part of the time interval, and then

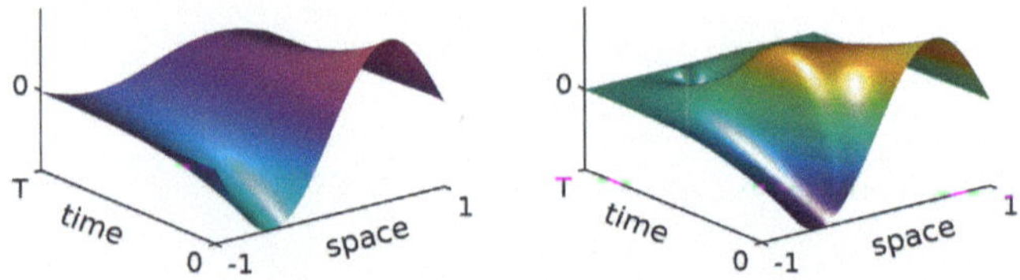

FIGURE 8 Free (left) and controlled (right) solution of (3.1) with $s = 0.8$ at time $T = 0.3$.

experiences large oscillations in the proximity of the final time. This fact, related with the characterization of the control as restrictions of solutions of the adjoint system, is in accordance with the *lazy* behavior (observed by Glowinski and Lions, 1994) of controls for the local heat equation which, at the very beginning, leave the solution evolve under the dissipative effect of the heat semi-group and, only when approaching the final controllability time, inject energy into the system in order to match the desired configuration.

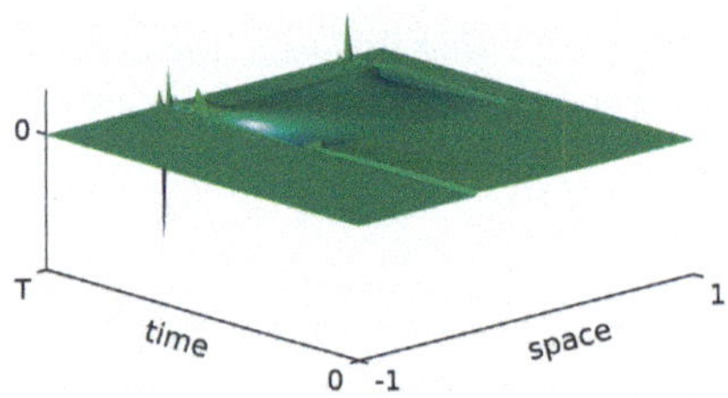

FIGURE 9 Behavior of the control for the fractional heat equation (3.1) with $s = 0.8$.

For completeness, we shall mention that it is by now well-known (see Porretta and Zuazua, 2013) that when the optimal control is chosen differently (in particular, including a *track term* in the cost functional) then the controls are better behaved and in particular exhibit the *turnpike* phenomenon. Nevertheless, this issue is beyond the interest of the present chapter and we will not discuss it further.

3.3.2 The 2-D case

Let us now consider the fractional heat equation (3.1) in space dimension $d = 2$. In this case, we know that null controllability holds in any time $T > 0$, acting from a neighborhood of the boundary of the domain Ω, once again if and only if $s \in (1/2, 1)$. On the other hand, if $s \in (0, 1/2]$, we only have approximate controllability.

For our simulations, we chose the domain $\Omega = B_0(1)$ as the unit ball centered at zero, while the control region is the ring $\omega = B_0(1) \setminus B_0(4/5)$. Moreover, we considered $y_0(x, y) = \sin(\pi x)\sin(\pi y)$ and the time horizon $T = 0.3$. The control for (3.1) has been computed once again by applying the penalized HUM of Section 3.2. Nevertheless, now we are considering a two dimensional dynamics described by a full stiffness matrix, whose associated optimal control problem may be computationally very heavy. To alleviate this computational effort and efficiently minimize the discrete functionals (3.14), we have relied on the expert interior-point optimization routine `IpOpt` (Wächter and Biegler, 2006), with the help of `CasADi` open-source tool for algorithmic differentiation (Andersson et al., 2019). We start by considering $s = 0.2$ and displaying in Fig. 10 the computed values as $h \to 0$ of the cost of control, the optimal energy and the discrete L^2 norm of $y(\cdot, T)$. We can see that:

- The discrete L^2 norm of the solution at time T decreases with h, thus confirming the approximate controllability of (3.1) according to the first part of Theorem 3.3.
- The control cost and the optimal energy both increase as $h \to 0$, thus violating (3.4). Hence, Theorem 3.3 yields the failure of the null controllability property.

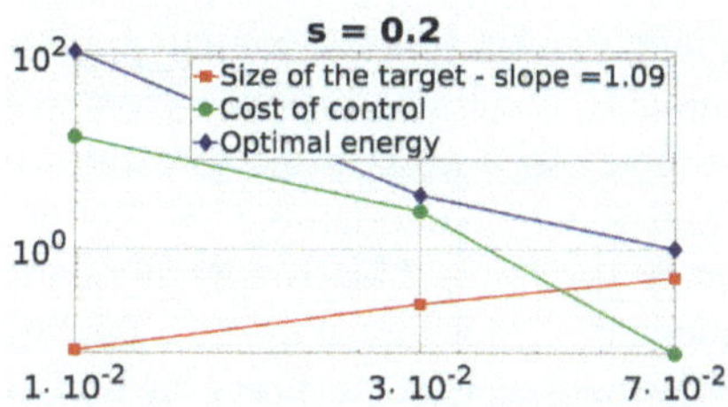

FIGURE 10 Behavior with respect to the mesh size h of the cost of control, optimal energy and size of the solution to (3.12) in space dimension $d = 2$ at time T when $s = 0.2$.

Finally, let us consider the case $s = 0.8$. This time, we can appreciate in Fig. 11 how the control cost and the optimal energy remain bounded as $h \to 0$. Besides, the discrete L^2 norm of $y(\cdot, T)$ decreases with rate h, which is the expected one according to (3.5b). Hence, Theorem 3.3 confirms once again the theoretical controllability properties of Theorem 3.1, assuring that the fractional heat equation in 2-D is null-controllable at time T.

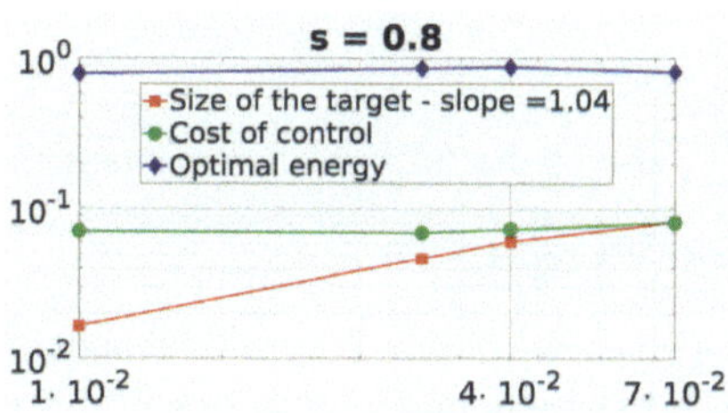

FIGURE 11 Behavior with respect to the mesh size h of the cost of control, optimal energy and size of the solution to (3.12) in space dimension $d = 2$ at time T when $s = 0.8$.

3.3.3 The constrained controllability case

In this section, we present some numerical evidences of the constrained controllability properties we obtained for (3.1) in Theorem 3.2. In particular, we focus on the controllability to trajectories of (3.1) by means of non-negative controls. To this end, we choose the initial datum $y_0(x) = \sin(\pi x)$ and we set as a target $\widehat{y}(\cdot, T)$ the solution at time T of (3.1) with initial datum $\widehat{y}_0(x) = 0.5\cos(\pi x/2)$ and right-hand side $\widehat{u} \equiv 0.02$ a.e. in $(-1, 1) \times (0, T)$. Moreover, we choose $\omega = (-0.3, 0.5) \subset (-1, 1)$ as the control region and set $s = 0.8$. Hence, Theorem 3.2 ensures the existence of a strictly positive controllability time $T_{\min}$ and a non-negative control function $u \in L^\infty(\omega \times (0, T))$ such that if $T \geq T_{\min}$ the solution of (3.1) is controllable to the trajectory $\widehat{y}(\cdot, T)$.

At this regard, we stress that we do not have analytical bounds for this minimal controllability time. As a matter of fact, the techniques developed in Loheac et al. (2017); Pighin and Zuazua (2018) to obtain these bounds for the local heat equation are not immediately extendable to our fractional context. We refer to Biccari et al. (2020b, Section 4.4) for a detailed discussion about this specific issue. Notwithstanding that, in what follows, we will provide numerical estimates of $T_{\min}$ by solving a suitable constrained optimization problem, which will also give us the minimal-time control $u_{\min}$ in the form of a linear combination of Dirac deltas, in accordance with Theorem 3.2. Secondly, we will consider the controllability problem in time $T > T_{\min}$. As our simulations will show, also in that case the target trajectory is matched at time T, although the control loses its atomic nature. Finally, we will consider a short time horizon $T < T_{\min}$, in which the controllability to trajectories is not achieved.

3.3.3.1 Numerical approximation of the minimal controllability time

We start by estimating numerically the minimal controllability time $T_{\min}$. To this end, we employ `IpOpt` and `CasADi` to solve the following constrained optimization problem:

$$\text{minimize } T \tag{3.17}$$

subject to the constraints

$$\begin{aligned} & T > 0, \\ & (y_h^m)_{m=1}^M \text{ solves the fully-discrete dynamics (3.12)} \\ & (y_h^m)_{m=1}^M, (u_h^m)_{m=1}^M \geq 0. \end{aligned} \tag{3.18}$$

By solving (3.17)–(3.18), we obtain $T_{\min} \simeq 0.68$ and the control $u_{\min}$. Fig. 12 shows the solution to (3.1) which, under the action of this minimal-time control, is steered from the initial datum y_0 to the desired target. Moreover, we can also see that the minimal-time control $u_{\min}$ is localized in certain specific points of the domain and time instants. This is in accordance with Theorem 3.2 which states that the minimal-time control is a Radon measure (in particular, a linear combination of Dirac masses).

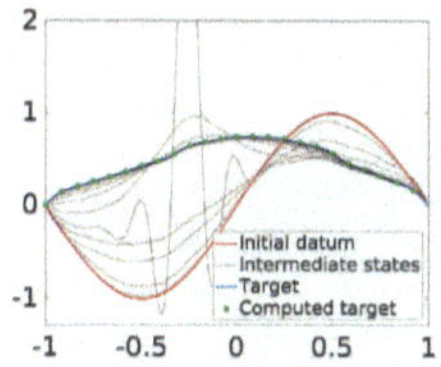

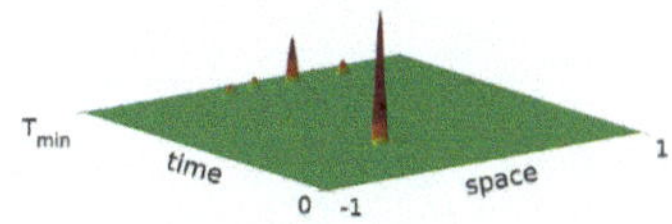

FIGURE 12 Evolution in the time horizon $T = T_{\min}$ of the solution of (3.1) with $s = 0.8$ (left) and of the control $u_{\min}$ (right). The blue curve is the target we want to reach while the green bullets indicate the final state of the computed numerical solution.

3.3.3.2 Constrained controllability in large time

The atomic behavior of the control is lost when extending the time horizon beyond $T_{\min}$. In Fig. 13, we show the evolution of the solution of the fractional heat equation (3.1) from the initial datum y_0 to the target $\widehat{y}(\cdot, T)$ in the time horizon $T = 1$, and the corresponding optimal control that we computed minimizing in `IpOpt` the cost functional

$$F_{\beta,\infty}(u) = \frac{1}{2}\|u\|_{L^\infty(\omega\times(0,T))} + \frac{1}{2\beta}\|y(\cdot, T) - \widehat{y}(\cdot, T)\|^2_{L^2(-1,1)} \tag{3.19}$$

with the constraints (3.18). As we can observe, in accordance with our theoretical results, the equation is still controllable in time T. Nevertheless, the control has lost its atomic nature and its action is now more distributed in ω.

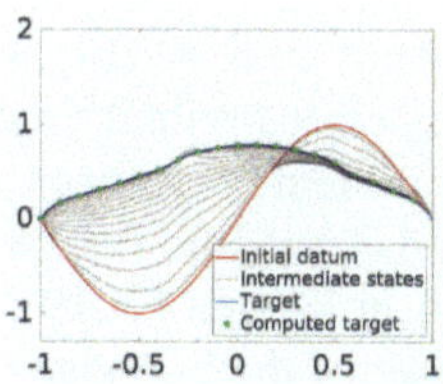

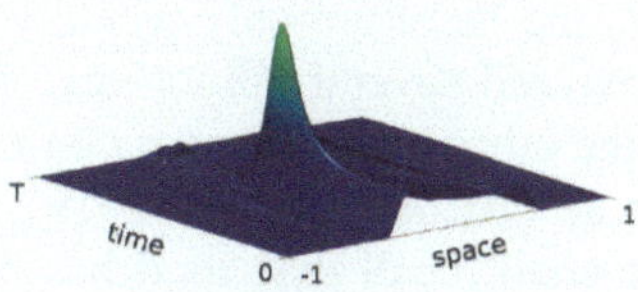

FIGURE 13 Evolution in the time horizon $T = 1 > T_{\min}$ of the solution of (3.1) with $s = 0.8$ (left) and of the control u (right), under the constraint $u \geq 0$. The atomic nature of the control is lost.

3.3.3.3 Lack of controllability in short time

To conclude this section, let us now consider the controllability problem in a short time horizon $T < T_{\min}$ (more specifically, $T = 0.25$). Once again, we have employed `IpOpt` combined with `CasADi` to minimize the functional (3.19) giving us the optimal control.

Our simulations displayed in Fig. 14 show that the solution of (3.1) fails to be controlled. In particular, the numerical solution at time T does not match with the trajectory in blue we want to reach. This is in accordance with the lack of constrained controllability in short time as proved in Theorem 3.2.

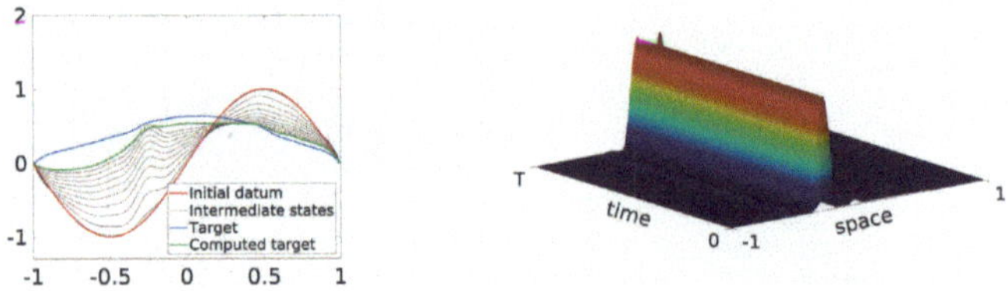

FIGURE 14 Evolution in the time horizon $T = 0.25 < T_{\text{min}}$ of the solution of (3.1) with $s = 0.8$ (left) and of the control u (right), under the constraint $u \geq 0$. The equation is not controllable.

4 Exterior controllability properties of the fractional heat equation

In this section, we discuss the exterior controllability properties of the one-dimensional fractional heat equation. That is, for any $y_0 \in L^2(-1, 1)$ and $\mathcal{O}$ bounded, nonempty and open subset of $(-1, 1)^c$, we are going to consider the following control system

$$\begin{cases} y_t + (-\Delta)^s y = 0 & \text{in } (-1,1) \times (0,T), \\ y = g\chi_{\mathcal{O}} & \text{in } (-1,1)^c \times (0,T), \\ y(\cdot, 0) = y_0 & \text{in } (-1,1). \end{cases} \tag{4.1}$$

Here g is the control function belonging to some functional space to be specified later.

The concept of exterior controllability for fractional models has been introduced in the literature only recently. First, in Antil et al. (2019, 2020b), the authors analyzed several elliptic and parabolic exterior optimal control problems. On the other hand, exterior controllability problems have been addressed in Antil et al. (2020a); Warma (2019); Warma and Zamorano (2020b). At this regard, we shall recall that, as it has been shown in Warma (2019), a boundary control (that is, a control g localized in a subset of the boundary) does not make sense in the presence of a fractional Laplacian. This is because of the non-locality of the operator and the fact that fractional models with standard boundary conditions (Dirichlet, Neumann, or Robin) are ill-posed. For problems involving the fractional Laplacian the correct notion of a boundary controllability is actually the exterior one, requiring the control function to be localized outside the domain where the PDE is satisfied, as in (4.1). Let us mention that exterior control problems also appear in many realistic applications, such as for instance:

1. Magnetic drug delivery: the drug with ferromagnetic particles is injected in the body and an external magnetic field is used to steer it to a desired location.

2. Acoustic testing: the aerospace structures are subjected to the sound from the loudspeakers.

We refer to Antil et al. (2019, 2020b) and the references therein for a further discussion and the derivation of the exterior control.

In this section, we will give a broad panorama of the exterior controllability problem (4.1). We will start by recalling in Subsection 4.1 the theoretical controllability results presently available in the literature, and commenting important aspects on the numerical approximation of exterior controls. Secondly, in Subsection 4.2, we will present our numerical experiments.

4.1 Review of theoretical controllability results

We summarize the theoretical controllability results which can be currently found in the literature for the fractional heat equation (4.1).

Theorem 4.1. *For the fractional heat equation* (4.1), *the following results hold.*

1. ***Approximate controllability**. Let $\mathcal{O} \subset (-1,1)^c$ be any nonempty and open subset of $(-1,1)^c$ and $s \in (0,1)$. For any $T > 0$, $y_0, y_T \in L^2(-1,1)$ and $\varepsilon > 0$, there exists a control function $g \in \mathcal{D}((-1,1)^c \times (0,T))$ such that the unique solution y of* (4.1) *satisfies $\|y(\cdot,T) - y_T\|_{L^2(-1,1)} \leq \varepsilon$.*
2. ***Null controllability**. Let $\mathcal{O} \subset (-1,1)^c$ be any nonempty and open subset of $(-1,1)^c$. For any $T > 0$ and $y_0 \in L^2(-1,1)$, there exists a control function $g \in L^2(0,T; H^s((-1,1)^c))$ such that the unique solution y of* (4.1) *satisfies $y(\cdot,T) = 0$ a.e. in $(-1,1)$, if and only if $s \in (1/2,1)$.*
3. ***L^∞ null controllability**. Let $\mathcal{O} \subset (-1,1)^c$ be any nonempty and open subset of $(-1,1)^c$. For any $T > 0$ and $y_0 \in L^2(-1,1)$, there exists a control function $g \in L^\infty(\mathcal{O} \times (0,T))$ such that the unique solution y of* (4.1) *satisfies $y(\cdot,T) = 0$ a.e. in $(-1,1)$, if and only if $s \in (1/2,1)$.*

For completeness, we shall mention that in Warma (2019, Theorem 6) the approximate controllability of (4.1) has actually been obtained in any space dimension $d \geq 1$. In Theorem 4.1, we have stated this result in the one-dimensional context to remain consistent with the presentation of this section.

4.1.1 Remarks on the theoretical controllability results

This section is devoted to some remarks and comments on the theoretical controllability results of Theorem 4.1. These remarks will be at the basis of the methodology we will adopt in our forthcoming numerical simulations.

First of all, for any function $\zeta \in H^s(\mathbb{R})$, let us denote with $\mathcal{N}_s\zeta$ the non-local normal derivative defined as (see (B.2))

$$\mathcal{N}_s\zeta(x) := C_s \int_{-1}^{1} \frac{\zeta(x) - \zeta(z)}{|x - z|^{1+2s}}\, dz, \quad x \in (-1,1)^c$$

Moreover, for any $p_T \in L^2(-1,1)$, let $p \in L^2(0,T; H_0^s(\Omega))$ be the solution of the adjoint equation

$$\begin{cases} p_t + (-\Delta)^s p = 0 & \text{in } (-1,1) \times (0,T), \\ p = 0 & \text{in } (-1,1)^c \times (0,T), \\ p(\cdot,T) = p_T & \text{in } (-1,1). \end{cases} \tag{4.2}$$

Then multiplying (4.1) by p and integrating over $(-1,1) \times (0,T)$ using the integration by parts formula given in Proposition B.2, it is easy to see that

$$y(\cdot,T) = 0 \text{ a.e. in } (-1,1) \text{ if and only if } \int_{-1}^{1} y_0 p(\cdot,0)\,dx = \int_0^T \int_{\mathcal{O}} g\mathcal{N}_s p\,dx. \tag{4.3}$$

Furthermore, the characterization (4.3) yields that (4.1) is null controllable at time $T > 0$ with $g \in L^2(0,T; H^s(\mathcal{O}))$ if and only if the following observability inequality for the adjoint system (4.2) holds (see Warma and Zamorano, 2020b, Lemma 2):

$$\|p(\cdot,0)\|^2_{L^2(-1,1)} \le C \int_0^T \|\mathcal{N}_s p(t)\|^2_{L^2(\mathcal{O})}\,dt.$$

This observability result has been proved in Warma and Zamorano (2020b, Theorem 1) and follows by employing spectral techniques and parabolic Ingham inequalities, taking into account the Fourier decomposition for the solution of (4.1) given in Theorem B.8.

Finally, in view of the above considerations, we can see that the exterior control for (4.1) can be obtained from the following optimal control problem:

$$g_\beta = \min_{g \in L^2((0,T);H^s(\mathcal{O}))} F_\beta^{ext}(g) \tag{4.4}$$

$$F_\beta^{ext}(g) := \frac{1}{2}\int_0^T \|g\|^2_{H^s(\mathcal{O})}\,dt + \frac{1}{2\beta}\|y(\cdot,T)\|^2_{L^2(-1,1)}.$$

Notice that the optimization process (4.4) has to be solved under the constraints given by the dynamics (4.1). As we commented in Section 2.1.3, there are several different possibility to discretize this dynamics. Here, we will use the approach presented in Antil et al. (2019, 2020b), consisting in approximating (4.1) through the following exterior Robin problem

$$\begin{cases} y_t^n + (-\Delta)^s y^n = 0 & \text{in } (-1,1) \times (0,T), \\ \mathcal{N}_s y^n + n\kappa y^n = n\kappa g & \text{in } (-1,1)^c \times (0,T), \\ y^n(\cdot,0) = y_0 & \text{in } (-1,1), \end{cases} \tag{4.5}$$

where $n \in \mathbb{N}$ is a fixed natural number and $\kappa \in L^1((-1,1)^c) \cap L^\infty((-1,1)^c)$ is a non-negative function.

The fact that (4.5) indeed approximates (4.1) has been proved in Antil et al. (2020b). As a matter of fact, we have the following result.

Theorem 4.2 (Antil et al., 2020b, Theorem 5.3). *Let* $y_0 \in L^2(-1,1)$, $g \in L^2((0,T); H^s(-1,1)^c)$, $\kappa \in L^1((-1,1)^c) \cap L^\infty((-1,1)^c)$ *non-negative and*

$$y^n \in L^2((0,T); H^s_\kappa(-1,1)) \cap H^1((0,T); H^{-s}_\kappa(-1,1))$$

be the weak solution of (4.5) *according to Definition B.9. Let* $y \in L^2((0,T); H^s(\mathbb{R}))$ *be the weak solution of* (4.1). *There is a constant* $C > 0$, *independent of* n, *such that*

$$\|y - y^n\|_{L^2(\mathbb{R}\times(0,T))} \leq \frac{C}{n}\|y\|_{L^2((0,T);H^s(\mathbb{R}))}. \tag{4.6}$$

In particular, y^n *converges strongly to* y *in* $L^2((-1,1)\times(0,T))$ *as* $n \to +\infty$.

We refer to Appendix B for the definition of the Hilbert space $H^s_\kappa(-1,1)$ and of the weak solutions to (4.5). At this regard, we shall stress that, at the numerical level, passing to the limit rigorously as $n \to +\infty$ in (4.5) would require a better understanding of the problems mentioned in Section 2.1.3 about the FE treatment of the exterior problem in the elliptic case. We shall comment more on this aspect in Section 6.

Moreover, also in the case of the Robin problem (4.5) controllability can be characterized through a dual argument. To this end, for any $p^n_T \in L^2(-1,1)$ we denote with p^n the solution of the following adjoint problem with Robin exterior conditions

$$\begin{cases} -p^n_t + (-\Delta)^s p^n = 0 & \text{in } (-1,1)\times(0,T), \\ \mathcal{N}_s p^n + n\kappa p^n = 0 & \text{in } (-1,1)^c \times (0,T), \\ p^n(\cdot,T) = p^n_T & \text{in } (-1,1). \end{cases} \tag{4.7}$$

It can be readily checked that

$$y^n(\cdot,T) = 0 \text{ a.e. in } (-1,1) \text{ if and only if}$$
$$\int_{-1}^{1} y_0 p^n(\cdot,0)\,dx + n\int_0^T\int_{\mathcal{O}} p^n \kappa g\,dx = 0.$$

In view of that, the controllability of (4.5) g replaced with $g\chi_{\mathcal{O}}$, where the control region $\mathcal{O} \subset (-1,1)^c$ is an arbitrary nonempty open set, is equivalent to the following observability inequality for (4.7)

$$\|p^n(\cdot,0)\|^2_{L^2(-1,1)} \leq C\int_0^T \|p^n(t)\|^2_{L^2(\mathcal{O})}\,dt,$$

and the exterior control can be obtained from the following optimal control problem:

$$g_\beta = \min_{g \in L^2(\mathcal{O}\times(0,T))} G_\beta^{ext}(g) \tag{4.8}$$

$$G_\beta^{ext}(g) := \frac{1}{2}\int_0^T \|g\|^2_{L^2(\mathcal{O})}\,dt + \frac{1}{2\beta}\left\|y^n(\cdot,T)\right\|^2_{L^2(-1,1)}.$$

To conclude this section, let us stress that the considerations which led to (4.8) are only formal, and currently not supported by rigorous mathematical results. We will comment more on this issue in Section 6.

4.2 Numerical experiments

Let us present our numerical simulations for the exterior control problem. At this regard, we recall that, as discussed in the previous section and supported by Theorem 4.2, instead of the Dirichlet problem (4.1) we will consider the Robin problem (4.5). The computation of the exterior control will be done through the optimization process (4.8).

For completeness, we stress that replacing our original dynamics (4.1) with (4.5) introduces an approximation error associated with the parameter $n \in \mathbb{N}$. Nevertheless, we shall also notice that this approximation error is of the order of n^{-1} in the L^2-norm, and can be kept small (in comparison with the error introduced by the space/time discretization of the dynamics) by selecting n large enough. In particular, we will take $n = 10^9$.

To approximate (4.5), we consider the interval $\mathcal{I} = (-2,2) \supset (-1,1)$ and assume that the control function g is supported in $\mathcal{O} \subset ((-2,2) \setminus (-1,1))$. Notice that, in this case, the regularity required in Theorem 4.2 for the function κ, namely $\kappa \in L^1((-2,2)\setminus(-1,1)) \cap L^\infty((-2,2)\setminus(-1,1))$, simply reduces to $\kappa \in L^\infty((-2,2)\setminus(-1,1))$ and is fulfilled by considering κ to be constant. For simplicity, we take $\kappa = 1$. In other words, we approximate the following problem:

$$\begin{cases} y_t^n + (-\Delta)^s y^n = 0 & \text{in } (-1,1)\times(0,T), \\ \mathcal{N}_s y^n + n y^n = n g \chi_{\mathcal{O}\times(0,T)} & \text{in } ((-2,2)\setminus(-1,1))\times(0,T), \\ y^n(\cdot,0) = y_0 & \text{in } (-1,1). \end{cases} \tag{4.9}$$

To discretize (4.9) in space, we introduce a uniform N-points mesh on $(-2,2)$ with mesh-size h and we use a globally continuous piece-wise linear FE scheme based on the following variational formulation (see Definition B.9): for all $v \in H^s_\kappa(-1,1)$,

$$n\int_{\mathcal{O}} \kappa g v\,dx\,dt = \int_{-1}^{1} y_t^n v\,dx\,dt + \mathcal{F}(y^n, v) + n\int_{(-1,1)^c} \kappa y^n v\,dx\,dt. \tag{4.10}$$

Notice that in (4.10) the non-homogeneous datum g enters in the integral on the left-hand side and not in the bilinear form.

For the time discretization, instead, we apply implicit Euler on a uniform M-points grid discretizing $[0, T]$.

To discuss the numerical controllability problem, in what follows, we adopt again the methodology of Section 3 and use Theorem 3.3 to find numerical evidences of the control results of Theorem 4.1. We do this by minimizing the functional (4.8) on several uniform meshes with decreasing mesh-size $h \to 0$ to compute the optimal control g_β, and by analyzing the behavior with respect to $\beta(h)$ (chosen as in (3.16)) of:

- The cost of control $\|g_\beta\|_{L^2((0,T);H^s(\mathcal{O}))}$.
- The optimal energy $G_\beta^{ext}(g_\beta)$.
- The L^2-norm of the corresponding solution at time T.

We begin by considering $s = 0.2$, for which we know that (4.1) is only approximately controllable. We set $\mathcal{O} = (1.7, 1.9)$, $T = 0.4$ and $y_0(x) = \cos(\pi x/2)$. We then employ IpOpt and CasADi to solve (4.8). The results of these numerical experiments are displayed in Fig. 15.

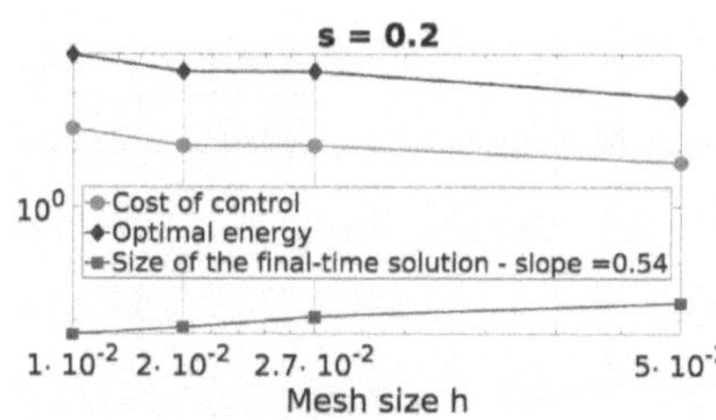

FIGURE 15 Behavior with respect to the mesh size h of the cost of control, optimal energy and size of the solution to (4.9) at time T when $s = 0.2$.

We observe that the L^2 norm of the final state tends to zero as $h \to 0$, confirming computationally the approximate controllability of (4.1). Notwithstanding that, we can also see that the cost of the control and the optimal energy slightly increase as $h \to 0$. Therefore, according to Theorem 3.3, the null controllability of (4.1) is not fulfilled.

Let us now take $s = 0.8$. In this case, as we can see in Fig. 16, the situation changes. Indeed, we can observe that this time the control cost and the optimal energy remain bounded as $h \to 0$. Furthermore, we also see that the L^2 norm of the solution at time T decreases with the expected rate $h = \sqrt{\beta}$ (see (3.16)). According to Theorem 3.3, all these facts confirm that, for $s = 0.8$, (4.1) is indeed null controllable.

This positive controllability result is also appreciated in Fig. 17, where we can clearly see that the uncontrolled solution diffuses under the action of the fractional heat semi-group, but does not reach zero at time T. On the other

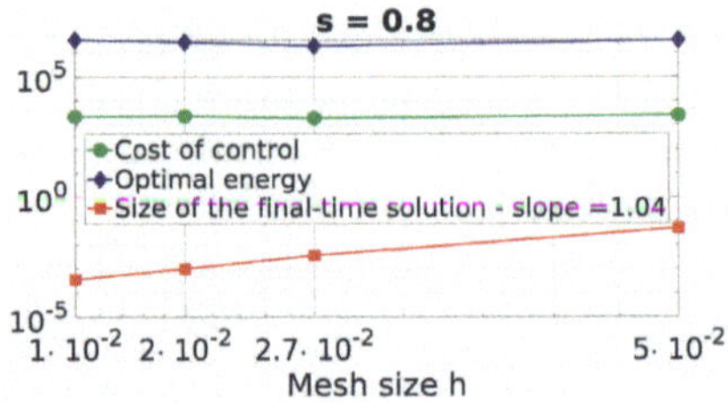

FIGURE 16 Behavior with respect to the mesh size h of the cost of control, optimal energy and size of the solution to (4.9) at time T when $s = 0.8$.

hand, the introduction of the computed optimal control modifies the dynamical behavior of y in such a way that we achieve $y(\cdot, T) = 0$.

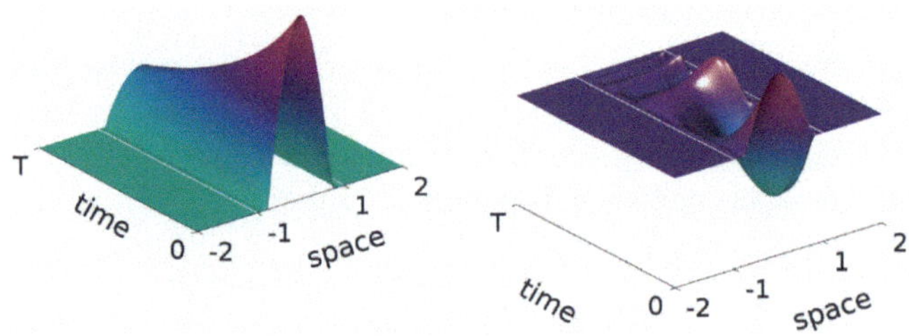

FIGURE 17 Free (left) and controlled (right) solution of (4.1) with $s = 0.8$ at time $T = 0.4$.

Finally, in Fig. 18, we can see the time evolution of the control function, which is only acting outside the interval $(-1, 1)$ where the dynamics evolves. Specifically, the control is active only on the right part of the exterior domain, since we have chosen $\mathcal{O} = (1.7, 1.9)$ as the control region.

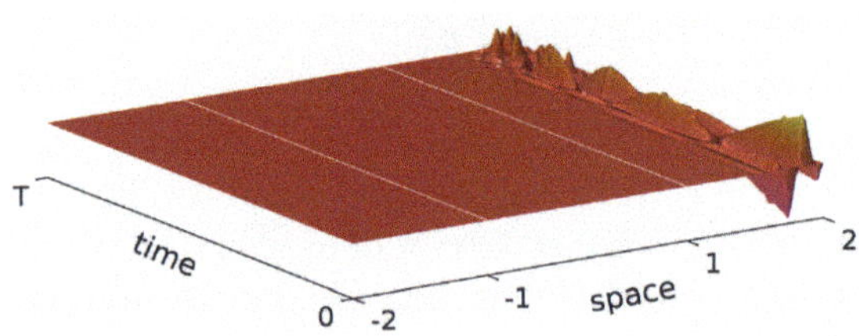

FIGURE 18 Exterior control obtained via the minimization of the functional (4.8).

Remark 4.3 (Constrained controllability problem). For completeness, we shall mention that the study of constrained controllability has been recently extended in Antil et al. (2020a) to the exterior control problem (4.1). In particular, in the same spirit of the interior control problem discussed in Section 3, the following results have been established in Antil et al. (2020a, Theorems 2.3, 2.4, and 2.5).

1. For all $s \in (1/2, 1)$, $y_0 \in L^2(-1, 1)$, and any positive trajectory $\widehat{y}$, there exists a minimal strictly positive controllability time $T_{\min} > 0$ such that, for all

$T > T_{\min}$, we can find a non-negative control $g \in L^\infty(\mathcal{O} \times (0, T))$ whose corresponding solution y of (4.1) satisfies $y(\cdot, T) = \widehat{y}(\cdot, T)$ a.e. in $(-1, 1)$. Moreover, if $y_0 \geq 0$, then $y(x, t) \geq 0$ for every $(x, t) \in (-1, 1) \times (0, T)$.

2. For $T = T_{\min}$, the above constrained controllability result holds with controls $g \in \mathcal{M}(\mathcal{O} \times (0, T_{\min}))$, the space of Radon measures on $\mathcal{O} \times (0, T_{\min})$.

Furthermore, the numerical computation of exterior non-negative controls has also been addressed in Antil et al. (2020a, Section 5), by combining the techniques developed for the interior control problem (3.1) (and described in Section 3) with the methodology we presented in Section 4.1.1 for the numerical resolution of (4.1).

An exhaustive discussion of the constrained exterior controllability properties of (4.1) (both at the continuous and discrete level) is omitted here for the sake of brevity, and can be found in Antil et al. (2020a).

5 Simultaneous control of parameter-dependent fractional heat equations

The concept of simultaneous control arises naturally in many contexts including parameter-dependent models, transmission problems, PDE on graphs, synchronization or multi-agent systems with applications, for instance, to robot coordination. The aim is to design a unique control policy which is independent of the model's changes and robust in a variegated spectrum of different realizations.

In this section, we discuss the simultaneous control problem in the context of parameter-dependent fractional heat equations.

One of the key issues when designing a control strategy for a dynamical system is the efficient computation of the control. This becomes an even more critical aspect in simultaneous control, where the corresponding optimization problem typically depends on a large amount of input data. For this reason, many analytical and computational techniques have been developed in the past years in order to speed up the simulation of parameterized control problems. Among others, we can mention Proper Orthogonal Decomposition, other more general Reduced Basis approaches, or the so-called *greedy methodology* (Hernández-Santamaría et al., 2019; Lazar and Lohéac, 2020; Lazar and Zuazua, 2016).

Here, we propose an alternative approach to simultaneous control, based on the employment of stochastic optimization techniques. We will start by presenting in Subsection 5.1 the problem we are going to address. Secondly, in Subsection 5.2, we will give an overview of several deterministic and stochastic optimization algorithms adapted to the computation of simultaneous controls, and we will discuss their convergence properties and computational cost. Finally, in Subsection 5.4 we will present some numerical experiments to compare the efficiency of deterministic and stochastic optimization in the framework of simultaneous control.

5.1 Problem formulation

Let $y_0 \in L^2(-1,1)$, $\mathcal{K} = \{s_1, s_2, \ldots, s_{|\mathcal{K}|}\} \subset (1/2, 1)$ a finite set of cardinality $|\mathcal{K}|$, and $s_\ell \in \mathcal{K}$ a random parameter following a uniform probability law μ, that is,

$$\mu(\mathcal{K}) = 1 \qquad \text{and} \qquad \mu(s_\ell) = |\mathcal{K}|^{-1}, \quad \text{for all } \ell = 1, \ldots, |\mathcal{K}|.$$

We consider the parameter-dependent fractional heat equation

$$\begin{cases} y_{s_\ell,t} + (-\Delta)^{s_\ell} y_{s_\ell} = u\chi_\omega & \text{in } (-1,1) \times (0,T), \\ y_{s_\ell} = 0 & \text{in } (-1,1)^c \times (0,T), \\ y_{s_\ell}(\cdot, 0) = y_0 & \text{in } (-1,1). \end{cases} \tag{5.1}$$

We are interested in the simultaneous control of (5.1), that is, we want to find a unique *parameter-independent* control u such that, at time $T > 0$, the solution y_{s_ℓ} satisfies

$$y_{s_\ell}(\cdot, T) = 0, \quad s_\ell \in \mathcal{K} \quad \mu - \text{a.e.} \tag{5.2}$$

In what follows, to simplify the notation, we will drop the sub-index ℓ and we will simply denote by s any parameter in the set $\mathcal{K}$. The corresponding solution of (5.1) will then be denoted by y_s.

Simultaneous control problems are typically very difficult to be tackled. In particular, to determine whether a parameter-dependent system is simultaneous controllable is not a trivial task.

For linear finite-dimensional (ODE) models, the question has been addressed and solved in Lohéac and Zuazua (2016). Nevertheless, to the best of our knowledge, in the infinite-dimensional PDE setting the problem is still poorly understood. In particular, there are no results of simultaneous control in the sense of (5.2). We will present a more complete discussion on this point in Section 6.

In view of this fact, in what follows, we will relax the problem to simply consider the numerical implementation of simultaneous controls for (5.1) via the minimization of a suitable cost function, leaving open the rigorous mathematical justification of the simulation evidences we will obtain. At this regard, we shall remark that, as pointed out in Lohéac and Zuazua (2016, Remark 1.1-5), the simultaneous control property (5.2) is guaranteed by the fact that

$$\mathbb{E}\left[\|y_s(\cdot, T)\|^2_{L^2(-1,1)}\right] = \frac{1}{|\mathcal{K}|}\sum_{s\in\mathcal{K}} \|y_s(\cdot, T)\|^2_{L^2(-1,1)} = 0,$$

where $\mathbb{E}[\cdot]$ denotes the expectation and we took into account the uniform probability distribution of the parameters. Hence, it is natural to address the simultaneous control of (5.1) by solving the optimization problem

$$\begin{aligned} u^* &= \min_{u\in L^2(\omega\times(0,T))} F_s(u) \\ F_s(u) &:= \frac{1}{2}\int_0^T \|u(t)\|^2_{L^2(\omega)}\,dt + \frac{1}{2\beta}\mathbb{E}\left[\|y_s(\cdot,T)\|^2_{L^2(-1,1)}\right]. \end{aligned} \tag{5.3}$$

Notice that, since $\mathbb{E}[\cdot]$ is convex, the functional F_s is convex as well.

A classical way to address the minimization problem (5.3) would be to use the Gradient Descent (GD) or the Conjugate Gradient (CG) algorithm. Nevertheless, when applied to parameter-dependent problems, these approaches have a main drawback. Indeed, their implementation requires, in each iteration, to solve the state equation (5.1) and the corresponding adjoint equation for all parameter values. This may rapidly increase the computational cost, especially when the dimension of $\mathcal{K}$ is large.

To bypass this issue, a possible approach is to employ a stochastic algorithm to reduce the number of gradient calculations and, consequently, the total computational complexity. Here we will consider the well-known Stochastic Gradient Descent (SGD) algorithm (see Robbins and Monro, 1951). Our main goal in this section will be to analyze to which extent this stochastic approach may be successfully applied for solving (5.3).

5.2 GD and SGD approaches

In this section, we give an overall description of GD, CG, and SGD for the resolution of (5.3). In particular, we recall the main results about the convergence of these methods and their computational complexity.

5.2.1 The GD approach

Let us start with the well-known GD procedure, consisting in finding the minimizer $\widehat{u}$ in (5.3) as the limit $k \to +\infty$ of the following iterative process

$$u^{k+1} = u^k - \eta_k \nabla F_s(u^k), \tag{5.4}$$

where $\eta_k > 0$ is called the *step-size* or *learning rate*. The selection of a correct learning rate is crucial for the algorithm performances. As a matter of fact, if η_k is not properly chosen, (5.4) may actually not converge to the minimum of F_s. See e.g. Boyd and Vandenberghe (2004, Section 9.2) or Nesterov (2004, Section 2.1.5) for more details.

For what concerns the gradient ∇F_s, this can be easily computed by means of a standard adjoint methodology. In our case, we can readily check that

$$\nabla F_s(u) = u - \frac{1}{\beta}\mathbb{E}[p_s\chi_\omega] = u - \frac{1}{\beta|\mathcal{K}|}\sum_{s\in\mathcal{K}} p_s\chi_\omega,$$

where p_s is the solution of the backward problem

$$\begin{cases} -p_{s,t} + (-\Delta)^s p_s = 0 & \text{in } (-1,1) \times (0,T), \\ p_s = 0 & \text{in } (-1,1)^c \times (0,T), \\ p_s(\cdot,T) = -y_s(\cdot,T) & \text{in } (-1,1). \end{cases} \tag{5.5}$$

Consequently, the GD scheme to solve the optimization problem (5.3) becomes

$$\textbf{GD:} \quad u^{k+1} = u^k - \eta_k \left(u^k - \frac{1}{\beta|\mathcal{K}|} \sum_{s \in \mathcal{K}} p_s^k \chi_\omega \right). \tag{5.6}$$

Hence, applying (5.6) for minimizing $F_s(u)$ requires to solve at each iteration $|\mathcal{K}|$ times the system

$$\begin{cases} y_{s,t} + (-\Delta)^s y_s = u\chi_\omega & \text{in } (-1,1) \times (0,T), \\ -p_{s,t} + (-\Delta)^s p_s = 0 & \text{in } (-1,1) \times (0,T), \\ y_s = p_s = 0 & \text{in } (-1,1)^c \times (0,T), \\ y_s(\cdot,0) = y_0, \; p_s(\cdot,T) = -y_s(\cdot,T) & \text{in } (-1,1). \end{cases} \tag{5.7}$$

Concerning now the convergence rate, since the functional F_s is convex, it is known (see, e.g., Nesterov, 2004, Theorem 2.1.15 or Nocedal and Wright, 2006, Theorem 3.3) that if we take η_k constant small enough we have that

$$\left\| u^k - u^* \right\|^2_{L^2(\omega\times(0,T))} \leq \left\| u^0 - u^* \right\|^2_{L^2(\omega\times(0,T))} e^{-2\mathcal{C}_{GD}k}, \tag{5.8}$$

where the positive constant $\mathcal{C}_{GD}$ is given by

$$\mathcal{C}_{GD} = \ln\left(\frac{\rho+1}{\rho-1}\right), \tag{5.9}$$

ρ denoting the problem's conditioning number.

As for the computational effort, let us denote with $\mathfrak{C}$ the cost of solving (5.7) once. Then, the per-iteration cost of GD will be $\mathfrak{C}|\mathcal{K}|$, due to the need of solving (5.7) for all $s \in \mathcal{K}$ to compute $\nabla F_s(u^k)$. Combining this with (5.8), we then conclude that, for computing the control $\widehat{u}$ up to some given tolerance $\varepsilon > 0$, i.e.

$$\left\| u^k - u^* \right\|^2_{L^2(\omega\times(0,T))} < \varepsilon,$$

the cost of the GD algorithm will be

$$cost_{GD} = \mathcal{O}\left(\frac{\mathfrak{C}|\mathcal{K}|\ln(\varepsilon^{-1})}{\mathcal{C}_{GD}}\right). \tag{5.10}$$

5.2.2 *The CG approach*

Let us now describe the CG approach and comment its convergence properties.

CG is an efficient algorithm to solve linear systems (see Saad, 2003). To apply it for minimizing F_s, the starting point is to notice that the gradient ∇F_s can be written as

$$\nabla F_s(u) = \underbrace{\left(I + \Lambda\right)}_{\mathbb{A}} u + \underbrace{\frac{1}{\beta}\mathbb{E}[q_s \chi_\omega]}_{-b} = \mathbb{A}u - b, \tag{5.11}$$

where

1. The operator Λ is defined as $\Lambda u = \mathbb{E}[p_s \chi_\omega]$ with p_s computed solving

$$\begin{cases} z_{s,t} + (-\Delta)^s z_s = u\chi_\omega & \text{in } (-1,1)\times(0,T), \\ -p_{s,t} + (-\Delta)^s p_s = 0 & \text{in } (-1,1)\times(0,T), \\ z_s = p_s = 0 & \text{in } (-1,1)^c \times (0,T), \\ z_s(\cdot,0) = 0, \;\; p_s(\cdot,T) = z_s(\cdot,T) & \text{in } (-1,1). \end{cases}$$

2. The function q_s is obtained solving the system

$$\begin{cases} \zeta_{s,t} + (-\Delta)^s \zeta_s = 0 & \text{in } (-1,1)\times(0,T), \\ -q_{s,t} + (-\Delta)^s q_s = 0 & \text{in } (-1,1)\times(0,T), \\ \zeta_s = q_s = 0 & \text{in } (-1,1)^c \times (0,T), \\ \zeta_s(\cdot,0) = y_0, \;\; q_s(\cdot,T) = \zeta_s(\cdot,T) & \text{in } (-1,1). \end{cases}$$

Since, clearly, the minimizer $\widehat{u}$ of F_s has to satisfy $\nabla F_s(\widehat{u}) = 0$, we see from (5.11) that computing $\widehat{u}$ is equivalent to solve the linear system $\mathbb{A}u = b$, for which purpose we can use the CG algorithm. Nevertheless, we immediately see from the above discussion that this requires to solve at each iteration the state and adjoint equations $|\mathcal{K}|$ times. Hence, as for GD before, if $|\mathcal{K}|$ is large to employ CG to compute the simultaneous control may become a very demanding task. Concerning now the convergence rate, we know from Saad (2003, Theorem 6.29, Equation 6.128) that

$$\left\|u^k - \widehat{u}\right\|^2_{L^2(\omega\times(0,T)} \leq 4\left\|u^0 - \widehat{u}\right\|^2_{L^2(\omega\times(0,T))} e^{-2\mathcal{C}_{CG}k}, \tag{5.12}$$

where the positive constant $\mathcal{C}_{CG}$ is given by

$$\mathcal{C}_{CG} = \ln\left(\frac{\sqrt{\rho}+1}{\sqrt{\rho}-1}\right). \tag{5.13}$$

From (5.12) we get that, for achieving ε-optimality, the cost of the CG algorithm will be

$$cost_{CG} = \mathcal{O}\left(\frac{\mathfrak{C}|\mathcal{K}|\ln(\varepsilon^{-1})}{\mathcal{C}_{CG}}\right). \tag{5.14}$$

Finally, let us stress that the convergence properties of CG are known to be better than the GD ones. This is due to two main reasons. First, since by definition of conditioning number we have $\rho > 1$, the constant $\mathcal{C}_{CG}$ in (5.13) is larger than $\mathcal{C}_{GD}$ given in (5.9). Hence, even if both GD and CG algorithms converge exponentially, this convergence will actually be faster for CG. Moreover, CG is known to enjoy the so-called *finite termination* property (see, e.g., Glowinski, 2015, Remark 2.4). This means that, when solving a N-dimensional problem, the algorithm will converge in at most N iterations. Practical implementations of CG may partially lose this finite termination property due to round-off errors. Nevertheless, this iterative method still provides monotonically improving approximations to the exact solution, which usually reach the required tolerance after a small (compared to N) number of iterations. See Saad (2003, Section 6.11.3) for more details.

5.2.3 The SGD approach

Let us now describe the SGD algorithm. The main difference with respect to GD and CG is that, in this iterative scheme, we do not employ all the components of $\nabla F_s(u)$. Instead, we pick a parameter s_k i.i.d. from $\mathcal{K}$ (here the sub-index k refers to the k-th iteration of the algorithm) with respect to the uniform probability μ and we use the corresponding gradient as descent direction. Hence, the SGD recursion process is given by

$$u^{k+1} = u^k - \eta_k \nabla F_{s_k}(u^k),$$

where $(\eta_k)_{k\geq 1}$ is a deterministic sequence of positive scalars which we still refer to as the *learning rates sequence*. Moreover, in view of the computations we already presented for GD, the descent direction ∇F_{s_k} is given by

$$\nabla F_{s_k}(u^k) = u^k - \frac{1}{\beta} p^k_{s_k} \chi_\omega,$$

with $p^k_{s_k}$ solution of the adjoint equation (5.5). Hence, the complete SGD scheme to solve the optimization problem (5.3) is given by

$$\textbf{SGD:} \quad u^{k+1} = u^k - \eta_k \left(u^k - \frac{1}{\beta} p^k_{s_k} \chi_\omega\right). \tag{5.15}$$

We then see that applying (5.15) for minimizing the functional $F_s(u)$ requires, at each iteration k, only one resolution of the system (5.7). Because of that, each iteration of this stochastic approach is very cheap.

Concerning now the convergence properties of SGD, some preliminary observations have to be made:

1. First of all, in the SGD method the iterate sequence $(u^k)_{k\geq 1}$ is a stochastic process whose behavior is determined by the random sequence $(s_k)_{k\geq 1} \subset \mathcal{K}$. In particular, this implies that the convergence properties of the algorithm have to be defined in terms of stochastic quantities, namely

$$\mathbb{E}\Big[\Big\| u^k - u^* \Big\|^2_{L^2(\omega\times(0,T))}\Big]$$

(see, e.g., Bach and Moulines, 2011; Bottou et al., 2018), or in the context of *almost sure convergence* (Bottou, 1998).

2. As for the deterministic case, the choice of a good learning rate is crucial for the performances of the algorithm. In the stochastic framework, this is a delicate issue. At this regard, let us stress that choosing a constant learning rate is not a viable option for SGD. Indeed, if SGD is run with a fixed step-size $\eta_k = \overline{\eta}$, even if $\overline{\eta}$ is small we may not reach convergence (see Bottou et al., 2018, Theorem 4.6). This is essentially due to the noise introduced by the randomness of the stochastic process defined by (5.15), as it has been exhaustively discussed for instance in Bottou et al. (2018, Section 4.2). Because of this noise, the convergence of SGD is guaranteed as long one is able to maintain the second moment $\mathbb{E}\Big[\big\|\nabla F_{s_k}(u^k)\big\|^2\Big]$ bounded above by a deterministic quantity, namely

$$\mathbb{E}\left[\Big\|\nabla F_{s_k}(u^k)\Big\|^2\right] \leq \sigma^2, \quad 0 < \sigma \in \mathbb{R}.$$

This may be achieved by properly reducing the value of η_k at each iteration. See, for instance, Bach and Moulines (2011); Bottou et al. (2018); Robbins and Monro (1951) for a complete discussion on this issue.

3. If the learning rate is properly chosen, by means of martingale techniques (see Bottou, 1998, Section 4.5) we can show that SGD converges almost surely

$$u^k \xrightarrow{a.s} \widehat{u}, \qquad \text{as } k \to +\infty.$$

In practice, this means that for solving (5.3) it is enough to run the SGD algorithm only once and we will have probability one of converging to the minimum u^*.

As for the convergence rate of SGD, if we choose the steps-size η_k reducing as $\eta_k = k^{-\alpha}$, $\alpha \in (0,1)$, it has been proved in Bach and Moulines (2011, Theorem 1) that

$$\mathbb{E}\left[\Big\|u^k - \widehat{u}\Big\|^2_{L^2(\omega\times(0,T))}\right] = \mathcal{O}\left(\sigma^2 k^{-\alpha}\right). \tag{5.16}$$

We then immediately understand the importance of keeping σ (which, we recall, is associated with the noise of the stochastic process) small, in order to have better convergence behavior. Moreover, confronting (5.16) with (5.8) and (5.12), we see that SGD converges slower than GD and CG to the minimum u^*. Nevertheless, it is crucial to recall that, at each iteration k, SGD only requires to approximate the dynamics (5.7) once instead of $|\mathcal{K}|$ times, as it was for the deterministic algorithms. Hence we can conclude that for computing the control $\widehat{u}$ up to some given tolerance $\varepsilon > 0$, i.e.

$$\mathbb{E}\left[\left\|u^k - u^*\right\|^2_{L^2(\omega\times(0,T))}\right] < \varepsilon,$$

the cost of the SGD algorithm with step-size $\eta_k = k^{-\alpha}$, $\alpha \in (0, 1)$, will be

$$cost_{SGD} = \mathcal{O}\left(\mathfrak{C}\sigma^2\varepsilon^{-1}\right). \tag{5.17}$$

Of course, (5.17) is larger than (5.10) and (5.14) for $|\mathcal{K}|$ small, while the comparison favors SGD when $|\mathcal{K}|$ is large. This suggests that, when $|\mathcal{K}|$ is small, GD and CG are expected to perform better than SGD due to the lower amount of iterations they require to converge. On the other hand, in case of a large parameter set for which a single iteration of GD and CG becomes very costly, a stochastic approach will be more efficient than a deterministic one for solving (5.3). Our numerical experiments in Section 5.4 will confirm this behavior.

As a final consideration, let us mention that the scheme (5.15) is the most basic version of a SGD algorithm, as it was introduced in the original paper Robbins and Monro (1951). In recent years, more sophisticated SGD-based algorithms have been proposed, to reduce the noise in the stochastic process and therefore achieve better convergence and stability properties. One of the most widely used nowadays is the so-called *Adam* scheme (see Kingma and Ba, 2015), which combines the momentum approach proposed in Polyak (1964) with a proper reduction of the step-size. In more detail, the Adam scheme consists in the following steps:

$$\begin{aligned} u^{k+1} &= u^k - \frac{\eta}{\sqrt{\widetilde{v}^k}+\delta}\widetilde{m}^k \\ \widetilde{m}^k &= \frac{m^k}{1-\gamma_1}, \qquad m^k = \gamma_1 m^{k-1} + (1-\gamma_1)\nabla F_{s_k}(u^k) \\ \widetilde{v}^k &= \frac{v^k}{1-\gamma_2}, \qquad v^k = \gamma_1 v^{k-1} + (1-\gamma_1)|\nabla F_{s_k}(u^k)|^2, \end{aligned} \tag{5.18}$$

where η, γ_1 and γ_2 are suitable chosen parameters and $0 < \delta \ll 1$ is introduced to avoid division by zero. This is the scheme that we will use for the simulations in Section 5.4.

5.3 Practical considerations on the implementation of GD and CG

In (5.10) and (5.14), we gave convergence rates for GD and CG in terms of the explicit constants $\mathcal{C}_{GD}$ and $\mathcal{C}_{CG}$ (see (5.9) and (5.13)), which depend on the conditioning ρ of the problem we are solving. Nevertheless, if we analyze the behavior of these constants with respect to ρ, since by definition $\rho > 1$, we immediately notice that both $\mathcal{C}_{GD}$ and $\mathcal{C}_{CG}$ are positive decreasing functions of ρ and they converge to zero as $\rho \to +\infty$. This implies that a bad conditioning in a minimization problem affects the actual convergence of GD and CG, which may deteriorate and violate (5.8) and (5.12).

This is a well-known computational limitation of gradient optimization methods. In particular, the GD algorithm is very sensitive to the problem conditioning and, if ρ is large, the convergence properties may deteriorate up to a linear rate. An illustrative example of this phenomenon is provided in Meza (2010).

The situation is less critical for CG, because of the considerations we presented in the previous section. In particular, CG is less sensible to the conditioning of the problem since the constant $\mathcal{C}_{CG}$ depends on $\sqrt{\rho}$ instead of ρ (see (5.13)). Furthermore, recall that CG enjoys the finite termination property, which helps in achieving convergence in a relatively small number of iterations.

In addition to that, let us mention that for a control problem ρ is typically very large (see Boyer, 2013, Remark 4.2 for some explicit estimates in the CG context). As a consequence of this bad conditioning, and according to the discussion above, we will see in our numerical simulations that the convergence rate of GD is significantly reduced with respect to the expected one given by (5.8), thus making this algorithm not very efficient in the context of our simultaneous control problem.

As a final remark, we stress that the considerations of this subsection concerning the conditioning of gradient methodologies apply also to SGD. Nevertheless, in this case, other key factors such as for instance the noise σ come into play, thus making more delicate the precise quantification of the problem's conditioning and of its influence on the overall optimization problem.

5.4 Numerical experiments

This section is devoted to some numerical experiments. The main goal is to confirm our previous discussion by comparing GD, CG, and SGD for the simultaneous control of the linear parameter-dependent model (5.1).

We have chosen the initial state $y_0 = \sin(\pi x)$, the time horizon $T = 0.4$ and the control region $\omega = (-0.5, 0.8)$. The parameter set $\mathcal{K}$ is a $|\mathcal{K}|$-points discretization of the interval $(0.6, 0.9)$, in which we know that each fractional heat equation (5.1) is null-controllable. The tolerance is set to $\varepsilon = 10^{-4}$. To test the efficiency of each algorithm, we have performed simulations for increasing values of $|\mathcal{K}|$. For SGD, we have implemented the Adam scheme (5.18) with $\eta = 10^{-3}$, $\gamma_1 = 0.9$, $\gamma_2 = 0.999$, and $\delta = 10^{-8}$.

Before comparing the performances of the three algorithms, let us show that the optimization problem (5.3) indeed provides an effective simultaneous control for (5.1). To this end, in Fig. 19, we display the final state $y_s(\cdot, T)$ for the free (that is, when $u = 0$) and controlled dynamics associated to (5.1). In order to increase the visibility of our plots, we consider here only the case of $|\mathcal{K}| = 10$ parameters in our system. Moreover, the simulations displayed in Fig. 19 have been performed with the CG algorithm, although the other two approaches that we described provide the same result.

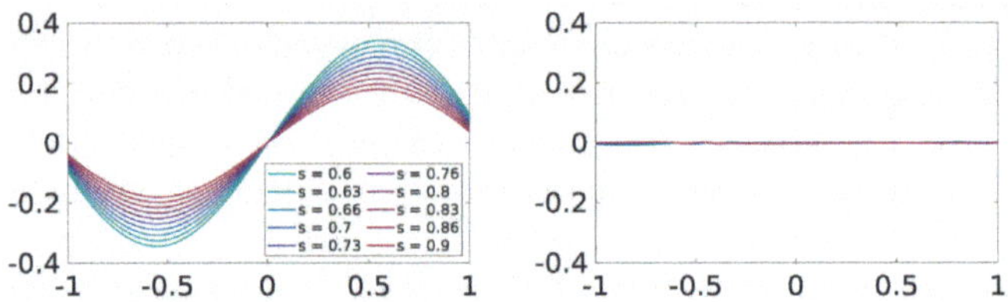

FIGURE 19 Final state $y_s(\cdot, T)$ of the free (left) and controlled (right) dynamics associated to (5.1) for different values of $s \in (0.6, 0.9)$.

We see how, while the free dynamics simply dissipates under the action of the heat semi-group without reaching zero at the final time, introducing a control allows us to steer all the realizations of (5.1) to zero at time T. For completeness, we shall stress that in the experiment displayed in Fig. 19 the initial datum y_0 is the same for all the realization of the parameter s. In principle, parameter-dependent initial condition could also be considered, and the sensitivity of the computed results with respect to these initial conditions could be analyzed. This kind of questions, although interesting, are beyond the interest of the present work and will not be discussed here.

Let us now analyze and discuss the behavior of GD, CG, and SGD with respect to the amount of parameters included in our model. To this end, we have run simulations for increasing values of the cardinality of $\mathcal{K}$, namely $|\mathcal{K}| = 2, 10, 100, 250, 500$. The results of our numerical experiments are collected in Table 1 and displayed in Fig. 20.

We notice that the number of iterations for GD and CG remains stable with respect to $|\mathcal{K}|$, while it grows for SGD. We believe that this is due to the noise introduced by the stochastic approach, although we do not have a rigorous mathematical justification of this phenomenon.

Nevertheless, despite this fact, we can see that our simulations confirm the behavior we expected from the discussion in Section 5.2. In particular, we can make the following observations:

1. GD is the worst performing algorithm. For $|\mathcal{K}| = 2$, its computational time is comparable with SGD. Nevertheless, already for $|\mathcal{K}| = 10$, its computational effort becomes considerable. This is a consequence of the bad conditioning of control problems, as we commented in Section 5.3.

TABLE 1 Number of iterations and computational time to converge to $\varepsilon = 10^{-4}$ for GD, CG and SGD applied to (5.3) with increasing values of $|\mathcal{K}|$.

	GD		CG		SGD	
$\lvert\mathcal{K}\rvert$	Iter.	Time (sec)	Iter.	Time (sec)	Iter.	Time (sec)
2	1424	11.5	35	0.4	3563	16.2
10	1363	51.5	25	1.3	3987	19.4
100	1343	507.6	25	11.8	4110	20.1
250	1341	1169.3	25	28.3	4019	18.6
500	1341	2659.9	25	55.39	4485	21.4

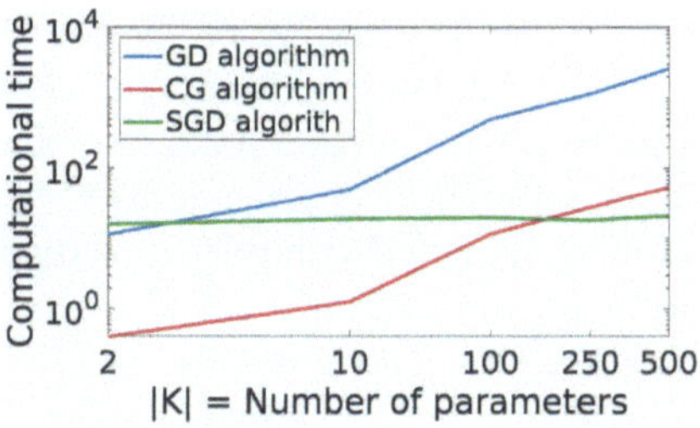

FIGURE 20 Computational time in logarithmic scale to converge to $\varepsilon = 10^{-4}$ for GD, CG, and SGD applied to (5.3) with increasing values of $|\mathcal{K}|$.

2. CG is the algorithm requiring the lowest number of iterations to converge. On the one hand, this confirms that CG is less sensible to the conditioning of the problem. On the other hand, this implies that CG is the best approach when dealing with a low and moderate amount of parameters, since when $|\mathcal{K}|$ is not too large the algorithm is capable to compensate the per-iteration cost with the very limited amount of iterations it requires to achieve ε-optimality.
3. SGD is the algorithm requiring the highest numbers of iterations to converge. This is in line with the fact that, for this algorithm, only a linear or sub-linear convergence rate is expected (see (5.16)), while for GD and CG this rate is expected to be exponential (see (5.8) and (5.12)). Notwithstanding that, SGD appears to be insensitive to the cardinality of $\mathcal{K}$. This is not surprising if we consider that, no matter how many parameters enter in our control problem, with SGD each iteration of the optimization process always requires only one resolution of the coupled system (5.7). Due to this low per-iteration cost, despite of the slower convergence rate, when $|\mathcal{K}|$ is large the total computational time for SGD is actually smaller than GD and CG. As a matter of fact, we can see in Fig. 20 that for a parameter set of cardinality around $|\mathcal{K}| = 150$ SGD starts outperforming CG.

All these considerations are aligned with our previous discussion and corroborate the fact that, when $|\mathcal{K}|$ is large, the SGD approach is preferable to the GD and CG ones to address the simultaneous control of (5.1).

6 Conclusion and open problems

In this chapter, we have addressed several aspects of the numerical control of non-local diffusive models involving the fractional Laplacian. Our discussion embraced a wide spectrum of situations: we considered both interior and exterior control problems, possibly under positivity constraints, as well as the relevant issue of simultaneous control. After having recalled the existing results in this fractional diffusive setting, we have shown how to compute the numerical controls by combining FE approximations of the fractional Laplacian with efficient optimization tools. Our simulations, on the one hand, have provided a numerical representation and validation of many known control properties for fractional diffusive processes. On the other hand, they have highlighted some analogies and differences between local and non-local heat-like models when facing the approximation of numerical controls. Nevertheless, many key questions related to our work remain currently unaddressed. We present below a short collection of them, which may be of interest for future investigation.

1. **Control of fully-discrete fractional problems.** Despite of the numerical evidences we presented in Sections 3 and 4 for the controllability properties of fully-discrete fractional heat equations, the complete analysis and proofs of these results are open. As a matter of fact, the controllability of fully-discrete problems has been considered only in a few research works and many important issues still remain unanswered. In Ervedoza and Valein (2010), the authors proved in a quite general framework that any parabolic equation is null-controllable after time discretization by applying an adequate filtering of high frequencies, under the condition that the space semi-discrete approximation schemes are uniformly observable with respect to the mesh size parameters. We remark that, in the context of the discrete Laplace operator, the uniform controllability of semi-discretized solutions has been considered in López and Zuazua (1998); Zuazua (2006) by means of spectral techniques and a careful analysis of the eigenvalues. Notice, however, that this analysis is not extendable to the discrete fractional Laplacian, due to the lack of an explicit knowledge of the spectrum of this operator. On the other hand, in Boyer et al. (2011), the controllability of fully-discrete finite-difference heat equations has been studied by adopting a Lebeau-Robbiano strategy, based on discrete spectral inequalities. Finally, in Hernández-Santamaría and Casanova (2020), similar results have been obtained by means of a Carleman approach. To the best of the authors' knowledge, there are no results currently available apart from the aforementioned ones. In particular, the study of control properties for the fully-discrete approximation of (3.1) is still completely open.

2. **Exterior control of the fractional heat equation.** In Section 4, we considered the exterior control problem for the fractional heat equation and discussed its theoretical and numerical aspects. Nevertheless, several important issues are still only partially understood. On the one hand, from the theoretical viewpoint, the only existing controllability results for (4.1) have been obtained in space dimension $d = 1$, while the multi-dimensional problem is completely open. At this regard, it would be interesting to analyze whether the techniques we recently developed in Biccari et al. (2021) for the interior control problem in multi-D are applicable also in the exterior control setting or, instead, new methodologies need to be derived. On the other hand, also the numerical control of (4.1) requires further investigation. In Section 4.1.1, we have proposed to compute the exterior controls passing through the Robin problem (4.5) since, as we discussed throughout this chapter, the numerical approximation of the exterior Dirichlet problem is still not completely clarified. Our simulations have shown the efficacy of this approach. Nevertheless, to completely justify this procedure, we shall rigorously analyze the control properties of the fully-discrete approximation of (4.5), in the same spirit of the discussion in point 1 above. Finally, also the Fenchel-Rockafellar duality has not been developed yet int the context of the exterior control problem, and it would be considered in a future work.
3. **Simultaneous control of fractional PDE.** In Section 5, we addressed the simultaneous control of parameter-dependent fractional heat equations and we proposed the use of stochastic optimization techniques for an efficient computation of numerical controls. Our discussion focused mainly on the implementation details, and the existence of a simultaneous control for the problem we considered has been addressed only at a numerical level. The reason behind this choice is that, at present time, the theory for simultaneous control of PDE is not sufficiently developed to tackle fractional models. As a matter of fact, existing techniques available in the local setting do not seem to be applicable in the non-local one. Just to give an example, in Zuazua (2016), we have proposed a methodology for the stable observation of additive superpositions of heat and wave equations which is applicable to the simultaneous control of such models. This methodology is rather systematic and easy to apply. It consists in observing only one component of the system and considering the others as unknown perturbations. One then composes all but one (the observable one) PDE operators so to reduce the problem to the consideration of a single equation and applying its known observability properties. Notwithstanding that, this approach is hardly extendable in the context of the fractional heat equation we have considered, since the commutation of different powers of the fractional Laplacian leads to non-local lower order perturbations which cannot be handled in terms of observability. It would then be of interest to provide a solid mathematical background to simultaneous control for non-local and fractional models, supporting the numerical evidences we displayed in Section 5. In addition to that, it would

be worth to consider the extension of Fenchel-Rockafellar duality in this simultaneous control framework.

4. **Wave-type equations.** Our discussion has focused only on the parabolic setting. Nevertheless, in recent years, several results have been obtained on the controllability properties of hyperbolic (wave) and dispersive (Schrödinger) models involving the fractional Laplacian. The interested reader may refer, for instance, to Biccari (2021); Biccari and Warma (2020); Warma and Zamorano (2020a). As for the numerical approximation of wave-type models, this issue is known to be quite delicate. As a matter of fact, already in the local framework, numerical high-frequency solutions of the wave equation can exhibit pathological behaviors such as lack of propagation in space or the so-called *rodeo effect*, i.e. waves that are trapped by the numerical grid in closed loops (see, e.g., Biccari et al., 2020a; Zuazua, 2005). When dealing with control and inversion problems, these behaviors then yield to the necessity of filtering high-frequency numerical components, to cope with the loss of uniform observability properties through numerical discretization. To the authors' knowledge, a rigorous analysis of how these mentioned pathologies transfer to the non-local setting of the fractional Laplacian has yet to be developed, and would be a very interesting question to be considered.
5. **Variable-order fractional Laplacian.** In some recent contributions (Antil and Rautenberg, 2019; Bahrouni and RĂdulescu, 2018; Rahmoune and Biccari, 2021; Xiang et al., 2019), elliptic problems involving a variable-order fractional Laplacian (that is, with $s = s(x) : \Omega \to (0, 1)$) have been considered, analyzing the existence of (possible several) solutions and some optimal control issues applied to image denoising. It would be of interest to investigate parabolic problems (which, as far as we can tell, are still unaddressed) and associated theoretical and numerical control problems.
6. **Memory-type equations.** Evolution equations involving memory terms appear in several different applications, for modeling natural and social phenomena which, apart from their current state, are influenced also by their history. Some classical examples are viscoelasticity or non-Fickian diffusion. In recent years, this class of models has got the attention of the control community (see for instance Biccari and Micu, 2019; Chaves-Silva et al., 2017; Lü et al., 2017, or Biccari and Warma, 2020 for models involving the fractional Laplacian). In particular, it has been observed that these models can be cast as coupled PDE-ODE systems, in which the ODE component introduces non-propagation effects similar to those produced by a low-order ($s \leq 1/2$) fractional Laplacian. This motivated the introduction of a *moving control* strategy to obtain positive controllability results. An illustrative example of this approach can be found in Oroya and Biccari (2020). Nevertheless, as far as the authors know, the controllability of memory-type equation has never been considered under a numerical perspective. In particular, the implementation of the moving control strategy in the hyperbolic setting, which already

requires a special treatment of high-frequency solutions (see Point 4 above) would be an important issue to be further investigated.

7. **Non-linear fractional control** The control of non-linear dynamical systems is most often a very difficult task, since a non-linear dynamics, in contrast with much simpler linear systems, may appear chaotic, unpredictable, or counter-intuitive. In the last decades, different techniques have been developed to control non-linear models, such as the employment of linearization and fixed-point arguments, or the so-called return method (see Coron, 2007). Nevertheless, these techniques are most of the time of difficult application to fractional models, whose complex and pathological behaviors are typically enhanced in a non-linear framework. As for the numerical implementation of the controls, also this is a very difficult question. In some simple situation (see e.g. Boyer, 2013, Section 4.3.2) the numerical control problem for semi-linear models has been tackled by combining the HUM approach we presented in Section 3.2 with some fixed-point strategy. However, the extension to general non-linear (possibly fractional) problems is, to the best of our knowledge, still missing. All this renders non-linear fractional control an almost unexplored yet very interesting field within control theory.

Acknowledgments

This project has received funding from the European Research Council (ERC) under the European Union's Horizon 2020 research and innovation programme (grant agreement NO: 694126-DyCon). The work of the three authors is supported by the Air Force Office of Scientific Research (AFOSR) under Award NO: FA9550-18-1-0242. The work of MW is supported by the US Army Research Office (ARO) under Award NO: W911NF-20-1-0115. The work of U.B. and E.Z. is supported by Grant MTM2017-92996-C2-1-R COSNET of MINECO (Spain) and by the Elkartek grant KK-2020/00091 CONVADP of the Basque government. The work of E.Z. is funded by the Alexander von Humboldt-Professorship program, the European Unions Horizon 2020 research and innovation programme under the Marie Sklodowska-Curie grant agreement No.765579-ConFlex and the Transregio 154 Project "Mathematical Modelling, Simulation and Optimization Using the Example of Gas Networks", project C08, of the German DFG.

Appendix A Fractional order Sobolev spaces and the fractional Laplacian

In this Appendix, we introduce the appropriate functional framework to work with the fractional Laplacian and present some technical results.

We start by giving a rigorous definition of the fractional Laplace operator. To this end, for any $0 < s < 1$, we consider the space

$$\mathcal{L}^s(\mathbb{R}^d) := \left\{ u : \mathbb{R}^d \to \mathbb{R} \text{ measurable} : \int_\Omega \frac{|u(x)|}{(1+|x|)^{d+2s}}\, dx < +\infty \right\}.$$

For $u \in \mathcal{L}^s(\mathbb{R}^d)$ and $\varepsilon > 0$, we let

$$(-\Delta)^s_\varepsilon u(x) := C_{d,s} \int_{\{y \in \mathbb{R}^d,\ |x-y| \geq \varepsilon\}} \frac{u(x)-u(y)}{|x-y|^{d+2s}}\, dy, \ \ x \in \mathbb{R}^d,$$

where the normalization constant is given by

$$C_{d,s} := \frac{s 2^{2s} \Gamma\left(\frac{2s+d}{2}\right)}{\pi^{\frac{d}{2}} \Gamma(1-s)},$$

and Γ is the Euler Gamma function. The fractional Laplace operator $(-\Delta)^s$ is then defined for every $u \in \mathcal{L}^s(\mathbb{R}^d)$ by the formula

$$(-\Delta)^s u(x) := C_{d,s} \text{P.V.} \int_{\mathbb{R}^d} \frac{u(x)-u(y)}{|x-y|^{d+2s}}\, dy = \lim_{\varepsilon \downarrow 0} (-\Delta)^s_\varepsilon u(x), \quad x \in \mathbb{R}^d, \tag{A.1}$$

provided that the limit exists for a.e. $x \in \mathbb{R}^d$.

Let us now introduce the appropriate function spaces needed to work with the fractional Laplacian, that is, the fractional order Sobolev spaces. In what follows, we will only provide the definitions and some relevant properties. More complete presentations can be found in several references, including but not limited to Adams and Fournier (2003); Di Nezza et al. (2012); Grisvard (2011); Lions and Magenes (1968); Warma (2015).

Let $\Omega \subset \mathbb{R}^d$ $(d \geq 1)$ be a bounded open set. For any $0 < s < 1$, we define the fractional order Sobolev space

$$H^s(\Omega) := \left\{ u \in L^2(\Omega) : \int_\Omega \int_\Omega \frac{|u(x)-u(y)|^2}{|x-y|^{d+2s}}\, dx dy < +\infty \right\}$$

and we endow it with the norm given by

$$\|u\|_{H^s(\Omega)} := \left(\int_\Omega |u|^2\, dx + \int_\Omega \int_\Omega \frac{|u(x)-u(y)|^2}{|x-y|^{d+2s}}\, dx dy \right)^{\frac{1}{2}}.$$

We let

$$H^s_0(\Omega) := \left\{ u \in H^s(\mathbb{R}^d) :\ u = 0 \ \text{ in } \Omega^c \right\} = \left\{ u \in H^s(\mathbb{R}^d) : \text{supp}[u] \subset \overline{\Omega} \right\}$$

and notice that if $0 < s \neq 1/2 < 1$, and $\partial\Omega$ is Lipschitz, then by Grisvard (2011, Chapter 1),

$$H^s_0(\Omega) = \overline{\mathcal{D}(\Omega)}^{H^s(\Omega)},$$

where $\mathcal{D}(\Omega)$ denotes the space of all continuous infinitely differentiable functions with compact support in Ω. In that case, we endow $H_0^s(\Omega)$ with the $H^s(\Omega)$-norm. Otherwise one uses the norm of $H^s(\mathbb{R}^d)$.

We denote by $H^{-s}(\Omega) := (H_0^s(\Omega))^\star$ the dual space of $H_0^s(\Omega)$ with respect to the pivot space $L^2(\Omega)$ so that the following continuous embeddings hold:

$$H_0^s(\Omega) \hookrightarrow L^2(\Omega) \hookrightarrow H^{-s}(\Omega).$$

Besides, let $\langle \cdot, \cdot \rangle_{-s,s}$ be the duality pairing between $H^{-s}(\Omega)$ and $H_0^s(\Omega)$. Finally, we denote by $H_{\text{loc}}^s(\Omega)$ the space defined by

$$H_{\text{loc}}^s(\Omega) := \Big\{ u \in L_{\text{loc}}^2(\Omega) \; : \; uv \in H^s(\Omega) \text{ for all } v \in \mathcal{D}(\Omega) \Big\}.$$

Let us now introduce the following Dirichlet problem associated with the fractional Laplace operator:

$$\begin{cases} (-\Delta)^s u = f & \text{in } \Omega, \\ u = 0 & \text{in } \Omega^c. \end{cases} \tag{A.2}$$

We have the following definition of weak solutions.

Definition A.1. Let $f \in H^{-s}(\Omega)$. A function $u \in H_0^s(\Omega)$ is said to be a weak solution of the Dirichlet problem (A.2) if the equality

$$\frac{C_{d,s}}{2} \int_{\mathbb{R}^d} \int_{\mathbb{R}^d} \frac{(u(x) - u(y))(v(x) - v(y))}{|x - y|^{d+2s}} \, dxdy = \langle f, v \rangle_{-s,s},$$

holds for every $v \in H_0^s(\Omega)$.

The existence and uniqueness of weak solutions to (A.2) is a direct consequence of the classical Lax-Milgram Theorem. In particular, we have the following result (see e.g. Biccari et al., 2017, Proposition 2.1 or Leonori et al., 2015, Theorem 12).

Proposition A.2. *Let $\Omega \subset \mathbb{R}^d$ be a bounded open set and $0 < s < 1$. Then for every $f \in H^{-s}(\Omega)$, the Dirichlet problem* (A.2) *has a unique weak solution $u \in H_0^s(\Omega)$ in the sense of Definition A.1. In addition, there is a constant $C > 0$ such that*

$$\|u\|_{H_0^s(\Omega)} \leq C \|f\|_{H^{-s}(\Omega)}.$$

We have the following maximal local regularity result for weak solutions of (A.2), which has been proved in Biccari et al. (2017, Theorem 1.3) (see also Grubb, 2015).

Proposition A.3. *Let $\Omega \subset \mathbb{R}^d$ be a bounded open set and $0 < s < 1$. Let $f \in L^2(\Omega)$ and $u \in H_0^s(\Omega)$ the unique weak solution of* (A.2) *in the sense of Definition A.1. Then $u \in H_{\text{loc}}^{2s}(\Omega)$.*

Next, we introduce the realization in $L^2(\Omega)$ of the fractional Laplacian with the zero Dirichlet exterior condition, that is, the operator

$$D((-\Delta)^s_D) := \left\{u \in H^s_0(\Omega) : ((-\Delta)^s u)|_\Omega \in L^2(\Omega)\right\}$$
$$(-\Delta)^s_D u = ((-\Delta)^s u)|_\Omega \text{ a.e. in } \Omega. \tag{A.3}$$

It is well-known (see, e.g., Claus and Warma, 2020) that $-(-\Delta)^s_D$ generates a strongly continuous sub-markovian semi-group on $L^2(\Omega)$. Moreover, it has been shown in Servadei and Valdinoci (2014) that $(-\Delta)^s_D$ has a compact resolvent, hence, it has a discrete spectrum which is formed with eigenvalues $(\lambda_j)_{j\in\mathbb{N}}$ satisfying

$$0 < \lambda_1 \leq \lambda_2 \leq \cdots \leq \lambda_j \leq \cdots \text{ and } \lim_{j\to+\infty} \lambda_j = +\infty.$$

We denote by $(\phi_j)_{j\in\mathbb{N}}$ the normalized eigenfunctions associated with the eigenvalues $(\lambda_j)_{j\in\mathbb{N}}$, i.e., the solutions of the Dirichlet problem

$$\begin{cases} (-\Delta)^s_D \phi_j = \lambda_j \phi_j & \text{in } \Omega, \\ \phi_j = 0 & \text{in } \Omega^c. \end{cases}$$

Finally, let us give our notion of solutions to the heat equation associated with the fractional Laplacian.

Definition A.4. Let $y_0 \in L^2(\Omega)$, $f \in L^2((0,T); H^{-s}(\Omega))$, and consider the following parabolic system

$$\begin{cases} y_t + (-\Delta)^s y = f & \text{in } \Omega \times (0,T), \\ y \equiv 0 & \text{in } \Omega^c \times (0,T), \\ y(\cdot,0) = y_0 & \text{in } \Omega. \end{cases} \tag{A.4}$$

We say that $y \in C([0,T]; L^2(\Omega)) \cap L^2((0,T); H^s_0(\Omega)) \cap H^1((0,T); H^{-s}(\Omega))$ is a finite energy solution to (A.4) if $y(\cdot,0) = y_0$ a.e. in Ω and the identity

$$\int_0^T \langle f, w\rangle_{-s,s}\, dt = \int_0^T \langle y_t, w\rangle_{-s,s}\, dt + \frac{C_{d,s}}{2}\int_0^T \int_{\mathbb{R}^d}\int_{\mathbb{R}^d} \frac{(y(x,t)-y(z,t))(w(x)-w(z))}{|x-z|^{d+2s}}\, dxdzdt$$

holds, for every $w \in H^s_0(\Omega)$.

We have the following well-posedness result (see Brezis, 2010, Chapter 10, Theorem 10.9 or Leonori et al., 2015, Theorem 26).

Proposition A.5. *Assume that $f \in L^2((0,T); H^{-s}(\Omega))$. Then, for every initial datum $y_0 \in L^2(\Omega)$ the fractional heat equation* (A.4) *has a unique finite energy solution y given by*

$$y(\cdot,t) = e^{-(-\Delta)_D^s t} y_0 + \int_0^t e^{-(-\Delta)_D^s (t-\tau)} f(\cdot,\tau)\, d\tau,$$

where $e^{-(-\Delta)_D^s t}$, $t \geq 0$, is the strongly continuous semi-group on $H^{-s}(\Omega)$ generated by the operator $-(-\Delta)_D^s$.

Appendix B The fractional Laplace operator with exterior conditions

We present here a general overview of elliptic and parabolic problems associated with the fractional Laplace operator on a bounded domain $\Omega \subset \mathbb{R}^d$ with non-zero exterior data. We first consider the following elliptic Dirichlet problem:

$$\begin{cases} (-\Delta)^s u = f & \text{in } \Omega \\ u = g & \text{in } \Omega^c. \end{cases} \tag{B.1}$$

Let us define our notion of solutions for (B.1). To this end, we need first to introduce the non-local normal derivative $\mathcal{N}_s$, defined for all $u \in H^s(\mathbb{R}^d)$ as

$$\mathcal{N}_s u(x) := C_{d,s} \int_\Omega \frac{u(x) - u(y)}{|x-y|^{d+2s}}\, dy, \quad x \in \mathbb{R}^d \setminus \overline{\Omega}. \tag{B.2}$$

Clearly, $\mathcal{N}_s$ is a non-local operator. Moreover, it is well defined on $H^s(\mathbb{R}^d)$ as the following result shows (see Ghosh et al., 2020, Lemma A.2 for the proof).

Lemma B.1. *The non-local normal derivative $\mathcal{N}_s$ maps $H^s(\mathbb{R}^d)$ continuously into $H^s_{\text{loc}}(\Omega^c) \subset L^2_{\text{loc}}(\Omega^c)$.*

Even if $\mathcal{N}_s$ is defined on the unbounded domain $\mathbb{R}^d \setminus \overline{\Omega}$, it is still known as the *normal* derivative. This is due to its similarity with the classical normal derivative as the following result taken from Dipierro et al. (2017) shows.

Proposition B.2. *The following assertions hold.*

1. ***Divergence theorem.*** *Let $u \in C^2(\mathbb{R}^d)$ vanishing at $\pm\infty$. Then*

$$\int_\Omega (-\Delta)^s u\, dx = -\int_{\Omega^c} \mathcal{N}_s u\, dx.$$

2. ***Integration by parts formula.*** *Let $u \in H^s(\mathbb{R}^d)$ be such that $(-\Delta)^s u \in L^2(\Omega)$ and $\mathcal{N}_s u \in L^2(\Omega^c)$. Then, for every $v \in H^s(\mathbb{R}^d)$ we have*

$$\int_\Omega v(-\Delta)^s u\, dx = \frac{C_{d,s}}{2} \int\int_{\mathbb{R}^{2d} \setminus (\Omega^c)^2} \frac{(u(x)-u(y))(v(x)-v(y))}{|x-y|^{d+2s}}\, dxdy$$

$$-\int_{\Omega^c} v\mathcal{N}_s u\,dx,$$

where $\mathbb{R}^{2d} \setminus (\Omega^c)^2 = (\Omega \times \Omega) \cup (\Omega \times \Omega^c) \cup (\Omega^c \times \Omega)$.

3. ***Limit as*** $s \uparrow 1^-$. *Let* $u, v \in C^2(\mathbb{R}^d)$ *vanishing at* $\pm\infty$. *Then*

$$\lim_{s\uparrow 1^-} \int_{\Omega^c} v\mathcal{N}_s u\,dx = \int_{\partial\Omega} v\frac{\partial u}{\partial \nu}\,d\sigma.$$

We are now ready to introduce the notion of transposition solutions to the Dirichlet problem (B.1).

Definition B.3. Let $g \in L^2(\Omega^c)$ and $f \in H^{-s}(\overline{\Omega})$. A function $u \in L^2(\mathbb{R}^d)$ is said to be a solution by transposition to (B.1) if the identity

$$\int_{\Omega} u(-\Delta)^s v\,dx = \langle f, v\rangle_{-s,s} - \int_{\Omega^c} g\mathcal{N}_s v\,dx, \tag{B.3}$$

holds for every $v \in V := \left\{v \in H_0^s(\Omega) \,:\, (-\Delta)^s v \in L^2(\Omega)\right\}$.

Moreover, we have the following existence and uniqueness result of solutions by transposition. We refer to Antil et al. (2019, Theorem 3.5) for the proof.

Theorem B.4. *Let* $f \in H^{-s}(\Omega)$ *and* $g \in L^2(\Omega^c)$. *Then,* (B.1) *has a unique solution by transposition* $u \in L^2(\mathbb{R}^d)$, *and there is a constant* $C > 0$ *such that*

$$\|u\|_{L^2(\mathbb{R}^N)} \le C\left(\|f\|_{H^{-s}(\overline{\Omega})} + \|g\|_{L^2(\Omega^c)}\right).$$

We now move to the parabolic problem associated to the fractional Laplacian with exterior condition.

$$\begin{cases} y_t + (-\Delta)^s y = 0 & \text{in } \Omega \times (0,T) \\ y = g & \text{in } \Omega^c \times (0,T) \\ y(\cdot,0) = y_0 & \text{in } \Omega. \end{cases} \tag{B.4}$$

Let $y_0 \in L^2(\Omega)$, $g \in L^2((0,T); H^s(\Omega^c))$ and consider the following two systems:

$$\begin{cases} \xi_t + (-\Delta)^s \xi = 0 & \text{in } \Omega \times (0,T) \\ \xi = 0 & \text{in } \Omega^c \times (0,T) \\ \xi(\cdot,0) = y_0 & \text{in } \Omega, \end{cases} \tag{B.5}$$

and

$$\begin{cases} z_t + (-\Delta)^s z = 0 & \text{in } \Omega \times (0,T) \\ z = g & \text{in } \Omega^c \times (0,T) \\ z(\cdot,0) = 0 & \text{in } \Omega. \end{cases} \tag{B.6}$$

Then, the solution of (B.4) is given by $y = \xi + z$. Moreover, noticing that (B.5) can be cast as a Cauchy problem for the operator $(-\Delta)^s_D$ we introduced in (A.3), using semi-group theory and the spectral theorem, one has the following result.

Theorem B.5. *Let $(\phi_k)_{k\in\mathbb{N}}$ be the normalized eigenfunctions of the operator $(-\Delta)^s_D$ associated with the eigenvalues $(\lambda_k)_{k\in\mathbb{N}}$. For every $y_0 \in L^2(\Omega)$, define $y_{0,k} := \langle y_0, \phi_k\rangle_{L^2(\Omega)}$. Then, there is a unique function*

$$\xi \in C([0,T]; L^2(\Omega)) \cap L^2((0,T); H^s_0(\Omega)) \cap H^1((0,T); H^{-s}(\Omega))$$

satisfying (B.5) *which is given for a.e. $x \in \Omega$ and every $t \in [0,T]$ by*

$$\xi(x,t) = \sum_{j\geq 1} y_{0,k} e^{-\lambda_k t}\phi_k(x).$$

We now consider the non-homogeneous exterior problem (B.6), for which we introduce the following notion of weak solution.

Definition B.6. Let $g \in L^2((0,T); H^s(\Omega^c))$. By a weak solution of (B.6) we mean a function $z \in L^2((0,T); H^s(\mathbb{R}))$ such that $z = g$ a.e. in $\Omega^c \times (0,T)$ and the identity

$$\int_0^T \langle -w_t + (-\Delta)^s w, z\rangle_{-s,s}\, dt = \int_\Omega z(x,T)w(x,T)\, dx + \int_0^T \int_{\Omega^c} g\mathcal{N}_s w\, dx dt \tag{B.7}$$

holds for every $w \in C([0,T]; L^2(\Omega)) \cap L^2((0,T); H^s_0(\Omega)) \cap H^1((0,T); H^{-s}(\Omega))$ with $\mathcal{N}_s w \in L^2((0,T)\times\Omega^c)$.

We then have the following existence result (see Warma, 2019).

Theorem B.7. *For every $g \in L^2((0,T); H^s(\Omega^c))$, the system* (B.6) *has a unique weak solution $z \in L^2((0,T); H^s(\mathbb{R}))$ given by*

$$z(x,t) = \sum_{k\geq 1}\left(\int_0^t \left(g(\cdot, t-\tau), \mathcal{N}_s\phi_k\right)_{L^2(\Omega^c)} e^{-\lambda_k \tau}\, d\tau\right)\phi_k(x).$$

Hence, from Theorems B.5 and B.7, we finally have the following result of existence and series representation of solutions to (B.4).

Theorem B.8. *For every $y_0 \in L^2(\Omega)$ and $g \in L^2((0,T); H^s(\Omega^c))$, the system* (B.4) *has a unique weak solution $y \in L^2((0,T)\times\mathbb{R}^d)$ given by*

$$\begin{aligned} y(x,t) = &\sum_{k\geq 1} y_{0,k} e^{-\lambda_k t}\phi_k \\ &+ \sum_{k\geq 1}\left(\int_0^t \left(g(\cdot, t-\tau), \mathcal{N}_s\phi_k\right)_{L^2(\Omega^c)} e^{-\lambda_k \tau}\, d\tau\right)\phi_k(x). \end{aligned}$$

Let us conclude this appendix by quickly discussing the following fractional heat equation with exterior Robin conditions

$$\begin{cases} y_t + (-\Delta)^s y = 0 & \text{in } \Omega \times (0,T) \\ \mathcal{N}_s y + \kappa y = \kappa g & \text{in } \Omega^c \times (0,T) \\ y(\cdot,0) = y_0 & \text{in } \Omega \end{cases} \tag{B.8}$$

which plays a fundamental role in the numerical simulations of Section 4. In (B.8), $\kappa \in L^1(\Omega^c) \cap L^\infty(\Omega^c)$ is a non-negative function.

To define the notion of solutions for the Robin problem (B.8), we first need to introduce the Sobolev space

$$H^s_\kappa(\Omega) := \left\{ y : \mathbb{R}^d \to \mathbb{R} \text{ measurable }, \; \|y\|_{H^s_\kappa(\Omega)} < +\infty \right\},$$

where

$$\|y\|_{H^s_\kappa(\Omega)} := \left(\|y\|^2_{L^2(\Omega)} + \left\| \kappa^{\frac{1}{2}} y \right\|^2_{L^2(\Omega^c)} + \int_{\mathbb{R}^{2d}} \frac{|y(x)-y(z)|}{|x-z|^{d+2s}}\, dxdz \right)^{\frac{1}{2}}.$$

We know from Dipierro et al. (2017, Proposition 3.1) that $H^s_\kappa(\Omega)$ is a Hilbert space. We denote with $H^{-s}_\kappa(\Omega) = (H^s_\kappa(\Omega))^\star$ its dual with respect to the pivot space $L^2(\Omega)$.

Definition B.9. Let $y_0 \in L^2(\Omega)$, $g \in L^2(\Omega^c \times (0,T))$ and $\kappa \in L^1(\Omega^c) \cap L^\infty(\Omega^c)$ non-negative. A function $y \in L^2((0,T); H^s_\kappa(\Omega)) \cap H^1((0,T); H^{-s}_\kappa(\Omega))$ is said to be a weak solution of (B.8) if $y(\cdot,0) = y_0$ a.e. in Ω and the identity

$$\int_0^T \langle y_t, v\rangle_{-s,s}\, dt + \int_0^T \mathcal{F}(y,v)\, dt + \int_0^T \int_{\Omega^c} \kappa y v\, dxdt = \int_0^T \int_{\Omega^c} \kappa g v\, dxdt, \tag{B.9}$$

holds for every $v \in H^s_\kappa(\Omega)$, where $\mathcal{F}(y,v)$ denotes the bilinear form

$$\mathcal{F}(y,v) := \int_{\mathbb{R}^{2d}\setminus(\Omega^c)^2} \frac{(y(x)-y(z))(v(x)-v(z))}{|x-z|^{d+2s}}\, dxdz.$$

Finally, we have the following existence result for (B.8) (see Antil et al., 2020b, Theorem 3.11).

Theorem B.10. *Let $\kappa \in L^1(\Omega^c) \cap L^\infty(\Omega^c)$ be non-negative. Then for every $y_0 \in L^2(\Omega)$ and $g \in L^2(\Omega^c \times (0,T))$, there exists a unique weak solution $y \in L^2((0,T); H^s_\kappa(\Omega)) \cap H^1((0,T); H^{-s}_\kappa(\Omega))$ of* (B.8).

References

Acosta, G., Bersetche, F.M., Borthagaray, J.P., 2017. A short FE implementation for a 2d homogeneous Dirichlet problem of a fractional Laplacian. Comput. Math. Appl. 74 (4), 784–816.

Acosta, G., Borthagaray, J.P., 2017. A fractional Laplace equation: regularity of solutions and finite element approximations. SIAM J. Numer. Anal. 55 (2), 472–495.

Adams, R.A., Fournier, J.J.F., 2003. Sobolev Spaces. Elsevier.

Andersson, J.A., Gillis, J., Horn, G., Rawlings, J.B., Diehl, M., 2019. CasADi: a software framework for nonlinear optimization and optimal control. Math. Program. Comput. 11 (1), 1–36.

Antil, H., Biccari, U., Ponce, R., Warma, M., Zamorano, S., 2020a. Controllability properties from the exterior under positivity constraints for a 1-d fractional heat equation. arXiv preprint. arXiv: 1910.14529.

Antil, H., Khatri, R., Warma, M., 2019. External optimal control of nonlocal PDEs. Inverse Probl. 35 (8), 084003.

Antil, H., Rautenberg, C., 2019. Sobolev spaces with non-Muckenhoupt weights, fractional elliptic operators, and applications. SIAM J. Math. Anal. 51 (3), 2479–2503.

Antil, H., Verma, D., Warma, M., 2020b. External optimal control of fractional parabolic PDEs. ESAIM Control Optim. Calc. Var. 26 (20).

Bach, F.R., Moulines, E., 2011. Non-asymptotic analysis of stochastic approximation algorithms for machine learning. In: Advances in Neural Information Processing Systems, pp. 451–459.

Bahrouni, A., RĂdulescu, V.D., 2018. On a new fractional Sobolev space and applications to non-local variational problems with variable exponent. Discrete Contin. Dyn. Syst. 11 (3), 379.

Bakunin, O.G., 2008. Turbulence and Diffusion: Scaling Versus Equations. Springer Science & Business Media.

Biccari, U., 2021. Internal control for a non-local Schrödinger equation involving the fractional Laplace operator. Evol. Equ. Control Theory 11 (1), 301–324.

Biccari, U., Hernández-Santamaría, V., 2018. The Poisson equation from non-local to local. Electron. J. Differ. Equ. 2018 (145), 1–13.

Biccari, U., Hernández-Santamaría, V., 2019a. Controllability of a one-dimensional fractional heat equation: theoretical and numerical aspects. IMA J. Math. Control Inf. 36 (4), 1199–1235.

Biccari, U., Hernández-Santamaría, V., 2019b. Null controllability of linear and semilinear nonlocal heat equations with an additive integral kernel. SIAM J. Control Optim. 57 (4), 2924–2938.

Biccari, U., Marica, A., Zuazua, E., 2020a. Propagation of one- and two-dimensional discrete waves under finite difference approximation. Found. Comput. Math. 20 (6), 1401–1438.

Biccari, U., Micu, S., 2019. Null-controllability properties of the wave equation with a second order memory term. J. Differ. Equ. 267 (2), 1376–1422.

Biccari, U., Warma, M., 2020. Null-controllability properties of a fractional wave equation with a memory term. Evol. Equ. Control Theory 9 (2), 399–430.

Biccari, U., Warma, M., Zuazua, E., 2017. Local elliptic regularity for the Dirichlet fractional Laplacian. Adv. Nonlinear Stud. 17 (2), 387–409.

Biccari, U., Warma, M., Zuazua, E., 2018. Local regularity for fractional heat equations. In: Recent Advances in PDEs: Analysis, Numerics and Control. Springer, pp. 233–249.

Biccari, U., Warma, M., Zuazua, E., 2020b. Controllability of the one-dimensional fractional heat equation under positivity constraints. Commun. Pure Appl. Anal. 19 (4), 1949–1978.

Biccari, U., Warma, M., Zuazua, E., 2021. Null control of multi-d fractional heat equations. Preprint.

Bonito, A., Lei, W., Pasciak, J.E., 2019. Numerical approximation of the integral fractional Laplacian. Numer. Math. 142 (2), 235–278.

Borthagaray, J.P., 2017. Laplaciano fraccionario: regularidad de soluciones y aproximaciones por elementos finitos. PhD thesis. Universidad de Buenos Aires, Buenos Aires, Argentina.

Bottou, L., 1998. Online learning and stochastic approximations. In: On-Line Learning in Neural Networks, Vol. 17(9), p. 142.

Bottou, L., Curtis, F.E., Nocedal, J., 2018. Optimization methods for large-scale machine learning. SIAM Rev. 60 (2), 223–311.

Boyd, S., Vandenberghe, L., 2004. Convex Optimization. Cambridge University Press, Cambridge.

Boyer, F., 2013. On the penalised HUM approach and its applications to the numerical approximation of null-controls for parabolic problems. In: ESAIM: Proc., Vol. 41. EDP Sciences, pp. 15–58.

Boyer, F., Hubert, F., Le Rousseau, J., 2011. Uniform controllability properties for space/time-discretized parabolic equations. Numer. Math. 118 (4), 601–661.

Brezis, H., 2010. Functional Analysis, Sobolev Spaces and Partial Differential Equations. Springer Science & Business Media.

Caffarelli, L., Silvestre, L., 2007. An extension problem related to the fractional Laplacian. Commun. Partial Differ. Equ. 32 (8), 1245–1260.

Caffarelli, L., Silvestre, L., 2009. Regularity theory for fully nonlinear integro-differential equations. Commun. Pure Appl. Math. 62 (5), 597–638.

Chaves-Silva, F.W., Zhang, X., Zuazua, E., 2017. Controllability of evolution equations with memory. SIAM J. Control Optim. 55 (4), 2437–2459.

Claus, B., Warma, M., 2020. Realization of the fractional Laplacian with nonlocal exterior conditions via forms method. J. Evol. Equ., 1–35.

Coron, J.-M., 2007. Control and Nonlinearity, Vol. 136. American Mathematical Soc.

De Roos, A.M., Persson, L., 2002. Size-dependent life-history traits promote catastrophic collapses of top predators. Proc. Natl. Acad. Sci. 99 (20), 12907–12912.

Di Nezza, E., Palatucci, G., Valdinoci, E., 2012. Hitchhiker's guide to the fractional Sobolev spaces. Bull. Sci. Math. 136 (5), 521–573.

Dipierro, S., Ros-Oton, X., Valdinoci, E., 2017. Nonlocal problems with Neumann boundary conditions. Rev. Mat. Iberoam. 33 (2), 377–416.

Dubkov, A.A., Spagnolo, B., Uchaikin, V.V., 2008. Lévy flight superdiffusion: an introduction. Int. J. Bifurc. Chaos Appl. Sci. Eng. 18 (9), 2649–2672.

Ekeland, I., Temam, R., 1999. Convex Analysis and Variational Problems, Vol. 28. SIAM.

Ervedoza, S., Valein, J., 2010. On the observability of abstract time-discrete linear parabolic equations. Rev. Mat. Complut. 23 (1), 163.

Fall, M.M., Felli, V., 2014. Unique continuation property and local asymptotics of solutions to fractional elliptic equations. Commun. Partial Differ. Equ. 39 (2), 354–397.

Fernández-Cara, E., Lü, Q., Zuazua, E., 2016. Null controllability of linear heat and wave equations with nonlocal spatial terms. SIAM J. Control Optim. 54 (4), 2009–2019.

Getoor, R., 1961. First passage times for symmetric stable processes in space. Trans. Am. Math. Soc. 101 (1), 75–90.

Ghosh, T., Salo, M., Uhlmann, G., 2020. The Calderón problem for the fractional Schrödinger equation. Anal. PDE 13 (2), 455–475.

Gilboa, G., Osher, S., 2008. Nonlocal operators with applications to image processing. Multiscale Model. Simul. 7 (3), 1005–1028.

Glowinski, R., 2015. Variational Methods for the Numerical Solution of Nonlinear Elliptic Problems. SIAM.

Glowinski, R., Lions, J.-L., 1994. Exact and approximate controllability for distributed parameter systems. Acta Numer. 3, 269–378.

Glowinski, R., Lions, J.-L., He, J., 2008. Exact and Approximate Controllability for Distributed Parameter Systems: a Numerical Approach. Cambridge University Press.

Gorenflo, R., Mainardi, F., Vivoli, A., 2007. Continuous-time random walk and parametric subordination in fractional diffusion. Chaos Solitons Fractals 34 (1), 87–103.

Grisvard, P., 2011. Elliptic Problems in Nonsmooth Domains. Classics in Applied Mathematics, vol. 69. Society for Industrial and Applied Mathematics (SIAM), Philadelphia, PA.

Grubb, G., 2015. Fractional Laplacians on domains, a development of Hörmander's theory of μ-transmission pseudodifferential operators. Adv. Math. 268, 478–528.

Hernández-Santamaría, V., Casanova, P.G., 2020. Carleman estimates and controllability results for fully-discrete approximations of 1-d parabolic equations. arXiv preprint. arXiv:2012.02156.

Hernández-Santamaría, V., Lazar, M., Zuazua, E., 2019. Greedy optimal control for elliptic problems and its application to turnpike problems. Numer. Math. 141 (2), 455–493.

Huang, Y., Oberman, A., 2014. Numerical methods for the fractional Laplacian: a finite difference-quadrature approach. SIAM J. Numer. Anal. 52 (6), 3056–3084.

Kingma, D.P., Ba, J.L., 2015. Adam: a method for stochastic optimization. In: Conference Paper, ICLR.

Lazar, M., Lohéac, J., 2020. Control of parameter dependent systems. Preprint.

Lazar, M., Zuazua, E., 2016. Greedy controllability of finite dimensional linear systems. Automatica 74, 327–340.

Leonori, T., Peral, I., Primo, A., Soria, F., 2015. Basic estimates for solutions of a class of nonlocal elliptic and parabolic equations. Discrete Contin. Dyn. Syst. 35 (12), 6031–6068.

Levendorskii, S.Z., 2004. Pricing of the American put under Lévy processes. Int. J. Theor. Appl. Finance 7 (03), 303–335.

Lions, J.-L., Magenes, E., 1968. Problèmes aux limites non homogènes et applications. Dunod.

Loheac, J., Trélat, E., Zuazua, E., 2017. Minimal controllability time for the heat equation under unilateral state or control constraints. Math. Models Methods Appl. Sci. 27 (09), 1587–1644.

Lohéac, J., Zuazua, E., 2016. From averaged to simultaneous controllability. Ann. Fac. Sci. Toulouse Math. 25 (4), 785–828.

Longhi, S., 2015. Fractional Schrödinger equation in optics. Opt. Lett. 40 (6), 1117–1120.

López, A., Zuazua, E., 1998. Some new results related to the null controllability of the $1-d$ heat equation. In: Séminaire Équations aux dérivées partielles (Polytechnique), pp. 1–22.

Lü, Q., Zhang, X., Zuazua, E., 2017. Null controllability for wave equations with memory. J. Math. Pures Appl. 108 (4), 500–531.

Mandelbrot, B.B., Van Ness, J.W., 1968. Fractional Brownian motions, fractional noises and applications. SIAM Rev. 10, 422–437.

Meza, J.C., 2010. Steepest descent. WIREs: Comput. Stat. 2 (6), 719–722.

Micu, S., Zuazua, E., 2006. On the controllability of a fractional order parabolic equation. SIAM J. Control Optim. 44 (6), 1950–1972.

Miller, L., 2006. On the controllability of anomalous diffusions generated by the fractional Laplacian. Math. Control Signals Syst. 18 (3), 260–271.

Nesterov, Y., 2004. Introductory Lectures on Convex Optimization: A Basic Course. Springer Science + Business Media, New York.

Nocedal, J., Wright, S., 2006. Numerical Optimization. Springer Science & Business Media.

Nochetto, R.H., Otárola, E., Salgado, A.J., 2015. A PDE approach to fractional diffusion in general domains: a priori error analysis. Found. Comput. Math. 15 (3), 733–791.

Oroya, J., Biccari, U., 2020. Moving control strategy for memory-type equations. DyCon Blog. https://deustotech.github.io/DyCon-Blog/tutorial/wp05/P0011.

Pham, H., 1997. Optimal stopping, free boundary, and American option in a jump-diffusion model. Appl. Math. Optim. 35 (2), 145–164.

Pighin, D., Zuazua, E., 2018. Controllability under positivity constraints of semilinear heat equations. Math. Control Relat. Fields 8 (3, 4), 935–964.

Polyak, B.T., 1964. Some methods of speeding up the convergence of iteration methods. USSR Comput. Math. Math. Phys. 4 (5), 1–17.

Porretta, A., Zuazua, E., 2013. Long time versus steady state optimal control. SIAM J. Control Optim. 51 (6), 4242–4273.

Rahmoune, A., Biccari, U., 2021. Multiplicity of solutions for fractional $q(.)$-Laplacian equations. Preprint. arXiv:2103.12600.

Robbins, H., Monro, S., 1951. A stochastic approximation method. Ann. Math. Stat. 22 (3), 400–407.

Ros-Oton, X., Serra, J., 2014. The Dirichlet problem for the fractional Laplacian: regularity up to the boundary. J. Math. Pures Appl. 101 (3), 275–302.

Saad, Y., 2003. Iterative Methods for Sparse Linear Systems, Vol. 82. SIAM.

Sauter, S.A., Schwab, C., 2010. Boundary element methods. In: Boundary Element Methods. Springer, pp. 183–287.

Schneider, W.R., 1990. Grey noise. In: Stochastic Processes, Physics and Geometry. Ascona and Locarno, 1988. World Sci. Publ., Teaneck, NJ, pp. 676–681.

Servadei, R., Valdinoci, E., 2012. Mountain pass solutions for non-local elliptic operators. J. Math. Anal. Appl. 389 (2), 887–898.

Servadei, R., Valdinoci, E., 2013. Variational methods for non-local operators of elliptic type. Discrete Contin. Dyn. Syst. 33 (5), 2105–2137.

Servadei, R., Valdinoci, E., 2014. On the spectrum of two different fractional operators. Proc. R. Soc. Edinb., Sect. A, Math. 144 (4), 831–855.

Vázquez, J.L., 2012. Nonlinear diffusion with fractional Laplacian operators. In: Nonlinear Partial Differential Equations. Springer, pp. 271–298.

Wächter, A., Biegler, L.T., 2006. On the implementation of an interior-point filter line-search algorithm for large-scale nonlinear programming. Math. Program. 106 (1), 25–57.

Warma, M., 2015. The fractional relative capacity and the fractional Laplacian with Neumann and Robin boundary conditions on open sets. Potential Anal. 42 (2), 499–547.

Warma, M., 2019. Approximate controllability from the exterior of space-time fractional diffusive equations. SIAM J. Control Optim. 57 (3), 2037–2063.

Warma, M., Zamorano, S., 2020a. Analysis of the controllability from the exterior of strong damping nonlocal wave equations. ESAIM Control Optim. Calc. Var. 26, 42.

Warma, M., Zamorano, S., 2020b. Null controllability from the exterior of a one-dimensional nonlocal heat equation. Control Cybern. 48 (3), 417–438.

Weiss, C., van Bloemen Waanders, B., Antil, H., 2020. Fractional operators applied to geophysical electromagnetics. Geophys. J. Int. 220 (2), 1242–1259.

Xiang, M., Zhang, B., Yang, D., 2019. Multiplicity results for variable-order fractional Laplacian equations with variable growth. Nonlinear Anal. 178, 190–204.

Zuazua, E., 2005. Propagation, observation, control and numerical approximation of waves. SIAM Rev. 47 (2), 197–243.

Zuazua, E., 2006. Control and numerical approximation of the wave and heat equations. In: International Congress of Mathematicians, Vol. 3. Madrid, Spain, pp. 1389–1417.

Zuazua, E., 2016. Stable observation of additive superpositions of Partial Differential Equations. Syst. Control Lett. 93, 21–29.

Chapter 2

Modeling, control, and numerics of gas networks

Martin Gugat[a,*,c] and Michael Herty[b,d]

[a]*FAU Erlangen-Nürnberg, Lehrstuhl für Dynamics, Control and Numerics (Alexander von Humboldt-Professorship), Erlangen, Germany,* [b]*RWTH Aachen University, Institut für Geometrie und Praktische Mathematik, Aachen, Germany*

**Corresponding author: e-mail address: martin.gugat@fau.de*

Contents

Abstract

In this chapter we survey recent progress on mathematical results on gas flow in pipe networks with a special focus on questions of control and stabilization. We briefly present the modeling of gas flow and coupling conditions for flow through vertices of a network. Our main focus is on gas models for spatially one-dimensional flow governed by hyperbolic balance laws. We survey results on classical solutions as well as weak solutions. We present results on well-posedness, controllability, feedback stabilization, the inclusion of uncertainty in the models and numerical methods.

[c] The author acknowledges support by DFG TRR154 Project C03 and C05.

[d] The author acknowledges support by BMBF ENets 05M18PAA, 320021702/GRK2326, 333849990/IRTG-2379, DFG EXC-2023 Internet of Production - 390621612, HIDSS-004, DFG 18,19-1.

Keywords
Hyperbolic balance laws, Stabilization, Exact controllability, Modeling of gas flow, Finite-volume schemes, Optimal control, Uncertainty

MSC Codes
35L65, 76N15, 82C40

1 Introduction

Gas transport on networks has been an active research topic for the past decade due to an increased demand in the sustainable use of natural resources and deregulation of energy markets. A particular focus has been on improved mathematical models of the physical flow for efficient simulation and for control purposes. In this chapter we contribute to the discussion by reviewing mathematical results on modeling, control, and numerical methods for partial differential equations. Continuum mechanical models for gas flow in a single pipe are typically of the following type

$$\partial_t \begin{pmatrix} \rho \\ \rho v \end{pmatrix} + \partial_x \begin{pmatrix} \rho v \\ \rho v^2 + p(\rho) \end{pmatrix} = - \begin{pmatrix} 0 \\ f\ \rho v|v| + g \sin(\alpha)\rho \end{pmatrix} \tag{1}$$

where ρ is the gas density, v the gas velocity, f is a friction parameter, $p(\rho)$ denotes the pressure as a function of ρ, g is the gravitational constant and α is the slope of the pipe and where a spatially one-dimensional model is chosen due to the particularities of pipe flow with low Mach number. Those also imply that temperature can be neglected. However, the model is capable of capturing transient phenomena driven by the need of simulating transient gas flow. Those flow patterns appear e.g. in the case of starting up gas power plants. If one is interested in average states of the gas network simplified models might be sufficient. The model (1) is posed on a single pipe and it is coupled by suitable transmission conditions to a flow model on networked pipes. The coupling typically leads to boundary conditions for the hyperbolic partial differential equation. Valves, compressor or generator stations are actuators of the control system. Those act pointwise (not distributed) in space and they are modeled by control through the boundary conditions and the corresponding coupling conditions in the networked system. Regarding detailed flow models, numerical challenges and further results, we cannot provide a complete list of reference at this point but refer to Benner et al. (2019); Sundar and Zlotnik (2018); Zlotnik et al. (2015); Brouwer et al. (2011); Bales et al. (2009); Osiadacz (1984), the references therein and the forthcoming sections. In the mathematical literature the study of hyperbolic balance laws, like (1), on metric graphs or networks has been studied over the past decade and we refer to the articles (Bressan et al., 2014; Čanić et al., 2017; Seo et al., 2017; Garavello, 2010), and references therein for further details. It is appropriate to mention also the early contributions on models for traffic flow on networks (Holden and Risebro, 1995) and (Coclite et al., 2005).

2 Modeling of gas flow

We are interested in the evolution equation (2) below on a network (or metric graph), consisting of a collection of one-dimensional manifolds connected at nodes. A general network is represented by a directed graph $\mathcal{G} = (\mathcal{E}, \mathcal{V})$ composed by a finite number of edges $\mathcal{E}$ connected by vertices or junctions. We use a directed graph to model the network but the discussion is not limited to this. For the sake of simplicity, we restrict the attention to a network that is composed by n edges labeled by j and modeled by the real interval $I_j := (0, L_j)$ where we could include the case $L_j = \infty$. We remark that this simplification does not imply a loss of generality due to the fact that the hyperbolic system (2) has the property that waves propagate with finite speed. The density and the flux on edge $j \in \mathcal{E}$ will be denoted ρ_j and $q_j = (\rho v)_j$. Without loss of generality the vertex is situated at $x_* = 0$ for all $j \in \mathcal{E}$. Note that in this particular network $v_j > 0$ implies the flow is emerging from the vertex. For a large, general network in order to achieve this property it might be necessary to transform the orientation of the flow at a vertex of the graph. This is achieved on the adjacent edge j using the transformation $x \to -x$ and $q_j(t,x) \to -q_j(t,x)$. Note that this does not change the balance law (2).

2.1 Models for gas flow on edges

For readability we skip the edge index in this section. Different scales might be relevant when modeling gas flow in pipe networks and we refer e.g. to Bales et al. (2009); Brouwer et al. (2011); Herty et al. (2010) for a detailed analysis of the involved scales as well as a corresponding hierarchy of suitable models. Here, we focus on a fine scale model for gas flow as given by the isentropic Euler equations

$$\partial_t \begin{pmatrix} \rho \\ \rho v \end{pmatrix} + \partial_x \begin{pmatrix} \rho v \\ \rho v^2 + p(\rho) \end{pmatrix} = - \begin{pmatrix} 0 \\ f\, \rho v|v| + g \sin(\alpha)\rho \end{pmatrix} \tag{2}$$

for the gas density $\rho = \rho(t,x)$ and the gas velocity $v = v(t,x)$. The gas flux is defined by $q(t,x) = (\rho v)(t,x)$ and $t \geq 0, x \in I$. The parameters f and $\alpha = \alpha(x)$ are the friction coefficient and the slope of the pipe. The number g denotes the gravitational force. It has been shown that often it is not necessary to consider temperature variations (Osiadacz, 1984; Osiadacz and Chaczykowski, 2001). Recent publication focuses on realistic pressure laws extending the isentropic choice

$$p(\rho) = \kappa \rho^{\gamma} \tag{3}$$

for $\kappa > 0$ and $\gamma \in [1, 3]$ towards

$$p(\rho) = z(p)\rho^{\gamma} \tag{4}$$

for some polynomial $\rho \to z(\rho)$ of order at most two (Grundel and Jansen, 2015; Gugat et al., 2018b). For the z-factor, often an affine linear model of the form

$$z(p) = R\Theta\,(1 + \tilde{\alpha}\,p) \tag{5}$$

is used, where $\tilde{\alpha} \in (-0.9, 0)$. Eq. (5) is sufficiently accurate within the network operating range, see Gugat et al. (2018b); Gugat and Wintergerst (2018). Often ideal gas corresponding to $\gamma = 1$ is considered for control and stabilization. Eq. (2) is accompanied by initial data

$$\rho(0, x) = \rho_0(x) \text{ and } q(0, x) = q_0(x), \tag{6}$$

and boundary data at $x = 0$ obtained through coupling conditions discussed in detail in the forthcoming section. In the case of $f = \alpha = 0$ the system (2) can be obtained formally and rigorously (Lions et al., 1994; Bouchut, 1999; Holle, 2020) through kinetic relaxation. For $f = f(t, x, \xi) \in \mathbb{R}^2$ with $\xi \in \mathbb{R}$ the BGK model for a fixed parameter $\epsilon > 0$ reads

$$\partial_t f(t, x, \xi) + \partial_x \xi f(t, x, \xi) = \frac{1}{\epsilon}\,(M[f](t, x, \xi) - f(t, x, \xi)). \tag{7}$$

The Maxwellian $M[f] = (M_1, M_2) \in \mathbb{R}^2$ is given by

$$\begin{aligned} M_1(t, x, \xi) &= \chi(\rho(t, x), \xi - v(t, x)), \\ M_2(t, x, \xi) &= \chi(\rho(t, x), \xi - v(t, x))\,((1 - \theta)v + \theta\xi) \end{aligned} \tag{8}$$

where χ is a rational polynomial and (ρ, v) are the moments of the kinetic distributions, i.e.,

$$\rho(t, x) = \int f_1(t, x, \xi)d\xi, \ (\rho v)(t, x) = \int f_2(t, x, \xi)d\xi. \tag{9}$$

In Holle (2020) it has been shown that for fixed $\epsilon > 0$ there exists a solution $f = f_\epsilon$ to Eq. (7) subject to suitable initial and boundary conditions. Furthermore, the sequence of associated densities and fluxes given by Eq. (9) converges towards a solution to Eq. (2) also in the presence of coupling conditions (19). For further details we refer to Holle (2020, Theorem 2.1).

In engineering application and connected with discrete decisions often stationary solutions to Eq. (2) and (4) are considered. The existence of steady states in the general case has been established in Gugat et al. (2015, 2018b). Those stationary solutions are important for stabilization of gas flow. Furthermore, they also appear in the design of optimal controls due to the turnpike phenomenon. This will be discussed in detail in the section below.

In the domain of operation of the gas pipeline networks (low Mach numbers and large pressure) the flow model can also be modeled by a degenerate parabolic model as studied e.g. by Bamberger (1977); Leugering and Mophou (2018). Formally, the model is obtained from Eq. (2) by neglecting the momentum term ρv^2 and the time derivative $\partial_t(\rho v)$. After small computations one obtains a degenerate parabolic equation in the pressure p in the isentropic case.

For further models we refer to the literature, e.g. Bales et al. (2009); Brouwer et al. (2011).

2.2 Models for the flow through vertices

The modeling of gas flow in pipes as spatially one-dimensional flow has severe implications on the modeling of the dynamics at the vertex. In particular, the modeling solely relies on the traces of the gas density $\rho_j(t, 0+)$ and gas flux $q_j(t, 0+)$ for all adjacent edge j and possibly a control input $u(t)$. In general the modeling of the vertex is hence given by a nonlinear function $\Psi : \mathbb{R}^+ \times (\mathbb{R}^2)^n \to \mathbb{R}^n$

$$\Psi\left(t, \rho_1(t, 0+), q_1(t, 0+), \ldots, \rho_n(t, 0+), q_n(t, 0+)\right) = u(t) \text{ a.e. } t \geq 0. \quad (10)$$

Thus the one-dimensional framework allows a rather simple modeling structure given by the form of Ψ, but very fast numerical integration. While a model in *one* space dimension well describes the dynamics within a pipe, it hardly covers geometry effects at a junction, which is clearly an intimately 3D phenomenon. As a consequence, the literature offers several different choices for a coupling or nodal condition Ψ and Eq. (10), depending on the specific needs of each particular situation. In engineering literature, the nodal conditions Ψ are typically accompanied by parameters whose values are empirically justified. A numerical study of one- and two-dimensional situations can be found e.g. in Banda et al. (2006b); Herty (2007a); Herty and Seaïd (2013).

A detailed well-posedness analysis of Eqs. (10) and (2) is deferred to the next section. However, it is important to notice that the condition (10) implicitly describes possible boundary conditions for the hyperbolic differential equation (2). This has implications on the modeling of suitable functions Ψ as seen below.

A further important aspect in the modeling of flows through vertices is the control action, here denoted by a given function $u(t)$. Those functions may model the closure of valves or the supplied power for compressor stations (Herty, 2007b). Most examples in the literature consider explicit control actions as shown in Eq. (10) even so an implicit dependence on u is possible. Additionally, the coupling condition might dependent explicitly on time $t \geq 0$ when fatigue of material is of importance.

In the following we turn to typical examples for Ψ. First, consider the case of subsonic data, i.e.,

$$\lambda_1(\rho_j(t, 0+), q_j(t, 0+)) < 0 < \lambda_2(\rho_j(t, 0+), q_j(t, 0+)). \quad (11)$$

Here, $\lambda_k(\rho, q)$ are the $k = 1, 2$ eigenvalues of the Jacobian of the flux function f. In the subsonic case we have therefore locally in time and phase space n boundary conditions to be determined by Eq. (10). A similar consideration in the supersonic case is possible and has been investigated in Gugat et al. (2017a). In general, the boundary conditions are not explicit but rather obtained by physical modeling considerations.

The first element of Ψ typically models the conservation of mass through the vertex and henceforth reads

$$\Psi_1(t, \rho_1, q_1, \ldots, \rho_n, q_n) = \sum_{j=1}^{n} q_j. \tag{12}$$

Except for a coupling in the gas-to-power setting (Herty et al., 2019) there is no control active at the first component of Ψ. The other components of Ψ may impose different physically desirable properties.

A condition typically used in the engineering community (Osiadacz, 1984; Banda et al., 2006a) is to assume equal pressure p at the vertex, i.e.,

$$\Psi_j(t, \rho_1, q_1, \ldots, \rho_n, q_n) = p(\rho_j) - p(\rho_1) \qquad j = 2, \ldots, n. \tag{13}$$

An alternative condition proposed in Banda et al. (2006a); Colombo and Garavello (2006) is to assume the continuity of the dynamic pressure or equality of momentum flux, i.e.,

$$\Psi_j(t, \rho_1, q_1, \ldots, \rho_n, q_n) = \left(\frac{{q_j}^2}{\rho_j} + p(\rho_j)\right) - \left(\frac{{q_1}^2}{\rho_1} + p(\rho_1)\right) \qquad j = 2, \ldots, n. \tag{14}$$

In Colombo and Garavello (2006) also geometric information has been included in the previous two conditions and an analytical comparison of qualitative properties has been conducted in Colombo and Garavello (2017). The conditions (13) and (14) might lead to a production of energy at the vertex as observed in Reigstad et al. (2015); Mindt et al. (2019). A condition that preserves the energy is to assume the equality of stagnation enthalpy or equality of Bernoulli invariant

$$\Psi_j(t, \rho_1, q_1, \ldots, \rho_n, q_n) = \frac{1}{2}\left(\frac{q_j}{\rho_j}\right)^2 + p'(\rho_j) - \frac{1}{2}\left(\frac{q_1}{\rho_1}\right)^2 + p'(\rho_1) \qquad j = 2, \ldots, n. \tag{15}$$

This condition implies the conservation of energy at the vertex. As noted in Holle et al. (2020), Eq. (2) itself dissipates energy and this might be also a desirable property of the coupling condition. An implicit condition ensuring this property has been introduced in Holle et al. (2020, Definition 1). The proposed condition is implicit in the sense that only the resulting boundary values $(\rho_k, q_k)(t, 0+)$ are given but *not* necessarily an explicit function Ψ. Hence, the starting points are constant initial data $(\hat{\rho}_k, \hat{q}_k)$ for $k = 1, \ldots, n$ adjacent to the vertex. Then, there exist boundary values $(\rho_k, q_k)(t, 0+)$ with the following properties.

- Mass is conserved at the junction:

$$\sum_{k=1}^{d} q_k(t, 0+) = 0, \ \ a.e. \, t \geq 0. \tag{16}$$

- There exists $\rho_* \geq 0$ such that for each fixed $k = 1, \ldots, n$ the boundary values $(\rho_k, q_k)(t, 0+)$ are equal to the restriction to $x > 0$ of the (unique) weak entropy solution of Eq. (2) in the sense of Lax with initial condition

$$(\rho, q)(0+, x) = \begin{cases} (\hat{\rho}_k, \hat{q}_k), & x > 0, \\ (\rho_*, 0), & x < 0. \end{cases} \tag{17}$$

The existence of $\rho_* > 0$ is proven in Holle et al. (2020, Lemma 1), the boundary values dependent continuously on ρ_* (Holle et al., 2020, Proposition 2) and *no* assumption on subsonic initial data is required. Further, the previous construction can be shown to be entropy decreasing at the vertex for a large class of symmetric entropies including the physical energy. The condition (17) can also be derived by using the kinetic formulation of the isentropic Euler equations (7). Coupling conditions for $f_k = f_k(t, x, \xi)$ where f_k is the kinetic particle density on edge $k = 1, \ldots, n$ can be formulated using a coupling condition of similar type as (10), i.e.,

$$\overline{\Psi}(t, f_1(t, 0+, \cdot), \ldots, f_n(t, 0+, \cdot))(\xi) = u(t) \, a.e.t \geq 0, \xi \in \mathbb{R}. \tag{18}$$

It has been shown in Holle (2020, Theorem 1) that among all coupling conditions $\overline{\Psi}$ that conserve the total mass, the condition that dissipates the most energy is given by

$$\overline{\Psi}_k(t, f_1(t, 0+, \cdot), \ldots, f_n(t, 0+, \cdot)) = M(\rho_*(t), 0, \xi), k = 1, \ldots, n. \tag{19}$$

Hence, in the formal limit $\epsilon \to 0$, we expect that at the vertex a state with zero velocity and some (unknown) density ρ_* prevails. This is precisely the condition imposed by Eq. (17).

In the uncontrolled case $u \equiv 0$ a qualitative and quantitative comparison of the previous conditions has been shown in Holle et al. (2020, Section 7). Conditions (13), (14) lead to boundary values resulting in an energy decay of the order of 10^{-2}, while condition (15) conserves the energy and condition (17) leads to a decay of the order of 10^{-1}. In the case of condition (13) no waves emerge from the vertex while for all other conditions except for (17) a (nonphysical) shock wave on $k = 2, 3$ appears. On the other hand, rarefaction waves on edges $k = 2, 3$ are observed for condition (17).

Next, we turn to the modeling of controls. The main control is due to compressor stations in gas networks. In terms of the formulation of Eq. (10) they are modeled by considering $n = 2$, Eq. (12) and

$$\Psi_2(t, \rho_1, q_1, \rho_2, q_2) = q_2 \left(\left(\frac{p(\rho_2)}{p(\rho_1)} - 1 \right)^{(\gamma-1)/\gamma} \right) \tag{20}$$

for some $\gamma \in (1, 3)$. Furthermore, $u(t) = (0, \Pi(t))$ where $\Pi(t)$ is the supplied compressor energy at time t, see e.g. Menon (2005); Herty and Sachers (2007); Colombo et al. (2009); Gugat and Herty (2011); Schmidt et al. (2016). This framework naturally leads to various control problems, where the open-loop control has to be chosen to satisfy suitable optimality criteria that will be discussed in the forthcoming sections. In contrast to the compressor control, control of valves can not add energy to the system and only one-way flow is possible through valves. In Corli and Rosini (2019) a mathematical model for this situation has been proposed and it amounts to prescribe a desired flow q_*. The control u acts in such way that the valve keeps the flow at a constant value q_* if possible; otherwise it is closed. Other types of valves where bi-directional flow is possible have also been considered e.g. in Corli et al. (2018).

Finally and for sake of completeness, we recall that the above conditions have been partly extended to the case of the full 3×3 system of Euler equations in Colombo and Marcellini (2010); Colombo and MauriI (2008).

3 Well-posedness of mathematical models for fixed control action

In this section we consider the case of a fixed given control action $u = u(t)$ and recall well-posedness results for classical and weak solutions.

3.1 Classical solutions

Tatsien Li and his collaborators have been very active in the study of semi-global classical solutions to the mixed initial-boundary value problem of one-dimensional quasilinear hyperbolic systems and the investigation of exact controllability in this framework, see for example Li and Jin (2001); Li and Rao (2019). As an example for a quasilinear initial boundary value problem, let us consider a density-velocity system that includes (1) as a special case:

$$\partial_t \begin{pmatrix} \rho \\ v \end{pmatrix} + \begin{pmatrix} v & \rho \\ \frac{(c(\rho))^2}{\rho} & v \end{pmatrix} \partial_x \begin{pmatrix} \rho \\ v \end{pmatrix} = - \begin{pmatrix} 0 \\ F(\rho, v) \end{pmatrix} \tag{21}$$

where the space interval is $[0, L]$, $\rho : [0, \infty) \times [0, L] \to (0, \infty)$, $v : [0, \infty) \times [0, L] \to \mathbb{R}$, $c \in C^2((0, \infty); (0, \infty))$, $F \in C^1((0, \infty) \times \mathbb{R}; \mathbb{R})$. The initial conditions are of the form

$$\rho(0, x) = \rho_0(x), \; v(0, x) = v_0(x), \; x \in [0, L] \tag{22}$$

where $\rho_0 \in C^1([0, L])$ and $v_0 \in C^1([0, L])$ are given initial data. As an example for boundary conditions on a time interval $[0, T]$, let us consider the following DIRICHLET boundary conditions for subsonic flow, where $|v| < c(\rho)$:

$$v(t, 0) = u_0(t), \; \rho(t, L) = u_L(t) \tag{23}$$

where $u_0, u_L \in C^1([0, T])$ are given control functions. Classical solutions can only exist if the initial conditions (22) and the boundary conditions (23) are C^1-compatible in the sense that

$$u_0(0) = v_0(0), \; u_L(0) = \rho_0(L) \tag{24}$$

and they satisfy the compatibility conditions implied by (21), that is

$$\begin{aligned} u_0'(0) &= -\frac{(c(\rho_0(0)))^2}{\rho_0(0)} \rho_0'(0) - v_0(0) v_0'(0) - F(\rho_0(0), v_0(0)), \\ u_L'(0) &= -v_0(L)\, \rho_0'(L) - \rho_0(L)\, v_0'(L). \end{aligned} \tag{25}$$

A typical existence result for semi-global classical solutions has the following structure:

Theorem 3.1. *Let a finite (arbitrarily large) time $T > 0$ be given. Then there exists a number $\varepsilon(T) > 0$ such that for all initial data and boundary data for which the maximal C^1-norm is less than $\varepsilon(T)$ and that satisfy the corresponding C^1-compatibility conditions there exists a unique classical solution on the time-interval $[0, T]$, i.e. a continuously differentiable function that satisfies the initial conditions, the boundary conditions and the partial differential equations. Moreover, there exists a constant $C_0(T) > 0$ such that the C^1-norm of the solution is bounded a priori by the product of $C_0(T)$ and the maximal C^1-norm of the initial data and the boundary data.*

The semi-global existence results are proved by rewriting the problem in terms of integral equations along the characteristic curves (see John, 1982) whose slopes are given by the eigenvalues of the system matrix. In the case of (21) the eigenvalues are $v \pm c(\rho)$. Using the notation $c^2 = \frac{\partial p}{\partial \rho}$, the eigenvalues for (1) can also be written as $\lambda_\pm = v \pm c$.

Starting from this formulation, similar as in the Picard iteration a map is defined such that every fixed point of this map solves the initial boundary value problem. Then the convergence of the fixed point iteration is shown and the assertion follows. The a priori bound is shown using Gronwall's Lemma.

Note that the fixed point iteration also allows to consider solutions that are only required to have Lipschitz regularity, see Gugat and Ulbrich (2018). The characteristic curves are often referred to only as characteristics. The method of characteristics is a classical method for the solution of one-dimensional quasi-linear hyperbolic systems. In order to obtain a well-posed problem, the boundary conditions have to be chosen according to the signs of the slopes of the characteristics. In the operation of gas networks, subsonic flow occurs, that is the velocity of the gas is smaller than the sound speed in the gas. This implies that one of the eigenvalues is positive in each point and the other one is negative in each point. Hence one family of characteristic curves travels from the left-hand side to the right-hand side and the other family of characteristics travels from the right-hand side to the left hand-side. Therefore and due to the structure of the Riemann invariants, for subsonic flow on a single pipe, at each end the value of one physical variable can be prescribed by the boundary conditions.

For the study of control problems, it is often useful to work in an Hilbert space. This is one reason why solutions of H^2-regularity are of interest. These solutions have a higher regularity than classical solutions. Solutions of this type have been studied in Bastin and Coron (2016) where an existence result is given in Appendix B. It has the same structure as Theorem 3.1 but with the C^1-norm replaced with the H^2-norm. In particular, the compatibility conditions remain unchanged.

Note that the existence result for semi-global classical solutions Theorem 3.1 can be extended to the case of networked systems that are defined on finite graphs if the node conditions uniquely define the necessary boundary input data for each adjacent pipe, see Gugat and Ulbrich (2018).

3.2 Weak solutions

Even for smooth initial data (ρ_0, q_0) (6) it is known that there may exist a time $t > 0$ such that solution $(\rho(t, \cdot), q(t, \cdot)$ to (2) may develop discontinuities (Dafermos, 2016). Hence, the notion of weak solutions has been introduced to treat solutions of lower regularity compared with solutions in the previous section. We refer to Dafermos (2016); Bressan (2009) for details on solutions to the Cauchy problem (2), (6) of systems of conservation and balance laws. Over several publications those results have extended to networked systems (2), (6), and (10) and refer to Bressan et al. (2014) for an overview and to for 2×2 hyperbolic balance laws with fixed control action to Colombo et al. (2009). Next, we briefly recall the basic definition of weak solution and well-posedness following Colombo and Garavello (2006); Colombo et al. (2008, 2009).

In order to present the notion of weak solutions for a single vertex located at $x = 0$ we define $\mathbf{u}_j = (\rho_j, q_j)$ as density and flux on the adjacent edge $j = 1, \dots, n$. Further, we define $f(\mathbf{u}) = (q, \frac{q^2}{\rho} + p(\rho))$ with $p(\rho)$ given by Eq. (3). Further, we consider a source term $g(t, x, \mathbf{u}) = (0, -f\rho v|v| - g\sin(\alpha)\rho)$. Then,

(2), (6), (10) reads

$$\begin{cases} \partial_t \mathbf{u}_1 + \partial_x f(\mathbf{u}_1) = g(t, x, \mathbf{u}_1) \\ \vdots \\ \partial_t \mathbf{u}_n + \partial_x f(\mathbf{u}_n) = g(t, x, \mathbf{u}_n) \end{cases} \tag{26}$$

coupled through the nodal condition

$$\Psi\left(\mathbf{u}_1(t, 0+), \cdots, \mathbf{u}_n(t, 0+)\right) = u(t). \tag{27}$$

Here, for every $j \in \{1, \ldots, n\}$, $\mathbf{u}_j : [0, T) \times I \to \Omega_j$, $T \in (0, +\infty]$, $I = (0, \infty)$, and Ω_j is a subset of $\mathbb{R}^2$. We supplement (26) and (27) with the initial condition

$$\begin{cases} \mathbf{u}_1(0, x) = \mathbf{u}_{1,0}(x), & x > 0, \\ \vdots \\ \mathbf{u}_n(0, x) = \mathbf{u}_{n,0}(x), & x > 0, \end{cases} \tag{28}$$

where, for every $j \in \{1, \ldots, n\}$, $\mathbf{u}_{j,0} : I_j \to \Omega_j$ are given functions. For brevity we introduce the notation

$$\vec{\mathbf{u}} = \begin{bmatrix} \mathbf{u}_1 \\ \cdots \\ \mathbf{u}_n \end{bmatrix}, \qquad \vec{f}(\mathbf{u}) = \begin{bmatrix} f(\mathbf{u}_1) \\ \cdots \\ f(\mathbf{u}_n) \end{bmatrix}, \qquad \vec{g}(t, x, \mathbf{u}) = \begin{bmatrix} g(t, x, \mathbf{u}_1) \\ \cdots \\ g(t, x, \mathbf{u}_n) \end{bmatrix}, \tag{29}$$

and we rewrite (26)-(27)-(28) in the form

$$\partial_t \vec{\mathbf{u}} + \partial_x \vec{f}(\mathbf{u}) = \vec{g}(t, x, \vec{\mathbf{u}}), \; \Psi\left(\vec{\mathbf{u}}(t, 0+)\right) = u(t), \; \vec{\mathbf{u}}(0, x) = \vec{\mathbf{u}}_0. \tag{30}$$

Let $L^1(\mathbb{R}^+; \Omega^n)$ denote the Lebesgue-integrable functions on $\mathbb{R}^+$ with values in Ω^n that have finite L^1-norm.

Definition 3.2. Fix $\hat{\mathbf{u}} = \left(\hat{\mathbf{u}}_1, \ldots, \hat{\mathbf{u}}_n\right) \in \prod_{j=1}^n \Omega_j$ and $T \in]0, +\infty]$. Assume $u \in BV(\mathbb{R}^+; \mathbb{R}^n)$. A weak solution to the Cauchy problem (30) on $[0, T)$ is a function $\vec{\mathbf{u}} \in C^0\left([0, T); \hat{\mathbf{u}} + L^1(\mathbb{R}^+; \Omega^n)\right)$ such that the following conditions hold.

1. For all $\phi \in C_c^\infty\left(]-\infty, T[\times \mathbb{R}^+; \mathbb{R}\right)$ and for $j \in \{1, \ldots, n\}$

$$\int_0^T \int_{\mathbb{R}^+} \left(\mathbf{u}_j \partial_t \varphi - g(t, x, \mathbf{u}_j)\varphi + f(\mathbf{u}_j)\partial_x \varphi\right) dx\, dt + \int_{\mathbb{R}^+} \mathbf{u}_{j,o}(x)\varphi(0, x) dx = 0. \tag{31}$$

2. For a.e. $t \in (0, T)$, $\Psi\left(\vec{\mathbf{u}}(t, 0+)\right) = u(t)$.

The weak solution $\vec{\mathbf{u}}$ is an entropy solution if for any convex entropy-entropy flux pair (η_j, Q_j), for all $\phi \in C_c^\infty\left(]-\infty, T[\times\mathbb{R}^+; \mathbb{R}\right)$ and for all $j = 1, \ldots, n$

$$\int_0^T \int_{\mathbb{R}^+} \left(\eta_j(\mathbf{u}_j)\partial_t\varphi - g(t, x, \mathbf{u}_j)D\eta_j(\mathbf{u}_j)\varphi + Q(\mathbf{u}_j)\partial_x\varphi\right) dx\, dt \geq 0. \quad (32)$$

Well-posedness of Eq. (30) is obtained for initial data having small total variation and under suitable assumptions on $\vec{f}$ and $\vec{g}$ as given by Colombo et al. (2009, Assumption F and G). Those assumptions are precisely as in the case of the Cauchy problem. In the following we only recall the additional assumptions on Ψ specific to the case of a network. We consider solutions in the space

$$X = \left(\vec{\mathbf{u}}_* + L^1(\mathbb{R}^+; \Omega), u_* + L^1(\mathbb{R}^+; \mathbb{R}^n)\right), \; \Omega = \prod_{j=1}^n \Omega_j, \quad (33)$$

where the constant states $(\vec{u}_*, u_*)$ are such that they fulfill the coupling condition

$$\Psi(\vec{u}_*) = u_*. \quad (34)$$

A solution is then obtained for data $(\vec{u}_0, u) = (\vec{u}_*, u_*) + (L^1(\mathbb{R}^+; \Omega), L^1(\mathbb{R}^+; \mathbb{R}^n))$ having small norm

$$TV(\vec{\mathbf{u}}_0) + TV(u) + \|\Psi(\vec{\mathbf{u}}_0(0+)) - u(0+)\|_{L^1(\mathbb{R}^+;\mathbb{R}^n)} \leq \delta, \quad (35)$$

for some possibly small $\delta > 0$. In the following we denote by $r_2(\mathbf{u})$ the right eigenvector of $Df(\mathbf{u})$ corresponding to the second characteristic family. Then, the assumption on Ψ is as follows: Assume $\Psi \in C^1(\Omega; \mathbb{R}^n)$ such that

$$\det\left[D_1\Psi(\vec{\mathbf{u}}_*)r_2(\mathbf{u}_{1,*})\; D_2\Psi(\vec{\mathbf{u}}_*)r_2(\mathbf{u}_{2,*}) \ldots D_n\Psi(\vec{\mathbf{u}}_*)r_2(\mathbf{u}_{n,*})\right] \neq 0. \quad (36)$$

Under the assumptions (35) and (36) the result (Colombo et al., 2009, Theorem 2.3) yields existence of a weak solution in the sense of Definition 3.2.

Some remarks are in order. The boundary traces of $\vec{u}(t, \cdot)$ are well-defined, since in fact we have $\mathbf{u}_j(t, \cdot) \in BV(\mathbb{R}^+; \Omega_j)$. However, the known result is only a perturbation result around the state $(\vec{\mathbf{u}}_*, u_*)$. Theorem 2.3 (Colombo et al., 2009) also shows continuous dependence of $\vec{\mathbf{u}}$ with respect to the initial data $(\vec{\mathbf{u}}_0, u)$. This leads to an existence result for suitable optimal control problems shown in the next section. It is important to mention that the stated nodal conditions in the previous section **all** fulfill condition (36) if $\vec{\mathbf{u}}_0(x)$ is subsonic for all $x \in \mathbb{R}^+$.

4 Control and controllability

4.1 Optimal control

In the context of weak solutions (3.2) existence of optimal controls has been established in Colombo et al. (2009). The result essentially follows by the continuous dependence of solutions $\vec{\mathbf{u}} \in X$ on $(\vec{\mathbf{u}}_0, u)$ where the space X is given by Eq. (33).

Theorem 4.1. *Let $n \in \mathbb{N}$, $n \geq 2$. Assume that $\vec{f}$ satisfies **(F)** at $\vec{\mathbf{u}}_*$ and $\vec{G}$ satisfies **(G)**. Fix a map $\Psi \in C^1(\Omega; \mathbb{R}^n)$ satisfying* (36) *and let $u_* = \Psi(\vec{\mathbf{u}}_*)$. For a fixed $\vec{\mathbf{u}}_o \in \mathcal{U}_\delta$ assume that*

$$J_o : \left\{ u_{|[0,T]} : u \in \left(u_* + \mathbf{L}^1(\mathbb{R}^+; \mathbb{R}^n)\right) \text{ and } (\mathbf{u}_o, u) \in \mathcal{D}^\delta \right\} \to \mathbb{R},$$
$$\text{and } J_1 : \mathcal{D}^\delta \to \mathbb{R}$$

are non-negative and lower semicontinuous with respect to the $\mathbf{L}^1$ norm. Then, the cost functional

$$\mathcal{J}(u) = J_o(u) + \int_0^T J_1\left(\vec{\mathbf{u}}(\tau, \cdot))\right) d\tau \tag{37}$$

admits a minimum on $\left\{ u_{|[0,T]} : u \in \left(u_* + \mathbf{L}^1(\mathbb{R}^+; \mathbb{R}^n)\right) \text{ and } (\vec{\mathbf{u}}_o, u) \in \mathcal{D}_0 \right\}$.

Here, we denote by $\vec{\mathbf{u}}(t, \cdot)$ the weak solution in the sense of Definition (3.2) with data $(\vec{\mathbf{u}}_0, u)$. The definition of the sets $\mathcal{D}_0, \mathcal{U}_\delta$, and $\mathcal{D}_\delta$ are given in Colombo et al. (2009). In particular, the sets $\mathcal{U}_\delta$ and $\mathcal{D}_\delta$ involve the assumption (35) on sufficiently small TV norm. The assumption **F** and **G** on flux and source term are the same as for the existence of a solution to a Cauchy problem and omitted here.

In the context of gas networks a typical cost functional $\mathcal{J}$ measures the distance to a given desired pressure $\bar{p}$ on a certain part $I_i = [x_1, x_2]$ of pipe i. If also oscillations in the control $u(\cdot)$ should be penalized the resulting functional reads

$$\mathcal{J}(u) = TV(u) + \int_0^T \int_{I_i} |p(\rho_i(t, x)) - \bar{p}| dx dt. \tag{38}$$

This functional fulfills the assumptions of Theorem 4.1. Note that as a possible substitute for the tracking term for the pressure in the objective function, also box constraints for the pressure are of interest.

Since the mapping $\vec{\mathbf{u}}_0 \to \vec{\mathbf{u}}$ is non-differentiable in any L^p, $p \geq 1$, optimality conditions are not straightforward. An alternative notion of differentiability has been introduced in Bressan and Lewicka (1998); Bressan and Marson (1995); Ulbrich (2003a). However, the extension to boundary control is still subject to active research (Pfaff and Ulbrich, 2017).

In Gugat et al. (2018a), the instantaneous control of mixed-integer PDE-constrained gas transport problems has been studied. Zero-one decisions occur in a natural way in the operation of gas networks in decisions as whether to switch on or off a compressor or whether to open or close a valve. As a first step towards the solution of transient optimal control problems with decisions of this type, an instantaneous control approach is suggested, where in the time-discrete problem, in each time step the control is chosen in such a way that the integrand in the objective function for the next time step is minimized. Approximation of the nonlinearities by piecewise linear functions leads to large mixed integer linear optimization problems where a solution close to global optimality is possible.

4.2 Controllability

The results about the exact boundary controllability of general one-dimensional quasilinear hyperbolic systems that have been obtained in the framework of classical solutions (see e.g. Li and Rao, 2019) can be applied to (1). It is typical for hyperbolic systems, that due to the finite speed of information flow exact controllability is only possible after a sufficiently large set-up time, see Hu (2015). In a nutshell, exact controllability is only possible after the flow of information from the boundary input has reached each point of the space interval on both families of characteristics. Exact boundary controllability on tree-like networks has also been studied in Gu and Li (2011a). Note that in general systems of conservation laws are not controllable in finite time in the framework of weak entropy solutions, see Bressan and Coclite (2002).

In the context of gas flow through pipelines, apart from the classical exact controllability, also the controllability of nodal profiles is of importance, since in the operation of gas pipeline networks, customer satisfaction is achieved by generating the desired pressure and flow rate profiles at the nodes where the customers are located. Stated as a controllability problem, the problem of the system operators is to steer the system in such a way that at the boundary nodes, after a finite time the desired nodal profiles are reached exactly during a certain time interval.

Note that in this notion of controllability of nodal profiles, in contrast to the classical notion of exact controllability, not the full state is prescribed at a fixed time, but instead the boundary trace of the state is prescribed on a certain time interval. Of course one of the variables can simply be prescribed by the boundary conditions, so the task is to drive simultaneously also the other variable to the values of the desired nodal profile. This notion of exact controllability of nodal profiles has been discussed in Gugat et al. (2011b) in the framework of classical solutions. The constructions in the proofs are based upon the exchange of the roles of time and space, that allows that the desired nodal profiles can play the role of virtual initial conditions. Recently, there has been a lot of research activity in the analysis of the exact controllability of nodal profiles in the framework of classical solutions of general networked quasilinear hyperbolic systems, see Li (2010); Gu and Li (2011b); Li (2019). For quasilinear wave equations it has been studied in Wang and Wang (2020) and for the Saint Venant system in Zhuang et al. (2019).

4.3 Feedback stabilization

If a desired stationary state is known, the problem arises, whether the system state in the gas network can be stabilized towards the desired state exponentially fast by suitably chosen boundary feedback laws. The corresponding analysis can be based upon suitably chosen Lyapunov functions. In particular Lyapunov functions in terms of Riemann invariants with exponential weights have been used successfully, see e.g. Bastin et al. (2009) in order to show the exponential decay of the L^2-norm for a linearized system with linear feedback laws in terms of the Riemann invariants. The corresponding analysis for the quasilinear system on a star-shaped network is given in Gugat et al. (2011a).

Extensions to the case of time delay in the feedback control have been presented in Gugat and Dick (2011).

The boundary feedback stabilization of the Saint-Venant system by a proportional feedback control is studied in Hayat and Shang (2019b) using a Lyapunov function in physical coordinates. A similar analysis for general hyperbolic density velocity systems (21) that also include (1) is presented in Bastin and Coron (2017). Here linear boundary conditions of the form

$$v(t,\, 0) = k_0\, \rho(t,\, 0),\; v(t,\, L) = k_L\, \rho(t,\, L) \tag{39}$$

with real feedback parameters k_0, k_L are considered. In Bastin and Coron (2017), intervals are defined in terms of the desired state for which the feedback law (39) leads to exponential decay of the L^2-norm of the distance between the current and the desired state. Note that in order to extend the semi-global solutions to global solutions, it is useful to work with solutions with H^2 regularity and to show that the H^2-norm of the solutions decays exponentially fast. This is the reason why in Bastin and Coron (2017); Hayat and Shang (2019b) also H^2 Lyapunov functions are presented and the exponential decay with respect to the H^2-norm is shown.

A similar analysis with a stabilizing Neumann feedback of the form

$$v_x(t,0) = v_x^*(0) + k_0 v_t(t,0),\; v(t,L) = v^*(L)$$

for the quasilinear wave equation

$$v_{tt} = (c^2 - v^2)\, v_{xx} - 2\, v \left(v_{tx} + (v_x)^2 \right) - 2\, v_t\, v_x - 2\, f\, |v| \left(v_t + \frac{3}{2}\, v\, v_x \right) \tag{40}$$

for the velocity that is derived from (1) is given in Gugat et al. (2017b). Here v^* is a desired stationary state for (40) and $k_0 > 0$ is a feedback parameter. The Lyapunov function that is used to show the exponential decay has the form $\mathcal{L}(t) = \mathcal{L}_1(t) + \mathcal{L}_2(t)$ with

$$\mathcal{L}_1(t) = \int_0^L k \left[\left(c^2 - v^2 \right) (v - v^*)_x^2 + v_t^2 \right]$$

$$
\begin{aligned}
&\qquad - 2\exp\left(-\frac{x}{L}\right)\left[v\,(v-v^*)_x^2 + v_t\,(v-v^*)_x\right]dx,\\
\mathcal{L}_2(t) &= \int_0^L k\left[\left(c^2-v^2\right)\,(v-v^*)_{xx}^2 + v_{tx}^2\right]\\
&\qquad - 2\exp\left(-\frac{x}{L}\right)\left[v\,(v-v^*)_{xx}^2 + v_{tx}\,(v-v^*)_{xx}\right]dx.
\end{aligned}
$$

A typical result then yields under suitable assumptions on k_0 the following estimate

$$
\mathcal{L}(t) \le \mathcal{L}(0)\exp(-\nu t),\ t \ge 0. \tag{41}
$$

Here, the constant $\nu > 0$ is typically not known explicitly and may depend among others on L. We refer to the section on numerical results for estimates of ν in the discrete case.

The boundary feedback stabilization by proportional-integral (PI) control is analyzed in Dos Santos et al. (2008), Xu and Sallet (2014), Bastin and Coron (2019), Hayat (2019) using a Lyapunov function in physical coordinates. The considered boundary conditions have the differential form

$$
\begin{aligned}
\rho(t,0)\,v(t,0) &= Q_0(t),\ \rho(t,L)\,v(t,L) = \kappa_L((1+k_L)\rho(t,L) - Z(t)),\\
Z'(t) &= \alpha_L(\rho^*(L) - \rho(t,L))
\end{aligned} \tag{42}
$$

where ρ^* denotes the desired stationary state.

Other techniques applied to stabilize linear hyperbolic balance rely on the backstepping technique and we refer e.g. to Anfinsen and Aamo (2018); Krstic and Smyshlyaev (2008) for further references and details. When using the backstepping technique an important aspect is the design of observers in order to determine the feedback law. State estimation and observer design have been studied in the context of system of hyperbolic equations in Anfinsen and Aamo (2016b); Bin and Di Meglio (2017); Anfinsen and Aamo (2016a); Anfinsen et al. (2016). Similar techniques as for the stabilization of gas dynamics have been used to stabilize St. Venant flow on general networks (Prieur and Winkin, 2018; Prieur et al., 2008; Hayat and Shang, 2019b). The partial differential equations (2) have a similar structure, except that the pressure is given by

$$
p(\rho) = \frac{g}{2}\rho^2 \tag{43}
$$

and that there are several models available for the friction term (Bastin and Coron, 2019; de Halleux et al., 2001). The system (2) can also be written in quasi-linear form using the variables (ρ, v) instead of (ρ, q). For classical solutions both systems are equivalent. Stabilization in terms of the variables (ρ, v) has been discussed recently e.g. in Hayat and Shang (2019a).

The boundary feedback stabilization for the degenerate parabolic model from Bamberger (1977) is studied in Gugat et al. (2020a). Also in this contribution the analysis is based upon a suitably chosen Lyapunov function. The suggested feedback law has the form

$$p(t, 0) = p_0, \; q(t, L) = \kappa_L \, p(t, L) \tag{44}$$

where $p_0 > 0$ is a desired pressure value and κ_L is a feedback parameter.

Note that in the analysis of the closed loop systems presented in this section, smallness assumptions for the initial state are used. In the analysis, these assumptions imply that no shocks are generated in the system. In order to take this into account, in the practical application of the feedback laws, it is important to keep in mind that it is often not clear whether these smallness assumptions are satisfied for the given initial states.

We want to mention that overviews on input-to-state stability can be found in Mironchenko and Prieur (2020); Sontag (2008).

5 Uncertainty quantification

In the operation of gas networks, the treatment of uncertainties plays a decisive role, since customer demands are uncertain. In practice, in a procedure with several steps the customers first buy the option to book their gas consumption later within a certain defined range. Then in a second step, precise quantities are nominated on a day-ahead market. A detailed description of the European entry-exit gas market is given in Grimm et al. (2019).

In order to take into account the uncertainty, random boundary data are included in the model. In Gotzes et al. (2016), a method for the computation of the probability of feasible load constellations in a stationary gas network with uncertain demand is given. A network with a single entry and several exits with uncertain loads is studied. The feasible flows have to satisfy given pressure bounds in the pipes.

The numerical method is based upon a spherical radial decomposition that is used both for the computation of the probabilities and the corresponding derivatives with respect to the control. Gradient formulae for nonlinear probabilistic constraints with non-convex quadratic forms are presented in van Ackooij and Pérez-Aros (2020).

In order to include the information on uncertainty in the optimization problems, probabilistic constraints of the form

$$\mathbb{P}(g(x, \omega) \geq 0) \geq p \tag{45}$$

are useful. Here the parameter $p \in (0, 1)$ is a probability threshold that can be chosen by the decision maker a priori according to his preferences, x denotes the decision variable and ω is a random variable. The deterministic form of (45) is a classical inequality constraint $g(x) \leq 0$.

For optimization problems, the structure of the corresponding set of feasible controls is relevant. In general, for the problems with probabilistic constraints this is not a convex set. However, in Gugat et al. (2020b) it has been shown that under weak assumptions the feasible set is star-shaped, which is an important result that implies that in the spherical radial decomposition on each ray at most one interval has to be considered in the computation of the probability. In Schuster et al. this computational approach is compared with a more general collocation method that is based upon kernel density estimators.

In Adelhütte et al. (2017), the approach with probabilistic constraints is generalized to a setting that also allows to take the dynamic case into account. In this paper, for a decision variable x, a set $\mathcal{U}$ (this could for example be a time interval) a desired probability threshold $p \in (0, 1)$ and a random variable ω probabilistic constraints of the form

$$\mathbb{P}\,(g(x, \omega, y) \geq 0\, \forall y \in \mathcal{U}) \geq p \tag{46}$$

are considered and referred to as *probust* constraints. This is useful for example to define a model where the probability that the state satisfies the pressure bounds throughout the time-interval is required to be at least p. Note that (46) is a stronger requirement than the constraint

$$\mathbb{P}\,(g(x, \omega, y) \geq 0) \geq p\, \forall y \in \mathcal{U} \tag{47}$$

which does not guarantee that for a feasible decision x the constraint is satisfied with probability p uniformly for all $y \in \mathcal{U}$.

In Farshbaf-Shaker et al. (2020), as a step towards the treatment of the full transient gas pipeline network flow using probabilistic constraints, the optimal Neumann boundary control of a vibrating string with uncertain initial data and probabilistic terminal constraints is analyzed and a numerical method is provided.

In Gugat and Schultz (2018), the quasilinear wave equation (40) is considered with the feedback law at $x = 0$ and uncertain boundary data $v^{\omega}(t)$ at $x = L$, that is with boundary conditions of the form

$$v_x(t, 0) = v_x^*(0) + k_0 v_t(t, 0),\; v(t, L) = v^*(L) + v^{\omega}(t).$$

It is shown that if the noise v^{ω} decays exponentially fast, the H^1-norm of the difference between v and the desired stationary state v^* decays exponentially fast.

In contrast to the probabilistic approach to robust control that we have presented above, the more conservative classical approach in robust optimization is to consider a certain range (called the uncertainty-set) for the uncertain parameters and optimize subject to the constraint that all elements of the uncertainty set are feasible (i.e. in particular the worst case scenario). For gas networks, this approach has been studied in Aßmann et al. (2019). In order to make the approach less costly, in Aßmann et al. (2019) a two-stage approach is proposed. For the two-stage model, the problem variables are classified as here-and-now variables that have to be decided at once before the uncertainty is realized and wait-and-see or adjustable variables whose values can be chosen later after the uncertainty is realized.

Another aspect of uncertainty is that also physical parameters that appear in the partial differential equations are uncertain. As a contribution to this topic, in Hintermüller and Strogies (2020) the problem to identify uncertain friction parameters that vary along the pipes is studied. The consequences of uncertain friction for the transport of natural gas through passive networks of pipelines have also been studied in Heitsch and Strogies (2019).

6 Numerical methods for simulation and control

The type of Eq. (2) is a hyperbolic balance law in one-spatial dimension. Hence, there is a vast literature on possible numerical schemes to discretize system (2) and we refer e.g. to LeVeque (2006) for an overview. The network structure imposes few particularities that will be reviewed in this section with particular focus on finite volume schemes. For simplicity we also use a regular spatial grid. In order to present numerical discretization in compact notation we introduce $\mathbf{u} = (\rho, q)$ and $g = (0, -f\rho v|v| - g\sin(\alpha)\rho)$.

6.1 Discretization of coupling conditions for finite volume schemes

Eq. (2) are approximated numerically using a finite volume method with numerical cell size $x_{i+1} - x_i = \Delta x > 0$ and time step $t^{m+1} - t^m = \Delta t$, chosen such the CFL condition (Lax, 2010) $\lambda_{\max}\Delta t \le \Delta x$ is satisfied. Here, $\lambda_{\max}$ is the maximum absolute value of the eigenvalues of Jacobian of f. The following discretization is done for each component $\mathbf{u} = (\mathbf{u}_1, \mathbf{u}_2)$ separately. Within a finite volume method the cell average $\overline{U}^m_{j,i}$ of $\mathbf{u}_j$ for $j = 1, 2$ of cell i at time t^m is given by

$$\overline{U}^m_{j,i} := \frac{1}{\Delta x}\int_{x_{i-\frac{1}{2}}}^{x_{i+\frac{1}{2}}} \mathbf{u}_j(x, t^m)dx.$$

The evolution of the cell average over time Δt is

$$\overline{U}^{m+1}_{j,i} = \overline{U}^m_{j,i} - \frac{\Delta t}{\Delta x}\left((\mathcal{F}_j)^m_{i+\frac{1}{2}} - (\mathcal{F}_j)^m_{i-\frac{1}{2}}\right) + \overline{G}^m_{j,i}, \tag{48}$$

where in Godunov's method (Godunov, 1959) $(\mathcal{F}_j)^m_{i+\frac{1}{2}} = \mathcal{F}_j(\overline{U}^m_{j,i}, \overline{U}^m_{j,i+1})$ denotes the numerical flux of component $u_j(t, x)$, $j = 1, 2$ through the boundary of the cells i and $i+1$. Further, $\overline{G}^n_{j,i}$ is an approximation to $\frac{1}{\Delta x}\int_{t^m}^{t^{m+1}}\int_{x_{i-\frac{1}{2}}}^{x_{i+\frac{1}{2}}} g_j(t, x, u_j(x, t))dxdt$ obtained by a suitable quadrature rule. Within Godunov's method the exact solution to a Riemann problem posed at the cell boundary $i + \frac{1}{2}$ is used to define the numerical flux $(\mathcal{F}_j)^m_{i+\frac{1}{2}}$. Many other numerical fluxes have been proposed and we refer to the literature (LeVeque, 2006; Toro, 2009) for further details. Here, we focus on Godunov's method as a basic first-order method. Therein, an approximation to $\mathbf{u}_j(t,x)$ is then obtained by the piecewise constant reconstruction

$$\mathbf{u}_j(t,x) = \sum_i \sum_m \overline{U}^m_{j,i}\, \chi_{[x_{i-\frac{1}{2}}, x_{i+\frac{1}{2}}]\times[t^m, t^{m+1}]}(t,x). \tag{49}$$

As discussed in the previous sections the coupling condition (10) induces boundary conditions for Eq. (2). Numerically, the corresponding discrete boundary conditions at $x = 0+$ can be obtained by the following method. Assume at time t^m the cell averages in the first cell $i = 0$ corresponding to $x = 0$ of the connected edges are $\overline{U}^m_{j,0}$ for $j = 1, \ldots, n$. Denote by $\sigma \to \mathcal{L}_\kappa(\mathbf{u}_o, \sigma)$ the κ-th Lax curve through the state $\mathbf{u}_o$ for $\kappa = 1, 2$. Then, we solve for $(\sigma_1^*, \ldots, \sigma_n^*)$ using Newton's method the nonlinear system

$$\Psi\left(t^m, \mathcal{L}_2(\overline{U}^m_{1,0}, \sigma_1), \ldots, \mathcal{L}_n(\overline{U}^m_{n,0}, \sigma_n)\right) = u(t^m), \tag{50}$$

where $u(t^m)$ denotes the discretized control at time t^m. Under suitable assumptions on Ψ a unique solution $(\sigma_1^*, \ldots, \sigma_n^*)$ to Eq. (50) exists. Then, a boundary value $\overline{U}^{m+1}_{j,0}$ at time t^{m+1} is given by Eq. (48) at $i = 0$

$$\overline{U}^m_{j,-1} := \mathcal{L}_2(\overline{U}^m_{j,0}, \sigma_j^*). \tag{51}$$

The given construction yields a first-order approximation to the coupling condition (10).

For finite-volume schemes of higher-order additional values at the boundary are required. In all recent publications (Banda et al., 2016; Borsche, 2016; Briani et al., 2016; Naumann et al., 2018) the additional information required for reconstruction is obtained using a Lax-Wendroff type approach. This amounts to differentiate condition (50) with respect to time and solve the additional equations for information on the slope of a reconstructed solution $t \to \mathbf{u}(t, 0+)$. It can be shown (Banda et al., 2016) that this construction allows to preserve the desired order.

Recently in Mantri et al. (2019) the interplay of the discretization order of the numerical flux $\mathcal{F}$ and the source term $\overline{G}$ has been investigated. In the case of spatially one-dimensional flow a numerical discretization has been proposed that allows to obtain steady-states up to machine precision even for large spatial grids Δx. This technique has been known as well-balanced schemes for the Cauchy problem but could be extended to the case of network equations.

6.2 Discretization of stabilization problems

The theoretical decay rates established in the previous section will be complemented by corresponding numerical results following Banda and Herty (2013).

Therein, a discrete stabilization result for a first-order spatial discretization of a quasi-linear system for $\mathbf{u} \in \mathbb{R}^d$

$$\partial_t \mathbf{u} + Df(\mathbf{u})\partial_x \mathbf{u} = 0, \ \mathbf{u}(x,0) = \mathbf{u}_0(x), \ \mathbf{u}(t,0) = K\mathbf{u}(t,1) \tag{52}$$

for $Df(\mathbf{u})_{i,j} = \delta_{i,j}\lambda_j(\mathbf{u}) > 0$ and $K_{ij} = \delta_{i,j}\kappa_i > 0$ has been established. A finite-volume discretization of the quasi-linear equation (52) is given by (48) where the flux $\mathcal{F}$ is chosen as Upwind flux due to the fixed direction $\lambda_j(\mathbf{u}) > 0$. This leads to the following discretization for $j = 1, \ldots, d$, $i = 0, \ldots, N$, and $m = 0, \ldots, K$

$$\overline{U}^{m+1}_{j,i} = \overline{U}^{m}_{j,i} - \frac{\Delta t}{\Delta x}\lambda_j(\overline{U}^{m}_{1,i}, \ldots, \overline{U}^{m}_{d,i})\left(\overline{U}^{m}_{j,i} - \overline{U}^{m}_{j,i-1}\right), \tag{53}$$

and for $j = 1, \ldots, d$, $i = 0, \ldots, N$, and $m \geq 0$ the initial boundary conditions are given by

$$\overline{U}^{0}_{j,i} = \frac{1}{\Delta x}\int_{I_i} \mathbf{u}_{j,0}(x)dx \text{ and } \overline{U}^{m}_{j,-1} = \kappa_j \overline{U}^{m}_{j,N}. \tag{54}$$

We further require that initial and boundary conditions are compatible, i.e.,

$$\overline{U}^{0}_{j,-1} = \kappa_j \overline{U}^{0}_{j,N}, \tag{55}$$

and that in the vicinity of $\mathbf{u} \in B_\delta(0) \subset \mathbb{R}^d$ the CFL condition (56) holds. For given $\delta > 0$, Δt is chosen such that

$$\frac{\Delta t}{\Delta x} \max_{j=1,\ldots,d} \max_{\mathbf{u} \in B_\delta(0)} |\lambda_j(\mathbf{u})| \leq 1. \tag{56}$$

For discrete initial data $\overline{U}^{0}_{j,i} \in B_\delta(0)$, having small and bounded discrete gradients discrete exponential stability of solutions to (53) has been established in Banda and Herty (2013, Theorem 2). Under suitable assumptions we obtain for discrete Lyapunov function

$$L^m = \Delta x \sum_{i=0}^{N}\sum_{j=1}^{d}\left(\overline{U}^{m}_{j,i}\right)^2 \exp(-\mu_j x_i), \ m \geq 0 \tag{57}$$

exponential decay in time

$$L^m \leq \exp(-\nu t^m) L^0, \ m \geq 0. \tag{58}$$

We do not review all required assumptions but recall that the exponential decay only holds provided that

$$0 < \kappa_j^2 \leq \frac{D_j^{\min}}{D_j^{\max}} \text{ and } \mu_j \leq \log\left(\sqrt{\kappa_j \frac{D_j^{\min}}{D_j^{\max}}}^{-2}\right), \tag{59}$$

where for $j = 1, \ldots, d$ we define $0 < D_j^{\min} := \min_{\mathbf{u} \in B_\delta(0)} \frac{\Delta t}{\Delta x} \lambda_j(\mathbf{u}) \leq \max_{\mathbf{u} \in B_\delta(0)} \frac{\Delta t}{\Delta x} \lambda_j(\mathbf{u}) \leq D_j^{\max} \leq 1$. The corresponding decay rate ν is given by

$$\nu = \min_{j=1,\ldots,d} \left(D_j^{\min} \exp(-\mu_j \Delta x) \mu_j \frac{\Delta x}{2\Delta t} \right). \tag{60}$$

The discrete scheme allows for explicit decay rate ν which is also independent of the grid since $\frac{\Delta t}{\Delta x}$ is fixed. Further, several extensions e.g. to $\lambda_j(\mathbf{u}) < 0$ and other boundary conditions are possible, see Banda and Herty (2013). In particular, in the linear case, $\lambda_j(\mathbf{u}) = \lambda_j$, the constant $D_j^{\min} = D_j^{\max}$. In this case $\nu = \min_{j=1,\ldots,d} \left(\frac{1}{2} \lambda_j \mu_j \right)$ for $\Delta x \to 0$.

The results on discrete L^2-stability have been extended to H^s-norm for any $s \geq 2$ in the case of linear flux $Df(u) = Au$ and linear source terms $g(\mathbf{u}) = G\mathbf{u}$ in the recent paper (Gerster and Herty, 2019).

6.3 Numerical methods for optimal control problems

As mentioned in the theoretical results, in the case of optimal boundary control problems optimality conditions are not straightforward to obtain. The main reason is the lack of differentiability in L^1 of the control to state mapping. A remedy in the case of systems has been introduced in Bressan and Marson (1995); Bressan and Shen (2007) and for scalar equations in Ulbrich (2003a,b) using a novel differential. A numerical implementation of those conditions remains challenging due to the required resolution of the fine structure of the solution to the system (2), (6), and (10). However, several approximations of the optimality system have been studied recently in the literature and we refer to Gugat et al. (2006); Giles and Ulbrich (2010b,a); Chertock et al. (2014); Herty et al. (2014, 2018); Herty and Piccoli (2016); Bardos and Pironneau (2002); Castro et al. (2008) for further details. For classical solutions, the evaluation of derivatives in the optimal nodal control of networked hyperbolic systems has been studied in Gugat (2005).

7 Open problems

In this section we briefly mention some open problems that might be relevant for the future development of modeling, control, and numerics for gas networks.

From a modeling perspective it would be interesting to extend the current models in at least two directions. It has been observed that in the operation of gas networks the quality of the injected gas varies. This leads to the transport of a mixture of different gases and the development of suitable models for gas mixtures is certainly interesting and a current research topic. On the other hand, future simulations and control of interconnected energy systems like e.g. coupled gas and electricity networks will be of importance. From a modeling point of view this requires suitable model couplings, from a numerical point of view this requires to treat multi-scale phenomena due to the different involved time scales. Preliminary results in this direction have been obtained but the full control of coupled infrastructure is still at large. This might also possibly require the development of novel control paradigms as well as reduced models.

From a control point of view a recent topic of interest for the operation of large-scale gas networks is the turnpike phenomenon. Turnpike results for the complex system dynamics can justify a combination of steady state models with dynamic models. A survey on large time horizon control and turnpike properties for wave equations is presented in Zuazua (2017). Due to the turnpike phenomenon, often the control time can be divided into dynamic phases at the beginning and the end of the time interval and a static phase between them. In the dynamic phase one goal is to develop novel efficient feedback controls. A second aspect is the development of suitable Lyapunov functions and feedback controls for weak solutions, like BV functions, in order to treat effects like closing valves and shock waves. Moreover, it would be of interest to have feedback laws that can stabilize the system for a large set of initial states, that is to weaken the smallness assumptions for the initial state. A further aspect in the control of complex systems is to include reinforcement learning of dynamics that does not require any mathematical model. An application of those techniques towards gas networks is still an open problem.

For numerical computations novel methods to treat large-scale networks might need to be developed in order to obtain practical relevance. Here, tools like model-order reduction or suitable adaptive schemes might be of importance. See Benner et al. (2019) for a model with a differential-algebraic equation. For the efficient computation of gradients on a network an adjoint calculus is desirable. While formally this system (Herty and Sachers, 2007) can be derived major obstacles appear due to the non-differentiability of the control to state mapping. Here, efficient schemes would be desirable.

Finally, results obtained for energy networks might also lead to new insights for networks appearing in different transport processes like e.g. blood flow, vehicular traffic flow or production engineering.

References

Adelhütte, D., Aßmann, D., Grandòn, T.G., Gugat, M., Heitsch, H., Liers, F., Henrion, R., Nitsche, S., Schultz, R., Stingl, M., Wintergerst, D., 2017. Joint model of probabilistic/robust (probust) constraints applied to gas network optimization.

Anfinsen, H., Aamo, O.M., 2016a. Boundary parameter and state estimation in 2x2 linear hyperbolic PDEs using adaptive backstepping. In: 2016 IEEE 55th Conference on Decision and Control, CDC 2016, pp. 2054–2060.

Anfinsen, H., Aamo, O.M., 2016b. Stabilization of a linear 2x2 hyperbolic system with uncertain parameters from anti-collocated sensing and control. In: 2016 20th International Conference on System Theory, Control and Computing, ICSTCC 2016 - Joint Conference of SINTES 20, SACCS 16, SIMSIS 20 - Proceedings, pp. 55–60.

Anfinsen, H., Aamo, O.M., 2018. Adaptive output-feedback stabilization of 2x2 linear hyperbolic PDEs with actuator and sensor delay. In: MED 2018 - 26th Mediterranean Conference on Control and Automation. Institute of Electrical and Electronics Engineers Inc., pp. 517–522.

Anfinsen, H., Di Meglio, F., Aamo, O.M., 2016. Estimating the left boundary condition of coupled 1-D linear hyperbolic PDEs from right boundary sensing. In: 2016 European Control Conference, ECC 2016, pp. 2179–2184.

Aßmann, D., Liers, F., Stingl, M., 2019. Decomposable robust two-stage optimization: an application to gas network operations under uncertainty. Networks 74, 40–61.

Bales, P., Kolb, O., Lang, J., 2009. Hierarchical Modelling and Model Adaptivity for Gas Flow on Networks. Lecture Notes in Computer Science (including subseries Lecture Notes in Artificial Intelligence and Lecture Notes in Bioinformatics), vol. 5544, pp. 337–346.

Bamberger, A., 1977. Étude D'Une Équation Doublement Non Linéaire. Journal of Functional Analysis 24, 148–155.

Banda, M.K., Häck, A.S., Herty, M., 2016. Numerical discretization of coupling conditions by high-order schemes. Journal of Scientific Computing 69, 122–145.

Banda, M.K., Herty, M., 2013. Numerical discretization of stabilization problems with boundary controls for systems of hyperbolic conservation laws. Mathematical Control and Related Fields 3, 121–142.

Banda, M.K., Herty, M., Klar, A., 2006a. Coupling conditions for gas networks governed by the isothermal Euler equations. Networks and Heterogeneous Media 1, 295–314.

Banda, M.K., Herty, M., Klar, A., 2006b. Gas flow in pipeline networks. Networks and Heterogeneous Media 1, 41–56.

Bardos, C., Pironneau, O., 2002. Un formalisme pour la dérivation des lois de conservations. Comptes Rendus. Mathématique 335, 839–845.

Bastin, G., Coron, J.-M., 2016. Stability and Boundary Stabilization of 1-D Hyperbolic Systems, Vol. 88. Birkhäuser/Springer, Basel.

Bastin, G., Coron, J.-M., 2017. A quadratic Lyapunov function for hyperbolic density-velocity systems with nonuniform steady states. Systems & Control Letters 104, 66–71.

Bastin, G., Coron, J.-M., 2019. Exponential stability of PI control for Saint-Venant equations with a friction term. Methods and Applications of Analysis 26, 101–112.

Bastin, G., Coron, J.-M., d'Andréa-Novel, B., 2009. On Lyapunov stability of linearised Saint-Venant equations for a sloping channel. Networks and Heterogeneous Media 4, 177–187.

Benner, P., Grundel, S., Himpe, C., Huck, C., Streubel, T., Tischendorf, C., 2019. Gas network benchmark models. In: Applications of Differential-Algebraic Equations: Examples and Benchmarks. Springer, Cham, pp. 171–197.

Bin, M., Di Meglio, F., 2017. Boundary estimation of boundary parameters for linear hyperbolic PDEs. IEEE Transactions on Automatic Control 62, 3890–3904.

Borsche, R., 2016. Numerical schemes for networks of hyperbolic conservation laws. Applied Numerical Mathematics 108, 157–170.

Bouchut, F., 1999. Construction of BGK models with a family of kinetic entropies for a given system of conservation laws. Journal of Statistical Physics 95, 113–170.

Bressan, A., 2009. Lecture Notes on Hyperbolic Conservation Laws. Department of Mathematics, Penn State University, …, pp. 1–85.

Bressan, A., Čanić, S., Garavello, M., Herty, M., Piccoli, B., 2014. Flows on networks: recent results and perspectives. EMS Surveys in Mathematical Sciences 1, 47–111.

Bressan, A., Coclite, G.M., 2002. On the boundary control of systems of conservation laws. SIAM Journal on Control and Optimization 41, 607–622.

Bressan, A., Lewicka, M., 1998. Shift Differentials of Maps in BV Spaces. SISSA Ref., 59/98/M. pp. 1–15.

Bressan, A., Marson, A., 1995. A variational calculus for discontinuous solutions of systems of conservation laws. Communications in Partial Differential Equations 20, 1491–1552.

Bressan, A., Shen, W., 2007. Optimality conditions for solutions to hyperbolic balance laws, pp. 129–152.

Briani, M., Piccoli, B., Qiu, J.M., 2016. Notes on RKDG methods for shallow-water equations in canal networks. Journal of Scientific Computing 68, 1101–1123.

Brouwer, J., Gasser, I., Herty, M., 2011. Gas pipeline models revisited: model hierarchies, nonisothermal models, and simulations of networks. Multiscale Modeling & Simulation 9, 601–623.

Čanić, S., Delle Monache, M.L., Piccoli, B., Qiu, J.M., Tambača, J., 2017. Numerical methods for hyperbolic nets and networks. Handbook of Numerical Analysis 18, 435–463.

Castro, C., Palacios, F., Zuazua, E., 2008. An alternating descent method for the optimal control of the inviscid Burgers equation in the presence of shocks. Mathematical Models and Methods in Applied Sciences 18, 369–416.

Chertock, A., Herty, M., Kurganov, A., 2014. An Eulerian–Lagrangian method for optimization problems governed by multidimensional nonlinear hyperbolic PDEs. Computational Optimization and Applications 59, 689–724.

Coclite, G.M., Garavello, M., Piccoli, B., 2005. Traffic flow on a road network. SIAM Journal on Mathematical Analysis 36, 1862–1886.

Colombo, R.M., Garavello, M., 2006. A well posed Riemann problem for the p–system at a junction. Networks and Heterogeneous Media 1, 495–511.

Colombo, R.M., Garavello, M., 2017. On the 1D modeling of fluid flowing through a Junction. Discrete and Continuous Dynamical Systems. Series B 22, 1–13.

Colombo, R.M., Guerra, G., Herty, M., Schleper, V., 2009. Optimal control in networks of pipes and canals. SIAM Journal on Control and Optimization 48, 2032–2050.

Colombo, R.M., Herty, M., Sachers, V., 2008. On 2x2 conservation laws at a junction. SIAM Journal on Mathematical Analysis 40, 605–622.

Colombo, R.M., Marcellini, F., 2010. Coupling conditions for the 3x3 Euler system. Networks and Heterogeneous Media 5, 675–690.

Colombo, R.M., Mauri, C., 2008. Euler system for compressible fluids at a junction. Journal of Hyperbolic Differential Equations 5, 547–567.

Corli, A., Figiel, M., Futa, A., Rosini, M.D., 2018. Coupling conditions for isothermal gas flow and applications to valves. Nonlinear Analysis: Real World Applications 40, 403–427.

Corli, A., Rosini, M.D., 2019. Coherence and chattering of a one-way valve. Zeitschrift für Angewandte Mathematik und Mechanik 99.

Dafermos, C., 2016. Hyperbolic Conservation Laws in Continuum Physics. Grundlehren der Mathematischen Wissenschaften, vol. 325.

de Halleux, J., Bastin, G., D'Andréa-Novel, B., Coron, J., 2001. A Lyapunov approach for the control of multi reach channels modelled by Saint-Venant equations. IFAC Proceedings Volumes 34, 1429–1434.

Dos Santos, V., Bastin, G., Coron, J.-M., D'Andréa-Novel, B., 2008. Boundary control with integral action for hyperbolic systems of conservation laws: stability and experiments. Automatica 44, 1310–1318.

Farshbaf-Shaker, M.H., Gugat, M., Heitsch, H., Henrion, R., 2020. Optimal Neumann boundary control of a vibrating string with uncertain initial data and probabilistic terminal constraints. SIAM Journal on Control and Optimization 58, 2288–2311.

Garavello, M., 2010. A review of conservation laws on networks.

Gerster, S., Herty, M., 2019. Discretized feedback control for systems of linearized hyperbolic balance laws. Mathematical Control and Related Fields 9, 517–539.

Giles, M., Ulbrich, S., 2010a. Convergence of linearized and adjoint approximations for discontinuous solutions of conservation laws. Part 1: linearized approximations and linearized output functionals. SIAM Journal on Numerical Analysis.

Giles, M., Ulbrich, S., 2010b. Convergence of linearized and adjoint approximations for discontinuous solutions of conservation laws. Part 2: adjoint approximations and extensions. SIAM Journal on Numerical Analysis 48, 905–921.

Godunov, S., 1959. Finite difference method for numerical computation of discontinuous solutions of the equations of fluid dynamics. Matematicheskii Sbornik 47 (89), 271–306.

Gotzes, C., Heitsch, H., Henrion, R., Schultz, R., 2016. On the quantification of nomination feasibility in stationary gas networks with random load. Mathematical Methods of Operations Research 84, 427–457.

Grimm, V., Schewe, L., Schmidt, M., Zöttl, G., 2019. A multilevel model of the European entry-exit gas market. Mathematical Methods of Operations Research 89, 223–255.

Grundel, S., Jansen, L., 2015. Efficient Simulation of Transient Gas Networks Using IMEX Integration Schemes and MOR Methods. In: Proceedings of the IEEE Conference on Decision and Control, Vol. 54. IEEE, pp. 4579–4584.

Gu, Q., Li, T., 2011a. Exact boundary controllability for quasilinear hyperbolic systems on a tree-like network and its applications. SIAM Journal on Control and Optimization 49, 1404–1421.

Gu, Q., Li, T., 2011b. Exact boundary controllability of nodal profile for quasilinear hyperbolic systems in a tree-like network. Mathematical Methods in the Applied Sciences 34, 911–928.

Gugat, M., 2005. Optimal nodal control of networked hyperbolic systems: evaluation of derivatives. Advanced Modelling and Optimization 7, 9–37.

Gugat, M., Dick, M., 2011. Time-delayed boundary feedback stabilization of the isothermal Euler equations with friction. Mathematical Control and Related Fields 1, 469–491.

Gugat, M., Dick, M., Leugering, G., 2011a. Gas flow in fan-shaped networks: classical solutions and feedback stabilization. SIAM Journal on Control and Optimization 49, 2101–2117.

Gugat, M., Hante, F., Jin, L., 2020a. Closed loop control of gas flow in a pipe: stability for a transient model. Automatisierungstechnik 68, 1001–1010.

Gugat, M., Hante, F.M., Hirsch-Dick, M., Leugering, G., 2015. Stationary states in gas networks. Networks and Heterogeneous Media 10, 295–320.

Gugat, M., Herty, M., 2011. Existence of classical solutions and feedback stabilization for the flow in gas networks. ESAIM. Control, Optimisation and Calculus of Variations 17, 28–51.

Gugat, M., Herty, M., Klar, A., Leugering, G., 2006. Conservation law constrained optimization based upon Front-Tracking. Mathematical Modelling and Numerical Analysis 40, 939–960.

Gugat, M., Herty, M., Müller, S., 2017a. Coupling conditions for the transition from supersonic to subsonic Fluid states. Networks and Heterogeneous Media 12, 371–380.

Gugat, M., Herty, M., Schleper, V., 2011b. Flow control in gas networks: exact controllability to a given demand. Mathematical Methods in the Applied Sciences 34, 745–757.

Gugat, M., Leugering, G., Martin, A., Schmidt, M., Sirvent, M., Wintergerst, D., 2018a. Mip-based instantaneous control of mixed-integer PDE-constrained gas transport problems. Computational Optimization and Applications 70, 267–294.

Gugat, M., Leugering, G., Wang, K., 2017b. Neumann boundary feedback stabilization for a nonlinear wave equation: a strict H^2-Lyapunov function. Mathematical Control and Related Fields 7, 419–448.

Gugat, M., Schultz, R., 2018. Boundary feedback stabilization of the isothermal Euler equations with uncertain boundary data. SIAM Journal on Control and Optimization 56, 1491–1507.

Gugat, M., Schultz, R., Schuster, M., 2020b. Convexity and starshapedness of feasible sets in stationary flow networks. Networks and Heterogeneous Media 15, 171–195.

Gugat, M., Schultz, R., Wintergerst, D., 2018b. Networks of pipelines for gas with nonconstant compressibility factor: stationary states. Computational & Applied Mathematics 37, 1066–1097.

Gugat, M., Ulbrich, S., 2018. Lipschitz solutions of initial boundary value problems for balance laws. Mathematical Models and Methods in Applied Sciences 28, 921–951.

Gugat, M., Wintergerst, D., 2018. Transient flow in gas networks: traveling waves. International Journal of Applied Mathematics and Computer Science 28, 341–348.

Hayat, A., 2019. PI controller for the general Saint-Venant equations. Preprint.

Hayat, A., Shang, P., 2019a. Exponential stability of density-velocity systems with boundary conditions and source term for the H 2 norm. Tech. Rep.

Hayat, A., Shang, P., 2019b. A quadratic Lyapunov function for Saint-Venant equations with arbitrary friction and space-varying slope. Automatica 100, 52–60.

Heitsch, H., Strogies, N., 2019. Consequences of uncertain friction for the transport of natural gas through passive networks of pipelines.

Herty, M., 2007a. Coupling conditions for networked systems of Euler equations. SIAM Journal on Scientific Computing 30, 1596–1612.

Herty, M., 2007b. Modeling, simulation and optimization of gas networks with compressors. Networks and Heterogeneous Media 2, 81–97.

Herty, M., Kurganov, A., Kurochkin, D., 2014. Numerical method for optimal control problems governed by nonlinear hyperbolic systems of PDEs. Communications in Mathematical Sciences 13, 15–48.

Herty, M., Kurganov, A., Kurochkin, D., 2018. On convergence of numerical methods for optimization problems governed by scalar hyperbolic conservation laws. Springer Proceedings in Mathematics & Statistics 236, 691–706.

Herty, M., Mohringb, J., Sachersa, V., 2010. A new model for gas flow in pipe networks. Mathematical Methods in the Applied Sciences 33, 845–855.

Herty, M., Müller, S., Sikstel, A., 2019. Coupling of compressible Euler equations. Vietnam Journal of Mathematics 47, 769–792.

Herty, M., Piccoli, B., 2016. A numerical method for the computation of tangent vectors to 2x2 hyperbolic systems of conservation laws. Communications in Mathematical Sciences 14, 683–704.

Herty, M., Sachers, V., 2007. Adjoint calculus for optimization of gas networks. Networks and Heterogeneous Media 2, 733–750.

Herty, M., Seaïd, M., 2013. Assessment of coupling conditions in water way intersections. International Journal for Numerical Methods in Fluids 71, 1438–1460.

Hintermüller, M., Strogies, N., 2020. Identification of the friction function in a semilinear system for gas transport through a network. Optimization Methods & Software 35, 576–617.

Holden, H., Risebro, N.H., 1995. A mathematical model of traffic flow on a network of unidirectional roads. SIAM Journal on Mathematical Analysis 26, 999–1017.

Holle, Y., 2020. Kinetic relaxation to entropy based coupling conditions for isentropic flow on networks. Journal of Differential Equations 269, 1192–1225.

Holle, Y., Herty, M., Westdickenberg, M., 2020. New coupling conditions for isentropic flow on networks. Networks and Heterogeneous Media 15, 605–631.

Hu, L., 2015. Sharp time estimates for exact boundary controllability of quasilinear hyperbolic systems. SIAM Journal on Control and Optimization 53, 3383–3410.

John, F., 1982. Partial Differential Equations, Vol. 1, 4th ed. Springer, New York, NY.

Krstic, M., Smyshlyaev, A., 2008. Backstepping boundary control for first-order hyperbolic PDEs and application to systems with actuator and sensor delays. Systems & Control Letters 57, 750–758.

Lax, P.D., 2010. Hyperbolic difference equations: a review of the Courant-Friedrichs-Lewy paper in the light of recent developments. IBM Journal of Research and Development 11, 235–238.

Leugering, G., Mophou, G., 2018. Instantaneous optimal control of friction dominated flow in a gas-network. International Series of Numerical Mathematics 169, 75–88.

LeVeque, R.J., 2006. Nonlinear conservation laws and finite volume methods. In: Computational Methods for Astrophysical Fluid Flow, Vol. M, pp. 1–159.

Li, T., 2010. Exact boundary controllability of nodal profile for quasilinear hyperbolic systems. Mathematical Methods in the Applied Sciences 33, 2101–2106.

Li, T., 2019. Exact boundary controllability of nodal profile for hyperbolic systems. Journal of Mathematical Research with Applications 39, 554–562.

Li, T., Jin, Y., 2001. Semi-global C^1 solution to the mixed initial-boundary value problem for quasilinear hyperbolic systems. Chinese Annals of Mathematics, Series B 22, 325–336.

Li, T., Rao, B., 2019. Exact boundary controllability for one-dimensional quasilinear hyperbolic systems. In: One-Dimensional Hyperbolic Conservation Laws and Their Applications. Lecture Notes for the LIASFMA Shanghai Summer School. Shanghai, China, August 16–27, 2015. World Scientific/Higher Education Press, Hackensack, NJ/Beijing, pp. 1–86.

Lions, P.L., Perthame, B., Tadmor, E., 1994. Kinetic formulation of the isentropic gas dynamics and p-systems. Communications in Mathematical Physics 163, 415–431.

Mantri, Y., Herty, M., Noelle, S., 2019. Well-balanced scheme for gas-flow in pipeline networks. Networks and Heterogeneous Media 14, 659–676.
Menon, E.S., 2005. Gas Pipeline Hydraulics.
Mindt, P., Lang, J., Domschke, P., 2019. Entropy-preserving coupling of hierarchical gas models. SIAM Journal on Mathematical Analysis 51, 4754–4775.
Mironchenko, A., Prieur, C., 2020. Input-to-state stability of infinite-dimensional systems: recent results and open questions. SIAM Review 62, 529–614.
Naumann, A., Kolb, O., Semplice, M., 2018. On a third order CWENO boundary treatment with application to networks of hyperbolic conservation laws. Applied Mathematics and Computation 325, 252–270.
Osiadacz, A., 1984. Simulation of transient gas flows in networks. International Journal for Numerical Methods in Fluids 4, 13–24.
Osiadacz, A.J., Chaczykowski, M., 2001. Comparison of isothermal and non-isothermal pipeline gas flow models. Chemical Engineering Journal 81, 41–51.
Pfaff, S., Ulbrich, S., 2017. Optimal control of scalar conservation laws by on/off-switching. Optimization Methods & Software 32, 904–939.
Prieur, C., Winkin, J., Bastin, G., 2008. Robust boundary control of systems of conservation laws. Mathematics of Control, Signals and Systems 20, 173–197.
Prieur, C., Winkin, J.J., 2018. Boundary feedback control of linear hyperbolic systems: application to the Saint-Venant–Exner equations. Automatica 89, 44–51.
Reigstad, G.A., Flatten, T., Haugen, N.E., Ytrehus, T., 2015. Coupling constants and the generalized Riemann problem for isothermal junction flow. Journal of Hyperbolic Differential Equations 12, 37–59.
Schmidt, M., Steinbach, M.C., Willert, B.M., 2016. High detail stationary optimization models for gas networks: validation and results. Optimization and Engineering 17, 437–472.
Schuster, M., Strauch, E., Gugat, M., Lang, J. Probabilistic constrained optimization on flow networks.
Seo, T., Bayen, A.M., Kusakabe, T., Asakura, Y., 2017. Traffic state estimation on highway: a comprehensive survey.
Sontag, E.D., 2008. Input to state stability: basic concepts and results. In: Nonlinear and Optimal Control Theory. Lectures Given at the C.I.M.E. Summer School. Cetraro, Italy, June 19–29, 2004. Springer, Berlin, pp. 163–220.
Sundar, K., Zlotnik, A., 2018. State and parameter estimation for natural gas pipeline networks using transient state data. IEEE Transactions on Control Systems Technology.
Toro, E.F., 2009. Riemann Solvers and Numerical Methods for Fluid Dynamics.
Ulbrich, S., 2003a. A sensitivity and adjoint calculus for discontinuous solutions of hyperbolic conservation laws with source terms. SIAM Journal on Control and Optimization 41, 740–797.
Ulbrich, S., 2003b. Adjoint-based derivative computations for the optimal control of discontinuous solutions of hyperbolic conservation laws. Systems & Control Letters 48, 313–328.
van Ackooij, W., Pérez-Aros, P., 2020. Gradient formulae for nonlinear probabilistic constraints with non-convex quadratic forms. Journal of Optimization Theory and Applications 185, 239–269.
Wang, K., Wang, L., 2020. Asymptotic stability of the exact boundary controllability of nodal profile for 1d quasi-linear wave equations. Mathematical Methods in the Applied Sciences 43, 1500–1515.
Xu, C.-Z., Sallet, G., 2014. Multivariable boundary PI control and regulation of a fluid flow system. Mathematical Control and Related Fields 4, 501–520.
Zhuang, K., Leugering, G., Li, T., 2019. Exact boundary controllability of nodal profile for Saint-Venant system on a network with loops. Journal de Mathématiques Pures et Appliquées (9) 129, 34–60.
Zlotnik, A., Chertkov, M., Backhaus, S., 2015. Optimal control of transient flow in natural gas networks. In: Proceedings of the IEEE Conference on Decision and Control, Vol. 54. IEEE, pp. 4563–4570.
Zuazua, E., 2017. Large time control and turnpike properties for wave equations. Annual Reviews in Control 44, 199–210.

Chapter 3

Optimal control, numerics, and applications of fractional PDEs

Harbir Antil[a,*], Thomas Brown[a], Ratna Khatri[b], Akwum Onwunta[c], Deepanshu Verma[d], and Mahamadi Warma[a]

[a]*Dept. of Mathematical Sciences, Center for Mathematics and Artificial Intelligence, George Mason University, Fairfax, VA, United States,* [b]*Optical Sciences Division, U.S. Naval Research Laboratory, Washington, DC, United States,* [c]*Department of Industrial and Systems Engineering, Lehigh University, Bethlehem, PA, United States,* [d]*Department of Mathematics, Emory University, Atlanta, GA, United States*

**Corresponding author: e-mail address: hantil@gmu.edu*

Contents

Abstract

This chapter provides a brief review of recent developments on two nonlocal operators: fractional Laplacian and fractional time derivative. We start by accounting for several applications of these operators in imaging science, geophysics, harmonic maps, and deep (machine) learning. Various notions of solutions to linear fractional elliptic equations are provided and numerical schemes for fractional Laplacian and fractional time derivative are discussed. Special emphasis is given to exterior optimal control problems with a linear elliptic equation as constraints. In addition, optimal control problems with interior control and state constraints are considered. We also provide a discussion on fractional

Handbook of Numerical Analysis, Volume 23, ISSN 1570-8659, https://doi.org/10.1016/bs.hna.2021.12.003

deep neural networks, which is shown to be a minimization problem with fractional in time ordinary differential equation as constraint. The paper concludes with a discussion on several open problems.

Keywords
Imaging, Geophysics, Harmonic maps, Quantum spin chains, Deep learning, Nonlocal operators, Fractional Laplacian, Caputo fractional derivative, Exterior optimal control, State constraints, Open problems

MSC Codes
49J20, 49K20, 35S15, 65R20

1 Introduction and applications of fractional operators

Fractional (nonlocal) models have received a significant amount of attention during the recent years. This can be attributed to their ability to account for long range interactions and their flexibility in being applicable to non-smooth functions. Motivated by these facts and several applications, this research area has attracted a significant amount of attention on both theoretical and computational fronts. The goal of this chapter is to discuss some of these developments. This chapter is meant to provide an overview, largely motivated by authors' own research, and it is not meant to be completely exhaustive. We begin by discussing several applications of fractional operators in data science, physics, and machine learning.

Fractional Laplacian in imaging: Image denoising problems have received a lot of attention over last several decades. A major breakthrough occurred when the article (Rudin et al., 1992) considered the Total Variation (TV) regularization to capture the jumps in an image reconstruction, during the noise removal process. Ever since, this approach has attracted a lot of attention from researchers interested in both theory and numerical methods. However, it is a still quite a challenging problem because of the non-smooth nature of TV and the fact that when we formally write the Euler-Lagrange equations, they are nonlinear and degenerate. Recently in Antil and Bartels (2017), the authors replaced the total variation semi-norm by fractional Laplacian regularization with fractional Laplacian $(-\Delta)^s$ of order s with constant $s \in (0, 1)$. For a mathematical justification, we refer to Bergounioux and Trélat (2010, Subsection 2.3) and Antil and Bartels (2017). The fractional variational model is given by

$$\min_u \frac{1}{2} \int_\Omega |(-\Delta)^{\frac{s}{2}} u|^2 \,\mathrm{d}x + \frac{\lambda}{2} \|u - f\|^2_{L^2(\Omega)} \tag{1.1}$$

where f is the noisy image, λ is the regularization parameter, and Ω is the image domain. Moreover, periodic boundary conditions are assumed, see Antil and Bartels (2017) for the definition of $(-\Delta)^s$. Notice that in this paper we will focus on the so-called integral fractional Laplacian, these two definitions coincide in case of periodic boundary conditions (Abatangelo and Valdinoci, 2019, Eq. (2.53)). The key advantage is that the Euler-Lagrange equation in the case of (1.1) is a linear elliptic fractional partial differential equation (PDE) of the type

$$(-\Delta)^s u + \lambda u = \lambda f \quad \text{in } \Omega \tag{1.2}$$

which can be easily solved in the case of periodic boundary conditions by using Fourier series (Antil and Bartels, 2017), or for general zero exterior conditions by using the method of bilinear forms. From left to right, Fig. 1 shows the original image, noisy image, denoised image using fractional model with fractional exponent of $s = 0.42$ and $\lambda = 10$. The rightmost image has been obtained using TV regularization approach (Chambolle, 2004; Burger et al., 2007), where we have even optimized over λ. It is clear that the reconstruction quality in the fractional case is comparable to the TV case even without optimizing over λ, however, it is significantly cheaper than TV. We also refer to Antil et al. (2020a) for the extension of this work to tomographic reconstruction problems where the second term in (1.1) has been replaced by $\|Ku - f\|^2_{L^2(\Omega)}$, with K denoting a linear operator. The article (Antil et al., 2020a) also discusses a bilevel optimization and machine learning based strategy to identify parameters such as λ and s. See also D'Elia et al. (2021) for another recent work on bilevel strategies for nonlocal problems. We remark that the total variation regularization is known to promote sparsity. However this sparsity behavior has also been expected in fractional Sobolev space case, under the strict condition $s < 1$, see Bergounioux and Trélat (2010, Subsection 2.3) and also Abraham et al. (2008); Bergounioux et al. (2018).

FIGURE 1 From left to right, original image, noisy image, denoised image using fractional regularization with randomly selected regularization parameter and denoised image using TV regularization with optimized regularization parameter. The figures have been reproduced from Antil and Bartels (2017).

The article (Antil and Rautenberg, 2019) introduces a new model with variable order $s(x) \in [0, 1]$ with $x \in \Omega$, i.e., s is not a constant any longer but instead, it is a function. Moreover, s is allowed to take the extreme values of 0 and 1. This can be quite beneficial for applications (e.g. imaging) where it is critical to capture the jump across the interfaces (e.g., phase field models). Fig. 2 shows a comparison between TV regularization and the variable order approach. A strategy to identify the function s is also discussed in Antil and

Rautenberg (2019). We clearly notice that TV is rounding up the corners in the middle panel, while the variable s based approach in the right panel leads to an almost perfect reconstruction. Notice that the regularization parameter has been optimized in TV case and has been chosen randomly in the variable s case. We further emphasize that it might be possible to further improve TV results, for instance, by using variable order λ regularizer (Calatroni et al., 2017). However, this approach is still challenging to implement in comparison to the single, linear PDE solve done above, in the case of variable order fractional Laplacian.

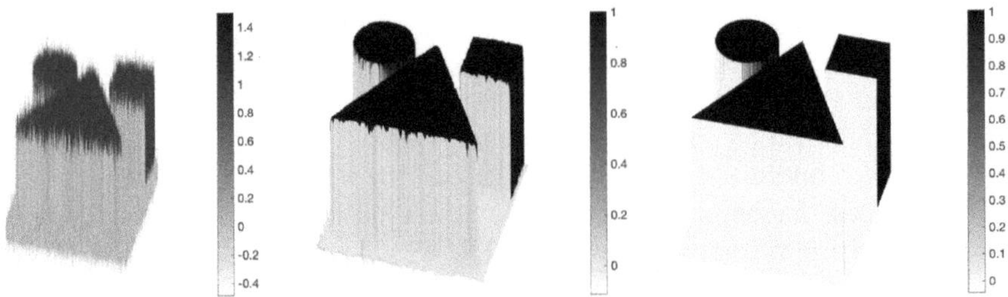

FIGURE 2 *Left:* Noisy image. *Middle:* reconstruction using TV regularization where the regularization parameter has been optimized. *Right:* Reconstruction using $s(x)$ approach where the regularization parameter λ has been chosen randomly. Notice that the reconstruction using $s(x)$ approach is almost perfect. The figures have been reproduced from Antil and Rautenberg (2019).

Fractional Laplacian in geophysics: The article (Weiss et al., 2020) has recently derived the fractional Helmholtz equation using the first principle arguments in conjunction with a constitutive relationship. A finite element based numerical method motivated by Bonito and Pasciak (2015) has been introduced to solve the problem. These concepts are applied to the scalar Helmholtz equation and its use in electromagnetic interrogation of Earth's interior through the magnetotelluric method. For the magnetotelluric problem, several interesting features are observable in the field data: long-range correlation of the predicted electromagnetic fields; a power-law relationship between the squared impedance amplitude and squared wavenumber whose slope is a function of the fractional exponent within the governing Helmholtz equation; and, a non-constant apparent resistivity spectrum whose variability arises solely from the fractional exponent. The latter can be seen in Fig. 3 (left). The figure on the right is the apparent resistivity data from USArray MT station for KSP34 located NW of Kansas City, KS, USA from the US Array. Notice that the fractional model provides an excellent qualitative match.

Other applications of fractional Laplacian include, *harmonic maps* (Antil et al., 2021a) and *quantum spin chains* (Haldane, 1986) etc.

Fractional time derivative in machine learning: Dynamical system based deep neural networks (DNNs) are gaining popularity, as they allow us to model connectivity among the network layers. Recently in Antil et al. (2020b), we have considered a new framework for classification problems and it has been further

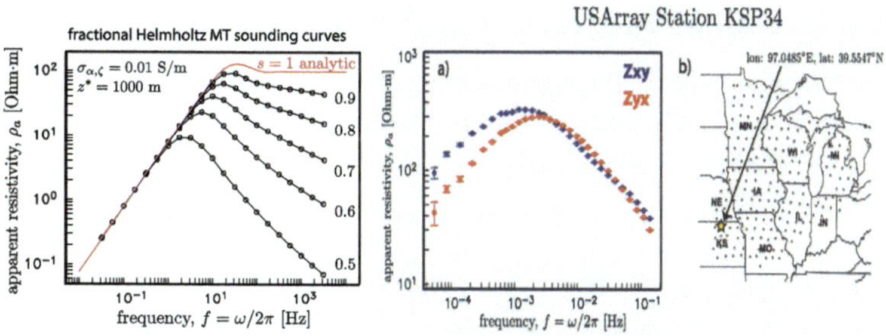

FIGURE 3 *Left:* Results from numerical simulations. *Right:* The apparent resistivity data from US-Array MT station for KSP34 located NW of Kansas City, KS, USA from the US Array.

extended to parameterized partial differential equations in Antil et al. (2021d). These works are motivated by Ruthotto and Haber (2019). These articles attempt to develop mathematical models for the analysis and understanding of DNNs. The idea is to consider DNNs as dynamical systems. In particular, in Benning et al. (2019); Günther et al. (2020), the process of training a DNN is thought of as an optimization problem constrained by a discrete ordinary differential equation. Designing DNN algorithms at the continuous level has the appealing advantage of architecture independence; in other words, the number of optimization iterations remains stable as the number of layers are increased. The articles (Antil et al., 2020b, 2021d) specifically consider fractional ODE as constraints. This fractional DNN allows the network to access historic information of input and gradients across antecedent layers since each layer is connected to all previous layers unlike the standard DNN, where each layer is connected only to a single previous layer.

Outline: In view of the above motivations, this chapter focuses on several optimization problems constrained by fractional partial differential equations. We will use the term "optimal control" to describe such problems. In Section 2 we begin by defining two fractional operators, i.e., fractional Laplacian and fractional time derivative. Section 3 focuses on various notions of solutions to a linear elliptic fractional diffusion equation. We discuss various notions of solutions, including weak and very-weak solutions under various regularity conditions on data. We also provide a discussion on various approaches to approximate the fractional Laplacian and fractional time derivative. Section 4 focuses on a new notion of optimal control, i.e., exterior optimal control. Moreover, Section 5 focuses on elliptic fractional state constraint problems where we also discuss a Moreau-Yosida based algorithm to solve the problem. Section 6 is devoted to fractional deep neural networks which is also an optimal control problem governed by time fractional ordinary differential equation. We conclude by stating various open problems in Section 7. We emphasize that this is a review article and the results have been previously published in the authors' other research papers.

2 Two fractional operators and their properties

Unless otherwise stated, throughout the paper, we assume that $\Omega \subset \mathbb{R}^N$ is a bounded open Lipschitz domain with boundary $\partial\Omega$. Notice that many of our results hold without the Lipschitz assumption, but we do not get into those details here in this review.

2.1 Fractional Laplacian $(-\Delta)^s$ and nonlocal normal derivative $\mathcal{N}_s$

To define the integral fractional Laplace operator we consider the weighted Lebesgue space

$$\mathbb{L}_s^1(\mathbb{R}^N) = \left\{ f : \mathbb{R}^N \to \mathbb{R} \text{ measurable and } \int_{\mathbb{R}^N} \frac{|f(x)|}{(1+|x|)^{N+2s}}\,\mathrm{d}x < \infty \right\},$$

and first define for $f \in \mathbb{L}_s^1(\mathbb{R}^N)$, $\varepsilon > 0$, and $x \in \mathbb{R}^N$ the quantity

$$(-\Delta)_\varepsilon^s f(x) = C_{N,s} \int_{\{y\in\mathbb{R}^N, |y-x|>\varepsilon\}} \frac{f(x)-f(y)}{|x-y|^{N+2s}}\,\mathrm{d}y,$$

where $C_{N,s} = \left(s2^{2s}\Gamma\left(\frac{2s+N}{2}\right)\right)/\left(\pi^{\frac{N}{2}}\Gamma(1-s)\right)$ is a normalizing constant and Γ is the Euler Gamma function. We then define the *integral version* of the fractional Laplace operator for $s \in (0, 1)$ via a limit passage for $\varepsilon \to 0$, i.e.,

$$(-\Delta)^s f(x) = C_{N,s}\text{P.V.} \int_{\mathbb{R}^N} \frac{f(x)-f(y)}{|x-y|^{N+2s}}\,\mathrm{d}y = \lim_{\varepsilon\to 0} (-\Delta)_\varepsilon^s f(x), \tag{2.1}$$

where P.V. indicates the Cauchy principal value. Note that this definition for the full space $\mathbb{R}^N$ coincides with the spectral definition of the fractional Laplacian obtained using Fourier transform (Di Nezza et al., 2012, Proposition 3.4), see also Caffarelli and Silvestre (2007). Such an equivalence also holds in the case of periodic boundary conditions (Abatangelo and Valdinoci, 2019, Eq. (2.53)).

Fractional order Sobolev spaces: Next we introduce the fractional order Sobolev space $H^s(\mathbb{R}^N)$ for $s \in (0, 1)$ by setting

$$[f]_{H^s(\mathbb{R}^N)} = \|(-\Delta)^{\frac{s}{2}} f\|_{L^2(\mathbb{R}^N)} = C_{N,s}\left(\int_{\mathbb{R}^N}\int_{\mathbb{R}^N} \frac{|f(x)-f(y)|^2}{|x-y|^{N+2s}}\,\mathrm{d}y\,\mathrm{d}x\right)^{\frac{1}{2}},$$

$$\|f\|_{H^s(\mathbb{R}^N)} = \|f\|_{L^2(\mathbb{R}^N)} + [f]_{H^s(\mathbb{R}^N)}.$$

Then, the Sobolev space $H^s(\mathbb{R}^N)$ is defined as

$$H^s(\mathbb{R}^N) = \left\{ f \in L^2(\mathbb{R}^N) : \|f\|_{H^s(\mathbb{R}^N)} < +\infty \right\}$$

is a Hilbert space. We denote the dual of $H^s(\mathbb{R}^N)$ by $H^{-s}(\mathbb{R}^N)$ and $\langle\cdot,\cdot\rangle$ the duality pairing between $H^s(\mathbb{R}^N)$ and $H^{-s}(\mathbb{R}^N)$. We mention that the notation

$\|(-\Delta)^{\frac{s}{2}} f\|_{L^2(\mathbb{R}^N)}$ is not just formal. Since we are in all $\mathbb{R}^N$, the notation has been justified in Di Nezza et al. (2012, Proposition 4.4) by using the Fourier transform.

For bounded open sets $\Omega \subset \mathbb{R}^N$ and parameters $s \in (0, 1)$ we define Sobolev spaces $\widetilde{H}^s(\Omega)$ by considering trivial extensions to $\mathbb{R}^N$, i.e., we set

$$\widetilde{H}^s(\Omega) = \{f \in L^2(\mathbb{R}^N) : (-\Delta)^{\frac{s}{2}} f \in L^2(\mathbb{R}^N), \quad f \equiv 0 \text{ in } \mathbb{R}^N \setminus \Omega\}.$$

We recall the following density result for $s \in (0, 1]$ and bounded domains $\Omega \subset \mathbb{R}^d$ with a Lipschitz continuous boundary (Fiscella et al., 2015)

$$\widetilde{H}^s(\Omega) = \overline{\mathcal{D}(\Omega)}^{\|\cdot\|_{\widetilde{H}^s(\Omega)}},$$

and by a Poincaré type inequality, which is a consequence of Hölder's inequality and Sobolev embedding theorem (Adams and Hedberg, 1996, Theorem 3.1.4.) a norm on $\widetilde{H}^s(\Omega)$ is given by

$$\|f\|_{\widetilde{H}^s(\Omega)} = \|(-\Delta)^{\frac{s}{2}} f\|_{L^2(\mathbb{R}^N)}.$$

Other fractional operators: Next we define the operator $(-\Delta)^s_D$ in $L^2(\Omega)$ as follows:

$$\begin{aligned} D((-\Delta)^s_D) &:= \left\{u|_\Omega,\ u \in \widetilde{H}^s(\Omega) : (-\Delta)^s u \in L^2(\Omega)\right\}, \\ (-\Delta)^s_D(u|_\Omega) &:= (-\Delta)^s u \ \text{ a.e. in } \Omega. \end{aligned} \tag{2.2}$$

Notice that $(-\Delta)^s_D$ is the realization of the fractional Laplace operator $(-\Delta)^s$ in $L^2(\Omega)$ with the zero Dirichlet exterior condition $u = 0$ in $\mathbb{R}^N \setminus \Omega$.

For $u \in H^s(\mathbb{R}^N)$, we define the *nonlocal normal derivative* $\mathcal{N}_s$ as:

$$\mathcal{N}_s u(x) := C_{N,s} \int_\Omega \frac{u(x) - u(y)}{|x - y|^{N+2s}}\, \mathrm{d}y, \quad x \in \mathbb{R}^N \setminus \overline{\Omega},$$

which maps continuously $H^s(\mathbb{R}^N)$ into $H^s_{\text{loc}}(\mathbb{R}^N \setminus \Omega)$. As a result, if $u \in H^s(\mathbb{R}^N)$, then $\mathcal{N}_s u \in L^2(\mathbb{R}^N \setminus \Omega)$. See Antil et al. (2019, Lemma 2.1) for details.

Moreover, the following *integration-by-parts formula* can be found in Antil et al. (2019, Proposition 2.2). If $u \in H^s(\mathbb{R}^N)$ is such that $(-\Delta)^s u \in L^2(\Omega)$, then for every $v \in H^s(\mathbb{R}^N)$ we have that

$$\begin{aligned} \frac{C_{N,s}}{2} \iint_{\mathbb{R}^{2N} \setminus (\mathbb{R}^N \setminus \Omega)^2} \frac{(u(x) - u(y))(v(x) - v(y))}{|x - y|^{N+2s}}\, dx dy &= \int_\Omega v (-\Delta)^s u \, dx \\ &+ \int_{\mathbb{R}^N \setminus \Omega} v \mathcal{N}_s u \, dx, \end{aligned} \tag{2.3}$$

where $\mathbb{R}^{2N} \setminus (\mathbb{R}^N \setminus \Omega)^2 := (\Omega \times \Omega) \cup (\Omega \times (\mathbb{R}^N \setminus \Omega)) \cup ((\mathbb{R}^N \setminus \Omega) \times \Omega)$.

2.2 Fractional time derivative: ${}_C\partial_t^\gamma$

The focus of this section is to define the (strong) fractional order Caputo derivative and to introduce the integration-by-parts formula. The latter is critical to derive the optimality conditions.

Definition 2.1 (Strong Caputo fractional derivative). Let $u \in W^{1,1}([0,T];X)$, with X denoting a Banach space. The (strong) Caputo fractional derivative of order $\gamma \in (0,1)$ is given by

$$
{}_C\partial_t^\gamma u(t) = \frac{1}{\Gamma(1-\gamma)} \int_0^t \frac{u'(r)}{(t-r)^\gamma}\, \mathrm{d}r. \tag{2.4}
$$

The following integration-by-parts formula holds (Agrawal, 2007) under the assumptions of Definition 2.1

$$
\int_0^T v(t)\, {}_C\partial_t^\gamma u(t)\, \mathrm{d}t = \int_0^T \partial_{t,T}^\gamma v(t) u(t)\, \mathrm{d}t + \left[(I_{t,T}^{1-\gamma} v)(t) u(t) \right]_{t=0}^{t=T}, \tag{2.5}
$$

provided the expressions on the left and right hand sides make sense. Here, $\partial_{t,T}^\gamma$ is the right Riemann-Liouville fractional derivative

$$
\partial_{t,T}^\gamma u(t) = -\frac{d}{dt} (I_{t,T}^{1-\gamma} u)(t), \quad \text{where} \quad I_{t,T}^\gamma u(t) := \frac{1}{\Gamma(\gamma)} \int_t^T (r-t)^{\gamma-1} u(r)\, \mathrm{d}r
$$

denotes the right Riemann-Liouville fractional integral of order γ. We refer to Samko et al. (1993, pg. 76) for further details. We notice that if the space X has the Radon-Nikodym property, then $W^{1,1}([0,T];X)$ is the optimal space for which (2.5) makes sense (see e.g. the monograph (Gal and Warma, 2020) for more details).

3 Fractional diffusion equation: analysis and numerical approximation

3.1 Non-homogeneous diffusion equations: analytic results

The focus of this section is on stating well-posedness results for the linear elliptic fractional equation

$$
\begin{cases} (-\Delta)^s u = f & \text{in } \Omega, \\ u = g & \text{in } \mathbb{R}^N \setminus \Omega, \end{cases} \tag{3.1}
$$

under various general assumptions on the data f and g. In particular, we first establish the notion of weak solution in a Sobolev space when g is an appropriate Sobolev function itself. When g is only square integrable, we establish the notion of very weak solution in L^2. Next for $g \equiv 0$, we provide conditions on f

and Ω which leads to boundedness and continuity of u. This is followed by the notion of very weak solution when f is a Radon measure. We will denote the space of Radon measures by $\mathcal{M}(\Omega)$. For parabolic versions of these results, we refer to Antil et al. (2020c,d). We further refer to Antil and Warma (2020) for semilinear problems, Antil and Warma (2019) for quasi-linear problems, Antil and Rautenberg (2018); Antil et al. (2021e) for variational and quasi-variational inequalities. Similar to the classical case of $s = 1$, the problem (3.1) serves as a fundamental model problem in the nonlocal case. It also serves as a model problem for some of the applications mentioned in Section 1.

We begin by stating the various notions of solutions to (3.1). This will be followed by several well-posedness results.

Definition 3.1 (Solutions to elliptic Dirichlet problem). We define various notions of solutions to (3.1):

(i) Weak solution to non-homogeneous problem: Let $f \in \widetilde{H}^{-s}(\Omega)$, $g \in H^s(\mathbb{R}^N \setminus \Omega)$, and $\mathcal{Z} \in H^s(\mathbb{R}^N)$ be such that $\mathcal{Z}|_{\mathbb{R}^N\setminus\Omega} = g$. A function $u \in H^s(\mathbb{R}^N)$ is said to be a weak solution to (3.1) if $u - \mathcal{Z} \in \widetilde{H}^s(\Omega)$ and

$$\frac{C_{N,s}}{2}\int_{\mathbb{R}^N}\int_{\mathbb{R}^N}\frac{(u(x)-u(y))(v(x)-v(y))}{|x-y|^{N+2s}}\,dx\,dy = \langle f, v\rangle, \quad \forall v \in \widetilde{H}^s(\Omega). \tag{3.2}$$

(ii) Very-weak solution to non-homogeneous Dirichlet problem: Let $g \in L^2(\mathbb{R}^N \setminus \Omega)$ and $f \in \widetilde{H}^{-s}(\Omega)$. A function $u \in L^2(\mathbb{R}^N)$ is said to be a very-weak solution to (3.1) if the identity

$$\int_\Omega u(-\Delta)^s v\,dx = \langle f, v\rangle - \int_{\mathbb{R}^N\setminus\Omega} g\mathcal{N}_s v\,dx, \tag{3.3}$$

holds for every $v \in V := \{v \in \widetilde{H}^s(\Omega) : (-\Delta)^s v \in L^2(\Omega)\}$.

(iii) Very-weak solution to homogeneous Dirichlet problem with measure data: Let p satisfy

$$p > \frac{N}{2s} \text{ if } N > 2s, \quad p > 1 \text{ if } N = 2s, \quad p = 1 \text{ if } N < 2s, \tag{3.4}$$

and $\frac{1}{p} + \frac{1}{p'} = 1$. Let $f \in \mathcal{M}(\Omega)$. A function $u \in L^{p'}(\Omega)$ is said to be a very-weak solution to (3.1), if for every $v \in V := \{v \in C_0(\Omega) \cap \widetilde{H}^s(\Omega) : (-\Delta)^s v \in L^p(\Omega)\}$ we have

$$\int_\Omega u(-\Delta)^s v\,dx = \int_\Omega v\,df. \tag{3.5}$$

Here $C_0(\Omega)$ is the space of all continuous functions in $\overline{\Omega}$ that vanish on $\partial\Omega$.

Theorem 3.2. *The following results hold for non-homogeneous Dirichlet boundary value problems:*

(i) ***Weak solution:*** *Given* $f \in \widetilde{H}^{-s}(\Omega)$ *and* $g \in H^s(\mathbb{R}^N \setminus \Omega)$ *there exists a unique weak solution to* (3.1) *according to Definition (3.1) (i). Also there is a constant* $C > 0$ *such that*

$$\|u\|_{H^s(\mathbb{R}^N)} \le C\left(\|f\|_{\widetilde{H}^{-s}(\Omega)} + \|g\|_{H^s(\mathbb{R}^N\setminus\Omega)}\right). \tag{3.6}$$

(ii) ***Very-weak solution:*** *Given* $f \in \widetilde{H}^{-s}(\Omega)$, $g \in L^2(\mathbb{R}^N \setminus \Omega)$ *then there exists a unique very-weak solution* u *to* (3.1) *according to Definition (3.1) (ii) that fulfills*

$$\|u\|_{L^2(\Omega)} \le C\left(\|f\|_{\widetilde{H}^{-s}(\Omega)} + \|g\|_{L^2(\mathbb{R}^N\setminus\Omega)}\right), \tag{3.7}$$

and the following assertions hold: (a) Every weak solution of (3.1) *is also a very-weak solution. (b) Every very-weak solution of* (3.1) *that belongs to* $H^s(\mathbb{R}^N)$ *is also a weak solution.*

In addition, the following holds true when $g \equiv 0$*:*

(iii) u ***is bounded in*** $L^\infty(\Omega)$*: Let* $f \in L^p(\Omega)$ *with* p *as in* (3.4), *then* $u \in L^\infty(\Omega) \cap \widetilde{H}^s(\Omega)$. *Moreover, if* Ω *fulfills the exterior cone condition, then the weak solution* $u \in C_0(\Omega)$ *and there is a* $C = C(N, s, p, \Omega) > 0$ *such that* $\|u\|_{C_0(\Omega)} \le C\|f\|_{L^p(\Omega)}$.

(iv) ***Very-weak solution with measure*** f*: Let* $f \in \mathcal{M}(\Omega)$, p *as in* (3.4), *and* Ω *fulfills the exterior cone condition, then there exists a unique* $u \in L^{p'}(\Omega)$ *solving* (3.1) *according to Definition (3.1) (iii), and there is a constant* $C = C(N, s, p, \Omega) > 0$ *such that* $\|u\|_{L^{p'}(\Omega)} \le C\|f\|_{\mathcal{M}(\Omega)}$.

3.2 Non-homogeneous diffusion equations: numerical approximation

Approximation of fractional Laplacian is a challenging topic and has received a tremendous amount of attention recently. However, most of the focus has been on the homogeneous problem. There exist several efficient works in 1D but for $N > 1$ the problem is very challenging and the number of works are limited. The main difficulty is the fact that the bilinear form (3.2) contains a singular integral and the traditional finite element approaches are not amenable to this. The first work that rigorously provides approximation to the weak solutions is by Acosta and Borthagaray (2017), see also Acosta et al. (2017). However, implementations in these works have been limited to $N = 2$ dimensions. We also refer to another finite element approach by Bonito et al. (2019), which also works for $N = 3$.

In contrast, in Antil et al. (2021c) the authors have introduced an efficient spectral method which works in arbitrary Lipschitz domains. Under this method, the evaluation of the fractional Laplacian and its application onto a vector has complexity of $\mathcal{O}(M \log(M))$ where M is the number of unknowns. For instance, for exponent $s = 1/4$ the 3D implementation can solve the Dirichlet problem with $5 \cdot 10^6$ unknowns under 2 hours on a standard office workstation.

The spectral method presented in Antil et al. (2021c) also extends to the nonhomogeneous Dirichlet problem. But below, instead, we briefly discuss a finite element method based on Antil et al. (2019, 2020c). We refer to Acosta et al. (2019) for an alternative approach. The main idea in Antil et al. (2019, 2020c) is to approximate the solutions to a Dirichlet problem by solutions to a Robin problem. Some of the key challenges for the Dirichlet problem are:

- Since we are interested in optimal control problems with L^2-exterior controls, the correct notion of solution is as given in (3.3). Numerically, this will require approximating $\mathcal{N}_s$.
- The control equation, in addition, will require approximation of $\mathcal{N}_s$ applied to the adjoint variable.

In other words, we have to approximate $\mathcal{N}_s$ in addition to approximating $(-\Delta)^s$. Notice that approximation of $\mathcal{N}_s$ using a finite element method will require triangulation of unbounded domain $\mathbb{R}^N \setminus \Omega$. A typical finite element is not well-suited for this. We further emphasize that even though $(-\Delta)^s$ is also defined on an unbounded domain, but its approximation using finite element method is by now well-established (Acosta et al., 2017). Approximating the Dirichlet problem by the Robin problem helps overcome the $\mathcal{N}_s$ computational challenge. To define the Robin problem, we consider the Sobolev space introduced in Dipierro et al. (2017). Let $\kappa \in L^1(\mathbb{R}^N \setminus \Omega)$ be fixed and set

$$W^{s,2}_{\Omega,\kappa} := \left\{ u : \mathbb{R}^N \to \mathbb{R} \text{ measurable and } \|u\|_{W^{s,2}_{\Omega,\kappa}} < \infty \right\},$$

where

$$\|u\|_{W^{s,2}_{\Omega,\kappa}} := \Bigg(\|u\|^2_{L^2(\Omega)} + \|u\|^2_{L^2(\mathbb{R}^N \setminus \Omega, \mu)} + \int\int_{\mathbb{R}^{2N} \setminus (\mathbb{R}^N \setminus \Omega)^2} \frac{|u(x) - u(y)|^2}{|x-y|^{N+2s}} dxdy \Bigg)^{\frac{1}{2}}, \tag{3.8}$$

and the measure μ on $\mathbb{R}^N \setminus \Omega$ is given by $d\mu = |\kappa| \, dx$.

The article (Dipierro et al., 2017, Proposition 3.1) has shown that for $\kappa \in L^1(\mathbb{R}^N \setminus \Omega)$, $W^{s,2}_{\Omega,\kappa}$ is a Hilbert space. Now for every $n \in \mathbb{N}$, we define the generalized Robin problem

$$\begin{cases} (-\Delta)^s u = f & \text{in } \Omega, \\ \mathcal{N}_s u + n\kappa u = n\kappa z & \text{in } \mathbb{R}^N \setminus \Omega. \end{cases} \tag{3.9}$$

Then the following result holds (see Antil et al., 2019, Proposition 3.9)

Proposition 3.3. *Let $n \in \mathbb{N}$ and $\kappa \in L^1(\mathbb{R}^N \setminus \Omega) \cap L^\infty(\mathbb{R}^N \setminus \Omega)$. Then for every $z \in L^2(\mathbb{R}^N \setminus \Omega, \mu)$ and $f \in (W^{s,2}_{\Omega,\kappa})^\star$, there exists a weak solution $u \in W^{s,2}_{\Omega,\kappa}$ of*

(3.9) *in the following sense:*

$$\frac{C_{N,s}}{2}\int\int_{\mathbb{R}^{2N}\setminus(\mathbb{R}^N\setminus\Omega)^2}\frac{(u(x)-u(y))(v(x)-v(y))}{|x-y|^{N+2s}}\,dxdy+n\int_{\mathbb{R}^N\setminus\Omega}\kappa uv\,dx$$
$$=\langle f,v\rangle_{(W^{s,2}_{\Omega,\kappa})^\star,W^{s,2}_{\Omega,\kappa}}+n\int_{\mathbb{R}^N\setminus\Omega}\kappa zv\,dx, \tag{3.10}$$

for all $v\in W^{s,2}_{\Omega,\kappa}$.

Let us mention that the motivation of using κ in the second equation of (3.9) is that with respect to the measure μ, the set $\mathbb{R}^N\setminus\Omega$ is bounded, that is, $\mu(\mathbb{R}^N\setminus\Omega)=\int_{\mathbb{R}^N\setminus\Omega}|\kappa|\,dx<\infty$, since $\kappa\in L^1(\mathbb{R}^N\setminus\Omega)$. This is a huge computational advantage over the use of unbounded set $\mathbb{R}^N\setminus\Omega$ with respect to the classical N-dimensional Lebesgue measure. However, we emphasize that all the results can be obtained by dropping out κ in (3.9) and using another fractional order Sobolev space. More precisely, in that case one has to use the following space:

$$W:=\Bigg\{u:\mathbb{R}^N\to\mathbb{R}\text{ measurable}:$$
$$\|u\|^2_{L^2(\mathbb{R}^N)}+\int\int_{\mathbb{R}^{2N}\setminus(\mathbb{R}^N\setminus\Omega)^2}\frac{|u(x)-u(y)|^2}{|x-y|^{N+2s}}dxdy<\infty\Bigg\}.$$

Next we make the following assumption.

Assumption 3.4. *We assume that* $\kappa\in L^1(\mathbb{R}^N\setminus\Omega)\cap L^\infty(\mathbb{R}^N\setminus\Omega)$ *and satisfies* $\kappa>0$ *almost everywhere in* $K:=supp[\kappa]\subset\mathbb{R}^N\setminus\Omega$, *where the Lebesgue measure* $|K|>0$.

Under this assumption, the solution to (3.10) belongs to $W^{s,2}_{\Omega,\kappa}\cap L^2(\mathbb{R}^N\setminus\Omega)$ (Antil et al., 2019, Lemma 6.2). Moreover, the following approximation result holds:

Theorem 3.5 (Approximation of Dirichlet solution by Robin solution). *Under Assumption 3.4, the following assertions hold:*

(a) *Let* $z\in H^s(\mathbb{R}^N\setminus\Omega)$ *and* $u_n\in W^{s,2}_{\Omega,\kappa}\cap L^2(\mathbb{R}^N\setminus\Omega)$ *be the weak solution of* (3.10). *Let* $u\in H^s(\mathbb{R}^N)$ *be the weak solution to the state equation* (3.2). *Then there is a constant* $C>0$ *(independent of* n*) such that*

$$\|u-u_n\|_{L^2(\mathbb{R}^N)}\le\frac{C}{n}\|u\|_{H^s(\mathbb{R}^N)}. \tag{3.11}$$

In particular u_n *converges strongly to* u *in* $L^2(\mathbb{R}^N)$ *as* $n\to\infty$.

(b) *Let* $z\in L^2(\mathbb{R}^N\setminus\Omega)$ *and* $u_n\in W^{s,2}_{\Omega,\kappa}\cap L^2(\mathbb{R}^N\setminus\Omega)$ *be the weak solution of* (3.10). *Then there exist a subsequence that we still denote by* $\{u_n\}_{n\in\mathbb{N}}$

and a function $\tilde{u} \in L^2(\mathbb{R}^N)$ such that $u_n \rightharpoonup \tilde{u}$ in $L^2(\mathbb{R}^N)$ as $n \to \infty$, and $\tilde{u}$ satisfies

$$\int_\Omega \tilde{u}(-\Delta)^s v \, dx = -\int_{\mathbb{R}^N \setminus \Omega} \tilde{u} \mathcal{N}_s v \, dx, \tag{3.12}$$

for all $v \in V := \{v \in \widetilde{H}^s(\Omega) : (-\Delta)^s v \in L^2(\Omega)\}$.

Next, we introduce a discrete scheme to approximate (3.10). Let $\widetilde{\Omega}$ be an open bounded set that contains Ω. We consider a conforming simplicial triangulation of Ω and $\widetilde{\Omega} \setminus \Omega$ such that the partition remains admissible in the classical sense (see Ciarlet, 2002, (FEM 1), pg. 38). We assume that the support of z and κ is contained in $\widetilde{\Omega} \setminus \Omega$. We let the finite element space $\mathbb{V}_h$ on $\widetilde{\Omega}$ to be the set of continuous piecewise linear functions. Then it remains to approximate the weak form (3.10). All other terms are standard and can be done using any standard finite element code and quadrature rule, except the stiffness matrix. We assemble the latter using (Acosta et al., 2017) and the details of this discretization can be found in another chapter in this handbook. All other matrices are computed by using quadrature which is accurate for polynomials of degree less than or equal to 4.

We consider an example taken from Acosta et al. (2019). Let $\Omega = B_0(1/2) \subset \mathbb{R}^2$, i.e., a ball centered at 0 with radius 1/2 and $\widetilde{\Omega} = B_0(3/2)$. Our goal is to find u solving $(-\Delta)^s u = 2$ in Ω and $u(\cdot) = \frac{2^{-2s}}{\Gamma(1+s)^2}(1 - |\cdot|)_+^s$ in $\mathbb{R}^N \setminus \Omega$. Fig. 4 shows our results and confirms our theoretical findings in Theorem 3.5.

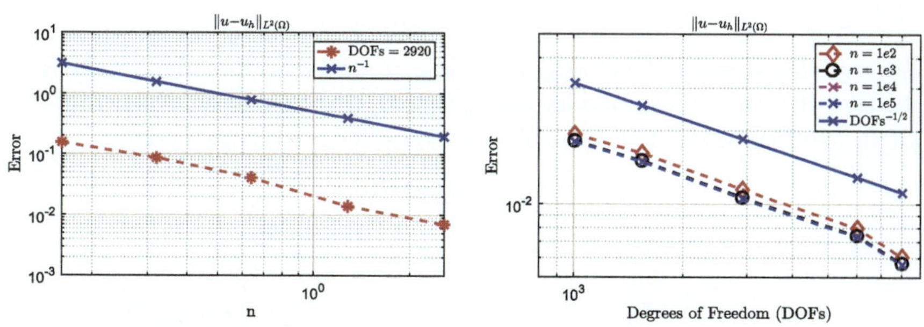

FIGURE 4 Left panel: Let $s = 0.5$ and DoFs $= 2920$ be fixed. We let $\kappa = 1$ and consider the L^2-error between the actual solution u to the Dirichlet problem and its approximation u_h which solves the Robin problem. We have plotted the error with respect to n. The solid line denotes a reference line and the actual error. We observe a rate of $1/n$ which confirms our theoretical result (3.11). Right panel: Let $s = 0.5$ be fixed. For each $n = 1e2, 1e3, 1e4, 1e5$ we have plotted the L^2-error with respect to the degrees of freedom (DOFs) as we refine the mesh. Notice that the error is stable with respect to n. Moreover, the observed rate of convergence is $(\text{DoFs})^{-\frac{1}{2}}$ and is independent of n.

3.3 Fractional time derivative: numerical approximation

We consider an Euler scheme to approximate ${}_C\partial_t$ and we state the result with the help of a nonlinear ODE (Lin and Xu, 2007; Lin et al., 2011; Nochetto et al., 2016)

$$d_t^{\gamma} u(t) = f(u(t)), \quad u(0) = u_0. \tag{3.13}$$

Consider the time discretization of the interval $[0, T]$ uniformly with step size τ. We let $0 = t_0 < t_1 < \cdots < t_{j+1} < \cdots < t_N = T$, where $t_j = j\tau$.

Then, using an Euler-scheme, the discretization of (3.13) is given by: for $j = 0, ..., N-1$,

$$u(t_{j+1}) = u(t_j) - \sum_{k=0}^{j-1} a_{j-k}\,(u(t_{k+1}) - u(t_k)) + \tau^{\gamma}\Gamma(2-\gamma) f(u(t_j)), \tag{3.14}$$

where the coefficients a_{j-k} are given by,

$$a_{j-k} = (j+1-k)^{1-\gamma} - (j-k)^{1-\gamma}. \tag{3.15}$$

Next, we present an example illustrating this approach. Consider the differential equation:

$$d_t^{\frac{1}{2}} u(t) = -4u(t), \quad u(0) = 0.5. \tag{3.16}$$

Then, the exact solution to (3.16) is given by, see Samko et al. (1993, Section 42), also Podlubny (1999, Section 1.2), $u(t) = 0.5\, E_{0.5}(-4t^{0.5})$, where E_{α}, with $0 < \alpha \in \mathbb{R}$, is the Mittag Leffler function, see Podlubny (1999, Pg. 17), defined by $E_{\alpha}(z) = E_{\alpha,1}(z) = \sum_0^{\infty} \frac{z^k}{\Gamma(\alpha k+1)}$. Fig. 5 depicts the true solution and the numerical solution using the discretization (3.14) for the above example with uniform step size $\tau = 0.005$ and final time, $T = 1$. For numerical analysis of partial differential equation with strong Caputo derivatives, see Nochetto et al. (2016); Antil et al. (2016); Jin et al. (2019) and the references therein.

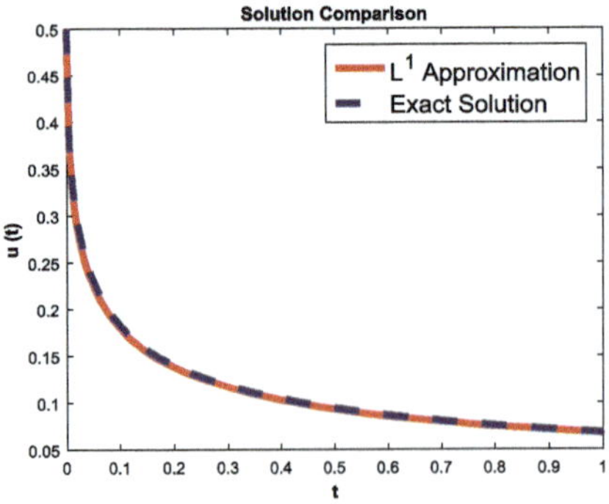

FIGURE 5 Comparison of exact solution of (3.16) and its approximation using an Euler-scheme.

4 Exterior optimal control of fractional parabolic PDEs with control constraints

This section provides details on the novel exterior optimal control problem introduced in Antil et al. (2019, 2020c). Under classical setting, the control either is on the boundary or in the interior, but this new framework allows the control placement away from the domain. This is depicted in Fig. 6. Such a situation can potentially arise in various scenarios, for instance, in magnetic drug targeting (Antil et al., 2018b,a) where the drug with ferromagnetic particles is injected into the body and an external magnetic field is used to steer it to a desired location.

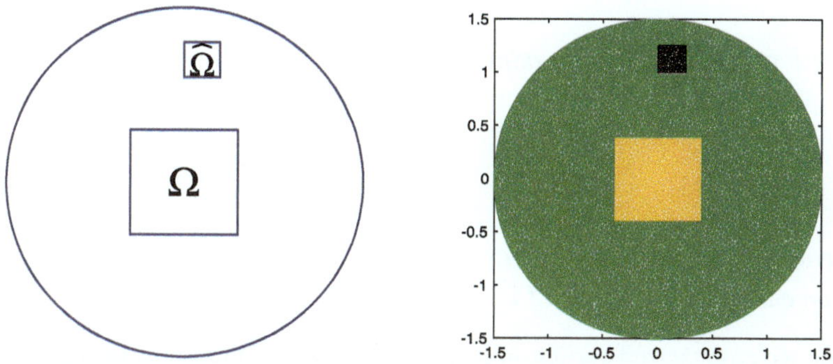

FIGURE 6 *Left:* External optimal control setup for a diffusion process in Ω with control supported in $\widehat{\Omega}$, disjoint from Ω. This is different than the classical control approaches where the control must be either inside Ω or on the boundary of Ω. *Right:* A finite element mesh.

Let $Z_D := L^2(\mathbb{R}^N \setminus \Omega)$, $U_D := L^2(\Omega)$, and $\lambda \geq 0$ be a constant penalty parameter. Then the fundamental optimal control problem amounts to:

$$\min_{(u,z)\in(U_D,Z_D)} J(u) + \frac{\lambda}{2}\|z\|^2_{Z_D}, \tag{4.1a}$$

subject to the fractional Dirichlet exterior value problem: Find $u \in U_D$ solving

$$\begin{cases} (-\Delta)^s u &= 0 \quad \text{in } \Omega, \\ u &= z \quad \text{in } \mathbb{R}^N \setminus \Omega, \end{cases} \tag{4.1b}$$

and the control constraints

$$z \in Z_{ad,D}, \tag{4.1c}$$

with $Z_{ad,D} \subset Z_D$ being a closed and convex subset. Well-posedness of this problem follows by the standard direct method of calculus of variations (Antil et al., 2019, Theorem 4.1).

Theorem 4.1. *Let $Z_{ad,D}$ be a closed and convex subset of Z_D. Let either $\lambda > 0$ or $Z_{ad,D}$ bounded and let $J : U_D \to \mathbb{R}$ be weakly lower-semicontinuous. Then* *there exists a solution $\bar{z}$ to* (4.1). *If either J is convex and $\lambda > 0$ or J is strictly convex and $\lambda \geq 0$, then $\bar{z}$ is unique.*

Moreover, the following first order optimality conditions hold (Antil et al., 2019, Theorem 4.3).

Theorem 4.2. *Let the assumptions of Theorem 4.1 hold. Let $\mathcal{Z}$ be an open set in Z_D such that $Z_{ad,D} \subset \mathcal{Z}$. Let $u \mapsto J(u) : U_D \to \mathbb{R}$ be continuously Fréchet differentiable with $J'(u) \in U_D$. If $\bar{z}$ is a minimizer of* (4.1) *over $Z_{ad,D}$, then the first order necessary optimality conditions are given by*

$$(-\mathcal{N}_s \bar{p} + \lambda \bar{z}, z - \bar{z})_{L^2(\mathbb{R}^N \setminus \Omega)} \geq 0, \quad \forall z \in Z_{ad,D}, \tag{4.2}$$

where $\bar{p} \in \widetilde{H}^s(\Omega)$ solves the adjoint equation

$$\begin{cases} (-\Delta)^s \bar{p} = J'(\bar{u}) & \text{in } \Omega, \\ \bar{p} = 0 & \text{in } \mathbb{R}^N \setminus \Omega. \end{cases} \tag{4.3}$$

Equivalently we can write (4.2) *as $\bar{z} = \mathcal{P}_{Z_{ad,D}}\left(\frac{1}{\lambda}\mathcal{N}_s \bar{p}\right)$, where $\mathcal{P}_{Z_{ad,D}}$ is the projection onto the set $Z_{ad,D}$. If J is convex, then* (4.2) *is also a sufficient condition.*

As we emphasized earlier, in its current form, (4.1) requires the approximation of $\mathcal{N}_s$ twice. First to compute the very-weak solution (Theorem 3.2(ii)) and second to evaluate the optimality condition (4.2). In order to overcome this issue in the state equation, recall that we had introduced the Robin problem (3.9). Also, recall the approximation of the Dirichlet solution by the Robin solution from Theorem 3.5. In fact, one can approximate the Dirichlet optimal control problem by the Robin optimal control problem: for $\lambda \geq 0$ the fractional Robin control problem is given by:

$$\min_{u \in U_R, z \in Z_R} J(u) + \frac{\lambda}{2}\|z\|^2_{L^2(\mathbb{R}^N \setminus \Omega)}, \tag{4.4a}$$

subject to the regularized exterior value problem (Robin problem): Find $u_n \in U_R$ solving

$$\begin{cases} (-\Delta)^s u & = 0 \quad \text{in } \Omega \\ \mathcal{N}_s u + n\kappa u & = n\kappa z \quad \text{in } \mathbb{R}^N \setminus \Omega, \end{cases} \tag{4.4b}$$

and the control constraints

$$z \in Z_{ad,R}. \tag{4.4c}$$

Here $Z_R := L^2(\mathbb{R}^N \setminus \Omega)$, $Z_{ad,R}$ is a closed and convex subset of Z_R and $U_R := W^{s,2}_{\Omega,\kappa} \cap L^2(\mathbb{R}^N \setminus \Omega)$. Then as $n \to \infty$ the Robin problem (4.4) approximates the Dirichlet problem (4.1), see Antil et al. (2019, Theorem 6.5) for details. We conclude this section by providing a numerical example, where we solve (4.1) by approximating it with (4.4).

We choose our objective function as

$$j(u,z) = J(u) + \frac{\lambda}{2}\|z\|^2_{L^2(\mathbb{R}^N\setminus\Omega)}, \quad \text{with} \quad J(u) := \frac{1}{2}\|u - u_d\|^2_{L^2(\Omega)},$$

and we let $Z_{ad,R} := \{z \in L^2(\mathbb{R}^N \setminus \Omega) \; : \; z \geq 0, \text{ a.e. in } \widehat{\Omega}\}$ where $\widehat{\Omega}$ is the support set of the control z that is contained in $\widetilde{\Omega} \setminus \Omega$. Moreover $u_d : L^2(\Omega) \to \mathbb{R}$ is the given data (observations). All the optimization problems below are solved using the projected-BFGS method with Armijo line search.

Our computational setup is shown in Fig. 6. The centered square region is $\Omega = (-0.4, 0.4)^2$ and $\widetilde{\Omega} = B_0(3/2)$. The smaller square inside $\widetilde{\Omega} \setminus \Omega$ is $\widehat{\Omega}$ which is the support of the source/control. The right panel in Fig. 6 shows a finite element mesh with DoFs $= 6103$.

We define u_d as follows. For $z = 1$, we first solve the state equation for $\tilde{u}$ (4.4b). To $\tilde{u}$, we then add a normally distributed random noise with mean zero and standard deviation 0.02 to $\tilde{u}$. We call the resulting expression as u_d. Furthermore, we set $\kappa = 1$, and $n = 1e5$.

Our goal is then to identify the source/control $\bar{z}_h$. For a fixed $\lambda = 1e\text{-}8$, Fig. 7 shows the optimal $\bar{z}_h$ for $s = 0.1$ (4), 0.7 (2), 0.9 (2). The numbers in the parenthesis denote the total number of iterations that BFGS has taken to achieve a stopping tolerance (for the projected gradient) of $1e\text{-}7$. Notice that the Armijo line search has remained inactive in these cases. We notice that for large s, $\bar{z}_h \equiv 0$. This is expected as larger the s is, the closer we are to the classical Poisson problem case and we know that we cannot impose the external condition in that case.

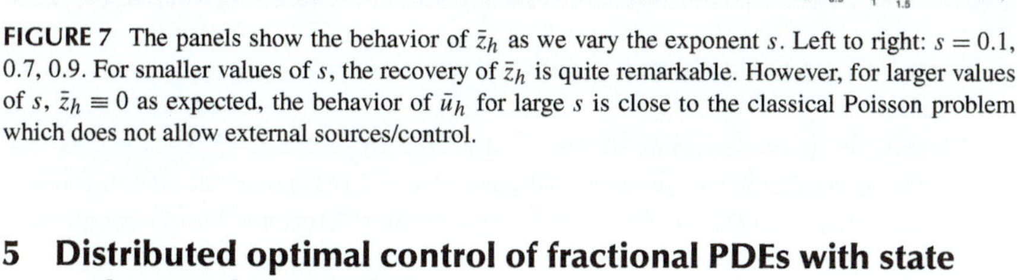

FIGURE 7 The panels show the behavior of $\bar{z}_h$ as we vary the exponent s. Left to right: $s = 0.1$, 0.7, 0.9. For smaller values of s, the recovery of $\bar{z}_h$ is quite remarkable. However, for larger values of s, $\bar{z}_h \equiv 0$ as expected, the behavior of $\bar{u}_h$ for large s is close to the classical Poisson problem which does not allow external sources/control.

5 Distributed optimal control of fractional PDEs with state and control constraints

This section is based on the article (Antil et al., 2021b). Given $u_d \in L^2(\Omega)$ and regularization parameter $\lambda > 0$, consider the state constrained optimal control problem

$$\min_{(u,z)\in(U,Z)} \frac{1}{2}\|u-u_d\|^2_{L^2(\Omega)} + \frac{\lambda}{2}\|z\|^2_{L^2(\Omega)}$$

subject to: (5.1)

$$(-\Delta)^s_D u = z \quad \text{in } \Omega, \qquad u|_\Omega \in \mathcal{K} \quad \text{and} \quad z \in Z_{ad}.$$

Recall the definition of $(-\Delta)^s_D$ from (2.2). Next, we introduce the relevant function spaces. We let

$$Z := L^p(\Omega) \text{ and } U := \left\{ u \in \widetilde{H}^s(\Omega) : ((-\Delta)^s_D)u|_\Omega \in L^p(\Omega) \right\}.$$

Then, U is a Banach space with the graph norm $\|u\|_U := \|u\|_{\widetilde{H}^s(\Omega)} + \|u\|_{C_0(\Omega)} + \|(-\Delta)^s_D u\|_{L^p(\Omega)}$. Here p is as in (3.4), but in addition we assume that $2 \le p < \infty$. We let $Z_{ad} \subset Z$ a nonempty, closed, and convex set and $\mathcal{K}$ defined as

$$\mathcal{K} := \left\{ w \in C_0(\Omega) \, : \, w(x) \le u_b(x), \quad \forall x \in \overline{\Omega} \right\}. \tag{5.2}$$

Recall that $C_0(\Omega)$ is the space of all continuous functions in $\overline{\Omega}$ that vanish on $\partial\Omega$. Moreover, $u_b \in C(\overline{\Omega})$ such that $u_b \ge 0$ on $\partial\Omega$. Existence of solution to (5.1) then follows from the direct method of calculus of variations (Antil et al., 2020d, Theorem 5.1).

In addition, under the Slater's constraint qualification (Antil et al., 2020d, Assumption 5.2) we can derive the first order optimality conditions: Let $(\bar{u}, \bar{z})$ be a solution to the optimization problem (4.1). Then, there exist a Lagrange multiplier $\bar{\mu} \in (C_0(\Omega))^\star = \mathcal{M}(\Omega)$ and an adjoint variable $\bar{\vartheta} \in L^{p'}(\Omega)$ such that

$$(-\Delta)^s_D \bar{u} = \bar{z} \quad \text{in } \Omega, \tag{5.3a}$$

$$\langle \bar{\vartheta}, (-\Delta)^s_D v \rangle_{L^{p'}(\Omega), L^p(\Omega)} = (\bar{u} - u_d, v)_{L^2(\Omega)} + \int_\Omega v \, d\bar{\mu}, \quad \forall\, v \in U \tag{5.3b}$$

$$\langle \bar{\vartheta} + J_z(\bar{u}, \bar{z}), z - \bar{z} \rangle_{L^{p'}(\Omega), L^p(\Omega)} \ge 0, \qquad \forall\, z \in Z_{ad} \tag{5.3c}$$

$$\bar{\mu} \ge 0, \quad \bar{u}(x) \le u_b(x) \text{ in } \Omega, \quad \text{and} \quad \int_\Omega (u_b - \bar{u}) \, d\mu = 0. \tag{5.3d}$$

Clearly, it is challenging to try to directly implement (5.3). Instead, we follow the approach from the classical case of $s = 1$ (Hinze et al., 2009; Hintermüller and Hinze, 2009) and consider the so-called Moreau-Yosida regularization. The Moreau-Yosida regularized optimal control problem is given by

$$\min J^\gamma(u, z) := \frac{1}{2}\|u - u_d\|^2_{L^2(Q)} + \frac{\lambda}{2}\|z\|^2_{L^2(Q)} + \frac{1}{2\gamma}\|(\hat{\mu} + \gamma(u - u_b))_+\|^2_{L^2(Q)}, \tag{5.4a}$$

subject to

$$(-\Delta)^s_D u = z \quad \text{in } \Omega \quad \text{and} \qquad z \in Z_{ad}, \tag{5.4b}$$

where $0 \le \hat{\mu} \in L^2(Q)$ is a shift parameter that can be taken to be zero, and $\gamma > 0$ denotes the regularization parameter. Here, $(\cdot)_+$ denotes $\max\{0, \cdot\}$. More information about this can be found in Ito and Kunisch (2008). Existence and uniqueness to (5.4) again follows by the standard direct method of calculus of variations. Moreover, we have the following first order optimality conditions:

Theorem 5.1. *Let $(\bar{u}^\gamma, \bar{z}^\gamma)$ be a solution to the regularized optimization problem (5.4). Then there exists a Lagrange multiplier $\bar{\vartheta}^\gamma \in \widetilde{H}^s(\Omega)$ such that*

$$(-\Delta)^s_D \bar{u}^\gamma = \bar{z}^\gamma, \qquad \textit{in } \Omega, \tag{5.5a}$$

$$(-\Delta)^s_D \bar{\vartheta}^\gamma = \bar{u}^\gamma - u_d + (\hat{\mu} + \gamma(\bar{u}^\gamma - u_b))_+, \qquad \textit{in } \Omega, \tag{5.5b}$$

$$(\bar{\vartheta}^\gamma + \lambda \bar{z}^\gamma, z - \bar{z}^\gamma)_{L^2(\Omega)} \ge 0, \qquad \forall\, z \in Z_{ad}. \tag{5.5c}$$

In addition, the following approximation result holds, see Antil et al. (2021b, Theorem 5.2) for the proof.

Proposition 5.2. *Let $(\bar{u}, \bar{z})$ solve (5.1) and $(\bar{u}^\gamma, \bar{z}^\gamma)$ solve (5.4). Then as $\gamma \to \infty$, we have $\bar{z}^\gamma \to \bar{z}$ in $L^2(\Omega)$ and $\|(\bar{u}^\gamma - u_b)_+\|_{L^2(\Omega)} = \mathcal{O}(\gamma^{-1/2})$. Moreover if $u_b \equiv 0$ in Ω, then under the additional assumption that $u_d \ge 0$, we obtain that $\|(\bar{u}^\gamma)_+\|_{L^2(\Omega)} = \mathcal{O}(\gamma^{-1})$ as $\gamma \to \infty$.*

For simplicity of presentation, next we consider the case where the control admissible set is

$$Z_{ad} = \{z \in Z : a \le z \le b \quad a.e. \text{ in } \Omega\}, \tag{5.6}$$

where $a < b$ are constants. Next, we state the regularity of optimal solutions to (5.4). See Antil et al. (2021b, Proposition 6.2) for the proof.

Proposition 5.3. *Let Ω be a bounded Lipschitz domain and $(\bar{u}^\gamma, \bar{\vartheta}^\gamma, \bar{z}^\gamma)$ be the solution to (5.5) for a fixed γ, then we have*

$$\bar{u}^\gamma \in H^{\sigma-\varepsilon}(\Omega), \qquad \bar{\vartheta}^\gamma \in H^{\sigma-\varepsilon}(\Omega), \qquad \bar{z}^\gamma \in H^{\tau}(\Omega),$$

for every $\varepsilon > 0$ where $\sigma = \min\{2s, s + 1/2\}$ and $\tau = \min\{1, 2s - \varepsilon\}$.

We discretize the state and adjoint pair using piecewise linear, globally continuous finite elements for a triangulation $\mathcal{T}_h$ of $\widetilde{\Omega}$ vanishing in the exterior $\widetilde{\Omega} \setminus \Omega$. Here we have assumed that $\overline{\Omega} \subset \widetilde{\Omega}$. We discretize the control using piecewise constants, i.e., the control space is $Z_h := \{z_h \in Z : z_h|_T \in \mathcal{P}_0, \ \forall\, T \in \mathcal{T}_h\}$, where $\mathcal{P}_0$ denotes the space of piecewise constants on the triangulation $\mathcal{T}_h$.

Then based on the regularity result in Proposition 5.3, the following result holds.

Corollary 5.4. *Let $\bar{z}^\gamma$ and $\bar{z}_h^\gamma$ denote the continuous and discrete optimal controls. Then the following result holds:*

$$\|\bar{z}^\gamma - \bar{z}_h^\gamma\|_{L^2(\Omega)} \le \frac{C}{\lambda}\Big((h^{2\beta}|\log h|^{2(1+\kappa)} + h^{\beta+s-\varepsilon})(\|z^\gamma\|_{L^2(\Omega)} + \|u_d\|_{L^2(\Omega)} + \|\hat{\mu}\|_{L^2(\Omega)}) + h^{2\beta}|\log h|^{2(1+\kappa)}(1+\gamma)\|\bar{z}^\gamma\|_{L^2(\Omega)} + h^\tau(1+\gamma+\lambda)|\bar{z}^\gamma|_{H^\tau(\Omega)}\Big),$$

for every $\varepsilon > 0$, where $\beta = \min\{s, 1/2\}$, $\tau = \min\{1, 2s-\varepsilon\}$, $\kappa = 1$ if $s = 1/2$ and zero otherwise.

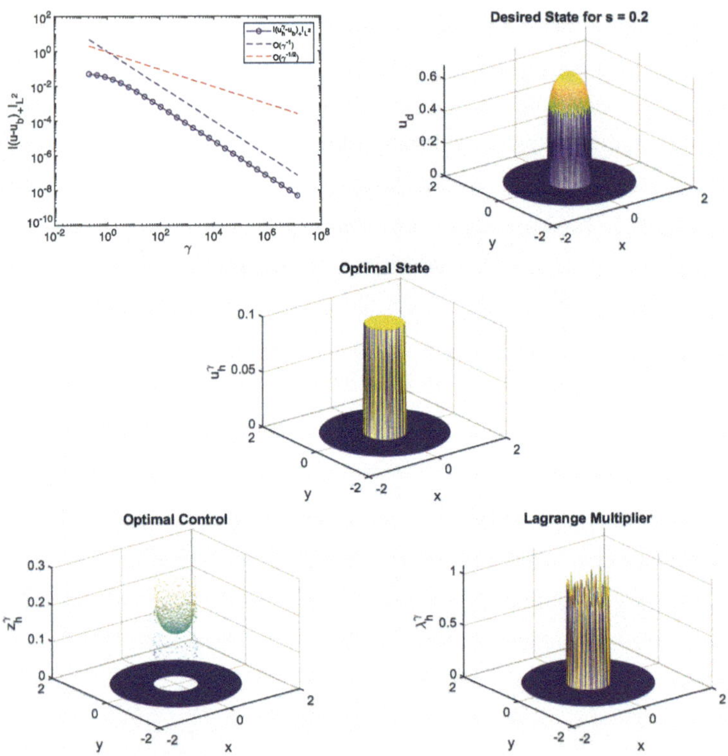

FIGURE 8 Convergence of $\|(u-u_b)_+\|_{L^2(\Omega)}$ as γ increases (top left). The desired state (top right), optimal state (middle), control (bottom left) and Lagrange multiplier (bottom right) for $s = 0.2$ and $\gamma = 419430.4$.

Next, we provide a numerical example. We let $u_d(x,y) = \frac{2^{-2s}}{\Gamma(1+s)^2}(1/4 - (x^2+y^2)_+)^s$, and we let $u_b = 0.1$. In the left panel in Fig. 8 we show the convergence of $\|(u-u_b)_+\|_{L^2(\Omega)}$ for $s = 0.4$ on a mesh with 24155 number of nodes and 48468 number of elements as γ increases. We observe a convergence of $\mathcal{O}(\gamma^{-1})$ which is better than expected. However, this has also been documented in the literature (when $s=1$) and it can be rigorously established when $u_b=0$ and $u_d \geq 0$, see Hintermüller and Hinze (2009).

In Fig. 8 we also show the optimal state, control, and Lagrange multiplier for $s=0.2$ and $\gamma=419430.4$. We note that the control is a piecewise constant on the mesh. The optimal state in Fig. 8 appears to be cleanly cut off at $u_b=0.1$, complying with the state constraint and resulting in a cylindrical profile. Moreover, notice that the Lagrange multiplier corresponding to the inequality constraint is a measure (bottom right panel) as expected.

6 Fractional deep neural networks – FDNNs

This section is based on the article (Antil et al., 2021d). Consider a generic (discrete) parameterized PDE

$$F(\mathbf{u}(\boldsymbol{\xi});\boldsymbol{\xi})=0, \tag{6.1}$$

where $\boldsymbol{\xi} \in \mathcal{P} \subset \mathbb{R}^{N_{\xi}}$ and $\mathbf{u}(\boldsymbol{\xi}) \in \mathcal{U} \subset \mathbb{R}^{N_x}$ represent a fixed parameter and the corresponding solution of the PDE, respectively. Moreover, $\mathcal{P}$ denotes the parameter domain and $\mathcal{U}$ the solution manifold. For a fixed parameter $\boldsymbol{\xi} \in \mathcal{P}$, we seek the solution $\mathbf{u}(\boldsymbol{\xi}) \in \mathcal{U}$. In other words, we have the functional relation given by the parameter-to-solution map

$$\boldsymbol{\xi} \mapsto \Phi(\boldsymbol{\xi}) \equiv \mathbf{u}(\boldsymbol{\xi}). \tag{6.2}$$

Parameter-dependent PDEs of the form (6.1) arise in several areas of computational sciences and engineering. Typical examples include Navier-Stokes equations (with Reynolds number as the parameter) (Elman et al., 2005), and Boussinesq equations (with Grashof or Prandtl numbers as the parameters) (Quarteroni et al., 2016), etc.

In real-world applications, solutions of (6.1) are required for many parameter values and $N_{\boldsymbol{\xi}}$ is often very large; thus, the associated computational complexity is enormous. Besides, a relatively large N_x (due to fine mesh in the discretization of the PDE) yields large (nonlinear) algebraic systems which are computationally expensive to solve and may also lead to huge storage requirements. This is, for instance, the case in Bayesian inverse problems governed by PDEs where several forward solves are required to adequately sample posterior distributions through MCMC-type schemes (Beskos et al., 2017; Martin et al., 2012). Due to the aforementioned challenges, it is a reasonable computational practice to replace the high-fidelity model by a surrogate model which is relatively easy to evaluate.

There are two major classes of surrogate models in the literature: the reduced-order models (ROMs), see e.g., Barrault et al. (2004); Hesthaven et al. (2016); Quarteroni et al. (2016); Antil et al. (2014) and the deep neural network models (DNNs), see e.g., Bengio et al. (2017); Raissi et al. (2019). A key feature of the ROMs is that they use the so-called *offline-online paradigm*. The main thrust of the offline step is the construction of a low-dimensional approximation to the solution space; this approximation is generally known as the *reduced basis*. Depending on the problem under consideration, the offline step can be computationally demanding, although the expense incurred is a one time cost. In the online step, one then uses the reduced basis to solve a smaller reduced problem. The resulting reduced solution accurately approximates the solution of the original problem. Typical examples of ROMs include the reduced basis method (Quarteroni et al., 2016), proper orthogonal decomposition (Hesthaven et al., 2016; Quarteroni et al., 2016), and the discrete empirical interpolation method (DEIM) and its variants (Barrault et al., 2004; Chaturantabut and Sorensen, 2010; Antil et al., 2014).

Deep neural network (DNN) models constitute another class of surrogate models which are well-known for their high approximation capabilities. The basic idea of DNNs is to approximate multivariate functions through a set of layers of increasing complexity (Bengio et al., 2017). As in the case of ROMs, we also note here that using DNNs also involves some offline cost, which is incurred in training the network. Examples of DNNs for surrogate modeling include Residual Neural Networks (e.g. ResNet) (He et al., 2016; Haber and Ruthotto, 2018), physics-informed neural network (PINNs) (Raissi et al., 2019), and fractional DNN (Antil et al., 2020b).

Despite the fact that, in the online phase of ROMs, one essentially has to do reduced system solves, note that these systems can still be ill-conditioned and highly intrusive especially for nonlinear problems. On the other hand, the DNN approach can be fully non-intrusive, which is essential for legacy codes. Undoubtedly, rigorous error estimates for ROMs under various assumptions have been well studied (Quarteroni et al., 2016); however, we like the advantage of DNN being nonintrusive, but recognize that error analysis is not yet as strong (Daubechies et al., 2021).

Next, we provide a description of the fractional derivative based DNN approach introduced in Antil et al. (2021d) and apply it to a Bayesian inverse problem. The idea of a DNN is to approximate the input-to-output map $\Phi : \mathbb{R}^{N_\xi} \to \mathbb{R}^{N_x}$ by a surrogate $\widehat{\Phi}$ which is the output of a DNN.

For sufficiently large $N_s \in \mathbb{N}$, suppose that $\mathbb{E} := \{\boldsymbol{\xi}_1, \boldsymbol{\xi}_2, \cdots, \boldsymbol{\xi}_{N_s}\}$ is a set of parameter samples with $\boldsymbol{\xi}_i \in \mathbb{R}^{N_\xi}$, and $\mathbb{S} := \{\mathbf{u}(\boldsymbol{\xi}_1), \mathbf{u}(\boldsymbol{\xi}_2), \cdots, \mathbf{u}(\boldsymbol{\xi}_{N_s})\}$ the corresponding snapshots (solutions of the model (6.1), with $\mathbf{u}_i := \mathbf{u}(\boldsymbol{\xi}_i) \in \mathbb{R}^{N_x}$). Here, we assume that $\text{span}\{\mathbf{u}(\boldsymbol{\xi}_1), \mathbf{u}(\boldsymbol{\xi}_2), \cdots, \mathbf{u}(\boldsymbol{\xi}_{N_s})\}$ sufficiently approximates the space of all possible solutions of (6.1). We refer to Quarteroni et al. (2016), for instance, for a description of "sufficient approximation" in the context of ROMs.

The idea of a DNN is to use the parameters $\boldsymbol{\xi}_j$ as an input to the DNN and try to match the output of DNN $\widehat{\Phi}(\boldsymbol{\xi}_j; \boldsymbol{\theta})$ with the vectors $\mathbf{u}_j$. Moreover, $\boldsymbol{\theta} = \{W_j, \mathbf{b}_j\}$, are the unknown parameters in the DNN that need to be learned.

This learning problem can be cast as an optimal control problem

$$\min_{\boldsymbol{\theta}=\{W_j,b_j\}} \mathcal{J}(\boldsymbol{\theta};\boldsymbol{\xi},\mathbf{u}) = \frac{1}{2N_s}\sum_{j=1}^{N_s} \|\hat{\Phi}(\boldsymbol{\xi}_j;\boldsymbol{\theta}) - \mathbf{u}_j\|_2^2 + \frac{\lambda}{2}\|\boldsymbol{\theta}\|_2^2, \tag{6.3}$$

subject to the fractional Deep Neural Network as constraint

$$\begin{aligned}
&\phi_0 = \boldsymbol{\xi},\\
&\phi_1 = \sigma(W_0\phi_0 + \mathbf{b}_0),\\
&\phi_j = \phi_{j-1} - \sum_{k=0}^{j-2}\Big(a_{j-1-k}\,(\phi_{k+1}-\phi_k)\\
&\qquad + \tau^{\gamma}\Gamma(2-\gamma)\sigma(W_{j-1}\phi_{j-1}+\mathbf{b}_{j-1})\Big), \quad 2\le j\le L-1,\\
&(\widehat{\Phi} :=)\ \phi_L = W_{L-1}\phi_{L-1},
\end{aligned} \tag{6.4}$$

where, recall that, the middle equation in (6.4) is the Euler-time discretization of a fractional differential equation (cf. (2.4) and (3.14)). For the definition of coefficients a_{j-1-k}, see (3.15). As pointed out in Antil et al. (2020b), designing the DNN solution algorithms at the continuous level has the appealing advantage of architecture independence; in other words, the number of optimization iterations will not significantly deteriorate even if the number of layers is increased. This comment is motivated by the literature on PDE constrained optimization where such mesh-independence has been extensively studied (Hintermüller and Ulbrich, 2004; Hintermüller et al., 2008).

Also, as noted in Antil et al. (2021d), this fractional DNN allows connectivity among all the layers unlike standard DNNs. This passage of historic information of input and gradients across all the subsequent layers allows one to overcome the vanishing gradient issue as discussed in Antil et al. (2020b) for classification problems. However, we emphasize that currently, only numerical evidence is available.

The learning problem (6.4) can be solved using adjoint based approach as discussed in the previous sections and by applying gradient based optimization methods, such as BFGS. We refer to Antil et al. (2020b, 2021d) for the details. We close this section by providing a numerical example.

Thermal fin problem: Consider the following parameterized system

$$\begin{aligned}
-\mathrm{div}\,(e^{\boldsymbol{\xi}(x)}\nabla u) &= 0, && \text{in } \Omega,\\
(e^{\boldsymbol{\xi}(x)}\nabla u)\cdot \boldsymbol{\nu} + 0.1u &= 0, && \text{on } \partial\Omega\setminus\Gamma,\\
(e^{\boldsymbol{\xi}(x)}\nabla u)\cdot \mathbf{n} &= 1, && \text{on } \Gamma = (-0.5, 0.5)\times\{0\},
\end{aligned} \tag{6.5}$$

which represents a forward model for heat conduction over the non-convex domain Ω as shown in Fig. 9 (left). When the heat conductivity function $e^{\boldsymbol{\xi}(\mathbf{x})}$ is known, the forward problem can be solved for u. The goal of this example is to infer 100 unknown parameters $\boldsymbol{\xi}$ from 262 noisy observations of u. Fig. 9 (left) also shows the location of the observations on the boundary $\partial\Omega \setminus \Gamma$.

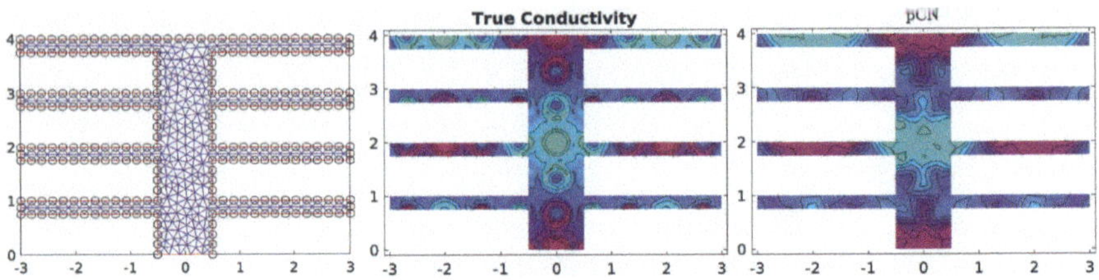

FIGURE 9 *Thermal fin problem:* Given a diffusion equation with coefficient $e^{\boldsymbol{\xi}(x)}$, the goal is to reconstruct u, from noisy measurements on the boundary (marked by circles). *Middle:* True conductivity field. *Right:* The posterior mean estimates obtained by preconditioned Crank-Nicolson (pCN) MCMC method using the fractional DNN as surrogate. The acceptance rate is comparable to Beskos et al. (2017). The figure has been reproduced from Antil et al. (2021d).

The fDNN is trained by first computing $N_s = 900$ solution snapshots corresponding to 900 parameters $\boldsymbol{\xi} \in \mathbb{R}^{100}$, drawn using Latin hypercube sampling. Next we compute the SVD of $\mathbb{S}$ and train the network on this reduced space. Fig. 9 shows the true conductivity (middle) and the posterior mean estimates obtained using standard MCMC approach.

Table 1 shows the average acceptance rates for these models and the computational times required to perform the MCMC simulation for different variants of the MCMC algorithm (preconditioned Crank-Nicolson method (pCN), the infinite variants of Riemannian manifold Metropolis-adjusted Langevin algorithm (∞-MALA) and Hamiltonian Monte-Carlo (∞-HMC) algorithms as presented in Beskos et al. (2017)), using both fDNN surrogate computations and full-order discrete PDE solution. The acceptance rate is clearly comparable for the fDNN surrogate model and the full order forward discrete PDE solver. In addition, a reduction of 90% in computational time is observed for the fDNN approach. The costs of the offline computations for fDNN were 63.2 seconds to generate the data used for the fractional DNN models and 33.5 seconds to train the network (with 1600 BFGS iterations); a total of 99.7 seconds.

TABLE 1 Acceptance rates and computational times needed to solve the inverse problem by pCN, ∞-MALA and ∞-HMC algorithms together with fDNN and full forward models.

Model	pCN	∞-MALA	∞-HMC
Acc. Rate (fDNN)	0.67	0.67	0.79
Acc. Rate (Full)	0.66	0.70	0.75
CPU time (fDNN)	16.46	98.0	228.4
CPU time (Full)	157.8	958.9	2585.3

7 Some open problems

There are still many open questions in this paradigm. We list a few of them below:

1. Theorem 3.2 in part (iv) assumed the domain Ω to fulfill exterior cone condition. It remains open if such a result holds true in case Ω is only Lipschitz continuous. Also, note that (iii)-(iv) shows boundedness and continuity of solution when the exterior datum $g \equiv 0$. However, it remains open to prove these results when $g \neq 0$. That is, given a function $g \in C(\mathbb{R}^N \setminus \Omega)$ and $f \equiv 0$ in Ω, under what conditions on Ω, there would exist an s-harmonic function $u \in C(\mathbb{R}^N)$ satisfying (3.1). The classical case $s = 1$ has been resolved by Wiener several decades ago, but the fractional case is more challenging and remains an open problem.
2. Section 3.2 discusses a numerical method to approximate the non-homogeneous exterior Dirichlet problem by the Robin problem. But a complete numerical analysis of this method is still open. In addition, a complete numerical analysis of the Robin problem in the full generality considered here is still open as well.
3. Section 4 discusses exterior optimal control problems, but numerical analysis of these problems are fully open.
4. Section 5 discusses state constrained problems and provides finite element γ-dependent error estimates in Corollary 5.4. Such estimates which are γ-independent are still open. Note that these questions are not just limited to elliptic problems, but are equally relevant to parabolic problems as well.
5. The article (Antil et al., 2020b) numerically showed that the fractional time derivative can help overcome the vanishing gradient problem. However, a theoretical justification is still an open problem.

Acknowledgments

This work is partially supported by NSF grants DMS-2110263, DMS-1913004 and the Air Force Office of Scientific Research under Award NO: FA9550-19-1-0036. The work of MW is partially supported by the AFOSR under Award NO: FA9550-18-1-0242 and by US Army Research Office (ARO) under Award NO: W911NF-20-1-0115.

DISTRIBUTION STATEMENT A. Approved for public release. Distribution is unlimited.

References

Abatangelo, N., Valdinoci, E., 2019. Getting acquainted with the fractional Laplacian. In: Contemporary Research in Elliptic PDEs and Related Topics. In: Springer INdAM Ser., vol. 33. Springer, Cham, pp. 1–105.

Abraham, R., Bergounioux, M., Trélat, E., 2008. A penalization approach for tomographic reconstruction of binary axially symmetric objects. Appl. Math. Optim. 58 (3), 345–371.

Acosta, G., Bersetche, F.M., Borthagaray, J.P., 2017. A short FE implementation for a 2d homogeneous Dirichlet problem of a fractional Laplacian. Comput. Math. Appl. 74 (4), 784–816.

Acosta, G., Borthagaray, J.P., 2017. A fractional Laplace equation: regularity of solutions and finite element approximations. SIAM J. Numer. Anal. 55 (2), 472–495.

Acosta, G., Borthagaray, J.P., Heuer, N., 2019. Finite element approximations of the nonhomogeneous fractional Dirichlet problem. IMA J. Numer. Anal. 39 (3), 1471–1501.

Adams, D.R., Hedberg, L.I., 1996. Function Spaces and Potential Theory. Grundlehren der Mathematischen Wissenschaften (Fundamental Principles of Mathematical Sciences), vol. 314. Springer-Verlag, Berlin.

Agrawal, O.P., 2007. Fractional variational calculus in terms of Riesz fractional derivatives. J. Phys. A 40 (24), 6287–6303.

Antil, H., Bartels, S., 2017. Spectral approximation of fractional PDEs in image processing and phase field modeling. Comput. Methods Appl. Math. 17 (4), 661–678.

Antil, H., Bartels, S., Schikorra, A., 2021a. Approximation of fractional harmonic maps. arXiv preprint. arXiv:2104.10049.

Antil, H., Brown, T.S., Verma, D., Warma, M., 2021b. Optimal Control of Fractional PDEs with State and Control Constraints. Pure Appl. Funct. Anal. In press.

Antil, H., Di, Z., Khatri, R., 2020a. Bilevel optimization, deep learning and fractional Laplacian regularization with applications in tomography. Inverse Probl. 36 (6), 064001.

Antil, H., Dondl, P., Striet, L., 2021c. Approximation of integral fractional Laplacian and fractional pdes via sinc-basis. SIAM J. Sci. Comput. 43 (4), A2897–A2922.

Antil, H., Elman, H.C., Onwunta, A., Verma, D., 2021d. Novel deep neural networks for solving Bayesian statistical inverse problems. arXiv preprint. arXiv:2102.03974.

Antil, H., Heinkenschloss, M., Sorensen, D.C., 2014. Application of the Discrete Empirical Interpolation Method to Reduced Order Modeling of Nonlinear and Parametric Systems. Springer International Publishing, Cham, pp. 101–136.

Antil, H., Khatri, R., Löhner, R., Verma, D., 2020b. Fractional deep neural network via constrained optimization. Mach. Learn.: Sci. Technol. 2 (1), 015003.

Antil, H., Khatri, R., Warma, M., 2019. External optimal control of nonlocal PDEs. Inverse Probl. 35 (8), 084003.

Antil, H., Nochetto, R.H., Venegas, P., 2018a. Controlling the Kelvin force: basic strategies and applications to magnetic drug targeting. Optim. Eng. 19 (3), 559–589.

Antil, H., Nochetto, R.H., Venegas, P., 2018b. Optimizing the Kelvin force in a moving target subdomain. Math. Models Methods Appl. Sci. 28 (1), 95–130.

Antil, H., Otárola, E., Salgado, A.J., 2016. A space-time fractional optimal control problem: analysis and discretization. SIAM J. Control Optim. 54 (3), 1295–1328.

Antil, H., Rautenberg, C.N., 2018. Fractional elliptic quasi-variational inequalities: theory and numerics. Interfaces Free Bound. 20 (1), 1–24.

Antil, H., Rautenberg, C.N., 2019. Sobolev spaces with non-Muckenhoupt weights, fractional elliptic operators, and applications. SIAM J. Math. Anal. 51 (3), 2479–2503.

Antil, H., Rautenberg, C.N., Schikorra, A., 2021e. On a fractional version of a Murat compactness result and applications. SIAM J. Math. Anal. 53 (3), 3158–3187.

Antil, H., Verma, D., Warma, M., 2020c. External optimal control of fractional parabolic PDEs. ESAIM Control Optim. Calc. Var. 26.

Antil, H., Verma, D., Warma, M., 2020d. Optimal control of fractional elliptic pdes with state constraints and characterization of the dual of fractional order Sobolev spaces. J. Optim. Theory Appl. 186 (1), 1–23.

Antil, H., Warma, M., 2019. Optimal control of the coefficient for the regional fractional p-Laplace equation: approximation and convergence. Math. Control Relat. Fields 9 (1), 1–38.

Antil, H., Warma, M., 2020. Optimal control of fractional semilinear PDEs. ESAIM Control Optim. Calc. Var. 26, 5.

Barrault, M., Maday, Y., Nguyen, N.C., Patera, A.T., 2004. An 'empirical interpolation' method: application to efficient reduced-basis discretization of partial differential equations. C. R. Math. Acad. Sci. Paris 339 (9), 667–672.

Bengio, Y., Goodfellow, I., Courville, A., 2017. Deep Learning, Vol. 1. MIT Press, Massachusetts, USA.

Benning, M., Celledoni, E., Ehrhardt, M., Owren, B., Schönlieb, C-B., 2019. Deep learning as optimal control problems: models and numerical methods. J. Comput. Dyn. 6, 171.
Bergounioux, M., Abraham, I., Abraham, R., Carlier, G., Le Pennec, E., Trélat, E., 2018. Variational methods for tomographic reconstruction with few views. Milan J. Math. 86 (2), 157–200.
Bergounioux, M., Trélat, E., 2010. A variational method using fractional order Hilbert spaces for tomographic reconstruction of blurred and noised binary images. J. Funct. Anal. 259 (9), 2296–2332.
Beskos, A., Girolami, M., Lan, S., Farrell, P.E., Stuart, A.M., 2017. Geometric MCMC for infinite-dimensional inverse problems. J. Comput. Phys. 335, 327–351.
Bonito, A., Lei, W., Pasciak, J.E., 2019. Numerical approximation of the integral fractional Laplacian. Numer. Math. 142 (2), 235–278.
Bonito, A., Pasciak, J.E., 2015. Numerical approximation of fractional powers of elliptic operators. Math. Comput. 84 (295), 2083–2110.
Burger, M., Frick, K., Osher, S.J., Scherzer, O., 2007. Inverse total variation flow. Multiscale Model. Simul. 6 (2), 365–395.
Caffarelli, L., Silvestre, L., 2007. An extension problem related to the fractional Laplacian. Commun. Partial Differ. Equ. 32 (7–9), 1245–1260.
Calatroni, L., Cao, C., De los Reyes, J.C., Schönlieb, C-B., Valkonen, T., 2017. Bilevel approaches for learning of variational imaging models. In: Variational Methods. In: Radon Ser. Comput. Appl. Math., vol. 18. De Gruyter, Berlin, pp. 252–290.
Chambolle, A., 2004. An algorithm for total variation minimization and applications. In: Special Issue on Mathematics and Image Analysis. J. Math. Imaging Vis. 20 (1–2), 89–97.
Chaturantabut, S., Sorensen, D.C., 2010. Nonlinear model reduction via discrete empirical interpolation. SIAM J. Sci. Comput. 32 (5), 2737–2764.
Ciarlet, P.G., 2002. The Finite Element Method for Elliptic Problems. Classics in Applied Mathematics, vol. 40. Society for Industrial and Applied Mathematics (SIAM), Philadelphia, PA. Reprint of the 1978 original [North-Holland, Amsterdam; MR0520174 (58 #25001)].
Daubechies, I., DeVore, R., Foucart, S., Hanin, B., Petrova, G., 2021. Nonlinear approximation and (deep) relu networks. Constr. Approx.
D'Elia, M., De Los Reyes, J.C., Miniguano-Trujillo, A., 2021. Bilevel parameter learning for nonlocal image denoising models. J. Math. Imaging Vis., 1–23.
Di Nezza, E., Palatucci, G., Valdinoci, E., 2012. Hitchhiker's guide to the fractional Sobolev spaces. Bull. Sci. Math. 136 (5), 521–573.
Dipierro, S., Ros-Oton, X., Valdinoci, E., 2017. Nonlocal problems with Neumann boundary conditions. Rev. Mat. Iberoam. 33 (2), 377–416.
Elman, H.C., Silvester, D.J., Wathen, A.J., 2005. Finite Elements and Fast Iterative Solvers: with Applications in Incompressible Fluid Dynamics. Numerical Mathematics and Scientific Computation. Oxford University Press, New York.
Fiscella, A., Servadei, R., Valdinoci, E., 2015. Density properties for fractional Sobolev spaces. Ann. Acad. Sci. Fenn., Math. 40 (1), 235–253.
Gal, C.G., Warma, M., 2020. Fractional-in-Time Semilinear Parabolic Equations and Applications. Mathématiques & Applications (Berlin) (Mathematics & Applications), vol. 84. Springer, Cham. ©2020.
Günther, S., Ruthotto, L., Schroder, J., Cyr, E., Gauger, N., 2020. Layer-parallel training of deep residual neural networks. SIAM J. Math. Data Sci. 2 (1–23).
Haber, E., Ruthotto, L., 2018. Stable architectures for deep neural networks. Inverse Probl. 34 (1), 014004.
Haldane, F.D.M., 1986. Geometrical interpretation of momentum and crystal momentum of classical and quantum ferromagnetic Heisenberg chains. Phys. Rev. Lett. 57 (12), 1488.
He, K., Zhang, X., Ren, S., Sun, J., 2016. Deep residual learning for image recognition. In: Proceedings of the IEEE Conference on Computer Vision and Pattern Recognition, pp. 770–778.
Hesthaven, J.S., Rozza, G., Stamm, B., 2016. Certified Reduced Basis Methods for Parametrized Partial Differential Equations. SpringerBriefs in Mathematics. Springer/BCAM Basque Center for Applied Mathematics, Cham/Bilbao. BCAM SpringerBriefs.

Hintermüller, M., Hinze, M., 2009. Moreau-Yosida regularization in state constrained elliptic control problems: error estimates and parameter adjustment. SIAM J. Numer. Anal. 47 (3), 1666–1683.

Hintermüller, M., Tröltzsch, F., Yousept, I., 2008. Mesh-independence of semismooth Newton methods for Lavrentiev-regularized state constrained nonlinear optimal control problems. Numer. Math. 108 (4), 571–603.

Hintermüller, M., Ulbrich, M., 2004. A mesh-independence result for semismooth Newton methods. Math. Program. 101 (1, Ser. B), 151–184.

Hinze, M., Pinnau, R., Ulbrich, M., Ulbrich, S., 2009. Optimization with PDE Constraints. Mathematical Modelling: Theory and Applications, vol. 23. Springer, New York.

Ito, K., Kunisch, K., 2008. Lagrange Multiplier Approach to Variational Problems and Applications. Advances in Design and Control, vol. 15. Society for Industrial and Applied Mathematics (SIAM), Philadelphia, PA.

Jin, B., Lazarov, R., Zhou, Z., 2019. Numerical methods for time-fractional evolution equations with nonsmooth data: a concise overview. Comput. Methods Appl. Mech. Eng. 346, 332–358.

Lin, Y., Li, X., Xu, C., 2011. Finite difference/spectral approximations for the fractional cable equation. Math. Comput. 80 (275), 1369–1396.

Lin, Y., Xu, C., 2007. Finite difference/spectral approximations for the time-fractional diffusion equation. J. Comput. Phys. 225 (2), 1533–1552.

Martin, J., Wilcox, L.C., Burstedde, C., Ghattas, O., 2012. A stochastic Newton MCMC method for large-scale statistical inverse problems with application to seismic inversion. SIAM J. Sci. Comput. 34 (3), A1460–A1487.

Nochetto, R.H., Otárola, E., Salgado, A.J., 2016. A PDE approach to space-time fractional parabolic problems. SIAM J. Numer. Anal. 54 (2), 848–873.

Podlubny, I., 1999. Fractional Differential Equations: an Introduction to Fractional Derivatives, Fractional Differential Equations, to Methods of Their Solution and Some of Their Applications. Mathematics in Science and Engineering. Academic Press, London.

Quarteroni, A., Manzoni, A., Negri, F., 2016. Reduced Basis Methods for Partial Differential Equations. An Introduction. Unitext, vol. 92. Springer, Cham. La Matematica per il, vol. 3+2.

Raissi, M., Perdikaris, P., Karniadakis, G.E., 2019. Physics-informed neural networks: a deep learning framework for solving forward and inverse problems involving nonlinear partial differential equations. J. Comput. Phys. 378, 686–707.

Rudin, L.I., Osher, S., Fatemi, E., 1992. Nonlinear total variation based noise removal algorithms. Phys. D: Nonlinear Phenom. 60 (1–4), 259–268.

Ruthotto, L., Haber, E., 2019. Deep neural networks motivated by partial differential equations. J. Math. Imaging Vis.

Samko, S.G., Kilbas, A.A., Marichev, O.I., 1993. Fractional Integrals and Derivatives. Gordon and Breach Science Publishers, Yverdon.

Weiss, C.J., Van Bloemen Waanders, B.G., Antil, H., 2020. Fractional operators applied to geophysical electromagnetics. Geophys. J. Int. 220 (2), 1242–1259.

Chapter 4

Optimal control of PDEs and FE-approximation

Eduardo Casas[a,*,d], Karl Kunisch[b,e], and Fredi Tröltzsch[c]
[a]*Department of Applied Mathematics and Computer Science, University of Cantabria, Santander, Spain,* [b]*Institute for Mathematics and Scientific Computing, University of Graz, Graz, Austria,* [c]*Institute for Mathematics, Technical University of Berlin, Berlin, Germany*
**Corresponding author: e-mail address: eduardo.casas@unican.es*

Contents

Abstract
Optimal control problems of partial differential equations are studied. Though the focus lies on elliptic partial differential equations, similar methods can be used for the analysis of control problems associated to evolution equations. In the first section, the state equation is semilinear. The existence of solutions is studied, first and second order optimality conditions are derived, the sparsity properties of the optimal controls are investigated, and error estimates for the numerical discretization of the control problem are provided. This analysis includes the case of sparse controls. In a second section, the controls are chosen as measures. Due to the difficulty of controlling a nonlinear system by measures, the state equation is assumed to be linear in this case. Existence of optimal controls,

[d] The first author was supported by MCIN/ AEI/10.13039/501100011033/ under research projects MTM2017-83185-P and PID2020-114837GB-I00.
[e] This author was supported by the ERC advanced grant 668998 (OCLOC) under the EU's H2020 research program.

sparsity and numerical approximation by finite element methods are studied. In the last section of the chapter, related topics are addressed and additional bibliographical data are provided.

Keywords
Optimal control, Sparse control, First and second order optimality conditions, Finite element approximation, Error estimates

MSC Codes
49-00, 49-01, 49J20, 49J52, 49K20, 49M25, 49N60

Introduction

The mathematical field of optimal control theory for partial differential equations (PDEs) emerged a bit later than the control theory for ordinary differential equations that had a first culmination point by the book (Pontryagin et al., 1962). Associated research was very much inspired and activated by the celebrated monograph (Lions, 1968), where basic principles and results for convex optimal control problems with linear elliptic, parabolic, and hyperbolic equations were established. In particular, the existence of optimal controls and first-order optimality conditions were discussed as well as some concepts for feedback control. Since that time, an extensive number of papers and books on PDE control was published, some of them mentioned in Section 3 of this chapter.

Optimal control of PDEs has become a very broad field so that a chapter on it has to concentrate on a part that is sufficiently small but yet provides some insight in main ideas of the associated theory. In view of this, we concentrate on principles of numerical analysis for the optimal control of a class of semilinear elliptic equations and pointwise constraints on the control. Concerning the control cost, the most frequent choice is a quadratic one, which in certain cases can be interpreted as a physical energy. It is also convenient since it entails simple expressions when computing derivatives of the cost functional with respect to the control. On certain occasions the choice of the cost functional can be dictated by well-posedness and regularity considerations of the controlled system. It can also be chosen to accommodate to certain control objectives. For the optimal actuator placement problem, for instance, L^1-type costs functionals fit particularly well. These functionals are sparsity enhancing and as a consequence a natural placement for actuators is the support of the optimal control (Stadler, 2009; Clason and Kunisch, 2011; Casas, 2017).

The plan of this chapter is as follows: In Section 1, we discuss problems where the control is a measurable function acting in a given spatial domain. Existence of optimal controls, the associated first- and second-order optimality conditions, and the application of these results for finite element based approximation and error analysis are addressed. In Section 2, the controls are regular Borel measures acting in the domain. Here we restrict ourselves to linear elliptic equations. The reader is referred to (Casas and Kunisch, 2014) and (Ponce and Wilmet, 2018) for the case of semilinear elliptic equations. The need for treating measure-valued controls originates from the fact that L^1 as the space of controls does not allow an existence theory for optimal control problems.

The reader will have observed that many important issues will not be addressed. The literature offers the extension of many of these results to other equations including equations of parabolic or hyperbolic type. A large diversity of different nonlinearities has also been investigated. Differently for the presented setting, the control can appear on the boundary or – for evolution equations – at any instant of time. Objective functionals may have various forms. Pointwise or functional constraints on the state might be given. Another important topics that we cannot address here include the algorithmic treatment of the resulting optimization problems, and different techniques of model order reduction. These and additional issues are briefly sketched in Section 3, where selected references are listed to guide the reader to further reading.

1 The L^2 framework

1.1 The control problem

In the first section of this chapter we study the optimal control problem

$$(\mathrm{P}) \qquad \min_{u \in \mathbf{U}_{\mathrm{ad}}} J(u)$$

with

$$J(u) := \frac{1}{2}\int_\Omega (y_u(x) - y_\Omega(x))^2\,dx + \frac{\nu}{2}\int_\Omega u^2(x)\,dx + \kappa \int_\Omega |u(x)|\,dx,$$

where y_u denotes the solution of the semilinear elliptic state equation

$$\begin{cases} Ay(x) + f(y(x)) = u(x) & \text{in } \Omega, \\ y(x) = 0 & \text{on } \Gamma. \end{cases} \tag{1.1.1}$$

This problem is selected to discuss main questions of optimal control theory, since its analysis is not too technical and yet sufficiently rich.

Above, Ω denotes an open and connected bounded subset of $\mathbb{R}^n$, $n \geq 2$, and Γ is its boundary, which is Lipschitz; see, for instance, (Nečas, 1967, Section 1.1.3). In Ω, A stands for the operator

$$Ay = -\sum_{i,j=1}^{n} \partial_{x_j}(a_{ij}(x)\partial_{x_i} y) + a_0(x)y \tag{1.1.2}$$

with $a_{ij}, a_0 \in L^\infty(\Omega)$, $1 \le i, j \le n$, such that $a_0(x) \ge 0$ a.e. in Ω and

$$\exists \Lambda > 0 : \Lambda |\xi|^2 \le \sum_{i,j=1}^{n} a_{ij}(x)\xi_i\xi_j \quad \text{for a.a. } x \in \Omega \text{ and } \forall \xi \in \mathbb{R}^n. \tag{1.1.3}$$

The function $f : \mathbb{R} \longrightarrow \mathbb{R}$ is monotone non-decreasing and of class C^2. In the cost functional, we have a fixed element $y_\Omega \in L^2(\Omega)$ and two real parameters $\nu \ge 0$ and $\kappa \ge 0$. Finally, the set of admissible controls

$$\mathbf{U}_{\text{ad}} = \{u \in L^\infty(\Omega) : \alpha \le u(x) \le \beta \text{ for a.a. } x \in \Omega\}$$

is given, where $-\infty \le \alpha < \beta \le +\infty$.

Remark 1.1.1. **(i)** More general functionals can be considered, for instance

$$J(u) = \int_\Omega L(x, y_u(x), u(x))\, dx \quad \text{with} \quad L : \Omega \times \mathbb{R} \times \mathbb{R} \longrightarrow \mathbb{R}.$$

Our functional is included as particular case by taking $L(x, y, u) = \frac{1}{2}(y - y_\Omega(x))^2 + \frac{\nu}{2}u^2 + \kappa|u|$. We preferred the above quadratic cost functional, since it is a standard one that is used for many applications. Moreover, we want to emphasize the role of the important parameters ν and κ. If $\kappa > 0$, then the optimal control has a sparse structure, which means that its support is small. Roughly speaking, the bigger κ is the smaller the support of the optimal control becomes. On the other hand, if $\nu > 0$, then the corresponding term, that may express the cost of the control, leads to a more regular solution of the optimal control problem. Therefore, ν is sometimes also called regularization parameter. For $\nu = 0$, the optimal control is essentially discontinuous, typically bang-bang. The second order analysis is very different in both cases. The reader is referred to Sections 1.2 and 1.3 for these issues.

(ii) For the nonlinear term f in the state equation, we could consider a more general function $f : \Omega \times \mathbb{R} \longrightarrow \mathbb{R}$, measurable with respect to the first variable and of class C^2 and non-decreasing with respect to the second. In this case, some additional assumptions should be imposed on f; see e.g. (Casas and Tröltzsch, 2009).

(iii) In many cases we are not free to place controls in the whole domain Ω. Often, the control can only act in a subdomain $\omega \subset \Omega$. In this case, we take $\mathbf{U}_{\text{ad}} \subset L^\infty(\omega)$ and write the right-hand side of (1.1.1) as $\chi_\omega u$, where χ_ω is the characteristic function of ω. Then, all the following results remain valid with obvious changes.

A first step in the analysis of the control problem is the discussion of the well-posedness of the state equation (1.1.1). We have the following result.

Theorem 1.1.2. *Assume that $p > \frac{n}{2}$. Then, the following statements hold:*

1. *For every* $u \in L^p(\Omega)$, *there exists a unique solution* $y_u \in H_0^1(\Omega) \cap C^{0,\mu}(\bar{\Omega})$ *of* (1.1.1) *for some* $\mu \in (0,1)$. *Moreover, the inequality*

$$\|y_u\|_{H_0^1(\Omega)} + \|y_u\|_{C^{0,\mu}(\bar{\Omega})} \leq C_{\Omega,A,p}(|f(0)| + \|u\|_{L^p(\Omega)}) \tag{1.1.4}$$

holds for a constant $C_{\Omega,A,p}$ *only depending on* Ω, A *and* p.

2. *The mapping* $G : L^p(\Omega) \longrightarrow H_0^1(\Omega) \cap C^{0,\mu}(\bar{\Omega})$ *defined by* $G(u) = y_u$ *is of class* C^2. *Further, for every* $u, v \in L^p(\Omega)$, $z_v = G'(u)v$ *is the unique solution to*

$$\begin{cases} Az(x) + f'(y_u(x))z(x) = v(x) & \text{in } \Omega, \\ z(x) = 0 & \text{on } \Gamma, \end{cases} \tag{1.1.5}$$

and, given $v_1, v_2 \in L^p(\Omega)$, $z_{v_1,v_2} = G''(u)(v_1, v_2)$ *is the solution of*

$$\begin{cases} Az(x) + f'(y_u(x))z(x) = -f''(y_u(x))z_{v_1}(x)z_{v_2}(x) & \text{in } \Omega, \\ z(x) = 0 & \text{on } \Gamma, \end{cases} \tag{1.1.6}$$

where $z_{v_i} = G'(u)v_i$ *for* $i = 1, 2$.

Proof. The proof of part 1 is classical, we only sketch it. First, we observe that $L^p(\Omega) \subset H^{-1}(\Omega)$ for every $p > \frac{n}{2}$, the embedding being even compact. For every integer $k \geq 1$, we define $f_k(t) = f(\text{Proj}_{[-k,+k]}(t))$, where $\text{Proj}_{[-k,+k]}(t) = \max\{-k, \min\{+k, t\}\}$. A direct application of Schauder's fixed point theorem allows to deduce the existence of a function $y_k \in H_0^1(\Omega)$ such that $Ay_k + f_k(y_k) = u$.

Now, the monotonicity of f yields

$$[f_k(y_k(x)) - f(0)]y_k(x) = [f_k(y_k(x)) - f_k(0)]y_k(x) \geq 0 \text{ for a.a. } x \in \Omega.$$

Then, testing the equation $Ay_k + f_k(y_k) = u$ with y_k, we infer that $\|y_k\|_{H_0^1(\Omega)} \leq C_1(|f(0)| + \|u\|_{L^p(\Omega)})$. Moreover, following the approach by (Stampacchia, 1965, Section 4), we deduce that $\|y_k\|_{L^\infty(\Omega)} \leq C_2$. The constants C_1 and C_2 are independent of f and k, they only depend on A, Ω and p. In addition, it is obvious that $f_k(y_k(x)) = f(y_k(x))$ for every $k > C_2$ and almost all $x \in \Omega$. Hence, $y_k \in H_0^1(\Omega) \cap L^\infty(\Omega)$ solves (1.1.1) for every $k > C_2$.

The uniqueness of a solution y_u in $H_0^1(\Omega) \cap L^\infty(\Omega)$ is a straightforward consequence of the ellipticity of A and the monotonicity of f. Finally, since $Ay_u = u - f(y_u) \in L^p(\Omega)$ with $p > \frac{n}{2}$, the Hölder regularity of y_u is a standard result, and the estimate $\|y_u\|_{C^{0,\mu}(\bar{\Omega})} \leq C_3(|f(0)| + \|u\|_{L^p(\Omega)})$ is fulfilled by a constant only depending on A, Ω and p; see, for instance, (Gilbarg and Trudinger, 1983, Theorem 8.29). Then, the inequality (1.1.4) is satisfied with $C_{\Omega,A,p} = \max\{C_1, C_2, C_3\}$.

To prove the differentiability of the control-to-state mapping G stated in part 2, we apply the implicit function theorem. To this end, we define the vector space

$$Y = \{y \in H_0^1(\Omega) \cap C^{0,\mu}(\bar{\Omega}) : Ay \in L^p(\Omega)\}$$

endowed with the norm

$$\|y\|_Y = \|y\|_{H_0^1(\Omega)} + \|y\|_{C^{0,\mu}(\bar{\Omega})} + \|Ay\|_{L^p(\Omega)}.$$

It is obvious that $(Y, \|\cdot\|_Y)$ is a Banach space. The first part of the theorem implies that $y_u \in Y$ for every $u \in L^p(\Omega)$. Now, we consider the mapping

$$\mathcal{F} : Y \times L^p(\Omega) \longrightarrow L^p(\Omega), \quad \mathcal{F}(y,u) = Au + f(y) - u.$$

Then, $\mathcal{F}$ is a C^2 function and $\mathcal{F}(G(u), u) = \mathcal{F}(y_u, u) = 0$. Moreover, the partial derivative

$$\frac{\partial \mathcal{F}}{\partial y}(y,u) : Y \longrightarrow L^p(\Omega), \quad \frac{\partial \mathcal{F}}{\partial y}(y,u)z = Az + f'(y)z,$$

is an isomorphism. This is a consequence from the fact that Eq. (1.1.5) has a unique solution $z \in Y$ for every $u \in L^p(\Omega)$, which follows from the first part of this theorem. Then, the implicit function theorem implies that G is of class C^2. Finally, the equations for $z_v = G'(u)v$ and $z_{v_1,v_2} = G''(u)(v_1, v_2)$ follow by differentiation of the identity $\mathcal{F}(G(u), u) = 0$. □

Remark 1.1.3. A look at the above proof reveals that the first part of the theorem only makes use of the continuity and monotonicity of f. The differentiability properties of f are only necessary to establish the differentiability of G.

We conclude this subsection by analyzing the cost functional J. In this functional, we distinguish two terms and write $J(u) = F(u) + \kappa j(u)$, where F and j are given by

$$F(u) = \frac{1}{2}\int_\Omega (y_u(x) - y_\Omega(x))^2\,dx + \frac{\nu}{2}\int_\Omega u^2(x)\,dx, \quad j(u) = \int_\Omega |u(x)|\,dx.$$

The following result is an immediate consequence of Theorem 1.1.2 and the chain rule.

Corollary 1.1.4. *Assume that $p > \frac{n}{2}$ and $p \geq 2$. Then, $F : L^p(\Omega) \longrightarrow \mathbb{R}$ is of class C^2, and its derivatives are given by*

$$F'(u)v = \int_\Omega (\varphi_u + \nu u)v\,dx, \tag{1.1.7}$$

$$F''(u)(v_1, v_2) = \int_\Omega \{[1 - f''(y_u)\varphi_u]z_{v_1} z_{v_2} + \nu v_1 v_2\}\,dx, \tag{1.1.8}$$

where $\varphi_u \in H_0^1(\Omega)$ is the adjoint state associated with u, solution of the equation

$$\begin{cases} A^*\varphi(x) + f'(y_u(x))\varphi(x) = y_u(x) - y_\Omega(x) & \text{in } \Omega, \\ \varphi(x) = 0 & \text{on } \Gamma, \end{cases} \tag{1.1.9}$$

and A^* denotes the adjoint operator of A:

$$A^*y = -\sum_{i,j=1}^{n} \partial_{x_j}(a_{ji}(x)\partial_{x_i} y) + a_0(x)y.$$

Furthermore, if $y_\Omega \in L^p(\Omega)$, then $\varphi \in C^{0,\mu_p}(\bar{\Omega})$ for some $\mu_p \in (0,1)$ independent of u.

The functional $j : L^1(\Omega) \longrightarrow \mathbb{R}$ is the non-differentiable part of J, it is convex and Lipschitz. Hence, there exists the convex subdifferential $\partial j(u)$ at every element $u \in L^1(\Omega)$. Moreover, a simple computation shows that $\lambda \in \partial j(u)$ if and only if

$$\begin{cases} \lambda(x) = +1 & \text{if } u(x) > 0, \\ \lambda(x) = -1 & \text{if } u(x) < 0, \\ \lambda(x) \in [-1, +1] & \text{if } u(x) = 0 \end{cases} \tag{1.1.10}$$

holds a.e. in Ω. In addition, j has a directional derivative at u given by

$$j'(u; v) = \int_{\Omega_u^+} v\,dx - \int_{\Omega_u^-} v\,dx + \int_{\Omega_u^0} |v|\,dx, \tag{1.1.11}$$

for $u, v \in L^1(\Omega)$, where Ω_u^+, Ω_u^- and Ω_u^0 represent the sets of points where u is positive, negative, or zero, respectively. Finally, the following relation holds

$$\max_{\lambda \in \partial j(u)} \int_\Omega \lambda\, v\,dx = j'(u; v) \le \frac{j(u + \rho\, v) - j(u)}{\rho} \quad \forall \rho \in (0, 1]. \tag{1.1.12}$$

We refer to (Bonnans and Shapiro, 2000, Section 2.4.3) and (Clarke, 1983, Chapter 2) for more details.

1.2 Existence of solutions and first order optimality conditions

Our analysis should start with proving the existence of a solution to (P). Using Theorem 1.1.2, this can be done by the classical approach of the Calculus of Variations if $\mathbf{U}_{\text{ad}}$ is bounded or if $\nu > 0$ and $n \le 3$. However, for unbounded $\mathbf{U}_{\text{ad}}$ and $n > 3$, we cannot apply Theorem 1.1.2 to deduce the continuity of J in $L^2(\Omega)$, unless f is affine: $f(t) = at + b$ with $a, b \in \mathbb{R}$ and $a \ge 0$.

Nevertheless, for any non-decreasing continuous functions f, we still can prove the existence of a solution for an unbounded control set $\mathbf{U}_{\text{ad}}$ and arbitrary dimension n, provided that $\nu > 0$. We will use first order optimality conditions to perform this task. For this reason, we start this subsection proving these conditions and postpone the existence theorem to the second part of the subsection. In the last part of the subsection, we study sparsity properties of locally optimal controls.

We remark that (P) is a strictly convex problem if f is affine. Hence, in this case, (P) has a unique solution if $\nu > 0$ or $\mathbf{U}_{\rm ad}$ is bounded. When f is not affine, the reader is referred to (Pighin, 2022) for some examples where (P) is not convex and the solution is not unique. We mention also (Ali et al., 2016) and (Porretta and Zuazua, 2016, §3.2) for some sufficient conditions for uniqueness of the solution.

Along this subsection, we will make use of the following assumption:

Assumption 1.2.1. *There exists $p > \frac{n}{2}$ with $p \geq 2$ such that $y_\Omega \in L^p(\Omega)$.*

Since (P) is not convex in general, we have to consider local solutions. Unlike the finite dimensional case, in infinite dimension several notions of local minima are defined depending on the topology under consideration. Let us introduce some different concepts of local solutions that are standard for control problems.

Definition 1.2.2. We say that $\bar u$ is an $L^r(\Omega)$-local minimizer (or solution) of (P) with $r \in [1, +\infty]$ if $\bar u \in \mathbf{U}_{\rm ad}$ and there exists some $\varepsilon > 0$ such that

$$J(\bar u) \leq J(u) \quad \text{for all } u \in \mathbf{U}_{\rm ad} \text{ with } \|\bar u - u\|_{L^r(\Omega)} \leq \varepsilon.$$

An element $\bar u \in \mathbf{U}_{\rm ad}$ is said to be a strong local minimizer (or solution) of (P) if there exists some $\varepsilon > 0$ such that

$$J(\bar u) \leq J(u) \text{ for all } u \in \mathbf{U}_{\rm ad} \text{ with } \|y_{\bar u} - y_u\|_{L^\infty(\Omega)} \leq \varepsilon.$$

We say that $\bar u \in \mathbf{U}_{\rm ad}$ is a strict (weak or strong) local minimizer (or solution) if the above inequalities are strict for $u \neq \bar u$.

The notions of weak and strong local solutions are well known in calculus of variations and control theory of ordinary differential equations. To our best knowledge, in the framework of optimal control theory for partial differential equations, the notion of strong local solutions was introduced in (Bayen et al., 2014) for the first time; see also (Bayen and Silva, 2016). We analyze the relationships among these concepts in the following proposition.

Proposition 1.2.3. *The following properties hold:*

If $\mathbf{U}_{\rm ad}$ is bounded in $L^\infty(\Omega)$, then

1. *$\bar u$ is an $L^1(\Omega)$-local minimizer of* (P) *if and only if it is an $L^r(\Omega)$-local minimizer of* (P) *for every $r \in (1, +\infty)$. Moreover, each $L^1(\Omega)$-local minimizer is also an $L^\infty(\Omega)$-local minimizer.*

2. *Any strong local minimizer of* (P) *is an* $L^1(\Omega)$*-local minimizer.*
3. *If in addition* $\nu > 0$, *then each* $L^1(\Omega)$*-local minimizer of* (P) *is a strong local minimizer.*

If $\mathbf{U}_{\text{ad}}$ *is not bounded in* $L^\infty(\Omega)$ *and* $\nu > 0$, *then*

1. *If* $1 \le p < \infty$, *then any* $L^p(\Omega)$*-local minimizer of* (P) *is an* $L^q(\Omega)$*-local minimizer for every* $p < q \le \infty$. *For* $p = 2$ *it is an* $L^1(\Omega)$*-local minimizer if* $n \le 3$.
2. *Each strong local minimizer of* (P) *is an* $L^p(\Omega)$ *local solution for every* $p \in (\frac{n}{2}, \infty]$.
3. *Any* $L^2(\Omega)$*-local minimizer of* (P) *is a strong local minimizer.*

The reader is referred to (Casas and Mateos, 2020), (Casas et al., 2020), and (Casas and Tröltzsch, 2020) for the proof of this proposition. We observe that the concepts of $L^r(\Omega)$-local solution for $r \in [1, \infty)$ and strong local solution are equivalent if $\mathbf{U}_{\text{ad}}$ is bounded and $\nu > 0$. If $\nu = 0$ and $\mathbf{U}_{\text{ad}}$ is bounded, then the strong local solution is the strongest concept in the sense that if $\bar{u}$ is a strong local solution, then $\bar{u}$ is an $L^r(\Omega)$-local solution for every $r \in [1, \infty]$. However, if $\nu > 0$ then the strongest concept of local solution corresponds to an $L^2(\Omega)$-local solution. In any case, the weakest concept always corresponds to $L^\infty(\Omega)$-local solution.

The different concepts of local solutions play an important role in sufficient second order optimality conditions. However, they do not influence the necessary conditions.

Theorem 1.2.4. *Assume that* $\bar{u}$ *is a local minimizer of* (P) *in any of the senses established in Definition 1.2.2. Then there exist* $\bar{y} \in H_0^1(\Omega) \cap C^{0,\mu}(\bar{\Omega})$ *with* μ *independent of* $\bar{u}$, $\bar{\varphi} \in H_0^1(\Omega)$, *and* $\bar{\lambda} \in \partial j(\bar{u})$ *that obey the optimality system*

$$\begin{cases} A\bar{y}(x) + f(\bar{y}(x)) = \bar{u}(x) & \text{in } \Omega, \\ \bar{y}(x) = 0 & \text{on } \Gamma, \end{cases} \tag{1.2.1}$$

$$\begin{cases} A^*\bar{\varphi}(x) + f'(\bar{y}(x))\bar{\varphi}(x) = \bar{y}(x) - y_\Omega(x) & \text{in } \Omega, \\ \bar{\varphi}(x) = 0 & \text{on } \Gamma, \end{cases} \tag{1.2.2}$$

$$\int_\Omega (\bar{\varphi} + \nu\bar{u} + \kappa\bar{\lambda})(u - \bar{u})\,dx \ge 0 \quad \forall u \in \mathbf{U}_{\text{ad}}. \tag{1.2.3}$$

Moreover, if Assumption 1.2.1 holds, then $\bar{\varphi} \in C^{0,\mu_p}(\bar{\Omega})$ *for some* $\mu_p \in (0, 1)$ *depending on* p.

Proof. Let $u \in \mathbf{U}_{\text{ad}}$ be arbitrary. Since $\bar{u}$ is a local minimizer, there exists $\rho_0 > 0$ such that $J(\bar{u} + \rho(u - \bar{u})) \ge J(\bar{u})$ $\forall \rho \in (0, \rho_0)$. Taking ρ in this interval and using the convexity of j, we get

$$0 \le \frac{F(\bar{u} + \rho(u - \bar{u})) - F(\bar{u})}{\rho} + \kappa\frac{j(\bar{u} + \rho(u - \bar{u})) - j(\bar{u})}{\rho}$$

$$\leq \frac{F(\bar{u}+\rho(u-\bar{u}))-F(\bar{u})}{\rho}+\kappa[j(u)-j(\bar{u})].$$

Passing to the limit as $\rho \to 0$ and using (1.1.7), we infer

$$\int_\Omega (\bar{\varphi}+\nu\bar{u})(u-\bar{u})\,dx+\kappa[j(u)-j(\bar{u})]\geq 0 \quad \forall u\in \mathbf{U}_{\text{ad}},$$

where $\bar{\varphi}$ is the solution of (1.2.2). Let us set

$$I(u)=\int_\Omega (\bar{\varphi}+\nu\bar{u})u\,dx+\kappa j(u)+I_{\mathbf{U}_{\text{ad}}}(u),$$

where $I_{\mathbf{U}_{\text{ad}}}$ is the indicator function of $\mathbf{U}_{\text{ad}}$, i.e. it takes the values 0 or ∞ if $u \in \mathbf{U}_{\text{ad}}$ or $u \notin \mathbf{U}_{\text{ad}}$, respectively. We infer from the above inequality that I achieves the minimum value at $\bar{u}$. Hence, using subdifferential calculus (Ekeland and Temam, 1976, Proposition I-5.6), we have that

$$0\in \partial I(\bar{u})=\bar{\varphi}+\nu\bar{u}+\kappa\partial j(\bar{u})+\partial I_{\mathbf{U}_{\text{ad}}}(\bar{u}).$$

This implies the existence of $\bar{\lambda}\in\partial j(\bar{u})$ such that inequality (1.2.3) holds for all $u\in\mathbf{U}_{\text{ad}}$. Finally, the regularity of $\bar{\varphi}$ follows from Theorem 1.1.2 and Assumption 1.2.1. □

Any function $\bar{u}\in\mathbf{U}_{\text{ad}}$ satisfying, along with $\bar{\lambda}\in\partial j(\bar{u})$, $\bar{y}$ and $\bar{\varphi}$, the optimality conditions (1.2.1)–(1.2.3), is called a stationary control for (P).

We should mention that the necessary optimality conditions (1.2.1)–(1.2.3) are also sufficient for global optimality if the term f in the state equation is affine. This is an obvious consequence of the convexity of (P) in this case.

Remark 1.2.5. We mention that (1.2.3) and (1.1.12) yield

$$\begin{aligned}0&\leq \int_\Omega (\bar{\varphi}+\nu\bar{u}+\kappa\bar{\lambda})(u-\bar{u})\,dx\leq F'(\bar{u})(u-\bar{u})+\kappa j'(\bar{u};u-\bar{u})\\&\leq F'(\bar{u})(u-\bar{u})+\kappa[j(u)-j(\bar{u})]\quad \forall u\in\mathbf{U}_{\text{ad}}.\end{aligned} \tag{1.2.4}$$

Other important and immediate consequences of (1.2.3) are the following:

$$\text{If } \nu>0, \text{ then } \bar{u}(x)=\text{Proj}_{[\alpha,\beta]}\Big(-\frac{1}{\nu}(\bar{\varphi}(x)+\kappa\bar{\lambda}(x))\Big), \tag{1.2.5}$$

$$\bar{\varphi}(x)+\nu\bar{u}(x)+\kappa\bar{\lambda}(x)\begin{cases}\geq 0 & \text{if } \bar{u}(x)=\alpha,\\ =0 & \text{if } \alpha<\bar{u}(x)<\beta,\\ \leq 0 & \text{if } \bar{u}(x)=\beta.\end{cases} \tag{1.2.6}$$

Let us observe that $\bar{\varphi}+\nu\,\bar{u}+\kappa\bar{\lambda}=0$ if there are no control constraints, i.e. if $\alpha=-\infty$ and $\beta=+\infty$.

Finally, if $\nu = \kappa = 0$ and $-\infty < \alpha < \beta < +\infty$, then

$$\bar{u}(x) = \begin{cases} \alpha & \text{if } \bar{\varphi}(x) > 0, \\ \beta & \text{if } \bar{\varphi}(x) < 0. \end{cases} \tag{1.2.7}$$

As a consequence of this property, if $|\{x \in \Omega : \bar{\varphi}(x) = 0\}| = 0$, then $\bar{u}$ only takes the two values α and β. Here $|\cdot|$ denotes the Lebesgue measure. In this case, $\bar{u}$ is said to be a bang-bang control.

Now, we address the issue of existence of a solution for (P).

Theorem 1.2.6. *Problem* (P) *has at least one solution $\bar{u}$ under each of the following conditions: (i) $-\infty < \alpha < \beta < +\infty$ or (ii) $\nu > 0$ and Assumption 1.2.1 holds.*

Proof. We analyze first the case (i). Here, the proof follows the direct method of the Calculus of Variations. Let $\{u_k\}_{k=1}^{\infty} \subset \mathbf{U}_{\text{ad}}$ be a minimizing sequence for (P), i.e. $J(u_k) \searrow \inf(\text{P})$. From the boundedness of $\mathbf{U}_{\text{ad}}$, we infer the existence of a subsequence, denoted in the same way, such that $u_k \stackrel{*}{\rightharpoonup} \bar{u}$ in $L^\infty(\Omega)$. Since $\mathbf{U}_{\text{ad}}$ is weakly* closed in $L^\infty(\Omega)$ we get that $\bar{u} \in \mathbf{U}_{\text{ad}}$. Moreover, from (1.1.4) it follows that $y_{u_k} \rightharpoonup y_{\bar{u}}$ in $H_0^1(\Omega)$ and $y_{u_k} \to y_{\bar{u}}$ in $C(\bar{\Omega})$. Finally, using the lower weak semicontinuity of the functional $L^\infty(\Omega) \ni u \mapsto \frac{\nu}{2}\|u\|_{L^2(\Omega)}^2 + \|u\|_{L^1(\Omega)}$, we conclude that $J(\bar{u}) \le \lim_{k\to\infty} J(u_k) = \inf(\text{P})$. Hence, $\bar{u}$ is a solution of (P).

Now we assume that (i) does not hold, but (ii) is fulfilled. If $\alpha > -\infty$, for any $M > |\alpha|$ we define $\alpha_M = \alpha$ and $\beta_M = M$. If $\beta < +\infty$, then for $M > |\beta|$ we take $\alpha_M = -M$ and $\beta_M = \beta$. If $\alpha = -\infty$ and $\beta = +\infty$, we set $\alpha_M = -M$ and $\beta_M = +M$ for any $M > 0$. Then, we consider the control problems

$$(\text{P}_M) \quad \min_{u \in \mathbf{U}_\text{M}} J(u)$$

where $\mathbf{U}_\text{M} \subset \mathbf{U}_{\text{ad}}$ is given by

$$\mathbf{U}_\text{M} = \{u \in L^\infty(\Omega) : \alpha_M \le u(x) \le \beta_M \text{ a.e. in } \Omega\}.$$

We will prove that, for M big enough, any solution of (P_M) also solves (P). First, we observe that (P_M) satisfies condition (i), hence there exists at least one solution u_M of (P_M). Lemma 1.2.7 below implies the existence of a constant K_∞ independent of M such that $\|u_M\|_{L^\infty(\Omega)} \le K_\infty$ for all M. Take $M_\infty = K_\infty$ and denote by $\bar{u} = u_{M_\infty}$ a solution of (P_{M_∞}). Then, $\bar{u}$ is a solution of (P). Indeed, let $u \in \mathbf{U}_{\text{ad}}$ be arbitrary. We set $M = \|u\|_{L^\infty(\Omega)}$. If $M \le M_\infty$, then $u \in \mathbf{U}_{M_\infty}$ and, consequently, $J(\bar{u}) = J(u_{M_\infty}) \le J(u)$ holds. If $M > M_\infty$, let u_M be a solution of (P_M). Then, u_M satisfies $\|u_M\|_{L^\infty(\Omega)} \le M_\infty$, hence $u_M \in \mathbf{U}_{M_\infty}$. Therefore, from the optimality of $\bar{u}$ and u_M in $\mathbf{U}_{M_\infty}$ and $\mathbf{U}_\text{M}$, respectively, we get $J(\bar{u}) = J(u_{M_\infty}) \le J(u_M) \le J(u)$, which proves the optimality of $\bar{u}$. □

Lemma 1.2.7. *Under Assumption (ii) of Theorem 1.2.6 and with the previous notation, there exists a constant* $K_\infty > 0$ *independent of* M *such that* $\|u_M\|_{L^\infty(\Omega)} \leq K_\infty$ *holds for all selected* M.

Proof. Given a solution u_M of (P_M), we denote by y_M, φ_M, and $\lambda_M \in \partial j(u_M)$ the functions satisfying the optimality conditions (1.2.1)–(1.2.3) along with u_M. Let us fix $u_0 = \alpha$ if $\alpha > -\infty$ or $u_0 = \beta$ if $\beta < +\infty$; otherwise we take $u_0 = 0$. Then, we have that $u_0 \in \mathbf{U}_\text{M}$ for all M and, consequently

$$\|u_M\|_{L^2(\Omega)} \leq \sqrt{\frac{2}{\nu}J(u_0)} \quad \forall M.$$

Now, from the state equation (1.1.1), using the monotonicity of f and (1.1.3), we get

$$\|y_M\|_{H_0^1\Omega)} \leq C_\Lambda\Big(\|u_M\|_{L^2(\Omega)} + |f(0)|\Big) \leq C_\Lambda\Big(\sqrt{\frac{2}{\nu}J(u_0)} + |f(0)|\Big) \ \forall M.$$

For $n = 2$ we have the Sobolev embedding $H_0^1(\Omega) \subset L^q(\Omega)$ for every $q < \infty$. If $n > 2$ we know that $H_0^1(\Omega) \subset L^{\frac{2n}{n-2}}(\Omega)$. A simple computation shows that $\frac{2n}{n-2} > \frac{n}{2}$ if $n < 6$. Since $y_\Omega \in L^p(\Omega)$ with $p > \frac{n}{2}$, for $n < 6$ we deduce the existence of some $r \in (\frac{n}{2}, p]$ such that $\|y_M - y_\Omega\|_{L^r(\Omega)} \leq C_r$ with C_r independent of M. Now, from the adjoint state equation (1.2.2) and the estimates of (Stampacchia, 1965, Theorem 4.2) we infer

$$\|\varphi_M\|_{L^\infty(\Omega)} \leq K_r\|y_M - y_\Omega\|_{L^r(\Omega)} \leq K_r C_r,$$

where K_r is independent of M. This estimate, (1.2.5), and (1.1.10) imply

$$\|u_M\|_{L^\infty(\Omega)} \leq \frac{1}{\nu}\Big(\|\varphi_M\|_{L^\infty(\Omega)} + \kappa\|\lambda_M\|_{L^\infty(\Omega)}\Big) \leq \frac{1}{\nu}(K_r C_r + \kappa) = K_\infty$$

if $\alpha = -\infty$ and $\beta = +\infty$. If $0 < \alpha < \infty$ or $-\infty < \beta < 0$, then we have to set $K_\infty = \max\{\alpha, \frac{1}{\nu}(K_r C_r + \kappa)\}$ or $K_\infty = \max\{-\beta, \frac{1}{\nu}(K_r C_r + \kappa)\}$, respectively.

If $n = 6$, we know that $y_M - y_\Omega \in L^{\frac{2n}{n-2}}(\Omega) = L^3(\Omega)$. With (Brezis and Strauss, 1973, Theorem 18), it follows that

$$\|\varphi_M\|_{L^p(\Omega)} \leq C\|y_M - y_\Omega\|_{L^3(\Omega)} \leq C_{\varphi,6} \ \forall M$$

with $C_{\varphi,6}$ independent of M.

If $n > 6$, we use $y_M - y_\Omega \in L^{\frac{2n}{n-2}}(\Omega)$. Then, we apply again (Brezis and Strauss, 1973, Theorem 18) to obtain that

$$\|\varphi_M\|_{L^{\frac{2n}{n-6}}(\Omega)} \leq C\|y_M - y_\Omega\|_{L^{\frac{2n}{n-2}}(\Omega)} \leq C_{\varphi,n}.$$

Invoking again (1.2.5), we deduce the existence of a constant independent of M such that $\|u_M\|_{L^p(\Omega)} \le C_{u,6}$ if $n = 6$ with $p > \frac{n}{2}$, and $\|u_M\|_{L^{\frac{2n}{n-6}}(\Omega)} \le C_{u,n}$ if $n > 6$. If $n < 10$, then $\frac{2n}{n-6} > \frac{n}{2}$ holds. Therefore, from (1.1.4) the inequality $\|y_M\|_{L^\infty(\Omega)} \le K_{y,\infty}$ follows for $n < 10$. From the adjoint state equation, we get that $\|\varphi_M\|_{L^\infty(\Omega)} \le K_{\varphi,\infty}$, and (1.2.5) leads to $\|u_M\|_{L^\infty(\Omega)} \le K_\infty$, with K_∞ independent of M.

For $n = 10$ and $n > 10$ we repeat the arguments used for $n = 6$ and $n > 6$ to deduce that $\|u_M\|_{L^\infty(\Omega)} \le K_\infty$ if $n < 14$. Hence, following this bootstrapping argument until the dimension of Ω is reached, we conclude the proof of lemma. □

The next theorem establishes some sparsity properties of a local minimizer $\bar{u}$; see (Casas, 2012) and (Casas et al., 2012a) for the proof.

Theorem 1.2.8. *Assume that $\alpha < 0 < \beta$ and $\kappa > 0$. Let $(\bar{u}, \bar{y}, \bar{\varphi}, \bar{\lambda}) \in \mathbf{U}_{\rm ad} \times [H_0^1(\Omega) \cap C^{0,\mu}(\bar{\Omega})] \times H_0^1(\Omega) \times \partial j(\bar{u})$ satisfy the optimality system* (1.2.1)–(1.2.3). *Then, the following properties are fulfilled for a.a. $x \in \Omega$:*

$$\bar{\lambda}(x) = \operatorname{Proj}_{[-1,+1]}\Big(-\frac{1}{\kappa}\bar{\varphi}(x)\Big), \tag{1.2.8}$$

$$\text{If } \nu > 0, \text{ then } \bar{u}(x) = 0 \quad \Leftrightarrow \quad |\bar{\varphi}(x)| \le \kappa, \tag{1.2.9}$$

$$\text{if } \nu = 0 \text{ and } -\infty < \alpha < \beta < +\infty, \text{ then}$$

$$\bar{u}(x) = \begin{cases} \alpha & \text{if } \bar{\varphi}(x) > +\kappa, \\ 0 & \text{if } |\bar{\varphi}(x)| < \kappa, \\ \beta & \text{if } \bar{\varphi}(x) < -\kappa. \end{cases} \tag{1.2.10}$$

Moreover, the regularity $\bar{\lambda} \in H_0^1(\Omega)$ holds. In addition, if Assumption 1.2.1 is fulfilled, we also have that $\bar{\varphi}, \bar{\lambda} \in C^{0,\mu_p}(\bar{\Omega})$ for some $\mu_p \in (0,1)$ independent of $\bar{u}$. Finally, if $\nu > 0$ then $\bar{u}$ enjoys the same regularity as $\bar{\lambda}$.

Remark 1.2.9. (i) If $\bar{u}$ is a local solution of (P), then formula (1.2.6) implies the *uniqueness* of the element $\bar{\lambda} \in \partial j(\bar{u})$ such that (1.2.1)–(1.2.3) hold.
(ii) If $\nu = 0$ and $|\{x \in \Omega : |\bar{\varphi}(x)| = \kappa\}| = 0$, then $\bar{u}(x)$ can take only the values α, β, and 0. We call controls of this form bang-bang-bang or bang-bang-off.

Theorem 1.2.10. *Suppose that $\alpha < 0 < \beta$ and Assumption 1.2.1 are satisfied. In addition, assume that one of the following conditions holds: (i) $-\infty < \alpha < \beta < +\infty$; (ii) $n \le 3$, (iii) $n < 6$ and $\nu > 0$. Then, there exists $\kappa_0 \in (0, \infty)$ such that $\bar{u} = 0$ is the only global solution of* (P) *if $\kappa > \kappa_0$.*

Proof. First, we assume condition (i). Let $\bar{u}$ be a global solution of (P) with associated state $\bar{y}$. Under the first assumption, Theorem 1.1.2 implies the existence of a constant $K_{\mathbf{U}_{\rm ad}}$ such that $\|y_u\|_{L^\infty(\Omega)} \le K_{\mathbf{U}_{\rm ad}}$ for every $u \in \mathbf{U}_{\rm ad}$. Therefore, we have

$$\|\bar{y} - y_\Omega\|_{L^p(\Omega)} \le K_{\mathbf{U}_{\rm ad}}|\Omega|^{\frac{1}{p}} + \|y_\Omega\|_{L^p(\Omega)}.$$

The adjoint state equation yields

$$\|\bar\varphi\|_{L^\infty(\Omega)} \le K_p\|\bar y - y_\Omega\|_{L^p(\Omega)} \le \kappa_0 = K_p(K_{\mathbf{U}_{\rm ad}}|\Omega|^{\frac{1}{p}} + \|y_\Omega\|_{L^p(\Omega)}).$$

If $\kappa > \kappa_0$, then $\|\bar\varphi\|_{L^\infty(\Omega)} < \kappa$ and with (1.2.9) and (1.2.10) we get $\bar u = 0$.

Let us now assume that (ii) holds and denote by y_0 the solution of (1.1.1) associated with the control $u = 0$. Since $u \in \mathbf{U}_{\rm ad}$, we have

$$\frac{1}{2}\|\bar y - y_\Omega\|^2_{L^2(\Omega)} \le J(\bar u) \le J(0) = \frac{1}{2}\|y_0 - y_\Omega\|^2_{L^2(\Omega)}$$

and hence $\|\bar y - y_\Omega\|_{L^2(\Omega)} \le \|y_0 - y_\Omega\|_{L^2(\Omega)}$. Then, (1.2.2) and Theorem 1.1.2 lead to the existence of a constant $K_2 > 0$ independent of $\bar u$ and κ such that

$$\|\bar\varphi\|_{L^\infty(\Omega)} \le K_2\|\bar y - y_\Omega\|_{L^2(\Omega)} \le \kappa_0 = K_2\|y_0 - y_\Omega\|_{L^2(\Omega)}$$

and the theorem follows arguing as before.

Finally, we assume that (iii) is fulfilled. Now, we use that

$$\|\bar u\|_{L^2(\Omega)} \le \frac{1}{\sqrt{\nu}}\|y_0 - y_\Omega\|_{L^2(\Omega)}.$$

Proceeding as in the first part of the proof of Lemma 1.2.7, we deduce the boundedness of $\bar\varphi$ by a constant κ_0 independent of κ. □

1.3 Second order optimality conditions

In this subsection, we carry out the second order analysis for (P). Since second order conditions are not needed for convex optimization problems, this subsection is only relevant when f is not affine. First, we derive the second order necessary optimality conditions to be satisfied by a local minimizer. Then, distinguishing the cases $\nu > 0$ and $\nu = 0$, we discuss second order sufficient optimality conditions. Along this subsection, we suppose that Assumption 1.2.1 is satisfied.

Let $\bar u \in \mathbf{U}_{\rm ad}$ be a stationary control for (P). Then, under Assumption 1.2.1, the linear and bilinear continuous forms $F'(\bar u) : L^p(\Omega) \longrightarrow \mathbb{R}$ and $F''(\bar u) : L^p(\Omega) \times L^p(\Omega) \longrightarrow \mathbb{R}$ can be extended to linear and bilinear continuous forms on $L^2(\Omega)$. This follows from the expressions (1.1.7) and (1.1.8); observe that Theorem 1.2.4 implies $\bar\varphi + \nu\bar u + \kappa\bar\lambda \in L^\infty(\Omega)$ and $f''(\bar y)\bar\varphi \in L^\infty(\Omega)$. Moreover, for every $v \in L^2(\Omega)$, the solution z_v of (1.1.5) is a well defined element of $H^1_0(\Omega)$, and $\|z_v\|_{H^1_0(\Omega)} \le C_{\Lambda,\Omega}\|v\|_{L^2(\Omega)}$ for some constant only depending on Ω and Λ. These extensions of $F'(\bar u)$ and $F''(\bar u)$ will be used in this subsection.

To set up the second-order optimality conditions, we need to introduce the critical cone. Given a control $\bar u \in \mathbf{U}_{\rm ad}$ satisfying (1.2.1)–(1.2.3) along with $\bar\lambda \in \partial j(\bar u)$ and $\bar y, \bar\varphi \in H^1_0(\Omega) \cap C^{0,\mu}(\bar\Omega)$ for some $\mu \in (0,1)$, we define

$$C_{\bar u} = \{v \in L^2(\Omega) \text{ satisfying (1.3.2) and } F'(\bar u)v + \kappa j'(\bar u; v) = 0\} \qquad (1.3.1)$$

with

$$\begin{cases} v(x) \geq 0 & \text{if } \bar{u}(x) = \alpha, \\ v(x) \leq 0 & \text{if } \bar{u}(x) = \beta \end{cases} \tag{1.3.2}$$

for a.a. $x \in \Omega$. For the proof of the next two results, the reader is referred to (Casas et al., 2012a).

Proposition 1.3.1. *The set $C_{\bar{u}}$ is a closed, convex cone in $L^2(\Omega)$.*

Theorem 1.3.2. *Assume that $\bar{u}$ is a local minimizer of* (P) *in any of the senses given in Definition 1.2.2. Then $F''(\bar{u})v^2 \geq 0$ is fulfilled for every $v \in C_{\bar{u}}$.*

1.3.1 Sufficient second order conditions. Case $\nu > 0$

The hypothesis $\nu > 0$ is assumed as standing assumption in this subsection. We also recall that Assumption 1.2.1 is assumed in the whole Section 1.3. Here we formulate a second order sufficient optimality condition with minimal gap with respect to the second order necessary condition. First, we mention the following two results whose proof can be found in (Casas et al., 2012a); see also (Casas, 2012). Though the dimension $n \leq 3$ is assumed in these papers, the same proof is valid for an arbitrary dimension.

Theorem 1.3.3. *Let $\bar{u} \in \mathbf{U}_{\text{ad}}$ satisfy the first order optimality conditions* (1.2.1)–(1.2.3). *Then the following statements are equivalent:*

1. *$F''(\bar{u})v^2 > 0$ for all $v \in C_{\bar{u}} \setminus \{0\}$,*
2. *There exist $\tau > 0$ and $\delta > 0$ such that $F''(\bar{u})v^2 \geq \delta \|v\|^2_{L^2(\Omega)}$ $\forall v \in S^\tau_{\bar{u}}$,*
3. *There exist $\tau > 0$ and $\delta > 0$ such that $F''(\bar{u})v^2 \geq \delta \|z_v\|^2_{L^2(\Omega)}$ $\forall v \in E^\tau_{\bar{u}}$, where $z_v = G'(\bar{u})v$,*

$$S^\tau_{\bar{u}} = \{v \in L^2(\Omega) \text{ satisfying (1.3.2) and } F'(\bar{u})v + \kappa j'(\bar{u}; v) \leq \tau \|v\|_{L^2(\Omega)}\},$$
$$E^\tau_{\bar{u}} = \{v \in L^2(\Omega) \text{ satisfying (1.3.2) and } F'(\bar{u})v + \kappa j'(\bar{u}; v) \leq \tau \|z_v\|_{L^2(\Omega)}\}.$$

The next theorem on second order sufficient optimality conditions is the main result of this subsection.

Theorem 1.3.4. *Assume that $\mathbf{U}_{\text{ad}}$ is bounded in $L^\infty(\Omega)$ or $n \leq 3$. Let $\bar{u} \in \mathbf{U}_{\text{ad}}$ satisfy the first order optimality conditions* (1.3.1)–(1.3.2). *If the second order condition*

$$F''(\bar{u})v^2 > 0 \quad \forall v \in C_{\bar{u}} \setminus \{0\} \tag{1.3.3}$$

holds, then there exist $\varepsilon > 0$ and $\delta > 0$ such that

$$J(\bar{u}) + \frac{\delta}{2}\|u - \bar{u}\|^2_{L^2(\Omega)} \leq J(u) \quad \forall u \in \mathbf{U}_{\text{ad}} \cap \bar{B}_\varepsilon(\bar{u}), \tag{1.3.4}$$

where $\bar{B}_\varepsilon(\bar{u})$ is the closed $L^2(\Omega)$-ball with radius ε centered at $\bar{u}$. Therefore, $\bar{u}$ is a strict $L^2(\Omega)$-local solution of (P).

In the above theorem we can remove the assumption $n \leq 3$ if we take the ball $\bar{B}_\varepsilon(\bar{u})$ in $L^p(\Omega)$ for some $p > \frac{n}{2}$.

According to Proposition 1.2.3, case $\nu > 0$, $\bar{u}$ is also a strong local minimizer. Consequently, it is natural to wonder if the quadratic growth condition (1.3.4) holds assuming that $\|y_u - \bar{y}\|_{L^\infty(\Omega)} \leq \varepsilon$ instead of taking $u \in \bar{B}_\varepsilon(\bar{u})$. The following corollary answers this question.

Corollary 1.3.5. *Under the assumptions of Theorem 1.3.4, there exist $\varepsilon > 0$ and $\delta > 0$ such that*

$$J(\bar{u}) + \frac{\delta}{2}\|u - \bar{u}\|^2_{L^2(\Omega)} \leq J(u) \; \forall u \in \mathbf{U}_{\rm ad} : \|y_u - \bar{y}\|_{L^\infty(\Omega)} \leq \varepsilon. \tag{1.3.5}$$

Proof. We proceed by contradiction and assume that (1.3.5) fails. Then, there exists a sequence of controls $\{u_k\}_{k=1}^\infty \subset \mathbf{U}_{\rm ad}$ with associated states $\{y_k\}_{k=1}^\infty \subset H_0^1(\Omega)$ such that

$$\|y_k - \bar{y}\|_{L^\infty(\Omega)} < \frac{1}{k}, \; J(u_k) < J(\bar{u}) + \frac{1}{2k}\|u_k - \bar{u}\|^2_{L^2(\Omega)}, \; \forall k \geq 1. \tag{1.3.6}$$

We are going to prove that $u_k \to \bar{u}$ strongly in $L^2(\Omega)$, then the second inequality of (1.3.6) contradicts (1.3.4) and the corollary is proved. First, we observe that $\{u_k\}_{k=1}^\infty$ is a bounded sequence of $L^2(\Omega)$. Indeed, we infer from (1.3.6)

$$\frac{\nu}{2}\|u_k\|^2_{L^2(\Omega)} \leq J(u_k) < J(\bar{u}) + \frac{1}{k}\|\bar{u}\|^2_{L^2(\Omega)} + \frac{1}{k}\|u_k\|^2_{L^2(\Omega)},$$

which obviously implies the mentioned boundedness. By taking a subsequence that we denote in the same way, we deduce the existence of an element $\tilde{u} \in \mathbf{U}_{\rm ad}$ such that $u_k \rightharpoonup \tilde{u}$ in $L^2(\Omega)$. Moreover, the boundedness of $\{u_k\}_{k=1}^\infty$ and (1.1.4) imply the boundedness of $\{y_k\}_{k=1}^\infty$ in $H_0^1(\Omega)$. Therefore, by (1.3.4), $y_k \to \bar{y}$ strongly in $L^\infty(\Omega)$ and $y_k \rightharpoonup \bar{y}$ in $H_0^1(\Omega)$ is fulfilled. Then, it is easy to pass to the limit in the equation satisfied by y_k to get

$$\bar{u} = A\bar{y} + f(\bar{y}) = w-\lim_{k\to\infty} Ay_k + \lim_{k\to\infty} f(y_k) = w-\lim_{k\to\infty} u_k = \tilde{u},$$

where $w-\lim$ denotes the weak limit in $H^{-1}(\Omega)$. Thus, we have $\bar{u} = \tilde{u}$ and, consequently, $u_k \rightharpoonup \bar{u}$ in $L^2(\Omega)$. Using (1.3.6) we get

$$\begin{aligned} J(\bar{u}) &\leq \liminf_{k\to\infty} J(u_k) \leq \limsup_{k\to\infty} J(u_k) \\ &\leq \limsup_{k\to\infty} \left\{ J(\bar{u}) + \frac{1}{k}\|\bar{u}\|^2_{L^2(\Omega)} + \frac{1}{k}\|u_k\|^2_{L^2(\Omega)} \right\} = J(\bar{u}). \end{aligned}$$

Finally, the convergences $J(u_k) \to J(\bar{u})$ and $\|y_k - \bar{y}\|_{L^2(\Omega)} \to 0$, deduced from the inequalities above and (1.3.6), imply that

$$\lim_{k\to\infty} \left\{ \frac{\nu}{2}\|u_k\|^2_{L^2(\Omega)} + \kappa\|u_k\|_{L^1(\Omega)} \right\} = \frac{\nu}{2}\|\bar{u}\|^2_{L^2(\Omega)} + \kappa\|\bar{u}\|_{L^1(\Omega)}.$$

Using this fact we get

$$
\begin{aligned}
&\frac{\nu}{2}\|\bar u\|^2_{L^2(\Omega)} \le \liminf_{k\to\infty}\frac{\nu}{2}\|u_k\|^2_{L^2(\Omega)} \le \limsup_{k\to\infty}\frac{\nu}{2}\|u_k\|^2_{L^2(\Omega)}\\
&\le \limsup_{k\to\infty}\left\{\frac{\nu}{2}\|u_k\|^2_{L^2(\Omega)} + \kappa\|u_k\|_{L^1(\Omega)}\right\} - \liminf_{k\to\infty}\kappa\|u_k\|_{L^1(\Omega)} \le \frac{\nu}{2}\|\bar u\|^2_{L^2(\Omega)}.
\end{aligned}
$$

From here we easily obtain $\|u_k\|_{L^2(\Omega)} \to \|\bar u\|_{L^2(\Omega)}$ and, consequently, $u_k \to \bar u$ in $L^2(\Omega)$ holds. □

The next corollary shows that $\bar u$ is the only stationary point of (P) in a neighborhood of $\bar u$. This was proved in (Casas and Tröltzsch, 2012) when $\kappa = 0$. The reader can easily modify the proof for the case $\kappa > 0$. Such a property excludes that $\bar u$ is an accumulation point of different local minima. In this case, numerical algorithms might fail to converge. This is one aspect that emphasizes the importance of second order conditions for the numerical analysis.

Corollary 1.3.6. *Under the assumptions of Theorem 1.3.4, there exists $\varepsilon_0 > 0$ such that $\bar u$ is the only control in $\mathbf{U}_{\rm ad}\cap \bar B_{\varepsilon_0}(\bar u)$ satisfying the first order necessary optimality conditions* (1.2.1)–(1.2.3).

1.3.2 Sufficient second order conditions. Case $\nu = 0$

In this subsection, we require the following standing assumption:

Assumption 1.3.7. *The conditions $\nu = 0$ and $-\infty < \alpha < \beta < +\infty$ are fulfilled.*

It is well known that in finite dimensional optimization the critical cone used to establish second order necessary optimality conditions is the same as the one used for second order sufficient conditions. However, this is not in general true for infinite-dimensional optimization problems, as an example by (Dunn, 1998) shows.

It holds for our control problem (P) under the assumption $\nu > 0$, as we have checked in Theorem 1.3.4. But, it does not work for $\nu = 0$. Then an extension of the cone of critical directions $C_{\bar u}$ is the standard approach to deal with second order sufficient conditions. Extensions were suggested by (Dunn, 1998), (Maurer and Zowe, 1979) and others.

As far as we know, two ways of enlarging the critical cone were proposed in the literature. In the context of abstract optimization problems, following (Maurer and Zowe, 1979), one could replace the cone $C_{\bar u}$ by $S^\tau_{\bar u}$ or, even better, by $E^\tau_{\bar u}$ for some small threshold $\tau > 0$; see Theorem 1.3.3. The latter extension was introduced in (Casas, 2012). However, also the following alternative extension was considered in the literature: In view of Remark 1.2.5, it is easy to check that the cone $C_{\bar u}$ can be written as follows

$$
\text{If } \kappa = 0,\ C_{\bar u} = \{v \in L^2(\Omega) \text{ satisfying (1.3.2) and } v(x) = 0 \text{ if } |\bar\varphi(x)| > 0\},
$$

$$\text{If } \kappa > 0,\ C_{\bar u} = \Big\{ v \in L^2(Q) \text{ satisfying (1.3.2) and } v(x) \begin{cases} \geq 0 & \text{if } \bar\varphi(x) = -\kappa \text{ and } \bar u(x) = 0, \\ \leq 0 & \text{if } \bar\varphi(x) = +\kappa \text{ and } \bar u(x) = 0, \\ = 0 & \text{if } \Big| |\bar\varphi(x)| - \kappa \Big| > 0 \end{cases} \Big\}.$$

As a consequence, the following extensions by some threshold $\tau > 0$ can be considered:

$$\text{If } \kappa = 0,\ G^\tau_{\bar u} = \{ v \in L^2(\Omega) \text{ satisfying (1.3.2) and } v(x) = 0 \text{ if } |\bar\varphi(x)| > \tau \},$$

$$\text{If } \kappa > 0,\ G^\tau_{\bar u} = \Big\{ v \in L^2(Q) \text{ satisfying (1.3.2) and } v(x) \begin{cases} \geq 0 & \text{if } \bar\varphi(x) = -\kappa \text{ and } \bar u(x) = 0, \\ \leq 0 & \text{if } \bar\varphi(x) = +\kappa \text{ and } \bar u(x) = 0, \\ = 0 & \text{if } \Big| |\bar\varphi(x)| - \kappa \Big| > \tau \end{cases} \Big\}.$$

Obviously, both cones $E^\tau_{\bar u}$ and $G^\tau_{\bar u}$ contain $C_{\bar u}$ for every $\tau > 0$, thus they are extensions of $C_{\bar u}$. For the use of the cones $E^\tau_{\bar u}$ and $G^\tau_{\bar u}$ in second order sufficient optimality conditions and their application to the stability analysis of the control problem, the reader is referred to (Casas et al., 2015). It is proved there, that a second order sufficient condition based on $G^\tau_{\bar u}$ leads to an $L^2(Q)$-weak local minimizer, while the same condition – but based on $E^\tau_{\bar u}$ – implies that $\bar u$ is a strong local minimizer.

Which of the two cones is better? There is not an easy answer. Of course, it is desirable to select a small cone such that the gap between second order necessary and sufficient conditions is as small as possible. Therefore, it is a natural idea to use the intersection $C^\tau_{\bar u} = E^\tau_{\bar u} \cap G^\tau_{\bar u}$. In (Casas and Mateos, 2020), it was proved that a second order sufficient condition based on $C^\tau_{\bar u}$ yields a strong local minimizer.

The reader might be tempted to assume that a natural coercivity condition is

$$F''(\bar u)v^2 \geq \delta \|v\|^2_{L^2(\Omega)} \quad \forall v \in C^\tau_{\bar u}.$$

However, an inspection of the expression for $F''(\bar u)$ with $\nu = 0$ reveals that the L^2-norm of v does not explicitly occur, hence this condition is not natural. Actually, it was proved in (Casas, 2012) that it is satisfied only in a few extreme cases. The expression of $F''(\bar u)$ suggests the second order condition

$$\exists \tau > 0 \text{ and } \exists \delta > 0 \text{ such that } F''(\bar u)v^2 \geq \delta \|z_v\|^2_{L^2(\Omega)} \quad \forall v \in C^\tau_{\bar u}, \tag{1.3.7}$$

where $z_v \in H^1_0(\Omega)$ is the solution of (1.1.5) with $y_u = \bar y$. Indeed, the following result can be shown:

Theorem 1.3.8. *Let $\bar{u} \in \mathbf{U}_{\text{ad}}$ satisfy the first order optimality conditions* (1.2.1)–(1.2.3). *Suppose that* (1.3.7) *and Assumptions 1.2.1 and 1.3.7 are fulfilled. Then, there exist $\varepsilon > 0$ and $\delta > 0$ such that*

$$J(\bar{u}) + \frac{\delta}{2}\|y_u - \bar{y}\|^2_{L^2(\Omega)} \le J(u) \ \forall u \in \mathbf{U}_{\text{ad}} : \|y_u - \bar{y}\|_{L^\infty(\Omega)} < \varepsilon. \tag{1.3.8}$$

This theorem was proved in (Casas and Mateos, 2021). The reader has surely realized the difference between (1.3.8) or (1.3.4) and (1.3.5). While (1.3.4) and (1.3.5) allow to prove stability of the solution $\bar{u}$ of (P) with respect to perturbations of the data and, as a consequence, stability of the state, (1.3.8) only ensures the stability of the state.

In order to get some extra information on the control, we need an additional assumption. Given a stationary control of (P), $\bar{u} \in \mathbf{U}_{\text{ad}}$, we impose the following assumption on $\bar{u}$, more precisely, on the associated adjoint state:

Assumption 1.3.9. *There exist $K > 0$ and $\gamma \in (0, 1]$ such that*

$$\left|\left\{x \in \Omega : \left||\bar{\varphi}(x)| - \kappa\right| \le \varepsilon\right\}\right| \le K\varepsilon^\gamma \quad \forall \varepsilon > 0. \tag{1.3.9}$$

Under condition (1.3.9) with $\gamma = 1$ and $\kappa = 0$, a deep second order analysis of (P) was carried out in (Casas et al., 2017). Moreover, for the same values of γ and κ, condition (1.3.9) was used in (Deckelnick and Hinze, 2012) to estimate the error for numerical discretizations of (P). In (Wachsmuth and Wachsmuth, 2011), it was applied to obtain regularization error estimates for bang-bang controls.

The property (1.3.9) is fulfilled with $\gamma = 1$ if $\bar{\varphi} \in C^1(\bar{\Omega})$ and there exists a constant $C > 0$ such that $|\nabla\bar{\varphi}(x)| \ge C \ \forall x \in \bar{\Omega}$ with $|\bar{\varphi}(x)| = \kappa$; see (Deckelnick and Hinze, 2012) for the case $\kappa = 0$. Nevertheless, (1.3.9) with $\gamma = 1$ may not hold if $\kappa = 0$ and Ω is polygonal or polyhedral; see (Qui and Wachsmuth, 2018, Eq. (3.5)).

As an immediate consequence of (1.3.9), we get

$$|\{x \in \Omega : |\bar{\varphi}(x)| = \kappa\}| = 0.$$

Therefore, (1.3.9) implies that $\bar{u}$ is bang-bang if $\kappa = 0$ or bang-bang-bang if $\kappa > 0$. The following result was proved in (Casas and Mateos, 2021).

Theorem 1.3.10. *Let $\bar{u}$ be a solution of* (P) *satisfying the first order optimality conditions* (1.2.1)–(1.2.3), *the second order sufficient condition* (1.3.7), *and the structural Assumption 1.3.9. Then, there exist $\varepsilon > 0$ and $\sigma > 0$ such that for all $u \in \mathbf{U}_{\text{ad}}$ with $\|y_u - \bar{y}\|_{L^\infty(\Omega)} \le \varepsilon$ the following inequality holds:*

$$J(\bar{u}) + \frac{\sigma}{2}\|u - \bar{u}\|^{1+\frac{1}{\gamma}}_{L^1(\Omega)} + \frac{\sigma}{2}\|y_u - \bar{y}\|^2_{L^2(\Omega)} \le J(u). \tag{1.3.10}$$

Remark 1.3.11. If $\bar{u}$ is not a bang-bang control, if $\kappa = 0$, or a bang-bang-bang control, if $\kappa > 0$, then the inequality

$$J(\bar{u}) + \sigma \|u - \bar{u}\|^{\rho}_{L^q(\Omega)} < J(u) \;\; \forall u \in \mathbf{U}_{\rm ad} : \|u - \bar{u}\|_{L^1(\Omega)} \le \varepsilon$$

never holds for any $q \geq 1$, and σ, ρ and ε strictly positive; see (Casas et al., 2018).

1.4 Finite element approximation of (P)

Numerical solutions of (P) are based on suitable discretizations of the problem. In this subsection, we consider such discretizations based on a finite element approximation. The goal is to prove convergence of the discretized solutions and to derive error estimates for the difference between continuous (exact) locally optimal controls/states and the associated discretized controls/states. To this end we impose the following assumption:

Assumption 1.4.1.

1. *Ω is a convex subset of $\mathbb{R}^n$ with $n = 2$ or 3.*
2. *The coefficients a_{ij} of the operator A are Lipschitz functions in $\bar{\Omega}$ for every $1 \leq i, j \leq n$.*

See Remark 1.4.2 for a discussion of this assumption.

Given $u \in L^2(\Omega)$, we know from Theorem 1.1.2 that there exists a unique solution $y_u \in H^1_0(\Omega) \cap C^{0,\mu}(\bar{\Omega})$ of (1.1.1). Due to Assumption 1.4.1, using (Grisvard, 1985, Chapter 3), we infer with equality $Ay_u = u - f(y_u) \in L^2(\Omega)$ that $y_u \in H^2(\Omega)$. Moreover, thanks to (1.1.4), we get

$$\|y_u\|_{H^2(\Omega)} \leq K_{A,\Omega,f}\Big(|f(0)| + \|u\|_{L^2(\Omega)}\Big) \quad \forall u \in L^2(\Omega) \tag{1.4.1}$$

for some constant $K_{A,\Omega,f}$ depending on A, Ω, and f. In addition, if Γ is of class $C^{1,1}$, we have that $y_u \in W^{2,p}(\Omega)$ for every $u \in L^p(\Omega)$ with $p \in (2, \infty)$, and the estimate

$$\|y_u\|_{W^{2,p}(\Omega)} \leq K_{A,\Omega,f}\, p\Big(|f(0)| + \|u\|_{L^p(\Omega)}\Big) \quad \forall u \in L^p(\Omega) \tag{1.4.2}$$

holds with $K_{A,\Omega,f}$ independent of p. The reader is referred to (Grisvard, 1985, Chapter 2) or (Gilbarg and Trudinger, 1983, Section 9.5) for the proof of the $W^{2,p}(\Omega)$-regularity. Though the dependence of p in the estimate (1.4.2) is very well known, the authors are not aware of a published reference where the proof is included. However, we mention two unpublished works where the explicit dependence of p is shown: (Cameron, 2010) and (Wahlbin).

Now, we consider a quasi-uniform family of triangulations $\{\mathcal{T}_h\}_{h>0}$ of $\bar{\Omega}$ with mesh size $h > 0$, cf. (Brenner and Scott, 2008, Definition (4.4.13)), and define $\bar{\Omega}_h = \cup_{T \in \mathcal{T}_h} T$. We assume that every boundary node of Ω_h is a point of

Γ. Additionally, we suppose that there exists a constant $C_\Gamma > 0$ independent of h such that the distance $d_\Gamma(x)$ of x to Γ obeys $d_\Gamma(x) \le C_\Gamma h^2$ for every $x \in \Gamma_h = \partial\Omega_h$. This is always satisfied if $n = 2$ and Γ is of class C^2; see, for instance, (Raviart and Thomas, 1983, Section 5.2). Under this assumption, there exists a constant $C_\Omega > 0$ independent of h such that

$$|\Omega \setminus \Omega_h| \le C_\Omega h^2. \tag{1.4.3}$$

Remark 1.4.2. The $H^2(\Omega)$-regularity of the states is crucial to obtain error estimates in the numerical discretization of (P). This regularity follows from Assumption 1.4.1. Indeed, Lipschitz continuity of the coefficients a_{ij} of A and some regularity of Ω are necessary to deduce the regularity of the states. It is possible to remove the convexity assumption on Ω if we impose that Γ is of class $C^{1,1}$ to achieve even more regularity as pointed out above. However, if Ω is not convex, the boundary triangles may not be contained in Ω, hence $\Omega_h \not\subset \Omega$ happens in this case. Then, the analysis is much more technical; see (Ciarlet, 1978, Chapter VI).

To address the approximation of the state equation by a finite element scheme we define the finite dimensional space

$$Y_h = \{y_h \in C(\bar{\Omega}) : \ y_{h|T} \in P_1(T)\ \forall T \in \mathcal{T}_h \text{ and } y_h \equiv 0 \text{ on } \Omega \setminus \Omega_h\},$$

where $P_1(T)$ denotes the space of polynomials in T of degree ≤ 1.

For every $u \in L^2(\Omega)$, we define its associated discrete state as the unique element $y_h(u) \in Y_h$ satisfying

$$a(y_h, z_h) + \int_{\Omega_h} f(y_h(x)) z_h dx = \int_{\Omega_h} u z_h dx \quad \forall z_h \in Y_h, \tag{1.4.4}$$

where $a : H^1(\Omega) \times H^1(\Omega) \longrightarrow \mathbb{R}$ is the bilinear form

$$a(y, z) = \sum_{i,j=1}^{n} \int_\Omega a_{ij} \partial_{x_i} y \partial_{x_j} z dx + \int_\Omega a_0 y z dx \quad \forall y, z \in H^1(\Omega).$$

The proof of the existence and uniqueness of a solution for (1.4.4) is standard. The existence follows from an easy application of Brouwer's fixed point theorem. Uniqueness is an immediate consequence of the monotonicity of f and the coercivity of a. The following theorem states the error estimates for the numerical discretization of the state equation.

Theorem 1.4.3. *Suppose that Assumption 1.4.1 holds. Let y_u and $y_h(u)$ be the solutions of* (1.1.1) *and* (1.4.4)*, respectively, corresponding to the control $u \in L^2(\Omega)$. Then, there exists a constant C independent of u and h such that*

$$\|y_u - y_h(u)\|_{H_0^1(\Omega)} \le Ch \|y_u\|_{H^2(\Omega)}, \tag{1.4.5}$$

$$\|y_u - y_h(u)\|_{L^2(\Omega)} \le Ch^2 \|y_u\|_{H^2(\Omega)}, \tag{1.4.6}$$

$$\|y_u - y_h(u)\|_{L^\infty(\Omega)} \le Ch^{2-\frac{n}{2}} \|y_u\|_{H^2(\Omega)}. \tag{1.4.7}$$

Moreover, if Γ is of class $C^{1,1}$ and u belongs to $L^\infty(\Omega)$, then we have the estimate

$$\|y_u - y_h(u)\|_{L^\infty(\Omega)} \le Ch^2 |\log h|^2 \Big(|f(0)| + \|u\|_{L^\infty(\Omega)} \Big). \tag{1.4.8}$$

Proof. The estimates (1.4.5)–(1.4.7) are proved in (Casas and Mateos, 2002). To prove (1.4.8) we introduce the function $y^h \in W^{2,p}(\Omega) \cap H_0^1(\Omega)$ for every $p < \infty$ as the solution of

$$\begin{cases} Ay(x) = u(x) - f(y_h(u)(x)) & \text{in } \Omega \\ y(x) = 0 & \text{on } \Gamma. \end{cases}$$

Applying the estimates (1.1.4). (1.4.1) and (1.4.7) to the previous linear equation and using (1.4.6), we infer for constants C_Ω, $K_{A,\Omega}$, and $C_{A,\Omega,f}$ only depending on the indicated subscripts

$$\begin{aligned} \|y_u - y^h\|_{L^\infty(\Omega)} &\le C_{\Omega,A} \|f(y_u) - f(y_h(u))\|_{L^2(\Omega)} \\ &\le C_{A,\Omega,f} \|y_u - y_h(u)\|_{L^2(\Omega)} \le C_{A,\Omega,f} Ch^2 \|y_u\|_{H^2(\Omega)} \\ &\le C_{A,\Omega,f} C K_{A,\Omega,f} h^2 \Big(|f(0)| + \|u\|_{L^2(\Omega)} \Big). \end{aligned} \tag{1.4.9}$$

From the classical $L^\infty(\Omega)$ estimates for linear equations, see (Rannacher, 1976), (Schatz, 1998), (Schatz and Wahlbin, 1982), we get with the bounds for the approximation error for the states (Brenner and Scott, 2008, Theorem 4.4.20), (1.4.2), and (1.4.7)

$$\begin{aligned} \|y^h - y_h(u)\|_{L^\infty(\Omega)} &\le M_{A,\Omega,f} h |\log h| \inf_{z_h \in Y_h} \|y^h - z_h\|_{W^{1,\infty}(\Omega)} \\ &\le M_{A,\Omega,f} K_\Omega h^{2-\frac{n}{p}} |\log h| \, \|y^h\|_{W^{2,p}(\Omega)} \\ &\le M_{A,\Omega,f} K_\Omega K_{A,\Omega,f}\, p h^{2-\frac{n}{p}} |\log h| \Big(|f(0)| + \|u\|_{L^\infty(\Omega)} \Big). \end{aligned}$$

Taking $p = |\log h|$ in the above estimate, we deduce

$$\|y^h - y_h(u)\|_{L^\infty(\Omega)} \le C'_{A,\Omega,f} h^2 |\log h|^2 \Big(|f(0)| + \|u\|_{L^\infty(\Omega)} \Big). \tag{1.4.10}$$

Combining (1.4.9) and (1.4.10), estimate (1.4.8) follows. □

Next, we address the numerical approximation of the optimal control problem (P). We denote by $\mathcal{U}_h$ the following space:

$$\mathcal{U}_h = \{u_h \in L^\infty(\Omega_h) : u_{h|T} \text{ is constant in all } T \in \mathcal{T}_h\}.$$

Every element $u_h \in \mathcal{U}_h$ can be written as

$$u_h = \sum_{T \in \mathcal{T}_h} u_T \chi_T,$$

where χ_T denotes the characteristic function of T and $u_T \in \mathbb{R}$. We set

$$\mathbf{U}_{\mathrm{h,ad}} = \{u_h \in \mathcal{U}_h : \alpha \le u_T \le \beta \ \forall T \in \mathcal{T}_h\}.$$

The discretized problem of (P) is

$$(\mathrm{P}_h) \qquad \min_{u_h \in \mathbf{U}_{\mathrm{h,ad}}} J_h(u_h)$$

with $J_h(u_h) := F_h(u_h) + \kappa j_h(u_h)$, where the functions $F_h : L^2(\Omega_h) \longrightarrow \mathbb{R}$ and $j_h : L^1(\Omega_h) \longrightarrow \mathbb{R}$ are defined by

$$F_h(u) = \frac{1}{2}\int_{\Omega_h} (y_h(u)(x) - y_\Omega(x))^2\,dx + \frac{\nu}{2}\int_{\Omega_h} u^2(x)\,dx$$

$$\text{and} \quad j_h(u) = \int_{\Omega_h} |u(x)|\,dx.$$

For $u_h \in \mathcal{U}_h$ we obtain

$$j_h(u_h) = \sum_{T \in \mathcal{T}_h} |T||u_T|.$$

Above, $y_h(u) \in Y_h$ denotes the discretized state associated with u, i.e. the solution of (1.4.4).

As in (1.1.10), a simple computation shows that $\lambda_h \in \partial j_h(u_h) \subset \mathcal{U}_h$ if and only if

$$\forall T \in \mathcal{T}_h \quad \begin{cases} \lambda_T = +1 & \text{if } u_T > 0, \\ \lambda_T = -1 & \text{if } u_T < 0, \\ \lambda_T \in [-1, +1] & \text{if } u_T = 0. \end{cases}$$

We also have

$$j_h'(u_h; v_h) = \sum_{T \in \mathcal{T}_h^+(u_h)} |T| v_T - \sum_{T \in \mathcal{T}_h^-(u_h)} |T| v_T + \sum_{T \in \mathcal{T}_h^0(u_h)} |T||v_T|,$$

where $\mathcal{T}_h^+(u_h)$, $\mathcal{T}_h^-(u_h)$ and $\mathcal{T}_h^0(u_h)$ denote the family of triangles T where u_T is positive, negative, or zero, respectively.

The optimality conditions for a local solution of (P_h) can be obtained analogously as for problem (P).

Theorem 1.4.4. *Let $\bar{u}_h$ be a local minimizer of* (P_h). *Then there exist $\bar{y}_h, \bar{\varphi}_h \in Y_h$, and $\bar{\lambda}_h \in \partial j_h(\bar{u}_h)$ such that*

$$a(\bar{y}_h, z_h) + \int_{\Omega_h} f(\bar{y}_h(x)) z_h dx dt = \int_{\Omega_h} \bar{u}_h z_h dx \quad \forall z_h \in Y_h, \tag{1.4.11}$$

$$a(z_h, \bar{\varphi}_h) + \int_{\Omega_h} f'(\bar{y}_h(x)) \bar{\varphi}_h z_h dx = \int_{\Omega_h} (\bar{y}_h - y_\Omega) z_h dx \quad \forall z_h \in Y_h, \tag{1.4.12}$$

$$\int_{\Omega_h} (\bar{\varphi}_h + \nu \bar{u}_h + \kappa \bar{\lambda}_h)(u_h - \bar{u}_h)\, dx \geq 0 \quad \forall u_h \in \mathbf{U}_{\mathrm{h,ad}}. \tag{1.4.13}$$

Remark 1.4.5. **(i)** Analogously to the continuous case, we deduce from (1.4.13) that

$$0 \leq \int_{\Omega_h} (\bar{\varphi}_h + \nu \bar{u}_h + \kappa \bar{\lambda}_h)(u_h - \bar{u}_h)\, dx \leq F_h'(\bar{u}_h)(u_h - \bar{u}_h) + \kappa j_h'(\bar{u}_h; u_h - \bar{u}_h)$$
$$\leq F_h'(\bar{u}_h)(u_h - \bar{u}_h) + \kappa [j_h(u_h) - j_h(\bar{u}_h)]. \tag{1.4.14}$$

(ii) Inequality (1.4.13) can also be written as

$$\Big[\sum_{T \in \mathcal{T}_h} \Big(\int_T \bar{\varphi}_h(x)\, dx + |T|(\nu \bar{u}_T + \kappa \bar{\lambda}_T)\Big)\Big](u_T - \bar{u}_T) \ \forall u_h \in \mathbf{U}_{\mathrm{h,ad}}.$$

From here we infer

$$\text{If } \nu > 0, \text{ then } \bar{u}_T = \mathrm{Proj}_{[\alpha,\beta]}\Big(-\frac{1}{\nu}\Big[\frac{1}{|T|}\int_T \bar{\varphi}_h(x)\, dx + \kappa \bar{\lambda}_T\Big]\Big), \tag{1.4.15}$$

$$\int_T \bar{\varphi}_h(x)\, dx + |T|(\nu \bar{u}_T + \kappa \bar{\lambda}_T) \begin{cases} \geq 0 & \text{if } \bar{u}_T = \alpha, \\ = 0 & \text{if } \alpha < \bar{u}_T < \beta, \\ \leq 0 & \text{if } \bar{u}_T = \beta. \end{cases} \tag{1.4.16}$$

(iii) As in Theorem 1.2.8, if $\alpha < 0 < \beta$ and $\kappa > 0$ we have, for all $T \in \mathcal{T}_h$,

$$\bar{\lambda}_T = \mathrm{Proj}_{[-1,+1]}\Big(-\frac{1}{\kappa}\frac{1}{|T|}\int_T \bar{\varphi}_h(x)\, dx\Big), \tag{1.4.17}$$

$$\text{If } \nu > 0, \text{ then } \bar{u}_T = 0 \Leftrightarrow \frac{1}{|T|}\Big|\int_T \bar{\varphi}_h(x)\, dx\Big| \leq \kappa, \tag{1.4.18}$$

If $\nu = 0$ and $-\infty < \alpha < 0 < \beta < +\infty$, then

$$\bar{u}_T = \begin{cases} \alpha & \text{if } \frac{1}{|T|}\int_T \bar{\varphi}_h(x)\, dx > +\kappa, \\ 0 & \text{if } \frac{1}{|T|}\Big|\int_T \bar{\varphi}_h(x)\, dx\Big| < \kappa, \\ \beta & \text{if } \frac{1}{|T|}\int_T \bar{\varphi}_h(x)\, dx < -\kappa. \end{cases} \tag{1.4.19}$$

For the analysis of the numerical approximations of (P), we use the hypothesis

Assumption 1.4.6. *At least one of the following conditions holds: (i)* $\mathbf{U}_{\rm ad}$ *is bounded in* $L^\infty(\Omega)$*; (ii)* $\nu > 0$.

In this numerical analysis we distinguish between two cases: $\nu > 0$ and $\nu = 0$.

1.4.1 Numerical analysis of (P)*. Case* $\nu > 0$

Assumption 1.4.1 and $\nu > 0$ are supposed in the whole subsection. The first step in this analysis is to prove the existence of a solution to problem (P_h) and the convergence of discretized solutions to a solution of (P). To this end, we set

$$u_0 = \max\{\alpha, \min\{0, \beta\}\} \text{ and } M_0 = \sup_{h\le h_0} \sqrt{\frac{2}{\nu} J_h(u_0)}, \tag{1.4.20}$$

where $h_0 > 0$ is arbitrarily fixed. Since $J_h(u_0) \to J(u_0)$ as $h \to 0$, we have $M_0 < \infty$. It is obvious that $u_0 \in \mathbf{U}_{\rm h,ad}$.

In the sequel, we will say that a sequence $\{u_{h_k}\}_{k=1}^\infty$ with $h_k \to 0$ converges weakly (weakly*) to u in $L^p(\Omega)$ with $1 \le p < \infty$ $(p = \infty)$ if

$$\lim_{k\to\infty} \int_{\Omega_h} w u_{h_k}\,dx = \int_\Omega wu\,dx \quad \forall w \in L^{p'}(\Omega).$$

We also say that $\{u_{h_k}\}_{k=1}^\infty$ converges in $L^p(\Omega)$ if $\|u_{h_k} - u\|_{L^p(\Omega_{h_k})} \to 0$.

Given an arbitrary element $v \in L^p(\Omega)$, we define the extensions

$$\tilde{u}_{h_k}(x) = \begin{cases} u_{h_k}(x) & \text{if } x \in \Omega_{h_k}, \\ v(x) & \text{if } x \in \Omega \setminus \Omega_{h_k}. \end{cases}$$

Then, the above definition of convergence is equivalent to saying that $\{\tilde{u}_{h_k}\}_{k=1}^\infty$ converges to u.

Theorem 1.4.7. *For every* $h > 0$*, the discrete problem* (P_h) *has at least one solution* $\bar{u}_h$*. Every family of solutions* $\{\bar{u}_h\}_{h\le h_0}$ *of the problems* (P_h) *is bounded in* $L^2(\Omega)$ *by* M_0*. If* $\{\bar{u}_{h_k}\}_{k=1}^\infty$ *with* $h_k \to 0$ *is a sequence of solutions to* (P_{h_k}) *converging weakly in* $L^2(\Omega)$ *to some function* $\bar{u}$*, then* $\bar{u}$ *is a solution of* (P)*,* $\bar{u}_{h_k} \to \bar{u}$ *strongly in* $L^\infty(\Omega)$*, and* $(\bar{y}_{h_k}, \bar{\varphi}_{h_k}) \to (\bar{y}, \bar{\varphi})$ *strongly in* $H_0^1(\Omega) \cap L^\infty(\Omega)$*, where* $\bar{y}_{h_k} = y_{h_k}(\bar{u}_{h_k})$*,* $\bar{y} = y_{\bar{u}}$*,* $\bar{\varphi}_{h_k}$ *is the discrete adjoint state associated with* $\bar{u}_{h_k}$*, and* $\bar{\varphi}$ *is the continuous adjoint state corresponding to* $\bar{u}$*. Moreover, if* $\kappa > 0$*, then* $\bar{\lambda}_{h_k} \to \bar{\lambda}$ *in* $L^\infty(\Omega)$ *holds, where* $\bar{\lambda} \in \partial j(\bar{u})$ *and* $\bar{\lambda}_{h_k} \in \partial j_{h_k}(\bar{u}_{h_k})$ *satisfy the optimality conditions* (1.2.1)–(1.2.3) *and* (1.4.11)–(1.4.13).

Proof. The existence of a solution to (P_h) follows from continuity and coercivity of J_h, and the fact that $\mathbf{U}_{\rm h,ad}$ is closed. Moreover, the inequality $\frac{\nu}{2}\|\bar{u}_h\|^2_{L^2(\Omega)} \le J_h(\bar{u}_h) \le J_h(u_0)$ implies that $\|\bar{u}_h\|_{L^2(\Omega)} \le M_0$ $\forall h \le h_0$. Therefore, we can extract a sequence $\{\bar{u}_{h_k}\}_{k=1}^\infty$ with $h_k \to 0$ weakly converging in $L^2(\Omega)$ to some element $\bar{u}$. We extend the elements $\bar{u}_{h_k}$ from Ω_{h_k} to Ω by u_0, keeping the same notation for the extensions. Then, these extensions are contained in $\mathbf{U}_{\mathrm{ad}}$, hence $\bar{u}$ belongs to $\mathbf{U}_{\mathrm{ad}}$ as well. Moreover, for $\bar{y}_{h_k} = y_{h_k}(\bar{u}_{h_k})$ and $\bar{y} = y_{\bar{u}}$, we infer from Theorem 1.1.2 and (1.4.7)

$$\|\bar{y} - \bar{y}_{h_k}\|_{L^\infty(\Omega)} \le \|\bar{y} - y_{u_{h_k}}\|_{L^\infty(\Omega)} + \|y_{u_{h_k}} - \bar{y}_{h_k}\|_{L^\infty(\Omega)} \to 0 \text{ as } k \to \infty.$$

Using (1.4.5) instead of (1.4.7), we also deduce the strong convergence $\bar{y}_{h_k} \to \bar{y}$ in $H_0^1(\Omega)$. By a similar argument, $\bar{\varphi}_{h_k} \to \bar{\varphi}$ converges strongly in $H_0^1(\Omega) \cap L^\infty(\Omega)$, where $\bar{\varphi}$ is the adjoint state associated with $\bar{u}$. Let us prove that $\bar{u}$ is a solution of (P). We take any solution $\tilde{u}$ of (P), whose existence follows from Theorem 1.2.6. Let $u_h \in \mathbf{U}_{\mathrm{h,ad}}$ be the $L^2(\Omega)$ projection of $\tilde{u}$ on $\mathcal{U}_h$. We have that $u_h \in \mathbf{U}_{\mathrm{h,ad}}$. Indeed, this follows from the representation

$$u_h = \sum_{T\in\mathcal{T}_h} u_T \chi_T \quad \text{with} \quad u_T = \frac{1}{T}\int_T \tilde{u}(x)\,dx.$$

Taking into account the convergence of the states $\bar{y}_{h_k} \to \bar{y}$, we get

$$J(\bar{u}) \le \liminf_{k\to\infty} J_{h_k}(\bar{u}_{h_k}) \le \liminf_{k\to\infty} J_{h_k}(u_{h_k}) = J(\tilde{u});$$

whence $\bar{u}$ is a solution of (P). Finally, it is enough to use the formulae (1.2.8) and (1.4.17), (1.2.5) and (1.4.15), to deduce the strong convergence of $\{\bar{\lambda}_{h_k}\}_{k=1}^\infty$ and $\{\bar{u}_{h_k}\}_{k=1}^\infty$ to $\bar{\lambda}$ and $\bar{u}$, respectively, in $L^\infty(\Omega)$. □

The next result is a kind of converse statement of the previous one.

Theorem 1.4.8. *If $\bar{u}$ is an $L^2(\Omega)$-strict local solution of* (P)*, then there exist $h_{\bar{u}} > 0$ and an $L^2(\Omega)$-local solution $\bar{u}_h$ of* (P$_h$) *for every $h \le h_{\bar{u}}$ such that $(\bar{u}_h, \bar{y}_h, \bar{\varphi}_h, \bar{\lambda}_h) \to (\bar{u}, \bar{y}, \bar{\varphi}, \bar{\lambda})$ in $L^\infty(\Omega)^4$.*

Proof. Since $\bar{u}$ is an $L^2(\Omega)$-strict local solution of (P), there exists $r > 0$ such that $\bar{u}$ is the only solution of the control problem

$$(\mathrm{P}_r) \qquad \min_{u\in \mathbf{U}_{\mathrm{ad}}\cap \bar{B}_r(\bar{u})} J(u).$$

In this proof, all the elements $v_h \in \mathcal{U}_h$ are extended from Ω_h to Ω by setting $v_h(x) = \bar{u}(x)$ for $x \in \Omega \setminus \bar{\Omega}_h$.

As in the proof of Theorem 1.4.7, for every $h > 0$ we can find some $u_h \in \mathbf{U}_{\mathrm{h,ad}}$ such that $\lim_{h\to 0}\|u_h - \bar{u}\|_{L^2(\Omega)} = 0$. This convergence yields the existence of $\bar{h} > 0$ such that $u_h \in \bar{B}_r(\bar{u})$ for every $h \le \bar{h}$. For every $h \le \bar{h}$ we define the problems

$$(\mathrm{P}_{r,h}) \qquad \min_{u_h\in \mathbf{U}_{\mathrm{h,ad}}\cap \bar{B}_r(\bar{u})} J_h(u_h).$$

Since $\mathbf{U}_{\mathrm{h,ad}} \cap \bar{B}_r(\bar{u})$ is a nonempty compact subset of $\mathcal{U}_h$ and J_h is continuous, problem $(\mathrm{P}_{r,h})$ has at least one solution $\bar{u}_h$. Each family of solutions $\{\bar{u}_h\}_{h\leq\bar{h}}$ is contained in the ball $\bar{B}_r(\bar{u})$. Therefore, there exist subsequences $\{\bar{u}_{h_k}\}_{k=1}^{\infty}$ with $h_k \to 0$ converging weakly to a control $\tilde{u} \in \mathbf{U}_{\mathrm{ad}} \cap \bar{B}_r(\bar{u})$. Arguing again as in the proof of Theorem 1.4.7, we confirm that $\tilde{u}$ is a solution of (P_r), hence $\tilde{u} = \bar{u}$ and $(\bar{u}_{h_k}, \bar{y}_{h_k}, \bar{\varphi}_{h_k}, \bar{\lambda}_{h_k}) \to (\bar{u}, \bar{y}, \bar{\varphi}, \bar{\lambda})$ in $L^\infty(\Omega)$. Since all the subsequences converge to $(\bar{u}, \bar{y}, \bar{\varphi}, \bar{\lambda})$ when $h_k \to 0$, we infer that $(\bar{u}_h, \bar{y}_h, \bar{\varphi}_h, \bar{\lambda}_h) \to (\bar{u}, \bar{y}, \bar{\varphi}, \bar{\lambda})$ strongly in $L^\infty(\Omega)^4$ when $h \to 0$. From this convergence we deduce the existence of $h_{\bar{u}} \leq \bar{h}$ such that $\bar{u}_h \in B_r(\bar{u})$, which means that $\bar{u}_h$ is an $L^2(\Omega)$-local minimizer of (P_h). □

Next, $\bar{u}$ will denote an $L^2(\Omega)$-local minimizer of (P) satisfying the sufficient second order condition (1.3.3). Hence, $\bar{u}$ is a strict local minimizer and Theorem 1.4.8 allows to deduce the existence of local minima $\{\bar{u}_h\}_{h\leq h_{\bar{u}}}$ of the problems (P_h) such that $\lim_{h\to 0} \|\bar{u}_h - \bar{u}\|_{L^\infty(\Omega_h)} = 0$. For convenience, we extend every control $\bar{u}_h$ from Ω_h to Ω by setting $\bar{u}_h(x) = \bar{u}(x)$ if $x \in \Omega \setminus \Omega_h$.

Theorem 1.4.9. *With the above notation, if $\bar{u}$ is an $L^2(\Omega)$-local minimizer of* (P) *satisfying* (1.3.3)*, then there exists a constant $C > 0$ such that*

$$\|\bar{u}_h - \bar{u}\|_{L^\infty(\Omega_h)} + \|\bar{y}_h - \bar{y}\|_{L^\infty(\Omega_h)} + \|\bar{\varphi}_h - \bar{\varphi}\|_{L^\infty(\Omega_h)} \leq Ch \quad \forall h \leq h_{\bar{u}}. \tag{1.4.21}$$

In addition, if $\kappa > 0$, then $\|\bar{\lambda}_h - \bar{\lambda}\|_{L^\infty(\Omega_h)} \leq Ch$ is fulfilled as well.

Proof. A detailed proof of this theorem can be found in (Casas et al., 2012a); see also (Arada et al., 2002) for the case $\kappa = 0$. Here we point out the lines of a slightly different proof. Since $\bar{u}_h(x) = \bar{u}(x)$ for $x \in \Omega \setminus \Omega_h$, we have that $\bar{u}_h \in \mathbf{U}_{\mathrm{ad}}$. Therefore, from (1.2.4) we get

$$F'(\bar{u})(\bar{u}_h - \bar{u}) + \kappa[j(\bar{u}_h) - j(\bar{u})] \geq 0 \quad \forall h \leq h_{\bar{u}}. \tag{1.4.22}$$

Now, let u_h be the $L^2(\Omega)$ projection of $\bar{u}$ on $\mathcal{U}_h$, extended to Ω by $\bar{u}$ as well. Then, $u_h \in \mathbf{U}_{\mathrm{h,ad}}$ and (1.4.14) yields

$$F_h'(\bar{u}_h)(u_h - \bar{u}_h) + \kappa[j_h(u_h) - j_h(\bar{u}_h)] \geq 0 \quad \forall h \leq h_{\bar{u}}. \tag{1.4.23}$$

Since $j(\bar{u}_h) - j(\bar{u}) = j_h(\bar{u}_h) - j_h(\bar{u})$ and $j_h(u_h) \leq j_h(\bar{u})$, adding (1.4.22) and (1.4.23) we infer

$$F'(\bar{u})(\bar{u}_h - \bar{u}) + F_h'(\bar{u}_h)(u_h - \bar{u}_h) \geq 0.$$

From here, we obtain

$$\begin{aligned}
[F'(\bar{u}_h) - F'(\bar{u})](\bar{u}_h - \bar{u}) &\leq [F_h'(\bar{u}_h) - F'(\bar{u}_h)](\bar{u} - \bar{u}_h) \\
&+ [F_h'(\bar{u}_h) - F'(\bar{u})](u_h - \bar{u}) + F'(\bar{u})(u_h - \bar{u}).
\end{aligned} \tag{1.4.24}$$

According to Theorem 1.3.3, the second order condition (1.3.3) is equivalent to the existence of $\tau > 0$ and $\delta > 0$ such that $F''(\bar u)v^2 \ge \delta\|v\|^2_{L^2(\Omega)}$ $\forall v \in S^\tau_{\bar u}$. In (Casas et al., 2012a, Lemma 4.6), the existence of $h_\delta > 0$ is proved such that for every $h \le h_\delta$ we have $\bar u_h - \bar u \in S^\tau_{\bar u}$ and

$$
\begin{aligned}
[F'(\bar u_h) - F'(\bar u)](\bar u_h - \bar u) &= F''(\bar u + \theta_h(\bar u_h - \bar u))(\bar u_h - \bar u)^2 \\
&\ge \frac{\delta}{2}\|\bar u_h - \bar u\|^2_{L^2(\Omega_h)} \qquad (1.4.25)
\end{aligned}
$$

with some $\theta_h \in (0, 1)$. On the other hand, using error estimates for the finite element approximation of elliptic equations similar to those of Theorem 1.4.3 applied to the adjoint and state equation, we obtain

$$
\begin{aligned}
&|[F'_h(\bar u_h) - F'(\bar u_h)](\bar u - \bar u_h)| \le \int_\Omega |\bar\varphi_h - \varphi_{\bar u_h}||\bar u_h - \bar u|\,dx \\
&\le \|\bar\varphi_h - \varphi_{\bar u_h}\|_{L^2(\Omega_h)}\|\bar u_h - \bar u\|_{L^2(\Omega)} \le C_1 h^2\|\bar u_h - \bar u\|_{L^2(\Omega)}. \qquad (1.4.26)
\end{aligned}
$$

The second term of the right hand side of (1.4.24) is estimated by

$$
\begin{aligned}
&|[F'_h(\bar u_h) - F'(\bar u)](u_h - \bar u)| \le \int_{\Omega_h} |\bar\varphi_h - \bar\varphi||u_h - \bar u|\,dx \\
&\le \Big(\|\bar\varphi_h - \varphi_{\bar u_h}\|_{L^2(\Omega_h)} + \|\varphi_{\bar u_h} - \bar\varphi\|_{L^2(\Omega_h)}\Big)\|u_h - \bar u\|_{L^2(\Omega_h)} \\
&\le \Big(C_1 h^2 + C_2\|\bar u_h - \bar u\|_{L^2(\Omega_h)}\Big)C_3 h, \qquad (1.4.27)
\end{aligned}
$$

where we have used that $\|u_h - \bar u\|_{L^2(\Omega_h)} \le Ch\|\bar u\|_{H^1(\Omega)}$ (notice that u_h is the $L^2(\Omega)$-projection of $\bar u$). Finally, for the last term of (1.4.24) we proceed as follows

$$
|F'(\bar u)(u_h - \bar u)| \le \|\bar\varphi + \nu\bar u\|_{H^1(\Omega)}\|u_h - \bar u\|_{H^1(\Omega)^*} \le C_4 h^2. \qquad (1.4.28)
$$

This estimate is a well known property of projections. Now, (1.4.24)–(1.4.28) imply that $\|\bar u_h - \bar u\|_{L^2(\Omega_h)} \le Ch$. Finally, arguing similarly as at the end of the proof of Theorem 1.4.7 we deduce (1.4.21) as well as the estimate for $\bar\lambda_h - \bar\lambda$ if $\kappa > 0$. □

Remark 1.4.10. (i) *Piecewise linear approximations.* Instead of taking piecewise constant functions to approximate the controls, we can consider piecewise linear and continuous approximations. This was investigated in (Casas et al., 2012b). In that paper, the authors proved the same order of convergence for a piecewise linear and continuous control discretization, which requires fewer degrees of freedom than the discretization by piecewise constant functions. A key idea to achieve this result is to use mass lumping for the following quadrature formulae to discretize the $L^2(\Omega)$ and $L^1(\Omega)$ norms in the functional J:

$$\sum_j u_j^2 \int_\Omega e_j\,dx \quad \text{and} \quad \sum_j |u_j| \int_\Omega e_j\,dx, \tag{1.4.29}$$

where $\{e_j\}$ are the usual piecewise linear nodal basis functions. These formulae allow a coefficient-wise representation of the control and of the subdifferential of the discretized $L^1(\Omega)$ norm analogous to the ones proved for the continuous control problem, which significantly facilitates the error analysis. Moreover, (1.4.29) is also advantageous from an algorithmic point of view.

In the case $\kappa = 0$, in (Casas, 2007) the order of convergence $o(h)$ was proved for piecewise linear approximations. Under an additional assumption on the optimal control that is non-restrictive in practice, the order $O(h^{\frac{3}{2}})$ is satisfied; see (Casas and Mateos, 2017).

(ii) *Variational discretization.* Assuming again that $\kappa = 0$, numerical computations to solve (P_h) can be performed without discretization of the control by taking $\mathcal{U}_h = L^\infty(\Omega)$, hence $\mathbf{U}_{\text{h,ad}} = \mathbf{U}_{\text{ad}}$. Therefore, in the proof of Theorem 1.4.9 we can replace u_h by $\bar{u}$. Then, inequality (1.4.24) simplifies to

$$[F'(\bar{u}_h) - F'(\bar{u})](\bar{u}_h - \bar{u}) \le [F_h'(\bar{u}_h) - F'(\bar{u}_h)](\bar{u} - \bar{u}_h).$$

Thus, using (1.4.25) and (1.4.26) we infer that $\|\bar{u}_h - \bar{u}\|_{L^2(\Omega)} \le Ch^2$. Then, we can proceed as in the proof of Theorem 1.4.9 to deduce (1.4.21) with h replaced by h^2. Since the local or global solutions of (P_h) are also piecewise linear and continuous, which follows from (1.4.13), a numerical implementation is possible. In the case of pointwise control constraints, the numerical computations are more involved because the linear structure of $\bar{u}_h$ is not supported on the grid defined by the nodes of the triangulation. In case of no constraints, we have $\bar{u} = -\frac{1}{\nu}\bar{\varphi}$, see Remark 1.2.5, and thus it is natural to choose the same discretizations for the controls as for the adjoint states. In this case, the variational discretization approach, which was introduced in (Hinze, 2005), is equivalent to discretizing the controls in the same space as the adjoint states.

(iii) *Superconvergence and postprocessing step.* A different approach to obtain the convergence order $O(h^2)$ is due to (Meyer and Rösch, 2004). In this paper, the authors explained the following order of convergence that was observed numerically:

$$\|\bar{y} - \bar{y}_h\|_{L^2(\Omega_h)} + \|\bar{\varphi} - \bar{\varphi}_h\|_{L^2(\Omega_h)} \le Ch^2.$$

The proof given in (Meyer and Rösch, 2004) considers problems governed by linear equations. Later, it was proved in (Krumbiegel and Pfefferer, 2015) that the result is valid for Neumann control of a semilinear elliptic equation. Their technique can also be applied to distributed problems as ours. The main idea is to construct a better approximation of the optimal discretized control $\bar{u}_h$, namely

$$\tilde{u}_h(x) = \text{Proj}_{[\alpha,\beta]}\Big(-\frac{1}{N}\bar{\varphi}_h(x)\Big) \text{ a.e. in } \Omega_h$$

that satisfies $\|\bar{u} - \tilde{u}_h\|_{L^2(\Omega)} \le Ch^2$.

1.4.2 Numerical analysis of (P). Case $\nu = 0$

Since $\nu = 0$, Assumption 1.4.6 implies that $-\infty < \alpha < \beta < +\infty$. Therefore, the existence of a solution for (P_h) follows from the compactness of $\mathbf{U}_{h,ad}$ and the continuity of J_h.

For the proofs of the next Theorems 1.4.11, 1.4.12, and 1.4.15 the reader is referred to (Casas and Mateos, 2021).

First, we have the following counterpart of Theorems 1.4.7 and 1.4.8.

Theorem 1.4.11. *Let $\bar{u}$ be a strict strong local minimizer for* (P)*, i.e.,*

$$\exists \rho > 0 : J(\bar{u}) < J(u) \quad \forall u \in \mathbf{U}_{ad} \setminus \{\bar{u}\} : \|y_u - \bar{y}\|_{L^\infty(\Omega)} \leq \rho. \tag{1.4.30}$$

Then, there exists a sequence $\{\bar{u}_h\}_h$ of local minimizers of (P_h) *such that $\bar{u}_h \stackrel{*}{\rightharpoonup} \bar{u}$ in $L^\infty(\Omega)$. Moreover, there exists $h_{\bar{u}} > 0$ such that $\forall h \leq h_{\bar{u}}$*

$$J_h(\bar{u}_h) \leq J_h(u_h) \; \forall u_h \in \mathbf{U}_{h,ad} \text{ with } \|y_h(u_h) - \bar{y}_h\|_{L^\infty(\Omega_h)} \leq \frac{\rho}{2}. \tag{1.4.31}$$

Conversely, let $\{\bar{u}_h\}_{h>0}$ be a family of local minimizers of (P_h) *satisfying* (1.4.31) *for some given $\rho > 0$ and such that $\bar{u}_h \stackrel{*}{\rightharpoonup} \bar{u}$ in $L^\infty(\Omega)$ as $h \to 0$. Then $\bar{u}$ is a strong local solution of* (P) *satisfying*

$$J(\bar{u}) \leq J(u) \quad \forall u \in \mathbf{U}_{ad} : \|y_u - \bar{y}\|_{L^\infty(\Omega)} < \frac{\rho}{2}. \tag{1.4.32}$$

Now, we establish the error estimates for the difference of the discretized and continuous states.

Theorem 1.4.12. *Let $\bar{u}$ be a strong local solution of* (P) *satisfying* (1.3.7) *and assume that $y_d \in L^\infty(\Omega)$. Let $\{\bar{u}_h\}_{h>0}$ be a family of local minima of* (P_h) *such that* (1.4.31) *holds for some $\rho > 0$ and $\bar{u}_h \stackrel{*}{\rightharpoonup} \bar{u}$ in $L^\infty(\Omega)$. Then, there exists a constant $C > 0$ independent of h such that*

$$\|\bar{y}_h - \bar{y}\|_{L^2(\Omega)} \leq C\sqrt{h}.$$

Corollary 1.4.13. *Under the assumptions and notations of Theorem 1.4.12, we have*

$$\|\bar{\varphi}_h - \bar{\varphi}\|_{L^\infty(\Omega)} \leq C\sqrt{h} \text{ and, if } \kappa > 0, \text{ then } \|\bar{\lambda}_h - \bar{\lambda}\|_{L^\infty(\Omega)} \leq C\sqrt{h}.$$

The proof of Theorem 1.4.12 is based on inequality (1.3.8). The corollary follows from the adjoint state equations and the representation formulae (1.2.5) and (1.4.17).

Regarding the convergence of the controls $\{\bar{u}_h\}_{h>0}$, we only have proved weak* convergence in $L^\infty(\Omega)$. However, the next theorem shows that strong convergence happens in some common cases.

Theorem 1.4.14. *Assume that $\bar{u} \in \mathbf{U}_{\mathrm{ad}}$, $\bar{u}_h \in \mathbf{U}_{\mathrm{h,ad}}$ for every h, and $\bar{u}_h \rightharpoonup \bar{u}$ as $h \to 0$ in $L^1(\Omega)$. Then, $\lim_{h\to 0} \|\bar{u}_h - \bar{u}\|_{L^q(\Omega_h)} = 0 \ \forall q \in [1, \infty)$ holds if one of the following conditions is fulfilled:*

1. *$\kappa = 0$ and $\bar{u}$ is bang-bang,*
2. *$\kappa > 0$, $\bar{u}$ is bang-bang-bang, and $\bar{u}$ and $\bar{u}_h$ satisfy the optimality conditions* (1.2.1)–(1.2.3) *and* (1.4.11)–(1.4.13), *respectively.*

Proof. We extend $\bar{u}_h$ to Ω by setting $\bar{u}_h(x) = \bar{u}(x)$ for $x \in \Omega \setminus \Omega_h$. Let us define

$$\Omega_\alpha = \{x \in \Omega : \bar{u}(x) = \alpha\} \quad \text{and} \quad \Omega_\beta = \{x \in \Omega : \bar{u}(x) = \beta\}.$$

If $\bar{u}$ is bang-bang then the weak convergence in $L^1(\Omega)$ implies

$$\lim_{h\to 0} \|\bar{u}_h - \bar{u}\|_{L^1(\Omega_h)} = \lim_{h\to 0} \left\{ \int_{\Omega_\alpha} (\bar{u}_h - \bar{u})\, dx + \int_{\Omega_\beta} (\bar{u} - \bar{u}_h)\, dx \right\} = 0.$$

The strong convergence in $L^1(\Omega)$ and the boundedness of $\{\bar{u}_h\}_{h>0}$ in $L^\infty(\Omega)$ yield the strong convergence in $L^q(\Omega)$ for every $q \in [1, \infty)$.

Now, we consider the second assumption. Since $-\infty < \alpha < \beta < +\infty$, the weak convergence in $L^1(\Omega)$ implies the weak* convergence in $L^\infty(\Omega)$. Then, writing $\bar{y}_h - \bar{y} = (\bar{y}_h - y_{\bar{u}_h}) + (y_{\bar{u}_h} - \bar{y})$, and using Theorem 1.1.2 and estimates (1.4.7) and (1.4.1), we deduce the convergence $\lim_{h\to 0} \|\bar{y}_h - \bar{y}\|_{L^\infty(\Omega)} = 0$. Taking advantage of this fact and arguing in a similar way with the adjoint states, we infer $\lim_{h\to 0} \|\bar{\varphi}_h - \bar{\varphi}\|_{L^\infty(\Omega)} = 0$. With the representation formulae (1.2.5) and (1.4.17) we get that $\lim_{h\to 0} \|\bar{\lambda}_h - \bar{\lambda}\|_{L^\infty(\Omega)} = 0$.

Let us consider the previously defined sets Ω_α and Ω_β, and put $\Omega_0 = \{x \in \Omega : \bar{u}(x) = 0\}$. Then, with the $L^1(\Omega)$ weak convergence, the properties of $\bar{\lambda}_h$, and the convergence $\bar{\lambda}_h \to \bar{\lambda}$ in $L^\infty(\Omega)$, we obtain

$$\begin{aligned}
&\lim_{h\to 0} \|\bar{u}_h - \bar{u}\|_{L^1(\Omega_h)} \\
&= \lim_{h\to 0} \left(\int_{\Omega_\alpha} (\bar{u}_h - \bar{u})\, dx + \int_{\Omega_\beta} (\bar{u} - \bar{u}_h)\, dx + \int_{\Omega_0} |\bar{u}_h|\, dx \right) \\
&= \lim_{h\to 0} \int_{\Omega_0} |\bar{u}_h|\, dx = \lim_{h\to 0} \int_{\Omega_0} \bar{\lambda}_h \bar{u}_h\, dx = \int_{\Omega_0} \bar{\lambda}\bar{u}\, dx = 0.
\end{aligned}$$

Once again $L^1(\Omega)$ convergence leads to $L^q(\Omega)$ convergence. □

We observe that, due to the absence of the quadratic regularization term in the cost functional ($\nu = 0$), it is not possible to prove error estimates for the optimal controls unless some extra assumption is made. This is stated in the following theorem.

Theorem 1.4.15. *Suppose that Γ is of class $C^{1,1}$ and y_d belongs to $L^\infty(\Omega)$. Let $\bar{u}$ be a solution of* (P) *satisfying the second order sufficient condition* (1.3.7) *and the structural assumption* (1.3.9). *Let* $\{\bar{u}_h\}_{h>0}$ *be a family of local minima of* (P_h) *such that* (1.4.31) *holds and* $\bar{u}_h \overset{*}{\rightharpoonup} \bar{u}$ *in* $L^\infty(\Omega)$ *as* $h \to 0$. *Then, there exists a constant* $C > 0$ *independent of* h *such that*

$$\|\bar{u}_h - \bar{u}\|_{L^1(\Omega_h)} \le Ch^{\gamma^2} \quad \text{and} \quad \|\bar{y}_h - \bar{y}\|_{L^2(\Omega)} \le Ch^{\frac{\gamma(\gamma+1)}{2}}.$$

The above estimates are deduced from the inequality (1.3.10) and utilize the adjoint equation. To this purpose, the regularity of Γ is needed to obtain an estimate for the adjoint states analogous to (1.4.8).

Corollary 1.4.16. *Under the assumptions and notations of Theorem 1.4.15, we have*

$$\|\bar{\varphi}_h - \bar{\varphi}\|_{L^\infty(\Omega)} \le Ch^{\frac{\gamma(\gamma+1)}{2}} \quad \text{and, if } \kappa > 0, \text{ then } \|\bar{\lambda}_h - \bar{\lambda}\|_{L^\infty(\Omega)} \le Ch^{\frac{\gamma(\gamma+1)}{2}}.$$

In (Casas and Mateos, 2021), the estimate

$$\|\bar{u}_h - \bar{u}\|_{L^1(\Omega_h)} \le C(h|\log h|)^{2\gamma}$$

was proved for the variational discretization. This order of convergence was obtained for problems governed by linear equations in (Deckelnick and Hinze, 2012).

2 Controlling with measures

2.1 The control problem

The second section of this chapter is dedicated to the analysis of optimal control problems where the controls are chosen as Borel measures. The necessity of allowing measure-valued controls in the context of problem (P) results from the fact that for $\nu = 0$, $\kappa > 0$ and no constraints on the controls, existence of a minimizer in $L^1(\Omega)$ cannot be guaranteed. Enlarging $L^1(\Omega)$ to the space of Borel measures, and replacing the $L^1(\Omega)$ cost by the measure-norm of u, allows to obtain existence of optimal controls and optimality conditions which exhibit their sparsity structure, with classical techniques from the calculus of variations and convex analysis. Beyond the mathematical interest, the use of measure controls allows controls with zero Lebesgue measure support which can be of practical relevance.

$$\text{(P)} \qquad \min_{u \in \mathcal{M}(\Omega)} J(u) = \frac{1}{2}\|y_u - y_d\|^2_{L^2(\Omega)} + \kappa\|u\|_{\mathcal{M}(\Omega)},$$

where y_u is the unique solution to the Dirichlet problem

$$\begin{cases} Ay = u & \text{in } \Omega, \\ y = 0 & \text{on } \Gamma, \end{cases} \tag{2.1.1}$$

A being defined by (1.1.2). Regarding the cost functional, we consider $y_d \in L^2(\Omega)$ and $\kappa > 0$. Here Ω is supposed to be a bounded domain of $\mathbb{R}^n$ with $n = 2$ or 3, with a Lipschitz boundary Γ. The controls are taken in the space of regular Borel measures $\mathcal{M}(\Omega)$. As usual, $\mathcal{M}(\Omega)$ is identified by the Riesz theorem with the dual space of $C_0(\Omega)$ – consisting of the continuous functions in $\bar{\Omega}$ vanishing on Γ – endowed with the norm

$$\|u\|_{\mathcal{M}(\Omega)} = \sup_{\|z\|_{C_0(\Omega)} \le 1} \langle u, z\rangle = \sup_{\|z\|_{C_0(\Omega)} \le 1} \int_\Omega z(x)\, du,$$

which is equivalent to the total variation of u. It will be convenient to recall the continuous embedding $W_0^{1,p'}(\Omega) \subset C_0(\Omega)$ if $p' = \frac{p}{p-1} \in (n, \infty)$, and consequently $\mathcal{M}(\Omega) \subset W^{-1,p}(\Omega)$ for $p \in (1, \frac{n}{n-1})$.

Given $u \in \mathcal{M}(\Omega)$, we say that $y_u \in L^2(\Omega)$ is a solution to (2.1.1) if

$$\int_\Omega y_u\, A^* z\, dx = \int_\Omega z\, du \qquad \text{for all } z \in \mathcal{Z}, \tag{2.1.2}$$

where

$$\mathcal{Z} = \{z \in H_0^1(\Omega) \cap C_0(\Omega) : A^* z \in L^2(\Omega)\}.$$

It is well known, see for instance (Stampacchia, 1965, Theorem 9.1), that there exists a unique solution to (2.1.1) satisfying $y_u \in W_0^{1,p}(\Omega)$ for all $1 \le p < \frac{n}{n-1}$ and

$$\|y_u\|_{W_0^{1,p}(\Omega)} \le C_p \|u\|_{\mathcal{M}(\Omega)}, \tag{2.1.3}$$

with C_p independent of $u \in \mathcal{M}(\Omega)$. We refer to (Casas and Kunisch, 2014) for a treatment of (P) with semilinear elliptic equations as constraint. This requires additional analytical techniques which go beyond this presentation. Let us note in passing that the equation $-\Delta y + y^3 = \delta_{x_0}$ in Ω, together with homogeneous Dirichlet boundary conditions does not admit a solution for $n = 3$; see (Benilan and Brezis, 2004). Here δ_{x_0} denotes the Dirac delta distribution located in $x_0 \in \Omega$.

Since $W_0^{1,p}(\Omega) \subset L^2(\Omega)$ for very $\frac{2n}{n+2} \le p < \frac{n}{n-1}$, the cost functional J is well defined on $\mathcal{M}(\Omega)$. Furthermore, the control-to-state mapping is injective, and thus J is strictly convex. Existence of a unique optimal solution $\bar{u}$ with an associated state $\bar{y}$ for problem (P) can be argued with standard variational techniques; see (Clason and Kunisch, 2011) for details. Henceforth we express J as

$$J(u) = F(u) + \kappa\, j(u), \text{ where } F(u) = \frac{1}{2}\|y_u - y_d\|^2_{L^2(\Omega)},\ j(u) = \|u\|_{\mathcal{M}(\Omega)}.$$

Introducing the adjoint state and using convex analysis techniques we obtain the first order optimality condition.

Theorem 2.1.1. *Let $\bar{u}$ denote the solution to* (P). *Then there exists a unique adjoint state $\bar{\varphi} \in H_0^1(\Omega) \cap C^{0,\mu}(\bar{\Omega})$ for some $\mu \in (0,1)$ satisfying $-\frac{1}{\kappa}\bar{\varphi} \in \partial j(\bar{u})$,*

$$\begin{cases} A^*\bar{\varphi} = \bar{y} - y_d & \text{in } \Omega, \\ \bar{\varphi} = 0 & \text{on } \Gamma, \end{cases} \tag{2.1.4}$$

$$\kappa \|\bar{u}\|_{\mathcal{M}(\Omega)} + \int_\Omega \bar{\varphi}\, d\bar{u} = 0, \tag{2.1.5}$$

$$\|\bar{\varphi}\|_{C_0(\bar{\Omega})} \begin{cases} = \kappa & \text{if } \bar{u} \neq 0, \\ \leq \kappa & \text{if } \bar{u} = 0. \end{cases} \tag{2.1.6}$$

Proof. Since $\bar{y} - y_d \in L^2(\Omega)$ there exists a unique solution $\bar{\varphi} \in H_0^1(\Omega) \cap C^{0,\mu}(\bar{\Omega})$ by Corollary 1.1.4. Now, we have with (2.1.4) and (2.1.2)

$$F'(\bar{u})u = \int_\Omega (\bar{y} - y_d) y_u \, dx = \int_\Omega y_u A^* \bar{\varphi}\, dx = \int_\Omega \bar{\varphi}\, du.$$

By convexity of j, we obtain for every $u \in \mathcal{M}(\Omega)$

$$0 \leq \lim_{\rho \searrow 0} \frac{J(\bar{u} + \rho(u - \bar{u})) - J(\bar{u})}{\rho} \leq F'(\bar{u})(u - \bar{u}) + \kappa(j(u) - j(\bar{u}))$$
$$= \int_\Omega \bar{\varphi}\, d(u - \bar{u}) + \kappa(j(u) - j(\bar{u})).$$

It follows that

$$\int_\Omega \frac{-1}{\kappa} \bar{\varphi}\, d(u - \bar{u}) + j(\bar{u}) \leq j(u) \quad \forall u \in \mathcal{M}(\Omega),$$

which by definition of the subdifferential implies that $\lambda := \frac{-1}{\kappa}\bar{\varphi} \in \partial j(\bar{u})$, where $\lambda \in C_0(\bar{\Omega})$. Hence, the above inequality can be written as

$$\langle u - \bar{u}, \lambda \rangle + \|\bar{u}\|_{\mathcal{M}(\Omega)} \leq \|u\|_{\mathcal{M}(\Omega)} \quad \forall u \in \mathcal{M}(\Omega). \tag{2.1.7}$$

Taking $u = 2\bar{u}$ and $u = 0$, we obtain the two inequalities

$$\langle \bar{u}, \lambda \rangle \leq \|\bar{u}\|_{\mathcal{M}(\Omega)} \leq \langle \bar{u}, \lambda \rangle$$

and hence (2.1.5) follows by the definition of λ. Inserting $\lambda = \frac{-1}{\kappa}\bar{\varphi}$ into (2.1.7) and using (2.1.5) yields $\langle u, \bar{\varphi} \rangle \leq \kappa \|u\|_{\mathcal{M}(\Omega)}$. This inequality together with (2.1.5) imply (2.1.6). □

Exploiting (2.1.5) and (2.1.6), additional insight into the sparsity structure of $\bar{u}$ can be obtained. Let $\bar{u} = \bar{u}^+ - \bar{u}^-$ denote the Jordan decomposition of $\bar{u}$. Then we have the following result.

Theorem 2.1.2 (Casas et al., 2013, Theorem 3.3). *Let $\bar{u} \neq 0$ denote the solution to* (P). *Then we have*

$$Supp\ (\bar{u}^{+}) \subset \{x \in \Omega : \bar{\varphi}(x) = -\kappa\}, \tag{2.1.8}$$

$$Supp\ (\bar{u}^{-}) \subset \{x \in \Omega : \bar{\varphi}(x) = +\kappa\}. \tag{2.1.9}$$

2.2 Finite element approximation of measure-valued control problem

Here we address the finite element approximation of (P) subject to (2.1.1). Throughout Assumption (1.4.1) is supposed to hold and we put ourselves into the finite element framework specified at the beginning of Section 1.4. In particular we recall the triangulation $\{\mathcal{T}_h\}_{h>0}$ and the finite dimensional state space Y_h introduced there.

We need to introduce some additional notation. Let $\{x_j\}_{j=1}^{N(h)}$ denote the interior nodes of the triangulation $\mathcal{T}_h$. Associated to these nodes we consider the nodal basis of Y_h given by the functions $\{e_j\}_{j=1}^{N(h)}$ such that $e_j(x_i) = \delta_{ij}$ for every $1 \leq i, j \leq N(h)$. Then every element y_h of Y_h can be written in the form

$$y_h = \sum_{j=1}^{N(h)} y_j e_j, \qquad \text{where} \quad y_j = y_h(x_j), \quad 1 \leq j \leq N(h).$$

The discretized controls will be taken from the space

$$D_h = \left\{ u_h \in \mathcal{M}(\Omega) : u_h = \sum_{j=1}^{N(h)} \lambda_j \delta_{x_j}, \text{ where } \{\lambda_j\}_{j=1}^{N(h)} \subset \mathbb{R} \right\}.$$

Here δ_{x_j} denotes the Dirac measure centered at the node x_j. We note that D_h can be identified with the dual of Y_h through the duality relation

$$\langle u_h, y_h \rangle = \sum_{j=1}^{N(h)} \lambda_j y_j. \tag{2.2.1}$$

It will be convenient to define the linear operators $\Pi_h : C_0(\Omega) \to Y_h$ and $\Lambda_h : \mathcal{M}(\Omega) \to D_h$ by

$$\Pi_h y = \sum_{j=1}^{N(h)} y(x_j) e_j \quad \text{and} \quad \Lambda_h u = \sum_{j=1}^{N(h)} \langle u, e_j \rangle \delta_{x_j}.$$

The operator Π_h is the nodal interpolation operator on Y_h and Λ_h is its adjoint with respect to the duality relation (2.2.1), as will be asserted in Theorem 2.2.3 below.

For every $u \in \mathcal{M}(\Omega)$, we denote by y_h the unique element of Y_h satisfying

$$a(y_h, z_h) = \int_{\Omega_h} z_h \, du \quad \forall z_h \in Y_h, \tag{2.2.2}$$

where a was defined just below (1.4.4).

The approximation of the optimal control problem (P) is chosen as

$$(\mathrm{P}_h) \qquad \min_{u \in \mathcal{M}(\Omega)} J_h(u_h) = \frac{1}{2}\|y_h - y_d\|^2_{L^2(\Omega_h)} + \alpha \|u\|_{\mathcal{M}(\Omega)},$$

where y_h is the solution to (2.2.2). This corresponds to the variational discretization of the control as discussed in Remark 1.4.10-(ii).

Existence of a solution to (P_h) can easily be argued. Concerning its uniqueness, observe that the functional J_h is convex. However, as a consequence of the non-injectivity of the control-to-discrete-state mapping and the non-strict convexity of the norm of $\mathcal{M}(\Omega)$, the functional J_h is not strictly convex. Thus solutions to (P_h) are not unique. On the other hand, note that for $u_h \in D_h$, the mapping $u_h \mapsto y_h(u_h)$, which denotes the solution to (2.2.2) for $u = u_h$, is linear, injective and since $\dim D_h = \dim Y_h$, this mapping is bijective. Therefore, the cost functional J_h is strictly convex on D_h. Now, if $\tilde{u}_h$ is a solution to (P_h) and we take $\bar{u}_h = \Lambda_h \tilde{u}_h$, then statement *4* of Theorem 2.2.3 below and (2.2.6) imply that $J_h(\bar{u}_h) \le J_h(\tilde{u}_h)$. Hence $\bar{u}_h$ is also a solution to (P_h), and it is the unique solution in D_h. It will be denoted by $\bar{u}_h$ hereafter. We summarize this discussion in the following theorem.

Theorem 2.2.1. *Problem* (P_h) *admits at least one solution. Among them there exists a unique one belonging to D_h, denoted by $\bar{u}_h$. Moreover, any other solution $\tilde{u}_h \in \mathcal{M}(\Omega)$ of* (P_h) *satisfies $\Lambda_h \tilde{u}_h = \bar{u}_h$.*

To emphasize the practical relevance of this result recall that, as an element of D_h, $\bar{u}_h$ has a unique representation of the form

$$\bar{u}_h = \sum_{j=1}^{N(h)} \bar{\lambda}_j \delta_{x_j}. \tag{2.2.3}$$

Hence numerical computations of $\bar{u}_h$ are reduced to the computation of the coefficients $\{\bar{\lambda}_j\}_{j=1}^{N(h)}$. We next address the convergence of the solutions to (P_h) in D_h to the solution of (P) as $h \to 0$.

Theorem 2.2.2. *For every $h > 0$, let $\bar{u}_h$ be the unique solution to* (P_h) *belonging to D_h and let $\bar{u}$ be the solution to* (P). *Then the following convergence properties hold for $h \to 0$:*

$$\bar{u}_h \stackrel{*}{\rightharpoonup} u \text{ in } \mathcal{M}(\Omega), \ \|\bar{u}_h\|_{\mathcal{M}(\Omega)} \to \|\bar{u}\|_{\mathcal{M}(\Omega)},$$

$$\|\bar{y} - \bar{y}_h\|_{L^2(\Omega)} \to 0, \; J_h(\bar{u}_h) \to J(\bar{u}),$$

where $\bar{y}$ and $\bar{y}_h$ are the continuous and discrete states associated to $\bar{u}$ and $\bar{u}_h$.

For the proof we refer to (Casas et al., 2012, Theorem 3.1). It uses the compact embedding $\mathcal{M}(\Omega) \hookrightarrow W^{-1,p}(\Omega)$ for every $1 \le p < \frac{n}{n-1}$. We turn to summarize properties of the nodal interpolation operator Π_h. They will be essential for the convergence rate results in the following subsection.

Theorem 2.2.3. *The following properties hold.*

1. *For every $u \in \mathcal{M}(\Omega)$ and every $z \in C_0(\Omega)$ and $z_h \in Y_h$ we have*

$$\langle u, z_h \rangle = \langle \Lambda_h u, z_h \rangle, \tag{2.2.4}$$
$$\langle u, \Pi_h z \rangle = \langle \Lambda_h u, z \rangle. \tag{2.2.5}$$

2. *For every $u \in \mathcal{M}(\Omega)$ we have*

$$\|\Lambda_h u\|_{\mathcal{M}(\Omega)} \le \|u\|_{\mathcal{M}(\Omega)}, \tag{2.2.6}$$
$$\Lambda_h u \overset{*}{\rightharpoonup} u \text{ in } \mathcal{M}(\Omega) \text{ and } \|\Lambda_h u\|_{\mathcal{M}(\Omega)} \to \|u\|_{\mathcal{M}(\Omega)}. \tag{2.2.7}$$

3. *There exists a constant $C > 0$ such that for every $u \in \mathcal{M}(\Omega)$*

$$\|u - \Lambda_h u\|_{W^{-1,p}(\Omega)} \le C h^{1-n/p'} \|u\|_{\mathcal{M}(\Omega)}, \quad 1 < p < \frac{n}{n-1}, \tag{2.2.8}$$
$$\|u - \Lambda_h u\|_{W_0^{1,\infty}(\Omega)^*} \le C h \|u\|_{\mathcal{M}(\Omega)}, \tag{2.2.9}$$

   where p' is the conjugate of p.

4. *Given $u \in \mathcal{M}(\Omega)$, let y_h and $\tilde{y}_h$ be the solutions to (2.2.2) associated to the controls u and $\Lambda_h u$, respectively. Then the equality $y_h = \tilde{y}_h$ holds.*

Proof. For $z_h = \sum_{j=1}^{N(h)} z_j e_j$ we have

$$\langle u, z_h \rangle = \sum_{j=1}^{N(h)} z_j \langle u, e_j \rangle = \sum_{j=1}^{N(h)} \langle u, e_j \rangle \langle \delta_{x_j}, z_h \rangle = \langle \Lambda_h u, z_h \rangle,$$

which proves (2.2.4). For (2.2.5) we proceed as follows

$$\langle u, \Pi_h z \rangle = \sum_{j=1}^{N(h)} z(x_j) \langle u, e_j \rangle = \sum_{j=1}^{N(h)} \langle u, e_j \rangle \langle \delta_{x_j}, z \rangle = \langle \Lambda_h u, z \rangle.$$

To verify (2.2.6) we introduce the function $s_h \in Y_h$ by

$$s_h = \sum_{j=1}^{N(h)} s_j e_j, \quad \text{with} \quad s_j = \begin{cases} +1 & \text{if } \langle u, e_j \rangle > 0, \\ -1 & \text{if } \langle u, e_j \rangle < 0, \\ 0 & \text{otherwise.} \end{cases}$$

Then we have

$$\|\Lambda_h u\|_{\mathcal{M}(\Omega)} = \sum_{j=1}^{N(h)} |\langle u, e_j\rangle| = \sum_{j=1}^{N(h)} s_j \langle u, e_j\rangle = \langle u, s_h\rangle$$
$$\leq \|u\|_{\mathcal{M}(\Omega)} \|s_h\|_{C_0(\Omega)} = \|u\|_{\mathcal{M}(\Omega)}.$$

Let us prove (2.2.7). Since $\{\Lambda_h u\}_{h>0}$ is bounded in $\mathcal{M}(\Omega)$ there exists a subsequence, denoted in the same way, such that $\Lambda_h u \stackrel{*}{\rightharpoonup} v$ in $\mathcal{M}(\Omega)$. Using the convergence $\|\Pi_h z - z\|_{C_0(\Omega)} \to 0$ for all $z \in C_0(\Omega)$ and (2.2.6) we get that

$$|\langle \Lambda_h u, z - \Pi_h z\rangle| \leq \|\Lambda_h u\|_{\mathcal{M}(\Omega)} \|z - \Pi_h z\|_{C_0(\Omega)} \to 0 \text{ as } h \to 0.$$

With this property and (2.2.4) we infer for every $z \in C_0(\Omega)$

$$\langle v, z\rangle = \lim_{h\to 0} \langle \Lambda_h u, z\rangle = \lim_{h\to 0} \langle \Lambda_h u, \Pi_h z\rangle + \lim_{h\to 0} \langle \Lambda_h u, z - \Pi_h z\rangle$$
$$= \lim_{h\to 0} \langle u, \Pi_h z\rangle = \langle u, z\rangle,$$

therefore $u = v$. Since any subsequence converges to u, the whole sequence converges to u weakly* in $\mathcal{M}(\Omega)$. From this convergence and (2.2.6) we obtain

$$\|u\|_{\mathcal{M}(\Omega)} \leq \liminf_{h\to 0} \|\Lambda_h u\|_{\mathcal{M}(\Omega)} \leq \limsup_{h\to 0} \|\Lambda_h u\|_{\mathcal{M}(\Omega)} \leq \|u\|_{\mathcal{M}(\Omega)},$$

consequently (2.2.7) holds.

To prove (2.2.8) we take an arbitrary element $z \in W_0^{1,p'}(\Omega)$, with $1 \leq p < \frac{n}{n-1}$. Using (2.2.5) and the well known interpolation error estimates in Sobolev spaces (see, for instance, (Ciarlet, 1978, Chapter 3)) we obtain

$$\langle u - \Lambda_h u, z\rangle = \langle u, z - \Pi_h z\rangle \leq \|u\|_{\mathcal{M}(\Omega)} \|z - \Pi_h z\|_{C_0(\Omega)}$$
$$\leq C h^{1-n/p'} \|u\|_{\mathcal{M}(\Omega)} \|z\|_{W_0^{1,p'}(\Omega)}.$$

Since $W^{-1,p}(\Omega)$ is the dual of $W_0^{1,p'}(\Omega)$ for $1 < p < \frac{n}{n-1}$, (2.2.8) follows from the above inequalities. For $p = 1$, we have $p' = \infty$ and the above inequality can be expressed as

$$\langle u - \Lambda_h u, z\rangle \leq C h \|u\|_{\mathcal{M}(\Omega)} \|z\|_{W_0^{1,\infty}(\Omega)}, \quad \forall z \in W_0^{1,\infty}(\Omega).$$

Since $W^{-1,1}(\Omega)$ is not the dual space of $W_0^{1,\infty}(\Omega)$, from these inequalities we only get (2.2.9). The last statement of the theorem is an immediate consequence of (2.2.4). □

2.2.1 Convergence rates

We discuss error estimates for the optimal costs as well as for the optimal states. Note that since $\mathcal{M}(\Omega)$ is not a separable space, strong convergence in $\mathcal{M}(\Omega)$ of the discrete optimal controls is not possible. In addition to Assumption (1.4.1) we need to assume throughout that

$$y_d \in L^r(\Omega) \text{ with } r = \begin{cases} 4 & \text{if } n = 2, \\ \frac{8}{3} & \text{if } n = 3. \end{cases} \tag{2.2.10}$$

As before we denote by $\bar{y}$ and $\bar{y}_h$ the continuous and discrete states associated to the optimal controls $\bar{u}$ and $\bar{u}_h$, respectively.

Theorem 2.2.4. *There exists a constant $C > 0$ independent of h such that*

$$|J(\bar{u}) - J_h(\bar{u}_h)| \leq Ch^\sigma, \tag{2.2.11}$$

where $\sigma = 1$ if $n = 2$ and $\sigma = 1/2$ if $n = 3$.

Proof. We establish some preliminary estimates. Given $u \in \mathcal{M}(\Omega)$, with associated continuous and discrete states y and y_h, we know from (Casas, 1985) that

$$\|y - y_h\|_{L^2(\Omega_h)} \leq Ch^\sigma \|u\|_{\mathcal{M}(\Omega)}, \tag{2.2.12}$$

with σ defined as in the statement of the theorem.

Taking r as in (2.2.10) and using Hölder's inequality and (1.4.3), we deduce that for all $\phi \in L^r(\Omega)$,

$$\|\phi\|_{L^2(\Omega\setminus\Omega_h)} \leq \|\phi\|_{L^r(\Omega\setminus\Omega_h)} |\Omega \setminus \Omega_h|^{\frac{r-2}{2r}} \leq C_\Omega \|\phi\|_{L^r(\Omega\setminus\Omega_h)} h^{\frac{\sigma}{2}} \tag{2.2.13}$$

holds. As a consequence of (2.2.12) and (2.2.13), with $\phi = y - y_d$, we get

$$\begin{aligned}
&\left| \|y - y_d\|^2_{L^2(\Omega)} - \|y_h - y_d\|^2_{L^2(\Omega_h)} \right| \\
&\leq \|y - y_d\|^2_{L^2(\Omega\setminus\Omega_h)} + \left(\|y - y_d\|_{L^2(\Omega_h)} + \|y_h - y_d\|_{L^2(\Omega_h)}\right) \|y - y_h\|_{L^2(\Omega_h)} \\
&\leq C_\Omega \Big(\|y - y_d\|^2_{L^r(\Omega\setminus\Omega_h)} \\
&+ [\|y - y_d\|_{L^2(\Omega_h)} + \|y_h - y_d\|_{L^2(\Omega_h)}] \|u\|_{\mathcal{M}(\Omega)} \Big) h^\sigma.
\end{aligned} \tag{2.2.14}$$

Using the optimality of $\bar{u}$ and $\bar{u}_h$ we obtain

$$J(\bar{u}) - J_h(\bar{u}) \leq J(\bar{u}) - J_h(\bar{u}_h) \leq J(\bar{u}_h) - J_h(\bar{u}_h),$$

hence

$$|J(\bar{u}) - J_h(\bar{u}_h)| \leq \max\{|J(\bar{u}) - J_h(\bar{u})|, |J(\bar{u}_h) - J_h(\bar{u}_h)|\}. \tag{2.2.15}$$

Since $\|\bar{u}_h\|_{\mathcal{M}(\Omega)} \to \|\bar{u}\|_{\mathcal{M}(\Omega)}$ by Theorem 2.2.2, we deduce that $\{\bar{u}_h\}_{h>0}$ is bounded in $\mathcal{M}(\Omega)$. Therefore, (2.1.3) implies that the associated states $\{y_{\bar{u}_h}\}_{h>0}$ are bounded in $W_0^{1,p}(\Omega)$ for every $1 \le p < \frac{n}{n-1}$ and therefore also in $L^r(\Omega)$. We apply (2.2.14) with $u = \bar{u}_h$ and $u = \bar{u}$, respectively. Together with (2.2.15) this establishes (2.2.11). □

In the next theorem a convergence rate for the states is obtained.

Theorem 2.2.5. *There exists a constant $C > 0$ independent of h such that*

$$\|\bar{y} - \bar{y}_h\|_{L^2(\Omega)} \le Ch^{\frac{\sigma}{2}}, \tag{2.2.16}$$

with σ as defined in Theorem 2.2.4.

Proof. By the optimality of $\bar{u}$ we have for all $u \in \mathcal{M}(\Omega)$, that

$$(\bar{y} - y_d, y_u - \bar{y}) + \kappa[\|u\|_{\mathcal{M}(\Omega)} - \|\bar{u}\|_{\mathcal{M}(\Omega)}] \ge 0,$$

where $(\cdot,\cdot)$ denotes the scalar product in $L^2(\Omega)$. In particular, taking $u = \bar{u}_h$, we get

$$(\bar{y} - y_d, y_{\bar{u}_h} - \bar{y}) + \kappa[\|\bar{u}_h\|_{\mathcal{M}(\Omega)} - \|\bar{u}\|_{\mathcal{M}(\Omega)}] \ge 0. \tag{2.2.17}$$

Analogously, the optimality of $\bar{u}_h$ implies that

$$(\bar{y}_h - y_d, y_h(\bar{u}) - \bar{y}_h) + \kappa[\|\bar{u}\|_{\mathcal{M}(\Omega)} - \|\bar{u}_h\|_{\mathcal{M}(\Omega)}] \ge 0. \tag{2.2.18}$$

We point out that by definition of Y_h, we have $y_h(u) = 0$ in $\Omega \setminus \Omega_h$. Then, the scalar product above in $L^2(\Omega)$ coincides with that in $L^2(\Omega_h)$. Rearrange terms in (2.2.18) lead to:

$$\begin{aligned}(y_{\bar{u}_h} &- y_d, \bar{y} - y_{\bar{u}_h}) + (\bar{y}_h - y_{\bar{u}_h}, y_h(\bar{u}) - \bar{y}_h)\\ &+ (y_d, \bar{y} - y_h(\bar{u}) + \bar{y}_h - y_{\bar{u}_h}) + (y_{\bar{u}_h}, y_h(\bar{u}) - \bar{y} + y_{\bar{u}_h} - \bar{y}_h)\\ &+ \kappa[\|\bar{u}\|_{\mathcal{M}(\Omega)} - \|\bar{u}_h\|_{\mathcal{M}(\Omega)}] \ge 0.\end{aligned}$$

Adding this inequality and (2.2.17) we obtain

$$\begin{aligned}\|\bar{y} - \bar{y}_h\|^2_{L^2(\Omega)} &= (\bar{y} - \bar{y}_h, \bar{y} - \bar{y}_h)\\ &\le (\bar{y}_h - y_{\bar{u}_h}, y_h(\bar{u}) - \bar{y}_h)\\ &+ (y_d - y_{\bar{u}_h}, \bar{y} - y_h(\bar{u}) + \bar{y}_h - y_{\bar{u}_h}).\end{aligned} \tag{2.2.19}$$

Let us estimate the right hand terms. For the first one we apply the Cauchy–Schwarz inequality, exploit the fact $y_h(\bar{u}) - \bar{y}_h = 0$ in $\Omega \setminus \Omega_h$ and use (2.2.12), to deduce

$$(\bar{y}_h - y_{\bar{u}_h}, y_h(\bar{u}) - \bar{y}_h) \le \|\bar{y}_h - y_{\bar{u}_h}\|_{L^2(\Omega_h)}\|y_h(\bar{u}) - \bar{y}_h\|_{L^2(\Omega_h)} \le Ch^{\sigma}, \tag{2.2.20}$$

where we have used that $\{\bar{u}_h\}_{h>0}$, $\{y_h(\bar{u})\}_{h>0}$ and $\{\bar{y}_h\}_{h>0}$ are bounded as a consequence of Theorem 2.2.4 and (2.2.12). For the second term we use (2.2.13) and once again (2.2.12) as well as the fact that $y_h(u) = 0$ in $\Omega \setminus \Omega_h$, to obtain

$$\begin{aligned}
&(y_d - y_{\bar{u}_h}, \bar{y} - y_h(\bar{u}) + \bar{y}_h - y_{\bar{u}_h}) \\
&\quad \le \|y_d - y_{\bar{u}_h}\|_{L^2(\Omega\setminus\Omega_h)} \|\bar{y} - y_{\bar{u}_h}\|_{L^2(\Omega\setminus\Omega_h)} \\
&\qquad + \|y_d - y_{\bar{u}_h}\|_{L^2(\Omega_h)} \|y_{\bar{u}-\bar{u}_h} - y_h(\bar{u} - \bar{u}_h)\|_{L^2(\Omega_h)} \\
&\quad \le C \left(\|y_d - y_{\bar{u}_h}\|_{L^r(\Omega\setminus\Omega_h)} \|y_{\bar{u}-\bar{u}_h}\|_{L^r(\Omega\setminus\Omega_h)} + \|\bar{u} - \bar{u}_h\|_{\mathcal{M}(\Omega)} \right) h^{\sigma} \\
&\quad \le C h^{\sigma},
\end{aligned} \tag{2.2.21}$$

where we have also used that $y_d \in L^r(\Omega)$ and (2.1.3). Finally, (2.2.19), (2.2.20) and (2.2.21) prove (2.2.16). □

Let us observe that from (2.2.11) and (2.2.16) we can deduce that

$$\left| \|\bar{u}\|_{\mathcal{M}(\Omega)} - \|\bar{u}_h\|_{\mathcal{M}(\Omega)} \right| \le C h^{\frac{\sigma}{2}}$$

for some constant $C > 0$ independent of h.

Remark 2.2.6. In (Casas et al., 2012) results analogous to Theorems 2.2.4 and 2.2.5 were obtained for the case of measure valued Neumann boundary control problems. Further, in (Casas et al., 2013) finite element approximation of linear parabolic optimal control problems with controls in $L^2(0, T; \mathcal{M}(\Omega))$ were investigated.

The convergence results presented above were improved in (Pieper and Vexler, 2013), under slightly stronger integrability assumptions on y_d, by employing error estimates on the state in $L^p(\Omega)$ with $p \in (1, \frac{n}{n-2})$. There also the case with control domain Ω_c and the observation domain Ω_o different from Ω were investigated.

3 Related topics

In view of the limited number of pages, we concentrated on some specific problems and questions. Here, we briefly sketch some additional topics out of the very broad field of optimal control theory that might be of interest for the reader. To bound the number of references, we have to keep citations to a minimum.

- **Pontryagin's maximum principle.** In the necessary optimality condition (1.2.3) for problem (P), the term $\nu\bar{u} + \kappa\bar{\lambda}$ arises from a certain differentiation of the objective functional J with respect to the control u. This differentiation might be impossible or cause some loss of information and is avoided by the Pontryagin maximum principle for the optimal control of PDEs. We refer to the book by (Li and Yong, 1995) and to (Bonnans and Casas, 1991), (Casas, 1997), (Casas et al., 2000), or (Raymond and Zidani, 1998) Moreover, we mention the early contributions (Wolfersdorf, 1976, 1977).

- **Control of parabolic equations.** The optimal control theory of linear or nonlinear parabolic equations is also very rich; we only refer to the books by (Lions, 1968) (linear equations), (Tröltzsch, 2010) (semilinear equations), or (Hinze et al., 2009) (numerical aspects) and to the references therein.
  The problem to reach a certain ball of $L^2(\Omega)$ in shortest time, starting from an initial value function $y_0 \in L^2(\Omega)$ is a typical example of time-optimal control. Time optimal control problems are addressed in the books by (Fattorini, 1999), (Fattorini, 2005), and (Wang et al., 2018).
  In comparison with elliptic problems, much less work was devoted to error estimates for the numerical approximation of parabolic control problems. We only refer to (Meidner and Vexler, 2008a), (Meidner and Vexler, 2008b), where error estimates for discontinuous Galerkin approximations of the parabolic equation and suitable control approximations are derived.
- **Control of hyperbolic equations.** For optimal control of the wave equation we refer exemplarily to (Gugat et al., 2005), (Gugat et al., 2016), (Gugat and Leugering, 2017), (Mordukhovich and Raymond, 2004), and the references cited there. For time optimal control see for instance (Kunisch and Wachsmuth, 2013), and for optimal control of the energy critical wave equation (Kunisch and Meinlschmidt, 2020). For optimal control of nonlinear hyperbolic conservation laws we refer to (Ulbrich, 2002) and (Lagnese et al., 1994), (Lagnese and Leugering, 2004).
- **Boundary control.** We concentrated on elliptic problems with control distributed in the domain Ω. The control might also appear as boundary control on $\Gamma = \partial\Omega$, e.g. in a nonlinear Robin boundary condition $\partial_n y + f(y) = u$, see (Lions, 1968), (Hinze et al., 2009), and (Tröltzsch, 2010), or in a Dirichlet boundary condition such as $y = u$ on Γ, cf. e.g. (Casas and Raymond, 2006).
- **Pointwise state constraints.** In our model problem (P), pointwise constraints were imposed only on the control. In addition, pointwise constraints on the state, say $g(x, y(x)) \le 0$, mixed control-state constraints such as $g(u(x), y(x)) \le 0$ or functional constraints $F(y, u) \le 0$ can be posed. The number of associated contributions is huge. We only mention (Bonnans and Casas, 1995), (Casas, 1986), (Casas, 1997), (Raymond and Zidani, 1998) for problems with pointwise state constraints and (Arada and Raymond, 2000), (Rösch and Tröltzsch, 2007) for mixed pointwise control-state constraints.
- **A posteriori error estimates.** Theorems 1.4.9, 1.4.12, 1.4.15 provide a priori error estimates that quantify the distance of a discretized optimal solution to an associated exact one. Given a fixed arbitrary state or control, an important question is to quantify the distance to the nearest optimal state or control. This is a question of a posteriori error analysis that is in particular meaningful for mesh adaptivity or model order reduction. Out of the extensive literature, we exemplarily quote (Becker et al., 2000), (Benedix and Vexler, 2009), (Gong and Yan, 2017), (Hintermüller et al., 2008), (Rösch et al., 2017), and (Vexler and Wollner, 2008).

- **Numerical optimization methods.** For solving optimal control problems, many numerical methods of nonlinear optimization were transferred and adapted to PDE control. Among them, thanks to their easy implementation, gradient, conditioned gradient, or conjugate gradient methods are often applied. Moreover, higher order methods such as BFGS, interior point, and Newton type methods are frequently used, e.g. sequential quadratic programming (SQP) methods. The semismooth Newton method is very often applied to solving nonlinear optimal control problems. It is related to the primal-dual active set strategy that is also of interest for convex problems with linear equations, see (Hintermüller et al., 2003), (Ito and Kunisch, 2008), or (Ulbrich, 2011).
  There are various ways of discretizing given optimal control problems, some of them mentioned in the preceding subsections. In this respect, the numerical realization of the gradient is essential. Automatic differentiation can play an important role here (Griewank and Walther, 2008).
  The number of contributions to numerical methods for PDE control is very large. We only quote the books by (Borzì and Schulz, 2012), (Hinze et al., 2009).
- **Model order reduction.** The discretization of optimal control problems for PDEs leads to large scale optimization problems that can exceed the capacity of computers. Therefore, model order reduction techniques are developed that reduce the size of the problems and yet lead to quite precise results. Among them, proper orthogonal decomposition (POD), reduced basis methods, or tensor product approximations were extensively studied. We mention (Kunisch and Volkwein, 2002), (Kunisch and Volkwein, 2001), for model order reduction by POD, (Ito and Ravindran, 2001), (Dedè, 2012), and (Kärcher and Grepl, 2014) for the reduced basis method, and (Benner et al., 2005), for a survey on model order reduction.
  In recent years much effort has gone into the development of tensor-based reduction techniques which are necessitated to partially reduce the curse of dimensionality in the context of optimal control and uncertainty quantification, and for solving closed loop nonlinear optimal control problems on the basis of approximations to the Hamilton-Jacobi-Bellman equation.
- **Optimal control of more complex PDEs.** Triggered by various applications, the optimal control of many different linear and nonlinear types of PDEs was investigated. We can only refer to some of them: optimal control of Navier-Stokes equations, reaction diffusion equations such as Allen-Cahn, Nagumo, FitzHugh-Nagumo, and phase field equations; the control of Fokker-Planck equations, electromagnetic waves and Maxwell equations, monodomain or bidomain equations related to cardiac electrophysiology, and quasilinear equations to name only a few topics.
- **Optimal control of functional differential equations.** In the seventies-nineties of the previous century significant attention was paid to control and optimal control of differential equations with delay terms and more gener-

ally to functional differential equations, see e.g. (Banks and Manitius, 1974), (Delfour, 1986). Such type control problems appear for instance in modeling biological phenomena or for transmission lines.

Selected books and monographs. Various topics of optimal PDE control are addressed by the following books: *Main theoretical principles and optimality conditions:* (Ahmed and Teo, 1981), (Barbu, 1994), (Lions, 1968), (Lions, 1983), (Fursikov, 2000), (Neittaanmäki et al., 2006), (Neittaanmäki and Tiba, 1994), (Tiba, 1990), (Tröltzsch, 2010). *Theory of time optimal control:* (Fattorini, 1999), (Fattorini, 2005), (Wang et al., 2018). *Pontryagin maximum principle:* (Li and Yong, 1995). *Numerical methods:* (Borzì and Schulz, 2012), (Hinze et al., 2009). *Variational methods:* (Ito and Kunisch, 2008). *Linear control theory and semigroup methods:* (Bensoussan et al., 1992). Moreover, we mention (Butkovskiy, 1969) (*applied optimal PDE control*), (Gunzburger, 2003) (*flow control*), (Kogut and Leugering, 2011) (*control in reticulated domains*), (Lasiecka and Triggiani, 2000a), (Lasiecka and Triggiani, 2000b) (*feedback control and related topics for linear systems*).

References

Ahmed, N.U., Teo, K.L., 1981. Optimal Control of Distributed Parameter Systems. North-Holland Publishing Co., New York-Amsterdam.

Ali, A.A., Deckelnick, K., Hinze, M., 2016. Global minima for semilinear optimal control problems. Comput. Optim. Appl. 65, 261–288.

Arada, N., Casas, E., Tröltzsch, F., 2002. Error estimates for the numerical approximation of a semilinear elliptic control problem. Comput. Optim. Appl. 23, 201–229.

Arada, N., Raymond, J.-P., 2000. Optimal control problems with mixed control-state constraints. SIAM J. Control Optim. 39 (5), 1391–1407.

Banks, H.T., Manitius, A., 1974. Application of abstract variational theory to hereditary systems–a survey. IEEE Trans. Autom. Control AC-19, 524–533.

Barbu, V., 1994. Mathematical Methods in Optimization of Differential Systems. Mathematics and Its Applications, vol. 310. Kluwer Academic Publishers Group, Dordrecht. Translated and revised from the 1989 Romanian original.

Bayen, T., Bonnans, F., Silva, F., 2014. Characterization of local quadratic growth for strong minima in the optimal control of semi-linear elliptic equations. Trans. Am. Math. Soc. 366 (4), 2063–2087.

Bayen, T., Silva, F., 2016. Second order analysis for strong solutions in the optimal control of parabolic equations. SIAM J. Control Optim. 54 (2), 819–844.

Becker, R., Kapp, H., Rannacher, R., 2000. Adaptive finite element methods for optimal control of partial differential equations: basic concept. SIAM J. Control Optim. 39 (1), 113–132.

Benedix, O., Vexler, B., 2009. A posteriori error estimation and adaptivity for elliptic optimal control problems with state constraints. Comput. Optim. Appl. 44 (1), 3–25.

Benilan, Ph., Brezis, H., 2004. Nonlinear problems related to the Thomas-Fermi equation, dedicated to Philippe Benilan. J. Evol. Equ. 3, 673–770.

Benner, P., Mehrmann, V., Sorensen, D.C. (Eds.), 2005. Dimension Reduction of Large-Scale Systems. Lecture Notes in Computational Science and Engineering, vol. 45. Springer, Berlin.

Bensoussan, A., Da Prato, G., Delfour, M.C., Mitter, S.K., 1992. Representation and Control of Infinite Dimensional Systems, Vols. I and II. Birkhäuser Verlag, Basel, 1993.

Bonnans, F., Casas, E., 1995. An extension of Pontryagin's principle for state-constrained optimal control of semilinear elliptic equations and variational inequalities. SIAM J. Control Optim. 33 (1), 274–298.

Bonnans, J.F., Casas, E., 1991. Un principe de Pontryagine pour le contrôle des systémes semilinéaires elliptiques. J. Differ. Equ. 90 (2), 288–303.

Bonnans, J.F., Shapiro, A., 2000. Perturbation Analysis of Optimization Problems. Springer-Verlag, Berlin.

Borzì, A., Schulz, V., 2012. Computational Optimization of Systems Governed by Partial Differential Equations. Computational Science & Engineering, vol. 8. Society for Industrial and Applied Mathematics (SIAM), Philadelphia, PA.

Brenner, S.C., Scott, L.R., 2008. The Mathematical Theory of Finite Element Methods, third edition. Texts in Applied Mathematics, vol. 15. Springer, New York.

Brezis, H., Strauss, W., 1973. Semilinear second order elliptic equations in L^1. J. Math. Soc. Jpn. 25 (4), 565–590.

Butkovskiy, A.G., 1969. Distributed Control Systems. Translated from the Russian by Scripta Technica, Inc. Translation Editor: George M. Kranc. Modern Analytic and Computational Methods in Science and Mathematics, vol. 11. American Elsevier Publishing Co., Inc., New York.

Cameron, A., 2010. Estimates for solutions of elliptic partial differential equations with explicit constants and aspects of the finite element method for second-order equations. PhD thesis. Cornell University.

Casas, E., 1985. L^2 estimates for the finite element method for the Dirichlet problem with singular data. Numer. Math. 47, 627–632.

Casas, E., 1986. Control of an elliptic problem with pointwise state constraints. SIAM J. Control Optim. 24 (6), 1309–1318.

Casas, E., 1997. Pontryagin's principle for state-constrained boundary control problems of semilinear parabolic equations. SIAM J. Control Optim. 35 (4), 1297–1327.

Casas, E., 2007. Using piecewise linear functions in the numerical approximation of semilinear elliptic control problems. Adv. Comput. Math. 26, 137–153.

Casas, E., 2012. Second order analysis for bang-bang control problems of PDEs. SIAM J. Control Optim. 50 (4), 2355–2372.

Casas, E., 2017. A review on sparse solutions in optimal control of partial differential equations. SeMA J. 74 (3), 319–344.

Casas, E., Clason, C., Kunisch, K., 2012. Approximation of elliptic control problems in measure spaces with sparse solutions. SIAM J. Control Optim. 50 (4), 1735–1752.

Casas, E., Clason, C., Kunisch, K., 2013. Parabolic control problems in measure spaces with sparse solutions. SIAM J. Control Optim. 51 (1), 28–63.

Casas, E., Herzog, R., Wachsmuth, G., 2012a. Optimality conditions and error analysis of semilinear elliptic control problems with L^1 cost functional. SIAM J. Optim. 22 (3), 795–820.

Casas, E., Herzog, R., Wachsmuth, G., 2012b. Approximation of sparse controls in semilinear equations by piecewise linear functions. Numer. Math. 122, 645–669.

Casas, E., Kunisch, K., 2014. Optimal control of semilinear elliptic equations in measure spaces. SIAM J. Control Optim. 52 (1), 339–364.

Casas, E., Mateos, M., 2002. Uniform convergence of the FEM. Applications to state constrained control problems. Comput. Appl. Math. 21 (1), 67–100.

Casas, E., Mateos, M., 2017. Optimal control of partial differential equations. In: Mateos, M., Alonso, P. (Eds.), Computational Mathematics, Numerical Analysis and Applications. In: SEMA SIMAI Springer Series, vol. 13, pp. 3–59.

Casas, E., Mateos, M., 2020. Critical cones for sufficient second order conditions in pde constrained optimization. SIAM J. Optim. 30 (1), 585–603.

Casas, E., Mateos, M., 2021. State error estimates for the numerical approximation of sparse distributed control problems in the absence of Tikhonov regularization. Vietnam J. Math. 49, 713–738.

Casas, E., Mateos, M., Rösch, A., 2020. Analysis of control problems of nonmonotone semilinear elliptic equations. ESAIM Control Optim. Calc. Var. 26.
Casas, E., Raymond, J.-P., 2006. Error estimates for the numerical approximation of Dirichlet boundary control for semilinear elliptic equations. SIAM J. Control Optim. 45 (5), 1586–1611.
Casas, E., Raymond, J.-P., Zidani, H., 2000. Pontryagin's principle for local solutions of control problems with mixed control-state constraints. SIAM J. Control Optim. 39 (4), 1182–1203.
Casas, E., Ryll, C., Tröltzsch, F., 2015. Second order and stability analysis for optimal sparse control of the FitzHugh-Nagumo equation. SIAM J. Control Optim. 53 (4), 2168–2202.
Casas, E., Tröltzsch, F., 2009. Optimality conditions for a class of optimal control problems with quasilinear elliptic equations. SIAM J. Control Optim. 48 (2), 688–718.
Casas, E., Tröltzsch, F., 2012. Second order analysis for optimal control problems: improving results expected from abstract theory. SIAM J. Optim. 22 (1), 261–279.
Casas, E., Tröltzsch, F., 2020. State-constrained semilinear elliptic optimization problems with unrestricted sparse controls. Math. Control Relat. Fields 10 (3), 527–546.
Casas, E., Wachsmuth, D., Wachsmuth, G., 2017. Sufficient second-order conditions for bang-bang control problems. SIAM J. Control Optim. 55 (5), 3066–3090.
Casas, E., Wachsmuth, D., Wachsmuth, G., 2018. Second order analysis and numerical approximation for bang-bang bilinear control problems. SIAM J. Control Optim. 56 (6), 4203–4227.
Ciarlet, P.G., 1978. The Finite Element Method for Elliptic Problems. North-Holland, Amsterdam.
Clarke, F.H., 1983. Optimization and Nonsmooth Analysis. John Wiley & Sons, New York.
Clason, C., Kunisch, K., 2011. A duality-based approach to elliptic control problems in non-reflexive Banach spaces. ESAIM Control Optim. Calc. Var. 17 (1), 243–266.
Deckelnick, K., Hinze, M., 2012. A note on the approximation of elliptic control problems with bang-bang controls. Comput. Optim. Appl. 51, 931–939.
Dedè, L., 2012. Reduced basis method and error estimation for parametrized optimal control problems with control constraints. J. Sci. Comput. 50 (2), 287–305.
Delfour, M.C., 1986. The linear-quadratic optimal control problem with delays in state and control variables: a state space approach. SIAM J. Control Optim. 24 (5), 835–883.
Dunn, J.C., 1998. On second order sufficient conditions for structured nonlinear programs in infinite-dimensional function spaces. In: Fiacco, A. (Ed.), Mathematical Programming with Data Perturbations. Marcel Dekker, New York, pp. 83–107.
Ekeland, I., Temam, R., 1976. Convex Analysis and Variational Problems. North-Holland-Elsevier, New York.
Fattorini, H.O., 1999. Infinite Dimensional Optimization and Control Theory. Cambridge University Press, Cambridge.
Fattorini, H.O., 2005. Infinite Dimensional Linear Control Systems. The Time Optimal and Norm Optimal Problems. North-Holland Mathematics Studies, vol. 201. Elsevier Science B.V., Amsterdam.
Fursikov, A.V., 2000. Optimal Control of Distributed Systems. Theory and Applications. Translations of Mathematical Monographs, vol. 187. American Mathematical Society, Providence, RI. Translated from the 1999 Russian original by Tamara Rozhkovskaya.
Gilbarg, G., Trudinger, N.S., 1983. Elliptic Partial Differential Equations of Second Order. Springer-Verlag, Berlin Heidelberg.
Gong, W., Yan, N., 2017. Adaptive finite element method for elliptic optimal control problems: convergence and optimality. Numer. Math. 135 (4), 1121–1170.
Griewank, A., Walther, A., 2008. Evaluating Derivatives. Principles and Techniques of Algorithmic Differentiation, second edition. Society for Industrial and Applied Mathematics (SIAM), Philadelphia, PA.
Grisvard, P., 1985. Elliptic Problems in Nonsmooth Domains. Pitman, Boston-London-Melbourne.
Gugat, M., Leugering, G., 2017. Time delay in optimal control loops for wave equations. ESAIM Control Optim. Calc. Var. 23 (1), 13–37.
Gugat, M., Leugering, G., Sklyar, G., 2005. L^p-optimal boundary control for the wave equation. SIAM J. Control Optim. 44 (1), 49–74.

Gugat, M., Trélat, E., Zuazua, E., 2016. Optimal Neumann control for the 1D wave equation: finite horizon, infinite horizon, boundary tracking terms and the turnpike property. Syst. Control Lett. 90, 61–70.

Gunzburger, M.D., 2003. Perspectives in Flow Control and Optimization. Advances in Design and Control, vol. 5. Society for Industrial and Applied Mathematics (SIAM), Philadelphia, PA.

Hintermüller, M., Hoppe, R.H.W., Iliash, Y., Kieweg, M., 2008. An a posteriori error analysis of adaptive finite element methods for distributed elliptic control problems with control constraints. ESAIM Control Optim. Calc. Var. 14 (3), 540–560.

Hintermüller, M., Ito, K., Kunisch, K., 2003. The primal-dual active set strategy as a semismooth Newton method. SIAM J. Optim. 13 (3), 865–888. 2002.

Hinze, M., 2005. A variational discretization concept in control constrained optimization: the linear-quadratic case. Comput. Optim. Appl. 30, 45–61.

Hinze, M., Pinnau, R., Ulbrich, M., Ulbrich, S., 2009. Optimization with PDE Constraints. Mathematical Modelling: Theory and Applications, vol. 23. Springer, New York.

Ito, K., Kunisch, K., 2008. Lagrange Multiplier Approach to Variational Problems and Applications. Advances in Design and Control, vol. 15. Society for Industrial and Applied Mathematics (SIAM), Philadelphia, PA.

Ito, K., Ravindran, S.S., 2001. Reduced basis method for optimal control of unsteady viscous flows. Int. J. Comput. Fluid Dyn. 15 (2), 97–113.

Kärcher, M., Grepl, M.A., 2014. A certified reduced basis method for parametrized elliptic optimal control problems. ESAIM Control Optim. Calc. Var. 20 (2), 416–441.

Kogut, P.I., Leugering, G.R., 2011. Optimal Control Problems for Partial Differential Equations on Reticulated Domains. Approximation and Asymptotic Analysis. Systems & Control: Foundations & Applications. Birkhäuser/Springer, New York.

Krumbiegel, K., Pfefferer, J., 2015. Superconvergence for Neumann boundary control problems governed by semilinear elliptic equations. Comput. Optim. Appl. 61, 373–408.

Kunisch, K., Meinlschmidt, H., 2020. Optimal control of an energy-critical semilinear wave equation in 3D with spatially integrated control constraints. J. Math. Pures Appl. 9 (138), 46–87.

Kunisch, K., Volkwein, S., 2001. Galerkin proper orthogonal decomposition methods for parabolic problems. Numer. Math. 90 (1), 117–148.

Kunisch, K., Volkwein, S., 2002. Galerkin proper orthogonal decomposition methods for a general equation in fluid dynamics. SIAM J. Numer. Anal. 40 (2), 492–515.

Kunisch, K., Wachsmuth, D., 2013. On time optimal control of the wave equation, its regularization and optimality system. ESAIM Control Optim. Calc. Var. 19 (2), 317–336.

Lagnese, J.E., Leugering, G., 2004. Domain Decomposition Methods in Optimal Control of Partial Differential Equations. International Series of Numerical Mathematics, vol. 148. Birkhäuser Verlag, Basel.

Lagnese, J.E., Leugering, G., Schmidt, E.J.P.G., 1994. Modeling, Analysis and Control of Dynamic Elastic Multi-Link Structures. Systems & Control: Foundations & Applications. Birkhäuser Boston, Inc., Boston, MA.

Lasiecka, I., Triggiani, R., 2000a. Control Theory for Partial Differential Equations: Continuous and Approximation Theories. I. Abstract Parabolic Systems. Encyclopedia of Mathematics and Its Applications, vol. 74. Cambridge University Press, Cambridge.

Lasiecka, I., Triggiani, R., 2000b. Control Theory for Partial Differential Equations: Continuous and Approximation Theories. II. Abstract Hyperbolic-Like Systems over a Finite Time Horizon. Encyclopedia of Mathematics and Its Applications, vol. 75. Cambridge University Press, Cambridge.

Li, X.J., Yong, J., 1995. Optimal Control Theory for Infinite-Dimensional Systems. Systems & Control: Foundations & Applications. Birkhäuser Boston, Inc., Boston, MA.

Lions, J.-L., 1968. Contrôle optimal de systèmes gouvernés par des équations aux dérivées partielles. Avant propos de P. Lelong. Dunod/Gauthier-Villars, Paris/Paris.

Lions, J.-L., 1983. Contrôle des systèmes distribués singuliers. Méthodes Mathématiques de l'Informatique (Mathematical Methods of Information Science), vol. 13. Gauthier-Villars, Montrouge.

Maurer, H., Zowe, J., 1979. First and second order necessary and sufficient optimality conditions for infinite-dimensional programming problems. Math. Program. 16, 98–110.
Meidner, D., Vexler, B., 2008a. A priori error estimates for space-time finite element discretization of parabolic optimal control problems. I. Problems without control constraints. SIAM J. Control Optim. 47 (3), 1150–1177.
Meidner, D., Vexler, B., 2008b. A priori error estimates for space-time finite element discretization of parabolic optimal control problems. II. Problems with control constraints. SIAM J. Control Optim. 47 (3), 1301–1329.
Meyer, C., Rösch, A., 2004. Superconvergence properties of optimal control problems. SIAM J. Control Optim. 43 (3), 970–985.
Mordukhovich, B., Raymond, J.-P., 2004. Neumann boundary control of hyperbolic equations with pointwise state constraints. SIAM J. Control Optim. 43.
Nečas, J., 1967. Les Méthodes Directes en Théorie des Equations Elliptiques. Editeurs Academia.
Neittaanmäki, P., Sprekels, J., Tiba, D., 2006. Optimization of Elliptic Systems. Theory and Applications. Springer Monographs in Mathematics. Springer, New York.
Neittaanmäki, P., Tiba, D., 1994. Optimal Control of Nonlinear Parabolic Systems. Theory, Algorithms, and Applications. Monographs and Textbooks in Pure and Applied Mathematics, vol. 179. Marcel Dekker, Inc., New York.
Pieper, K., Vexler, B., 2013. A priori error analysis for discretization of sparse elliptic optimal control problems in measure space. SIAM J. Control Optim. 51 (4), 2788–2808.
Pighin, D., 2022. Nonuniqueness of minimizers for semilinear optimal control problems. J. Eur. Math. Soc. In press.
Ponce, A.C., Wilmet, N., 2018. Optimal control of nonlinear elliptic problems with sparsity. SIAM J. Control Optim. 56 (4), 2513–2535.
Pontryagin, L.S., Boltyanskii, V.G., Gamkrelidze, R.V., Mishchenko, E.F., 1962. The Mathematical Theory of Optimal Processes. Translated from the Russian by K.N. Trirogoff: edited by L.W. Neustadt. Interscience Publishers John Wiley & Sons, Inc., New York-London.
Porretta, A., Zuazua, E., 2016. Remarks on long time versus steady state optimal control. In: Mathematical Paradigms of Climate Science. In: Springer INdAM Ser., vol. 15. Springer, Cham, pp. 67–89.
Qui, N.T., Wachsmuth, D., 2018. Stability for bang-bang control problems of partial differential equations. Optimization 67 (12), 2157–2177.
Rannacher, R., 1976. Zur L^{∞}-Konvergenz linearer finiter Elemente beim Dirichlet-Problem. Math. Z. 149 (1), 69–77.
Raviart, P.A., Thomas, J.M., 1983. Introduction à L'analyse Numérique des Equations aux Dérivées Partielles. Masson, Paris.
Raymond, J.-P., Zidani, H., 1998. Pontryagin's principle for state-constrained control problems governed by parabolic equations with unbounded controls. SIAM J. Control Optim. 36, 1853–1879.
Rösch, A., Siebert, K.G., Steinig, S., 2017. Reliable a posteriori error estimation for state-constrained optimal control. Comput. Optim. Appl. 68 (1), 121–162.
Rösch, A., Tröltzsch, F., 2007. On regularity of solutions and Lagrange multipliers of optimal control problems for semilinear equations with mixed pointwise control-state constraints. SIAM J. Control Optim. 46 (3), 1098–1115.
Schatz, A.H., 1998. Pointwise error estimates and asymptotic error expansion inequalities for the finite element method on irregular grids: Part I. Global estimates. Math. Comput. 67 (223), 877–899.
Schatz, A.H., Wahlbin, L.B., 1982. On the quasi-optimality in L_{∞} of the $\overset{o}{H}{}^{1}$-projection into finite element spaces. Math. Comput. 38 (157), 1–22.
Stadler, G., 2009. Elliptic optimal control problems with L^1-control cost and applications for the placement of control devices. Comput. Optim. Appl. 44 (2), 159–181.
Stampacchia, G., 1965. Le problème de Dirichlet pour les équations elliptiques du second ordre à coefficients discontinus. Ann. Inst. Fourier (Grenoble) 15, 189–258.

Tiba, D., 1990. Optimal Control of Nonsmooth Distributed Parameter Systems. Lecture Notes in Mathematics, vol. 1459. Springer-Verlag, Berlin.

Tröltzsch, F., 2010. Optimal Control of Partial Differential Equations. Theory, Methods and Applications, Vol. 112. American Math. Society, Providence.

Ulbrich, M., 2011. Semismooth Newton Methods for Variational Inequalities and Constrained Optimization Problems in Function Spaces. MOS-SIAM Series on Optimization, vol. 11. Industrial and Applied Mathematics (SIAM)/Mathematical Optimization Society, Philadelphia, PA/Philadelphia, PA.

Ulbrich, S., 2002. A sensitivity and adjoint calculus for discontinuous solutions of hyperbolic conservation laws with source terms. SIAM J. Control Optim. 41 (3), 740–797.

Vexler, B., Wollner, W., 2008. Adaptive finite elements for elliptic optimization problems with control constraints. SIAM J. Control Optim. 47 (1), 509–534.

Wachsmuth, D., Wachsmuth, G., 2011. Convergence and regularisation results for optimal control problems with sparsity functional. ESAIM Control Optim. Calc. Var. 17 (3), 858–886.

Wahlbin, L. Exact constants in certain L_p estimates. Unpublished.

Wang, G., Wang, L., Xu, Y., Zhang, Y., 2018. Time Optimal Control of Evolution Equations. Progress in Nonlinear Differential Equations and Their Applications, vol. 92. Birkhäuser/Springer, Cham. Subseries in Control.

Wolfersdorf, L.v., 1976. Optimal control for processes governed by mildly nonlinear differential equations of parabolic type I. Z. Angew. Math. Mech. 56, 531–538.

Wolfersdorf, L.v., 1977. Optimal control for processes governed by mildly nonlinear differential equations of parabolic type II. Z. Angew. Math. Mech. 57, 11–17.

Chapter 5

Numerical solution of multi-objective optimal control and hierarchic controllability problems

Enrique Fernández-Cara
EDAN and IMUS, Universidad de Sevilla, Sevilla, Spain
e-mail address: cara@us.es

Contents

Abstract
We present several control problems involving multiple criteria. There, the state is governed by a partial differential equation, with special attention paid to the heat and wave equations. In the first part of the paper, we deal with multi-objective optimal control, where the goal is to find and characterize controls and associated states that "minimize" simultaneously several cost functionals (or, if this is impossible, furnish control-state pairs at equilibrium). In the second part, we consider hierarchical controllability problems. Here, the interest is to drive the state to a desired target following at the same time

Handbook of Numerical Analysis, Volume 23, ISSN 1570-8659, https://doi.org/10.1016/bs.hna.2021.12.005

appropriate strategies that, in some sense, minimize additional costs. In both parts, we describe related numerical methods and we illustrate the results with some experiments.

Keywords
Control of partial differential equations, Bi-objective optimal control, Nash and Pareto equilibria, Exact controllability, Hierarchic control, Stackelberg-Nash strategy, Numerical approximation, Space-time discretization

MSC Codes
35K58, 93A13, 93B05, 90C29

1 Introduction

In this review paper, we present several control problems, governed by partial differential equations, that involve multiple criteria. In most cases, the state will be a solution to a heat or wave equation.

We will first deal with *multi-objective* optimal control problems. This corresponds to a situation where we are interested in the simultaneous optimization of several criteria. Accordingly, the control(s) must be chosen in such a way that the associated states "minimize" simultaneously several cost functionals or, at least, provide appropriate equilibria. In fact, as explained below, one of the key points in this context is the precise definition of equilibrium.

Then, we will be concerned with hierarchical controllability. Here, the interest is to drive or steer the state to a desired target respecting appropriate strategies that minimize one or several cost functionals. Thus, this can be viewed as a kind of constrained controllability problem.

In both cases, our analysis will be illustrated with related numerical methods and the results of experiments. We will introduce to this purpose appropriate numerical approximations, continuous and piecewise polynomial in time and space, that solve finite dimensional versions of the considered control problems, see Section 2.2, 2.6, 3.2, and 3.3.

These problems are very important from the theoretical and numerical viewpoints and appear frequently in the applications. Thus, multicriteria optimal control problems for partial differential equations have been considered in several papers since the 90's. In particular, the existence and computation of Nash equilibria have been the subject of Ramos et al. (2002a,b); Ramos and Roubicek (2007). The techniques have been applied in many areas, such as, for instance, mathematical modeling and optimal management of urban road networks (Vázquez-Méndez et al., 2019) and food treatment process optimization (Ferrández et al., 2020). On the other hand, the analysis of hierarchic control problems has been first performed in Díaz (2002); Díaz and Lions (2004); Lions (1986, 1994) and, more recently, in Araruna et al. (2017, 2015); Guillén and Rojas-Medar (2013). Applications can be found in finance (Araruna et al., 2018a) and population dynamics (Mophou et al., 2020), among other fields.

These considerations indicate that the numerical analysis of the problems mentioned above is at the same time relevant and nontrivial and deserves attention.

For a better understanding of what follows, let us briefly present a biobjective optimal control problem and a hierarchic controllability problem in the context of the modeling of an epidemic. Thus, let us assume that, in a population of N individuals, a contagious disease is declared, starting at time $t = 0$. We assume that the disease propagation during the time interval $(0, T)$ is modeled by an improved version of the well known SIR model introduced by W.O. Kermack and A.G. McKendrick, where the effects of vaccination and quarantine are incorporated:

$$\begin{cases} Q_t = p(t)S - \lambda Q \\ S_t = -\beta \dfrac{I}{N} S - (p(t) + \rho(t))S + \lambda Q \\ I_t = \beta \dfrac{I}{N} S - \gamma I \\ R_t = \gamma I + \rho(t) S \end{cases} \tag{1}$$

Here, the variables, constants, and coefficients can be interpreted as follows:

- Q, S, I, and R respectively denote the number of confined, susceptible, infectious, and recovered individuals. Of course, $S + E + I + R \equiv N$.
- The positive constants p, β, and γ are respectively the percentage of quarantined susceptible individuals and the infection and recovering rates (for instance, $1/\gamma$ is the average time of recovering).
- $\lambda = \lambda(t)$ and $\rho = \rho(t)$ are respectively the instantaneous quarantine rate ($1/\lambda(t)$ is the average time of confinement) and the number of individuals vaccinated at time unit.

For more information, see for instance (Behncke, 2000; Grigorieva and Khailov, 2014).

Let us fix initial values of Q, S, I, and R

$$Q(0) = Q_0, \quad S(0) = S_0, \quad I(0) = I_0, \quad R(0) = R_0. \tag{2}$$

In (1)–(2), we must view λ and ρ as controls and (Q, S, I, R) as associated states. In a realistic situation, λ and ρ must be subject to appropriate constraints; moreover, in a simplified model, it can be assumed that a fixed percentage of the recovered population dies.

Several different goals make full sense: to minimize the number of infectious and/or recovered individuals, to drive (Q, S, I, R) to a desired final state at time T, etc. Thus, an interesting multi-criteria optimal control problem is to find a *Pareto equilibrium* (λ, ρ) of the functionals J_1 and J_2, with

$$J_1(\lambda, \rho) := \int_0^T I(t)\,dt + \int_0^T \rho(t)\,dt$$

and

$$J_2(\lambda, \rho) := |S(T) - S_T| \quad \text{(where } S_T \in (0, N) \text{ is given).}$$

Recall that, in this context, a Pareto equilibrium is a couple (λ, ρ) such that there is no other (λ', ρ') satisfying

$$J_1(\lambda', \rho') \leq J_1(\lambda, \rho) \ \text{ and } \ J_2(\lambda', \rho') \leq J_2(\lambda, \rho),$$

with at least one strict inequality. On the other hand, a rather natural hierarchic control problem is to first associate to each λ a control $\rho[\lambda]$ satisfying

$$J_1(\lambda, \rho[\lambda]) \leq J_1(\lambda, \rho') \ \text{ for all admissible } \rho'$$

and then find λ such that the state associated to $(\lambda, \rho[\lambda])$ satisfies $S(T) = S_T$.

The plan of the paper is the following. Section 2 is devoted to analyze bi-objective optimal control problems for systems governed by linear and semi-linear heat and wave equations. We prove the existence of Nash and Pareto equilibria and give appropriate characterizations (optimality systems) in terms of the controls, their associated states and the corresponding adjoint states. Also, we present the results of some numerical experiments. In Section 3, we consider hierarchic control problems. As already mentioned, we find here controls of two kinds, in charge of two different tasks: some of them are assumed to solve a mono-criterium or multi-criteria optimal control problem; another distinguished control tries lo lead the system to a desired state. We prove existence and characterization results and we illustrate the situation with numerical experiments.

2 Bi-objective control problems for heat and wave equations

As already mentioned, we will consider in this section optimal control problems where the choice of the control(s) is subject to more than one criteria.

In these problems, in contrast with the single-objective case, we can give several different definitions of "good" or "optimal" controls, depending on the objectives in mind. They lead to what we call *equilibria,* that can be cooperative (when the controls collaborate to achieve the goals) and noncooperative (in the opposite case).

Specifically, for an extremal problem with p cost functionals J_i to minimize, a Nash strategy reduces to the search of a set of p players or controls v_i, each of them optimizing J_i with respect to the i-th variable. If the J_i are regular enough and no constraint is imposed, the v_i can be characterized as follows:

$$D_i J_i(v_1, \ldots, v_p) = 0, \quad i = 1, \ldots, p.$$

On the other hand, a Pareto equilibrium is by definition a p-tuple $(v_1, \ldots, v_p)$ such that there is no other $(v_1', \ldots, v_p')$ such that $J_i(v_1', \ldots, v_p') \leq J_i(v_1, \ldots, v_p)$

for all i, with at least one strict inequality. For regular functions J_i, if no constraint is required, a Pareto equilibrium must satisfy

$$\sum_{i=1}^{p} \lambda_i D_i J_i(v_1, \ldots, v_p) = 0$$

for some $\lambda_i \geq 0$ with $\sum_{i=1}^{p} \lambda_i = 1$; although this is just a necessary optimality condition, not sufficient in general.

The characteristics and roles of Nash and Pareto equilibria are illustrated in Fig. 1 and 2 in an elementary case, with $p = 2$, $J_1(v_1, v_2) \equiv v_1^2 + v_2^2$ and $J_2(v_1, v_2) \equiv \frac{4}{9}(v_1 - 2)^2 + (v_2 - 2)^2$. We see there that they can be completely different. It is also found that Pareto equilibria provide a set of *optimal* nonscalar values, the so called *Pareto front,* that can be of great interest from the viewpoint of applications. In fact, it is common to find situations where there are "few" Nash equilibria and "many" Pareto equilibria, that can be written as the points of a family depending on a real parameter; for details, see for instance Ehrgott (2005).

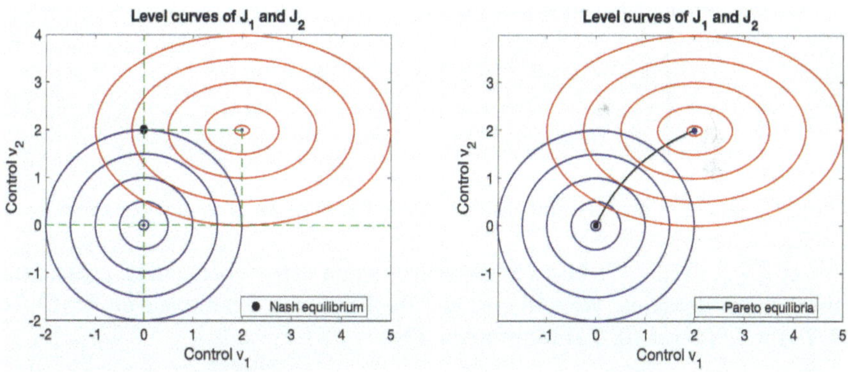

(a) The Nash equilibrium (b) The Pareto equilibria

FIGURE 1 The level curves of two different quadratic functions $J_i = J_i(v_1, v_2)$ in the plane and the associated Nash and Pareto equilibria.

In the sequel, we will present efficient strategies for the numerical solution of multi-objective control problems of these kinds (for simplicity with $p = 2$), governed by linear and semilinear partial differential equations.

2.1 Equilibria for linear heat equations

Let Ω be a bounded domain of $\mathbb{R}^d$ $(d \geq 1)$ with (at least) Lipschitz-continuous boundary and let $\omega \subset\subset \Omega$ be a non-empty open set. Let $T > 0$ be given and let us set $Q_T := \Omega \times (0, T)$, $q_T := \omega \times (0, T)$, and $\Sigma_T := \partial\Omega \times (0, T)$.

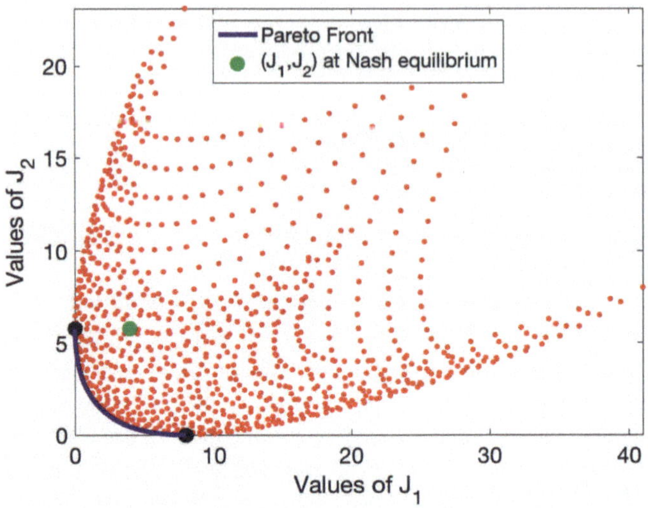

FIGURE 2 The Pareto front corresponding to the equilibria in Fig. 1(b).

We will consider the state system

$$\begin{cases} \partial_t y - \Delta y = f + v_1 1_{\mathcal{O}_1} + v_2 1_{\mathcal{O}_2} & \text{in } Q_T \\ y = 0 & \text{on } \Sigma_T, \\ y(x,0) = y_0(x) & \text{in } \Omega, \end{cases} \tag{3}$$

where $\mathcal{O}_1$ and $\mathcal{O}_2$ are disjoint non-empty subsets of Ω, $f \in L^2(Q_T)$ and $y_0 \in L^2(\Omega)$.

Let $\mathcal{O}_{1,d}, \mathcal{O}_{2,d} \subset \Omega$ be two non-empty open sets, representing prescribed observations domains. We will consider the following functionals on $L^2(\mathcal{O}_1 \times (0,T)) \times L^2(\mathcal{O}_2 \times (0,T))$ for problem (3):

$$J_i(v_1, v_2) := \frac{1}{2} \iint_{\mathcal{O}_{i,d} \times (0,T)} |y - y_{i,d}|^2 + \frac{\mu_i}{2} \iint_{\mathcal{O}_i \times (0,T)} |v_i|^2, \quad i = 1, 2, \tag{4}$$

where the μ_i are positive constants and the $y_{i,d} \in L^2(\mathcal{O}_{i,d} \times (0,T))$ are given functions.

For any (v_1, v_2), the real quantity $J_i(v_1, v_2)$ can be interpreted as the cost we have to pay to get a state y close to $y_{i,d}$ in $\mathcal{O}_{i,d} \times (0,T)$. The first and second term in (4) respectively account for the *success* (y is close to the target) and the *effort* in the process. Of course, μ_i indicates the relative relevance of these terms: a large μ_i is associated with the search of non-expensive controls, while a small μ_i corresponds to the search of controls that furnish states close to $y_{i,d}$. In the limit $\mu_i = 0$ and then we try to find controls that drive the state as close as possible to the target, no matter what we have to pay for.

The optimal control problems are the following:

1. NASH BI-OBJECTIVE PROBLEM: Find controls $v_i \in L^2(\mathcal{O}_i \times (0,T))$ $(i = 1,2)$ satisfying

$$\begin{cases} J_1(v_1, v_2) \le J_1(v_1', v_2) & \forall v_1' \in L^2(\mathcal{O}_1 \times (0,T)), \\ J_2(v_1, v_2) \le J_2(v_1, v_2') & \forall v_2' \in L^2(\mathcal{O}_2 \times (0,T)). \end{cases} \tag{5}$$

Note that, since the state equation (3) is linear, the J_i are strictly convex and consequently (v_1, v_2) is a Nash equilibrium if and only if it solves the linear system

$$D_1 J_1(v_1, v_2) = 0, \quad D_2 J_2(v_1, v_2) = 0. \tag{6}$$

2. PARETO BI-OBJECTIVE PROBLEM: Find controls $v_i \in L^2(\mathcal{O}_i \times (0,T))$ $(i = 1,2)$, such that there is no other (v_1', v_2') with $v_i' \in L^2(\mathcal{O}_i \times (0,T))$ satisfying

$$\begin{cases} J_1(v_1', v_2') \le J_1(v_1, v_2) \text{ and } J_2(v_1', v_2') \le J_2(v_1, v_2), \\ \text{with strict inequality for at least one } J_i. \end{cases} \tag{7}$$

Taking again into account the fact that (3) is linear, it is not difficult to prove that (v_1, v_2) is a Pareto equilibrium if and only if there exists $\lambda \in [0,1]$ with

$$(\lambda D_1 J_1 + (1-\lambda) D_2 J_2)(v_1, v_2) = 0, \quad i = 1, 2. \tag{8}$$

Indeed, if the couple (v_1, v_2) is a Pareto equilibrium, it minimizes J_i in the set of couples (v_1', v_2') satisfying $J_{3-i}(v_1', v_2') \le E_{3-i} := J_{3-i}(v_1, v_2)$ for $i = 1, 2$. From the well known *Karun-Kuhn-Tucker's Theorem,* this implies (8) for some $\lambda \in [0,1]$. Conversely, from the continuity and strict convexity of the J_i, we find that any solution to (8) is the (unique) minimizer of $\lambda J_1 + (1-\lambda) J_2$ and is therefore a Pareto equilibrium.

2.2 Computation of Nash equilibria

With the help of standard techniques in optimal control, it is not difficult to check that (6) can be equivalently rewritten as a coupled system for (v_1, v_2), the associated state y and the new variables ϕ_1 and ϕ_2 (the associated adjoint states). It is the following:

$$\begin{cases} \partial_t y - \Delta y = f - v_1 1_{\mathcal{O}_1} - v_2 1_{\mathcal{O}_2} & \text{in } Q_T, \\ y = 0 & \text{on } \Sigma_T, \\ y(x,0) = y_0(x) & \text{in } \Omega, \end{cases} \tag{9}$$

$$\begin{cases} -\partial_t \phi_i - \Delta \phi_i = (y - y_{i,d}) 1_{\mathcal{O}_{i,d}} & \text{in } Q_T, \\ \phi_i = 0 & \text{on } \Sigma_T, \\ \phi_i(x,T) = 0 & \text{in } \Omega, \end{cases} \tag{10}$$

$$v_i = -\frac{1}{\mu_i}\phi_i\big|_{\mathcal{O}_i\times(0,T)}, \quad i = 1, 2 \tag{11}$$

(see for instance Lions, 1971; Fursikov, 2000).

This is the optimality system we have to solve in order to compute a Nash equilibrium for (3) with respect to the functionals (4).

A short explanation of the equivalence of (5), (6), and (9)–(11) is as follows. For example, one has

$$\langle D_1 J_1(v_1, v_2), w_1\rangle = \iint_{\mathcal{O}_{1,d}\times(0,T)} (y - y_{1,d})z + \mu_1 \iint_{\mathcal{O}_1\times(0,T)} v_1 w_1, \tag{12}$$

where we have denoted by z the derivative of y with respect to v_1 in the direction w_1, that is, the solution to

$$\begin{cases} \partial_t z - \Delta z = w_1 1_{\mathcal{O}_1} & \text{in } Q_T, \\ z = 0 & \text{on } \Sigma_T, \\ z(x,0) = 0 & \text{in } \Omega, \end{cases} \tag{13}$$

Consequently, if (v_1, v_2) is a Nash equilibrium and we introduce ϕ_1 as in (10) with $i = 1$, we see that

$$0 = \langle D_1 J_1(v_1, v_2), w_1\rangle = \iint_{\mathcal{O}_1\times(0,T)} (\phi_1 + \mu_1 v_1) w_1$$

for all $w_1 \in V_1$. These identities, together with a similar argument for the other partial derivative in (6), yield (11). Therefore, (6) implies (9)–(11). The converse is also true and can be proved with a similar argument.

Let us introduce the notation $V_i := L^2(\mathcal{O}_i \times (0, T))$ and $V := V_1 \times V_2$ and let $(\cdot, \cdot)_V$ stand for the usual scalar product in V. Let us also introduce the bilinear form $a : V \times V \mapsto \mathbb{R}$, where

$$a((v_1, v_2), (w_1, w_2)) := \sum_{i=1}^{2} \iint_{\mathcal{O}_i\times(0,T)} \left(v_i + \frac{1}{\mu_i}\xi_i\right) w_i,$$

the ξ_i solve the systems

$$\begin{cases} -\partial_t \xi_i - \Delta \xi_i = h 1_{\mathcal{O}_{i,d}} & \text{in } Q_T, \\ \xi_i = 0 & \text{on } \Sigma_T, \\ \xi_i(x,T) = 0 & \text{in } \Omega \end{cases}$$

and h solves

$$\begin{cases} \partial_t h - \Delta h = v_1 1_{\mathcal{O}_1} + v_2 1_{\mathcal{O}_2} & \text{in } Q_T, \\ h = 0 & \text{on } \Sigma_T, \\ h(x,0) = 0 & \text{in } \Omega. \end{cases}$$

Also, let the linear form $\ell : V \mapsto \mathbb{R}$ be given by

$$\ell(w) := -\sum_{i=1}^{2} \frac{1}{\mu_i} \iint_{\mathcal{O}_i \times (0,T)} \eta_i w_i,$$

where

$$\begin{cases} -\partial_t \eta_i - \Delta \eta_i = (\overline{y} - y_{i,d}) 1_{\mathcal{O}_{i,d}} & \text{in } Q_T, \\ \eta_i = 0 & \text{on } \Sigma_T, \\ \eta_i(x,T) = 0 & \text{in } \Omega \end{cases}$$

and

$$\begin{cases} \partial_t \overline{y} - \Delta \overline{y} = f & \text{in } Q_T, \\ \overline{y} = 0 & \text{on } \Sigma_T, \\ \overline{y}(x,0) = y_0(x) & \text{in } \Omega. \end{cases}$$

Then, (6) can be written in the form

$$a(v, v') = \ell(v') \ \forall v' \in V, \ \ v \in V. \tag{14}$$

It is clear that $a(\cdot\,,\cdot)$ is continuous and symmetric and $\ell : V \mapsto \mathbb{R}$ is continuous. Furthermore, if μ_1 and μ_2 are sufficiently large, $a(\cdot\,,\cdot)$ is V-elliptic. Indeed, one has

$$\begin{aligned} a((v_1, v_2), (v_1, v_2)) &:= \sum_{i=1}^{2} \iint_{\mathcal{O}_i \times (0,T)} \left(v_i + \frac{1}{\mu_i} \xi_i \right) v_i \\ &\geq \sum_{i=1}^{2} \left(1 - \frac{C_0}{\mu_i} \right) \iint_{\mathcal{O}_i \times (0,T)} |v_i|^2, \end{aligned}$$

where C_0 only depends on Ω, the $\mathcal{O}_i$, the $\mathcal{O}_{i,d}$ and T. Therefore, as a consequence of the *Lax-Milgram Theorem,* (6) possesses a unique solution in this case.

In other words, we have proved the following:

Theorem 1. *There exists $\mu_N(\Omega, \mathcal{O}_i, \mathcal{O}_{i,d}, T) > 0$ such that, if $\mu_1, \mu_2 \geq \mu_N$, there exists exactly one Nash equilibrium for* (3), (4).

In the conditions of this theorem, the Nash equilibrium can be computed by the following algorithm of the fixed-point kind:

ALG 1

1 Choose $v^0 = (v_1^0, v_2^0) \in V$.

2 Then, for given $n \geq 0$ and $v^n = (v_1^n, v_2^n) \in V$, compute the solution y^n to (9) with controls $v_i = v_i^n$, then the solutions ϕ_i^n to the systems (10) with $y = y^n$ and finally the v_i^{n+1} from (11) with ϕ_i^n.

If μ_1 and μ_2 are sufficiently large, the iterates produced by **ALG 1** converge to the unique Nash equilibrium. Indeed, from the definitions of $a(\cdot, \cdot)$ and ℓ, it is easy to see that, if the μ_i are sufficiently large, **ALG 1** is the usual *method of successive approximations* for a contraction on V.

The reduction of (9)–(11) to finite dimension can be performed in two steps:

- STEP 1: APPROXIMATION IN TIME. We consider the time discretization step τ, defined by $\tau = T/M$, where M is a (large) positive integer. Then, if we set $t^m := m\tau$ for all m, we have:

$$0 = t^0 < t^1 < t^2 < \cdots < t^M = T.$$

We approximate V_1 and V_2 respectively by

$$V_1^\tau := \big(L^2(\mathcal{O}_1)\big)^M \quad \text{and} \quad V_2^\tau := \big(L^2(\mathcal{O}_2)\big)^M.$$

Thus, we can interpret the elements of V_i^τ as continuous and piecewise constant in time controls in V_i.

- STEP 2: APPROXIMATION IN SPACE. From now on, we will assume that Ω, the $\mathcal{O}_{i,d}$ and the $\mathcal{O}_i$ are polygonal. We introduce a triangulation $\mathcal{T}_h$ of Ω, where the $\mathcal{O}_{i,d}$ and $\mathcal{O}_i$ are unions of triangles. Next, we approximate the space $L^2\big(0, T; H^1(\Omega)\big)$ by W_h^τ, where

$$W_h^\tau := (W_h)^M, \quad W_h := \big\{z \in C^0(\bar{\Omega}) : z\big|_K \in \mathbb{P}_1(K) \ \forall K \in \mathcal{T}_h\big\}$$

and $\mathbb{P}_1(K)$ is the space of the polynomial functions on K of degree ≤ 1; thus, $\dim(\mathbb{P}_1(K)) = 3$ and $\dim(W_h) = N_h$, where N_h is the number of vertices of $\mathcal{T}_h$. In this second step, we first approximate V_i by the finite dimensional space $V_{i,h}^\tau$, defined as follows:

$$V_{i,h}^\tau := (V_{i,h})^M, \quad V_{i,h} := \big\{z \in C^0(\bar{\mathcal{O}}_i) : z\big|_K \in \mathbb{P}_1(K)) \ \forall K \in \mathcal{O}_i\big\}.$$

Then, we set $V_h^\tau := V_{1,h}^\tau \times V_{2,h}^\tau$. Finally, we consider the space

$$\big\{z \in L^2(0, T; H^1(\Omega)) : z(\cdot, t)\big|_{\Gamma_1} = 0 \text{ a.e. in } (0, T)\big\}$$

and its finite dimensional version $W_{h,0}^\tau$, where

$$W_{h,0}^\tau = (W_{h,0})^M, \quad W_{h,0} = \big\{z \in W_h : z\big|_{\Gamma_1} = 0\big\}.$$

The state equation (3) and the adjoint systems (10) can be approximated in time and space by incorporating (for instance) implicit Euler finite difference in time and finite element techniques in space; for details, see for instance Quarteroni and Valli (1994). This allows to compute a state and two adjoint states for each control pair $v = (v_1, v_2) \in V_h^\tau$. Accordingly, we can approximate (6) and it is meaningful to consider a fixed-point algorithm for the computation of a numerical Nash equilibrium.

Using standard results, it is not difficult to check that the fixed point of the numerical approximation of (9)–(11) converges, as $h, \tau \to 0$, to a control-state triplet (v_1, v_2, y) where (v_1, v_2) is a Nash equilibrium and y is the associated state.

2.3 Computation of Pareto equilibria

Now, we will solve (8) for any fixed $\lambda \in (0, 1)$. This will provide a whole family of equilibria where, eventually, we can perform a further choice. It is natural to exclude the values $\lambda = 1$ and $\lambda = 0$, since they respectively correspond to the minimizers of J_1 and J_2.

Again, applying standard arguments, we see that (8) is equivalent to the optimality system

$$\begin{cases} \partial_t y - \Delta y = f + v_1 1_{\mathcal{O}_1} + v_2 1_{\mathcal{O}_2} & \text{in } Q_T, \\ y = 0 & \text{on } \Sigma_T, \\ y(x,0) = y_0(x) & \text{in } \Omega, \end{cases} \tag{15}$$

$$\begin{cases} -\partial_t \phi_i - \Delta \phi_i = (y - y_{i,d}) 1_{\mathcal{O}_{i,d}} & \text{in } Q_T, \\ \phi_i = 0 & \text{on } \Sigma_T, \\ \phi_i(x,T) = 0 & \text{in } \Omega, \end{cases} \tag{16}$$

$$\begin{aligned} v_1 &= -\frac{1}{\mu_1}\Big(\phi_1 + \frac{1-\lambda}{\lambda}\phi_2\Big)\Big|_{\mathcal{O}_1 \times (0,T)} \quad \text{and} \\ v_2 &= -\frac{1}{\mu_2}\Big(\frac{\lambda}{1-\lambda}\phi_1 + \phi_2\Big)\Big|_{\mathcal{O}_2 \times (0,T)}. \end{aligned} \tag{17}$$

Let us fix $\lambda \in (0, 1)$. As in the case of Nash equilibria, we can rewrite (15)–(17) as a Lax-Milgram problem. For brevity, we leave the details to the reader.

If μ_1 and μ_2 are sufficiently large, we deduce again existence and uniqueness:

Theorem 2. *Let $\lambda \in (0, 1)$ be given. There exists $\mu_P(\Omega, \mathcal{O}_i, \mathcal{O}_{i,d}, T, \lambda) > 0$ such that, if $\mu_1, \mu_2 \geq \mu_P$, there exists exactly one Pareto equilibrium for* (3), (8).

In order to compute this solution, we can argue as before and use the following fixed-point algorithm:

ALG 2

1 Choose $v^0 = (v_1^0, v_2^0) \in V$.

2 Then, for given $n \geq 0$ and $v^n = (v_1^n, v_2^n) \in V$, compute the solution y^n to (15) with $v_i = v_i^n$, then the solutions ϕ_i^n to the adjoint systems (16) with $y = y^n$ and, finally, compute the v_i^{n+1} using (17) with $\phi_i = \phi_i^n$.

Again, the iterates converge provided μ_1 and μ_2 are large enough.

2.4 Nash and Pareto bi-objective optimal control for semilinear heat equations

We will consider in this section the system

$$\begin{cases} \partial_t y - \Delta y + F(y) = f + v_1 1_{\mathcal{O}_1} + v_2 1_{\mathcal{O}_2} & \text{in } Q_T, \\ y = 0 & \text{on } \Sigma_T, \\ y(x,0) = y_0(x) & \text{in } \Omega, \end{cases} \tag{18}$$

where $F : \mathbb{R} \mapsto \mathbb{R}$ is C^1 and F' is uniformly bounded.

Obviously, (6) is now a necessary condition (not sufficient in general) for (v_1, v_2) to be a Nash equilibrium for (18), (4). However, in practice it is many times sufficient for our purposes to solve this optimality system. The proof that (5) implies (6) is immediate. We only have to take into account that the functions $(v_1, v_2) \mapsto J_i(v_1, v_2)$ are C^1 in V.

On the other hand, as in the linear case, we can rewrite (6) equivalently as a coupled optimality system. This time, the following is found:

$$\begin{cases} \partial_t y - \Delta y + F(y) = f + v_1 1_{\mathcal{O}_1} + v_2 1_{\mathcal{O}_2} & \text{in } Q_T, \\ y = 0 & \text{on } \Sigma_T, \\ y(x,0) = y_0(x) & \text{in } \Omega, \end{cases} \tag{19}$$

$$\begin{cases} -\partial_t \phi_i - \Delta \phi_i + F'(y)\phi_i = (y - y_{i,d}) 1_{\mathcal{O}_{i,d}} & \text{in } Q_T, \\ \phi_i = 0 & \text{on } \Sigma_T, \\ \phi_i(x,T) = 0 & \text{in } \Omega, \end{cases} \tag{20}$$

$$v_i = -\frac{1}{\mu_i} \phi_i \Big|_{\mathcal{O}_i \times (0,T)}, \quad i = 1, 2. \tag{21}$$

Indeed, one has (12) where z is now the solution to

$$\begin{cases} \partial_t z - \Delta z + F'(y) z = w_1 1_{\mathcal{O}_1} & \text{in } Q_T, \\ z = 0 & \text{on } \Sigma_T, \\ z(x,0) = 0 & \text{in } \Omega \end{cases} \tag{22}$$

and a similar assertion can be established for $D_2 J_2$. Therefore, if (v_1, v_2) is a Nash equilibrium and the ϕ_i satisfy (20), we necessarily have (21).

Once more, let us assume that μ_1 and μ_2 are large enough. Then, (19)–(21) is equivalent to a fixed-point equation in V that possesses at least one solution. More precisely, it suffices to introduce the mapping $\mathcal{A} : V \mapsto V$, with

$$\mathcal{A}(v_1, v_2) = -(\mu_1^{-1} \phi_1|_{\mathcal{O}_1 \times (0,T)}, \mu_2^{-1} \phi_2|_{\mathcal{O}_2 \times (0,T)}) \quad \forall (v_1, v_2) \in V$$

and put (21) in the form $(v_1, v_2) = \mathcal{A}(v_1, v_2)$. Since $\mathcal{A}$ is continuous and compact and (if the μ_i are large) maps a ball in V into itself, we deduce from *Schauder's Theorem* that a fixed-point exists.

This argument leads to the following:

Theorem 3. *There exists* $\mu_N(\Omega, \mathcal{O}_i, \mathcal{O}_{i,d}, T, y_0, f, F) > 0$ *such that, if* $\mu_1, \mu_2 \geq \mu_N$, *there exists at least one Nash equilibrium for* (18), (4).

In the sequel, we will present several iterative algorithms for the solution of (19)–(21). We will introduce $G : \mathbb{R} \mapsto \mathbb{R}$, with

$$G(s) := \begin{cases} \dfrac{1}{s}(F(s) - F(0)) & \text{if } s \neq 0, \\ F'(0) & \text{otherwise} \end{cases}$$

and we will use that $F(s) \equiv G(s)s + F(0)$. Consequently, for any $(v_1, v_2) \in V$, the differential equation in (18) can be rewritten in the form

$$\partial_t y - \Delta y + G(y)y = f - F(0) + v_1 1_{\mathcal{O}_1} + v_2 1_{\mathcal{O}_2}.$$

The first two algorithms are of the fixed-point kind:

ALG 3

1 Choose $v^0 = (v_1^0, v_2^0) \in V$. Set $y^{-1}(x,t) \equiv y_0(x)$.
2 Then, for given $n \geq 0$, y^{n-1} and $v^n = (v_1^n, v_2^n) \in V$, do the following:
2.1 Set $y^{n,0} = y^{n-1}$.
2.2 Then, for given $\ell \geq 0$ and $y^{n,\ell}$, compute the solution $y^{n,\ell+1}$ to the system

$$\begin{cases} \partial_t y - \Delta y + G(y^{n,\ell})y = f - F(0) + v_1^n 1_{\mathcal{O}_1} + v_2^n 1_{\mathcal{O}_2} & \text{in } Q_T, \\ y = 0 & \text{on } \Sigma_T, \\ y(x,0) = y_0(x) & \text{in } \Omega. \end{cases} \tag{23}$$

After convergence, denote by y^n the last computed $y^{n,\ell+1}$.
3 Compute the adjoint states ϕ_i^n by solving the systems (20) with $y = y^n$.
4 Finally, set

$$v_i^{n+1} = -\frac{1}{\mu_i}\phi_i^n\Big|_{\mathcal{O}_i \times (0,T)}, \quad i = 1, 2. \tag{24}$$

Our second algorithm is a simplification of **ALG 3** where we simultaneously iterate to solve the semilinear equation in (19) and upgrade the controls:

ALG 4

1 Choose $v^0 = (v_1^0, v_2^0) \in V$ and set $y^{-1}(x,t) \equiv y_0(x)$.
2 Then, for given $n \geq 0$, y^{n-1} and $v^n = (v_1^n, v_2^n) \in V$, compute y^n by solving

$$\begin{cases} \partial_t y - \Delta y + G(y^{n-1})y = f - F(0) + v_1^n 1_{\mathcal{O}_1} + v_2^n 1_{\mathcal{O}_2} & \text{in } Q_T, \\ y = 0 & \text{on } \Sigma_T, \\ y(x,0) = y_0(x) & \text{in } \Omega, \end{cases} \tag{25}$$

compute the adjoint states ϕ_i^n by solving the systems (20) with $y = y^n$ and finally compute the v_i^{n+1} from (24).

The analysis of the convergence of **ALG 3** needs some work. It is expected that, at least when $F \in C_b^2(\mathbb{R})$ (that is, F is of class C^2 and its first and second derivatives are uniformly bounded) and μ_1 and μ_2 are sufficiently large, the iterates converge in $L^2(\mathcal{O}_1 \times (0,T)) \times L^2(\mathcal{O}_2 \times (0,T))$ to a solution to (19)–(21). However, up to our knowledge, this is unknown.

For our third algorithm in this section, we also assume that $F \in C_b^2(\mathbb{R})$. We solve (19)–(21) by Newton's method:

ALG 5

1 Choose $v^0 = (v_1^0, v_2^0) \in V$, set $y^{-1}(x,t) \equiv y_0(x)$ and compute y^0 and ϕ_1^0 and ϕ_2^0 (for example) by solving

$$\begin{cases} \partial_t y^0 - \Delta y^0 + F'(0)y^0 = f - F(0) + v_1^0 1_{\mathcal{O}_1} + v_2^0 1_{\mathcal{O}_2}, & \text{in } Q_T, \\ -\partial_t \phi_i^0 - \Delta \phi_i^0 + F'(0)\phi_i^0 = (y^0 - y_{i,d})1_{\mathcal{O}_{i,d}}, & \text{in } Q_T, \\ y^0 = 0, \quad \phi_i^0 = 0, & \text{on } \Sigma_T, \\ y^0(x,0) = y_0(x), \quad \phi_i^0(x,T) = 0, & \text{in } \Omega, \end{cases} \tag{26}$$

2 Then, for given $n \geq 0$, y^n, ϕ_i^n, and v_i^n, compute z^n, ψ_i^n, and w_i^n, with

$$\begin{cases} \partial_t z^n - \Delta z^n + F'(y^n)z^n = w_1^n 1_{\mathcal{O}_1} + w_2^n 1_{\mathcal{O}_2} + H^n & \text{in } Q_T, \\ -\partial_t \psi_i^n - \Delta \psi_i^n + F'(y^n)\psi_i^n = -F''(y^n)z^n \phi_i^n + z^n 1_{\mathcal{O}_{i,d}} + K^n & \text{in } Q_T, \\ z^n = 0, \quad \psi_i^n = 0 & \text{on } \Sigma_T, \\ z^n(x,0) = 0, \quad \psi_i^n(x,T) = 0 & \text{in } \Omega, \end{cases} \tag{27}$$

where $H^n := f - (\partial_t y^n - \Delta y^n + F(y^n) - v_1^n 1_{\mathcal{O}_1} - v_1^n 1_{\mathcal{O}_2})$, $K^n := (y^n - y_{i,d})1_{\mathcal{O}_{i,d}} - (\partial_t \phi_i^n - \Delta \phi_i^n + F'(y^n)\phi_i^n)$ and

$$w_i^n = -\frac{1}{\mu_i} \psi_i^n \Big|_{\mathcal{O}_i \times (0,T)}, \quad i = 1, 2 \tag{28}$$

and set $y^{n+1} = y^n + z^n$, $\phi_i^{n+1} = \phi_i^n + \psi_i^n$, and $v_i^{n+1} = v_i^n + w_i^n$ for $i = 1, 2$.

If y_0 is sufficiently small and μ_1 and μ_2 are sufficiently large, it is possible to establish a convergence result for **ALG 5**; to this respect, see for instance Argyros (2008).

In a similar way, we can try to solve optimality systems for Pareto equilibria.

If (v_1, v_2) is a Pareto equilibrium, there exists $\lambda \in [0,1]$ such that (8) holds. This is (again) a consequence of the regularity of the J_i and Karun-Kuhn-Tucker's Theorem. On the other hand, it is also true that if, for some $\lambda \in (0,1)$, the couple (v_1, v_2) minimizes $\lambda J_1 + (1-\lambda)J_2$, then (v_1, v_2) is a Pareto equilibrium. Accordingly, it seems reasonable to try to find a solution to (8) for any fixed $\lambda \in (0,1)$.

As before, we can get an equivalent formulation in terms of appropriate adjoint states:

$$
\begin{cases}
\partial_t y - \Delta y + F(y) = f + v_1 1_{\mathcal{O}_1} + v_2 1_{\mathcal{O}_2} & \text{in } Q_T, \\
y = 0 & \text{on } \Sigma_T, \\
y(x,0) = y_0(x) & \text{in } \Omega,
\end{cases}
\tag{29}
$$

$$
\begin{cases}
-\partial_t \phi_i - \Delta \phi_i + F'(y)\phi_i = [y - y_{i,d}] 1_{\mathcal{O}_{i,d}} & \text{in } Q_T, \\
\phi_i = 0 & \text{on } \Sigma_T, \\
\phi_i(x,T) = 0 & \text{in } \Omega,
\end{cases}
\tag{30}
$$

$$
\begin{aligned}
v_1 &= -\frac{1}{\mu_1}\Big(\phi_1 + \frac{1-\lambda}{\lambda}\phi_2\Big)\Big|_{\mathcal{O}_1\times(0,T)} \quad \text{and} \\
v_2 &= -\frac{1}{\mu_2}\Big(\frac{\lambda}{1-\lambda}\phi_1 + \phi_2\Big)\Big|_{\mathcal{O}_2\times(0,T)}.
\end{aligned}
\tag{31}
$$

That (8) can be written in the form (29)–(31) can be deduced arguing as before. Again, assuming that μ_1 and μ_2 are large enough, it is possible to rewrite (29)–(31) as a fixed-point equation in V that possesses at least one solution.

Of course, we can introduce fixed-point algorithms for the solution of (29)–(31) similar to **ALG 3** and **ALG 4** and a Newton algorithm similar to **ALG 5**. For brevity, we omit the details. Also, arguing as before, we can introduce approximations based on finite difference and finite element techniques and formulate finite dimensional versions of **ALG 2**, **ALG 3**, etc.

2.5 Equilibria for linear and semilinear wave equations

It makes sense to consider control problems like those above for systems governed by linear or semilinear wave equations completed with appropriate boundary and initial conditions, that is,

$$
\begin{cases}
\partial_{tt} y - \Delta y = f + v_1 1_{\mathcal{O}_1} + v_2 1_{\mathcal{O}_2} & \text{in } Q_T, \\
y = 0 & \text{on } \Sigma_T, \\
y(x,0) = y_0(x), \quad y_t(x,0) = y_1(x) & \text{in } \Omega
\end{cases}
\tag{32}
$$

or

$$
\begin{cases}
\partial_{tt} y - \Delta y + F(y) = f + v_1 1_{\mathcal{O}_1} + v_2 1_{\mathcal{O}_2} & \text{in } Q_T, \\
y = 0 & \text{on } \Sigma_T, \\
y(x,0) = y_0(x), \quad y_t(x,0) = y_1(x) & \text{in } \Omega.
\end{cases}
\tag{33}
$$

Again, we can prove existence and characterization results and we can formulate related iterative algorithms. For simplicity, we will only give the details of a gradient-like algorithm for the computation of a Nash equilibrium in the linear case, that is, for (32), (4).

Thus, let (v_1, v_2) be given in V. With standard techniques, it is straightforward to deduce that (v_1, v_2) is a Nash equilibrium for (32), (4) if and only if it satisfies, together with the associated state y and the adjoint states ϕ_1 and ϕ_2, the following:

$$\begin{cases} \partial_{tt} y - \Delta y = f - \mu_1^{-1}\phi_1 1_{\mathcal{O}_1} - \mu_2^{-1}\phi_2 1_{\mathcal{O}_2} & \text{in } Q_T, \\ y = 0 & \text{on } \Sigma_T, \\ y(x,0) = y_0(x), \quad y_t(x,0) = y_1(x) & \text{in } \Omega, \end{cases} \tag{34}$$

$$\begin{cases} \partial_{tt}\phi_i - \Delta\phi_i = (y - y_{i,d}) 1_{\mathcal{O}_{i,d}} & \text{in } Q_T, \\ \phi_i = 0 & \text{on } \Sigma_T, \\ \phi_i(x,T) = 0, \quad \phi_{i,t}(x,T) = 0 & \text{in } \Omega, \end{cases} \tag{35}$$

$$v_i = -\frac{1}{\mu_i}\phi_i\Big|_{\mathcal{O}_i \times (0,T)}, \quad i = 1, 2. \tag{36}$$

Note that $(D_1 J_1(v_1, v_2), D_2 J_2(v_1, v_2)) = (\mu_1 v_1 + \phi_1|_{\mathcal{O}_1 \times (0,T)}, \mu_2 v_2 + \phi_2|_{\mathcal{O}_2 \times (0,T)})$ for all $(v_1, v_2) \in V$, where each ϕ_i is the solution to the corresponding linear problem (35). Accordingly, the solution to (34)–(36) can be computed by a gradient-like method with "optimal" steps as follows:

ALG 6

1 Choose $v^0 = (v_1^0, v_2^0) \in V$.
2 For given $n \geq 0$ and $v^n = (v_1^n, v_2^n) \in V$, first compute the solution y^n to (34) with controls $v_i = v_i^n$ and then the solutions ϕ_i^n to the systems (35) with $y = y^n$.
3 Then, assuming that y^n, ϕ^n, v^n are known, compute the direction $r^n := (r_1^n, r_2^n)$, with

$$r^n = -(\mu_1 v_1^n + \phi_1^n|_{\mathcal{O}_1 \times (0,T)}, \mu_2 v_2^n + \phi_2^n|_{\mathcal{O}_2 \times (0,T)}).$$

4 Then, assuming that v^n and r^n are known, we compute the couple $\rho^n = (\rho_1^n, \rho_2^n)$ "minimizing" $J(v^n + \rho^n r^n)$, with $J = (J_1, J_2)$. This means the following:

$$\begin{cases} J_1(v_1^n + \rho_1^n r_1^n, v_2^n) \leq J_1(v_1^n + \rho_1 r_1^n, v_2^n) \ \forall \rho_1 \geq 0, \\ J_2(v_1^n, v_2^n + \rho_2^n r_2^n) \leq J_2(v_1^n, v_2^n + \rho_2 r_2^n) \ \forall \rho_2 \geq 0. \end{cases} \tag{37}$$

5 Finally, set $v_i^{n+1} = v_i^n + \rho_i^n r_i^n$ and $v^{n+1} = (v_1^{n+1}, v_2^{n+1})$.

The extremal problems in (37) can be solved very easily, through explicit formulæ for ρ_1^n and ρ_2^n. Indeed, note that for instance $\rho_1 \mapsto J_1(v_1^n + \rho_1 r_1^n, v_2^n)$ is a quadratic polynomial function. Moreover, if μ_1 and μ_2 are sufficiently large, it is reasonable to expect (and confirmed by experiments) that the previous iterates converge to the unique Nash equilibrium. But a rigorous proof of this is, to our knowledge, unknown at present.

A natural and obvious variant of **ALG 6** can be obtained by using conjugate gradient directions in step 3. This was done in Ramos et al. (2002a) and Ramos et al. (2002b) respectively for linear and nonlinear equations.

2.6 A numerical experiment concerning bi-objective optimal control

This section is devoted to present the results of some numerical experiments; they have been performed with the FreeFem++ library; see http://www.freefem.org/ff++. We have only considered the computation of Pareto equilibria for the semilinear heat equation (18) via **ALG 4**. With the exception of **ALG 5**, the computational effort and the performance of the other algorithms in the previous sections are similar. In the case of the Newton algorithm **ALG 5**, the situation is obviously different and deserves a particular study that is omitted for reasons of space.

The domain Ω and the sets $\mathcal{O}_i$ and $\mathcal{O}_{i,d}$ will be the following:

$$\Omega = B((0,0);6), \quad \mathcal{O}_1 = (3,4)\times(1,3), \quad \mathcal{O}_2 = (3,4)\times(-3,-1),$$
$$\mathcal{O}_{1,d} = \mathcal{O}_{2,d} = (1,2)\times(-3,3).$$

The sets Ω, $\mathcal{O}_i$ and $\mathcal{O}_{i,d}$ are divided in triangles and a working mesh is depicted in Fig. 3.

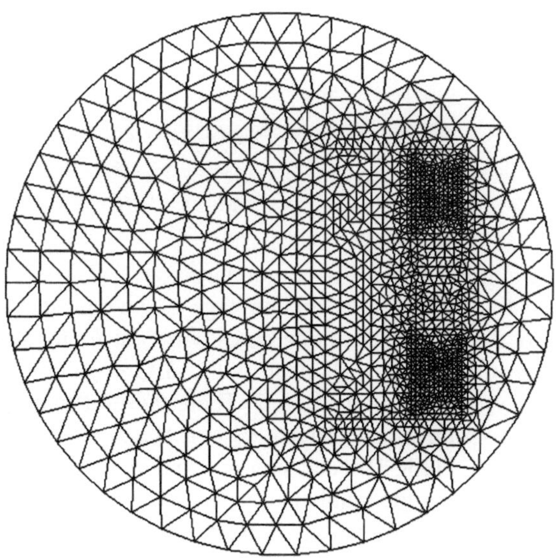

FIGURE 3 The domain and a mesh (1228 vertices, 2404 triangles).

The functions f, y_0, F, and $y_{i,d}$ are given by

$$\begin{cases} f(x,t) \equiv 0.5\, 1_{\mathcal{O}} \text{ with } \mathcal{O} = (1,2) \times (-3,3), \;\; y_0(x) \equiv 1, \;\; F(s) \equiv s(1 + \sin s), \\ y_{1,d}(x,t) \equiv x_1^2 - x_2^2, \quad y_{2,d}(x,t) \equiv x_1 x_2. \end{cases}$$

We have taken $T = 1.5$ and $\mu_1 = \mu_2$ (with this common value denoted by μ). The number of time steps was $M = 30$ (this corresponds to $\tau = 0.05$) and the stopping criteria have been

$$\frac{\|(v_1^{n+1}, v_2^{n+1}) - (v_1^n, v_2^n)\|}{\|(v_1^{n+1}, v_2^{n+1})\|} \leq 10^{-5}. \tag{38}$$

We present in Table 1 the results produced by **ALG 4** with several meshes; we have taken here $\lambda = 0.5$; NV and NT respectively denote the numbers of vertices and triangles; h is the typical diameter of a triangle; ITER is the number of iterates needed for convergence; COST and SN are respectively the final computed cost $\|(v_1, v_2)\|_V$ and state norm $\|y\|_{L^2(Q_T)}$; finally, REL ERR is the L^2 relative error respect to the reference solution (the last computed state, corresponding to NV = 24542 and NT = 48482; for that reason, the last relative error is zero).

TABLE 1 Pareto equilibria for semilinear heat equation. The behavior of ALG 4 for various meshes with $\mu_1 = \mu_2 = 10.5$ and $\lambda = 0.5$.

NV	NT	h	ITER	COST	SN	REL ERR
479	876	0.17	6	2.72298	2567.39	2.17×10^{-1}
1833	3504	0.09	6	2.85223	2609.25	3.77×10^{-2}
2785	5368	0.07	6	2.89361	2616.12	2.14×10^{-3}
11239	22076	0.04	6	2.98039	2628.33	6.55×10^{-4}
16285	32088	0.03	6	3.00195	2630.18	1.30×10^{-4}
24542	48482	0.02	6	3.01601	2631.91	—

Although the number of cases is not too large, several aspects seem clear. In particular, the cost and the state norm converge as we consider a finer mesh to finite values; also, the relative error decreases with h as h^a for some $a > 2$.

A comparison of the results obtained with **ALG 4** for several values of μ and various $(y_{1,d}, y_{2,d})$ is given in Table 2. There, ITER 1 indicates the number of iterates needed for convergence when $y_{1,d} \equiv y_{2,d} \equiv 1$. Similar explanations hold for ITER 2 (resp. ITER 3) when $y_{1,d} \equiv 1$ and $y_{2,d} \equiv -1$ (resp. $y_{1,d} \equiv -1$ and $y_{2,d} \equiv 1$). We see that the number of iterates is similar. On the other hand, as expected, as μ increases the required effort decreases, while letting $\mu \to 0^+$ makes the task much more difficult. This confirms what we said in Section 2.1 about the meaning of the terms in (4) and, in particular, the role of μ_i.

TABLE 2 Pareto equilibria for semilinear heat equation via **ALG 4**. Iterates vs. μ for various $y_{i,d}$.

μ	ITER 1	ITER 2	ITER 3
0.5	130	263	259
1.5	75	86	85
2.5	43	50	49
4.5	22	25	24
5.5	17	19	17
6.5	13	15	15
9.5	5	5	6
10.5	6	6	6

For fixed μ, we have also computed the solution for several values of λ, ranging from 0.05 to 0.95. The behavior of **ALG 4** is illustrated in Fig. 4 and 5.

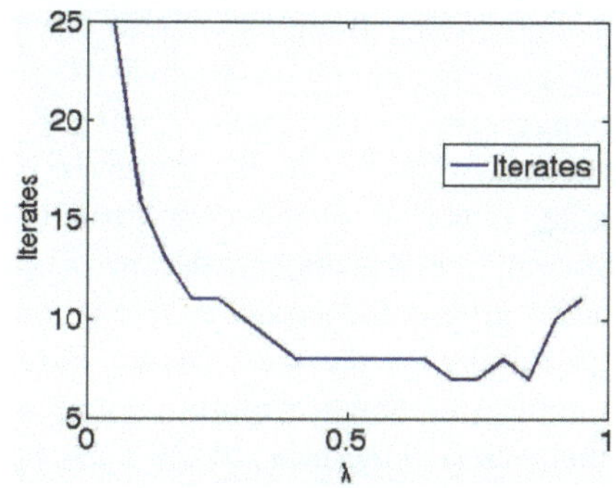

FIGURE 4 Pareto equilibria for semilinear heat equation via **ALG 4**. Iterates vs. λ.

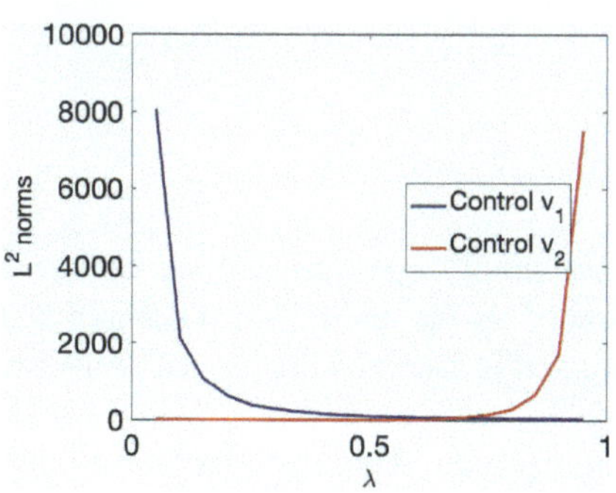

FIGURE 5 Pareto equilibria for semilinear heat equation via **ALG 4**. The norms of the controls vs. λ.

3 Stackelberg strategies and hierarchical control problems

This section deals with the solution of some hierarchic control problems. To this purpose, we will apply *Stackelberg-Nash* and *Stackelberg-Pareto* strategies, that can be briefly described as follows:

- We have controls of two kinds: *leaders* and *followers*. We associate to each leader a Nash or Pareto equilibrium pair, given (as in the previous sections) by the solution to a bi-objective optimal control problem and, then, the corresponding state.
- Then, we compute a leader whose action, added to the followers' action, produces a state that vanishes at final time. In fact, we choose the leader within the family of *null controls*, that is, controls that drive the states to zero, by minimizing a suitable functional.

If, for instance, we interpret that the controls are heat sources applied at different locations of a room and the state is the room temperature, it is completely meaningful to try to steer the temperature to zero at the end of the day and, additionally, keep the temperature "not too far" from prescribed values at some other domains. Other fields where these techniques can be of help are traffic flow, finance, production and marketing, economics of growth, etc.; see for instance Sethi (2019); Ungureanu (2018); Bauso (2016). It is shown in Araruna et al. (2017, 2015) that this is possible to achieve, even if the considered state equation is semilinear and under other more complicate circumstances.

Recall that, in general, the numerical solution of an exact controllability problem for a partial differential equation is not a simple task. We will follow here a strategy relying on the well known *Fursikov-Imanuvilov formulation* of the null control of linear evolution equations, see Fursikov and Imanuvilov (1996) (this was also the basis of the numerical controllability results in Fernández-Cara and Münch (2013, 2014); Fernández-Cara et al. (2017)).

For brevity, we will only consider controlled systems of the form

$$\begin{cases} \partial_t y - \Delta y = f 1_{\mathcal{O}} + v_1 1_{\mathcal{O}_1} + v_2 1_{\mathcal{O}_2}, & \text{in} \quad Q_T, \\ y = 0, & \text{on} \quad \Sigma_T, \\ y(\cdot\,,0) = y^0, & \text{in} \quad \Omega, \end{cases} \tag{39}$$

where $\mathcal{O} \subset \Omega$ is a non-empty open set and we conserve the notation of the previous sections. Now, $\mathcal{O}$ is the main control domain and $\mathcal{O}_1$ and $\mathcal{O}_2$ are secondary control domains; it is assumed that $\mathcal{O} \cap \mathcal{O}_i = \emptyset$ for $i = 1, 2$. We will set $\mathcal{H} := L^2(\mathcal{O} \times (0,T))$.

Let $\mathcal{O}_d \subset \Omega$ be another non-empty open set (an *observation* domain) and assume that the $\mu_i > 0$ and the functions $y_{i,d} \in L^2(\mathcal{O}_{i,d} \times (0,T))$ and the weights $\rho = \rho(x,t)$ and $\rho_0 = \rho_0(x,t)$ are given. The *Stackelberg-Nash* strategy for the null controllability and bi-objective optimal control problem is as follows:

- Consider the *secondary cost functionals*

$$J_i(f; v_1, v_2) := \frac{1}{2} \iint_{\mathcal{O}_d \times (0,T)} |y - y_{i,d}|^2 + \frac{\mu_i}{2} \iint_{\mathcal{O}_i \times (0,T)} |v_i|^2, \quad i = 1, 2, \tag{40}$$

where, for each (f, v_2, v_2), y is the unique solution to (39). Then, for each "leader" $f \in \mathcal{H}$, we search for a Nash equilibrium $(v_1(f), v_2(f))$ for J_1 and J_2. The secondary controls $v_1(f)$ and $v_2(f)$ will be called the "followers" associated to f. It is clear that $(v_1(f), v_2(f))$ is given by the optimality system

$$\begin{cases} \partial_t y - \Delta y = f\,1_{\mathcal{O}} - \displaystyle\sum_{i=1}^{2} \mu_i^{-1} \phi_i 1_{\mathcal{O}_i}, & \text{in} \quad Q_T, \\ -\partial_t \phi_i - \Delta \phi_i = (y - y_{i,d}) 1_{\mathcal{O}_d}, & \text{in} \quad Q_T, \\ y = 0, \quad \phi_i = 0, & \text{on} \quad \Sigma_T, \\ y(\cdot, 0) = y^0, \quad \phi_i(\cdot, T) = 0, & \text{in} \quad \Omega, \end{cases} \tag{41}$$

$$v_i(f) = -\frac{1}{\mu_i} \phi_i |_{\mathcal{O}_i \times (0,T)}, \quad i = 1, 2. \tag{42}$$

As before, there exists $\mu_N(\Omega, \mathcal{O}_i, \mathcal{O}, T) > 0$ such that, if $\mu_1, \mu_2 \geq \mu_N$, we can associate to each $f \in \mathcal{H}$ exactly one Nash equilibrium $(v_1(f), v_2(f))$.
- Let us also consider the *main cost functional*

$$J(f) := \frac{1}{2} \iint_{Q_T} \rho^2 |y|^2 + \frac{1}{2} \iint_{\mathcal{O} \times (0,T)} \rho_0^2 |f|^2, \tag{43}$$

where, for each $f \in \mathcal{H}$, y is the unique solution to (39) with $v_i = v_i(f)$. Then, we search for a minimizer f of J in the family of controls in $\mathcal{H}$ such that

$$y(x, T) = 0 \ \text{in} \ \Omega. \tag{44}$$

From classical results in control theory, we know that null controls for (41) exist and depend continuously on the data y_0 if and only if a suitable observability inequality is satisfied by the solutions to the associated adjoint system

$$\begin{cases} -\partial_t \psi - \Delta \psi = \displaystyle\sum_{i=1}^{2} \gamma_i 1_{\mathcal{O}_d}, & \text{in} \quad Q_T, \\ \partial_t \gamma_i - \Delta \gamma_i = -\mu_i^{-1} \psi 1_{\mathcal{O}_i}, & \text{in} \quad Q_T, \\ \psi = 0, \quad \gamma_i = 0, & \text{on} \quad \Sigma_T, \\ \psi(\cdot, T) = \psi^T, \quad \gamma_i(\cdot, 0) = 0, & \text{in} \quad \Omega. \end{cases} \tag{45}$$

More precisely, this happens if and only if there exist $C > 0$ and a suitable weight ρ_d such that

$$\|\psi(\cdot,0)\|^2 + \sum_{i=1}^{2} \iint_{\mathcal{O}\times(0,T)} \rho_d^{-2}|\gamma^i|^2 \le C \iint_{\mathcal{O}\times(0,T)} |\psi|^2 \qquad \forall \psi^T \in L^2(\Omega).$$

This is explained in detail in Araruna et al. (2015) and Araruna et al. (2017), where the existence of C is established for sufficiently large μ_1 and μ_2, provided $y_{1,d}$ and $y_{2.d}$ decay sufficiently fast to zero as $t \to T$.

More precisely, the following holds:

Theorem 4. *There exists $\mu_{SN}(\Omega, \mathcal{O}_i, \mathcal{O}_d, T, y_{1,d}, y_{2,d}) > 0$ and a positive function $\rho_d = \rho_d(x,t)$, blowing up as $t \to T$, with the following property: if $\mu_1, \mu_2 \ge \mu_{SN}$ and*

$$\iint_{\mathcal{O}_d\times(0,T)} \rho_d^2 |y_{i,d}|^2 < +\infty, \quad i = 1, 2, \tag{46}$$

for each $y^0 \in L^2(\Omega)$ there exist controls $f \in L^2(\mathcal{O} \times (0,T))$ and associated Nash equilibria (v_1, v_2) such that the corresponding solutions to (39) satisfy (44).

3.1 A strategy for the computation of null controls

We will present in this section a method that can be used in practice to compute null controls. There have been many contributions to this purpose the last years, based on various ideas, see for instance Carthel et al. (1994); Glowinski et al. (2008); Boyer et al. (2011); Boyer (2013); Ervedoza and Zuazua (2013). In most cases, the goal is to compute the minimal L^2 norm null control. Here, as in Fernández-Cara and Münch (2013), we will recall another strategy, relying on the so called Fursikov-Imanuvilov formulation, where we pay the prize of computing "larger" controls (in the sense of the L^2 norm) but, on the other hand, we get controls that go, as well as the states, exponentially to zero as $t \to T$.

In the sequel, we assume that $\mu_1, \mu_2 \ge \mu_{SN}$ and we consider the extremal problem

$$\begin{cases} \text{Minimize } K(y,\phi_1,\phi_2,f) = \dfrac{1}{2}\left(\displaystyle\iint_{Q_T} \rho^2|y|^2 + \iint_{\mathcal{O}\times(0,T)} \rho_0^2|f|^2\right) \\ \text{Subject to } (y,\phi_1,\phi_2,f) \in H(y_0,T), \end{cases} \tag{47}$$

where

$$H(y_0,T) := \{(y,\phi_1,\phi_2,f) : f \in \mathcal{U},\ (y,\phi_1,\phi_2,f) \text{ satisfies (41) and (44)}\}.$$

From now on, we take

$$\rho(x,t) := e^{\chi(x)/(T-t)}, \quad \chi(x) := K_1\left(e^{K_2} - e^{\chi_0(x)}\right),$$
$$\rho_0(x,t) := (T-t)^{3/2}\rho(x,t), \tag{48}$$

where K_1 and K_2 are sufficiently large positive constants (depending on T) and $\chi_0 = \chi_0(x)$ is a regular bounded function that is positive in Ω, vanishes on $\partial\Omega$ and satisfies $|\nabla\chi_0| > 0$ in $\overline{\Omega} \setminus \mathcal{O}$ (for a justification of the existence of χ_0, see Fursikov and Imanuvilov (1996)).

Let (y, ϕ_1, ϕ_2, f) be a solution to (47) and let us introduce the new adjoint system

$$\begin{cases} -\partial_t\psi - \Delta\psi = \sum_{i=1}^{2} \gamma_i 1_{\mathcal{O}_d} + \rho^2 y, & \text{in} \quad Q_T, \\ \partial_t\gamma_i - \Delta\gamma_i = -\mu_i^{-1}\psi 1_{\mathcal{O}_i} + \rho^2\phi_i, & \text{in} \quad Q_T, \\ \psi = 0, \quad \gamma_i = 0, & \text{on} \quad \Sigma_T, \\ \gamma_i(\cdot, 0) = 0, & \text{in} \quad \Omega \end{cases} \tag{49}$$

and the following notation

$$M(y,\phi_1,\phi_2) := (\partial_t y - \Delta y + \sum_{i=1}^{2}\mu_i^{-1}\phi_i 1_{\mathcal{O}_i}, -\phi_{1,t} - \Delta\phi_1 - y 1_{\mathcal{O}_{1,d}}, -\phi_{2,t} - \Delta\phi_2 - y 1_{\mathcal{O}_{2,d}})$$

and

$$M^*(\psi,\gamma_1,\gamma_2) := (-\partial_t\psi - \Delta\psi - \sum_{i=1}^{2}\gamma_i 1_{\mathcal{O}_d}, \gamma_{1,t} - \Delta\gamma_1 + \mu_1^{-1}\psi 1_{\mathcal{O}_1}, \gamma_{2,t} - \Delta\gamma_2 + \mu_2^{-1}\psi 1_{\mathcal{O}_2}).$$

Then, it is not difficult to check that, at least formally, one has

$$(y,\phi_1,\phi_2) = \rho^{-2}M^*(\psi,\gamma_1,\gamma_2) \text{ and } f = -\rho_0^{-2}\psi|_{\mathcal{O}\times(0,T)}, \tag{50}$$

where $(\psi, \gamma_1, \gamma_2)$ solves (in some sense) the system

$$\begin{cases} M(\rho^{-2}M^*(\psi,\gamma_1,\gamma_2)) + \rho_0^{-2}\psi 1_{\mathcal{O}} = 0, & \text{in} \quad Q_T, \\ (\psi,\gamma_1,\gamma_2) = (0,0,0), \quad \rho^{-2}M^*(\psi,\gamma_1,\gamma_2) = (0,0,0), & \text{on} \quad \Sigma_T, \\ \rho^{-2}M^*(\psi,\gamma_1,\gamma_2)_1(\cdot,0) = y_0, \quad \gamma_1(\cdot,0) = \gamma_2(\cdot,0) = 0, & \text{in} \quad \Omega, \\ \rho^{-2}M^*(\psi,\gamma_1,\gamma_2)(\cdot,T) = (0,0,0), & \text{in} \quad \Omega. \end{cases} \tag{51}$$

On the other hand, if $(\psi, \gamma_1, \gamma_2)$ is a solution to (51) and (y, ϕ_1, ϕ_2) and f are given by (50), then (y, ϕ_1, ϕ_2, f) solves (47). Also, it can be seen that (51) is well-posed in an appropriate space. Indeed, let us introduce

$$P_0 := \{(\psi, \gamma_1, \gamma_2) \in C^2(Q_T)^3 : (\psi, \gamma_1, \gamma_2) = (0, 0, 0) \text{ on } \Sigma_T,\\ \gamma_1(\cdot\,, 0) = \gamma_2(\cdot\,, 0) = 0\} \tag{52}$$

and the bilinear form

$$m((\psi, \gamma_1, \gamma_2), (\psi', \gamma_1', \gamma_2')) := \iint_{Q_T} \rho^{-2} M^*(\psi, \gamma_1, \gamma_2) \cdot M^*(\psi', \gamma_1', \gamma_2')\\ + \iint_{\mathcal{O}\times(0,T)} \rho_0^{-2} \psi \psi'.$$

Then, as a consequence of the Carleman estimates established in the following result, $m(\cdot\,, \cdot)$ is a scalar product in P_0:

Proposition 1. *Let us set $\rho_2 := (T-t)^{1/2}\rho$ and $\rho_1 := (T-t)^{-1/2}\rho$. There exists $C_0(\Omega, \mathcal{O}, T) > 0$ such that the following holds for any $(\psi, \gamma_1, \gamma_2) \in P_0$, with $h := \gamma_1 + \gamma_2$:*

$$\iint_{Q_T} \left[\rho_2^{-2}(|\partial_t \psi|^2 + |\Delta\psi|^2) + \rho_1^{-2}|\nabla\psi|^2 + \rho_0^{-2}|\psi|^2 + \rho^{-2}|h|^2 \right] dx\, dt\\ \le C_0\, m((\psi, \gamma_1, \gamma_2), (\psi, \gamma_1, \gamma_2)) \tag{53}$$

The proof of this result can be found in Araruna et al. (2015).

Let P be the completion of P_0 with respect to this scalar product. Then P is a Hilbert space and the functions $(\psi, \gamma_1, \gamma_2) \in P$ satisfy

$$\iint_{Q_T} \rho^{-2}|M^*(\psi, \gamma_1, \gamma_2)|^2 + \iint_{\mathcal{O}\times(0,T)} \rho_0^{-2}|\psi|^2 < +\infty. \tag{54}$$

From Proposition 1 and a standard density argument, we also have (53) for all $(\psi, \gamma_1, \gamma_2) \in P$. In particular, we see that for any $(\psi, \gamma_1, \gamma_2) \in P$ one has $\psi \in C^0([0, T-\delta]; H_0^1(\Omega))$ for all $\delta > 0$ and, moreover,

$$\|\psi(\cdot\,, 0)\|_{H_0^1}^2 \le C\, m((\psi, \gamma_1, \gamma_2), (\psi, \gamma_1, \gamma_2)) \quad \forall (\psi, \gamma_1, \gamma_2) \in P. \tag{55}$$

Let us introduce the linear form $\ell : P \mapsto \mathbb{R}$, with

$$\langle \ell, (\psi, \gamma_1, \gamma_2) \rangle := \int_\Omega y_0(x)\, \psi(x, 0)\, dx - \sum_{i=1}^{2} \iint_{\mathcal{O}_d l\times(0,T)} y_{i,d}\, \gamma_i \psi.$$

In view of (55), ℓ is well defined and continuous. Furthermore, from the previous considerations, the following result holds:

Theorem 5. *Let* (y, ϕ_1, ϕ_2, f) *be the unique solution to* (47). *Then one has* (50), *where* $(\psi, \gamma_1, \gamma_2)$ *is the unique solution to the following variational equality in the Hilbert space* P*:*

$$\begin{cases} m((\psi, \gamma_1, \gamma_2), (\psi', \gamma_1', \gamma_2')) = \langle \ell, (\psi', \gamma_1', \gamma_2') \rangle \\ \forall (\psi', \gamma_1', \gamma_2') \in P, \quad (\psi, \gamma_1, \gamma_2) \in P. \end{cases} \tag{56}$$

Obviously, (56) must be viewed as the weak formulation of the boundary-value problem (51), that is fourth-order in x and second-order in t. It is then clear that what we have to do in practice is to solve numerically (56). This will furnish (an approximation of) the triplet $(\psi, \gamma_1, \gamma_2)$. Then, we will use (50) and deduce that the controls are given by

$$f = -\rho_0^{-2}\psi|_{\mathcal{O}\times(0,T)}, \quad v_i = -\frac{1}{\mu_i}\phi_i|_{\mathcal{O}_i\times(0,T)}, \quad i = 1, 2,$$

where ϕ_1 and ϕ_2 are respectively the second and third component of $\rho^{-2}M^*(\psi, \gamma_1, \gamma_2)$.

Let P_h denote a finite dimensional subspace of P. Then, a completely natural approximation of (56) is the following:

$$\begin{cases} m((\psi_h, \gamma_{1,h}, \gamma_{2,h}), (\psi_h', \gamma_{1,h}', \gamma_{2,h}')) = \langle \ell, (\psi_h', \gamma_{1,h}', \gamma_{2,h}') \rangle \\ \forall (\psi_h', \gamma_{1,h}', \gamma_{2,h}') \in P_h, \quad (\psi_h, \gamma_{1,h}, \gamma_{2,h}) \in P_h. \end{cases} \tag{57}$$

Thus, it could seem that, in order to solve numerically the variational equality (56), it suffices to construct explicitly finite dimensional spaces $P_h \subset P$. Note however that this is possible but needs a lot of work and leads to expensive computations, specially in spatial dimensions $d \geq 2$. The reason is that, in order to get $(\psi_h, \gamma_{1h}, \gamma_{2h}) \in P$, we need $\psi_{h,t} + \Delta\psi_h \in L^2_{loc}(Q_T)$. In practice, this means that ψ_h must possess first-order time derivatives and up to second-order spatial derivatives in $L^2_{\text{loc}}(Q_T)$. Therefore, an approximation based on a standard triangulation of Q_T requires functions that must be C^0 in (x, t) and C^1 in x and this can be complicated and too expensive. Spaces of this kind are constructed for instance in Fernández-Cara and Münch (2013); Münch and Souza (2016) for $d = 1$. For $d \geq 2$, one has to consider other spaces, based on *reduced HTC*, *Bell*, or *Bogner-Fox-Schmidt* finite elements; see Ciarlet (2002).

Despite its complexity, the direct approximation of (57) has an advantage: it is possible to adapt the standard finite element theory to this framework and deduce strong convergence results for the numerical controls and states.

3.2 Mixed formulations and numerical approximations of Stackelberg-Nash control problems

The goal of this section is to present an efficient numerical method for the solution of (57). The approach will be different from the one proposed in Section 2.

In the biobjective part of the paper, the computational task was in practice (more or less) equivalent to solving a coupled parabolic system and we have indicated several ways to accomplish this. For hierarchical Stackelberg-Nash or Stackelberg-Pareto problems, this system may be viewed as a subsystem of the optimality characterization of the leader control and it seems much more appropriate to try to solve the whole system at once, via a direct method.

Let us introduce the Hilbert spaces $L^2(\rho^s; Q_T) := \{g : \rho^s g \in L^2(Q_T)\}$ (where $s \in \mathbb{R}$) and the new variable $Z = (z, z_1, z_2) := M^*(\psi, \gamma_1, \gamma_2)$. Then, $Z - M^*(\psi, \gamma_1, \gamma_2) = 0$ (an equality in $L^2(\rho^{-1}; Q_T)^3$) and a first mixed reformulation of (56) is the following:

$$\begin{cases} \alpha((Z,(\psi,\gamma_1,\gamma_2)),(Z',(\psi',\gamma_1',\gamma_2'))) + \beta((Z',(\psi',\gamma_1',\gamma_2')),\Lambda) \\ \qquad = \langle L,(Z,(\psi',\gamma_1',\gamma_2'))\rangle, \\ \beta((Z,(\psi,\gamma_1,\gamma_2)),\Lambda') = 0 \\ \forall (Z',(\psi',\gamma_1',\gamma_2'),\Lambda') \in X, \quad (Z,(\psi,\gamma_1,\gamma_2),\Lambda) \in X, \end{cases} \tag{58}$$

where we have used the notation $X := L^2(\rho^{-1}; Q_T)^3 \times P \times L^2(\rho; Q_T)^3$,

$$\begin{cases} \alpha((Z,(\psi,\gamma_1,\gamma_2)),(Z',(\psi',\gamma_1',\gamma_2'))) \\ \qquad := \iint_{Q_T} \left(\rho^{-2}(z\,z' + z_1\,z_1' + z_2\,z_2') + \rho_0^{-2}\psi\,\psi' 1_{\mathcal{O}}\right), \\ \beta((Z,(\psi,\gamma_1,\gamma_2)),\Lambda) := \iint_{Q_T} \left[Z - M^*(\psi,\gamma_1,\gamma_2)\right]\cdot \Lambda \, dx\, dt \quad \text{and} \\ \langle L,(Z,(\psi,\gamma_1,\gamma_2))\rangle := \int_\Omega y_0(x)\,\psi(x,0)\,dx - \sum_{i=1}^{2} \iint_{\mathcal{O}_d\times(0,T)} y_{i,d}\gamma_i. \end{cases} \tag{59}$$

It can be easily proved that there exists exactly one solution $(Z, (\psi, \gamma_1, \gamma_2), \Lambda)$ to (58) such that $(\psi, \gamma_1, \gamma_2)$ is the unique solution to (56). An advantage of (58) with respect to (56) is that the solution furnishes directly the null state-control couple: it suffices to take

$$y = \rho^{-2}M^*(\psi,\gamma_1,\gamma_2)_1, \quad \phi_i = \rho^{-2}M^*(\psi,\gamma_1,\gamma_2)_{i+1} \quad \text{and} \\ f = -\rho_0^{-2}\psi\big|_{\mathcal{O}\times(0,T)}. \tag{60}$$

However, we again find spatial second-order derivatives in the integrals in (58) and, consequently, a finite element approximation of (58) still needs C^1 in space subspaces.

Let us set $h := \gamma_1 + \gamma_2$ and let us introduce the spaces

$$\tilde{P} := \{(\psi,\gamma_1,\gamma_2) : \iint_{Q_T} \rho^{-2}[(T-t)|\partial_t\psi|^2 + (T-t)^{-1}|\nabla\psi|^2 \\ + (T-t)^{-3}|\psi|^2 + |h|^2] < +\infty,$$

$$\psi = \gamma_1 = \gamma_2 = 0 \text{ on } \Sigma_T\},$$
$$\tilde{Y} := \{\lambda : \iint_{Q_T} \left((T-t)^{-1}\rho^2|\lambda|^2 + (T-t)^{-1}\rho^2|\nabla\lambda^2\right) < +\infty,$$
$$\lambda = 0 \text{ on } \Sigma_T\}$$

and $\tilde{X} := L^2(\rho^{-1}; Q_T)^3 \times \tilde{P} \times \tilde{Y}^3$ and the bilinear form $\tilde{\beta}(\cdot,\cdot)$, with

$$\begin{aligned}\tilde{\beta}\Big((Z,(\psi,\gamma_1,\gamma_2)),\Lambda\Big) := &\iint_{Q_T}\left[\left(z + \partial_t\psi + \sum_{i=1}^{2}\gamma_i 1_{\mathcal{O}_{id}}\right)\cdot\lambda - \nabla\psi\cdot\nabla\lambda\right] \\ &+ \iint_{Q_T}\left[\left(z_1 - \partial_t\gamma^1 - \frac{1}{\mu_1}\psi 1_{\mathcal{O}_1}\right)\cdot\lambda_1 - \nabla\gamma_1\cdot\nabla\lambda_1\right] \\ &+ \iint_{Q_T}\left[\left(z_2 - \partial_t\gamma^2 - \frac{1}{\mu_2}\psi 1_{\mathcal{O}_2}\right)\cdot\lambda_2 - \nabla\gamma_2\cdot\nabla\lambda_2\right].\end{aligned} \tag{61}$$

Then $\tilde{\beta}(\cdot,\cdot)$ is well-defined and continuous on $(L^2(\rho^{-1}; Q_T)^3 \times \tilde{P}) \times \tilde{Y}^3$ and the linear form L is also continuous on $L^2(\rho^{-1}; Q_T)^3 \times \tilde{P}$. Let us consider the following mixed formulation (that is not exactly the same as before):

$$\begin{cases}\alpha((Z,(\psi,\gamma_1,\gamma_2)),(Z',(\psi',\gamma_1',\gamma_2'))) + \tilde{\beta}((Z',(\psi',\gamma_1',\gamma_2')),\Lambda) \\ \qquad\qquad = \langle L,(Z',(\psi',\gamma_1',\gamma_2'))\rangle, \\ \tilde{\beta}((Z,(\psi,\gamma_1,\gamma_2)),\Lambda') = 0 \\ \forall((Z',(\psi',\gamma_1',\gamma_2')),\Lambda') \in \tilde{X}, \quad ((Z,(\psi,\gamma_1,\gamma_2)),\Lambda) \in \tilde{X}.\end{cases} \tag{62}$$

It is easy to see that any possible solution to (62) also solves (58). Indeed, if $((Z,(\psi,\gamma_1,\gamma_2)),\Lambda)$ solves (62), then $Z = M^*(\psi,\gamma_1,\gamma_2)$ in the sense of $\mathcal{D}'(Q_T)^3$, whence $(\psi,\gamma_1,\gamma_2) \in P$; thus, the integration by parts with respect to the spatial variables in $\tilde{\beta}((Z,(\psi,\gamma_1,\gamma_2)),\Lambda')$ is fully justified and $(Z,(\psi,\gamma_1,\gamma_2),\Lambda)$ certainly solves (58).

Consequently, there exists at most one solution to (62). However a rigorous proof of the existence of a solution to (62) is, to our knowledge, unknown. In practice, what we would need to prove is that $\tilde{\beta}$ satisfies the "inf-sup" condition for $(Z,(\psi,\gamma_1,\gamma_2)) \in L^2(\rho^{-1}; Q_T)^3 \times \tilde{P}$ and $\Lambda \in \tilde{Y}$. But whether or not this is true is an open question. Unfortunately, for the finite dimensional approximations that follow, one has the same at present.

It is very convenient in practice to rewrite (62) by introducing new variables. Thus, let us set

$$\hat{Z} = (\hat{z},\hat{z}_1,\hat{z}_2) := \rho^{-1}Z, \quad \hat{M} = (\hat{\psi},\hat{\gamma}_1,\hat{\gamma}_2) := \rho_0^{-1}(\psi,\gamma_1,\gamma_2),$$
$$\hat{\Lambda} = (\hat{\lambda},\hat{\lambda}_1,\hat{\lambda}_2) := \rho\Lambda.$$

This will serve to improve the conditioning of the approximations given below.

The mixed problem (62) can be rewritten in the form

$$\begin{cases} \hat{\alpha}((\hat{Z},\hat{M}),(\hat{Z}',\hat{M}'))+\hat{\beta}\big((\hat{Z}',\hat{M}'),\hat{\Lambda}\big)=\langle\hat{L},(\hat{Z}',\hat{M}')\rangle, \\ \qquad\qquad\qquad\qquad \hat{\beta}\big((\hat{Z},\hat{M}),\hat{\Lambda}'\big)=0 \\ \forall(\hat{Z}',\hat{M}',\hat{\Lambda}')\in\hat{X},\quad (\hat{Z},\hat{M},\hat{\Lambda})\in\hat{X}, \end{cases} \tag{63}$$

where $\hat{X} := \{(\hat{Z},\hat{M},\hat{\Lambda}) : (\rho_0\hat{Z},\rho_0\hat{M},\rho^{-1}\hat{\Lambda}) \in \tilde{X}\}$ and the precise values of $\hat{\alpha}(\cdot\,,\cdot)$, $\hat{\beta}(\cdot\,,\cdot)$ and $\hat{L}$ are easily deduced from (59) and (61).

It is straightforward to reduce (63) to an approximation based on Lagrangian finite elements. Thus, let us assume for simplicity that $d \le 3$ and Ω, $\mathcal{O}$, $\mathcal{O}_1$, and $\mathcal{O}_2$ are intervals, polygonal domains, or polyhedrical domains. Let $\mathcal{T}_\kappa$ be a classical $(d-1)$-simplex triangulation of $\overline{\Omega}$ such that $\overline{\mathcal{O}} = \cup_{R\in\mathcal{T}_\kappa, R\subset\mathcal{O}} R$ and $\overline{\mathcal{O}_i} = \cup_{R\in\mathcal{T}_\kappa, R\subset\mathcal{O}_i} R$ (with $i=1,2$) and let $\mathcal{P}_\tau$ denote a partition of the time interval $[0,T]$ (here, κ and τ denote the respective mesh size parameters). We will use the notation $h := (\kappa,\tau)$ and we will denote by $\mathcal{Q}_h$ the family of all sets of the form $K = R\times[t_1,t_2]$ with $R\in\mathcal{T}_\kappa$ and $[t_1,t_2]\in\mathcal{P}_\tau$ and by $\mathcal{R}_h$ (resp. $\mathcal{R}_{i,h}$) the subfamily of the sets $K = R\times[t_1,t_2]\in\mathcal{Q}_h$ such that $R\subset\mathcal{O}$ (resp. $R\subset\mathcal{O}_i$).

We have then

$$\overline{Q_T} = \bigcup_{K\in\mathcal{Q}_h} K,\quad \overline{\mathcal{O}}\times[0,T] = \bigcup_{K\in\mathcal{R}_h} K \quad\text{and}\quad \overline{\mathcal{O}_i}\times[0,T] = \bigcup_{K\in\mathcal{R}_{i,h}} K.$$

For any couple of integers $m,n\ge 1$, let us introduce the finite dimensional spaces

$$\hat{E}_h(m,n) := \{\hat{Z}_h \in C^0(\overline{Q_T})^3 : \hat{Z}_h|_K \in (\mathbb{P}_{m,x}\otimes\mathbb{P}_{n,t})(K)^3 \ \ \forall K\in\mathcal{Q}_h\},$$
$$\hat{P}_h(m,n) := \{\hat{Z}_h \in \hat{E}_h(m,n) : Z_h = 0 \ \text{ on } \ \Sigma_T\}$$

and

$$\hat{Y}_h(m,n) := \{\hat{Z}_h \in \hat{E}_h(m,n) : Z_h(x,T) = 0 \ \text{ in } \ \Omega\},$$

where, $\mathbb{P}_{\ell,\xi}$ denotes the space of polynomial functions of order ℓ in the variable ξ and $\mathbb{P}_{m,x}\otimes\mathbb{P}_{n,t}$ stands for the usual tensorial product.

Then, for any given $m,n,m',n',m'',n''\ge 1$, we set

$$\hat{X}_h = \hat{X}_h(m,n,m',n',m'',n'') := \hat{E}_h(m,n)\times\hat{P}_h(m',n')\times\hat{Y}_h(m'',n'')$$

and, accordingly, the following approximation of (63) makes sense:

$$\begin{cases} \hat{\alpha}\big((\hat{Z}_h,\hat{M}_h),(\hat{Z}'_h,\hat{M}'_h)\big)+\hat{\beta}\big((\hat{Z}'_h,\hat{M}'_h),\hat{\Lambda}_h\big) = \langle\hat{L},(\hat{Z}'_h,\hat{M}'_h)\rangle, \\ \qquad\qquad\qquad\qquad \hat{\beta}((\hat{Z}_h,\hat{M}_h),\hat{\Lambda}'_h) = 0 \\ \forall(\hat{Z}'_h,\hat{M}'_h,\hat{\Lambda}'_h)\in\hat{X}_h,\quad (\hat{Z}_h,\hat{M}_h,\hat{\Lambda}_h)\in\hat{X}_h. \end{cases} \tag{64}$$

Obviously, (64) can be rewritten as a linear system of the form

$$\begin{bmatrix} \hat{A}_h & \hat{B}_h^T \\ -\hat{B}_h & 0 \end{bmatrix} \begin{bmatrix} p \\ q \end{bmatrix} = \begin{bmatrix} \hat{L}_h \\ 0 \end{bmatrix}, \tag{65}$$

where the matrix $\hat{A}_h$ is symmetric and positive semidefinite but not positive definite for all h. In the following section, we present the numerical results obtained by solving a "regularized" version of (65).

More precisely, taking into account well known techniques, instead of (65), the following system will be solved:

$$\begin{bmatrix} \hat{A}_h & \hat{B}_h^T \\ -\hat{B}_h & \epsilon\,\mathrm{Id.} \end{bmatrix} \begin{bmatrix} p \\ q \end{bmatrix} = \begin{bmatrix} \hat{L}_h \\ 0 \end{bmatrix}, \tag{66}$$

with $\epsilon = 10^{-8}$. This will serve to illustrate the approach.

3.3 A numerical experiment concerning Stackelberg-Nash control

This section is devoted to present the results of a numerical experiment; it deals with the computation of leaders and followers for linear heat equations. Again, it has been performed with the help of the FreeFem++ library; see http://www.freefem.org/ff++.

Here, we deal with the numerical solution of (63) with $\mathbb{P}_1$-Lagrange finite elements in (x, t) for all the variables. From the components $\hat{Z}_h$ and $\hat{M}_h$ of the solution to (64), we can obtain approximations of the controls

$$f_h = -\rho_0^{-1}\hat{\psi}_h 1_{\mathcal{O}} \text{ and } v_{i,h} = -\frac{1}{\mu_i}\rho^{-1}\hat{z}_{ih}\Big|_{\mathcal{O}_i\times(0,T)}.$$

We have taken $d = 1$, $\Omega = (0, 1)$, $T = 0.6$, $\mathcal{O}_1 = (0, 0.2)$, $\mathcal{O}_2 = (0.8, 1)$, $\mathcal{O}_d = (0, 1)$ and $y_{1,d} = y_{2,d} = 0$. We have also fixed $\mu_1 = \mu_2 = 80$ and the main control domain $\mathcal{O}$ has been $\mathcal{O} = (0.25, 0.75)$.

We have used mesh adaptation techniques based on the values of y_h; the final mesh in the computations has been displayed. In fact, the mesh refinement process has been performed with the help of the `adaptmesh` FreeFem utility (see for example Huang and Russell, 2011; Le Dret and Lucquin, 2016 for details on adaptation techniques).

The resolution of (66) has been achieved by applying the UMFPACK method, see Davis (2004). In view of the regularizing effect of the heat equation, the lack of compatibility of the initial and boundary data has no consequence. Indeed, it is seen below that the boundary conditions are satisfied as soon as $t > 0$.

For the reasons explained above, it is natural to have here a presentation of the numerical results that is different from what is shown in Section 2.6. Note that we are applying a direct method to the solution of the problem (essentially, what we have to do is to solve (66)), so convergence tables are meaningless here; on the other hand, since we are using adaptive mesh techniques, the number of unknowns, that is, the step sizes in time and space are not fixed a priori but given by the computed solution itself.

The initial state is given by $y_0(x) \equiv 100x^{0.2}(1-x)^{0.2}$ and $\mathcal{O} = (0.25, 0.75)$. We present the initial and final adapted meshes, the computed controls, the computed state and the time evolution of the L^2 norms in Fig. 6–10.

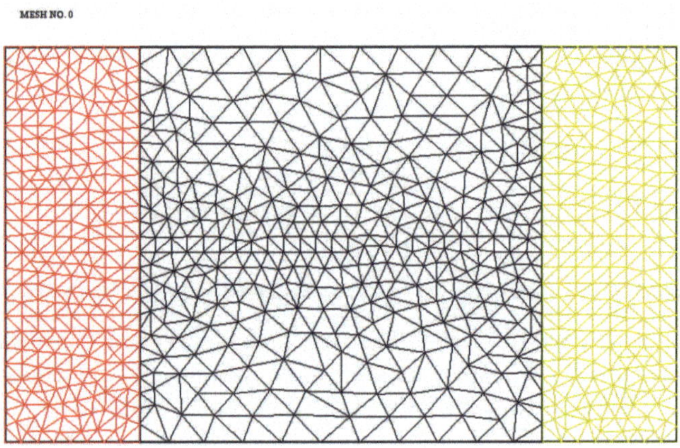

FIGURE 6 Stackelberg-Nash control of the heat equation. The initial mesh of $Q_T = \Omega \times (0, T)$. Number of elements: 1586. Number of vertices: 846. $\mathcal{O}_1 \times (0, T)$ and $\mathcal{O}_2 \times (0, T)$ respectively correspond to the red and yellow regions.

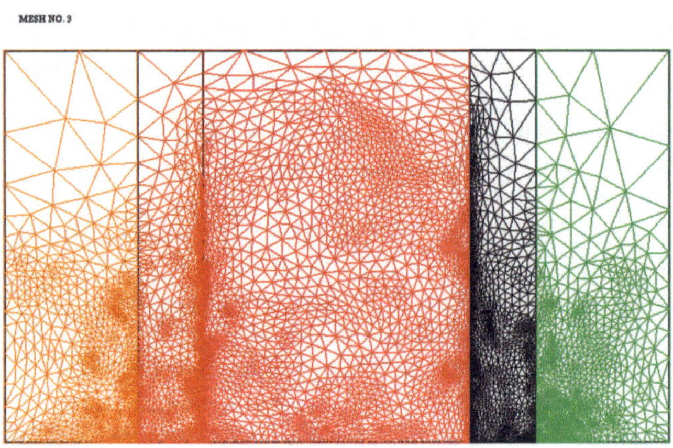

FIGURE 7 Stackelberg-Nash control of the heat equation. The final mesh. Number of vertices: 17608. Number of elements: 9000.

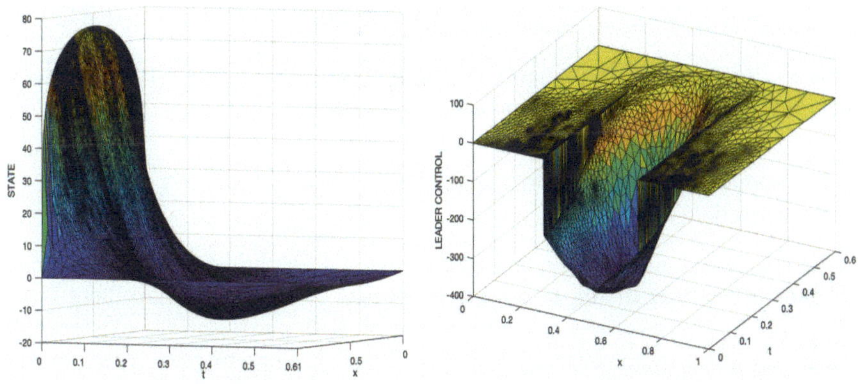

(a) The computed state y (b) The computed leader f

FIGURE 8 Stackelberg-Nash control of the heat equation.

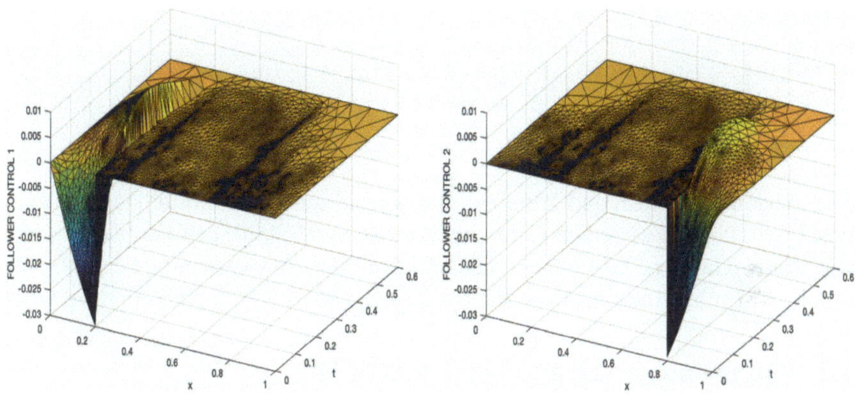

(a) The computed follower v_1 (b) The computed follower v_2

FIGURE 9 Stackelberg-Nash control of the heat equation.

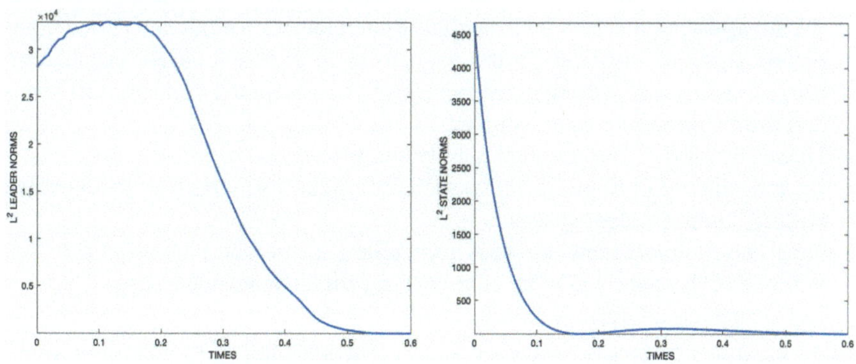

(a) The L^2 norm of the leader (b) The L^2 norm of the state

FIGURE 10 Test 3 – Evolution in time of the L^2 norms.

4 Additional comments and conclusions

This section is devoted to review some of the results in the previous sections and also present possible continuation and extensions.

4.1 Stackelberg-Nash controllability for semilinear and nonlinear equations

If the state equation is nonlinear, the convexity is lost in the functionals J_i and J and the analysis of the bi-objective optimal control and Stackelberg-Nash controllability problems is much more complex. In particular, the existence of Nash equilibria is not ensured and sometimes one has to work with weaker or generalized concepts. Of course, this introduces nontrivial difficulties at the numerical level.

An interesting situation appears when the state is a velocity-pressure couple (y, p) governed by the stationary Navier-Stokes equations:

$$\begin{cases} -\Delta y + (y \cdot \nabla) y + \nabla p = f + v_1 1_{\mathcal{O}_1} + v_2 1_{\mathcal{O}_2} & \text{in} \quad \Omega, \\ \nabla \cdot y = 0 & \text{in} \quad \Omega, \\ y = 0 & \text{on} \quad \partial\Omega. \end{cases}$$

The existence of Nash and Pareto equilibria and quasi-equilibria has been analyzed in detail in Fernández-Cara and Marín-Gayte (2020) and Fernández-Cara and Marín-Gayte (2021), where successful numerical experiments can also be found. A similar study for the evolution Navier-Stokes problem is in progress.

4.2 On the numerical approximation of (47)

The numerical approximation of (47) can be carried out at least in two ways:

- By working with C^1 finite element spaces.
  This is related to the Lax-Milgram formulation (56). The advantages are that we are applying more or less standard finite element techniques with a completely rigorous justification and we can, as in Fernández-Cara and Münch (2013), deduce the strong convergence of the computed controls and states and even appropriate error estimates.
  However, for $d \geq 2$, this is technically complicate.
- By introducing multipliers and associated mixed formulations and working with C^0 finite element spaces.
  Now, the finite elements are easy to handle and, in fact, numerical approximations of the state y and the control f are directly obtained.
  But a rigorous analysis is missing.

In Section 3.2, we have chosen the second approach. In a forthcoming paper, we will explore other methods that share advantages with both methodologies and, in some sense, can be viewed as intermediate.

4.3 More extensions

Of course, the same ideas and techniques can be applied to the computation of Stackelberg-Nash controls in many other similar situations: non-cylindrical control domains, similar boundary control problems, etc. In particular, the hierarchical exact controllability of linear and semilinear parabolic equations with distributed leaders and boundary followers (and also boundary leaders and distributed followers) has been investigated in Araruna et al. (2019).

It would also be interesting to extend and/or adapt these ideas to other multi-objective control problems, such as the computation of Stackelberg-Pareto equilibria. More precisely, we can try to establish a strategy of the following kind:

- For each fixed $f \in \mathcal{U}$, find the corresponding *Pareto Front* for (39), J_1 and J_2. In other words, find the family of couples $(v_1, v_2) \in \mathcal{H}$ with the following property: there exist no $(v_1', v_2') \in \mathcal{H}$ such that $J_i(v_1', v_2') \leq J_i(v_1, v_2)$ for $i = 1, 2$, at least one of these inequalities being strict.
- Then, find $f \in \mathcal{U}$ and an associated Pareto equilibrium $(v_1(f), v_2(f))$ such that (44) holds.

It can be proved that, for each leader f, there exists a whole family of Pareto equilibria

$$\{(v_1^\alpha(f), v_2^\alpha(f)) : \alpha \in (0, 1)\}.$$

The parameter α indicates how relevant is one of the functionals regarding the other one. For any fixed α, it is relatively simple to prove the existence of f such that the state corresponding to f and $(v_1^\alpha(f), v_2^\alpha(f))$ vanishes at time $t = T$. Moreover, the computation of f and the associated Pareto equilibrium can be achieved with the techniques in this paper.

However, it is much more difficult (and actually open) to find leader controls f such that the same holds simultaneously for a family of associated equilibria.

References

Araruna, F.D., Araújo, B.S.V., Fernández-Cara, E., 2018a. Stackelberg-Nash null controllability for some linear and semilinear degenerate parabolic equations. Math. Control Signals Syst. 30 (3), 14.

Araruna, F.D., Fernández-Cara, E., da Silva, L.C., 2019. Hierarchical exact controllability of semilinear parabolic equations with distributed and boundary controls. Commun. Contemp. Math., 1950034.

Araruna, F.D., Fernández-Cara, E., Guerrero, S., Santos, M.C., 2017. New results on the Stackelberg-Nash exact control of linear parabolic equations. Syst. Control Lett. 104, 78–85.

Araruna, F.D., Fernández-Cara, E., Santos, M.C., 2015. Stackelberg-Nash exact controllability for linear and semilinear parabolic equations. ESAIM Control Optim. Calc. Var. 21, 835–856.

Argyros, I.K., 2008. Convergence and Applications of Newton-Type Iterations. Springer, New York.

Bauso, D., 2016. Game Theory with Engineering Applications. Advances in Design and Control, vol. 30. Society for Industrial and Applied Mathematics (SIAM), Philadelphia, PA.

Behncke, H., 2000. Optimal control of deterministic epidemics. Optim. Control Appl. Methods 21 (6), 269–285.

Boyer, F., 2013. On the penalised HUM approach and its applications to the numerical approximation of null-controls for parabolic problems. In: CANUM 2012, 41e Congrès National d'Analyse Numérique. In: ESAIM Proc., vol. 41. EDP Sci., Les Ulis, pp. 15–58.

Boyer, F., Hubert, F., Le Rousseau, J., 2011. Uniform controllability properties for space/time-discretized parabolic equations. Numer. Math. 118 (4), 601–661.

Carthel, C., Glowinski, R., Lions, J.-L., 1994. On exact and approximate boundary controllability for the heat equation: a numerical approach. J. Optim. Theory Appl. 82 (3), 429–484.

Ciarlet, Ph.G., 2002. The Finite Element Methods for Elliptic Problems. Classics in Applied Mathematics, vol. 40. Society for Industrial and Applied Mathematics (SIAM), Philadelphia, PA. Reprint of the 1978 original [North-Holland, Amsterdam].

Davis, T.A., 2004. Algorithm 832: UMFPACK V4.3 – an unsymmetric-pattern multifrontal method. ACM Trans. Math. Softw. 30 (2), 196–199.

Díaz, J.I., 2002. On the von Neumann problem and the approximate controllability of Stackelberg-Nash strategies for some environmental problems. Rev. R. Acad. Cien., Ser. A Math. 96 (3), 343–356.

Díaz, J.I., Lions, J.-L., 2004. On the approximate controllability of Stackelberg-Nash strategies. In: Ocean Circulation and Pollution Control: a Mathematical and Numerical Investigation. Springer, Berlin, pp. 17–27.

Ehrgott, M., 2005. Multicriteria Optimization, second edition. Springer, Berlin.

Ervedoza, S., Zuazua, E., 2013. Numerical Approximation of Exact Controls for Waves. Springer Briefs in Mathematics. Springer, New York.

Fernández-Cara, E., Marín-Gayte, I., 2020. Theoretical and numerical results for some bi-objective optimal control problems. Commun. Pure Appl. Anal. 19 (4), 2101–2126.

Fernández-Cara, E., Marín-Gayte, I., 2021. Bi-objective optimal control of some PDEs: Nash equilibria and quasi-equilibria. ESAIM Control Optim. Calc. Var. 27, 50.

Fernández-Cara, E., Münch, A., 2013. Strong convergent approximations of null controls for the 1D heat equation. SeMA J. 61, 49–78.

Fernández-Cara, E., Münch, A., 2014. Numerical exact controllability of the 1-D heat equation: duality and Carleman weights. J. Optim. Theory Appl. 163, 253–285.

Fernández-Cara, E., Münch, A., Souza, D.A., 2017. On the numerical controllability of the two-dimensional heat, Stokes and Navier-Stokes equations. J. Sci. Comput. 70, 819–858.

Ferrández, M.R., Redondo, J.L., Ivorra, B., Ramos, A.M., Ortigosa, P.M., Paechter, B., 2020. Improving the performance of a preference-based multi-objective algorithm to optimize food treatment processes. Eng. Optim. 52 (5), 896–913.

Fursikov, A.V., 2000. Optimal Control of Distributed Systems. Theory and Applications. Translations of Mathematical Monographs, vol. 187. American Mathematical Society, Providence, RI.

Fursikov, A.V., Imanuvilov, O.Yu., 1996. Controllability of Evolution Equations. Lecture Notes Series, vol. 34. Seoul National University, Research Institute of Mathematics, Global Analysis Research Center, Seoul.

Glowinski, R., Lions, J.-L., He, J., 2008. Exact and Approximate Controllability for Distributed Parameter Systems. A Numerical Approach. Encyclopedia of Mathematics and Its Applications, vol. 117. Cambridge University Press, Cambridge.

Grigorieva, E.V., Khailov, E.N., 2014. Optimal vaccination, treatment, and preventive campaigns in regard to the SIR epidemic model. Math. Model. Nat. Phenom. 9 (4), 105–121.

Guillén, F., Rojas-Medar, M.A., 2013. On the approximate controllability of Stackelberg-Nash strategies for Stokes equations. Proc. Am. Math. Soc. 141 (5), 1759–1773.

Huang, W., Russell, R.D., 2011. Adaptive Moving Mesh Methods. Applied Mathematical Sciences, vol. 174. Springer, New York.

Le Dret, H., Lucquin, B., 2016. Partial Differential Equations: Modeling, Analysis and Numerical Approximation. International Series of Numerical Mathematics, vol. 168. Birkhäuser/Springer, Cham.

Lions, J.-L., 1971. Optimal Control of Systems Governed by Partial Differential Equations. Die Grundlehren der Mathematischen Wissenschaften, vol. 170. Springer-Verlag, New York-Berlin.

Lions, J.-L., 1986. Contrôle de Pareto de systèmes distribués. Le cas d'évolution. C. R. Acad. Sci. Paris, Ser. I 302 (11), 413–417.

Lions, J.-L., 1994. Some remarks on Stackelberg's optimization. Math. Models Methods Appl. Sci. 4, 477–487.

Mophou, G., Kéré, M., Landry Djomegne Njoukoué, L., 2020. Robust hierarchic control for a population dynamics model with missing birth rate. Math. Control Signals Syst. 32 (2), 209–239.

Münch, A., Souza, D.A., 2016. A mixed formulation for the direct approximation of L^2-weighted controls for the linear heat equation. Adv. Comput. Math. 42 (1), 85–125.

Quarteroni, A., Valli, A., 1994. Numerical Approximation of Partial Differential Equations. Springer Series in Computational Mathematics, vol. 23. Springer-Verlag, Berlin.

Ramos, A.M., Glowinski, R., Periaux, J., 2002a. Nash equilibria for the multi-objective control of linear partial partial differential equations. J. Optim. Theory Appl., 457–498.

Ramos, A.M., Glowinski, R., Periaux, J., 2002b. Pointwise control of the Burgers equation and related Nash equilibria problems: a computational approach. J. Optim. Theory Appl. 112, 499–516.

Ramos, A.M., Roubicek, T., 2007. Nash equilibria in noncooperative predator-prey games. Appl. Math. Optim. 56 (2), 211–241.

Sethi, S.P., 2019. Optimal Control Theory. Applications to Management Science and Economics, third edition. Springer, Cham.

Ungureanu, V., 2018. Pareto-Nash-Stackelberg Game and Control Theory. Intelligent Paradigms and Applications. Smart Innovation, Systems and Technologies, vol. 89. Springer, Cham.

Vázquez-Méndez, M.E., Alvarez-Vázquez, L.J., García-Chan, N., Martínez, A., 2019. Optimal management of an urban road network with an environmental perspective. Comput. Math. Appl. 77 (6), 1786–1797.

Chapter 6

Numerics for stochastic distributed parameter control systems: a finite transposition method

Qi Lü[a,d], Penghui Wang[b,e], Yanqing Wang[c,f], and Xu Zhang[a,*,g]

[a]*School of Mathematics, Sichuan University, Chengdu, China,* [b]*School of Mathematics, Shandong University, Jinan, China,* [c]*School of Mathematics and Statistics, Southwest University, Chongqing, China*

* *Corresponding author: e-mail address: zhang_xu@scu.edu.cn*

Contents

Abstract

In this chapter, we present some recent progresses on the numerics for stochastic distributed parameter control systems, based on the *finite transposition method* introduced in our previous works. We first explain how to reduce the numerics for these control problems to that for backward stochastic evolution equations. Then we present a method

[d] The research of this author is supported by the NSF of China under grants 12025105, 11971334 and 11931011, by the Chang Jiang Scholars Program from the Chinese Education Ministry, and by the Science Development Project of Sichuan University under grants 2020SCUNL201 and 2020SCUNL101.

[e] The research of this author is supported by the NSF of China under grant 11871308.

[f] The research of this author is supported by the NSF of China under grant 11801467, and by the Chongqing Natural Science Foundation under grant cstc2018jcyjAX0148.

[g] The research of this author is partially supported by the NSF of China under grants 11931011 and 11821001, and by the Science Development Project of Sichuan University under grant 2020SCUNL201.

Handbook of Numerical Analysis, Volume 23, ISSN 1570-8659, https://doi.org/10.1016/bs.hna.2021.12.006

to find finite transposition solutions to the latter equations. Also, we give an illuminating example.

Keywords
Numerics, Stochastic distributed parameter control system, Backward stochastic evolution equation, Transposition solution, Finite transposition method

MSC Codes
93E20, 60H15, 65M60

1 Introduction

In this chapter, we study the numerics for stochastic distributed parameter control systems, including particularly controlled stochastic partial differential equations. Although some pioneer works appeared in 1960s (e.g., Bensoussan, 1969; Kushner, 1968; Tzafestas and Nightingale, 1968), the theory for stochastic distributed parameter control systems is far from mature till now. Lots of interesting and important problems therein are not well studied. One of them is the numerical computation on such sort of control systems, for which one will meet many substantially difficulties:

- For stochastic control systems, numerical schemes should keep the adaptedness of states and controls with respect to the underlying filtration. This restricts the application of many implicit schemes.
- Unlike the deterministic setting, solutions to stochastic evolution equations (even in finite dimensions) are usually non-differentiable with respect to the time variable. This leads to some serious troubles in the analysis of convergence rates.
- In the continuous-time setting, similarly to the deterministic situation, a typical way to solve many stochastic control problems is to introduce the corresponding dual systems, which are however backward stochastic evolution equations usually (see Lü and Zhang, 2021, for more details). Numerically, it is very hard to compute the correction terms in such sort of backward equations since, so far, there is no efficient method to compute the conditional expectation of a general random variable.

Due to these difficulties, there are very few works on the numerical analysis for stochastic distributed parameter control systems (e.g., Dunst and Prohl, 2016; Prohl and Wang, 2021a,b; Li and Zhou, 2021).

Because of the importance of backward stochastic evolution equations in the study of stochastic distributed parameter control systems, in this chapter we shall focus on the numerical solutions to such equations and the related theoretical analysis.

Up to now, there exist quite a number of numerical schemes for backward stochastic differential equations, i.e. backward stochastic evolution equations in finite dimensions, such as the method of four-step scheme (e.g., Douglas et al., 1996), the quantization tree method (e.g., Bally and Pagès, 2003), the forward/backward Euler method (e.g., Zhang, 2004; Gobet et al., 2005; Bender and Denk, 2007), the Malliavin calculus based method (e.g., Bouchard and Touzi, 2004; Hu et al., 2011), the fully time-space discretization method (e.g., Zhao et al., 2006, 2014), the finite transposition method (Wang and Zhang, 2011), the Wiener chaos decomposition method (e.g., Briand and Labart, 2014), the machine learning method (E et al., 2019), etc. However, numerical schemes for backward stochastic evolution equations in infinite dimensions (including typically backward stochastic partial differential equations) are quite limited (Wang, 2013, 2016; Dunst and Prohl, 2016).

Based on the stochastic transposition method introduced in Lü and Zhang (2013, 2014), in this chapter, we propose a new numerical algorithm to solve backward stochastic evolution equations, which can be used to compute numerically the desired controls of some controllability/optimal control problems for stochastic distributed parameter systems.

To present the key idea in the simplest way, we do not pursue the full technical generality. Firstly, we consider only the simplest case of one dimensional standard Brownian motion (with respect to the time variable). It would not be difficult to extend the results to the case of cylindrical Brownian motions and other infinite-dimensional Brownian motions. Secondly, we impose considerably strong regularity and boundedness assumptions on the data appeared in the state equations and the cost functionals. Thirdly, we assume that the control operators are bounded linear ones, which cannot cover the case of stochastic partial differential equations with boundary/pointwise controls. All of these assumptions can be considerably relaxed. Fourthly, we only consider a null controllability and a linear quadratic optimal control problem. More general controllability and optimal control problems can be studied by the method presented in this chapter.

The rest of this chapter is organized as follows. In Section 2, we shall introduce the control problems and the corresponding backward stochastic evolution equations to be considered in this chapter. In Section 3, we firstly give an outline of the finite transposition method for solving backward stochastic evolution equations numerically, and then provide a choice of finite transposition spaces. In Section 4, based on the chosen finite transposition spaces, we define finite transposition solutions to backward stochastic evolution equations, and then present the existence and uniqueness of these solutions as well as their convergence rate. Finally, in Section 5, as an application of our finite transposition method, we propose a numerical scheme for a stochastic optimal control problem in infinite dimensions.

2 Dual equations for stochastic distributed parameter control problems

We begin with some notations and notions to be used later. Let $T > 0$ and $(\Omega, \mathcal{F}, \mathbf{F}, \mathbb{P})$ (with $\mathbf{F} = \{\mathcal{F}_t\}_{t\in[0,T]}$) be a filtered probability space so that

$(\Omega, \mathcal{F}, \mathbb{P})$ is complete, $\mathcal{F}_0$ contains all $\mathbb{P}$-null sets in $\mathcal{F}$, and $\mathbf{F}$ is right continuous. Let $\{W(t)\}_{t\in[0,T]}$ be an $\mathbf{F}$-adapted 1-dimensional standard Brownian motion on $(\Omega, \mathcal{F}, \mathbf{F}, \mathbb{P})$. Denote by $\mathbb{F}$ the progressive σ-field (in $[0,T]\times\Omega$) with respect to $\mathbf{F}$, by $\mathbb{E}\xi$ the (mathematical) expectation of an integrable random variable $\xi : (\Omega, \mathcal{F}, \mathbb{P}) \to \mathbb{R}$, and by $\mathcal{C}$ a generic positive constant, which may vary from one place to another.

For a Banach space $\mathcal{X}$, we write $\|\cdot\|_{\mathcal{X}}$ for its norm. For another Banach space $\mathcal{Y}$, denote by $\mathcal{L}(\mathcal{X};\mathcal{Y})$ the Banach space of all bounded linear operators from $\mathcal{X}$ to $\mathcal{Y}$, with the usual operator norm. When $\mathcal{X}=\mathcal{Y}$, we simply write $\mathcal{L}(\mathcal{X})$ instead of $\mathcal{L}(\mathcal{X};\mathcal{X})$. If $\mathcal{X}$ is a Hilbert space, we denote its inner product by $\langle\cdot,\cdot\rangle_{\mathcal{X}}$ (when $\mathcal{X}$ is a Euclidean space, for simplicity, we denote its norm and inner product by $|\cdot|$ and $\langle\cdot,\cdot\rangle$, respectively). For any filtration $\mathbf{G}=\{\mathcal{G}_t\}_{t\in[0,T]}\subseteq\mathbf{F}$, $p,q\in[1,\infty)$ and $t\in[0,T]$, let $L^p_{\mathcal{G}_t}(\Omega;\mathcal{X})\triangleq L^p(\Omega,\mathcal{G}_t,\mathbb{P};\mathcal{X})$, and

$$
\begin{aligned}
L^q_{\mathbb{F}}(0,T;L^p(\Omega;\mathcal{X})) &\triangleq \Big\{\varphi:(0,T)\times\Omega\to\mathcal{X}\;\Big|\;\varphi(\cdot)\text{ is }\mathbf{F}\text{-adapted}\\
&\qquad\text{and }\int_0^T\big|\mathbb{E}\|\varphi(t)\|^p_{\mathcal{X}}\big|^{\frac{q}{p}}dt<\infty\Big\},\\
L^p_{\mathbb{F}}(\Omega;C([0,T];\mathcal{X})) &\triangleq \Big\{\varphi:[0,T]\times\Omega\to\mathcal{X}\;\Big|\;\varphi(\cdot)\text{ is continuous,}\\
&\qquad\mathbf{F}\text{-adapted and }\mathbb{E}\big(\|\varphi(\cdot)\|^p_{C([0,T];\mathcal{X})}\big)<\infty\Big\},\\
C_{\mathbb{F}}([0,T];L^p(\Omega;\mathcal{X})) &\triangleq \Big\{\varphi:[0,T]\times\Omega\to\mathcal{X}\;\Big|\;\varphi(\cdot)\text{ is }\mathbf{F}\text{-adapted}\\
&\qquad\text{and }\varphi(\cdot):[0,T]\to L^p_{\mathcal{F}_T}(\Omega;\mathcal{X})\text{ is continuous}\Big\}.
\end{aligned}
$$

In a similar way, we may define $L^\infty_{\mathcal{G}_t}(\Omega;\mathcal{X})$ and $L^\infty_{\mathbb{F}}(0,T;L^\infty(\Omega;\mathcal{X}))$. One can show that, for $1\le p,q\le\infty$, $L^p_{\mathcal{G}_t}(\Omega;\mathcal{X})$ and $L^q_{\mathbb{F}}(0,T;L^p(\Omega;\mathcal{X}))$ are Banach spaces (with the canonical norms). Further, both $L^p_{\mathbb{F}}(\Omega;C([0,T];\mathcal{X}))$ and $C_{\mathbb{F}}([0,T];L^p(\Omega;\mathcal{X}))$ are Banach spaces with norms given by $\|\varphi(\cdot)\|_{L^p_{\mathbb{F}}(\Omega;C([0,T];\mathcal{X}))}=\big(\mathbb{E}(\|\varphi(\cdot)\|^p_{C([0,T];\mathcal{X})})\big)^{1/p}$ and $\|\varphi(\cdot)\|_{C_{\mathbb{F}}([0,T];L^p(\Omega;\mathcal{X}))}=\max_{t\in[0,T]}\big(\mathbb{E}(\|\varphi(t)\|^p_{\mathcal{X}})\big)^{1/p}$, respectively. Also, we write $D_{\mathbb{F}}([0,T];L^p(\Omega;\mathcal{X}))$ for the Banach space of all $L^p(\Omega;\mathcal{X})$-valued, $\mathbf{F}$-adapted, càdlàg stochastic processes $\varphi(\cdot)$ such that $\max_{t\in[0,T]}\big(\mathbb{E}(\|\varphi(t)\|^p_{\mathcal{X}})\big)^{1/p}<\infty$, with the canonical norm. In the sequel, we shall simply denote $L^p_{\mathbb{F}}(0,T;L^p(\Omega;\mathcal{X}))$ by $L^p_{\mathbb{F}}(0,T;\mathcal{X})$; and further simply denote $L^p_{\mathcal{G}_t}(\Omega;\mathbb{R})$, $L^p_{\mathbb{F}}(0,T;\mathbb{R})$ and $L^p_{\mathbb{F}}(\Omega;C([0,T];\mathbb{R}))$ by $L^p_{\mathcal{G}_t}(\Omega)$, $L^p_{\mathbb{F}}(0,T)$ and $L^p_{\mathbb{F}}(\Omega;C([0,T]))$, respectively.

In what follows, we assume that H and U are (real) separable Hilbert spaces, and that A is an unbounded linear operator (with domain $D(A)\subseteq H$), which generates a C_0-semigroup $\{e^{At}\}_{t\ge0}$ on H. Write A^* for the adjoint operator of A. For any $y_0\in H$, consider the following controlled linear stochastic evolution equation:

$$\begin{cases} dy(t) = \big(Ay(t) + Bu(t)\big)dt + \big(Cy(t) + Du(t)\big)dW(t), & t \in (0, T], \\ y(0) = y_0. \end{cases} \tag{2.1}$$

In (2.1), $C(\cdot) \in L^\infty_{\mathbb{F}}(0, T; \mathcal{L}(H))$, $B(\cdot), D(\cdot) \in L^\infty_{\mathbb{F}}(0, T; \mathcal{L}(U; H))$, $u(\cdot) \in L^2_{\mathbb{F}}(0, T; U)$ is the control variable, and $y(\cdot) = y(\cdot; y_0, u(\cdot))$ is the state variable. By the well-posedness result for stochastic evolution equations (e.g., Lü and Zhang, 2021, Section 3.2 of Chapter 3), the system (2.1) admits a unique mild solution $y(\cdot) \in C_{\mathbb{F}}([0, T]; L^2(\Omega; H))$.

Now, we recall the notion of null controllability of (2.1).

Definition 2.1. System (2.1) is called null controllable at time T if for any $y_0 \in H$, there exists $u(\cdot) \in L^2_{\mathbb{F}}(0, T; U)$ such that the corresponding mild solution to (2.1) satisfies $y(T) = 0$, a.s.

Next, we introduce the following quadratic cost functional

$$\begin{aligned} \mathcal{J}(y_0; u(\cdot)) = \frac{1}{2}\mathbb{E}\Big[\int_0^T \big(\langle Qy(t), y(t)\rangle_H + \langle Ru(t), u(t)\rangle_U\big)dt \\ + \langle Gy(T), y(T)\rangle_H\Big], \end{aligned} \tag{2.2}$$

where $Q(\cdot) \in L^\infty_{\mathbb{F}}(0, T; \mathcal{L}(H))$ and $R(\cdot) \in L^\infty_{\mathbb{F}}(0, T; \mathcal{L}(U))$ are self-adjoint operator-valued stochastic processes, while $G \in L^\infty_{\mathcal{F}_T}(\Omega; \mathcal{L}(H))$ is a self-adjoint operator-valued random variable. Consider the linear quadratic optimal control problem (SLQ problem for short) as follows:

Problem (SLQ): For each $y_0 \in H$, find a $\bar{u}(\cdot) \in L^2_{\mathbb{F}}(0, T; U)$ such that

$$\mathcal{J}(y_0; \bar{u}(\cdot)) = \inf_{u(\cdot) \in L^2_{\mathbb{F}}(0,T;U)} \mathcal{J}(y_0; u(\cdot)). \tag{2.3}$$

Any $\bar{u}(\cdot)$ satisfying (2.3) is called an *optimal control*, the corresponding state $\bar{y}(\cdot)$ is called an *optimal state*, and $(\bar{y}(\cdot), \bar{u}(\cdot))$ is called an *optimal pair*.

As in the deterministic case, in order to solve the above null controllability and optimal control problems, one may employ the duality argument. For this purpose, people introduce respectively the following two backward stochastic evolution equations:

$$\begin{cases} dz(t) = -\big(A^*z(t) + C^*Z(t)\big)dt + Z(t)dW(t), & t \in [0, T), \\ z(T) = z_T (\in L^2_{\mathcal{F}_T}(\Omega; H)), \end{cases} \tag{2.4}$$

and

$$\begin{cases} dz(t) = -\big(A^*z(t) - Qy(t) + C^*Z(t)\big)dt + Z(t)dW(t), & t \in [0, T), \\ z(T) = -Gy(T). \end{cases} \tag{2.5}$$

These two equations can be used to solve the aforementioned two stochastic control problems respectively (see Theorems 2.2 and 2.3 in the sequel).

As far as we know, the study of backward stochastic evolution equations is stimulated by the works Bensoussan (1983); Hu and Peng (1991). Now, backward stochastic evolution equations and its variants play fundamental roles in the theory of stochastic distributed parameter control systems (e.g., Lü and Zhang, 2021).

Since neither the usual natural filtration condition nor the quasi-left continuity is assumed for the filtration **F**, one cannot apply the existing results on infinite dimensional backward stochastic evolution equations (e.g., Hu and Peng, 1991; Al-Hussein, 2009) to obtain the well-posedness of Eqs. (2.4) and (2.5) in the sense of mild or weak solutions. Therefore, we shall employ the stochastic transposition method, developed first in Lü and Zhang (2013) for backward stochastic differential equations, then in Lü and Zhang (2014) for backward stochastic evolution equations, to study the well-posedness of Eqs. (2.4) and (2.5).

Consider the following backward stochastic evolution equation (in a general form):

$$\begin{cases} dz(t) = -A^*z(t)dt + F(t, z(t), Z(t))dt + Z(t)dW(t), & t \in [0, T), \\ z(T) = z_T. \end{cases} \tag{2.6}$$

Here $z_T \in L^2_{\mathcal{F}_T}(\Omega; H)$ and $F : [0, T] \times \Omega \times H \times H \to H$ is a given function satisfying that

$$\begin{cases} F(\cdot, 0, 0) \in L^1_{\mathbb{F}}(0, T; L^2(\Omega; H)), \\ \|F(t, y_1, z_1) - F(t, y_2, z_2)\|_H \le \mathcal{C}(\|y_1 - y_2\|_H + \|z_1 - z_2\|_H), \\ \qquad\qquad \forall\, y_1, y_2, z_1, z_2 \in H, \ \text{ a.e. } (t, \omega) \in [0, T] \times \Omega. \end{cases} \tag{2.7}$$

To define the transposition solution to (2.6), we introduce the following stochastic evolution equation:

$$\begin{cases} d\varphi(s) = \big(A\varphi(s) + v_1(s)\big)ds + v_2(s)dW(s), & s \in (t, T], \\ \varphi(t) = \eta, \end{cases} \tag{2.8}$$

where $t \in [0, T]$, $v_1(\cdot) \in L^1_{\mathbb{F}}(t, T; L^2(\Omega; H))$, $v_2(\cdot) \in L^2_{\mathbb{F}}(t, T; H)$ and $\eta \in L^2_{\mathcal{F}_t}(\Omega; H)$. By the classical well-posedness result for stochastic evolution equations (e.g., Lü and Zhang, 2021, Section 3.2 of Chapter 3), Eq. (2.8) admits a unique mild solution $\varphi(\cdot) \in C_{\mathbb{F}}([t, T]; L^2(\Omega; H))$, and

$$\begin{aligned} &\|\varphi(\cdot)\|_{C_{\mathbb{F}}([t,T];L^2(\Omega;H))} \\ &\le \mathcal{C}\big(\|\eta\|_{L^2_{\mathcal{F}_t}(\Omega;H)} + \|v_1(\cdot)\|_{L^1_{\mathbb{F}}(t,T;L^2(\Omega;H))} + \|v_2(\cdot)\|_{L^2_{\mathbb{F}}(t,T;H)}\big). \end{aligned}$$

If (2.6) admits a classical mild solution in Hu and Peng (1991); Al-Hussein (2009) (say, when **F** is the natural filtration of the Brownian motion $W(\cdot)$), then by Itô's formula, we have

$$\begin{aligned}&\mathbb{E}\big\langle\varphi(T),z_T\big\rangle_H-\mathbb{E}\int_t^T\big\langle\varphi(s),F(s,z(s),Z(s))\big\rangle_H ds\\&=\mathbb{E}\big\langle\eta,z(t)\big\rangle_H+\mathbb{E}\int_t^T\big\langle v_1(s),z(s)\big\rangle_H ds+\mathbb{E}\int_t^T\big\langle v_2(s),Z(s)\big\rangle_H ds.\end{aligned}\tag{2.9}$$

Motivated by this, we introduce the following notion.

Definition 2.2. We call $(z(\cdot),Z(\cdot))\in D_{\mathbb{F}}([0,T];L^2(\Omega;H))\times L^2_{\mathbb{F}}(0,T;H)$ a transposition solution to Eq. (2.6) if for any $t\in[0,T]$, $v_1(\cdot)\in L^1_{\mathbb{F}}(t,T;L^2(\Omega;H))$, $v_2(\cdot)\in L^2_{\mathbb{F}}(t,T;H)$, $\eta\in L^2_{\mathcal{F}_t}(\Omega;H)$ and the corresponding solution $\varphi(\cdot)\in C_{\mathbb{F}}([t,T];L^2(\Omega;H))$ to (2.8), the equality (2.9) holds.

We remark that, the space for the first component of the solution (to (2.6)) is chosen to be $D_{\mathbb{F}}([0,T];L^2(\Omega;H))$ rather than $C_{\mathbb{F}}([0,T];L^2(\Omega;H))$. This is quite natural because the filtration **F** is assumed only to be right-continuous.

It is easy to see that if (2.6) admits a mild solution, then such a solution is also a transposition solution to (2.6).

Remark 2.1. The above stochastic transposition method is stimulated by the classical transposition method to solve the non-homogeneous boundary value problems for deterministic partial differential equations (e.g., Lions and Magenes, 1972). This method is a variant of the standard duality method, and in some sense it provides a way to see something which is not easy to be detected directly. Specifically, for Eq. (2.6), the point of this method is to interpret the solution to a backward stochastic evolution equation in terms of a forward one which is well studied. The key tool to achieve this goal is a new Riesz-type representation theorem for $L^q_{\mathbb{F}}(0,T;L^p(\Omega;\mathcal{X}))$ $(1\le p,q<\infty)$ (see Lü et al., 2012, 2018). On the other hand, the equality (2.9) can be regarded as a variational formulation of Eq. (2.6). As pointed out in (Lü and Zhang, 2013, Remark 3.2), it provides a way to solve Eq. (2.6) numerically.

We have the following well-posedness result for Eq. (2.6).

Theorem 2.1 (Lü and Zhang (2014, Theorem 3.1)). *For any $z_T\in L^2_{\mathcal{F}_T}(\Omega;H)$, Eq. (2.6) admits a unique transposition solution $(z(\cdot),Z(\cdot))\in D_{\mathbb{F}}([0,T];L^2(\Omega;H))\times L^2_{\mathbb{F}}(0,T;H)$. Furthermore,*

$$\begin{aligned}&\|(z(\cdot),Z(\cdot))\|_{D_{\mathbb{F}}([0,T];L^2(\Omega;H))\times L^2_{\mathbb{F}}(0,T;H)}\\&\le C\big(\|z_T\|_{L^2_{\mathcal{F}_T}(\Omega;H)}+\|F(\cdot,0,0)\|_{L^1_{\mathbb{F}}(0,T;L^2(\Omega;H))}\big).\end{aligned}$$

From Theorem 2.1, it follows that both (2.4) and (2.5) are well-posed in the sense of transposition solution. Then we can study the null controllability and optimal control problems by (2.4) and (2.5), respectively.

We first consider the problem of finding the control which drives the state of the system (2.1) to 0 (at time T). To this end, we define a functional on $L^2_{\mathcal{F}_T}(\Omega; H)$ as follows:

$$
\begin{aligned}
&\mathfrak{J}(z_T) = \frac{1}{2}\mathbb{E}\int_0^T \|B^*z(t) + D^*Z(t)\|_U^2 dt + \mathbb{E}\langle y_0, z(0)\rangle_H, \\
&\forall\, z_T \in L^2_{\mathcal{F}_T}(\Omega; H),
\end{aligned}
\tag{2.10}
$$

where $(z(\cdot), Z(\cdot)) = (z(\cdot; z_T), Z(\cdot; z_T))$ is the transposition solution to Eq. (2.4) (corresponding to the final datum z_T).

The following result holds:

Theorem 2.2 (Lü and Zhang (2021, Theorems 7.17 and 7.28)). *If the system* (2.1) *is null controllable at time T, then, among all controls transferring the state of* (2.1) *from y_0 to 0 at time T, the one given by*

$$
u(t) = B^*z(t; \hat{z}_T) + D^*Z(t; \hat{z}_T), \quad a.e.\ (t,\omega) \in [0,T]\times\Omega \tag{2.11}
$$

has the minimal $L^2_{\mathbb{F}}(0,T;U)$-norm, where $\hat{z}_T \in L^2_{\mathcal{F}_T}(\Omega; H)$ is the minimizer of the functional $\mathfrak{J}(\cdot)$.

Remark 2.2. As in the context of deterministic setting, the null controllability of (2.1) can be reduced to a suitable observability estimate for Eq. (2.4) (e.g., Lü and Zhang, 2021, Theorem 7.17).

Theorem 2.2 provides an explicit formula of the desired control for the null controllability of (2.1). Therefore, to get such a control, one has to solve the following two problems:

(1) Finding the minimizer $\hat{z}_T$ of the functional $\mathfrak{J}(\cdot)$ (defined by (2.10));
(2) Solving Eq. (2.4) (with z_T replaced by $\hat{z}_T$) to obtain $B^*z(t;\hat{z}_T) + D^*Z(t;\hat{z}_T)$.

If, numerically, we know how to solve Eq. (2.4), then we can adopt the gradient method/Newton's method to find the minimizer of the functional $\mathfrak{J}(\cdot)$. Hence, the key point to compute the control (2.11) is how to find the numerical solutions to (2.4).

Next, we consider Problem (SLQ). With the aid of (2.5), we have the following result.

Theorem 2.3 (Lü and Zhang (2015, Theorem 5.2)). *Let $(\bar{y}(\cdot), \bar{u}(\cdot))$ be an optimal pair of Problem* (SLQ). *Then, for the transposition solution $(z(\cdot), Z(\cdot))$ to* (2.5) *with $y(\cdot)$ replaced by $\bar{y}(\cdot)$, it holds that*

$$
R(t)\bar{u}(t) - B(t)^*z(t) - D(t)^*Z(t) = 0, \quad a.e.\ (t,\omega)\in[0,T]\times\Omega. \tag{2.12}
$$

According to Theorem 2.3, by (2.1) and (2.5), when $R(\cdot)^{-1} \in L^\infty_{\mathbb{F}}(0,T;\mathcal{L}(U))$, in order for finding the optimal pair $(\bar{y}(\cdot),\bar{u}(\cdot))$ of Problem (SLQ), it suffices to solve the following coupled forward-backward stochastic evolution equation:

$$\begin{cases} d\bar{y}(t) = \big(A\bar{y}(t) + BR^{-1}B^*z(t) + BR^{-1}D^*Z(t)\big)dt \\ \qquad\quad +\big(C\bar{y}(t) + DR^{-1}B^*z(t) + DR^{-1}D^*Z(t)\big)dW(t), & t \in (0,T], \\ dz(t) = -\big(A^*z(t) - Q\bar{y}(t) + C^*Z(t)\big)dt + Z(t)dW(t), & t \in [0,T), \\ \bar{y}(0) = y_0, \qquad z(T) = -G\bar{y}(T). \end{cases} \tag{2.13}$$

Nevertheless, a key point to solve (2.13) is again to handle the backward equation (for $(z(\cdot), Z(\cdot))$) therein.

From the above discussion, we see that it is very important to solve backward stochastic evolution equations in the study of control problems for stochastic evolution equations. Note that, for almost all backward stochastic evolution equations, it is impossible to find an explicit formula of their solutions. Hence, for practical applications, it is crucial to find numerical solutions to these equations.

3 The space of finite transposition

In the following two sections, based on the transposition method, we shall present the *finite transposition method*, introduced in (Wang and Zhang, 2011; Wang et al., 2020) for backward stochastic differential equations, to find numerical solutions to backward stochastic evolution equations. To present this method clearly, we only consider the following backward stochastic heat equation:

$$\begin{cases} dz(t,x) = \big(-\Delta z(t,x) + F(t,x,z(t,x),Z(t,x))\big)dt \\ \qquad\quad +Z(t,x)dW(t), & (t,x) \in [0,T) \times D, \\ z(t,x) = 0, & (t,x) \in [0,T) \times \partial D, \\ z(T,x) = z_T(x), & x \in D, \end{cases} \tag{3.1}$$

where $D \subset \mathbb{R}^\ell$ (for some $\ell \in \mathbb{N}$) is a bounded domain with a C^2 boundary ∂D.

Let us first introduce the following two assumptions:

(A1) $\mathbf{F}$ is the natural filtration generated by $\{W(t)\}_{t\in[0,T]}$, $z_T \in L^2_{\mathcal{F}_T}(\Omega; H^1_0(D))$, and $F(\cdot,\cdot,\cdot,\cdot) : [0,T] \times D \times \mathbb{R} \times \mathbb{R} \to \mathbb{R}$ is $\frac{1}{2}$-Hölder continuous with respect to the first variable and globally Lipchitz continuous with respect to the third and fourth variables. Moreover, $F(\cdot,\cdot,0,0) \in L^2(0,T; H^1_0(D))$.

(A2) For the (equal time-interval) partition $\pi : 0 = t_0 < t_1 < \cdots < t_N = T$ with $N \in \mathbb{N}$, $\tau = \frac{T}{N}$ and $t_i = i\tau$ (for $i = 0, 1, \cdots, N$), there exists a constant $\mathcal{C} > 0$

such that the correction term Z in Eq. (3.1) satisfies

$$\sum_{n=0}^{N-1} \mathbb{E} \int_{t_n}^{t_{n+1}} \Big(\|Z(t) - Z(t_{n+1})\|_{L^2(D)}^2 + \|Z(t) - Z(t_n)\|_{L^2(D)}^2 \Big) \, dt \leq C\tau. \quad (3.2)$$

Remark 3.1. **(1)** The filtration $\mathbf{F}$ can be generalized to quasi-left continuous one, and $F(\cdot, \cdot, \cdot, \cdot)$ can be stochastic. But we do not consider these general case to avoid technical complexity.

(2) In this chapter, we adopt the equipartition for simplicity. Our method can also be applied to quasi-uniform partition, i.e.

$$\max_{k=0,1,\cdots,N-1} \{t_{k+1} - t_k\} \leq C \min_{k=0,1,\cdots,N-1} \{t_{k+1} - t_k\}.$$

(3) In Zhang (2004), the estimate in the form of (3.2) is called the L^2-regularity of $Z(\cdot)$. Under some suitable conditions, (3.2) can be guaranteed (see, e.g., Wang, 2016, 2020). When $Z(\cdot)$ appears in the drift term, (3.2) is crucial to prove rates of convergence for temporal discretization. If $Z(\cdot)$ does not appear in the drift term, when proving the rate for $z(\cdot)$, we do not need the condition (3.2) (see Prohl and Wang, 2021b).

We shall use the following notations: For $k = 0, 1, \cdots, N-1$, $t \in [t_k, t_{k+1})$,

$$\Delta_{k+1} W = W(t_{k+1}) - W(t_k), \quad I_k = [t_k, t_{k+1}), \quad (3.3)$$

$$\nu(t) = t_k, \quad \mu(t) = t_{k+1}, \quad \pi(t) = k, \quad (3.4)$$

and

$$\nu(T) = \mu(T) = T, \quad \pi(T) = N.$$

Since our numerical method is based on transposition solutions to backward stochastic differential equations (see, Wang and Zhang, 2011; Lü and Zhang, 2013; Wang et al., 2020), we call it the finite transposition method. Let us propose below an outline of the finite transposition method to solve Eq. (3.1):

(1) Determine a finite dimensional subspace $\mathbb{S}$ of $L^2(D)$.

(2) Choose a *finite transposition space* $\mathbb{H} = \text{span}\{e_i\}$, which is a finite dimensional subspace of $L^2_{\mathbb{F}}(0, T; \mathbb{S})$ (in analogy with the finite element spaces of the finite element method to solve partial differential equations).

(3) Introduce the following variational equation:

$$\begin{aligned} &\mathbb{E}\langle \varphi(T), z_T \rangle_{L^2(D)} \\ &= \mathbb{E} \int_0^T \Big[\langle \varphi(t), F(\mu(t), \cdot, z(t), Z(t)) \rangle_{L^2(D)} + \langle v_1(t), z(t) \rangle_{L^2(D)} \quad (3.5) \\ &\quad + \langle v_2(t), Z(t) \rangle_{L^2(D)} \Big] dt, \end{aligned}$$

where

$$\varphi(t)=\int_0^{\mu(t)}\big(\Delta\varphi(\nu(s))+v_1(s)\big)ds+\int_0^{\nu(t)}v_2(s)dW(s),$$

with $v_1(\cdot)$ and $v_2(\cdot)$ being stochastic processes in suitable finite dimensional subspaces $\mathbb{H}_1\subset L^1_{\mathbb{F}}(0,T;L^2(\Omega;\mathbb{S}))$ and $\mathbb{H}_2\subset L^2_{\mathbb{F}}(0,T;\mathbb{S})$, respectively, and $\nu(\cdot)$ and $\mu(\cdot)$ being piecewise constant functions defined in (3.4). Based on this variational equation, one can prove the existence and uniqueness of approximate solutions $(z(\cdot),Z(\cdot))$, called the *finite transposition solutions* to Eq. (3.1), in the following form

$$z(\cdot)=\sum_{i=1}^{\dim(\mathbb{H}_1)}\alpha_i e_i,\quad Z(\cdot)=\sum_{i=1}^{\dim(\mathbb{H}_2)}\beta_i e_i.$$

(4) Find coefficients α_i, β_i in the above *finite transposition solutions* via the variational equation (3.5), and prove the rate of convergence.

Remark 3.2. There is a useful method — the stochastic finite element method — to solve partial differential equations with random parameter (see, e.g., Ghanem and Spanos, 1991). A main ingredient in this method is the orthogonal expansions, such as polynomial chaos expansion, Karhunen-Loève expansion, etc. Let $\{\mathbf{e}_i\}_{i=1}^{\infty}$ be an orthogonal basis for $L^2_{\mathcal{F}_T}(\Omega;L^2(D))$. Then for any stochastic process $X(\cdot)$, there exists an expansion

$$X(\cdot)=\sum_{i=1}^{\infty}x_i(\cdot)\mathbf{e}_i.$$

Since the solutions $(z(\cdot),Z(\cdot))$ to backward stochastic evolution equations are progressively measurable stochastic processes, people need to find the orthogonal basis of $L^2_{\mathbb{F}}(0,T;L^2(D))$. However, it seems that there is no simple explicit orthogonal basis for such a Hilbert space. Note that $L^2_{\mathbb{F}}(0,T;L^2(D))$ is much more complicated than $L^2_{\mathcal{F}_T}(\Omega;L^2(D))$, and hence the classical "stochastic finite element method" does not work for our problem.

Now we give details for the above outline of the finite transposition method.

Firstly, we determine a finite dimensional subspace of $L^2(D)$ based on the classical Galerkin method. Define

$$A:D(A)=H_0^1(D)\cap H^2(D)\to L^2(D),\quad A\varphi=\Delta\varphi,\quad \forall\varphi\in D(A).\tag{3.6}$$

Let $\{(-\lambda_i,\phi_i)\}_{i=1}^{\infty}$ be the sequence of eigenvalues and eigenfunctions of operator A, such that $\|\phi_i\|_{L^2(D)}=1$ for $i\in\mathbb{N}$. Then $\{\phi_i\}_{i=1}^{\infty}$ constitutes an orthonormal basis of $L^2(D)$ and $\{\phi_i\}_{i=1}^{\infty}$ is also an orthogonal basis of $H_0^1(D)$. Take the subspaces

$$\mathbb{S}_m=\operatorname{span}\{\phi_1,\cdots,\phi_m\},\qquad m\in\mathbb{N}.\tag{3.7}$$

Secondly, we construct a *finite transposition space* by the Wiener chaos. To this end, we review some notations and results on Wiener chaos (see Nualart, 2006, for more details). Recall that $\mathcal{F}_T = \sigma\{W(t); 0 \le t \le T\}$. Define the Itô isometry $\mathbb{W} : L^2(0,T) \to L^2_{\mathcal{F}_T}(\Omega)$ by

$$\mathbb{W}(h) = \int_0^T h(t)dW(t).$$

For $x \in \mathbb{R}$, let

$$H_n(x) = \begin{cases} \dfrac{(-1)^n}{n!} e^{\frac{x^2}{2}} \dfrac{d^n}{dx^n}(e^{-\frac{x^2}{2}}), & n > 0, \\ 1, & n = 0 \end{cases}$$

be the n-th Hermite polynomial. Denote by $\mathcal{I}$ the set of all sequences $\alpha = (\alpha_1, \alpha_2, \cdots)$, $\alpha_i \in \mathbb{N}_0$, such that all the terms, except a finite number of them, vanish. For $k \in \mathbb{N}$, let $\mathcal{I}(k)$ be the set of all sequences $\alpha = (\alpha_1, \alpha_2, \cdots, \alpha_k)$, $\alpha_i \in \mathbb{N}_0$. For all $\alpha \in \mathcal{I}$ (*resp.* $\alpha \in \mathcal{I}(k)$), write

$$\alpha! \equiv \prod_{i=1}^{\infty} \alpha_i! \left(resp. \prod_{i=1}^{k} \alpha_i! \right), \quad \text{and} \quad |\alpha| \equiv \sum_{i=1}^{\infty} \alpha_i \left(resp. \sum_{i=1}^{k} \alpha_i \right),$$

and for $x = (x_1, x_2, \cdots) \in \mathbb{R}^\infty$ (defined as the set of all real sequences),

$$H_\alpha(x) = \prod_{i=1}^{\infty} H_{\alpha_i}(x_i), \quad \alpha \in \mathcal{I} \left(resp. \prod_{i=1}^{k} H_{\alpha_i}(x_i), \quad \alpha \in \mathcal{I}(k) \right),$$

which is called the generalized Hermite polynomial. Also, we put

$$g_i(\cdot) = \frac{\chi_{I_{i-1}}(\cdot)}{\sqrt{\tau}}, \qquad i = 1, 2, \cdots, N,$$

where $I_{i-1} = [t_{i-1}, t_i)$ (see (3.3)). Furthermore, for any $k = 1, 2, \cdots, N$, define the *Wiener chaos of order* n in a Hilbert space H as follows:

$$\mathcal{H}_n(k; H) \triangleq \text{span} \left\{ \prod_{i=1}^{k} H_{\alpha_i}(\mathbb{W}(g_i))\phi \,\middle|\, \alpha \in \mathcal{I}(k), \ |\alpha| = n, \ \phi \in H \right\}.$$

Set $\mathcal{H}_n(0; H) = H$. Take $\mathcal{G}_0 = \mathcal{B}(\mathbb{R})$, and for $k = 1, 2, \cdots, N$, define

$$\mathcal{G}_{t_k} \triangleq \sigma\{\Delta_1 W, \Delta_2 W, \cdots, \Delta_k W\},$$

where $\Delta_k W$ is defined in (3.3). By virtue of the Wiener chaos of order n, we have the following orthogonal decomposition result.

Theorem 3.1 (Nualart (2006, Theorem 1.1.1)). *For any $N, m \in \mathbb{N}$, and $k = 0, 1, \cdots, N$, it holds that*

$$L^2_{\mathcal{G}_{t_k}}(\Omega; \mathbb{S}_m) = \bigoplus_{n=0}^{\infty} \mathcal{H}_n(k; \mathbb{S}_m).$$

Based on Theorem 3.1, we set

$$\mathcal{H}^M(k; \mathbb{S}_m) \triangleq \bigoplus_{n=0}^{M} \mathcal{H}_n(k; \mathbb{S}_m),$$
$$\mathcal{H}^M_N(k; \mathbb{S}_m) \triangleq \text{span}\left\{\chi_{I_k}(\cdot)\xi \mid \xi \in \mathcal{H}^M(k; \mathbb{S}_m)\right\},$$

and the desired *finite transposition space* as follows:

$$\mathbb{H}_{N,M}(\mathbb{S}_m) \triangleq \bigoplus_{k=0}^{N-1} \mathcal{H}^M_N(k; \mathbb{S}_m) \subset L^2_{\mathbb{F}}(0, T; \mathbb{S}_m). \tag{3.8}$$

For simplicity, we denote $\mathcal{H}_n(k; \mathbb{R})$, $\mathcal{H}^M(k; \mathbb{R})$, $\mathcal{H}^M_N(k; \mathbb{R})$, and $\mathbb{H}_{N,M}(\mathbb{R})$ by $\mathcal{H}_n(k)$, $\mathcal{H}^M(k)$, $\mathcal{H}^M_N$, and $\mathbb{H}_{N,M}$, respectively. We write $\left(\mathcal{H}^M(k)\right)^{\oplus n}$ for the space whose elements have n $\mathcal{H}^M(k)$-valued components. Similarly, we can introduce the notation $\mathbb{H}^{\oplus n}_{N,M}$.

Remark 3.3. In Theorem 3.1, we only list a decomposition of $L^2_{\mathcal{G}_{t_k}}(\Omega; \mathbb{S}_m)$, which is a subspace of $L^2_{\mathcal{F}_T}(\Omega; L^2(D))$. For the latter space, by setting

$$\begin{aligned}
&\mathcal{H}_n(L^2(D)) \\
&\triangleq \text{span}\left\{\prod_{i=1}^{\infty} H_{\alpha_i}(\mathbb{W}(g_i))\phi_j \;\middle|\; \alpha \in \mathcal{I},\ |\alpha| = n,\ \{g_i\}_{i=1}^{\infty} \text{ is an orthonormal}\right. \\
&\qquad\left. \text{basis of } L^2(0, T),\ \{\phi_j\}_{j=1}^{\infty} \text{ is an orthonormal basis of } L^2(D)\right\},
\end{aligned}$$

we have

$$L^2_{\mathcal{F}_T}(\Omega; L^2(D)) = \bigoplus_{n=0}^{\infty} \mathcal{H}_n(L^2(D)).$$

For the projection operator

$$\Gamma_M : L^2_{\mathcal{F}_T}(\Omega; L^2(D)) \to \bigoplus_{n=0}^{M} \mathcal{H}_n(L^2(D)), \tag{3.9}$$

it is easy to check that $\Gamma_M\left(L^2_{\mathcal{G}_T}(\Omega; \mathbb{S}_m)\right) = \mathcal{H}^M(N; \mathbb{S}_m)$. This fact is crucial in the sequel.

Remark 3.4. By the construction of the finite transposition space $\mathbb{H}_{N,M}$, it is easy to see that the orthonormal basis of $\mathcal{H}^M(k)$ is

$$\{h_{k,i}\}_{i=1}^{N_{k,M}} = \left\{ \sqrt{\alpha!} \prod_{i=1}^{k} H_{\alpha_i}(\mathbb{W}(g_i)) \;\middle|\; \alpha \in \mathcal{I}(k), |\alpha| \le M \right\},$$

which is a finite dimensional subspace of $L^2_{\mathcal{G}_{t_k}}(\Omega)$. Note that $N_{k,M}$ depends on M. For example, it is easy to check that

$$N_{k,M} = \begin{cases} k+1, & M=1, \\ \dfrac{(k+1)(k+2)}{2}, & M=2, \\ \dfrac{k^3+8k^2+19k+6}{6}, & M=3. \end{cases}$$

Hence the orthonormal basis of $\mathbb{H}_{N,M}$ is

$$\{e_{k,i}(\cdot)\}_{k,i} = \left\{ \chi_{I_k}(\cdot)\frac{1}{\sqrt{\tau}}h_{k,i} \;\middle|\; 1 \le i \le N_{k,M}, 0 \le k \le N-1 \right\},$$

which is also a finite dimensional subspace of $L^2_{\mathbb{F}}(0,T)$.

By virtue of $\{h_{k,i}\}_{i=1}^{N_{k,M}}$, $\{e_{k,i}(\cdot)\}_{1\le i\le N_{k,M}, 0\le k\le N-1}$ and $\{\phi_j\}_{j=1}^m$, we can obtain the orthonormal basis of $\mathcal{H}^M(k;\mathbb{S}_m)$ and $\mathbb{H}_{N,M}(\mathbb{S}_m)$.

Remark 3.5. In this chapter, since we mainly focus on presenting the idea of handling the difficulties caused by the stochastic setting, we choose the finite dimensional subspaces of $L^2(D)$ given by (3.7). One can also replace that spaces as the ones in other numerical methods for solving deterministic partial differential equations, such as the finite element spaces.

For other choice of infinite dimensional spaces of $L^2_{\mathbb{F}}(0,T)$, we refer to Dai et al. (2017).

Remark 3.6. A Wiener chaos decomposition method is proposed in Briand and Labart (2014). It should be pointed out that, although in both Briand and Labart (2014) and the present chapter, some tools of Wiener chaos expansion are used, the numerical method in Briand and Labart (2014) (based on Malliavin analysis) is totally different from our finite transposition method in this chapter (based on the variational equation (3.5)).

4 Finite transposition method for backward stochastic evolution equations

Based on the finite transposition space $\mathbb{H}_{N,M}(\mathbb{S}_n)$, we give below the definition of finite transposition solutions to Eq. (3.1).

Definition 4.1. For any $N, M, n \in \mathbb{N}$, a couple $(z(\cdot), Z(\cdot)) \in \mathbb{H}_{N,M}(\mathbb{S}_n) \times \mathbb{H}_{N,M-1}(\mathbb{S}_n)$ is called a finite transposition solution to Eq. (3.1), if for any $v_1(\cdot) \in \mathbb{H}_{N,M}(\mathbb{S}_n)$ and $v_2(\cdot) \in \mathbb{H}_{N,M-1}(\mathbb{S}_n)$, the following variational equation holds

$$\begin{aligned}&\mathbb{E}\big\langle\varphi(T), z^{\pi}_{N,T}\big\rangle_{L^2(D)} \\ &= \mathbb{E}\int_0^T \Big[\langle\varphi(t), F(\mu(t), \cdot, z(t), Z(t))\rangle_{L^2(D)} + \langle v_1(t), z(t)\rangle_{L^2(D)} \\ &\quad + \langle v_2(t), Z(t)\rangle_{L^2(D)}\Big]dt.\end{aligned} \tag{4.1}$$

Here $z^{\pi}_{N,T} (\in \mathcal{H}^M(N; \mathbb{S}_n))$ is an approximation of z_T, and $\varphi(\cdot)$ solves the following forward equation:

$$\begin{cases}\varphi(t_k) = \displaystyle\int_0^{t_{k+1}} (A\varphi(\nu(t)) + v_1(t))\,dt \\ \qquad\qquad + \displaystyle\int_0^{t_k} v_2(t)dW(t), \quad k = 1, 2, \cdots, N-1, \\ \varphi(0) = 0, \\ \varphi(t) = \varphi(\nu(t)), \quad t \in (0, T],\end{cases} \tag{4.2}$$

where A is defined in (3.6).

The finite transposition method to solve Eq. (3.1) is an algorithm to obtain its finite transposition solutions, via the variational equation (4.1).

The rest of this section is divided into two subsections. In the first one, we discretize the original equation (3.1) with respect to space variables. In the second one, we discretize the obtained equation with respect to the time variable.

4.1 Spacial discretization

Denote by Π_n the orthogonal projection from $L^2(D)$ to $\mathbb{S}_n$ and define A_n by $A_n = A|_{\mathbb{S}_n}$. The semi-discrete problem in space corresponding to Eq. (3.1) is to find a pair $(z_n(\cdot), Z_n(\cdot)) \in L^2_{\mathbb{F}}(\Omega; C([0,T]; \mathbb{S}_n)) \times L^2_{\mathbb{F}}(0, T; \mathbb{S}_n)$ solving the following equation:

$$\begin{cases}dz_n(t) = \big(-A_n z_n(t) + \Pi_n F(t, z_n(t), Z_n(t))\big)dt \\ \qquad\qquad + Z_n(t)dW(t), \quad t \in [0, T], \\ z_n(T) = \Pi_n z_T.\end{cases} \tag{4.3}$$

The following result states the rate of convergence for the solution of Eq. (4.3) to the one of Eq. (3.1).

Theorem 4.1 (Wang (2013, Theorem 5.4), Wang (2016, Theorem 3.1)). *Suppose that* (A1) *holds. Then the following estimate holds*

$$\begin{aligned}
&\|z(\cdot)-z_n(\cdot)\|^2_{L^2_{\mathbb{F}}(\Omega;C([0,T];L^2(D)))\cap L^2_{\mathbb{F}}(0,T;H^1_0(D))}+\|Z(\cdot)-Z_n(\cdot)\|^2_{L^2_{\mathbb{F}}(0,T;L^2(D))}\\
&\le \frac{\mathcal{C}}{\lambda_{n+1}}\Big(\|z_T(\cdot)\|^2_{L^2_{\mathcal{F}_T}(\Omega;H^1_0(D))}+\|F(\cdot,\cdot,0,0)\|^2_{L^2(0,T;H^1_0(D))}\Big),
\end{aligned}$$

where $(z(\cdot), Z(\cdot))$ *and* $(z_n(\cdot), Z_n(\cdot))$ *are solutions to Eqs.* (3.1) *and* (4.3), *respectively.*

We recall that $\mathbb{S}_n$ in the finite transposition space $\mathbb{H}_{N,M}(\mathbb{S}_n)$ is constructed by the first n eigenfunctions of the operator A. As explained in Remark 3.5, one can also choose other finite element spaces $\mathbb{V}_h$, and construct the corresponding finite transposition spaces $\mathbb{H}_{N,M}(\mathbb{V}_h)$. The readers are referred to Dunst and Prohl (2016); Prohl and Wang (2021a,b) for error estimates for finite-element based spacial discretizations.

4.2 Temporal discretization

In this part, firstly we rewrite the definition of finite transposition solutions and the finite transposition method for the backward stochastic differential equation (4.3). Then we present the existence and uniqueness of these solutions. Finally, we give the rate of convergence.

Since $\mathbb{S}_n = \text{span}\{\phi_1, \cdots, \phi_n\}$, and the solution $(z_n(\cdot), Z_n(\cdot))$ to Eq. (4.3) is in the space $L^2_{\mathbb{F}}(\Omega; C([0, T]; \mathbb{S}_n)) \times L^2_{\mathbb{F}}(0, T; \mathbb{S}_n)$, we may take $(z_n(\cdot), Z_n(\cdot))$ to be the following form

$$z_n(\cdot)=\sum_{j=1}^n a_{n,j}(\cdot)\phi_j, \quad Z_n(\cdot)=\sum_{j=1}^n b_{n,j}(\cdot)\phi_j, \tag{4.4}$$

where $a_{n,j}(\cdot) \in L^2_{\mathbb{F}}(\Omega; C([0, T]))$ and $b_{n,j}(\cdot) \in L^2_{\mathbb{F}}(0, T)$, for $j = 1, 2, \cdots, n$. Set

$$a_n(\cdot)=\begin{pmatrix} a_{n,1}(\cdot)\\ a_{n,2}(\cdot)\\ \vdots\\ a_{n,n}(\cdot)\end{pmatrix},\quad b_n(\cdot)=\begin{pmatrix} b_{n,1}(\cdot)\\ b_{n,2}(\cdot)\\ \vdots\\ b_{n,n}(\cdot)\end{pmatrix},\quad \Lambda_n=\begin{pmatrix} -\lambda_1 & 0 & \cdots & 0\\ 0 & -\lambda_2 & \cdots & 0\\ \vdots & \vdots & \ddots & \vdots\\ 0 & 0 & \cdots & -\lambda_n\end{pmatrix},$$

and

$$F_n(\cdot,a_n(\cdot),b_n(\cdot))=\begin{pmatrix} \langle F(\cdot,\cdot,z_n(\cdot),Z_n(\cdot)),\phi_1\rangle_{L^2(D)}\\ \langle F(\cdot,\cdot,z_n(\cdot),Z_n(\cdot)),\phi_2\rangle_{L^2(D)}\\ \vdots\\ \langle F(\cdot,\cdot,z_n(\cdot),Z_n(\cdot)),\phi_n\rangle_{L^2(D)}\end{pmatrix}.$$

Then $(a_n(\cdot), b_n(\cdot))$ solves the following backward stochastic differential equation:

$$\begin{cases} da_n(t) = \big(-\Lambda_n a_n(t) + F_n(t, a_n(t), b_n(t))\big)dt + b_n(t)dW(t), & t \in [0,T], \\ a_n(T) = \big(\langle z_T, \phi_1\rangle_{L^2(D)}, \langle z_T, \phi_2\rangle_{L^2(D)}, \cdots, \langle z_T, \phi_n\rangle_{L^2(D)}\big)^\top. \end{cases} \tag{4.5}$$

Definition 4.2. For any $N, M, n \in \mathbb{N}$, a couple $(a_{n,N,M}(\cdot), b_{n,N,M-1}(\cdot)) \in \mathbb{H}_{N,M}^{\oplus n} \times \mathbb{H}_{N,M-1}^{\oplus n}$ is called a finite transposition solution to the backward stochastic differential equation (4.5) if for any $v_1(\cdot) \in \mathbb{H}_{N,M}^{\oplus n}$ and $v_2(\cdot) \in \mathbb{H}_{N,M-1}^{\oplus n}$, the following variational equation holds

$$\begin{aligned} \mathbb{E}\big\langle x(T), a_{n,T}^\pi\big\rangle = \mathbb{E}\int_0^T \Big[&\big\langle x(t), F_n(\mu(t), a_{n,N,M}(t), b_{n,N,M-1}(t))\big\rangle \\ &+ \big\langle v_1(t), a_{n,N,M}(t)\big\rangle + \big\langle v_2(t), b_{n,N,M-1}(t)\big\rangle\Big]dt, \end{aligned} \tag{4.6}$$

where $x(\cdot)$ is given by

$$\begin{cases} x(t_k) = \displaystyle\int_0^{t_{k+1}} (\Lambda_n x(\nu(t)) + v_1(t))\, dt \\ \qquad\quad + \displaystyle\int_0^{t_k} v_2(t) dW(t), \quad k = 1, \cdots, N-1, \\ x(0) = 0, \\ x(t) = x(\nu(t)), \quad t \in (0,T], \end{cases} \tag{4.7}$$

and $a_{n,T}^\pi \in \big(\mathcal{H}^M(N)\big)^{\oplus n}$ is an approximation of $a_n(T)$.

By choosing $a_{n,T}^\pi = \big(\langle z_{N,T}^\pi, \phi_1\rangle_{L^2(D)}, \langle z_{N,T}^\pi, \phi_2\rangle_{L^2(D)}, \cdots, \langle z_{N,T}^\pi, \phi_n\rangle_{L^2(D)}\big)^\top$, and combining with Definitions 4.1 and 4.2, we can see that the pair of stochastic processes $\big(\sum_{j=1}^n \big(a_{n,N,M}(\cdot)\big)_j \phi_j, \sum_{j=1}^n \big(b_{n,N,M-1}(\cdot)\big)_j \phi_j\big)$ is a finite transposition solution to (3.1), where $(a_{n,N,M}(\cdot))_j$, $(b_{n,N,M-1}(\cdot))_j$ are the j-th component of $a_{n,N,M}(\cdot)$, $b_{n,N,M-1}(\cdot)$ respectively. Based on this, from now on, we study (4.6) instead of (4.1).

By Remark 3.4, we can denote the orthonormal basis of $\mathbb{H}_{N,M}^{\oplus n}$ by

$$\Big\{e_{k,i}^\ell(\cdot) \;\Big|\; 1 \le \ell \le n,\ 1 \le i \le N_{k,M}, 0 \le k \le N-1\Big\}.$$

Here

$$e_{k,i}^\ell(\cdot) = (0, \cdots, 0, e_{k,i}(\cdot), 0, \cdots, 0)^\top,$$

the ℓ-th component of which is $e_{k,i}(\cdot)$. Similarly, we can define $h_{k,i}^\ell$.

The following result is addressed to the existence and uniqueness of finite transposition solutions to backward stochastic differential equations. The readers are referred to Wang et al. (2020) for a proof.

Theorem 4.2. *Under the assumptions* (A1) *and* (A2), *for any* $N, M, n \in \mathbb{N}$, *the backward stochastic differential equation* (4.5) *admits a unique finite transposition solution* $(a_{n,N,M}(\cdot), b_{n,N,M-1}(\cdot)) \in \mathbb{H}_{N,M}^{\oplus n} \times \mathbb{H}_{N,M-1}^{\oplus n}$.

Remark 4.1. By Definition 4.2, the variational equation (4.6) leads to an implicit algorithm. We can also provide other schemes based on the finite transposition method, such as the following explicit algorithm:

$$\begin{aligned}\mathbb{E}\big\langle x(T), a_{n,T}^{\pi}\big\rangle = \mathbb{E}\int_0^T \Big[&\big\langle x(t), F_n(\mu(t), a_{n,N,M}(\mu(t)), b_{n,N,M-1}(\mu(t)))\big\rangle \\ &+ \big\langle v_1(t), a_{n,N,M}(t)\big\rangle + \big\langle v_2(t), b_{n,N,M-1}(t)\big\rangle\Big]dt.\end{aligned}$$

By Remark 3.4, we see that $\{e_{k,i}(\cdot) \mid 1 \le i \le N_{k,M},\ 0 \le k \le N-1\}$ is an orthonormal basis of the finite transposition space $\mathbb{H}_{N,M}$. Hence, by Theorem 4.2, we may write

$$\begin{aligned}
&a_{n,N,M}(\cdot) = \sum_{\ell=1}^{n}\sum_{k=0}^{N-1}\sum_{i=1}^{N_{k,M}} \alpha_{k,i}^{\ell} e_{k,i}^{\ell}(\cdot), \\
&\quad \alpha_{k,i}^{\ell} \in \mathbb{R} \text{ for } i = 1, \cdots, N_{k,M},\ k = 0, \cdots, N-1,\ \ell = 1, \cdots, n, \\
&b_{n,N,M-1}(\cdot) = \sum_{\ell=1}^{n}\sum_{k=0}^{N-1}\sum_{i=1}^{N_{k,M-1}} \beta_{k,i}^{\ell} e_{k,i}^{\ell}(\cdot), \\
&\quad \beta_{k,i}^{\ell} \in \mathbb{R} \text{ for } i = 1, \cdots, N_{k,M-1},\ k = 0, \cdots, N-1,\ \ell = 1, \cdots, n.
\end{aligned} \tag{4.8}$$

In what follows, we shall apply the variational equation (4.6) to determine the coefficients of the finite transposition solution $(a_{n,N,M}(\cdot), b_{n,N,M-1}(\cdot))$.

By choosing $v_1(\cdot) = e_{k,i}^{\ell}(\cdot)$ and $v_2(\cdot) = 0$, we obtain

$$x(t) = \begin{cases} 0, & t \in [0, t_k), \\ \Lambda_0^{\pi(t)-k+1}\sqrt{\tau} h_{k,i}^{\ell}, & t \in [t_k, T]. \end{cases}$$

Here

$$\Lambda_0 = (\mathbb{I}_n - \Lambda_n \tau)^{-1},$$

where $\mathbb{I}_n$ is the identity matrix of $n \times n$. By virtue of (4.6), we have

$$\begin{aligned}&\mathbb{E}\langle \Lambda_0^{N-k}\sqrt{\tau} h_{k,i}^{\ell}, a_{n,T}^{\pi}\rangle \\ &= \mathbb{E}\int_{t_k}^{T} \Big\langle \Lambda_0^{\pi(t)-k+1}\sqrt{\tau} h_{k,i}^{\ell}, F_n\big(\mu(t), a_{n,N,M}(t), b_{n,N,M-1}(t)\big)\Big\rangle dt\end{aligned}$$

$$
\begin{aligned}
&+\frac{1}{\sqrt{\tau}}\mathbb{E}\int_{t_k}^{t_{k+1}}\Big\langle h_{k,i}^{\ell}, a_{n,N,M}(t)\Big\rangle dt\\
&=\mathbb{E}\int_{t_k}^{T}\Big\langle \Lambda_0^{\pi(t)-k+1}\sqrt{\tau}h_{k,i}^{\ell}, F_n\big(\mu(t), a_{n,N,M}(t), b_{n,N,M-1}(t)\big)\Big\rangle dt\\
&+\frac{1}{\tau}\mathbb{E}\int_{t_k}^{t_{k+1}}\Big\langle h_{k,i}^{\ell}, \sum_{m=1}^{n}\sum_{l=0}^{N-1}\sum_{j=1}^{N_{l,M}}\alpha_{l,j}^{m}\chi_{I_l}(t)h_{l,j}^{m}\Big\rangle dt\\
&=\mathbb{E}\int_{t_k}^{T}\Big\langle \Lambda_0^{\pi(t)-k+1}\sqrt{\tau}h_{k,i}^{\ell}, F_n\big(\mu(t), a_{n,N,M}(t), b_{n,N,M-1}(t)\big)\Big\rangle dt+\alpha_{k,i}^{\ell},
\end{aligned}
$$

which implies

$$
\begin{aligned}
\alpha_{k,i}^{\ell}=&\,\mathbb{E}\Big\langle \Lambda_0^{N-k}\sqrt{\tau}h_{k,i}^{\ell}, a_{n,T}^{\pi}\Big\rangle\\
&-\mathbb{E}\int_{t_k}^{T}\Big\langle \Lambda_0^{\pi(t)-k+1}\sqrt{\tau}h_{k,i}^{\ell}, F_n\big(\mu(t), a_{n,N,M}(t), b_{n,N,M-1}(t)\big)\Big\rangle dt.
\end{aligned}
\tag{4.9}
$$

Similarly, by taking $v_1=0$, $v_2=e_{k,i}^{\ell}$, we see that

$$
x(t)=\begin{cases}0, & t\in[0,t_{k+1}),\\ \Lambda_0^{\pi(t)-k}\dfrac{\Delta_{k+1}W}{\sqrt{\tau}}h_{k,i}^{\ell}, & t\in[t_{k+1},T],\end{cases}
$$

and

$$
\begin{aligned}
&\mathbb{E}\Big\langle \Lambda_0^{N-k-1}\frac{\Delta_{k+1}W}{\sqrt{\tau}}h_{k,i}^{\ell}, a_{n,T}^{\pi}\Big\rangle\\
&=\mathbb{E}\int_{t_{k+1}}^{T}\Big\langle \Lambda_0^{\pi(t)-k}\frac{\Delta_{k+1}W}{\sqrt{\tau}}h_{k,i}^{\ell}, F_n\big(\mu(t), a_{n,N,M}(t), b_{n,N,M-1}(t)\big)\Big\rangle dt\\
&+\frac{1}{\sqrt{\tau}}\mathbb{E}\int_{t_k}^{t_{k+1}}\Big\langle h_{k,i}^{\ell}, b_{n,N,M-1}(t)\Big\rangle dt\\
&=\mathbb{E}\int_{t_{k+1}}^{T}\Big\langle \Lambda_0^{\pi(t)-k}\frac{\Delta_{k+1}W}{\sqrt{\tau}}h_{k,i}^{\ell}, F_n\big(\mu(t), a_{n,N,M}(t), b_{n,N,M-1}(t)\big)\Big\rangle dt\\
&+\frac{1}{\tau}\mathbb{E}\int_{t_k}^{t_{k+1}}\Big\langle h_{k,i}^{\ell}, \sum_{m=1}^{n}\sum_{l=0}^{N-1}\sum_{j=1}^{N_{l,M-1}}\beta_{l,j}^{m}\chi_{I_l}(t)h_{l,j}^{m}\Big\rangle dt\\
&=\mathbb{E}\int_{t_{k+1}}^{T}\Big\langle \Lambda_0^{\pi(t)-k}\frac{\Delta_{k+1}W}{\sqrt{\tau}}h_{k,i}^{\ell}, F_n\big(\mu(t), a_{n,N,M}(t), b_{n,N,M-1}(t)\big)\Big\rangle dt+\beta_{k,i}^{\ell}.
\end{aligned}
$$

It follows that

$$\beta_{k,i}^{\ell} = \mathbb{E}\Big\langle \Lambda_0^{N-k-1} \frac{\Delta_{k+1} W}{\sqrt{\tau}} h_{k,i}^{\ell}, a_{n,T}^{\pi} \Big\rangle$$
$$- \mathbb{E} \int_{t_{k+1}}^{T} \Big\langle \Lambda_0^{\pi(t)-k} \frac{\Delta_{k+1} W}{\sqrt{\tau}} h_{k,i}^{\ell}, F_n\big(\mu(t), a_{n,N,M}(t), b_{n,N,M-1}(t)\big) \Big\rangle dt. \tag{4.10}$$

By virtue of (4.9) and (4.10), we obtain the finite transposition solution $(a_{n,N,M}(\cdot), b_{n,N,M-1}(\cdot))$ to the backward stochastic differential equation (4.5).

To give a convergence rate of the finite transposition method, we need the following space

$$D\big((-A)^{3/2}\big) = \Big\{ \varphi = \sum_{i=1}^{\infty} \varrho_i \phi_i \;\Big|\; \sum_{i=1}^{\infty} |\varrho_i|^2 \lambda_i^3 < \infty \Big\}.$$

We have the following result on the convergence rate, which also shows the relationship between the Euler method and the finite transposition one.

Theorem 4.3. *Assume* (A1) *and* (A2), *and that for given* $N, M, n \in \mathbb{N}$, $(a_n(\cdot), b_n(\cdot)) \in L_{\mathbb{F}}^2(\Omega; C([0,T]; \mathbb{R}^n)) \times L_{\mathbb{F}}^2(0,T; \mathbb{R}^n)$ *and* $(a_{n,N,M}(\cdot), b_{n,N,M-1}(\cdot)) \in \mathbb{H}_{N,M}^{\oplus n} \times \mathbb{H}_{N,M-1}^{\oplus n}$ *are adapted solution and finite transposition solution to the backward stochastic differential equation* (4.5), *respectively. Then,*

$$a_{n,N,M}(\cdot) = \sum_{k=0}^{N-1} \chi_{I_k}(\cdot) \Big[\mathbb{E}\Big(\Lambda_0^{N-k} a_{n,T}^{\pi} \;\Big|\; \mathcal{F}_{t_k} \Big)$$
$$- \mathbb{E}\Big(\int_{t_k}^{T} \Lambda_0^{\pi(t)-k+1} \Gamma_M F_n(\mu(t), a_{n,N,M}(t), b_{n,N,M-1}(t)) dt \;\Big|\; \mathcal{F}_{t_k} \Big) \Big], \tag{4.11}$$

$$b_{n,N,M-1}(\cdot) = \sum_{k=0}^{N-1} \chi_{I_k}(\cdot) \mathbb{E}\Big(\frac{\Delta_{k+1} W}{\tau} a_{n,N,M}(t_{k+1}) \;\Big|\; \mathcal{F}_{t_k} \Big), \tag{4.12}$$

where $\Lambda_0 = (\mathbb{I}_n - \Lambda_n \tau)^{-1}$. *Furthermore, the following rate of convergence holds true:*

$$\sup_{0 \le t \le T} \mathbb{E}|a_{n,N,M}(t) - a_n(t)|^2 + \mathbb{E}\int_0^T |b_{n,N,M-1}(t) - b_n(t)|^2 dt$$
$$\le C\Big[\mathbb{E}|a_n(T) - a_{n,T}^{\pi}|^2 + \lambda_n^2 \tau + \mathbb{E}\int_0^T |(\mathbb{I}_n - \Gamma_M) F_n(s, a_n(s), b_n(s))|^2 ds \Big], \tag{4.13}$$

where Γ_M *is defined in* (3.9). *Moreover, if* $F(\cdot,\cdot,\cdot,\cdot)$ *is linear with respect to the last two components, and* $z_T \in L_{\mathcal{F}_T}^2\big(\Omega; D\big((-A)^{3/2}\big)\big)$, $F(\cdot,\cdot,0,0) \in$

$L^2(0, T; H_0^1(D) \cap H^2(D))$, *then*

$$\sup_{0\le t\le T} \mathbb{E}|a_{n,N,M}(t) - a_n(t)|^2 + \mathbb{E}\int_0^T |b_{n,N,M-1}(t) - b_n(t)|^2 dt \le \mathcal{C}\Big[\mathbb{E}|a_n(T) - a_{n,T}^\pi|^2 + \tau\Big]. \tag{4.14}$$

Proof. The proof is long, and we carry out it by the following three steps.

Step 1. In this step, we prove (4.11). By (4.8) and (4.9), it follows that

$$\begin{aligned} a_{n,N,M}(\cdot) = \sum_{\ell=1}^{n}\sum_{k=0}^{N-1}\sum_{i=1}^{N_{k,M}} \Big[& \mathbb{E}\Big\langle \Lambda_0^{N-k}\sqrt{\tau} h_{k,i}^\ell, a_{n,T}^\pi \Big\rangle \\ & - \mathbb{E}\int_{t_k}^{T} \Big\langle \Lambda_0^{\pi(t)-k+1}\sqrt{\tau} h_{k,i}^\ell, F_n(\mu(t), a_{n,N,M}(t), b_{n,N,M-1}(t))\Big\rangle dt \Big] \\ & \times e_{k,i}^\ell(\cdot). \end{aligned}$$

For the first term on the right side of the above equality, we have that

$$\begin{aligned} \mathbb{E}\Big\langle \Lambda_0^{N-k}\sqrt{\tau} h_{k,i}^\ell, a_{n,T}^\pi \Big\rangle e_{k,i}^\ell(\cdot) &= \mathbb{E}\Big\langle \Lambda_0^{N-k}\sqrt{\tau} h_{k,i}^\ell, a_{n,T}^\pi \Big\rangle \chi_{I_k}(\cdot)\frac{1}{\sqrt{\tau}} h_{k,i}^\ell \\ &= \chi_{I_k}(\cdot)\mathbb{E}\Big\langle h_{k,i}^\ell, \Lambda_0^{N-k} a_{n,T}^\pi \Big\rangle h_{k,i}^\ell. \end{aligned} \tag{4.15}$$

By the definition of $L^2_{\mathcal{G}_{t_k}}(\Omega;\mathbb{R}^n)$, $\mathcal{H}^M(k;\mathbb{R}^n)$ and Remark 3.4, we can extend the orthonormal basis $\{h_{k,i}^\ell \mid 1\le \ell\le n,\ 1\le i\le N_{k,M}\}$ of $\big(\mathcal{H}^M(k)\big)^{\oplus n}$ to the orthonormal basis $\{h_{k,i} \mid 1\le k\le n,\ 1\le i\le\infty\}$ of $L^2_{\mathcal{F}_{t_k}}(\Omega;\mathbb{R}^n)$. Since for any $\eta\in L^2_{\mathcal{G}_T}(\Omega;\mathbb{R}^n)$, $\Gamma_M\eta\in\big(\mathcal{H}^M(N)\big)^{\oplus n}$,

$$\mathbb{E}(\Gamma_M\eta \mid \mathcal{F}_{t_k}) \in \Big(\mathcal{H}^M(N)\Big)^{\oplus n} \cap L^2_{\mathcal{G}_{t_k}}(\Omega;\mathbb{R}^n) = \Big(\mathcal{H}^M(k)\Big)^{\oplus n}. \tag{4.16}$$

By (4.15), (4.16) and the fact that $\{h_{k,i}^\ell \mid 1\le\ell\le n,\ 1\le i\le N_{k,M}\}$ is an orthonormal basis of $\big(\mathcal{H}^M(k)\big)^{\oplus n}$ and $a_{n,T}^\pi\in\big(\mathcal{H}^M(N)\big)^{\oplus n}$, we see that

$$\sum_{\ell=1}^{n}\sum_{k=0}^{N-1}\sum_{i=1}^{N_{k,M}} \mathbb{E}\Big\langle \Lambda_0^{N-k}\sqrt{\tau} h_{ki}^\ell, a_{n,T}^\pi \Big\rangle e_{k,i}^\ell(\cdot) = \sum_{k=0}^{N-1}\chi_{I_k}(\cdot)\Lambda_0^{N-k}\mathbb{E}\left(a_{n,T}^\pi \mid \mathcal{F}_{t_k}\right). \tag{4.17}$$

Similarly,

$$\sum_{\ell=1}^{n}\sum_{k=0}^{N-1}\sum_{i=1}^{N_{k,M}}\mathbb{E}$$
$$\times\int_{t_k}^{T}\Big\langle\Lambda_0^{\pi(t)-k+1}\sqrt{\tau}h_{k,i}^{\ell},F_n\big(\mu(t),a_{n,N,M}(t),b_{n,N,M-1}(t)\big)\Big\rangle dt\,e_{k,i}^{\ell}(\cdot)$$
$$=\sum_{\ell=1}^{n}\sum_{k=0}^{N-1}\chi_{I_k}(\cdot)$$
$$\times\sum_{i=1}^{N_{k,M}}\mathbb{E}\int_{t_k}^{T}\Big\langle h_{k,i}^{\ell},\Lambda_0^{\pi(t)-k+1}\Gamma_M F_n\big(\mu(t),a_{n,N,M}(t),b_{n,N,M-1}(t)\big)\Big\rangle dt\,h_{k,i}^{\ell}$$
$$=\sum_{k=0}^{N-1}\chi_{I_k}(\cdot)\mathbb{E}\left(\int_{t_k}^{T}\Lambda_0^{\pi(t)-k+1}\Gamma_M F_n\big(\mu(t),a_{n,N,M}(t),b_{n,N,M-1}(t)\big)dt\;\Big|\;\mathcal{F}_{t_k}\right).\tag{4.18}$$

Therefore, by (4.15), (4.17), and (4.18), we have (4.11).

Step 2. In this step, we prove (4.12). Noting that $a_{n,T}^{\pi}\in\big(\mathcal{H}^M(N)\big)^{\oplus n}$, by Remark 3.4, we obtain

$$a_{n,T}^{\pi}=\sum_{\ell=1}^{n}\sum_{m=0}^{M}\sum_{|\alpha|=m}d^{\alpha,\ell}\sqrt{\alpha!}\prod_{i=1}^{N}H_{\alpha_i}^{\ell}(\mathbb{W}(g_i)).$$

Recalling that $H_n(x)$ is the Hermite polynomial, we have

$$xH_n(x)=(n+1)H_{n+1}(x)-H_{n-1}(x).$$

Consequently,

$$\Delta_{k+1}Wa_{n,T}^{\pi}=\sum_{\ell=1}^{n}\sum_{m=0}^{M}\sum_{|\alpha|=m}d^{\alpha,\ell}\sqrt{\alpha!}\prod_{i=1}^{N}H_{\alpha_i}^{\ell}(\mathbb{W}(g_i))\sqrt{\tau}H_1(\mathbb{W}(g_{k+1}))$$
$$=\sqrt{\tau}\sum_{\ell=1}^{n}\sum_{m=0}^{M}\sum_{|\alpha|=m}d^{\alpha,\ell}\sqrt{\alpha!}$$
$$\times\prod_{i=1}^{N}\Big[(\alpha_{k+1}+1)H_{\bar{\alpha}_i}^{\ell}(\mathbb{W}(g_i))-H_{\hat{\alpha}_i}^{\ell}(\mathbb{W}(g_i))\Big],$$

where $\bar{\alpha}=\alpha+\gamma_{k+1}$ and $\hat{\alpha}=\alpha-\gamma_{k-1}$. Here the k-th component of $\gamma_k(\in\mathcal{I})$ is 1, and the others are 0. By the definition of $\bar{\alpha}$, we have $\bar{\alpha}_{k+1}\geq 1$. Therefore, by noting that $|\widehat{\alpha}| \le M-1$, we arrive at

$$
\begin{aligned}
&\mathbb{E}(\Delta_{k+1} W a^{\pi}_{n,T} \mid \mathcal{F}_{t_k}) \\
&= \sqrt{\tau} \sum_{\ell=1}^{n} \sum_{m=0}^{M} \sum_{|\alpha|=m} d^{\alpha,\ell} \sqrt{\alpha!} \\
&\quad \times \mathbb{E}\Big(\prod_{i=1}^{N} \Big[(\alpha_{k+1}+1) H^{\ell}_{\bar{\alpha}_i}(\mathbb{W}(g_i)) - H^{\ell}_{\hat{\alpha}_i}(\mathbb{W}(g_i)) \Big] \Big| \mathcal{F}_{t_k} \Big) \\
&= \sqrt{\tau} \sum_{\ell=1}^{n} \sum_{m=0}^{M} \sum_{|\alpha|=m} d^{\alpha,\ell} \sqrt{\alpha!} (\alpha_{k+1}+1) \\
&\quad \times \prod_{i=1}^{k} H^{\ell}_{\bar{\alpha}_i}(\mathbb{W}(g_i)) \mathbb{E}\Big(\prod_{i=k+1}^{N} H^{\ell}_{\bar{\alpha}_i}(\mathbb{W}(g_i)) \Big| \mathcal{F}_{t_k} \Big) \\
&\quad - \sqrt{\tau} \sum_{\ell=1}^{n} \sum_{m=0}^{M} \sum_{|\alpha|=m} d^{\alpha,\ell} \sqrt{\alpha!} \mathbb{E}\Big(\prod_{i=1}^{N} H^{\ell}_{\hat{\alpha}_i}(\mathbb{W}(g_i)) \Big| \mathcal{F}_{t_k} \Big) \\
&= -\sqrt{\tau} \sum_{\ell=1}^{n} \sum_{m=0}^{M} \sum_{|\alpha|=m} d^{\alpha,\ell} \sqrt{\alpha!} \mathbb{E}\Big(\prod_{i=1}^{N} H^{\ell}_{\hat{\alpha}_i}(\mathbb{W}(g_i)) \Big| \mathcal{F}_{t_k} \Big) \\
&\in \Big(\mathcal{H}^{M-1}(k) \Big)^{\oplus n}.
\end{aligned}
$$

Furthermore, we deduce that

$$
\begin{aligned}
&\sum_{\ell=1}^{n} \sum_{i=1}^{N_{k,M-1}} \mathbb{E}\Big\langle \frac{\Delta_{k+1} W}{\sqrt{\tau}} h^{\ell}_{k,i}, a^{\pi}_{n,T} \Big\rangle h^{\ell}_{k,i} \\
&= \sum_{\ell=1}^{n} \sum_{i=1}^{N_{k,M-1}} \mathbb{E}\Big\langle h^{\ell}_{k,i}, \mathbb{E}\Big(\frac{\Delta_{k+1} W}{\sqrt{\tau}} a^{\pi}_{n,T} \Big| \mathcal{F}_{t_k} \Big) \Big\rangle h^{\ell}_{k,i} = \mathbb{E}\Big(\frac{\Delta_{k+1} W}{\sqrt{\tau}} a^{\pi}_{n,T} \Big| \mathcal{F}_{t_k} \Big).
\end{aligned}
\tag{4.19}
$$

With the same procedure, we have

$$
\begin{aligned}
&\sum_{\ell=1}^{n} \sum_{k=0}^{N-1} \sum_{i=1}^{N_{k,M-1}} \mathbb{E} \\
&\times \int_{t_{k+1}}^{T} \Big\langle \frac{\Delta_{k+1} W}{\sqrt{\tau}} h^{\ell}_{k,i}, F_n\big(\mu(t), a_{n,N,M}(t), b_{n,N,M-1}(t)\big) \Big\rangle dt \, e^{\ell}_{k,i}(\cdot) \\
&= \sum_{k=0}^{N-1} \chi_{I_k}(\cdot)
\end{aligned}
$$

$$\times \sum_{\ell=1}^{n} \sum_{i=1}^{N_{k,M-1}} \mathbb{E}\Big\langle h_{k,i}^{\ell}, \int_{t_{k+1}}^{T} \frac{\Delta_{k+1} W}{\tau} \Gamma_M F_n\big(\mu(t), a_{n,N,M}(t), b_{n,N,M-1}(t)\big) dt \Big\rangle h_{k,i}^{\ell}$$

$$= \sum_{k=0}^{N-1} \chi_{I_k}(\cdot) \mathbb{E}\Big(\int_{t_{k+1}}^{T} \frac{\Delta_{k+1} W}{\tau} \Gamma_M F_n\big(\mu(t), a_{n,N,M}(t), b_{n,N,M-1}(t)\big) dt \,\Big|\, \mathcal{F}_{t_k} \Big). \tag{4.20}$$

Combining (4.8), (4.10), (4.19), and (4.20), we conclude that

$$\begin{aligned}
& b_{n,N,M-1}(\cdot) \\
&= \sum_{\ell=1}^{n} \sum_{k=0}^{N-1} \sum_{i=1}^{N_{k,M-1}} \Big[\mathbb{E}\Big\langle \Lambda_0^{N-k-1} \frac{\Delta_{k+1} W}{\sqrt{\tau}} h_{k,i}^{\ell}, a_{n,T}^{\pi} \Big\rangle \\
&- \mathbb{E} \int_{t_{k+1}}^{T} \Big\langle \Lambda_0^{\pi(t)-k} \frac{\Delta_{k+1} W}{\sqrt{\tau}} h_{k,i}^{\ell}, \Gamma_M F_n(\mu(t), a_{n,N,M}(t), b_{n,N,M-1}(t)) \Big\rangle dt \Big] e_{k,i}^{\ell}(\cdot) \\
&= \sum_{k=0}^{N-1} \chi_{I_k}(\cdot) \mathbb{E}\Big(\frac{\Delta_{k+1} W}{\tau} \Big[\Lambda_0^{N-k-1} a_{n,T}^{\pi} \\
&- \int_{t_{k+1}}^{T} \Lambda_0^{\pi(t)-k} \Gamma_M F_n(\mu(t), a_{n,N,M}(t), b_{n,N,M-1}(t)) dt \Big] \,\Big|\, \mathcal{F}_{t_k} \Big) \\
&= \sum_{k=0}^{N-1} \chi_{I_k}(\cdot) \mathbb{E}\Big(\frac{\Delta_{k+1} W}{\tau} a_{n,N,M}(t_{k+1}) \,\Big|\, \mathcal{F}_{t_k} \Big),
\end{aligned}$$

which is (4.12).

Step 3. In this step, by means of (4.11) and (4.12), we prove the error estimates for the finite transposition method, which is a slight variation of (Wang, 2016, Theorem 4.2). We provide here a sketch for the completeness.

Set $\bar{a}_n(t_k) = a_{n,N,M}(t_k) \in L^2_{\mathcal{F}_{t_k}}(\Omega; \mathbb{R}^n)$, for $k = 0, 1, \cdots, N$, and note that $\bar{\alpha}_n(T) = \alpha_{n,T}^{\pi}$. By the Martingale Representation Theorem, there exists a square integrable process $\bar{b}_n(\cdot)$, such that

$$\bar{a}_n(t_{k+1}) = \mathbb{E}(\bar{a}_n(t_{k+1}) \,|\, \mathcal{F}_{t_k}) + \int_{t_k}^{t_{k+1}} \bar{b}_n(s) dW(s), \quad k = 0, 1, \cdots, N-1.$$

For $t \in [t_k, t_{k+1})$, $k = 0, 1, \cdots, N-1$, define

$$\begin{aligned}
\bar{a}_n(t) = {} & (\mathbb{I}_n - \Lambda_n \tau) \bar{a}_n(t_k) + (t - t_k) \Gamma_M F_n(t_{k+1}, \bar{a}_n(t_k), b_{n,N,M-1}(t_k)) \\
& + \int_{t_k}^{t} \bar{b}_n(s) dW(s).
\end{aligned}$$

Subsequently, by the fact that $b_{n,N,M-1}(t_k)=\frac{1}{\tau}\mathbb{E}\big(\int_{t_k}^{t_{k+1}}\bar{b}_n(s)ds\,\big|\,\mathcal{F}_{t_k}\big)$, the difference between $(a_n(\cdot),b_n(\cdot))$ and $(\bar{a}_n(\cdot),\bar{b}_n(\cdot))$ can be estimated as follows:

$$
\begin{aligned}
&\mathbb{E}|\,(\mathbb{I}_n-\Lambda_n\tau)\,(a_n(t_k)-\bar{a}_n(t_k))|^2+\mathbb{E}\int_{t_k}^{t_{k+1}}|b_n(s)-\bar{b}_n(s)|^2ds\\
&=\mathbb{E}\Big|(a_n(t_k)-\bar{a}_n(t_k))\\
&\qquad-\int_{t_k}^{t_{k+1}}\big(F_n(s,a_n(s),b_n(s))-\Gamma_M F_n(t_{k+1},\bar{a}_n(t_k),b_{n,N,M-1}(t_k))\big)\,ds\Big|^2\\
&\le(1+C\tau)\mathbb{E}|a_n(t_{k+1})-\bar{a}_n(t_{k+1})|^2 \qquad\qquad (4.21)\\
&\quad+C\Big\{\mathbb{E}\int_{t_k}^{t_{k+1}}\Big[\big|\Lambda_n(a_n(t_j)-a_n(s))\big|^2+|(\mathbb{I}_n-\Gamma_M)F_n(s,a_n(s),b_n(s))|^2\Big]ds\\
&\quad+\mathbb{E}\int_{t_k}^{t_{k+1}}\big|\Gamma_M\big(F_n(t_{k+1},\bar{a}_n(t_k),b_{n,N,M-1}(t_k))-F_n(s,a_n(s),b_n(s))\big)\big|^2ds\Big\}.
\end{aligned}
$$

By Gronwall's inequality and the assumption (A2), we get that

$$
\begin{aligned}
&\max_{k=0,1,\cdots,N}\mathbb{E}|a_n(t_k)-a_{n,N,M}(t_k)|^2=\max_{k=0,1,\cdots,N}\mathbb{E}|a_n(t_k)-\bar{a}_n(t_k)|^2\\
&\le C\Big[\mathbb{E}|a_n(T)-\bar{a}_n(T)|^2+\tau\\
&\quad+\tau\mathbb{E}\int_0^T\lambda_n^2\Big(|\Lambda_n a_n(s)|^2+|b_n(s)|^2+|F_n(t,0,0)|^2\Big)\,ds\\
&\quad+\mathbb{E}\int_0^T|(\mathbb{I}_n-\Gamma_M)F_n(s,a_n(s),b_n(s))|^2ds\Big]\\
&\le C\Big[\mathbb{E}|a_n(T)-\bar{a}_n(T)|^2+\lambda_n^2\tau+\mathbb{E}\int_0^T|(\mathbb{I}_n-\Gamma_M)F_n(s,a_n(s),b_n(s))|^2ds\Big].
\end{aligned}
\qquad (4.22)
$$

Summing (4.21) from $k=0$ to $N-1$, applying (4.22) and the fact that $b_{n,N,M-1}(t_k)=\frac{1}{\tau}\mathbb{E}\big(\int_{t_k}^{t_{k+1}}\bar{b}_n(s)ds\,\big|\,\mathcal{F}_{t_k}\big)$, we obtain (4.13). In a similar way, we can prove (4.14). This completes the proof of Theorem 4.3. □

Remark 4.2. By the formulas (4.11) and (4.12), we may see that, the transposition method to solve a linear backward stochastic evolution equation is simply the Euler method under the $\mathcal{H}^M(N;\mathbb{S}_n)$-valued approximations of the terminal value. Nevertheless, to solve nonlinear equations, the finite transposition method is different from the Euler one. Also, for the linear equation, since the terminal value is approximated by $\mathcal{H}^M(N;\mathbb{S}_n)$-valued random variables, thanks to the variational equation (4.1), there is no need to calculate the conditional expectations.

5 Numerical method for optimal controls

In this section, we present an application of the finite transposition method. To avoid the technical complexity, we only consider here a simple SLQ problem. More precisely, we consider a cost functional

$$\mathcal{J}(y_0; u(\cdot)) = \frac{1}{2}\mathbb{E}\Big[\int_0^T \big(\|y(t)\|^2_{L^2(D)} + \|u(t)\|^2_{L^2(D)}\big)dt + \|y(T)\|^2_{L^2(D)}\Big]$$

subject to the (controlled forward) stochastic heat equation with an additive noise

$$\begin{cases} dy(t) = \big(Ay(t) + u(t)\big)dt + \sigma dW(t), & t \in (0, T], \\ y(0) = y_0. \end{cases} \tag{5.1}$$

Here A is defined in (3.6), and $y_0, \sigma \in H_0^1(D) \cap H^2(D)$. Let us now state an SLQ problem as follows:

Problem (SLQ). Find a $\bar{u}(\cdot) \in L^2_{\mathbb{F}}(0, T; L^2(D))$ such that

$$\mathcal{J}(y_0; \bar{u}(\cdot)) = \inf_{u(\cdot) \in L^2_{\mathbb{F}}(0,T;L^2(D))} \mathcal{J}(y_0; u(\cdot)).$$

By Lü and Zhang (2015), the solvability of Problem (SLQ) is equivalent to the solvability of the following forward-backward stochastic evolution equation:

$$\begin{cases} d\bar{y}(t) = \big(A\bar{y}(t) + \bar{u}(t)\big)dt + \sigma dW(t), & t \in (0, T], \\ dz(t) = \big(-Az(t) + \bar{y}(t)\big)dt + Z(t)dW(t), & t \in [0, T), \\ \bar{y}(0) = y_0, \qquad z(T) = -\bar{y}(T), \end{cases} \tag{5.2}$$

with the condition

$$\bar{u}(t) - z(t) = 0, \quad \text{a.e. } (t, \omega) \in [0, T] \times \Omega. \tag{5.3}$$

In the following, we mainly propose a discretization of (5.2)–(5.3). To be specific, for the forward equation $(5.2)_1$, we adopt the time-implicit Galerkin method (see, e.g., Grecksch and Kloeden, 1996), while for the backward one $(5.2)_2$, we apply the finite transposition method proposed in Section 4; see also Algorithm 1 below.

Since (5.2)–(5.3) is coupled, it is nontrivial to analyze the convergence of the above discretization strategy. In what follows, we adopt optimal control theory to deduce the convergence rates. To do this, we discretize Problem (SLQ) within two steps: firstly, we get the spacial semi-discretization (which is referred to as Problem $(\text{SLQ})_S$); secondly, we obtain a spatio-temporal discretization (which is referred to as Problem $(\text{SLQ})_{ST}$). Now, we present these two discretizations.

Problem (SLQ)$_S$. For a fixed $n \in \mathbb{N}$, minimize the following cost functional over $L^2_{\mathbb{F}}(0,T;\mathbb{S}_n)$:

$$\mathcal{J}(\Pi_n y_0; u(\cdot)) = \frac{1}{2}\mathbb{E}\Big[\int_0^T \big(\|y_n(t)\|^2_{L^2(D)} + \|u_n(t)\|^2_{L^2(D)}\big)dt + \|y_n(T)\|^2_{L^2(D)}\Big]$$

subject to the following $\mathbb{S}_n$-valued stochastic differential equation:

$$\begin{cases} dy_n(t) = \big(A_n y_n(t) + u_n(t)\big)dt + \Pi_n \sigma dW(t), & t \in [0,T], \\ y_n(0) = \Pi_n y_0 = \sum_{k=1}^{n} \langle y_0, \phi_k \rangle_{L^2(D)} \phi_k. \end{cases} \tag{5.4}$$

By Yong and Zhou (1999, Proposition 5.5 in Chapter 6), the solvability of Problem (SLQ)$_S$ is equivalent to the solvability of the following forward-backward stochastic differential equation:

$$\begin{cases} d\bar{y}_n(t) = \big(A_n \bar{y}_n(t) + \bar{u}_n(t)\big)dt + \Pi_n \sigma dW(t), & t \in (0,T], \\ dz_n(t) = \big(-A_n z_n(t) + \bar{y}_n(t)\big)dt + Z_n(t)dW(t), & t \in [0,T), \\ \bar{y}_n(0) = \Pi_n y_0, \qquad z_n(T) = -\bar{y}_n(T), \end{cases} \tag{5.5}$$

with the condition

$$\bar{u}_n(t) - z_n(t) = 0, \quad \text{a.e. } (t,\omega) \in [0,T] \times \Omega. \tag{5.6}$$

The following result is on the rate of convergence for spacial discretization of Problem (SLQ).

Theorem 5.1. *Let* $(\bar{y}(\cdot), \bar{u}(\cdot))$ *be the optimal pair of Problem* (SLQ) *and* $(\bar{y}_n(\cdot), \bar{u}_n(\cdot))$ *be the solution of Problem* (SLQ)$_S$ *for* $n \in \mathbb{N}$. *Then, there exists a constant* $\mathcal{C}$ *such that*

(i) $\mathbb{E}\big(\sup_{t\in[0,T]} \|\bar{u}(t) - \bar{u}_n(t)\|^2_{L^2(D)}\big) \le \dfrac{\mathcal{C}}{\lambda_{n+1}}$,

(ii) $\mathbb{E}\big(\sup_{t\in[0,T]} \|\bar{y}(t) - \bar{y}_n(t)\|^2_{L^2(D)}\big) + \mathbb{E}\displaystyle\int_0^T \|\bar{y}(t) - \bar{y}_n(t)\|^2_{H_0^1(D)} dt \le \dfrac{\mathcal{C}}{\lambda_{n+1}}$.

Proof. Applying Itô's formula to $\|\nabla \bar{y}(\cdot)\|^2_{L^2(D)} + \|\bar{y}(\cdot)\|^2_{L^2(D)}$, we obtain

$$\sup_{t\in[0,T]} \mathbb{E}\|\bar{y}(t)\|^2_{H_0^1(D)} \le \mathcal{C}\Big[\|y_0\|^2_{H_0^1(D)} + \mathbb{E}\int_0^T \Big(\|\bar{u}(t)\|^2_{L^2(D)} + \|\sigma\|^2_{L^2(D)}\Big)dt\Big].$$

On the other hand, Theorem 4.1 yields that

$$\begin{aligned}&\mathbb{E}\big(\sup_{t\in[0,T]}\|z(t)-z_n(t)\|^2_{L^2(D)}\big)\\&\le\frac{\mathcal{C}}{\lambda_{n+1}}\Big(\mathbb{E}\|\bar{y}(T)\|^2_{H_0^1(D)}+\mathbb{E}\int_0^T\|\bar{y}(t)\|^2_{H_0^1(D)}dt\Big).\end{aligned}$$

The maximum conditions (5.3) and (5.6) and above two inequalities lead to the assertion (i).

Itô's formula to $\|\bar{y}(\cdot)-\bar{y}_n(\cdot)\|^2_{L^2(D)}$, Gronwall's inequality and the assertion (i) lead to

$$\begin{aligned}&\sup_{t\in[0,T]}\mathbb{E}\|\bar{y}(t)-\bar{y}_n(t)\|^2_{L^2(D)}+\mathbb{E}\int_0^T\|\nabla(\bar{y}(s)-\bar{y}_n(s))\|^2_{L^2(D)}ds\\&\le\mathcal{C}\Big[\|y_0-\Pi_n y_0\|^2_{L^2(D)}+\mathbb{E}\int_0^T\big[\|\bar{u}(s)-\bar{u}_n(s)\|^2_{L^2(D)}+\|\sigma-\Pi_n\sigma\|^2_{L^2(D)}\big]ds\Big]\\&\le\frac{\mathcal{C}}{\lambda_{n+1}}.\end{aligned}\tag{5.7}$$

Applying Itô's formula to $\|\bar{y}(\cdot)-\bar{y}_n(\cdot)\|^2_{L^2(D)}$ again, then using Burkholder-Davis-Gundy inequality and (5.7), we can derive the assertion (ii). □

Before presenting the spatio-temporal discretization of Problem (SLQ), we introduce the following two spaces:

$$\begin{aligned}\mathbb{X}_\tau&\triangleq\Big\{x(\cdot)\in L^2_{\mathbb{F}}(0,T;\mathbb{S}_n)\;\big|\;x(t)=x(t_k),\ \forall t\in I_k,\ k=0,1,\cdots,N-1\Big\},\\\mathbb{U}_\tau&\triangleq\Big\{u(\cdot)\in L^2_{\mathbb{F}}(0,T;\mathbb{S}_n)\;\big|\;u(t)=u(t_k),\ \forall t\in I_k,\ k=0,1,\cdots,N-1\Big\},\end{aligned}$$

where I_k is defined in (3.3), and for any $x(\cdot)\in\mathbb{X}_\tau$ and $u(\cdot)\in\mathbb{U}_\tau$,

$$\begin{aligned}\|x(\cdot)\|_{\mathbb{X}_\tau}&\triangleq\Big(\tau\sum_{n=1}^{N}\mathbb{E}\|x(t_n)\|_{L^2(D)}\Big)^{1/2},\\\|u(\cdot)\|_{\mathbb{U}_\tau}&\triangleq\Big(\tau\sum_{n=0}^{N-1}\mathbb{E}\|u(t_n)\|_{L^2(D)}\Big)^{1/2}.\end{aligned}$$

Problem (SLQ)$_{ST}$. For fixed $N,n\in\mathbb{N}$, minimize the following functional over $\mathbb{U}_\tau$:

$$\mathcal{J}_{ST}(\Pi_n y_0;u_{nN}(\cdot))=\frac{1}{2}\Big(\|y_{nN}(\cdot)\|^2_{\mathbb{X}_\tau}+\|u_{nN}(\cdot)\|^2_{\mathbb{U}_\tau}\Big)+\frac{1}{2}\mathbb{E}\,\|y_{nN}(T)\|^2_{L^2(D)},$$

where $(y_{nN}(\cdot), u_{nN}(\cdot))$ satisfies

$$\begin{cases} y_{nN}(t_{k+1}) - y_{nN}(t_k) = \tau\big[A_n y_{nN}(t_{k+1}) + u_{nN}(t_k)\big] \\ \qquad\qquad\qquad\qquad + \Pi_n \sigma \Delta_{k+1} W, \qquad k = 0, 1, \cdots, N-1, \\ y_{nN}(0) = \Pi_n y_0. \end{cases} \tag{5.8}$$

The following result, which guarantees the solvability of Problem $(\mathrm{SLQ})_{ST}$, is a variant of Prohl and Wang (2021b, Theorem 4.2).

Theorem 5.2. *Problem* $(\mathrm{SLQ})_{ST}$ *admits a unique minimizer* $(\bar{y}_{nN}(\cdot), \bar{u}_{nN}(\cdot)) \in \mathbb{X}_\tau \times \mathbb{U}_\tau$, *which satisfies the following coupled equation:*

$$\begin{cases} \bar{y}_{nN}(t_{k+1}) = \Lambda_0 \bar{y}_{nN}(t_k) + \tau \Lambda_0 \bar{u}_{nN}(t_k) + \Lambda_0 \Pi_n \sigma \Delta_{k+1} W, \\ z_{nN}(t_k) = \Lambda_0 \mathbb{E}\left(z_{nN}(t_{k+1}) - \tau \bar{y}_{nN}(t_{k+1}) \,\middle|\, \mathcal{F}_{t_k}\right), \\ \bar{y}_{nN}(0) = y_{nN}(0), \quad z_{nN}(T) = -\bar{y}_{nN}(T), \end{cases} \tag{5.9}$$

together with the maximum condition

$$\bar{u}_{nN}(t_k) - z_{nN}(t_k) = 0, \qquad k = 0, 1, \cdots, N-1, \quad a.s.\,,$$

where $\Lambda_0 = (\mathbb{I}_n - \Lambda_n \tau)^{-1}$.

Based on Malanowski (1982), as well as the regularity of the optimal pairs $(\bar{y}_n(\cdot), \bar{u}_n(\cdot))$ and $(\bar{y}_{nN}(\cdot), \bar{u}_{nN}(\cdot))$, we can prove the following convergence rates (e.g., Prohl and Wang, 2021b, Theorem 4.3).

Theorem 5.3. *For any* $n, N \in \mathbb{N}$, *assume that* $(\bar{y}_n(\cdot), \bar{u}_n(\cdot))$ *and* $(\bar{y}_{nN}(\cdot), \bar{u}_{nN}(\cdot))$ *are the optimal pairs of Problem* $(\mathrm{SLQ})_S$ *and* $(\mathrm{SLQ})_{ST}$, *respectively. Then,*

(i) $\displaystyle\sum_{k=0}^{N-1} \mathbb{E} \int_{t_k}^{t_{k+1}} \|\bar{u}_n(t) - \bar{u}_{nN}(t_k)\|^2_{L^2(D)}\, dt \le \mathcal{C}\tau;$

(ii) $\displaystyle\max_{0\le k\le N} \mathbb{E}\,\|\bar{y}_n(t_k) - \bar{y}_{nN}(t_k)\|^2_{L^2(D)} + \tau\mathbb{E}\sum_{k=1}^{N} \|\bar{y}_n(t_k) - \bar{y}_{nN}(t_k)\|^2_{H_0^1(D)} \le \mathcal{C}\tau.$

Remark 5.1. Eq. (5.9) is the implicit Euler scheme for Eq. (5.5) and the Galerkin based implicit Euler method for (5.2). Notice that (5.2) and (5.5) are coupled forward-backward equations. Numerical methods for these coupled equations and the convergence analysis is highly nontrivial. To deduce the rates of convergence, additional conditions such as sufficiently small T, or equations' weak coupling, are needed (see, e.g., Bender and Zhang, 2008). In this section, we show an idea to solve these equations. To be specific, if a coupled equation can be transferred to an optimal control problem, one can try to utilize tools in optimal control theory.

Eq. (5.9) is a coupled equation and contains conditional expectations. There are some methods to deal with these conditional expectations. But for coupled forward-backward equation, even for that in finite dimensions, numerical schemes are rare and the convergence analysis is far from complete (see, e.g., Bender and Zhang, 2008). Here we can adopt the finite transposition method and gradient method to solve (5.2).

Algorithm 1: Solving Problem (SLQ) by the Galerkin based finite transposition method and the gradient method

Fix $N, M, n \in \mathbb{N}$, and $\kappa > 0$. Let $\widetilde{u}^{(0)} \in \mathbb{H}_{N,M}(\mathbb{S}_n)$. For any $\ell \in \mathbb{N}$, update $\widetilde{u}^{(\ell)} \in \mathbb{H}_{N,M}(\mathbb{S}_n)$ as follows:

1. Compute $\widetilde{y}^{(\ell)} \in \mathbb{H}_{N,M}(\mathbb{S}_n)$ via the time-implicit Galerkin method:

$$\begin{cases} \widetilde{y}^{(\ell)}(t_{n+1}) - \widetilde{y}^{(\ell)}(t_n) = \tau A_n \widetilde{y}^{(\ell)}(t_{n+1}) + \Pi_n \sigma \Delta_{n+1} W, & n = 0, 1, \cdots, N-1, \\ \widetilde{y}^{(\ell)}(0) = \Pi_n y_0. \end{cases}$$

2. Utilize $\widetilde{y}^{(\ell)} \in \mathbb{H}_{N,M}(\mathbb{S}_n)$ in $(5.2)_2$ to compute $\widetilde{z}^{(\ell)} \in \mathbb{H}_{N,M}(\mathbb{S}_n)$ via the finite transposition method.
3. Compute the update $\widetilde{u}^{(\ell+1)} \in \mathbb{H}_{N,M}(\mathbb{S}_n)$ via

$$\widetilde{u}^{(\ell+1)} = \left(1 - \frac{1}{\kappa}\right)\widetilde{u}^{(\ell)} + \frac{1}{\kappa}\widetilde{z}^{(\ell)}.$$

Finally, Algorithm 1 is used to show how to solve Problem (SLQ) numerically. In this algorithm, κ can be taken as $1 + T + T^2$. We can prove that $\big(\widetilde{u}^{(\ell)}(\cdot), \widetilde{y}^{(\ell)}(\cdot)\big)$ converges to $(\bar{u}(\cdot), \bar{y}(\cdot))$ in $\mathbb{X}_\tau \times \mathbb{U}_\tau$ as $\ell \to \infty$ (e.g., Prohl and Wang, 2021b, Section 5).

References

Al-Hussein, A., 2009. Backward stochastic partial differential equations driven by infinite dimensional martingales and applications. Stochastics 81 (6), 601–626.

Bally, V., Pagès, G., 2003. Error analysis of the optimal quantization algorithm for obstacle problems. Stoch. Process. Appl. 106 (1), 1–40.

Bender, C., Denk, R., 2007. A forward scheme for backward SDEs. Stoch. Process. Appl. 117 (12), 1793–1812.

Bender, C., Zhang, J., 2008. Time discretization and Markovian iteration for coupled FBSDEs. Ann. Appl. Probab. 18 (1), 143–177.

Bensoussan, A., 1969. Contrôle optimal stochastique de système gouvernés par des équations aux dérivées partielles de type parabolique. Rend. Mat. (6) 2, 135–173.

Bensoussan, A., 1983. Stochastic maximum principle for distributed parameter systems. J. Franklin Inst. 315 (5–6), 387–406.

Bouchard, B., Touzi, N., 2004. Discrete-time approximation and Monte-Carlo simulation of backward stochastic differential equations. Stoch. Process. Appl. 111 (2), 175–206.

Briand, P., Labart, C., 2014. Simulation of BSDEs by Wiener chaos expansion. Ann. Appl. Probab. 24 (3), 1129–1171.

Dai, L., Zhang, Y., Zou, J., 2017. Numerical schemes for forward-backward stochastic differential equations using transposition solutions. Preprint.

Douglas Jr., J., Ma, J., Protter, P., 1996. Numerical methods for forward-backward stochastic differential equations. Ann. Appl. Probab. 6 (3), 940–968.

Dunst, T., Prohl, A., 2016. The forward-backward stochastic heat equation: numerical analysis and simulation. SIAM J. Sci. Comput. 38 (5), A2725–A2755.

E, W., Hutzenthaler, M., Jentzen, A., Kruse, T., 2019. On multilevel Picard numerical approximations for high-dimensional nonlinear parabolic partial differential equations and high-dimensional nonlinear backward stochastic differential equations. J. Sci. Comput. 79 (3), 1534–1571.

Ghanem, R.G., Spanos, P.D., 1991. Stochastic Finite Elements: a Spectral Approach. Springer-Verlag, New York.

Gobet, E., Lemor, J.-P., Warin, X., 2005. A regression-based Monte Carlo method to solve backward stochastic differential equations. Ann. Appl. Probab. 15 (3), 2172–2202.

Grecksch, W., Kloeden, P.E., 1996. Time-discretised Galerkin approximations of parabolic stochastic PDEs. Bull. Aust. Math. Soc. 54 (1), 79–85.

Hu, Y., Nualart, D., Song, X., 2011. Malliavin calculus for backward stochastic differential equations and application to numerical solutions. Ann. Appl. Probab. 21 (6), 2379–2423.

Hu, Y., Peng, S.G., 1991. Adapted solution of a backward semilinear stochastic evolution equation. Stoch. Anal. Appl. 9 (4), 445–459.

Kushner, H.G., 1968. On the optimal control of a system governed by a linear parabolic equation with white noise inputs. SIAM J. Control 6, 596–614.

Li, B., Zhou, Q., 2021. Discretization of a distributed optimal control problem with a stochastic parabolic equation driven by multiplicative noise. J. Sci. Comput. 87, 68.

Lions, J.-L., Magenes, E., 1972. Non-homogeneous Boundary Value Problems and Applications. Vol. I. Springer-Verlag, New York-Heidelberg.

Lü, Q., Yong, J., Zhang, X., 2012. Representation of Itô integrals by Lebesgue/Bochner integrals. J. Eur. Math. Soc. 14 (6), 1795–1823.

Lü, Q., Yong, J., Zhang, X., 2018. Erratum to "Representation of Itô integrals by Lebesgue/Bochner integrals" (J. Eur. Math. Soc. 14, 1795-1823 (2012)). J. Eur. Math. Soc. 20 (1), 259–260.

Lü, Q., Zhang, X., 2013. Well-posedness of backward stochastic differential equations with general filtration. J. Differ. Equ. 254 (8), 3200–3227.

Lü, Q., Zhang, X., 2014. General Pontryagin-Type Stochastic Maximum Principle and Backward Stochastic Evolution Equations in Infinite Dimensions. SpringerBriefs in Mathematics. Springer, Cham.

Lü, Q., Zhang, X., 2015. Transposition method for backward stochastic evolution equations revisited, and its application. Math. Control Relat. Fields 5 (3), 529–555.

Lü, Q., Zhang, X., 2021. Mathematical Control Theory for Stochastic Partial Differential Equations. Probability Theory and Stochastic Modelling, vol. 101. Springer, Switzerland AG.

Malanowski, K., 1982. Convergence of approximations vs. regularity of solutions for convex, control-constrained optimal-control problems. Appl. Math. Optim. 8 (1), 69–95.

Nualart, D., 2006. The Malliavin Calculus and Related Topics, second ed. Springer-Verlag, Berlin.

Prohl, A., Wang, Y., 2021a. Strong error estimates for a space-time discretization of the linear-quadratic control problem with the stochastic heat equation with linear noise. IMA J. Numer. Anal. https://doi.org/10.1093/imanum/drab069. In press.

Prohl, A., Wang, Y., 2021b. Strong rates of convergence for a space-time discretization of the backward stochastic heat equation, and of a linear-quadratic control problem for the stochastic heat equation. ESAIM Control Optim. Calc. Var., 54.

Tzafestas, G., Nightingale, J.M., 1968. Optimal control of a class of linear stochastic distributed-parameter systems. Proc. Inst. Electr. Eng. 115, 1213–1220.

Wang, P., Wang, Y., Zhang, X., 2020. Error analysis of finite transposition method for solving backward stochastic differential equations. Preprint.

Wang, P., Zhang, X., 2011. Numerical solutions of backward stochastic differential equations: a finite transposition method. C. R. Math. Acad. Sci. Paris 349 (15–16), 901–903.

Wang, Y., 2013. Transposition Solutions of Backward Stochastic Differential Equations and Numerical Schemes. Ph. D. Thesis. Academy of Mathematics and Systems Science, Chinese Academy of Sciences.

Wang, Y., 2016. A semidiscrete Galerkin scheme for backward stochastic parabolic differential equations. Math. Control Relat. Fields 6 (3), 489–515.

Wang, Y., 2020. L^2-regularity of solutions to linear backward stochastic heat equations, and a numerical application. J. Math. Anal. Appl. 486 (1), 123870.

Yong, J., Zhou, X.Y., 1999. Stochastic Controls: Hamiltonian Systems and HJB Equations. Springer-Verlag, New York.

Zhang, J., 2004. A numerical scheme for BSDEs. Ann. Appl. Probab. 14 (1), 459–488.

Zhao, W., Chen, L., Peng, S., 2006. A new kind of accurate numerical method for backward stochastic differential equations. SIAM J. Sci. Comput. 28 (4), 1563–1581.

Zhao, W., Fu, Y., Zhou, T., 2014. New kinds of high-order multistep schemes for coupled forward backward stochastic differential equations. SIAM J. Sci. Comput. 36 (4), A1731–A1751.

Chapter 7

Numerical solutions of stochastic control problems: Markov chain approximation methods

Zhuo Jin[a,d], **Ky Tran**[b,e], **and George Yin**[c,*,f]
[a] *Department of Actuarial Studies and Business Analytics, Macquarie University, North Ryde, NSW, Australia,* [b] *Department of Applied Mathematics and Statistics, SUNY Korea, Incheon, Republic of Korea,* [c] *Department of Mathematics, University of Connecticut, Storrs, CT, United States*
**Corresponding author: e-mail address: gyin@uconn.edu*

Contents

Abstract

This chapter is devoted to Markov chain approximation methods for solving stochastic control problems numerically. While the underlying problems are in continuous time, the numerical methods are based on the construction of discrete-time controlled Markov chains that are locally consistent so that the local mean and covariance coincide with that of the continuous-time controlled stochastic systems. One of the main advantages of the Markov chain approximation method is that not much of *a priori* knowledge of the regularity of the related partial differential equations is needed.

Keywords

Controlled switching diffusion, Numerical method, Markov chain approximation

[d] The research of Z. Jin was supported in part by Research Grants Council of the Hong Kong Special Administrative Region (project No. 17304921).
[e] The research of K. Tran was supported in part by the National Research Foundation of Korea Grant funded by the Korea Government (MIST) NRF-2021R1F1A1062361.
[f] The research of G. Yin was supported in part by the US National Science Foundation under grant DMS-2114649.

MSC Codes
60J27, 65C99, 65Kxx, 93E20

This chapter begins with an introduction to controlled switching diffusions and regular and singular controls of the corresponding systems. Then Markov chain approximation methods are discussed together with several application examples. Because of the page limitation, proofs of convergence are not given. However, appropriate references are provided. The rest of the chapter is arranged as follows. Formulations of controlled switching diffusions are recalled next. Section 2 introduces the Markov chain approximation methods, followed by numerical methods for insurance problems in Section 3, and applications to mathematical biology in Section 4. Section 5 concludes the paper with further remarks on rates of convergence and the use of deep learning.

When the switching is absent, the systems reduce to controlled diffusions. The numerical methods of this chapter can be used to treat numerous controlled diffusions and controlled switching diffusions. Note that switching diffusions and diffusions may display very different behavior. A simple example shows that putting two well-behaved linear equations together with switching can produce a switching system with finite explosion time; see Wang et al. (1999) for details.

Further readings can be found in Fleming and Soner (2006) and Yong and Zhou (1999) for controlled diffusions, Yin and Zhu (2010) for switching diffusions, Lü and Zhang (2021) for controlled stochastic partial differential equations, Kushner (1977), Kushner (2002), and Kushner and Dupuis (2001) for numerical methods of controlled diffusions, Song et al. (2006), Song et al. (2008), and Song and Yin (2009) for numerical methods of controlled switching diffusions and games. While this chapter focuses on Markov chain approximation methods, there are many other numerical methods. An often used technique to prove the convergence of numerical schemes is the viscosity solution method of Barles and Souganidis (1991). For approximation to the solutions of Hamilton-Jacobi-Bellman (HJB) equations, we refer the reader to Falcone and Ferretti (2014), McEneaney (2007), Jiang and Dai (2004), Kalise et al. (2020), Fehrenbach and Mirebeau (2014), and references therein; see also Banks and Kunisch (1989) for related inverse problems and numerical methods.

1 Stochastic control problems

Throughout this chapter, $(\Omega, \mathcal{F}, \mathbb{F}, \mathbb{P})$ is a complete filtered probability space, where $\mathbb{F} := \{\mathcal{F}_t\}$ is a right-continuous, $\mathbb{P}$-complete filtration. Suppose that $W(\cdot)$ is a standard $\mathbb{P}$-Brownian motion with appropriate dimension and that $\Lambda(\cdot)$ is a continuous-time Markov chain taking values in $\mathcal{M} = \{1, 2, \ldots, m_0\}$ with a generator $Q = (q_{ij}) \in \mathbb{R}^{m_0 \times m_0}$. Moreover, $W(\cdot)$ and $\Lambda(\cdot)$ are adapted to $\{\mathcal{F}_t\}$ and are independent. For $x, y \in \mathbb{R}^d$, with $x = (x_1, \ldots, x_d)^\top$ and $y = (y_1, \ldots, y_d)^\top$, $x \cdot y = \sum_{i=1}^d x_i y_i$ denotes the scalar product, where $z^\top$ denotes the transpose of

z. For a real number s, we denote by $s^+ = \max\{s, 0\}$ and $s^- = \max\{-s, 0\}$. Denote by $\mathbf{e_i} \in \mathbb{R}^d$ the standard unit vector in the ith direction for $i = 1, 2, \ldots, d$.

Controlled switching diffusions. A switching diffusion has a continuous component representing the continuous dynamics and a discrete component delineating the discrete events. The continuous dynamics and discrete events coexist and interact. Suppose that $\mathcal{M} = \{1, \ldots, m_0\}$ represents the possible values of the discrete state (regimes). In this chapter, we assume that $\Lambda(\cdot)$ is a continuous-time Markov chain having state space $\mathcal{M}$ with a generator $Q = (q_{ij}) \in \mathbb{R}^{m_0 \times m_0}$ satisfying $q_{ij} \geq 0$ for $i \neq j$ and $\sum_j q_{ij} = 0$ for each $i \in \mathcal{M}$. By the Markov property, $\mathbb{P}(\Lambda(t+\Delta) = j | \Lambda(t) = i) = q_{ij}\Delta + o(\Delta)$ if $j \neq i$, and $= 1 + q_{ii}\Delta + o(\Delta)$ otherwise. The construction of $\Lambda(\cdot)$ can be summarized as follows. Let $\Lambda(0) = \alpha \in \mathcal{M}$. Without loss of generality, we suppose that $q_{\ell\ell} \neq 0$ for all $\ell \in \mathcal{M}$. Let $\tau_0 = 0$ and τ_1 be an exponential random variable with mean $-1/q_{\alpha\alpha}$. Then $\Lambda(t) = \alpha$ on the time interval $[\tau_0, \tau_1)$, and τ_1 is the first jump time of $\Lambda(\cdot)$ with the post-jump location $\Lambda(\cdot)$ given by $\mathbb{P}(\Lambda(\tau_1) = \ell) = -q_{\alpha\ell}/q_{\alpha\alpha}$, if $\ell \neq \alpha$. Let $\{\tau_n : n \geq 1\}$ be the sequence of jump times of $\Lambda(\cdot)$. In general, $\Lambda(t) = \Lambda(\tau_n)$ on the time interval $[\tau_n, \tau_{n+1})$. Moreover, given $\{\tau_k, \Lambda(\tau_k) : k \leq n\}$, $\tau_{n+1} - \tau_n$ is an exponential random variable with mean $-1/q_{\Lambda(\tau_n)\Lambda(\tau_n)}$. The post-jump location of $\Lambda(\cdot)$ satisfies $\mathbb{P}(\Lambda(\tau_{n+1}) = j \big| \tau_k, \Lambda(\tau_k) : k \leq n) = -q_{\Lambda(\tau_n)j}/q_{\Lambda(\tau_n)\Lambda(\tau_n)}$, for $j \neq \Lambda(\tau_n)$. The controlled dynamic system of interest is given by $X(0) = x$, $\Lambda(0) = \alpha$, and

$$dX(t) = b(X(t), \Lambda(t), u(t))dt + \sigma(X(t), \Lambda(t))dW(t), \tag{1}$$

where $u(\cdot)$ is a control taking values in a nonempty compact set $\mathcal{U} \subset \mathbb{R}^d$, and $b(\cdot) : \mathbb{R}^d \times \mathcal{M} \mapsto \mathbb{R}^d$ and $\sigma(\cdot) : \mathbb{R}^d \times \mathcal{M} \mapsto \mathbb{R}^{d\times d}$ are deterministic functions representing the drift and diffusion coefficients of the controlled switching diffusion, respectively; see Yin and Zhu (2010) for further discussion on switching diffusions and controlled switching diffusions. Let G be a nonempty bounded open subset of $\mathbb{R}^d$ and $\tau = \inf\{t \geq 0 : X(t) \notin G\}$ be the first exit time from G. With given initial data (x, α), consider the discounted payoff functional

$$J(x, \alpha, u(\cdot)) = \mathbb{E}_{x,\alpha}[\int_0^\tau e^{-rt} f(X(t), \Lambda(t), u(t))dt + e^{-r\tau} g(X(\tau), \Lambda(\tau))], \tag{2}$$

where $r > 0$ is the discount factor, $f : \mathbb{R}^d \times \mathcal{M} \times \mathcal{U} \mapsto \mathbb{R}$ and $g : \mathbb{R}^d \times \mathcal{M} \mapsto \mathbb{R}$ are appropriate deterministic functions representing the running payoff and terminal payoff, respectively.

The control $u(\cdot)$ is said to be *admissible* if it satisfies (a) $X(t) \in G$ for any $t < \tau$, where τ is the first exit time from G; (b) $u(\cdot)$ takes values in $\mathcal{U}$ and is adapted to $\sigma\{W(s), \Lambda(s) : 0 \leq s \leq t\}$ augmented by the $\mathbb{P}$-null sets; (c) $J(x, \alpha, u(\cdot)) < \infty$, where $J(x, \alpha, u(\cdot))$ is the functional defined in (2).

Let $\mathcal{A}_{x,\alpha}$ be the collection of all admissible controls with the initial data (x, α). The value function is then defined by

$$V(x, \alpha) := \sup_{u(\cdot)\in\mathcal{A}_{x,\alpha}} J(x, \alpha, u(\cdot)). \tag{3}$$

By (2) and (3), $V(x,\alpha) = g(x,\alpha)$ for $(x,\alpha) \notin G \times \mathcal{M}$. For a function $\Phi(\cdot,\cdot): \mathbb{R}^d \times \mathcal{M} \mapsto \mathbb{R}$ satisfying $\Phi(\cdot,\alpha) \in C^2(\mathbb{R}^d)$ for each $\alpha \in \mathcal{M}$, we define

$$\mathcal{L}^{\xi}\Phi(x,\alpha) = b(x,\alpha,\xi)\nabla_x\Phi(x,\alpha) + \frac{1}{2}\mathrm{tr}\big(\sigma(x,\alpha)\sigma^{\top}(x,\alpha)\nabla_x^2\Phi(x,\alpha)\big), \quad (4)$$

where $\nabla_x\Phi(\cdot,\alpha)$ and $\nabla_x^2\Phi(\cdot,\alpha)$ denote the gradient and Hessian matrix of $\Phi(\cdot,\alpha)$, respectively. Formally, by applying the dynamic programming principle, the value function $V(\cdot)$ satisfies the following system of HJB equations

$$\begin{aligned}
&\sup_{\xi\in\mathcal{U}}[(\mathcal{L}^{\xi} - r)V(x,\alpha) + f(x,\alpha,\xi)] + QV(x,\cdot)(\alpha) = 0, \ \ (x,\alpha) \in G \times \mathcal{M},\\
&V(x,\alpha) = g(x,\alpha) \ \text{ for } \ (x,\alpha) \notin G \times \mathcal{M}, \ \text{ with}\\
&QV(x,\cdot)(\alpha) := \sum_{j\in\mathcal{M}} q_{\alpha j}[V(x,j) - V(x,\alpha)].
\end{aligned} \quad (5)$$

As a standing assumption to be used in the chapter, we suppose that for each $\alpha \in \mathcal{M}$, the functions $f(\cdot,\alpha)$ and $g(\cdot,\alpha)$ are bounded and continuous, and the system coefficients $b(\cdot,\alpha)$ and $\sigma(\cdot,\alpha)$ are locally Lipschitz continuous. Then the value function $V(\cdot,\cdot)$ is bounded on $\mathbb{R}^d \times \mathcal{M}$.

Singular controls. Building on the controlled switching diffusion (1), we consider the model with $X(0^-) = x$, $\Lambda(0) = \alpha$, and

$$dX(t) = b(X(t),\Lambda(t),u(t))dt + \sigma(X(t),\Lambda(t))dW(t) - dZ(t), \quad (6)$$

where $u(\cdot)$, termed regular control, takes values in a nonempty compact set $\mathcal{U}$ and $Z(\cdot)$ is a singular stochastic control in the sense that it may not be absolutely continuous with respect to the Lebesgue measure of the time variable; see Jin et al. (2012) for details and see Pham (2009) for syntheses of singular control problems for diffusion processes. Let G be a nonempty subset of $\mathbb{R}_+^d$ and τ be defined by

$$\tau = \inf\{t \ge 0 : X(t) \notin G\}. \quad (7)$$

We assume the initial state is $X(0-) = x$ and initial regime is $\Lambda(0) = \alpha$. Hence $X(0)$ may not be equal to $X(0-)$ due to an instantaneous control $Z(0)$. We use the convention that $Z(0-) = 0$.

For a control policy $(u(\cdot), Z(\cdot))$, the accumulated discounted payoffs of the control $u(\cdot)$ and the singular control $Z(\cdot)$ are $\int_0^{\tau} e^{-rt} f(X(t),\Lambda(t),u(t))dt$ and $\int_0^{\tau} e^{-rt} c \cdot dZ(t)$, where $r > 0$ is the discounting factor, f is a function representing the running yield, $c = (c_1,\ldots,c_d)^{\top} \in \mathbb{R}^d$ and $c \cdot dZ(t) = \sum_{j=1}^{d} c_j dZ_j(t)$. Thus, the expected discounted payoff functional is given by

$$J(x,\alpha,u(\cdot),Z(\cdot)) = \mathbb{E}_{x,\alpha}[\int_0^{\tau} e^{-rt}(f(X(t),\Lambda(t),u(t))dt + c \cdot dZ(t))]. \quad (8)$$

Here c is a constant vector in $\mathbb{R}^d$. While the methods do not work if c is x-dependent, it seems that the case $c = c(\alpha)$ can be handled. However, since the proofs involve a time rescaling, more thoughts and investigations are needed. A control policy $(u(\cdot), Z(\cdot))$ is said to be *admissible* with the initial condition (x, α) if it satisfies (a) $X(t) \in G$ for any $t < \tau$, where τ is the stopping time defined in (7); (b) $u(\cdot)$ and $Z(\cdot)$ are adapted to $\sigma\{W(s), \Lambda(s) : 0 \le s \le t\}$ augmented by the $\mathbb{P}$-null sets; $u(\cdot)$ takes values in $\mathcal{U}$, $Z(\cdot)$ has nonnegative, component-wise non-decreasing, and right continuous sample paths; (c) $J(x, \alpha, u(\cdot), Z(\cdot)) < \infty$, where $J(x, \alpha, u(\cdot), Z(\cdot))$ is the functional defined in (8).

Let $\mathcal{A}_{x,\alpha}$ be the collection of all admissible controls with the initial condition (x, α). The value function is then defined by

$$V(x,\alpha) := \sup_{(u(\cdot),Z(\cdot))\in\mathcal{A}_{x,\alpha}} J(x,\alpha,u(\cdot),Z(\cdot)). \tag{9}$$

Formally, by applying the dynamic programming principle, the value function $V(\cdot)$ satisfies the following system of HJB variational inequalities

$$\begin{aligned} &\max\{\sup_{\xi\in\mathcal{U}}\{(\mathcal{L}^{\xi} - r)V(x,\alpha) + f(x,\alpha,\xi)\} + QV(x,\cdot)(\alpha) = 0,\\ &\qquad c - V_x(x,\alpha)\} = 0,\\ &V(x,\alpha) = 0, \quad (x,\alpha) \notin G \times \mathcal{M}, \end{aligned} \tag{10}$$

where $\mathcal{L}^{\xi}(\cdot)$ and $QV(x,\cdot)(\alpha)$ are defined in (4) and (5), respectively. $c - V_x(x,\alpha) = 0$ means that $c_i - V_{x_i}(x,\alpha) = 0$ for each $i = 1, 2, \ldots, d$. Intuitively, the set G can be divided into two regions: the regular-control region

$$\sup_{\xi\in\mathcal{U}} \left\{(\mathcal{L}^{\xi} - r)V(x,\alpha) + f(x,\alpha,\xi)\right\} + QV(x,\cdot)(\alpha) = 0, \quad c - V_x(x,\alpha) < 0,$$

and the singular-control region

$$\sup_{\xi\in\mathcal{U}} \left\{(\mathcal{L}^{\xi} - r)V(x,\alpha) + f(x,\alpha,\xi)\right\} + QV(x,\cdot)(\alpha) < 0, \quad c - V_x(x,\alpha) = 0.$$

In the regular-control region, the regular control is dominant while in the singular-control region, the singular control is dominant. We refer to (Song et al., 2011, Theorem 4.9) and (Hening et al., 2019, Theorem 2.9) for the derivation of the HJB variational inequalities.

As a standing assumption, we suppose that for each $\alpha \in \mathcal{M}$, the function $f(\cdot, \alpha)$ is bounded and continuous; the system coefficients $b(\cdot, \alpha)$ and $\sigma(\cdot, \alpha)$ are locally Lipschitz continuous. We can show that if (6) has a unique solution for each initial condition $(x, \alpha) \in G \times \mathcal{M}$ and there is a positive number N such that $b(x, \alpha, \xi) \le rx + N$ for any $(x, \alpha, \xi) \in G \times \mathcal{M} \times \mathcal{U}$, then $V(\cdot, \cdot)$ is finite on $\mathbb{R}^d \times \mathcal{M}$; see Hening et al. (2019).

2 Methods of Markov chain approximation

To find the value functions and the optimal controls, one needs to solve the associated HJB equations. However, for controlled switching diffusions, the HJB equation is, in fact a coupled system of nonlinear equations. A closed-form solution is virtually impossible to obtain. The Markov chain approximation methodology becomes a viable alternative. In this section, we introduce the Markov chain approximation for the controlled regime-switching diffusions. A comprehensive introduction of the Markov chain approximation method can be found in Kushner and Dupuis (2001). As pointed out in Kushner and Dupuis (2001), the probabilistic approach using the Markov chain approximation method for controlled diffusions has the following main advantages. First, the Markov chain approximation method allows one to use physical insights derived from the dynamics of the controlled switching diffusions to obtain a suitable approximation scheme. Second, the Markov chain approximation method does not require much regularity of the controlled processes (solutions of the controlled stochastic differential equations) nor does it rely on the properties of the associated HJB equations. This is particularly appealing when not much information concerning the regularity of the associated PDEs is known. In addition, we note that when a Markov chain is used in the systems, we need to treat the process $(X(t), \Lambda(t))$ jointly; see Song et al. (2006) and Song and Yin (2009). To use the Markov chain approximation method, our approximating Markov chain must have two components as well. One component delineates the diffusive behavior whereas the other keeps track of the regimes.

2.1 Controlled switching diffusions

Recall that G is a nonempty bounded open subset of $\mathbb{R}^d$. We construct a discrete-time, finite-state, controlled Markov chain to approximate the controlled switching diffusion process (1). Let $h > 0$ be a discretization parameter for the continuous state variable. For convenience, let $\widetilde{G}$ be a bounded open subset of $\mathbb{R}^d$ so that $\widetilde{G}$ contains the closure of G. Define $S_h := \{x = (k_1 h, \dots, k_d h)^\top : k_1, \dots, k_d \in \mathbb{Z}_+\} \cap \widetilde{G}$. Let $\{(X_n^h, \Lambda_n^h) : n \in \mathbb{Z}_+\}$ be a discrete-time controlled Markov chain with state space $S_h \times \mathcal{M}$. At any discrete-time step n, the magnitude of the control component u_n^h must be specified. The space of controls is $\mathcal{U}$. Let $u^h = \{u_n^h\}$ be a sequence of controls. We denote by $p^h((x, \alpha), (y, \ell)|\xi)$ the transition probability from state (x, i) to another state (y, ℓ) under the control ξ. Denote $\mathcal{F}_n^h = \sigma\{X_k^h, \Lambda_k^h, u_k^h, k \le n\}$.

The sequence u^h is said to be admissible if it satisfies the following conditions: (a) u^h is $\sigma\{X_0^h, \dots, X_n^h, \Lambda_0^h, \dots, \Lambda_n^h, u_0^h, \dots, u_{n-1}^h\}$-adapted, (b) For any $(x, \alpha) \in S_h \times \mathcal{M}$, we have

$$\begin{aligned}
&\mathbb{P}\{(X_{n+1}^h, \Lambda_{n+1}^h) = (x, \alpha)|\mathcal{F}_n^h\} \\
&\quad = \mathbb{P}\{(X_{n+1}^h, \Lambda_{n+1}^h) = (x, \alpha)|X_n^h, \Lambda_n^h, u_n^h\} = p^h((X_n^h, \Lambda_n^h), (x, \alpha)|u_n^h),
\end{aligned}$$

(c) $(X_n^h, \Lambda_n^h) \in S_h \times \mathcal{M}$ for all $n \in \mathbb{Z}_+$.

The collection of all admissible control sequences for initial state (x, α) will be denoted by $\mathcal{A}_{x,\alpha}^h$. For each $(x, \alpha, \xi) \in S_h \times \mathcal{M} \times \mathcal{U}$, we define a family of the interpolation intervals $\Delta t^h(x, \alpha, \xi)$ so that $\sup_{x,\alpha,\xi} \Delta t^h(x, \alpha, \xi) \to 0$ as $h \to 0$. The values of $\Delta t^h(x, \alpha, \xi)$ will be specified later. Next we define $t_0^h = 0$, $\Delta t_k^h = \Delta t^h(X_k^h, \Lambda_k^h, u_k^h)$, $t_n^h = \sum_{k=0}^{n-1} \Delta t_k^h$. For $(x, \alpha) \in S_h \times \mathcal{M}$ and $u^h \in \mathcal{A}_{x,\alpha}^h$, the payoff functional for the controlled Markov chain is defined as

$$J^h(x, \alpha, u^h) = \mathbb{E}\Big[\sum_{k=0}^{\eta_h - 1} e^{-rt_k^h} f(X_k^h, \Lambda_k^h, u_k^h)\Delta t_k^h + e^{-r\eta_h} g(X_{\eta_h}^h, \Lambda_{\eta_h}^h)\Big], \tag{11}$$

with $\eta_h = \inf\{n \geq 0 : X_n^h \notin G\}$. The value function of the controlled Markov chain is

$$V^h(x, \alpha) = \sup_{u^h \in \mathcal{A}_{x,\alpha}^h} J^h(x, \alpha, u^h). \tag{12}$$

In computation, we can compute $V^h(x, \alpha)$ by solving the corresponding dynamic programming equation using either value iteration or policy iteration methods (see Kushner and Dupuis (2001) and references therein); that is, we solve

$$\begin{aligned} V^h(x, \alpha) = \sup_{\xi \in \mathcal{U}} \Bigg[& e^{-r\Delta t^h(x,\alpha,\xi)} \sum_{(y,\ell) \in S_h \times \mathcal{M}} V^h(y, \ell) p^h\big((x, \alpha), (y, \ell) | \xi\big) \\ & + f(x, \alpha, \xi)\Delta t^h(x, \alpha, \xi) \Bigg], \end{aligned} \tag{13}$$

for any $(x, \alpha) \in G \times \mathcal{M}$. Note that $V^h(x, \alpha) = g(x, \alpha)$ for $(x, \alpha) \notin G \times \mathcal{M}$.

Transition probabilities and local consistency. To proceed, we denote $a(x, \alpha) = \big(a_{ij}(x, \alpha)\big) := \sigma(x, \alpha)\sigma^\top(x, \alpha)$ and assume that

$$a_{ii}(x, \alpha) - \sum_{j:j \neq i} |a_{ij}(x, \alpha)| \geq 0 \quad \text{for} \quad i = 1, \dots, d, x \in \overline{G}, \alpha \in \mathcal{M}, \tag{14}$$

which will be used to ensure the validity of transition probabilities $p^h(\cdot, \cdot | \xi)$. This assumption is motivated by Kushner (1976) and (Kushner, 1977, pp. 91-92). Under this assumption, the matrix $a(x, \alpha)$ is diagonally dominated. There are several alternatives to handle the non-diagonally dominated case; we refer to Kushner and Dupuis (2001) for a discussion on scaling. In general, $a(x, \alpha)$ might not be diagonally dominated even after scaling and one needs to employ an algorithmic procedure; see Bonnans et al. (2004) for $d = 2$ and Mirebeau (2014) for $d = 3$.

Let $\mathbb{E}^{h,\xi}_{x,\alpha,n}$, $\mathbb{C}\mathrm{ov}^{h,\xi}_{x,\alpha,n}$ denote the conditional expectation and covariance with given $\{X^h_k, \Lambda^h_k, u^h_k, k \le n, X^h_n = x, \Lambda^h_n = \alpha, u^h_n = \xi\}$, respectively. Define $\Delta X^h_n = X^h_{n+1} - X^h_n$. Our objective is to define transition probabilities $p^h((x,\alpha),(y,\ell)|\xi)$ so that the controlled Markov chain $\{(X^h_n, \alpha^h_n)\}$ is locally consistent with the controlled switching diffusion (1) if the following conditions hold:

$$
\begin{aligned}
&\mathbb{E}^{h,\xi}_{x,\alpha,n}\Delta X^h_n = b(x,\alpha,\xi)\Delta t^h(x,\alpha,\xi) + o(\Delta t^h(x,\alpha,\xi)),\\
&\mathbb{C}\mathrm{ov}^{h,\xi}_{x,\alpha,n}\Delta X^h_n = a(x,\alpha)\Delta t^h(x,\alpha,\xi) + o(\Delta t^h(x,\alpha,\xi)),\\
&\mathbb{P}^{h,\xi}_{x,\alpha,n}(\Lambda^h_{n+1} = \ell) = \Lambda_{\alpha\ell}\Delta t^h(x,\alpha,\xi) + o(\Delta t^h(x,\alpha,\xi)) \quad \text{for} \quad \alpha \ne \ell,\\
&\mathbb{P}^{h,\xi}_{x,\alpha,n}(\Lambda^h_{n+1} = \alpha) = 1 + \Lambda_{\alpha\alpha}\Delta t^h(x,\alpha,\xi) + o(\Delta t^h(x,\alpha,\xi)),\\
&\sup_{n,\,\omega}|\Delta X^h_n| \to 0 \quad \text{as} \quad h \to 0.
\end{aligned}
\tag{15}
$$

Any method of constructing transition probabilities $p^h(\cdot)$ for which (15) holds can be used. We present a finite-difference method, which uses finite-difference approximations to reflect the actual direction (or probability distribution of the direction) of motion of the given dynamical system. To proceed, we use the following approximation for the first and the second derivatives of $V(\cdot,\cdot)$

$$
\begin{aligned}
V(x,\alpha) &\to V^h(x,\alpha),\\
V_{x_i}(x,\alpha) &\to \frac{V^h(x+h\mathbf{e_i},\alpha) - V^h(x,\alpha)}{h} \qquad \text{if } b_i(x,\alpha,\xi) \ge 0,\\
V_{x_i}(x,\alpha) &\to \frac{V^h(x,\alpha) - V^h(x-h\mathbf{e_i},\alpha)}{h} \qquad \text{if } b_i(x,\alpha,\xi) < 0,\\
V_{x_ix_i}(x,\alpha) &\to \frac{V^h(x+h\mathbf{e_i},\alpha) - 2V^h(x,\alpha) + V^h(x-h\mathbf{e_i},\alpha)}{h^2},
\end{aligned}
$$

if $a_{ij}(x,\alpha) \ge 0$,

$$
\begin{aligned}
&V_{x_ix_j}(x,\alpha)\\
&\to [2V^h(x,\alpha) + V^h(x+h\mathbf{e_i}+h\mathbf{e_j},\alpha) + V^h(x-h\mathbf{e_i}-h\mathbf{e_j},\alpha)]/(2h^2)\\
&\quad - [V^h(x+h\mathbf{e_i},\alpha) + V^h(x-h\mathbf{e_i},\alpha) + V^h(x+h\mathbf{e_j},\alpha)\\
&\quad + V^h(x-h\mathbf{e_j},\alpha)]/(2h^2),
\end{aligned}
$$

and if $a_{ij}(x,\alpha) < 0$,

$$
\begin{aligned}
&V_{x_ix_j}(x,\alpha)\\
&\to -[2V^h(x,\alpha) + V^h(x+h\mathbf{e_i}-h\mathbf{e_j},\alpha) + V^h(x-h\mathbf{e_i}+h\mathbf{e_j},\alpha)]/(2h^2)\\
&\quad + [V^h(x+h\mathbf{e_i},\alpha) + V^h(x-h\mathbf{e_i},\alpha) + V^h(x+h\mathbf{e_j},\alpha)\\
&\quad + V^h(x-h\mathbf{e_j},\alpha)]/(2h^2).
\end{aligned}
$$

Then we substitute the finite-difference expressions above in $(\mathcal{L}^{\xi} - r)V(x,\alpha) + f(x,\alpha,\xi) + QV(x,\cdot)(\alpha) = 0$, which arises in (5). We multiply all terms by h^2 (to clear the denominator), and then divide all terms in the resulting expression by the coefficient of $V^h(x,\alpha)$ (which is $Q_h(x,\alpha,\xi)$). The $p^h\big((x,\alpha),(y,\ell)\big|\xi\big)$ and $\Delta t^h(x,\alpha,\xi)$ are coefficients of the resulting equation. Particularly, for $(x,\alpha) \in G \times \mathcal{M}$ and $\xi \in \mathcal{U}$, we define $Q_h(x,\alpha,\xi) = \sum_{i=1}^d a_{ii}(x,\alpha) - \sum\limits_{i,j:i\neq j} \frac{1}{2}|a_{ij}(x)| + h \sum\limits_{i=1}^{d} |b_i(x,\alpha,\xi)| - h^2 q_{\alpha\alpha} + \zeta(h)$. We set $\zeta(h) = h$. If, in addition, $a_{ii}(x,\alpha) - \sum\limits_{j:j\neq i} \big|a_{ij}(x,\alpha)\big| > 0$ for $i = 1,\dots,d$, $x \in \overline{G}$, $\alpha \in \mathcal{M}$, we can simply take $\zeta(h) = 0$.

Assumption (14) guarantees that the transition probabilities in (16) below are well defined. The definition of $p^h((x,\alpha),\cdot|\xi)$ for $(x,\alpha) \notin G \times \mathcal{M}$ is not important since in the analysis of the control problem, the chain will be stopped at the first time it exits $G \times \mathcal{M}$. Using the transition probabilities, we can check that the local-consistency conditions of $\{(X_n^h, \Lambda_n^h)\}$ in (15) are satisfied. We define

$$
\begin{aligned}
&p^h((x,\alpha),(x+h\mathbf{e_i},\alpha)|\xi) \\
&\qquad = \frac{a_{ii}(x,\alpha)/2 - \sum\limits_{j:j\neq i} |a_{ij}(x,\alpha)|/2 + \big(b_i(x,\alpha,\xi)\big)^+ h}{Q_h(x,\alpha,\xi)}, \\
&p^h((x,\alpha),(x-h\mathbf{e_i},\alpha)|\xi) \\
&\qquad = \frac{a_{ii}(x,\alpha)/2 - \sum\limits_{j:j\neq i} |a_{ij}(x,\alpha)|/2 + \big(b_i(x,\alpha,\xi)\big)^- h}{Q_h(x,\alpha,\xi)}, \\
&p^h\big((x,\alpha),(x+h\mathbf{e_i}+h\mathbf{e_j},\alpha)|\xi\big) \\
&\qquad = p^h\big((x,\alpha),(x-h\mathbf{e_i}-h\mathbf{e_j},\alpha)|\xi\big) = \frac{a_{ij}^+(x)}{2Q_h(x,\alpha,\xi)}, \\
&p^h\big((x,\alpha),(x+h\mathbf{e_i}-h\mathbf{e_j},\alpha)|\xi\big) \\
&\qquad = p^h\big((x,\alpha),(x-h\mathbf{e_i}+h\mathbf{e_j},\alpha)|\xi\big) = \frac{a_{ij}^-(x)}{2Q_h(x,\alpha,\xi)}, \\
&p^h((x,\alpha),(x,\ell)|\xi) = \frac{h^2 \Lambda_{\alpha\ell}}{Q_h(x,\alpha,\xi)} \quad \text{for } \alpha \neq \ell, \\
&p^h((x,\alpha),(x,\alpha)|\xi) = \frac{\zeta(h)}{Q_h(x,\alpha,\xi)}, \qquad \Delta t^h(x,\alpha,\xi) = \frac{h^2}{Q_h(x,\alpha,\xi)}, \\
&p^h((x,\alpha),(y,\ell)|\xi) = 0 \quad \text{for all non-listed values of } (y,\ell) \in S_h \times \mathcal{M}.
\end{aligned} \tag{16}
$$

Continuous-time interpolation. To proceed, we construct a continuous-time interpolation of the approximating chain. For this, we define $n^h(t) = \max\{n : t_n^h \le t\}$, $t \ge 0$. Let $\mathbb{E}_n^h$ denote the expectation given $\sigma\{X_k^h, \Lambda_k^h, u_k^h : k \le n\}$. The discrete time processes associated with the controlled Markov chain $\{(X_n^h, \Lambda_n^h)\}$ are defined as follows. Let

$$B_0^h = M_0^h = 0, \quad B_n^h = \sum_{k=0}^{n-1} \mathbb{E}_k^h \Delta X_k^h, \quad M_n^h = \sum_{k=0}^{n-1} (\Delta X_k^h - \mathbb{E}_k^h \Delta X_k^h), \quad n \geq 1. \tag{17}$$

The piecewise constant interpolation processes, denoted by $X^h(\cdot)$, $\Lambda^h(\cdot)$, $B^h(\cdot)$, $M^h(\cdot)$, $u^h(\cdot)$ are naturally defined as

$$\begin{aligned} X^h(t) &= X^h_{n^h(t)}, \quad \Lambda^h(t) = \Lambda^h_{n^h(t)}, \quad u^h(t) = u^h_{n^h(t)}, \\ B^h(t) &= B^h_{n^h(t)}, \quad M^h(t) = M^h_{n^h(t)}, \quad t \geq 0. \end{aligned} \tag{18}$$

Let $\tau_h = t^h_{\eta_h}$. The payoff functional from (11) can be rewritten as

$$\begin{aligned} J^h&(x, \alpha, u^h) \\ &= \mathbb{E}\Big[\int_0^{\tau_h} e^{-rt} f(X^h(t), \Lambda^h(t), u^h(t))dt + e^{-r\tau_h} g(X^h(\tau_h), \Lambda^h(\tau_h))\Big]. \end{aligned} \tag{19}$$

Relaxed controls representation. For our analysis, it is more convenient to use the notion of relaxed controls. We first briefly recall the notion of "relaxed control", which arises naturally in the weak convergence analysis for the approximation to the optimal control problems.

Definition 2.1. Let $\mathcal{B}(\mathcal{U} \times [0, \infty))$ be the σ-algebra of Borel subsets of $\mathcal{U} \times [0, \infty)$. An admissible relaxed control or simply a relaxed control $m(\cdot)$ is a measure on $\mathcal{B}(\mathcal{U} \times [0, \infty))$ such that $m(\mathcal{U} \times [0, t]) = t$ for all $t \geq 0$. Given a relaxed control $m(\cdot)$, there is a probability measure $m_t(\cdot)$ defined on the σ-algebra $\mathcal{B}(\mathcal{U})$ such that $m(d\xi dt) = m_t(d\xi)dt$.

With the given probability space, we say that $m(\cdot)$ is an admissible relaxed stochastic control for $(W(\cdot), \Lambda(\cdot))$ or $(m(\cdot), W(\cdot), \Lambda(\cdot))$ is admissible, if (i) for each fixed $t \geq 0$, $m(t, \cdot)$ is a random variable taking values in $\mathcal{R}(\mathcal{U} \times [0, \infty))$, and for each fixed ω, $m(\cdot, \omega)$ is a deterministic relaxed control; (ii) the function defined by $m(A \times [0, t])$ is $\mathcal{F}_t$-adapted for any $A \in \mathcal{B}(\mathcal{U})$. As a result, with probability one, there is a measure $m_t(\cdot, \omega)$ on the Borel σ-algebra $\mathcal{B}(\mathcal{U})$ such that $m(d\xi dt) = m_t(d\xi)dt$.

Remark 2.2. Let $\mathcal{R}(\mathcal{U} \times [0, \infty))$ denote the space of all relaxed controls on $\mathcal{U} \times [0, \infty)$. Then $\mathcal{R}(\mathcal{U} \times [0, \infty))$ can be metrized using the Prohorov metric in the usual way as in (Kushner and Dupuis, 2001, pp. 263-264). With the Prohorov metric, $\mathcal{R}(\mathcal{U} \times [0, \infty))$ is a compact space. It follows that any sequence of relaxed controls has a convergent subsequence. Moreover, a sequence

$\zeta_n \in \mathcal{R}(\mathcal{U} \times [0, \infty))$ converges to $\zeta \in \mathcal{R}(\mathcal{U} \times [0, \infty))$ if and only if for all continuous functions with compact support $\Psi(\cdot)$ on $\mathcal{U} \times [0, \infty)$,

$$\int_{\mathcal{U}\times[0,\infty)} \Psi(\xi, t)\zeta_n(d\xi, dt) \to \int_{\mathcal{U}\times[0,\infty)} \Psi(\xi, t)\zeta(d\xi, dt).$$

With the notion of relaxed control given above, we can write (19) as

$$J^h(x, \alpha, u^h) = J^h(x, \alpha, m^h(\cdot)) = \mathbb{E}\Big[\int_0^{\tau_h} e^{-rt} f(X^h(t), \Lambda^h(t), \xi) m_t^h(d\xi)dt + e^{-r\tau_h} g(X^h(\tau_h), \Lambda^h(\tau_h))\Big]. \tag{20}$$

Note also that the value function defined in (3) can be rewritten as

$$V(x, \alpha) = \sup\{J(x, \alpha, m(\cdot)) : m(\cdot) \text{ is an admissible relaxed control}\},$$

where

$$J(x, \alpha, m(\cdot)) := \mathbb{E}_{x,\alpha}\Big[\int_0^{\tau} e^{-rt} f\big(X(t), \Lambda(t), \xi\big) m_t(d\xi)dt + e^{-r\tau} g(X(\tau), \Lambda(\tau))\Big]. \tag{21}$$

Convergence results. Let $H^h(\cdot) = \big(X^h(\cdot), \Lambda^h(\cdot), B^h(\cdot), M^h(\cdot), m^h(\cdot), \tau_h\big)$. Then we can show that any weakly convergent subsequence of $H^h(\cdot)$ has the weak limit, denoted by $H(\cdot)$, which satisfies (1). Also, the associated payoff functionals $J^h(x, \alpha, m^h(\cdot))$ converge to $J(x, \alpha, m(\cdot))$ given by (21). Together with the tightness of $H^h(\cdot)$, we can show that $V^h(x, \alpha) \to V(x, \alpha)$ as $h \to 0$, where $V^h(x, \alpha)$ and $V(x, \alpha)$ are defined in (12) and (3), respectively. The main results are stated Theorem 2.3 and Theorem 2.4, whereas the proofs can be found in Song et al. (2006).

Theorem 2.3. *Let the chain $\{(X_n^h, \Lambda_n^h)\}$ be constructed with transition probabilities defined in* (16), *$X^h(\cdot)$, $\Lambda^h(\cdot)$, $B^h(\cdot)$, $M^h(\cdot)$ be the continuous-time interpolation defined in* (17)-(18), *$\{u_n^h\}$ be an admissible control, and $m^h(\cdot)$ be its relaxed control representation. Then the following assertions hold. (a) The sequence $H^h(\cdot) = \big(X^h(\cdot), \Lambda^h(\cdot), B^h(\cdot), M^h(\cdot), m^h(\cdot), \tau_h\big)$ is tight. As a result, $H^h(\cdot)$ has a weakly convergent subsequence with limit $H(\cdot) = \big(X(\cdot), \Lambda(\cdot), B(\cdot), M(\cdot), m(\cdot), \tau\big)$. (b) For any $t \geq 0$,*

$$B(t) = \int_0^t b(X(s), \Lambda(s), \xi) m_s(d\xi)ds. \tag{22}$$

(c) There is a filtration $\{\mathcal{F}(t)\}$ so that $M(\cdot)$ is a continuous $\mathcal{F}(t)$-martingale with quadratic variation $\int_0^t a(X(s), \Lambda(s))ds$. Thus, there is an $\mathcal{F}(t)$-standard

Brownian motion $W(t)$, in which is defined on the probability space (or an augmented probability space) satisfying

$$M(t) = \int_0^t \sigma(X(s), \Lambda(s))dW(s). \tag{23}$$

(d) The limit processes satisfy

$$X(t) = x + B(t) + M(t), \quad t \geq 0. \tag{24}$$

Theorem 2.4. *Let $V^h(x,\alpha)$ and $V(x,\alpha)$ be value functions defined in* (12) *and* (3), *respectively. Then $V^h(x,\alpha) \to V(x,\alpha)$ as $h \to 0$.*

Numerical methods. We use classical iterative methods to find the optimal control by solving the dynamic programming equation (13). A sketch of the algorithm is provided below, which is a generalization of (Kushner and Dupuis, 2001, Chapter 6) because a switching process is present. We refer to Song et al. (2006) for numerical experiments.

For $(x,\alpha) \in G \times \mathcal{M}$, we choose the initial control and the initial value $u_0(x,\alpha) \equiv \xi_0 \in \mathcal{U}$, $V_0^h(x,\alpha) = f(x,\alpha,\xi_0)\Delta t(x,\alpha,\xi_0)$. We proceed to find the sequence $V_n^h(\cdot)$ recursively as follows. For each $(x,\alpha) \in G \times \mathcal{M}$ and $\xi \in \mathcal{U}$, we compute

$$\begin{aligned} V_{n+1}^h(x,\alpha \mid \xi) = e^{-r\Delta t^h(x,\alpha,\xi)} \sum_{(y,\ell)\in S_h \times \mathcal{M}} V_n^h(y,\ell) p^h\big((x,\alpha),(y,\ell)|\xi\big) \\ + f(x,\alpha,\xi)\Delta t^h(x,\alpha,\xi). \end{aligned}$$

Note also that if $y \notin G$, then $V_n^h(y,\ell) = g(y,\ell)$ for any ℓ and n. We choose the control $u_{n+1}^h(x,\alpha)$ and record an improved value $V_{n+1}^h(x,\alpha)$ by

$$u_{n+1}^h(x,\alpha) = \text{argmax}_{\xi\in\mathcal{U}} V_{n+1}^h(x,\alpha \mid \xi),\ \ V_{n+1}^h(x,\alpha) = V_{n+1}^h(x,\alpha \mid u_{n+1}^h(x,\alpha)),$$

where the argmax on the right-hand side is computed row by row. The iterations stop when the increment $V_{n+1}^h(\cdot) - V_n^h(\cdot)$ reaches a predetermined tolerance level.

2.2 Singular control

As in the preceding formulation, we construct a discrete-time, finite-state, controlled Markov chain to approximate the controlled switching-diffusion process (6); see Jin et al. (2012) for details. Let G be a nonempty subset of $\mathbb{R}_+^d$. This consideration is motivated by the applications in finance and biology in which the processes of interest are nonnegative. Let U be an upper bound introduced for numerical computation purpose. The rationale is that in practical situations, the continuous state $X(\cdot)$ cannot be infinite. Moreover, it is frequently observed that under optimal or near-optimal control policies, $X(\cdot)$ lies in a bounded set in $\mathbb{R}^d$; see (Tran et al., Theorem 3.4) for a related result. Hence, we define

$$\begin{aligned}\mathcal{A}^U_{x,\alpha} &= \{(u, Z) \in \mathcal{A}_{x,\alpha} : X(\cdot) \in [0, U]^d\},\\ V^U(x,\alpha) &= \sup_{(u,Z)\in\mathcal{A}^U_{x,\alpha}} J(x,\alpha,u,Z).\end{aligned} \tag{25}$$

Let $h > 0$ be a discretization parameter for the continuous state variable. Define

$$S_h := \{x = (k_1 h, \ldots, k_d h)^\top : k_1, \ldots, k_d \in \mathbb{Z}_+\} \cap [0, U + h]^d.$$

Let $\{(X^h_n, \Lambda^h_n) : n \in \mathbb{Z}_+\}$ be a discrete-time controlled Markov chain with state space $S_h \times \mathcal{M}$. For each n, the increments of the chain $\Delta X^h_n = X^h_{n+1} - X^h_n$ approximate exactly one of the following quantities dynamically.

- Diffusion step: $b(X(t), \Lambda(t), u(t))dt + \sigma(X(t), \Lambda(t))dW(t)$.
- Singular control step on the ith component: $-dZ_i(t)$.

We use π^h_n to denote the type of step n. We set $\pi^h_n = 0$ if step n is a diffusion step and $\pi^h_n = i$ if step n is a singular control step on the ith component. Each of these steps is described precisely in what follows.

If $\pi^h_n = 0$, a regular control u^h_n is exercised so that $\Delta X^h_n = X^h_{n+1} - X^h_n$ is to behave like an increment of $\int b dt + \int \sigma dw$ over a small time interval. If $\pi^h_n = i$, a regular control u^h_n and a singular control $\Delta Z^h_n = h\mathbf{e_i}$ are applied so that the singular control term pushes the state process in the direction $-\mathbf{e_i}$. Thus, the amount of the singular control term at any step is $\Delta Z^h_n = h\mathbf{e_i}_{\{\pi^h_n = i\}}$.

Let $u^h = \{u^h_n\}$ be a sequence of regular controls. The space of regular controls is given by $\mathcal{U}$. Let $\pi^h = \{\pi^h_n\}_{n\geq 0}$ denote the sequence of control types. We denote by $p^h((x,\alpha),(y,\ell)|\pi,\xi)$ the transition probability from state (x,α) to another state (y,ℓ) under the control type π and regular control ξ. Denote $\mathcal{F}^h_n = \sigma\{X^h_k, \Lambda^h_k, \pi^h_k, u^h_k, k \leq n\}$.

The sequence $\{(\pi^h_n, u^h_n)\}$ is said to be admissible if it satisfies the following conditions: (a) π^h_n, u^h_n are $\sigma\{X^h_0, \ldots, X^h_n, \Lambda^h_0, \ldots, \Lambda^h_n, \pi^h_0, \ldots, \pi^h_{n-1}, u^h_1, \ldots, u^h_{n-1}\}$-adapted, (b) For any $x \in S_h$, we have

$$\begin{aligned}\mathbb{P}\{X^h_{n+1} = x, \Lambda^h_{n+1} = \alpha | \mathcal{F}^h_n\} &= \mathbb{P}\{X^h_{n+1} = x, \Lambda^h_{n+1} = \alpha | X^h_n, \Lambda^h_n, \pi^h_n, u^h_n\}\\ &= p^h\big((X^h_n, \Lambda^h_n), (x,\alpha) | \pi^h_n, u^h_n\big),\end{aligned}$$

(c) $\pi^h_n \in \{0, 1, \ldots, d\}$, $u^h_n \in \mathcal{U}$, $X^h_n \in S_h$ for all $n \in \mathbb{Z}_+$. The class of all admissible control sequences (π^h, u^h) for initial state (x,α) is denoted by $\mathcal{A}^h_{x,\alpha}$.

For each $(x,\alpha,i,\xi) \in S_h \times \mathcal{M} \times \{0,1,\ldots,d\} \times \mathcal{U}$, we define a family of the interpolation intervals $\Delta t^h(x,\alpha,i,\xi)$ so that $\sup_{x,\alpha,i,\xi} \Delta t^h(x,\alpha,i,\xi) \to 0$ as $h \to 0$. The values of $\Delta t^h(x,\alpha,i,\xi)$ will be specified later. Then we define

$$t^h_0 = 0,\ \Delta t^h_k = \Delta t^h(X^h_k, \Lambda^h_k, \pi^h_k, u^h_k),\ t^h_n = \sum_{k=0}^{n-1} \Delta t^h_k.$$

For $(x,\alpha) \in S_h \times \mathcal{M}$ and $(\pi^h, u^h) \in \mathcal{A}^h_{x,\alpha}$, the payoff for the controlled Markov chain is defined as

$$J^h(x,\alpha,\pi^h,u^h) = \mathbb{E}\sum_{k=0}^{\eta_h-1} e^{-rt_k^h}\Big\{f(X_k^h,\Lambda_k^h,u_k^h)\Delta t_k^h + c\cdot\Delta Z_k^h\Big\}, \tag{26}$$

with $\eta_h = \inf\{n \geq 0 : X_n^h \notin G\}$. The value function of the controlled Markov chain is

$$V^h(x,\alpha) = \sup_{(\pi^h,u^h)\in\mathcal{A}^h_{x,\alpha}} J^h(x,\alpha,\pi^h,u^h). \tag{27}$$

The corresponding dynamic programming equation for the discrete approximation is given by

$$\begin{aligned} V^h(x,\alpha) = \sup_{(i,\xi)} \Big\{ & e^{-r\Delta t^h(x,\alpha,i,\xi)} \sum_{(y,\ell)\in S_h} V^h(y,\ell)p^h\big((x,\alpha),(y,\ell)|i,\xi\big) \\ & + f(x,\alpha,\xi)\Delta t^h(x,\alpha,i,\xi), \big(V^h(x-h\mathbf{e_i},\alpha)+c_ih\big)I_{\{x-h\mathbf{e_i}\in G\}}\Big\}, \end{aligned} \tag{28}$$

for any $(x,\alpha) \in G \times \mathcal{M}$. Note that $V^h(x,\alpha) = 0$ for $(x,\alpha) \notin G \times \mathcal{M}$.

Transition probabilities and local consistency. To proceed, we define the transition probability function $p^h(\cdot)$ so that the evolution of $\{(X_n^h, \Lambda_n^h)\}$ well approximates the local behavior of $\{(X(t), \Lambda(t))\}$ given by (6). Under assumption (14), at the diffusion steps, $p^h(\cdot|0,\xi)$ and $\Delta t^h(\cdot,\cdot,0,\xi)$ are defined as in (16). As a singular control step of the ith component, we define

$$p^h\big((x,\alpha),(x-h\mathbf{e_i},\alpha)\,\big|i,\xi\,\big) = 1,\ \ \Delta t^h(x,\alpha,i,\xi) = 0.$$

The definition of $p^h\,((x,\alpha),\cdot|i,\xi)$ for $(x,\alpha)\notin G\times\mathcal{M}$ is not important since in the analysis of the control problem, the chain will be stopped the first time it exits $G \times \mathcal{M}$. Using the above transition probabilities, we can check that the local-consistency conditions of $\{(X_n^h, \Lambda_n^h)\}$ in (15) are satisfied.

Convergence results. The convergence result is based on a continuous-time interpolation of the chain and relaxed control representations as in the preceding case. In addition, a "stretched-out" time scale is introduced to overcome the possible non-tightness of the piecewise constant interpolation associated with Z^h (Kushner and Martins (1991)). Using weak convergence methods, we can obtain the convergence of the value function; see Jin et al. (2012).

Theorem 2.5. *Let $V^U(x,\alpha)$ and $V^h(x,\alpha)$ be value functions defined in* (25) *and* (27)*, respectively. Then $V^h(x,\alpha) \to V^U(x,\alpha)$ as $h \to 0$.*

Numerical methods. We use classical iterative methods to find the optimal control by solving the dynamic programming equation (28). A sketch of the algorithm is provided below. We refer to Jin et al. (2012) for numerical experiments.

For $(x,\alpha) \in G \times \mathcal{M}$, we choose the initial control and the initial value $u_0(x,\alpha) \equiv \xi_0 \in \mathcal{U}$, $\pi_0^h(x,\alpha) = 1$, $V_0^h(x,\alpha) = c \cdot x$. We proceed to find the sequence $V_n^h(\cdot)$ recursively as follows. For each $(x,\alpha) \in G \times \mathcal{M}$ and $(i,\xi) \in \{0,1,\dots,d\} \times \mathcal{U}$, we compute

$$\begin{aligned} V_{n+1}^h(x,\alpha \mid 0,\xi) &= e^{-r\Delta t^h(x,\alpha,0,\xi)} \sum_{(y,\ell)\in S_h\times\mathcal{M}} V_n^h(y,\ell) p^h\big((x,\alpha),(y,\ell)|0,\xi\big) \\ &\quad + f(x,\alpha,\xi)\Delta t^h(x,\alpha,0,\xi), \\ V_{n+1}^h(x,\alpha \mid i,\xi) &= \big(V^h(x-h\mathbf{e_i},\alpha) + c_i h\big) I_{\{x-h\mathbf{e_i}\in G\}}, \quad i = 1,2,\dots,d. \end{aligned}$$

Note also that if $y \notin G$, then $V_n^h(y,\ell) = 0$ for any ℓ and n. We choose the controls $\pi_{n+1}^h(x,\alpha)$, $u_{n+1}^h(x,\alpha)$, and record an improved value $V_{n+1}^h(x,\alpha)$ by

$$\begin{aligned} \big(\pi_{n+1}^h(x,\alpha), u_{n+1}^h(x,\alpha)\big) &= \operatorname{argmax}_{(i,\xi)} V_{n+1}^h(x,\alpha \mid i,\xi), \\ V_{n+1}^h(x,\alpha) &= V_{n+1}^h\big(x,\alpha \mid \pi_{n+1}^h(x,\alpha), u_{n+1}^h(x,\alpha)\big), \end{aligned}$$

where the argmax on the right-hand side is computed row by row. The iterations stop when the increment $V_{n+1}^h(\cdot) - V_n^h(\cdot)$ reaches a predetermined tolerance level.

3 Application to insurance

In this section, we consider several insurance applications. Due to the availability of versatile risk management tools in the insurance portfolio management, decision makers generally face to a high-dimensional stochastic optimization problem with multiple controls such as dividend payment, investment, reinsurance, etc. The numerical studies can be referred to a series of works in Jin et al. (2012), Jin et al. (2013a), Jin et al. (2013b) and Bui et al. (2019). In what follows, we will introduce two insurance games.

Reinsurance zero-sum games. Consider a reinsurance zero-sum game between two insurance companies (players). Each insurance company can adopt a corresponding reinsurance strategy to outperform its competitor. The reinsurance game formulation can be found in Jin et al. (2013b).

Let $x^{(1)}$ be the initial surplus of player 1 and $x^{(2)}$ be the initial surplus of player 2. Similarly, by considering a Markov regime-switching process $\Lambda(t)$, the surplus difference process between the player 1 and 2 satisfies the following stochastic differential equation

$$\begin{cases} dX(t) = [c^{(1)}(\Lambda(t)) - g(u_1) - c^{(2)}(\Lambda(t)) + g(u_2)]dt - u_1(t)\sigma_1(\Lambda(t))dW_1 \\ \qquad\qquad + u_2(t)\sigma_2(\Lambda(t))dW_2, \\ X(0) = x = x^{(1)} - x^{(2)}, \end{cases} \tag{29}$$

for all $t < \tau$, where $\tau = \inf\{t \geq 0 : X(t) \notin (a,b)\}$ represents the time of exiting the game, and a, b are constants satisfying $a < b$. For $k = 1, 2$, $c^{(k)}$ represents the primary insurance premium income of players 1 and 2, respectively. The u_1 and u_2 are the reinsurance strategies of players 1 and 2. The $g(\cdot)$ is the reinsurance premium principle; σ_i for $i = 1, 2$ represent the volatility of claims. The game will stop if the absolute value of the difference of the surpluses are too large. Either in positive or negative direction, the large difference means that one company dominates the market and wins the game.

Denote by $r > 0$ the discount factor. Let U_1 and U_2 be the collection of all reinsurance strategies, respectively, which are assumed to be compact sets. Let $u(\cdot) = (u_1(\cdot), u_2(\cdot))$ and $u(\cdot) \in U = U_1 \times U_2$. For an arbitrary admissible control u and each $\alpha \in \mathcal{M}$, the expected discounted payoff is

$$J(x, \alpha, u(\cdot)) = \mathbb{E}_{x,\alpha}\left[\int_0^{\tau} e^{-rt}\left[f(X(t), \Lambda(t), u(t))dt + l(X(\tau^-), \Lambda(\tau^-))\right]\right], \tag{30}$$

where $\mathbb{E}_{x,\alpha}$ denotes the expectation conditioned on $X(0) = x$ and $\Lambda(0) = \alpha$. The control $u(\cdot) = (u_1(\cdot), u_2(\cdot))$ is said to be *admissible* if $u_1(\cdot)$, $u_2(\cdot)$ satisfy (a) $u_1(t)$, $u_2(t)$ are nonnegative for any $t \geq 0$, (b) $X(t) \in (a,b)$, for any $t \leq \tau$, (c) both $u_1(\cdot)$, $u_2(\cdot)$ are adapted to $\mathcal{F}_t$ that contains at least $\sigma\{W(s), \Lambda(s), 0 \leq s \leq t\}$, (d) $J(x, \alpha, u(\cdot)) < \infty$ for any $x \in G$ and admissible pair $u(\cdot) = (u_1(\cdot), u_2(\cdot))$, where J is the functional defined in (30). Suppose that U is the collection of possible retention levels $u(t)$, where U is a given compact set. We further assume the function $f(\cdot)$ is concave-convex with respect to (u_1, u_2), that is, $f(\cdot)$ is concave with respect to u_1 for every (x, α, u_2) and convex with respect to u_2 for every (x, α, u_1). Assume that $l(\cdot)$ is a continuous function. For an arbitrary $u \in \mathcal{A}$, and $V(\cdot, \alpha) \in C^2(\mathbb{R})$, define an operator $\mathcal{L}^u$ by

$$\begin{aligned} \mathcal{L}^u V(x,\alpha) = {} & V_x(x,\alpha)[c^{(1)}(\alpha) - g(u_1) - c^{(2)}(\alpha) + g(u_2)] \\ & + \frac{1}{2}(\sigma_1^2(\alpha) + \sigma_2^2(\alpha))V_{xx}(x,\alpha). \end{aligned} \tag{31}$$

Formally, V satisfies the following systems of HJI equations: for each $\alpha \in \mathcal{M}$,

$$\begin{cases} \inf\limits_{u_1 \in U_1} \sup\limits_{u_2 \in U_2} (\mathcal{L}^u V(x,\alpha) - rV(x,\alpha) + f(x,\alpha,u)) \\ \quad = \sup\limits_{u_2 \in U_2} \inf\limits_{u_1 \in U_1} (\mathcal{L}^u V(x,\alpha) - rV(x,\alpha) + f(x,\alpha,u)) = 0, \quad x \in (a,b) \\ V(x,\alpha) = l(x,\alpha), \quad x = a, b. \end{cases} \tag{32}$$

We can then define the upper value with $J^h(\cdot)$ and $\mathcal{U}^h$ as

$$V^+(x,\alpha,) = \inf_{u_1\in\mathcal{U}_1(1)} \sup_{u_2\in\mathcal{U}_2(2)} J^h(x,\alpha,u_1,u_2). \tag{33}$$

Analogously, the lower value is defined as

$$V^-(x,\alpha) = \sup_{u_2\in\mathcal{U}_2(1)} \inf_{u_1\in\mathcal{U}_1(2)} J^h(x,\alpha,u_1,u_2). \tag{34}$$

Recall that a saddle point exists, if

$$V(x,\alpha) = V^+(x,\alpha) = V^-(x,\alpha),\ \forall\alpha\in\mathcal{M}. \tag{35}$$

For computation, we try to find $V(x,\alpha)$ by solving the corresponding dynamic programming equation using either value iteration or policy iteration. In fact, we can use

$$\begin{aligned} V^+(x,\alpha) &= \inf_{u_1\in U_1}\{\sup_{u_2\in U_2}\sum_{y,\ell} e^{-r\Delta t(x,\alpha,u)} p((x,\alpha),(y,\ell))|u)V(y,\ell) \\ &\quad + f(x,\alpha,u)\Delta t(x,\alpha,u)\}, \\ V^-(x,\alpha) &= \sup_{u_2\in U_2}\{\inf_{u_1\in U_1}\sum_{y,\ell} e^{-r\Delta t(x,\alpha,u)} p((x,\alpha),(y,\ell))|u)V(y,\ell) \\ &\quad + f(x,\alpha,u)\Delta t(x,\alpha,u)\}. \end{aligned} \tag{36}$$

On the other hand, for $\alpha\in\mathcal{M}$, the saddle point $V(x,\alpha)$ satisfies the HJI equation

$$\begin{aligned} 0 &= V_x(x,\alpha)[c^{(1)}(\alpha) - g(u_1) - c^{(2)}(\alpha) + g(u_2)] \\ &\quad + \frac{1}{2}V_{xx}(x,\alpha)(\sigma_1^2(\alpha)+\sigma_2^2(\alpha)) - rV(x,\alpha). \end{aligned} \tag{37}$$

The transition probabilities are constructed as follows

$$\begin{aligned} & p^h(((x,\alpha),(x+h,\alpha))|u^h) \\ &= \frac{(\sigma_1^2(\alpha)+\sigma_2^2(\alpha))/2 + h[c^{(1)}(\alpha) - g(u_1^h) - c^{(2)}(\alpha) + g(u_2^h)]^+}{\widetilde{D} - rh^2}, \\ & p^h(((x,\alpha),(x-h,\alpha))|u^h)) \\ &= \frac{(\sigma_1^2(\alpha)+\sigma_2^2(\alpha))/2 + h[c^{(1)}(\alpha) - g(u_1^h) - c^{(2)}(\alpha) + g(u_2^h)]^-}{\widetilde{D} - rh^2}, \\ & \Delta t(x,\alpha,u^h) = \frac{h^2}{\widetilde{D}}, \end{aligned} \tag{38}$$

with $\widetilde{D} = \sigma_1^2(\alpha)+\sigma_2^2(\alpha)+h|c^{(1)}(\alpha) - g(u_1^h) - c^{(2)}(\alpha) + g(u_2^h)| + (r - q_{\alpha\alpha})h^2$ being well defined.

Furthermore, the system of dynamic programming equations is a modification of (36). That is,

$$V^h(x,\alpha)=\begin{cases}\sup\limits_{u_2^h\in U_2^h}\Big\{\inf\limits_{u_1^h\in U_1^h}\Big[e^{-r\Delta t^h(x,\alpha,u^h)}\sum\limits_{y,\ell}(p^h((x,\alpha),(y,\ell))|u^h)V^h(y,\ell)\\ \quad +f(x,\alpha,u^h)\Delta t^h(x,\alpha,u^h)\Big]\Big\}, \qquad \text{for } x\in(a,b),\\ l(x,\alpha), \qquad \text{for } x=a,b.\end{cases} \tag{39}$$

Reinsurance and investment non-zero-sum games. We are considering an insurance market including two competing insurance companies. Each of them adopts optimal investment and reinsurance strategies to manage the insurance portfolios. The reinsurance game formulation can be found in Bui et al. (2019).

The surplus process of each insurance company is subject to a financial market with a market index $Z(t)$, whose prices satisfy

$$dZ(t)=\mu_Z(t,Z(t))dt+\sigma_Z(t,Z(t))dW_Z(t), \tag{40}$$

where $W_Z(t)$ is a standard Brownian motion. Denote by $\{\mathcal{F}_t^{W_Z}\}$ the filtration generated by Brownian motion $\{W_Z(t)\}_{t\geq 0}$.

Following the classical Cramér-Lundberg process, we assume that $\widehat{X}_k(t)$, $k\in\{1,2\}$, the surplus of insurance company k without investment and reinsurance satisfies

$$\widehat{X}_k(t)=\hat{x}_k+\int_0^t c_k(\Lambda(s),Z(s))ds-Y_k(t),\quad t\geq 0, \tag{41}$$

where $\widehat{X}_k(0):=\hat{x}_k$ is the initial surplus, $c_k(\Lambda(t),Z(t))$ is the rate of premium, and $Y_k(t)=\sum_{i=1}^{N_k(t)}A_i^k$ is a compound Poisson process with the claim size A_i^k with $\{A_i^k : i>1\}$ being a sequence of positive, independent and identically distributed random variables.

Consider a Poisson measure in lieu of the traditionally used Poisson process. Suppose $\Theta\subset\mathbb{R}_+$ is a compact set and the function $q_k(\cdot)$ is the magnitude of the claim sizes.

$$N_k(t,H)=\text{number of claims on } [0,t] \text{ with claim size taking values in } H\subset\Theta, \tag{42}$$

counts the number of claims up to time t, which is a Poisson counting process. For $k=1,2$, $Y_k(t)$ is a jump process representing claims for each company with arrival rate λ_k. Note that claim frequencies depend on the economy and financial market states. The function $q_k(\Lambda(t),Z(t),\rho_k)$ is assumed to be the magnitude of the claim sizes, where ρ_k has distribution $\Pi_k(\cdot)$, and $q_k(\alpha,\cdot,\rho_k)$ is continuous for each ρ_k and each $\alpha\in\mathcal{M}$. At different regimes and financial market states, taking into consideration of random environment, the values of $q_k(\alpha,\cdot,\rho_k)$ could be much different. Then the Poisson measure

$N_k(\cdot,\cdot)$ has intensity $\lambda_k dt \times \Pi_k(d\rho_k)$ where $\Pi_k(d\rho_k) = f(\rho_k)d\rho_k$. Let $\nu_{k,n}$ denote the time of the n-th claim and $\zeta_{k,n} = \nu_{k,n+1} - \nu_{k,n}$. Let $\{\zeta_{k,v}, \rho_k, v \geq n\}$ be independent of $\{X_k(s), \Lambda(s), s \leq \nu_{k,n}, \zeta_{k,v}, \rho_k, v \leq n\}$. Then the n-th claim term is $q_k(\Lambda(\nu_{k,n}), Z(\nu_{k,n}), \rho_k)$, and the claim amount of Y_k can be written as $Y_k(t) = \sum_{\nu_{k,n} \leq t} q_k(\Lambda(\nu_{k,n}), Z(\nu_{k,n}), \rho_k)$.

Let $a_k(t)$ be an $\mathcal{F}_t$-progressively measurable process valued in [0, 1], an exogenous retention level, which is a control chosen by the insurance company representing the reinsurance policy and $g(a_k)$ is the reinsurance premium rate. Denote by $\mathcal{A}_k := \{a_k(t) : 0 \leq a_k(t) \leq 1, 0 \leq t \leq T\}$, the set of reinsurance strategies of insurer k. Recall that A_i^k is the size of the ith claim. Let $A_i^k(a_k)$ be the fraction of each claim paid by the primary insurance company. Then the aggregation claim amount paid by the primary insurance company is denoted as $Y_k^{a_k}(t)$.

The insurance companies invest in both risk-free assets $S_0(t)$ and risky assets $S(t)$ with prices satisfying

$$\begin{cases} \dfrac{dS_0(t)}{S_0(t)} = r(\Lambda(t))dt, \\ \dfrac{dS(t)}{S(t)} = \mu_S(\Lambda(t), Z(t))dt + \sigma_S(\Lambda(t), Z(t))dW_S(t), \end{cases} \tag{43}$$

where $r(\Lambda(t))$ and $\mu(\Lambda(t), Z(t))$ are the return rates of the risk-free and risky assets, respectively; $\sigma_S(\Lambda(t), Z(t))$ is the corresponding volatility; $W_S(t)$ is a standard Brownian motion independent of $W_Z(t)$. For $k = 1, 2$, the investment behavior of the insurer k is modeled as a portfolio process $b_k(t)$, where $b_k(t)$ is invested in the risky asset $S(t)$. Let $\mathcal{B}_k = \{b_k(t) : 0 \leq t \leq T\}$ denote the set of investment strategies of insurer k.

Combining the reinsurance and investment strategies, the surplus process of the insurance company k, denoted by $\widetilde{X}_k(t)$, follows

$$\begin{cases} d\widetilde{X}_k(t) = \big\{r(\Lambda(t))\widetilde{X}_k(t) + b_k(t)[\mu_S(\Lambda(t), Z(t)) - r(\Lambda(t))] \\ \qquad + c_k(\Lambda(t), Z(t)) - g(a_k(t))\big\}dt \\ \qquad + b_k(t)\sigma_S(\Lambda(t), Z(t))dW_S(t) - dY_k^{a_k}(t), \\ \widetilde{X}_k(0) = \widetilde{x}_k, \end{cases} \tag{44}$$

where $Y_k^{a_k}(t) = \int_0^t \int_{\mathbb{R}_+} \tilde{q}_k(q_k, a_k)N_k(dt, d\rho_k)$ and $\tilde{q}_k(q_k, a_k)$ is the magnitude of the claim sizes with respect to the surplus process.

We model the competition of the two insurance companies with investment and reinsurance schemes in finite time horizon using a game theoretic formulation. The performance of each company is measured by the relative performance of their surpluses against their competitor's. Thus, the competition between the two companies formulates a game with two players, each of which can adjust its reinsurance strategies based on the competitor's scheme. Let the relative surplus performance for insurance company k be $X_k(t) := \widetilde{X}_k(t) - \kappa_k\widetilde{X}_l(t)$, where

$l = 3 - k$. And $X_k(0) = x_k$. Hence, $X_k(t)$ is governed by the following dynamic system

$$
\begin{aligned}
dX_k(t) = \sum_{\alpha \in \mathcal{M}} I_{\{\Lambda(t)=\alpha\}} \Big\{ & r(\alpha)X_k(t) + (b_k(t) - \kappa_k b_l(t))[\mu_S(\alpha, Z(t)) - r(\alpha)] \\
& + c_k(\alpha, Z(t)) - \kappa_k c_l(\alpha, Z(t)) - (g(a_k(t)) - \kappa_k g(a_l(t)))dt + (b_k(t) \\
& - \kappa_k b_l(t))\sigma_S(\alpha, Z(t))dW_S(t) \Big\} - dY_k^{a_k}(t) + \kappa_k dY_l^{a_l}(t),
\end{aligned}
\tag{45}
$$

where

$$Y_k^{a_k}(t) = \int_0^t \int_{\mathbb{R}_+} \tilde{q}_k(q_k, a_k) N_k(dt, d\rho_k).$$

For $k = 1, 2$, insurer k has a utility function $U_k : \mathbb{R} \to \mathbb{R}$, where U_k is assumed to be increasing, strictly concave, and satisfies Inada conditions, i.e., $\partial_x U_k(-\infty) = +\infty$, $\partial_x U_k(+\infty) = 0$. The insurer k aims to maximize the expected utility of his relative performance at the terminal time T by adopting a pair of investment and reinsurance strategy $u_k = (a_k, b_k) \in \mathcal{A}_k \times \mathcal{B}_k$, denote $\mathcal{U}_k \triangleq \mathcal{A}_k \times \mathcal{B}_k$. For an arbitrary pair of admissible controls $u = (u_1, u_2) \in \mathcal{U} \triangleq \mathcal{U}_1 \times \mathcal{U}_2$, the objective function is

$$
\begin{aligned}
J^k(t, x_k, z, \alpha, u) &= \mathbb{E}\left[U_k\left((1-\kappa_k)\widetilde{X}_k(T) + \kappa_k(\widetilde{X}_k(T) - \widetilde{X}_l(T))\right)\right] \\
&= \mathbb{E}\left[U_k\left(\widetilde{X}_k(T) - \kappa_k \widetilde{X}_l(T)\right)\right],
\end{aligned}
\tag{46}
$$

for $k \neq l \in \{1, 2\}$. For $k = 1, 2$, κ_k measures the sensitivity of insurer k to the performance of his competitor.

A Nash equilibrium $u^* = (u_1^*, u_2^*) \in \mathcal{U}$ is achieved such that

$$
\begin{aligned}
\mathbb{E}[U_1(\widetilde{X}_1^{u_1}(T) - \kappa_1 \widetilde{X}_2^{u_2^*}(T))] &\leq \mathbb{E}[U_1(\widetilde{X}_1^{u_1^*}(T) - \kappa_1 \widetilde{X}_2^{u_2^*}(T))], \\
\mathbb{E}[U_2(\widetilde{X}_2^{u_2}(T) - \kappa_2 \widetilde{X}_1^{u_1^*}(T))] &\leq \mathbb{E}[U_2(\widetilde{X}_2^{u_2^*}(T) - \kappa_1 \widetilde{X}_1^{u_1^*}(T))].
\end{aligned}
\tag{47}
$$

For $\Lambda(t) = \alpha \in \mathcal{M}$, $Z(t) = z$, and $X_k(t) = x_k$, where $0 \leq t \leq T$ and $k \neq l \in \{1, 2\}$, the value function of insurance company k follows

$$V^k(t, x_k, z, \alpha) = \sup_{u_k \in \mathcal{H}_k} \mathbb{E}[U_k(\widetilde{X}_k^{u_k}(T) - \kappa_k \widetilde{X}_l^{u_l^*}(T))], \tag{48}$$

where $V^k(\cdot, \cdot, \cdot, \cdot)$ is the value function in $\mathbb{R}_+ \times \mathbb{R} \times \mathbb{R} \times \mathcal{M}$.

To obtain the system of Hamilton-Jacobi-Bellman (HJB) equations, we assume the existence of optimal control. For an arbitrary $u_k \in \mathcal{U}_k$, $\Lambda(t) = \alpha \in \mathcal{M}$, $k \neq l \in \{1, 2\}$, and $V^k(\cdot, \cdot, \cdot, \alpha) \in C^2(\mathbb{R}_+ \times \mathbb{R} \times \mathbb{R} \times \mathcal{M})$, define an integro-differential operator $\mathcal{L}^{u_k, u_l}$ by

$$
\begin{aligned}
&\mathcal{L}^{u_k,u_l} V^k(t,x_k,z,\alpha)\\
&= V^k_{x_k}(t,x_k,z,\alpha)\Big\{ r(\alpha)x_k + (b_k-\kappa_k b_l)[\mu_S(\alpha,z)-r(\alpha)] + c_k(\alpha,z)\\
&\quad - \kappa_k c_l(\alpha,z) - (g(a_k)-\kappa_k g(a_l))\Big\} + V^k_z(t,x_k,z,\alpha)\mu_Z(t,z)\\
&\quad + \frac{1}{2}(b_k-\kappa_k b_l)^2\sigma_S^2(\alpha,z)V_{x_k x_k}(t,x_k,z,\alpha) + \frac{1}{2}\sigma_Z^2(t,z)V_{zz}(t,x_k,z,\alpha) \quad (49)\\
&\quad - \lambda_k \int_{\mathbb{R}_+} [V^k(t,x_k-\tilde{q}(\alpha,z,\rho_k),z,\alpha) - V^k(t,x_k,z,\alpha)]f(\rho_k)d\rho_k\\
&\quad + \lambda_l \int_{\mathbb{R}_+} [V^k(t,x_k+\kappa_k\tilde{q}(\alpha,z,\rho_l),z,\alpha) - V^k(t,x_k,z,\alpha)]f(\rho_l)d\rho_l\\
&\quad + QV(t,x_k,z,\cdot)(\alpha),
\end{aligned}
$$

where

$$
QV(t,x_k,z,\cdot)(\alpha) = \sum_{j\neq\alpha} q_{\alpha j}(V(t,x_k,z,j) - V(t,x_k,z,\alpha)).
$$

Formally, for $k \neq l \in \{1,2\}$, we conclude that V^k satisfies the following system of integro-differential HJI equations: for each $\alpha \in \mathcal{M}$,

$$
\begin{cases}
V^k_t(t,x_k,z,\alpha) + \sup\limits_{u_k\in\mathcal{H}_k} \mathcal{L}^{u_k,u_l^*} V^k(t,x_k,z,\alpha) = 0,\\
V^k(T,x_k,z,\alpha) = U_k(x_k).
\end{cases} \quad (50)
$$

A numerical experiment. We present a numerical result for a high-dimensional reinsurance and investment non-zero-sum game where the associated system of PDEs has multiple states and controls. Both insurance companies are constant absolute risk aversion (CARA) agents. More precisely, the utility function of each insurer has the form

$$
U_k(X_k) = -\frac{1}{\eta_k}\exp(-\eta_k X_k), \quad \text{for } \eta_k > 0,\ k=1,2. \quad (51)
$$

Based on the algorithm constructed above, we carry out the computation by valuing iterations in a backward manner by time.

1. Set $t = T-\delta$ and $J^{k,h,\delta}(T,x,\alpha,u^{h,\delta}) = U_k(x_k)$, for each $k=1,2$.
2. We have

$$
\begin{aligned}
&J^{k,\delta}(t,x,\alpha,u^{h,\delta})\\
&= (1-\lambda\delta)\sum_{(y,j)} p_k^{h,\delta}\Big((x,\alpha),(y,j)|u^{h,\delta}\Big) J^{k,\delta}(t+\delta,y,j,u^{h,\delta})
\end{aligned}
$$

$$+\lambda_k\delta\int_{\mathbb{R}_+} J^{k,\delta}(t+\delta, x_k - \widetilde{q}_k^h(q_k(\alpha, z, \rho_k), a_k^{h,\delta}), z, \alpha, u^{h,\delta})\Pi_k(d\rho_k)$$

$$+\lambda_l\delta\int_{\mathbb{R}_+} J^{k,\delta}(t+\delta, x_k + \kappa_k\widetilde{q}_l^h(q_l(\alpha, z, \rho_l), a_l^{h,\delta}), z, \alpha, u^{h,\delta})\Pi_l(d\rho_l).$$

Find the pair $\{\hat{u}_k^{h,\delta}, k=1,2\}$ and record $u_k^{h,\delta}(t) = \hat{u}_k^{h,\delta}$ satisfying that for any $u_k^{h,\delta} \in \mathcal{U}_k$:

$$J^{1,h,\delta}(t, x, \alpha, u_1^{h,\delta}, \hat{u}_2^{h,\delta}) \le J^{1,h,\delta}(t, x, \alpha, \hat{u}_1^{h,\delta}, \hat{u}_2^{h,\delta})$$
$$J^{2,h,\delta}(t, x, \alpha, \hat{u}_1^{h,\delta}, u_2^{h,\delta}) \le J^{2,h,\delta}(t, x, \alpha, \hat{u}_1^{h,\delta}, \hat{u}_2^{h,\delta}).$$

3. Let $t = t - \delta$ and continue the procedure until $t = 0$. We consider the case in which the discrete event consists of two states, or equivalently, the Markov chain has two states with given claim size distributions. In addition, we assume that the claim size distributions are identical in each regime. By using the value iteration methods, we numerically solve the optimal control problems. The continuous-time Markov chain $\Lambda(t)$ representing the discrete event state has the generator $Q = \begin{pmatrix} -0.5 & 0.5 \\ 0.5 & -0.5 \end{pmatrix}$ and takes values in $\mathcal{M} = \{1, 2\}$.

The reinsurance premium rates are computed from expectation premium principle as $g_k(a_k) = (1 + l_k)(1 - a_k)\mathbb{E}[A_k]$. Details of the parameters values can be referred to Bui et al. (2019). State discretization follows $\delta = 0.04$ and $h = 0.2$. For reinsurance, we discretize the reinsurance rate into six levels uniformly located from 0 to 1. The investment amount is free from restrictions, and can vary from -3 to 3 with 0.2 increase. The trends of investment and reinsurance for varying relative surplus of each player are plotted in Fig. 1.

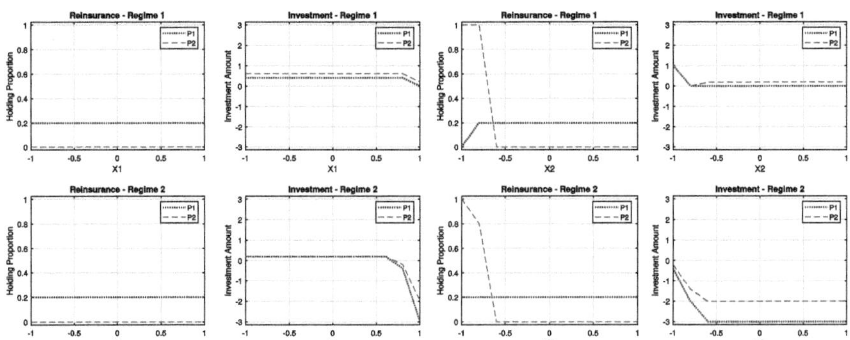

FIGURE 1 Controls for varying X_1 and X_2. Left 4 figures: $T = 0.08$, $Z = 0.81$, and $X_2 = 0.2$. Right 4 figures: $T = 0.08$, $Z = 0.81$, and $X_1 = 1$.

4 Application to mathematical biology

Mixed regular-singular control formulation. We consider d species interacting nonlinearly in a stochastic environment with harvesting; see Hening et al. (2019) for details and see also Alvarez and Shepp (1998); Lungu and Øksendal (1997); Song et al. (2011) and references therein for related formulations. Let $G = [0, \infty)^d$. Let $X_i(t)$ be the population size of species i at time t and $X(t) = (X_1(t), \ldots, X_d(t))^\top \in \mathbb{R}^d$. Let $Z_i(t)$ denote the amount of species i that has been harvested up to time t. We consider a model with a regular control $u(\cdot)$ as follows

$$dX(t) = \Big[b(X(t), \Lambda(t)) - u(t)\Big]dt + \sigma(X(t), \Lambda(t))dW(t) - dZ(t), \quad (52)$$

where $b, \sigma : [0, \infty)^d \times \mathcal{M} \mapsto \mathbb{R}^d$ are suitable functions. Suppose that $u(t) = (u_1(t), \ldots, u_d(t))^\top$ takes values in a nonempty compact set $\mathcal{U} \subset \mathbb{R}^d$ and $0 \in \mathcal{U}$. Let $f : [0, \infty)^d \times \mathcal{M} \times \mathcal{U} \to \mathbb{R}^d_+$ be the running yield associated with $u(\cdot)$. If $u(t) > 0$, $u(t)$ is the harvesting rate at time t. If $u(t) < 0$, $u(t)$ is the seeding rate at time t. The key generalization is making it possible to not only harvest, but also 'seed' individuals into the ecosystem. The payoff functional is defined by

$$J(x, u(\cdot), Z(\cdot)) := \mathbb{E}_{x,\alpha} \int_0^\infty e^{-rt}\Big[f(X(t), \Lambda(t), u(t))dt + c \cdot dZ(t)\Big], \quad (53)$$

where $c = (c_1, \ldots, c_d)^\top \in \mathbb{R}^d_+$ and $r > 0$ is the discounting factor. For each (x, α), the set of admissible controls $\mathcal{A}_{x,\alpha}$ can be defined as that for (6). The value function is defined by

$$V(x, \alpha) := \sup_{(u(\cdot), Z(\cdot)) \in \mathcal{A}_{x,\alpha}} J(x, \alpha, u(\cdot), Z(\cdot)). \quad (54)$$

Our objective is to maximize the payoff functional. This harvesting problem belongs to the class of optimal control problems described in Section 5. Hence, it can be solved by the Markov chain approximation method as in Section 2.2. We omit the details but refer the reader to Jin et al. (2012) for further reading. Note that we can consider various mixed regular-singular control formulations for optimal harvesting problems. In fact, each of harvesting and seeding components can be a singular control or a regular control. There are many possible combinations leading to different models. In a specific case, one model can be more suitable than the others. We refer to Tran et al. for a recent effort with a general formulation and state constraints.

A numerical experiment. Consider two species competing according to the following stochastic Lotka-Volterra system

$$
\begin{aligned}
dX_1(t) &= X_1(t)\Big(b_1(\Lambda(t)) - a_{11}(\Lambda(t))X_1(t) - a_{12}(\Lambda(t))X_2(t)\Big)dt - u_1(t)dt \\
&\quad + \sigma_1(\Lambda(t))X_1(t)dw_1(t) - dZ_1(t) \\
dX_2(t) &= X_2(t)\Big(b_2(\Lambda(t)) - a_{21}(\Lambda(t))X_1(t) - a_{22}(\Lambda(t))X_2(t)\Big)dt - u_2(t)dt \\
&\quad + \sigma_2(\Lambda(t))X_2(t)dw_2(t) - dZ_2(t).
\end{aligned}
\tag{55}
$$

Here $b_1(\alpha)$, $b_2(\alpha)$ are the per-capita growth rates, $a_{12}(\alpha)$, $a_{21}(\alpha)$ the per-capita interspecific competition rates, $a_{11}(\alpha)$, $a_{22}(\alpha)$ the per-capita intraspecific competition rates and $\sigma_1^2(\alpha)$, $\sigma_2^2(\alpha)$ the per-capita variances of the environmental fluctuations in regime α. In the model, $u(t) = (u_1(t), u_2(t))^\top$ is the regular control taking values in a nonempty compact set $\mathcal{U} \subset \mathbb{R}^2$. In addition, $Z_1(\cdot)$ and $Z_2(\cdot)$ are harvesting controls of species 1 and species 2, respectively. We set

$$
\begin{aligned}
&r = 0.05,\ \ c_1 = 1.6,\ \ c_2 = 1,\ \ \mathcal{U} = \{0, -0.5\} \times \{0, -0.5\},\ \ \mathcal{M} = \{1, 2\},\\
&q_{11} = -0.3,\ \ q_{12} = 0.3,\ \ q_{21} = 0.7,\ \ q_{22} = -0.7,\\
&f(x, 1, \xi_1, \xi_2) = 3\xi_1 + 3\xi_2,\ f(x, 2, \xi_1, \xi_2) = 3.5\xi_1 + 4\xi_2 \text{ for } (\xi_1, \xi_2)^\top \in \mathcal{U},\\
&b_1(1) = 2.5, b_1(2) = 0.05, b_2(1) = 2, b_2(2) = 0.03, a_{11}(\alpha) = 1.5,\\
&a_{12}(\alpha) = 1.5, \sigma_1(\alpha) = 2, a_{21}(\alpha) = 2.5, a_{22}(\alpha) = 1.5, \sigma_2(\alpha) = 2.25, \alpha \in \mathcal{M}.
\end{aligned}
$$

Let $U = 2$. Since $\mathcal{U} = \{0, -0.5\} \times \{0, -0.5\}$, it is possible to seed both species. Since $\xi_1 \le 0$ and $\xi_2 \le 0$ for any $(\xi_1, \xi_2)^\top \in \mathcal{U}$, $|u_1(t)|$ and $|u_2(t)|$ are seeding rates of species 1 and species 2 at time t, respectively. In Fig. 2 one can see the value function and the control type as functions of the population size $(x_1, x_2)^\top$. Here "1" denotes the harvesting of species 1, "2" the harvesting of species 2, and "0" the seeding (including seeding zero). In each regime α, there is a nonlinear curve Γ_α (see Fig. 3) such that one should harvest whenever the population size $(X_1(t), X_2(t))^\top$ lies above Γ_α. Seeding takes place only when $(X_1(t), X_2(t))^\top$ is in the domain close to the origin in $\mathbb{R}^2$ (green domain in the color print or domain with darker print in the black and white print). In particular, only species 1 should be seeded. This observation is well connected with the chosen system parameters which shows that the environment is more favorable to species 1 than to species 2.

Associated with the model (55), there are two subsystems that interact and switch back and forth. Using the same parameters as above, consider the harvesting problem for the subsystem in regime 2 with the value function $V_2(x)$. Since $b_1(2) \le r$ and $b_2(2) \le r$, it is known that the optimal policy is to drive the process instantaneously to $(0, 0)^\top$ (see (Alvarez and Shepp, 1998, Theorem 1)); that is, $V_2(x) = c_1x_1 + c_2x_2$ for $x = (x_1, x_2)^\top \in \mathbb{R}^2$ and $V_2(0, 0) = 0$. Meanwhile, for the switching model (55), the figures show that $V(x, 2) > c_1x_1 + c_2x_2$

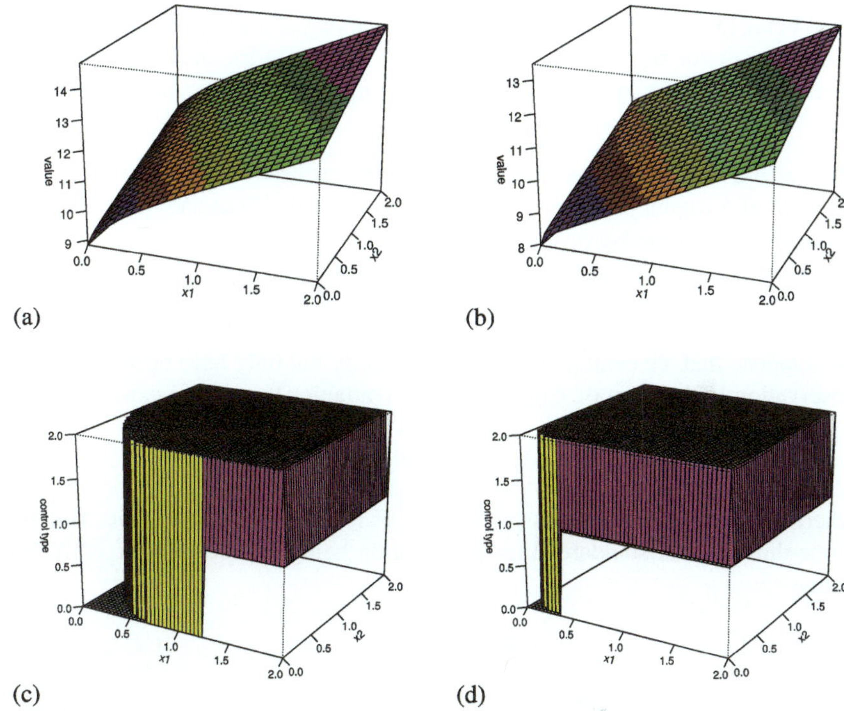

FIGURE 2 (a) The value function in regime 1; (b) the value function in regime 2; (c) the optimal control in regime 1; (d) the optimal control in regime 2 ("1": harvesting of species 1, "2": harvesting of species 2, "0": seeding).

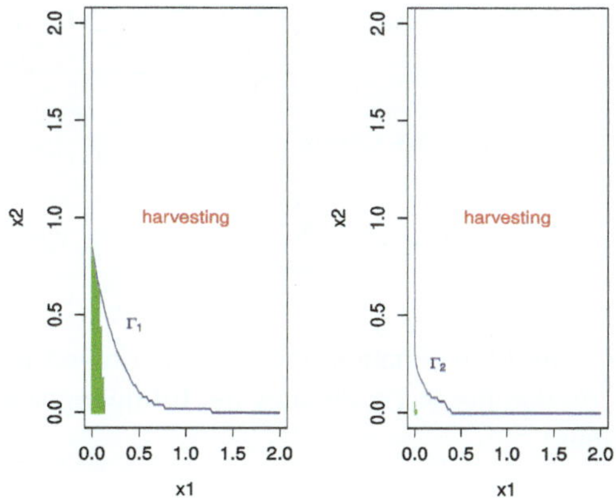

FIGURE 3 The optimal control in regime 1 (left) and regime 2 (right): if the population sizes lie in the domain close to the origin (green in the color print or darker in the black and white print), species 1 should be seeded.

and in particular $V((0,0),2) > 0$. Thus, in the presence of the switching component, being at regime 2, it is not necessary to harvest all instantaneously. Instead, one can wait for regime 1 which is more favorable to the development of the species. This shows the effects of the switching process to the harvesting problem.

Harvesting with a hidden Markov chain. We proceed to study the harvesting problem of a regime-switching population system; see Tran and Yin (2016) for details. Consider the harvested population system given by

$$dX(t) = b(X(t), \Lambda(t))dt + \sigma(X(t), \Lambda(t))dW(t) - dZ(t). \tag{56}$$

We suppose that we cannot observe $\Lambda(t)$ directly but only have noise-corrupted observation in the form of $\Lambda(t)$ plus noise available. That is, we can observe

$$dy(t) = g_0(\Lambda(t))dt + \sigma_0 dW_0(t), \quad y(0) = 0, \tag{57}$$

where σ_0 is a positive constant, $g_0 : \mathcal{M} \mapsto \mathbb{R}$ is a one-to-one function, $W_0(\cdot)$ is a one-dimensional standard Brownian motion being independent of $W(\cdot)$. We define

$$p_j(t) := \mathbb{1}_{\{\Lambda(t)=j\}}, \quad \phi_j(t) := \mathbb{P}\left(\Lambda(t) = j | y(s), 0 \le s \le t\right), \quad j = 1, \ldots, m_0. \tag{58}$$

Clearly, $\phi_j(t) \ge 0$ and $\sum_{j=1}^{m_0} \phi_j(t) = 1$. Then, it is sufficient to work with $\phi(t) := \left(\phi_1(t), \ldots, \phi_{m_0-1}(t)\right)^\top$. The actual state space for $\phi(t)$ is

$$S_{m_0-1} := \{\phi := (\phi_1, \ldots, \phi_{m_0-1})^\top \in \mathbb{R}^{m_0-1} : \phi_j \ge 0, \sum_{j=1}^{m_0-1} \phi_j \le 1\}.$$

For each $(x, \phi) \in \mathbb{R}^d \times S_{m_0-1}$ we define

$$\overline{b}(x,\phi) = \sum_{j=1}^{m_0} \phi_j b(x,j), \; \overline{\sigma}(x,\phi) = \sum_{j=1}^{m_0} \phi_j \sigma(x,j), \; \overline{g}_0(\phi) = \sum_{j=1}^{m_0} \phi_j g_0(j).$$

We recall some results of Wonham's filter. It was proved in Wonham (1965) that the posterior probability $\phi_j(\cdot)$ satisfies the following system of stochastic differential equations

$$\begin{aligned} d\phi_j(t) = & \Big[\sum_{k=1}^{m_0} q_{kj}\phi_k(t) - \sigma_0^{-2}\big(g_0(j) - \overline{g}_0(\phi(t))\big)\overline{g}_0(\phi(t))\phi_j(t)\Big]dt \\ & + \sigma_0^{-2}\big(g_0(j) - \overline{g}_0(\phi(t))\big)\phi_j(t)dy(t), \quad j = 1, \ldots, m_0, \end{aligned} \tag{59}$$

where $(\phi_1^0, \ldots, \phi_{m_0}^0)^\top$ is the initial distribution of $\Lambda(t)$. Introduce a one-dimensional innovation process

$$d\overline{W}(t) = \sigma_0^{-1}\big(dy(t) - \overline{g}_0(\phi(t))dt\big), \quad \overline{W}(0) = 0.$$

Then the first $m-1$ equations in (59) can be rewritten as

$$\begin{aligned} d\phi_j(t) &= \sum_{k=1}^{m_0} q_{kj}\phi_k(t)dt + \sigma_0^{-1}\phi_j(t)\big(g_0(j) - \overline{g}_0(\phi(t))\big)d\overline{W}(t), \\ j &= 1, \ldots, m_0 - 1. \end{aligned} \tag{60}$$

With the use of (58),

$$b(X(s), \alpha(s)) = \sum_{j=1}^{m_0} p_j(s) b(X(s), j), \; \sigma(X(s), \alpha(s)) = \sum_{j=1}^{m_0} p_j(s)\sigma(X(s), j).$$

By replacing the hidden state $p_j(t)$ with its estimate $\phi_j(t)$, we arrive at

$$X(t) = x + \int_0^t \overline{b}(X(s), \phi(s))ds + \int_0^t \overline{\sigma}(X(s), \phi(s))dW(s) - Z(t). \tag{61}$$

Hence (60) and (61) form a controlled process $(X(t), \phi(t)) \in \mathbb{R}^d \times S_{m_0-1}$ with complete observation and the initial condition (x, ϕ). The corresponding set of all admissible strategies is denoted by $\mathcal{A}_{x,\phi}$. Let $f_i(\cdot) : \mathcal{M} \mapsto (0, \infty)$ be the price of species i. Define $\overline{f}_i(\phi) = \sum_{j=1}^{m_0} \phi_j f_i(j)$ for each $\phi \in S_{m_0-1}$. Let $r > 0$ be the discounting factor and $\tau = \inf\{t \geq 0 : X_i(t) \leq 0, i = 1, 2, \ldots, d\}$. For a strategy $Z(\cdot) \in \mathcal{A}_{x,\phi}$, the payoff functional and the value function are

$$\begin{aligned} J(x, \phi, Z(\cdot)) &:= \sum_{i=1}^{d} \mathbb{E}_{x,\phi} \int_0^\tau e^{-rs}\overline{f}_i(\phi(s))dZ_i(s), \\ V(x, \phi) &:= \sup_{Z(\cdot)\in\mathcal{A}_{x,\phi}} J(x, \phi, Z(\cdot)), \end{aligned} \tag{62}$$

respectively. Problem (61)-(62) belongs to the class of optimal control problems described in Section 5. Hence, it can be solved by the Markov chain approximation method as in Section 2.2. We omit the details but refer the reader to Tran and Yin (2016) for further reading.

Controlled SIS models. A SIS system is an epidemic model, which subdivides the population into susceptible (S) and infected (I) classes. The word SIS is an abbreviation of susceptible-infected-susceptible. That is, a susceptible individual becomes infected, and later on becomes susceptible again; see Bolzoni et al. (2019); Grandits et al. (2019); Tran and Yin (2021) and references therein for detailed discussions. For such diseases, a promising model is the following classical deterministic SIS epidemic model

$$\begin{cases} dS(t) = \big[\mu N - \beta S(t)I(t) + \gamma I(t) - \mu S(t)\big]dt, & S(0) = s_0, \\ dI(t) = \big[\beta S(t)I(t) - (\mu + \gamma)I(t)\big]dt, & I(0) = i_0, \end{cases} \tag{63}$$

where the initial values $S(0) = s_0 > 0$ and $I(0) = i_0 > 0$ satisfy $s_0 + i_0 = N$, $S(t)$ and $I(t)$ are the numbers of susceptible and infected individuals at time t, respectively, and μ and γ^{-1} are the average death/birth rate and the average infectious period, respectively. The parameter β is the disease transmission coefficient with $\beta = \lambda/N$ and λ being the disease contact rate of an infective individual. It can be seen that $S(t) + I(t) = N$ for any $t \ge 0$. Hence, (63) amounts to a single ordinary differential equation. Started with the model given in (63), we normalize the population size to one by replacing S and I with S/N and I/N, respectively. We obtain

$$\begin{cases} dS(t) = \big[\mu - \lambda S(t)I(t) + \gamma I(t) - \mu S(t)\big]dt, & S(0) = s_0, \\ dI(t) = \big[\lambda S(t)I(t) - (\mu + \gamma)I(t)\big]dt, & I(0) = i_0, \end{cases} \tag{64}$$

where $S(t)$ and $I(t)$ are the fraction of susceptible and infected individuals at time t, respectively. Taking into account the environmental noise, the system parameters μ, λ, γ may experience abrupt changes, which is modeled by a Markov chain $\Lambda(\cdot)$ taking values in $\mathcal{M}$ as follows

$$\begin{cases} dS(t) = \big[\mu_{\Lambda(t)} - \lambda_{\Lambda(t)} S(t)I(t) + \gamma_{\Lambda(t)} I(t) - \mu_{\Lambda(t)} S(t)\big]dt, \\ dI(t) = \big[\lambda_{\Lambda(t)} S(t)I(t) - (\mu_{\Lambda(t)} + \gamma_{\Lambda(t)})I(t)\big]dt, \\ S(0) = s_0, \quad I(0) = i_0, \quad \Lambda(0) = \alpha_0. \end{cases} \tag{65}$$

A key parameter of the system is the average number of contacts per infected people per day $\lambda_{\Lambda(t)}$, which can be perturbed by white noise; that is, $\lambda_{\Lambda(t)} \to \lambda_{\Lambda(t)} + \sigma_{\Lambda(t)} \dot{W}(t)$. Thus,

$$\begin{cases} dS(t) = \big[\mu_{\Lambda(t)} - \lambda_{\Lambda(t)} S(t)I(t) + \gamma_{\Lambda(t)} I(t) - \mu_{\Lambda(t)} S(t)\big]dt \\ \qquad\qquad -\sigma_{\Lambda(t)} S(t)I(t)dW(t), \\ dI(t) = \big[\lambda_{\Lambda(t)} S(t)I(t) - (\mu_{\Lambda(t)} + \gamma_{\Lambda(t)})I(t)\big]dt + \sigma_{\Lambda(t)} S(t)I(t)dW(t), \\ S(0) = s_0, \quad I(0) = i_0, \quad \Lambda(0) = \alpha_0. \end{cases} \tag{66}$$

We suppose that the decision maker is able to control the magnitudes of the disease contact rate and the recovery rate to some extent. Thus, we introduce the disease contact rate control (or social distancing control) $u_1(\cdot)$ and the recovery rate control $u_2(t)$, which take values in the nonempty compact sets of real numbers $\mathcal{U}_1 \in [0, 1]$ and $\mathcal{U}_2$, respectively. The controlled version is given by

$$\begin{cases} dS(t) = \big[\mu_{\Lambda(t)} - u_1(t)\lambda_{\Lambda(t)}S(t)I(t) + \gamma_{\Lambda(t)}I(t) - \mu_{\Lambda(t)}S(t) + u_2(t)I(t)\big]dt \\ \qquad\qquad -\sigma_{\Lambda(t)}S(t)I(t)dw(t), \\ dI(t) = \big[u_1(t)\lambda_{\Lambda(t)}S(t)I(t) - \big(\mu_{\Lambda(t)} + \gamma_{\Lambda(t)} + u_2(t)\big)I(t)\big]dt \\ \qquad\qquad +\sigma_{\Lambda(t)}S(t)I(t)dw(t), \\ S(0) = s_0, \quad I(0) = i_0, \quad \Lambda(0) = \alpha_0. \end{cases} \tag{67}$$

Given that $I(t) + S(t) = 1$, $I(t)$, the fraction of infected individuals, obeys the stochastic differential equation with Markovian switching given by

$$\begin{aligned} dI(t) = I(t)\big[&u_1(t)\lambda_{\Lambda(t)} - \mu_{\Lambda(t)} - \gamma_{\Lambda(t)} - u_1(t)\lambda_{\Lambda(t)}I(t) \\ &- u_2(t)\big]dt + \sigma_{\Lambda(t)}I(t)\big[1 - I(t)\big]dw(t). \end{aligned} \tag{68}$$

We choose the eradication time τ as the terminal time of planning (that is, time horizon). Let ζ be a positive constant such that $\zeta < 1$, then $\tau = \inf\{t \ge 0 : I(t) \le \zeta\}$. Let $\mathcal{A}_{i,\alpha}$ denote the collection of all admissible controls with initial value $(i, \alpha) \in (0, 1) \times \mathcal{M}$. A strategy $u(\cdot) = (u_1(\cdot), u_2(\cdot))$ will be in $\mathcal{A}_{i,\alpha}$ if $u(\cdot)$ is $\sigma\{W(s), \alpha(s) : s \le t\}$-adapted and $u(t) \in \mathcal{U}_1 \times \mathcal{U}_2$ for any $t \ge 0$. We suppose that controlling the recovery rate and the disease contact rate is costly and the treatment of the infected individuals also create additional costs. The cost is described by the bounded cost function $f : [0, 1] \times \mathcal{M} \times \mathcal{U}_1 \times \mathcal{U}_2 \mapsto [0, \infty)$. For a control strategy $u(\cdot) \in \mathcal{A}_{i,\alpha}$. We define the cost functional as $J(i, \alpha, u(\cdot)) := \mathbb{E}_{i,\alpha} \int_0^\tau e^{-rt} f(I(t), \Lambda(t), u(t))dt$, where $r > 0$ is the discounting factor. The goal is to minimize the cost functional and find an optimal strategy $u^*(\cdot)$ such that $J(i, \alpha, u^*(\cdot)) = V(i, \alpha) := \inf_{u(\cdot)\in\mathcal{A}_{i,\alpha}} J(i, \alpha, u(\cdot))$. This control problem belongs to the class of optimal control problems described in Section 1. Hence, it can be solved by the Markov chain approximation method as in Section 2.1. We omit the details but refer the reader to Tran and Yin (2021) for further reading.

A numerical experiment. We present a numerical experiment for the optimal control problem of SIS models. We can treat a Markov chain with any number of states. However, for illustration, we choose a Markov chain with two states. Consider Eq. (68) with $r = 0.05$, $\mathcal{M} = \{1, 2\}$, $\zeta = 0.05$, $\mathcal{U}_1 = \{1\}$, and $\mathcal{U}_2 = \{k/5 : k \in \mathbb{Z}, 0 \le k \le 15\}$; that is, we focus on the recovery rate control $u_2(\cdot)$. The cost function is of the form $f(i, \alpha, \xi) = a_0 + a_1 i + a_2 i\xi^2$ for $(i, \alpha, \xi) \in [0, 1] \times \mathcal{M} \times \mathcal{U}_2$, where a_0, a_1, and a_2 represent the cost per unit time of the components $1, i, i\xi^2$, respectively. In particular, $\int_0^\tau a_0 dt$ is the cost due to the time period needed for outbreak eradication, while $\int_0^\tau a_1 I(t)dt$ is the cost that infected individual creates for the society due to lost working hours and standard medical care, not including the treatment $u_2(\cdot)$. $\int_0^\tau a_2 I(t)u_2^2(t)dt$ is the cost of treating infected individuals. We assume that the cost is proportional to the number of infected individuals, and the cost per each patient depends quadratically on the treatment effort $u_2(\cdot)$. The system parameters are given by $\mu_1 = 0.45$, $\mu_2 = 0.05$, $\gamma_1 = 0.35$, $\gamma_2 = 0.15$, $\lambda_1 = 2$, $\lambda_2 = 2.4$, $\sigma_1 = 1$, $\sigma_2 = 0$.

The generator Q of the Markov chain $\Lambda(\cdot)$ is given by $q_{11} = -1$, $q_{12} = 1$, $q_{21} = 1$, $q_{22} = -1$. Let $f(i, \alpha, \xi) = 2 + \alpha i + \alpha i \xi^2$. The results in Fig. 4 show that the approximating value function and approximating optimal control highly depends on the regime of the Markov chain.

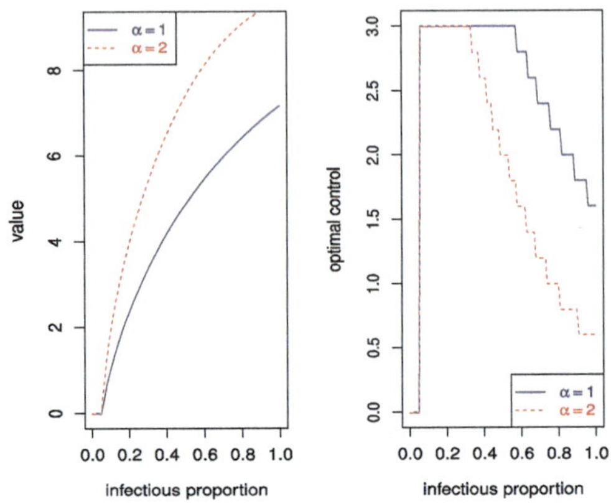

FIGURE 4 Value function (left) and optimal control (right).

In each regime α, there is a level $L(\alpha)$ such that the optimal control is the maximum control effort for $i < L(\alpha)$. The value function in regime 2 is higher than that of regime 1, while the optimal control in regime 2 is strictly smaller than that of regime 1 possibly because of the higher cost function in regime 2; that is $f(i, 2, \xi) > f(i, 1, \xi)$ for any $(i, \xi) \in (0, 1) \times \mathcal{U}_2$.

5 Final remarks

This chapter has been devoted to numerical methods for solving stochastic optimal control of switching diffusions. An important problem that is not discussed here is the rate of convergence. Early study can be found in Menaldi (1989). Then started in 1997, Krylov launched a series of study for analyzing the rates of convergence; see Krylov (2000). Both of the aforementioned works are analytic in nature by using mainly a nonlinear PDE approach, whereas Song and Yin (2009) used a probabilistic approach to treat the rates of convergence issues. We refer the reader to the cited references for further reading. Finally, we remark that there is a recent trend on using deep learning in conjunction with numerical approximation for stochastic control problems Han et al. (2018), which opens up new domains for further investigation with a wider range of applications.

References

Alvarez, L.H., Shepp, L., 1998. Optimal harvesting of stochastically fluctuating populations. J. Math. Biol. 37, 155–177.

Banks, H.T., Kunisch, K., 1989. Estimation Techniques for Distributed Parameter Systems. Systems Control: Foundations Appl., vol. 1. Birkhäuser, MA.

Barles, G., Souganidis, P.E., 1991. Convergence of approximation schemes for fully nonlinear second order equations. Asymptot. Anal. 4, 271–283.

Bolzoni, L., Bonacini, E., Marca, R.D., Groppi, M., 2019. Optimal control of epidemic size and duration with limited resources. Math. Biosci. 315, 108–232.

Bonnans, J.F., Ottenwaelter, E., Zidani, H., 2004. A fast algorithm for the two dimensional HJB equation of stochastic control. Math. Model. Numer. Anal. 38, 723–735.

Bui, T., Cheng, X., Jin, Z., Yin, G., 2019. Approximation of a class of non-zero-sum investment and reinsurance games for regime-switching jump-diffusion models. Nonlinear Anal. Hybrid Syst. 32, 276–293.

Falcone, M., Ferretti, R., 2014. Semi-Lagrangian Approximation Schemes for Linear and Hamilton-Jacobi Equations. SIAM, Philadelphia, PA.

Fehrenbach, J., Mirebeau, J.-M., 2014. Sparse non-negative stencils for anisotropic diffusion. J. Math. Imaging Vis. 49, 123–147.

Fleming, W., Soner, H., 2006. Controlled Markov Processes and Viscosity Solutions, 2nd ed. Springer-Verlag, New York, NY.

Grandits, P., Kovacevic, R.M., Veliov, V.M., 2019. Optimal control and the value of information for a stochastic epidemiological SIS-model. J. Math. Anal. Appl. 476, 665–695.

Han, J., Jentzen, A., Weinan, E., 2018. Solving high-dimensional partial differential equations using deep learning. Proc. Natl. Acad. Sci. 115 (34), 8505–8510.

Hening, A., Tran, K., Phan, T.T., Yin, G., 2019. Harvesting of interacting stochastic populations. J. Math. Biol. 79, 533–570.

Jiang, L., Dai, M., 2004. Convergence of binomial tree methods for European/American path-dependent options. SIAM J. Numer. Anal. 42, 1094–1109.

Jin, Z., Yang, H., Yin, G., 2013a. Numerical methods for optimal dividend payment and investment strategies of regime-switching jump diffusion models with capital injections. Automatica 49, 2317–2329.

Jin, Z., Yin, G., Wu, F., 2013b. Optimal reinsurance strategies in regime-switching jump diffusion models: stochastic differential game formulation and numerical methods. Insur. Math. Econ. 53, 733–746.

Jin, Z., Yin, G., Zhu, C., 2012. Numerical solutions of optimal risk control and dividend optimization policies under a generalized singular control formulation. Automatica 48, 1489–1501.

Kalise, D., Kundu, S., Kunisch, K., 2020. Robust feedback control of nonlinear PDEs by numerical approximation of high-dimensional Hamilton–Jacobi–Isaacs equations. SIAM J. Appl. Dyn. Syst. 19, 1496–1524.

Krylov, N.V., 2000. On the rate of convergence of finite-difference approximations for Bellman's equations with variable coefficients. Probab. Theory Relat. Fields 117, 1–16.

Kushner, H.J., 1976. A survey of some applications of probability and stochastic control theory to finite difference methods for degenerate elliptic and parabolic equations. SIAM Rev. 18, 545–577.

Kushner, H.J., 1977. Probability Methods for Approximations in Stochastic Control and for Elliptic Equations. Academic Press.

Kushner, H.J., 2002. Numerical approximations for stochastic differential games. SIAM J. Control Optim. 41, 457–486.

Kushner, H.J., Dupuis, P., 2001. Numerical Methods for Stochastic Control Problems in Continuous Time, 2nd ed. Springer, New York.

Kushner, H.J., Martins, L.F., 1991. Numerical methods for stochastic singular control problems. SIAM J. Control Optim. 29, 1443–1475.

Lü, Q., Zhang, X., 2021. Mathematical Control Theory for Stochastic Partial Differential Equations. Springer, Berlin.

Lungu, E.M., Øksendal, B., 1997. Optimal harvesting from a population in a stochastic crowded environment. Math. Biosci. 145, 47–75.

McEneaney, W.M., 2007. A curse-of-dimensionality-free numerical method for solution of certain HJB PDEs. SIAM J. Control Optim. 46, 1239–1276.

Menaldi, J.L., 1989. Some estimates for finite difference approximations. SIAM J. Control Optim. 27, 579–607.

Mirebeau, J.M., 2014. Anisotropic fast-marching on Cartesian grids using lattice basis reduction. SIAM J. Numer. Anal. 52, 1573–1599.

Pham, H., 2009. Continuous-Time Stochastic Control and Optimization with Financial Applications. Springer.

Song, Q.S., Stockbridge, R.H., Zhu, C., 2011. On optimal harvesting problems in random environments. SIAM J. Control Optim. 49, 859–889.

Song, Q.S., Yin, G., 2009. Rates of convergence of numerical methods for controlled regime-switching diffusions with stopping times in the costs. SIAM J. Control Optim. 48, 1831–1857.

Song, Q.S., Yin, G., Zhang, Z., 2006. Numerical method for controlled regime-switching diffusions and regime-switching jump diffusions. Automatica 42, 1147–1157.

Song, Q.S., Yin, G., Zhang, Z., 2008. Numerical solutions for stochastic differential games with regime switching. IEEE Trans. Autom. Control 53, 509–521.

Tran, K., Bich, L.T.N., Yin, G. Harvesting of a stochastic population under a mixed regular-singular control formulation. Preprint. arXiv:2110.14418.

Tran, K., Yin, G., 2016. Numerical methods for optimal harvesting strategies in random environments under partial observations. Automatica 70, 74–85.

Tran, K., Yin, G., 2021. Optimal control and numerical methods for hybrid stochastic SIS models. Nonlinear Anal. Hybrid Syst. 41, 101051.

Wang, L.Y., Khargonekar, P.P., Beydoun, A., 1999. Robust control of hybrid systems: performance guided strategies. In: Antsaklis, P., Kohn, W., Lemmon, M., Nerode, A., Sastry, S. (Eds.), Hybrid Systems V. Berlin. In: Lecture Notes in Computer Sci., vol. 1567, pp. 356–389.

Wonham, W.M., 1965. Some applications of stochastic differential equations to optimal nonlinear filtering. J. SIAM Control Ser. A 2, 347–369.

Yin, G., Zhu, C., 2010. Hybrid Switching Diffusions: Properties and Applications. Springer, New York.

Yong, J., Zhou, X.Y., 1999. Stochastic Controls: Hamiltonian Systems and HJB Equations. Springer-Verlag, New York.

Chapter 8

Control of parameter dependent systems

Martin Lazar[a,*] and Jérôme Lohéac[b]
[a]*Department of Electrical Engineering and Computing, University of Dubrovnik, Dubrovnik, Croatia,* [b]*Université de Lorraine, CNRS, CRAN, Nancy, France*
**Corresponding author: e-mail address: mlazar@unidu.hr*

Contents

Abstract

In this chapter we provide an overview of recent progress on the problem of controllability of parameter dependent systems. We explore different control notions successfully developed through the last decade. The aim of the control function is to steer the system to a state satisfying some properties prescribed either at some time instant $T > 0$ or during a given time interval. These properties may be separated with respect to parameter values and can refer just to a single system itself (e.g. greedy control), or may consider solutions corresponding to the whole parameter range (e.g. ensemble control, averaged control). In the latter case control functions are designed as parameter invariant, implying a same control is to be applied to the system independently of a particular realization of the parameter, while in the first case controls vary along with the parameter.
Beside the positive theoretical results, for each notion we provide a precise computational algorithm accompanied by a numerical example.

Handbook of Numerical Analysis, Volume 23, ISSN 1570-8659, https://doi.org/10.1016/bs.hna.2021.12.008

Keywords
Averaged control, Greedy control, Ensemble control, Parameter dependent systems, Reduced basis

MSC Codes
93C15, 49J55, 49M29, 93-08

1 Introduction

In this chapter, we consider parameter dependent control systems of the form,

$$\dot{x}_\theta = A_\theta x_\theta + B_\theta u_\theta, \qquad x_\theta(0) = \mathrm{x}^0_\theta, \tag{1.1}$$

where $x_\theta(t) \in \mathbb{R}^N$ is the state of the system at time $t > 0$, $u_\theta(t)$ is the control at time $t > 0$, and $\mathrm{x}^0_\theta \in \mathbb{R}^N$ is a given initial condition. The dynamics operator A_θ is an $N \times N$ matrix, the control operator B_θ is an $N \times M$ matrix, while θ is a parameter describing the system.

If not precised otherwise, the following assumptions will be used throughout the chapter. More precisely, this assumption will be replaced by Assumption 2, when considering averaged controllability, see Section 2.1.

Assumption 1. The parameter θ belongs to a compact set $\Theta \subset \mathbb{R}^d$, and the map $\theta \in \Theta \mapsto (A_\theta, B_\theta, \mathrm{x}^0_\theta) \in \mathbb{R}^{N\times N} \times \mathbb{R}^{N\times M} \times \mathbb{R}^N$ is Lipschitz continuous.

The aim of the control function is to steer the system to a state satisfying some properties prescribed either at some time instant $T > 0$ or during a given time interval. These properties may be separated with respect to parameter values and can refer just to a single system itself (e.g. greedy control), or may consider solutions corresponding to the whole parameter range (e.g. ensemble control, averaged control). In the latter case controls u_θ are designed as parameter invariant, implying a same control is to be applied to the system (1.1) independently of a particular realization of the parameter θ, while in the first case the controls u_θ vary with θ.

Analysis of systems of the form (1.1) is often related to infinite dimensional problems, as they result from semi discretization of partial differential equations (PDEs). This is important from the application point of view, as many physical problems are modeled by parameter dependent PDEs. Typical examples are the heat equation where the system operator depends on the thermal diffusivity parameter and control is provided through an external heat source (cf. Example 3.6 below), or the wave equation, with parameter standing for a damping constant and/or velocity of propagation, and controlled by external disturbances.

However, besides the connection to PDEs, the above system models directly physical phenomena governed by ordinary differential equations. An example is the Bloch equation that describes evolution of bulk magnetization of a sample of nuclear spins. The parameter included is the gyromagnetic ratio and control is provided through the magnetic field (cf. Beauchard et al., 2010). Another example is control of vibrational systems (e.g. a system of coupled oscillators) where parameters represent mass and stiffness of the system, while control is provided, similar as for the wave equation, through external disturbances (e.g. Nakić et al., 2019). More examples of finite dimensional, parameter dependent control problems with applications arising from different disciplines can be found for instance in Ackermann et al. (1993, Chapter 1).

The chapter is organized as follows. In the next subsection we present a brief introduction to controllability of finite dimensional systems. Section 2 deals with various types of parameter independent controls, and in Section 3, we present different methods for efficient design of (an approximation of) control u_θ for every parameter θ. Each presented control notion is accompanied by a precise algorithm, providing construction of an optimal control function, and a numerical example.

Let us precise the notations used in this paper. We define the classical sets: $\mathbb{R}_+$ (the set of nonnegative real numbers), $\mathbb{R}_+^* = \mathbb{R}_+ \setminus \{0\}$, $\mathbb{N}$ (the set of nonnegative integers), and $\mathbb{N}^* = \mathbb{N} \setminus \{0\}$. For $N, M \in \mathbb{N}^*$, we define by $\mathbb{R}^{N\times M}$ the set of $N \times M$ real matrices. For $A \in R^{N\times M}$, A^* is the adjoint of A, rank A is the rank of A, and for $N = M$, $\sigma(A)$ is the set of eigenvalues of A. We also define Span$\{x_1, \ldots, x_k\}$ the vector space generated by $x_1, \ldots, x_k \in \mathbb{R}^N$. For Θ a set of $\mathbb{R}^d$ (with $d \in \mathbb{N}^*$), we define $\mathcal{P}(\Theta) = \{E \subset \Theta\}$, and $\mathcal{B}(\Theta)$ the Lebesgue σ-algebra on Θ. The notations $|\cdot|$ and $\langle\cdot,\cdot\rangle$ are used to denote the Euclidean norm and dot product on $\mathbb{R}^d$. We also define for every $\theta \in \mathbb{R}^d$, $\mathrm{dist}(\theta, \Theta) = \inf_{\eta\in\Theta} |\theta - \eta|$. Finally, for $I \subset \mathbb{R}$ an interval and $p \in [1, \infty)$ (respectively $p = \infty$), we denote by $L^p(I)$ the set of p-integrable (respectively essentially bounded) real functions defined on I. We also denote by $\dot{f}$ the time derivative of the function f.

Recall on finite dimensional control systems. Here we present some classical results on control of a general finite dimensional system of a form as in (1.1), with dependence with respect to parameter being omitted. A more elaborated overview of the topic can be found in e.g. Tucsnak and Weiss (2009, Chapter 1) or Zuazua (2006).

A system

$$\dot{x} = Ax + Bu, \quad x(0) = \mathrm{x}^0, \tag{1.2}$$

is said to be *controllable* if every initial datum x^0 can be driven to any final datum $\mathrm{x}^1 \in \mathbb{R}^N$ in some (and consequently in any) time $T > 0$.

This controllability property can be fully characterized by the so called *Kalman rank condition*. It says that system (1.2) is controllable if and only if

$$\mathrm{rank}\left[B, AB, \ldots, A^{N-1}B\right] = N. \tag{1.3}$$

This condition is of purely algebraic nature, and also time independent. For this reason controllability in some time $T > 0$ implies controllability in any positive time, and it is easy to observe that there are infinitely many controls steering the trajectory from the initial datum to the prescribed final target.

Another characterization of the controllability property is provided by the dual problem of observability of the *adjoint system*

$$-\dot{\varphi} = A^*\varphi, \qquad \varphi(T) = \varphi^1. \tag{1.4}$$

More precisely the following result holds.

Proposition 1.1. *System* (1.2) *is controllable in time T if and only if the adjoint system* (1.4) *is observable in time T, i.e., if there exists a constant $C > 0$ such that, for all solution φ of* (1.4),

$$|\varphi^1|^2 \le C \int_0^T |B^*\varphi(t)|^2 \, \mathrm{d}t. \tag{1.5}$$

Both properties hold in all time $T > 0$ if and only if the Kalman rank condition (1.3) *is satisfied.*

Furthermore, this dual approach provides a control of minimal L^2-norm out of all admissible controls. More precisely, if $\widetilde{\varphi}^1$ is a minimizer of the quadratic functional $J : \mathbb{R}^N \to \mathbb{R}$:

$$J(\varphi^1) = \frac{1}{2} \int_0^T |B^*\varphi(t)|^2 \, \mathrm{d}t - \langle \mathrm{x}^1, \varphi^1 \rangle + \langle \mathrm{x}^0, \varphi(0) \rangle, \tag{1.6}$$

with φ solution of (1.4), then the control given by

$$u = B^*\widetilde{\varphi} \in L^2(0, T)^M,$$

where $\widetilde{\varphi}$ is the corresponding solution of (1.4) (with $\varphi^1 = \widetilde{\varphi}^1$), is the control of minimal L^2-norm steering the system from x^0 to x^1 in time T.

Note that the minimizer $\widetilde{\varphi}^1$ of the functional J exists since the latter is differentiable, convex and, by the observability inequality (1.5), coercive. In addition, it satisfies the requirement $DJ(\widetilde{\varphi}^1) = 0$ which can be equivalently restated as

$$\Lambda \widetilde{\varphi}^1 = \mathrm{x}^1 - e^{TA}\mathrm{x}^0. \tag{1.7}$$

Here the operator Λ stands for the (infinite time) controllability Gramian associated to the pair (A, B), given by the relation

$$\Lambda = \int_0^T e^{(T-t)A} B B^* e^{(T-t)A^*} \, \mathrm{d}t.$$

Thus solving the control problem reduces to finding the minimizer of the functional (1.6), which is equivalent to solving the system (1.7). In practice, this is done by exploring (conjugate) gradient or some other iterative method.

Note this is usually implemented without construction of the Gramian matrix Λ itself. Indeed, application of the Gramian operator to any vector φ^1 corresponds to the state of the system at time T driven by the control of the form $u = B^*\varphi$, where φ is the solution of the adjoint system (1.4). Thus, each iteration requires solving in turn the dual and the primal system, which for systems of a large dimension N represents a significant computational challenge, specially if it has to be resolved and applied in a real time.

For these reasons, when dealing with large systems that are in addition parameter dependent, it is of crucial importance to develop robust algorithms that provide (at least approximatively) desired control functions in a stable and computationally efficient way.

2 Parameter invariant controls

In this paragraph, we are looking for controls u_θ independent of the parameter θ, that is to say that we consider the system

$$\dot{x}_\theta = A_\theta x_\theta + B_\theta u, \qquad x_\theta(0) = \mathrm{x}_\theta^0. \tag{2.1}$$

More precisely, in order to define the controllability notion, we propose two different approaches. In the first notion, (averaged controllability), the goal is to steer the expectation of the system to the target, while for the second one (ensemble controllability), the aim is to steer every realization of the system to an arbitrarily small ball around the target. Of course for the averaged controllability notion, it is required that the parameter realization follows some probability law. One can see that ensemble controllability is a stronger notion, and it has been checked in Triggiani (1975) that ensemble controllability implies averaged controllability.

In each case, the major goal is to steer the parameter dependent solution to some target by a single control. The problem is relevant in applications in which the control has to be chosen in a robust manner, independently of a particular realization of parameters obeying some deterministic or stochastic law.

Let us make more precise these two notions. To this end, we denote by $x_\theta(t; x_\theta^0, u)$ the solution of (2.1) at time t.

Definition 1 (Averaged controllability). Assume that Assumption 2 holds, and let $(\Theta, \mathcal{F}, \mu)$ be a probability space (i.e., $\mathcal{F}$ is a σ-algebra on Θ and μ is a probability measure defined on $\mathcal{F}$). We say that the system (2.1) is controllable in average, with respect to the probability distribution μ, if for every μ-integrable function $\theta \in \Theta \mapsto \mathrm{x}_\theta^0 \in \mathbb{R}^N$, and every $\overline{\mathrm{x}}^1 \in \mathbb{R}^N$, there exist a time $T > 0$ and a control $u \in L^2(0, T)^M$ such that the solution of (2.1) satisfies

$$\int_\Theta x_\theta(T; \mathrm{x}_\theta^0, u)\, \mathrm{d}\mu = \overline{\mathrm{x}}^1. \tag{2.2}$$

Of course, for Definition 1 to make sense, one needs an integrability property of $\theta \mapsto x_\theta(T; \mathrm{x}^0_\theta, u)$. This can be ensured using Assumption 2. Note however that more general well-posedness conditions can be stated in terms of the adjoint state (see Zuazua, 2014).

Assumption 2 (Averaged controllability assumptions). $\theta \in \Theta \mapsto (A_\theta, B_\theta) \in \mathbb{R}^{N\times N} \times \mathbb{R}^{N\times M}$ is μ-measurable, and $\sup\{\mathrm{Re}(\lambda_\theta) \mid \lambda_\theta \in \sigma(A_\theta),\ \theta \in \Theta\} < \infty$.

Hence, for averaged controllability, i.e., in Section 2.1, only Assumption 2 will be assumed.

Definition 2 (Ensemble controllability). Assume that Assumption 1 holds. We say that the system (2.1) is ensemble controllable, if for every $\varepsilon > 0$, every continuous function $\theta \in \Theta \mapsto (\mathrm{x}^0_\theta, \mathrm{x}^1_\theta) \in \mathbb{R}^N \times \mathbb{R}^N$, there exist a time $T > 0$ and a control $u \in L^2(0, T)^M$ such that the solution of (2.1) satisfies

$$\sup_{\theta\in\Theta} \left| x_\theta(T; \mathrm{x}^0_\theta, u) - x^1_\theta \right| < \varepsilon. \tag{2.3}$$

By taking the formal limit $\varepsilon \to 0$ in Definition 2, we obtain the notion of simultaneous controllability (see e.q. Russell, 1986 or Lions, 1988, Chapter 5 for this notion). However, let us mention that if the cardinal of Θ is infinite, then simultaneous controllability cannot hold, see Triggiani (1975). In addition, when $\Theta = \{\theta_1, \ldots, \theta_K\}$, then the ensemble controllability of (2.1) coincides with the simultaneous controllability of (2.1), that is to say to the controllability of the pair $(\mathbb{A}, \mathbb{B})$, with $\mathbb{A} = \begin{pmatrix} A_{\theta_1} & & 0 \\ & \ddots & \\ 0 & & A_{\theta_K} \end{pmatrix} \in \mathbb{R}^{KN\times KN}$ and $\mathbb{B} = \begin{pmatrix} B_{\theta_1} \\ \vdots \\ B_{\theta_K} \end{pmatrix} \in \mathbb{R}^{KN\times M}$. This is the consequence of the well-known result stating that approximate controllability implies exact controllability in finite dimension.

Remark 2.1. Definition 2 is also called uniform ensemble controllability. Indeed, one can consider the more general notion of $L^q(\Theta, \mu)$-ensemble controllability, for $q \in (1, \infty)$ and some measure μ on Θ, by substituting (2.3) with $\left(\int_\Theta \left|x_\theta(T; \mathrm{x}^0_\theta, u) - x^1_\theta\right|^q \mathrm{d}\mu\right)^{1/q} < \varepsilon$. Note however that for a bounded measure μ, the ensemble controllability defined by Definition 2 implies $L^q(\Theta, \mu)$-ensemble controllability. This is an easy consequence of the continuous injection of $C^0(\Theta) \cap L^\infty(\Theta)$ in $L^\infty(\Theta; \mu)$ and of $L^\infty(\Theta; \mu)$ in $L^q(\Theta; \mu)$.

In the rest of this section, we consider averaged controllability in Section 2.1 and ensemble controllability in Section 2.2. For each of these notions, we recall some known results and propose numerical methods to compute a control realization.

2.1 Averaged controllability

2.1.1 Averaged controllability notion and results

The notion of averaged controllability has first been introduced in Zuazua (2014). Rapidly averaged controllability results have been obtained for different classes of PDE systems, see e.g. Asier Bárcena-Petisco and Zuazua (2020); Avetisyan and Khurshudyan (2019); Coulson et al. (2019); Klamka and Khurshudyan (2019); Lazar (2018); Lazar and Zuazua (2014); Lohéac and Zuazua (2017); Lü and Zuazua (2016); Marín et al. (2017). Links between averaged and simultaneous controllability have been obtained in Lohéac and Zuazua (2016) and a refined version of averaged controllability (the notion of long time averaged controllability has been performed in Lazar and Lohéac (2020)). Dealing with numerical aspects of averaged controllability, we refer to Abdelli and Castro (2020) for wave equations and to Marín et al. (2017, 2018) for Euler-Bernoulli plate equations.

In Zuazua (2014), it has been shown that the system (2.1) is controllable in average if and only if a Kalman type rank condition is satisfied. It has also been shown there, that the averaged controllability property is equivalent to the positivity of some controllability Gramian. In addition, if the system (2.1) is controllable in average, then one can build a control of minimal L^2-norm. These results are summarized in the following theorem.

Theorem 2.2 (Zuazua, 2014). *The following properties are equivalent.*

1. *The system* (2.1) *is controllable in average, with probability measure* μ.
2. $\operatorname{rank}\left[\int_\Theta B_\theta \, \mathrm{d}\mu, \int_\Theta A_\theta B_\theta \, \mathrm{d}\mu, \ldots, \int_\Theta A_\theta^k B_\theta \, \mathrm{d}\mu, \ldots\right] = N.$
3. *There exists* $T > 0$ *such that the* $N \times N$ *matrix*

$$\overline{\Lambda}(T) = \int_0^T \left(\int_\Theta e^{(T-t)A_\theta} B_\theta \, \mathrm{d}\mu \int_\Theta B_\theta^* e^{(T-t)A_\theta^*} \, \mathrm{d}\mu \right) \mathrm{d}t \tag{2.4}$$

   is positive, i.e., there exists a constant $c(T) > 0$ *such that* $\varphi^{1*}\overline{\Lambda}(T)\varphi^1 \geqslant c(T)|\varphi^1|$ *for every* $\varphi^1 \in \mathbb{R}^N$.
4. *For every* $T > 0$, *the matrix* $\overline{\Lambda}(T)$, *defined by* (2.4), *is positive.*

In addition, if the system (2.1) *is controllable in average, then for every* μ-*integrable* $(\mathrm{x}_\theta^0)_{\theta\in\theta} \in (\mathbb{R}^N)^\Theta$, *every* $\overline{\mathrm{x}}^1 \in \mathbb{R}^N$, *and every* $T > 0$, *then there exist a unique control of minimal* L^2-*norm steering the initial condition* $(\mathrm{x}_\theta^0)_\theta$ *to the averaged value* $\overline{\mathrm{x}}^1$ *in time* T, *and this control is given by*

$$u(t) = \int_\Theta B_\theta^* e^{(T-t)A_\theta^*} \overline{\Lambda}(T)^{-1} \left(\overline{\mathrm{x}}^1 - \int_\Theta e^{TA_\theta} \mathrm{x}_\theta^0 \, \mathrm{d}\mu \right) \mathrm{d}\mu \qquad (t \in (0, T)). \tag{2.5}$$

We trivially have the existence of a constant $C(T) > 0$ such that $\varphi^{1*}\overline{\Lambda}(T)\varphi^1 \leqslant C(T)|\varphi^1|^2$ for every $\varphi^1 \in \mathbb{R}^N$. In addition, the 3rd condition of Theorem 2.2 is also equivalent to the observability of the adjoint. These fact leads to this condition is equivalent to,

$$c(T)|\varphi^1|^2 \leqslant \int_0^T \left| \int_\Theta B_\theta^* \varphi_\theta(t)\, \mathrm{d}\mu \right|^2 \mathrm{d}t \leqslant C(T)|\varphi^1|^2 \qquad (\varphi^1 \in \mathbb{R}^N), \tag{2.6}$$

where φ_θ is solution of

$$-\dot{\varphi}_\theta = A_\theta^* \varphi_\theta, \qquad \varphi_\theta(T) = \varphi^1. \tag{2.7}$$

Note also that the control of minimal L^2-norm given in Theorem 2.2 is given by

$$u(t) = \int_\Theta B_\theta^* \varphi_\theta(t)\, \mathrm{d}\mu, \tag{2.8}$$

where φ_θ is solution of (2.7), with $\varphi^1 \in \mathbb{R}^N$ is the minimizer of

$$J(\varphi^1) = \frac{1}{2} \int_0^T \left| \int_\Theta B_\theta^* \varphi_\theta(t)\, \mathrm{d}\mu \right|^2 \mathrm{d}t - \left\langle \varphi^1, \overline{\mathrm{x}}^1 \right\rangle + \int_\Theta \left\langle \varphi_\theta(0), \mathrm{x}_\theta^0 \right\rangle \mathrm{d}\mu$$
$$(\varphi^1 \in \mathbb{R}^N).$$

This expression of the control is easier to consider, since we only have to solve a quadratic minimization problem in $\mathbb{R}^N$ to obtain φ^1. The associated minimal L^2-norm control is then obtained from (2.8). More precisely, in order to solve the quadratic problem, one can use for instance any gradient descent method. Then, once φ^1 has been found, we obtain $\varphi_\theta(t)$ by solving (2.7), and finally, using the expression (2.8), we get $u(t)$. Of course, in this procedure, one has to discretize time and integrals with respect to Θ. Note that the time discretization is rather standard and hence will not be developed here. The discretization with respect to θ, will be discussed in the rest of this section.

2.1.2 Numerical approximation

In this subsection we deal with the issue of constructing a (numerical approximation of) averaged control for a general system for which formulae (2.5) or (2.8) can not be calculated exactly, either due to computational complexity or to issues related to integration (with respect to θ).

The idea behind all the numerical methods presented here is to consider $K \in \mathbb{N}^*$ values $\theta_1^K, \ldots, \theta_K^K \in \Theta$ and K weights $m_1^K, \ldots, m_K^K \in \mathbb{R}_+$, so that $\sum_{k=1}^K m_k^K e^{tA_{\theta_k^K}} B_{\theta_k^K}$ converges for every $t \in [0, T]$ to $\int_\Theta e^{tA_\theta} B_\theta \, \mathrm{d}\mu$, as K goes to ∞. Once these parameters are given, we consider the matrices,

$$\mathbb{A}_K = \begin{pmatrix} A_{\theta_1^K} & & 0 \\ & \ddots & \\ 0 & & A_{\theta_K^K} \end{pmatrix} \in \mathbb{R}^{NK \times NK}, \quad \mathbb{B}_K = \begin{pmatrix} B_{\theta_1^K} \\ \vdots \\ B_{\theta_K^K} \end{pmatrix} \in \mathbb{R}^{NK \times M},$$

$$\mathbb{C}_K = \begin{pmatrix} m_1^K \mathrm{I}_n & \dots & m_K^K \mathrm{I}_n \end{pmatrix} \in \mathbb{R}^{N \times NK} \tag{2.9}$$

and the system

$$\dot{X}_K = \mathbb{A}_K X_K + \mathbb{B}_K u, \quad X_K(0) = X_K^0 := \begin{pmatrix} x_{\theta_1^K}^{0\ *} & \dots & x_{\theta_K^K}^{0\ *} \end{pmatrix}^* \in \mathbb{R}^{NK}, \tag{2.10a}$$

$$\overline{x}_K = \mathbb{C}_K X_K. \tag{2.10b}$$

Given some time $T > 0$, the problem is to find a control $u \in L^2(0,T)^M$ such that the output of the system (2.10) satisfies $\overline{x}_K(T) = \overline{\mathrm{x}}^1$.

Definition 3. We say that the system (2.10) is output controllable if for every $X^0 \in \mathbb{R}^{NK}$ and every $\overline{\mathrm{x}}^1 \in \mathbb{R}^N$, there exist a time $T > 0$ and a control $u \in L^2(0,T)^M$ such that the solution of (2.10) satisfies, $\overline{x}_K(T) = \overline{\mathrm{x}}^1$.

Remark 2.3. The notion of output controllability is more general than the averaged one, and can be applied for general systems $\dot{x} = Ax + Bu$, with output $y = Cx + Du$. We refer to Kreindler and Sarachik (1964) for this notion and corresponding controllability results.

Note that the output controllability of (2.10) coincides with the averaged controllability of (2.1) with the measure $\mu_K = \sum_{k=1}^N m_k^K \delta_{\theta_k^K}$, where δ_θ is the Dirac measure supported at $\theta \in \Theta$. As a consequence output controllability criteria for the system (2.10) can be directly obtained from Theorem 2.2. In particular, (2.10) is averaged controllable if and only if for every $T > 0$, the Gramian

$$\overline{\Lambda}_K(T) = \mathbb{C}_K \left(\int_0^T e^{(T-t)\mathbb{A}_K} \mathbb{B}_K \mathbb{B}_K^* e^{(T-t)\mathbb{A}_K^*} \, \mathrm{d}t \right) \mathbb{C}_K^* \tag{2.11}$$

is positive. In addition, if $\overline{\Lambda}_K(T)$ is positive, the minimal L^2-norm control in time $T > 0$ is given by

$$u_K(t) = \mathbb{B}_K^* \varphi_K(t) \tag{2.12}$$

with $\varphi_K(t)$ solution of

$$\dot{\varphi}_K = -\mathbb{A}_K^* \varphi_K, \qquad \varphi_K(T) = \mathbb{C}_K^* \varphi^1 \in \mathbb{R}^{NK}, \tag{2.13}$$

where φ_K^1 is the minimizer of

$$J(\varphi^1) = \frac{1}{2} \varphi^{1*} \overline{\Lambda}_K(T) \varphi^1 + \langle \varphi^1, \mathbb{C}_K e^{T\mathbb{A}_K} X_K^0 - \overline{\mathrm{x}}^1 \rangle \qquad (\varphi^1 \in \mathbb{R}^N). \tag{2.14}$$

Algorithm 1 below will be then used to compute numerically an averaged control for K evaluations of the system. In practice, given some error $\varepsilon > 0$, one wants to determine an integer $K = K(\varepsilon)$ such that $\|u_K - u\|_{L^2(0,T)^M} \leqslant \varepsilon$ (which in turn implies that $|\overline{x}(T; \mathrm{x}_\theta^0, u_K) - \mathrm{x}_\theta^1|$ is small as well), where u is the control of minimal L^2-norm such that $\overline{x}_\theta(T; \mathrm{x}_\theta^0, u) = \overline{\mathrm{x}}^1$. The difficulty is that, in practice, we know the convergence rate, but do not know the multiplicative constant in front of it, i.e., with the notations of Proposition 2.5, we know p_o, but not C in (2.15). To overcome this difficulty, one checks a posteriori that $|\overline{x}(T; \mathrm{x}_\theta^0, u_K) - \mathrm{x}_\theta^1| < \varepsilon$. This is the aim of Algorithm 2.

Remark 2.4. Obviously, to compute $|\overline{x}(T; \mathrm{x}_\theta^0, u_K) - \overline{x}_\theta^1|$ in Algorithm 2, one needs again to discretize $\int_\Theta x_\theta(T; \mathrm{x}_0^\theta, u_K)\, \mathrm{d}\mu$, this can be done with a finer discretization $\theta_1, \dots, \theta_L$ than the one $(\theta_1^K, \dots, \theta_K^K)$ used in Algorithm 1.

Algorithm 1 Output control with K evaluations of the system

Initialize: Given $\theta \mapsto \mathrm{x}_\theta^0, \overline{\mathrm{x}}^1, T > 0$,
$K \in \mathbb{N}^*, \theta_1^K, \dots, \theta_K^K \in \Theta$ and $m_1^K, \dots, m_K^K \in \mathbb{R}_+$.
1: Build $\mathbb{A}_K, \mathbb{B}_K, \mathbb{C}_K$ and X_K^0 from (2.9)–(2.10a).
2: **if** rank $\mathbb{C}_K \left[\mathbb{A}_K^0 \mathbb{B}_K, \dots, \mathbb{A}^{nK-1}\mathbb{B}_K\right] < N$ **then**
3: No control can be computed, **return** $\varnothing$.
4: **else**
5: Compute φ_K^1 the minimizer of (2.14).
6: Compute $\varphi_K(t)$ solution of (2.13).
7: **end if**
Return: $u_K(t)$ given by (2.12).

Algorithm 2 Averaged control computation

Initialize: Given $\epsilon > 0, \theta \mapsto \mathrm{x}_\theta^0, \overline{\mathrm{x}}^1, T > 0$.
1: $K \leftarrow 1$, and compute u_1 from Algorithm 1.
2: **while** u_K is not defined (Algorithm 1 returns $\varnothing$) **or** $|\overline{x}(T; \mathrm{x}_\theta^0, u_K) - \overline{x}_\theta^1| \geqslant \varepsilon$ **do**
3: $K \leftarrow K + 1$.
4: Compute u_K from Algorithm 1.
5: **end while**
Return: u_K.

Based on the expressions (2.11)–(2.14), and with well-chosen parameters $\theta_1^K, \dots, \theta_K^K$ and $m_1^K, \dots, m_K^K$, one can ensure that if K is large enough, then the system (2.10) is output controllable. In addition, the numerical control u_K given by Algorithm 1 converges to the exact control as $K \to \infty$.

Proposition 2.5. *Let $T > 0$, $\overline{\mathrm{x}}^1 \in \mathbb{R}^N$ and assume that (2.1) is μ-averaged controllable. Assume in addition that $\theta_1^K, \dots, \theta_K^K \in \Theta$, and $m_1^K, \dots, m_K^K \in \mathbb{R}_+$ are*

designed so that there exist $C_o > 0$ and $p_o > 0$ such that

$$\sup_{t\in[0,T]}\left|\int_\Theta e^{tA_\theta} B_\theta \,\mathrm{d}\mu - \sum_{k=1}^{K} m_k^K e^{tA_{\theta_k^K}} B_{\theta_k^K}\right| \leqslant C_o K^{-p_o} \ \textit{and}$$

$$\left|\int_\Theta e^{TA_\theta} \mathrm{x}_\theta^0 \,\mathrm{d}\mu - \sum_{k=1}^{K} m_k^K e^{TA_{\theta_k^K}} \mathrm{x}_{\theta_k^K}^0\right| \leqslant C_o K^{-p_o}.$$

Then there exists $K_0 \in \mathbb{N}^*$ such that the system (2.10) *is output controllable for every $K \geqslant K_0$, and there exist $C > 0$ such that*

$$\|u_K - u\|_{L^2(0,T)^M} \leqslant C K^{-p_o} \quad \textit{and}$$
$$\left|\int_\Theta x_\theta(T;\mathrm{x}_\theta^0, u_K)\,\mathrm{d}\mu - \overline{\mathrm{x}}^1\right| \leqslant C K^{-p_o} \qquad (K \geqslant K_0), \quad (2.15)$$

where u and u_K are the controls respectively given by (2.5) *and* (2.12), *and where $x_\theta(\cdot;\mathrm{x}_\theta^0, u_K)$ is the solution of* (2.1) *with control u_K.*

Proof. Let us first prove the existence of $K_0 \in \mathbb{N}$ such that for every $K \geqslant K_0$, the system (2.10) is output controllable. To this end, we define the quadrature error

$$E_K(t) = \sum_{k=1}^{K} m_k^K e^{tA_{\theta_k^K}} B_{\theta_k^K} - \int_\Theta e^{tA_\theta} B_\theta \,\mathrm{d}\mu \qquad (t \in [0,T]).$$

We observe that $\overline{\Lambda}_K(T)$ given by (2.11) can be expressed as

$$\begin{aligned}\overline{\Lambda}_K(T) = \overline{\Lambda}(T) &+ \int_0^T \left(\int_\Theta e^{tA_\theta} B_\theta \,\mathrm{d}\mu\right) E_K(t)^* \,\mathrm{d}t \\ &+ \int_0^T E_K(t)\left(\int_\Theta e^{tA_\theta} B_\theta \,\mathrm{d}\mu\right)^* \mathrm{d}t + \int_0^T E_K(t)E_K(t)^* \,\mathrm{d}t.\end{aligned}$$

Hence, for every $\varphi^1 \in \mathbb{R}^N$, we get,

$$\varphi^{1*}\overline{\Lambda}_K(T)\varphi^1 \geqslant \varphi^{1*}\overline{\Lambda}(T)\varphi^1 - 2\sqrt{TC(T)}|\varphi^1|^2 C_o K^{-p_o} - TC_0^2 K^{-2p_o}|\varphi^1|^2,$$

with $C(T)$ given in (2.6). Since the system (2.1) is averaged controllable, we have $\varphi^{1*}\overline{\Lambda}(T)\varphi^1 \geqslant c(T)|\varphi^1|^2$ (here again, $c(T)$ is defined in (2.6)). This ensures the existence of $K_0 \in \mathbb{N}^*$ such that for every $K \geqslant K_0$, we have,

$$\varphi^{1*}\overline{\Lambda}_K(T)\varphi^1 \geqslant c_K(T)|\varphi^1|^2 \qquad (\varphi^1 \in \mathbb{R}^N),$$

for some positive constant $c_K(T)$. That is to say that the system (2.10) is output controllable.

Proving the convergence rate of u_K to u is a classical exercise which is not developed here. This fact also leads to the convergence of $\int_\Theta x_\theta(T, u_K)\,\mathrm{d}\mu$ to $\overline{\mathrm{x}}^1$. □

In the rest of this section, we check this result in Example 2.11 with two different quadrature rules. The first one is based on the mid-point rule, while the second one on the Monte Carlo integration.

Mid-point quadrature rule. For this approach, we assume that the set of parameter Θ is some real interval. We will also assume[1] that $\mathrm{d}\mu = w(\theta)\mathrm{d}\theta$, where w is some positive function on Θ. Let us then define for every $\theta \in \Theta$, $W(\theta) = \mu(\Theta \cap (-\infty, \theta)) = \int_{\Theta\cap(-\infty,\theta)} w(s)\,\mathrm{d}s$. It is clear that W is an increasing function, and hence, $W : \Theta \to [0, 1]$ is a bijective map.

In order to apply Proposition 2.5, one needs to have a consistent numerical method for the integration with respect to the measure μ. This can be for instance done with the mid-point rule. To this end, we define for every $K \in \mathbb{N}^*$, $K + 1$ points $z_0^K, \dots, z_K^K \in [0, 1]$ satisfying $0 = z_0^K < z_1^K < \cdots < z_K^K = 1$, and set

$$\theta_k^K = W^{-1}\left(\tfrac{z_{k-1}^K + z_k^K}{2}\right) \quad \text{and} \quad m_k^K = z_k^K - z_{k-1}^K \qquad (k \in \{1, \dots, K\})$$

and for every $f \in C^0(\Theta)$, such that $\int_\Theta |f|\,\mathrm{d}\mu < \infty$, we numerically approach $\int_\Theta f\,\mathrm{d}\mu$ by $\sum_{k=1}^K m_k^K f(\theta_k^K)$.

Lemma 2.6. *In addition to the above notations and assumptions, let us assume that f and W are C^2 functions on Θ. Let us also set $g : \Theta \to \mathbb{R}_+$ a μ-integrable function such that $g \geqslant |f|$. Then we have*

$$\left| \int_\Theta f(\theta)\,\mathrm{d}\mu - \sum_{k=1}^{K} m_k^K f(\theta_k^K) \right| \leqslant \sum_{k=2}^{K-1} \frac{M_k}{24}(m_k^K)^3 + r_1 + r_K,$$

where we have set $M_k = \sup_{(W^{-1}(z_{k-1}^K), W^{-1}(z_k^K))} w^{-2}\left|f'' - f'\frac{w'}{w}\right|$, for $k \in \{1, \dots, K\}$), with the convention[2] that $W^{-1}(0) = \inf\Theta$ and $W^{-1}(1) = \sup\Theta$, and for $k \in \{1, K\}$, r_k is given by $(m_k^K)^3 M_k/24$ if $\lim_{z\to\epsilon_k} f \circ W^{-1}(z)$ exist (with $\epsilon_1 = 0$ and $\epsilon_K = 1$) or by $\int_{z_{k-1}^K}^{z_k^K} g \circ W^{-1}\,\mathrm{d}z + m_1^K g(\theta_k^K)$ otherwise.

The proof of this result is classical and postponed in Appendix A.1.

[1] The following description can also be adapted for more general probability measures (including the one having a countable number of atomic masses). However, we keep this assumption for the sake of simplicity.

[2] This convention is only relevant when Θ is unbounded.

Remark 2.7. In this paragraph, we have considered one of the most basic quadrature rule: the mid-point quadrature one. However, one can use some more sophisticated quadrature method, see e.g. Kronrod (1965) for some other quadrature rules. For instance for Example 2.11 below, one can consider Clenshaw-Curtis quadrature, see e.g. Trefethen (2000) or Martínez-Frutos and Periago Esparza (2018, §4.2), or Gauss-Laguerre quadrature, see e.g. Shao et al. (1964), and we refer to Trefethen (2008) for a comparison between these two methods.

Monte Carlo integration. For this approach, we consider $K \in \mathbb{N}^*$ realizations of the random variable θ, following the probability law μ. This leads to some parameters $\theta_1^K, \ldots, \theta_K^K \in \Theta$. We then apply the output controllability procedure to the system (2.10), with matrices $\mathbb{A}$, $\mathbb{B}$, and $\mathbb{C}$ given by (2.9), with $m_k^K = 1/K$.

More precisely, we approach $I(f) = \int_\Theta f(\theta)\, \mathrm{d}\mu$ by the random variable

$$I_K(f) = \frac{1}{K} \sum_{k=1}^{K} f(\theta_k^K), \tag{2.16}$$

where θ_k^K is a random variable following the probability law μ. Based on the central limit theorem, we know that the expectation of the error, $I(f) - I_K(f)$ is null, and furthermore, $\sqrt{K}(I(f) - I_K(f))/\sigma$ converges in law to a random variable following the probability law $\mathcal{N}(0, 1)$, where σ is the standard derivation of $f(\theta)$. More precisely, we have the following result.

Lemma 2.8. *Given $(\Theta, \mathcal{F}, \mu)$ a probability space and let $f : \Theta \to \mathbb{R}$ be a measurable function such that $\int_\Theta |f(\theta)|^2\, \mathrm{d}\mu < \infty$. We define $I(f) = \displaystyle\int_\Theta f(\theta)\, \mathrm{d}\mu$ and $\sigma = \sqrt{\displaystyle\int_\Theta (f(\theta) - I(f))^2\, \mathrm{d}\mu}$. Then for every sequence $(\theta_k)_{k \in \mathbb{N}^*}$ of independent and identically distributed random variables following the probability law μ, we have that $\sqrt{K}\,(I_K(f) - I(f))\,/\sigma$ (with $I_K(f)$ defined by* (2.16)*) converges in law to the normal distribution $\mathcal{N}(0, 1)$.*

Lemma 2.8 ensures in particular that for every $a > 0$, the probability that $|I_K(f) - I(f)| < a/\sqrt{K}$ converges to a constant as $K \to \infty$. That is to say that, when evaluating the random variable $\theta_1, \ldots, \theta_K$, it is expected that $|I(f) - I_K(f)|$ decays polynomially with rate $1/2$. This rate will be confirmed on Example 2.11.

Remark 2.9. Note that this method is very simple to implement in practice. Its convergence rate is rather small, but it is easy to extend this method for general parameter set Θ and general measure μ. In addition, the convergence rate of this method is independent of the dimension of Θ.

Remark 2.10. In this paragraph, we give the classical Monte-Carlo integration, and we refer to Robert and Casella (2004) for a general overview of Monte-Carlo methods. As for the previous paragraph, some more advanced numerical integration can be used. We refer for instance to Martínez-Frutos and Periago Esparza (2018, §5.3), and to Dick et al. (2013); Caflisch (1998) for quasi-Monte-Carlo integration.

Numerical example. To conclude this subsection, we consider both the mid-point quadrature rule and Monte Carlo integration approaches on an example.

Example 2.11. Let us consider the probability space $(\mathbb{R}_+, \mathcal{B}(\mathbb{R}_+), \mu)$, where the measure μ is given by $d\mu = w(\theta)d\theta$, with $w(\theta) = e^{-\theta}$. Let us also assume that $N = 2$ and $M = 1$. We then consider, for every $\theta \in \Theta$, the matrices $A_\theta = \begin{pmatrix} 0 & -\theta^2 \\ 1 & 0 \end{pmatrix}$ and $B_\theta = \begin{pmatrix} 0 \\ \theta - 1 \end{pmatrix}$. For this example, everything can be computed by hands. In particular, for $T = 1$, the controllability Gramian $\overline{\Lambda}(T)$ given in (2.4) is $\overline{\Lambda}(1) = \frac{1}{240}\begin{pmatrix} 32 + 15\pi & 30 \\ 30 & 15\pi - 20 \end{pmatrix}$. One can observe that $\overline{\Lambda}(1)$ is a positive matrix. Hence, the system considered in this example is controllable in average. In addition, we set as initial condition $x_\theta^0 = \begin{pmatrix} \theta & \sin\theta \end{pmatrix}^*$ and an averaged target condition $\overline{x}^1 = 0$. Using the relation (2.5), we can analytically compute the control u_{ex} of minimal L^2-norm[3] steering $(x_\theta^0)_\theta$ on the average $\overline{x}^1$ in time $T = 1$.

In order to apply Proposition 2.5, we have to consider the μ-integration of f, with f given by $f(\theta) = e^{TA_\theta}x_\theta^0$ or $f(\theta) = e^{tA_\theta}B_\theta$.

• **Mid-point quadrature rule.** Our aim is to apply Lemma 2.6 with well-chosen values z_k^K. For every $\varepsilon > 0$, we get (see Appendix A.2) the existence of $C_\varepsilon > 0$ such that

$$w(\theta)^{-2}\left|f''(\theta) - f'(\theta)\frac{w'(\theta)}{w(\theta)}\right| \leqslant C_\varepsilon e^{(2+\varepsilon)\theta} \qquad (\theta \in \mathbb{R}_+),$$

where the function f is either defined by $f(\theta) = e^{TA_\theta}x_\theta^0$, or by $f(\theta) = e^{tA_\theta}B_\theta$. In addition, we have $|f(\theta)| \leqslant g(\theta) := \theta + 1$ (see Appendix A.2). We can then apply Lemma 2.6 to obtain that the quadrature error $E_f^K = \left|\int_\Theta f(\theta)\,d\mu - \sum_{k=1}^K m_k^K f(\theta_k^K)\right|$ satisfies,

$$E_f^K \leqslant C_\varepsilon \sum_{k=1}^{K-1} e^{(2+\varepsilon)W^{-1}(z_k^K)}(z_k^K - z_{k-1}^K)^3 + \left(2 + W^{-1}(z_{K-1}^K)\right)e^{-W^{-1}(z_{K-1}^K)}$$

[3] The explicit expression of this control is:
$u_{ex}(t) = \frac{48(t-1)(144-90\pi+520t-156\pi t-188t^2+33\pi t^2-72t^3+45\pi t^3)}{5(45\pi^2+36\pi-308)(2-2t+t^2)^3}$ for every $t \in [0, 1]$.

$$+ (1 - z_{K-1}^K)\left(W^{-1}\left(\frac{z_{K-1}^K+1}{2}\right)+1\right).$$

Using the fact that $W^{-1}(z) = -\ln(1-z)$, and setting $z_k^K = k/K$ for every $k \in \{0, \ldots, K\}$, we finally get

$$E_f^K \leqslant \frac{1}{K^{1-\varepsilon}}\left(C_\varepsilon \frac{2+\varepsilon}{1+\varepsilon} + \frac{3+2\ln(2K)}{K^\varepsilon} - \frac{C_\varepsilon}{(1+\varepsilon)K^{1+\varepsilon}}\right).$$

That is to say that, for every $\varepsilon > 0$, the quadrature error decays polynomially with rate $1-\varepsilon$, and we can apply Proposition 2.5, with $C_o = C_\varepsilon$ and $p_o = 1-\varepsilon$, to obtain that the L^2-norm of $u_K - u_{ex}$ decays polynomially with rate $1-\varepsilon$. Of course the constant C_ε appearing in the previous computation increases as ε goes to 0. On Fig. 1a, we have computed the L^2-norm of $u_K - u_{ex}$ for $K \in \{10, \cdots, K_{\max}\}$, with $K_{\max} = 1000$, and numerically obtain a decay rate of It is expected that the decay rate, numerically computed, will slowly increase to 1 as $K_{\max}$ goes to ∞.

• **Monte Carlo integration.** As already noticed, in both cases we have $|f(\theta)| \leqslant \theta + 1$. This ensures that $\int_\Theta |f(\theta)|^2 \,\mathrm{d}\mu$ is finite. Hence, according to Lemma 2.8, the Monte Carlo method is convergent in expectation with rate $1/2$. Hence, by application of Proposition 2.5,[4] we obtain that $K \mapsto \|u_K - u_{ex}\|_{L^2(0,T)^M}$ decays polynomially in expectation with rate $1/2$. This fact is confirmed numerically on Fig. 1b, where we have run, for each value of $K \in \{10, \ldots, 1000\}$, Algorithm 1 with Monte Carlo integration 100 times, and displayed the obtained mean value of the L^2-error between the computed control u_K and the exact control u_{ex}. On this figure, we have also displayed the 90%-confidence interval for the error.

Note that in order to generate the sequence of random variables, we observe that there is a one-to-one correspondence between random variable $\theta \in (\mathbb{R}_+, \mathcal{B}(\mathbb{R}_+), \mu)$ and random variable $z \in ((0,1), \mathcal{B}((0,1)), \lambda)$ (with λ the Lebesgue measure). More precisely, if $z \in ((0,1), \mathcal{B}((0,1)), \lambda)$, then $\theta = W^{-1}(z) \in (\mathbb{R}_+, \mathcal{B}(\mathbb{R}_+), \mu)$, with $W(s) = \mu([0,s])$.

2.2 Ensemble controllability

2.2.1 Ensemble controllability notion and results

The notion of ensemble controllability has been at first stage introduced for Bolch equation, and we refer to Beauchard et al. (2010); Li and Khaneja (2009) for related articles. We also refer to Agrachev et al. (2016); Chen (2019) for ensemble controllability for more general nonlinear systems. Dealing with linear control system, this notion has been studied in Dirr and Schönlein (2020); Helmke and Schönlein (2014); Li (2011); Li et al. (2020); Li and Qi (2016); Zeng and Allgöwer (2016).

4 Strictly speaking, Proposition 2.5 cannot be used. Indeed, one shall rely on its randomized version, whose proof is similar to the original proof of Proposition 2.5.

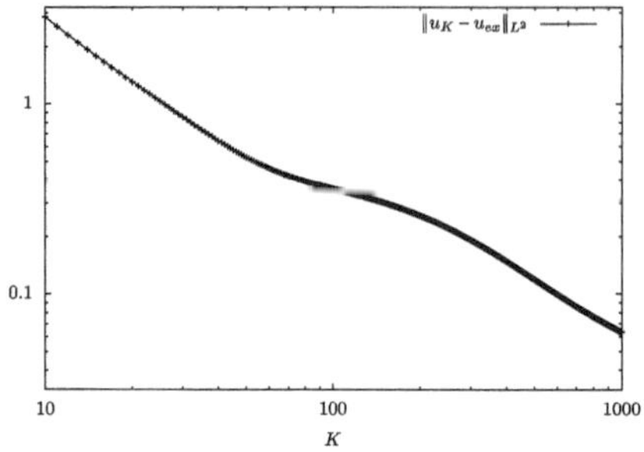

(a) For the mid-point quadrature rule, we numerically obtain polynomially decay rate equal to `0.761540`.

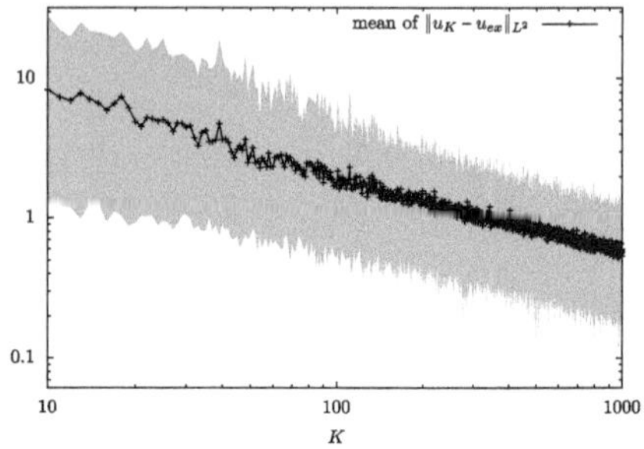

(b) For the Monte Carlo integration, we numerically obtain polynomially decay rate equal to `0.549031` In black, we have the mean error and in gray the 90%-confidence interval. These results have been obtained over 100 tests for each value of K.

FIGURE 1 Convergence as the number of evaluation points goes to ∞ for Example 2.11.

A necessary and sufficient condition for ensemble controllability can be stated.

Proposition 2.12. *(Triggiani, 1975, Theorem 3.1.1 or Dirr and Schönlein, 2020, Section 2) The system* (2.1) *is ensemble controllable if and only if* Span $\{\theta \in \Theta \mapsto A_\theta^k B_\theta \mathrm{e}_j \mid k \in \mathbb{N},\ j \in \{1, \dots, M\}\}$ *is dense in* $C^0(\Theta)^N$, *where* $\mathrm{e}_1, \dots, \mathrm{e}_M$ *denote vectors of the canonical basis of* $\mathbb{R}^M$.

Note that if the system (2.1) is ensemble controllable, then it is ensemble controllable in any time $T > 0$, see Dirr and Schönlein (2020).

The condition of Proposition 2.12 might be hard to check, and instead we shall employ the following sufficient condition.

Proposition 2.13. *(Dirr and Schönlein, 2020, Corollary 3) In addition to Assumption 1, let us assume that* $M = 1$ *(i.e.,* $B_\theta \in \mathbb{R}^N$*). The system* (2.1) *is ensemble controllable if the following conditions are satisfied:*

1. *the pair* (A_θ, B_θ) *is controllable for every* $\theta \in \Theta$;
2. *the eigenvalues of* A_θ *are simple for every* $\theta \in \Theta$;
3. *the map* $\theta \in \Theta \mapsto \sigma(A_\theta) \in \mathcal{P}(\mathbb{C})$ *is injective.*

2.2.2 Numerical approximation

In this paragraph, we propose an algorithm to compute, for a given precision $\varepsilon > 0$, given time $T > 0$, and given initial and target data $\theta \mapsto (\mathrm{x}_\theta^0, \mathrm{x}_\theta^1)$, a control u^ε such that (2.3) holds. To this end, we first state a regularity lemma.

Lemma 2.14. *Let us assume, in addition to Assumption 1 that* $\theta \mapsto \mathrm{x}_\theta^1$ *is Lipschitz continuous. Then, for every* $T > 0$ *and every* $u \in L^2(0, T)^M$, *the map* $\theta \in \Theta \mapsto \mathrm{x}_\theta^1 - x_\theta(T; \mathrm{x}_\theta^0, u) \in \mathbb{R}^N$ *is Lipschitz continuous with Lipschitz constant* $C_0 + C_1 \|u\|_{L^2(0,T)^M}$, *where* C_0 *and* C_1 *are nonnegative constants independent of* u.

Proof. Recall, that we have, $x_\theta(T; \mathrm{x}_\theta^0, u) = e^{TA_\theta}\mathrm{x}_\theta^0 + \int_0^T e^{(T-t)A_\theta} B_\theta u(t)\,\mathrm{d}t$. The conclusion follows by noticing that $\theta \mapsto e^{tA_\theta}$ is Lipschitz continuous with a Lipschitz constant independent of $t \in [0, T]$. □

Let us also state a simple lemma, whose proof is classical and postponed in Appendix A.3.

Lemma 2.15. *Assume that the system* (2.1) *is ensemble controllable, and let* $\theta \in \Theta \mapsto (\mathrm{x}_\theta^0, \mathrm{x}_\theta^1) \in \mathbb{R}^N \times \mathbb{R}^N$ *be any continuous function on* Θ*. Then for every* $T > 0$*, every* $\varepsilon > 0$*, every* $K \geqslant 2$ *and every set of distinct values* $\theta_1, \ldots, \theta_K \in \Theta$*, there exists a control* $u_K^\varepsilon \in L^2(0, T)^M$ *such that the solution* $x_\theta(\cdot; \mathrm{x}_\theta^0, u_K^\varepsilon)$ *of* (2.1) *satisfies*

$$\sup_{k\in\{1,\ldots,K\}} \left| \mathrm{x}_{\theta_k}^1 - x_{\theta_k}(T; \mathrm{x}_{\theta_k}^0, u_K^\varepsilon) \right| \leqslant \varepsilon. \tag{2.17}$$

Furthermore the control of minimal L^2*-norm such that* (2.17) *holds, is given by*

$$u_K^\varepsilon(t) = \sum_{k=1}^K B_{\theta_k}^* \varphi_k(t), \tag{2.18}$$

where we have set $\varphi_k(t)$ *the solution of,*

$$-\dot{\varphi}_k = A_{\theta_k}^* \varphi_k, \qquad \varphi_k(T) = \varphi_k^1, \tag{2.19}$$

with $(\varphi_1^1, \ldots, \varphi_K^1) \in (\mathbb{R}^N)^K$ *the minimizer of*

$$\begin{aligned} J(\varphi_1^1, \ldots, \varphi_K^1) = \frac{1}{2} \int_0^T \left| \sum_{k=1}^K B_{\theta_k}^* e^{(T-t)A_{\theta_k}^*} \varphi_k^1 \right|^2 \mathrm{d}t \\ + \sum_{k=1}^K \langle \varphi_k^1, e^{TA_{\theta_k}} \mathrm{x}_{\theta_k}^0 - \mathrm{x}_{\theta_k}^1 \rangle + \varepsilon \sum_{k=1}^K |\varphi_k^1|. \end{aligned} \tag{2.20}$$

Using Lemmas 2.14 and 2.15, we can deduce the following result.

Proposition 2.16. *Assume that the system* (2.1) *is ensemble controllable, and let* $\theta \in \Theta \mapsto (\mathrm{x}_\theta^0, \mathrm{x}_\theta^1) \in \mathbb{R}^N \times \mathbb{R}^N$ *be any Lipschitz continuous function on* Θ*. For every* $K \in \mathbb{N}^*$*, we consider a set* $\Theta_K \subset \Theta$ *of cardinal* K *such that* $\lim_{K\to\infty} \sup_{\theta\in\Theta} \operatorname{dist}(\theta, \Theta_K) = 0$*. Then for every* $T > 0$*, every* $\varepsilon > 0$*, every* $\varepsilon_0 \in (0, \varepsilon)$*, there exist* $K \in \mathbb{N}^*$ *such that the solution* $x_\theta(\cdot; \mathrm{x}_\theta^0, u_K^{\varepsilon_0})$ *of* (2.1) *satisfies* $\sup_{\theta\in\Theta} \left|\mathrm{x}_\theta^1 - x_\theta(T; \mathrm{x}_\theta^0, u_K^{\varepsilon_0})\right| \leqslant \varepsilon$*, where* $u_K^{\varepsilon_0}$ *is the control of minimal* L^2*-norm such that* (2.17) *holds, with* $\{\theta_1, \ldots, \theta_K\} = \Theta_K$*.*

Remark 2.17. Since Θ is a compact set of $\mathbb{R}^d$, one can check that there exist a constant $C > 0$ and a sequence $(\Theta_K)_K$ (so that the cardinal Θ_K equals K) such that for every $K \in \mathbb{N}$, we have $\sup_{\theta\in\Theta} \operatorname{dist}(\theta, \Theta_K) \leqslant CK^{-1/d}$.

Proof. By assumption, there exist $u^{\varepsilon_0} \in L^2(0,T)^M$ such that, $\sup_{\theta\in\Theta}\left|\mathrm{x}^1_\theta - x_\theta(T;\mathrm{x}^0_\theta, u^{\varepsilon_0})\right| \leqslant \varepsilon_0$. It is also easy to see that $\|u^{\varepsilon_0}_K\|_{L^2(0,T)^M} \leqslant \|u^{\varepsilon_0}\|_{L^2(0,T)^M}$, hence, by application of Lemma 2.14, we obtain,

$$\sup_{\theta\in\Theta}\left|\mathrm{x}^1_\theta - x_\theta(T;\mathrm{x}^0_\theta, u^{\varepsilon_0}_K)\right| \leqslant \varepsilon_0 + \delta_K\left(C_0 + C_1\|u^{\varepsilon_0}_K\|_{L^2(0,T)^M}\right)$$
$$\leqslant \varepsilon_0 + \delta_K\left(C_0 + C_1\|u^{\varepsilon_0}\|_{L^2(0,T)^M}\right), \qquad (2.21)$$

where C_0 and C_1 are defined by Lemma 2.14, and $\delta_K = \sup_{\theta\in\Theta}\mathrm{dist}(\theta,\Theta_K)$. It is then trivial to conclude the existence of $K(\varepsilon,\varepsilon_0) \in \mathbb{N}^*$ such that for every $K > K(\varepsilon,\varepsilon_0)$ we have, $\delta_K(C_0 + C_1\|u^{\varepsilon_0}\|_{L^2(0,T)}) \leqslant \varepsilon - \varepsilon_0$. □

Remark 2.18. Let us also mention that, up to an extraction of a subsequence, the sequence of controls $(u^{\varepsilon_0}_K)_K$, given in Proposition 2.16, converges in L^2 to the control u^{ε_0} of minimal L^2-norm such that $\sup_{\theta\in\Theta}|\mathrm{x}^1_\theta - x_\theta(T;\mathrm{x}^0_\theta, u^{\varepsilon_0})| < \varepsilon_0$.

Indeed, for every K, we have $\|u^{\varepsilon_0}_K\|_{L^2(0,T)^M} \leqslant \|u^{\varepsilon_0}\|_{L^2(0,T)^M}$. Hence, there exist $u^{\varepsilon_0}_* \in L^2(0,T)^M$ such that up to an extraction of a subsequence, the sequence $(u^{\varepsilon_0}_K)_K$ is weakly convergent to $u^{\varepsilon_0}_*$. In addition, we have $\|u^{\varepsilon_0}_*\|_{L^2(0,T)^M} \leqslant \|u^{\varepsilon_0}\|_{L^2(0,T)^M}$, and the weak convergence of $(u^{\varepsilon_0}_K)_K$ to $u^{\varepsilon_0}_*$ ensures that $\lim_{K\to\infty} x_\theta(T;\mathrm{x}^0_\theta, u^{\varepsilon_0}_K) = \mathrm{x}_\theta(T;\mathrm{x}^0_\theta, u^{\varepsilon_0}_*)$. Furthermore, the two maps $\theta \mapsto x_\theta(T;\mathrm{x}^0_\theta, u^{\varepsilon_0}_K)$ and $\theta \mapsto \mathrm{x}_\theta(T;\mathrm{x}^0_\theta, u^{\varepsilon_0}_*)$ are both Lipschitz with the same Lipschitz constant, $C_0 + C_1\|u^{\varepsilon_0}\|_{L^2(0,T)^M}$, with the notations of Lemma 2.14. This, together with the compactness assumption on Θ, ensures that $\sup_{\theta\in\Theta}|\mathrm{x}^1_\theta - x_\theta(T;\mathrm{x}^0_\theta, u^{\varepsilon_0}_*)| = \lim_{K\to\infty}\sup_{\theta\in\Theta}|\mathrm{x}^1_\theta - x_\theta(T;\mathrm{x}^0_\theta, u^{\varepsilon_0}_K)|$. This last fact, combined with the estimation (2.21), leads to $\sup_{\theta\in\Theta}|\mathrm{x}^1_\theta - x_\theta(T;\mathrm{x}^0_\theta, u^{\varepsilon_0}_*)| \leqslant \varepsilon_0$, and hence, the uniqueness of the minimal L^2-norm control implies $u^{\varepsilon_0}_* = u^{\varepsilon_0}$. Finally, since $\|u^{\varepsilon_0}_K\|_{L^2(0,T)^M} \leqslant \|u^{\varepsilon_0}\|_{L^2(0,T)^M}$, we also conclude that $\lim_{K\to\infty}\|u^{\varepsilon_0}_K\|_{L^2(0,T)^M} = \|u^{\varepsilon_0}\|_{L^2(0,T)^M}$, that is to say that $(u^{\varepsilon_0}_K)_K$ converges strongly in $L^2(0,T)^M$ to u^{ε_0}.

Proposition 2.16 directly leads to Algorithm 3 used to compute an appropriate control for ensemble controllability. In Algorithm 3, we use Algorithm 4, in order to compute the control of minimal L^2-norm such that (2.17) holds (with ε replaced by ε_0).

Remark 2.19. Obviously, in the 3rd step of Algorithm 3, it is not possible to numerically compute $\sup_{\theta\in\Theta}\left|\mathrm{x}^1_\theta - x_\theta(T;\mathrm{x}^0_\theta, u^{\varepsilon_0}_K)\right|$. Hence, one has to discretize Θ once again. This can be done, by picking some large number L and some values $\widetilde{\theta}_1,\ldots,\widetilde{\theta}_L \in \Theta$, and substituting $\sup_{\theta\in\Theta}\left|\mathrm{x}^1_\theta - x_\theta(T;\mathrm{x}^0_\theta, u^{\varepsilon_0}_K)\right|$ by $\sup_{l\in\{1,\ldots,L\}}\left|\mathrm{x}^1_{\widetilde{\theta}_l} - x_{\widetilde{\theta}_l}(T;\mathrm{x}^0_{\widetilde{\theta}_l}, u^{\varepsilon_0}_K)\right|$. Note that the choice of the values of $\widetilde{\theta}_l$ can be random.

To conclude this subsection, we consider the following example.

Algorithm 3 Ensemble control

Initialize: Given $\epsilon > 0$, $\varepsilon_0 \in (0, \varepsilon)$,
$T > 0$ and $\theta \mapsto (\mathrm{x}_\theta^0, \mathrm{x}_\theta^1)$ continuous.
1: $K \leftarrow 2$.
2: Compute $u_K^{\varepsilon_0}$ from Algorithm 4.
3: **while** $\sup_{\theta\in\Theta} \left|\mathrm{x}_\theta^1 - x_\theta(T; \mathrm{x}_\theta^0, u_K^{\varepsilon_0})\right| > \varepsilon$ **do**
4: $K \leftarrow K + 1$.
5: Compute $u_K^{\varepsilon_0}$ from Algorithm 4.
6: **end while**
Return: $u_K^{\varepsilon_0}$.

Algorithm 4 Approximate control for K values of θ

Initialize: Given $K \geqslant 2$, $\varepsilon_0 > 0$,
$T > 0$ and $\theta \mapsto (\mathrm{x}_\theta^0, \mathrm{x}_\theta^1)$.
1: Compute $\Theta_K = \{\theta_1, \ldots, \theta_K\} \subset \Theta$ (where $(\Theta_K)_K$ is a sequence satisfying assumptions of Proposition 2.16).
2: Compute $(\varphi_1^1, \ldots, \varphi_K^1) \in \mathbb{R}^N \times \cdots \times \mathbb{R}^N$, the minimizer of (2.20) (with ε replaced by ε_0).
3: Compute, for $k \in \{1, \ldots, K\}$, $\varphi_k(t)$, solution of (2.19).
Return: $u_K^{\varepsilon_0}(t)$ given by (2.18).

Example 2.20. Let us consider $\Theta = [1, 10]$, let us also assume that $N = 2$ and $M = 1$. We then consider for every $\theta \in \Theta$, the matrices $A_\theta = \begin{pmatrix} -1 & -\theta \\ \theta & -1 \end{pmatrix}$ and $B_\theta = \begin{pmatrix} 0 \\ 1 \end{pmatrix}$. With these matrices, we observe that the system (2.1) is ensemble controllable. Indeed, for every $\theta \in \Theta$, we have, $\sigma(A_\theta) = \{-1 - i\theta, -1 + i\theta\}$ and $\operatorname{rank}(B_\theta, A_\theta B_\theta) = \operatorname{rank} \begin{pmatrix} 0 & -\theta \\ 1 & -1 \end{pmatrix} = 2$. It is then easy to see that items 1–3 of Proposition 2.13 are fulfilled and hence, the system (2.1), with the parameter set Θ and the matrices A_θ and B_θ defined above, is ensemble controllable.

Let us now apply the result of Proposition 2.16. To this end, we consider the controllability time $T = 2$, and the initial and final conditions given by $\mathrm{x}_\theta^0 = (\theta \;\; \sin\theta)^*$ and $\mathrm{x}_\theta^1 = 0$. We then use Algorithm 4, with $\varepsilon_0 = 0.1$ and let $K \to \infty$. For step 1 of this algorithm, we chose $\Theta_K = \{1 + 9(k-1)/(K-1), k \in \{1, \ldots, K\}\}$. On Fig. 2, we display the L^2-norm of the control u_K and the maximum of $\theta \mapsto \mathrm{x}_\theta^1 - x_\theta(T; \mathrm{x}_\theta^0, u_K)$ for $K \in \{3, \ldots, 50\}$. To compute the maximum of $|\mathrm{x}_\theta^1 - x_\theta(T; \mathrm{x}_\theta^0, u_K)|$, we use Remark 2.19 with $L = 5.10^4$ uniformly distributed values of θ. We observe on this figure that $|\mathrm{x}_\theta^1 - x_\theta(T; \mathrm{x}_\theta^0, u_K)|$ decays to ε_0 as K increase, we also observe that

$\|u_K\|_{L^2(0,T)}$ is bounded. Note that, we cannot guarantee that $K \mapsto \|u_K\|_{L^2(0,T)}$ is increasing, and this is in general not the case, as we see on Fig. 2. However, we can ensure that $\|u_{2K-1}\|_{L^2(0,T)} \geqslant \|u_K\|_{L^2(0,T)}$, this is due to the fact that the sequence of interpolation points for $2K-1$ points contains the one for K points.

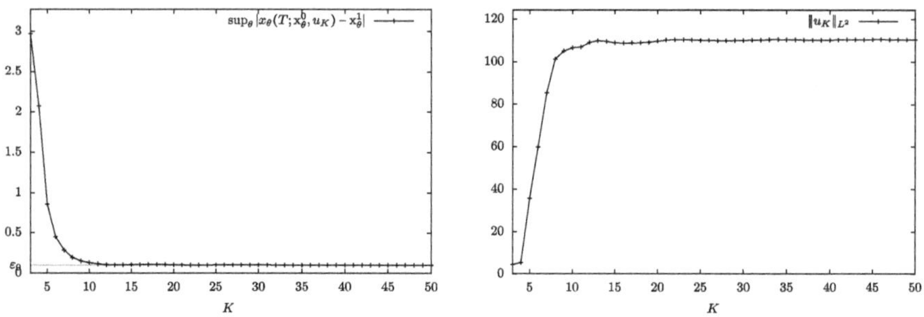

FIGURE 2 Display of $\|u_K^{\varepsilon_0}\|_{L^2(0,T)}$ and $\sup_{\theta\in\Theta}|x_\theta(T;\mathrm{x}_\theta^0,u_K^{\varepsilon_0}) - \mathrm{x}_\theta^1|$, for $K \in \{3,\ldots,50\}$, with u_K given by Algorithm 4 for the ensemble control of the system considered in Example 2.20.

3 Parameter dependent controls

Throughout this section we consider the system (1.1) accompanied by the control problem

$$x_\theta(T) = \mathrm{x}_\theta^1, \qquad (\theta \in \Theta), \tag{3.1}$$

where the prescribed target state in general depends on the parameter, while the control time $T > 0$, at which the system should reach the target, is θ independent. In addition, we suppose the system (1.1) is controllable for each value of the parameter $\theta \in \Theta$. In particular, it implies that for each given control time $T > 0$ and the target x_θ^1 there exists a (parameter dependent) control such that the state of the system satisfies $x_\theta(T) = \mathrm{x}_\theta^1$. As explained in the introduction, such control (with minimal L^2-norm) can be explicitly constructed by solving the adjoint equation starting from optimal datum at time T determined by Eq. (3.3).

However, the construction of control for a large dimensional system is an expansive, time and computation consuming process. When dealing with parameter dependent problems, if this process is to be repeated many times, with every change of parameter value, we need to seek alternatives for solving the full problem over and over. For these reasons one tries to develop an approximate model of lower complexity providing the solution to problem of interest within a prescribed error. Such model we refer to as a reduced basis model. We stress that such approach is relevant in case of large system dimensions, typically resulting from (semi) discretization of PDE control problems (cf. Example 3.6 below).

An underlying assumption in application of reduced model techniques is that the solution manifold can be well approximated by a linear space of low dimension. In the setting of this chapter, this manifold consists of all controls (of minimal L^2-norm) of the system (1.1) obtained through variation of the parameter. Their implementation generally follows the *offline-online* paradigm. The purpose of the *offline* phase can be described as follows: given a Banach space X and its compact subset $\mathcal{K}$, find a low-dimensional subspace $V \subset X$ such that V approximates $\mathcal{K}$ fairly well. Once a reduced basis (of V) is determined, the approximation of an arbitrary element from $\mathcal{K}$ is calculated *online* as a linear combination of the reduced basis vectors. Normally, the cost of the *offline* routine is much higher and it requires large computational and memory capacities, but it can be considered as a pre-processing step that is performed only once.

The construction of the reduced basis mainly explores a proper orthogonal decomposition (POD) or a greedy sampling procedure, or their combination (POD-greedy) in case of time-dependent problems (Haasdonk, 2013; Hesthaven et al., 2016). Here we also refer the reader to Quarteroni et al. (2016, Chapter 12) for a general introduction to reduced basis methods for parameter dependent control problems.

The former procedure consists of an explore-and-compress strategy, relying on a singular value decomposition of numerous snapshots (vectors selected from $\mathcal{K}$) and provide an optimal approximation space in the mean squared error sense. In the latter case the basis vectors coincide with the snapshots themselves, carefully selected through some optimality strategy, which significantly reduces the cost of the offline computation when compared to POD (Hesthaven et al., 2016; Maday, 2006). In addition, the greedy approach provides approximation rates comparable to the optimal ones, expressed through Kolmogorov widths.

By $u(\Theta) = \{u_\theta, \theta \in \Theta\}$ we denote the control manifold consisting of all optimal controls associated to problem (1.1) + (3.1) as the parameter ranges over the whole set Θ. As explained in the introduction, each such control can be uniquely determined by the relation

$$u_\theta = B_\theta^* e^{(T-t)A_\theta^*} \varphi_\theta^1,$$

where φ_θ^1 is the unique minimizer of the quadratic functional

$$J_\theta(\varphi^1) = \frac{1}{2} \int_0^T |B_\theta^* \varphi(t)|^2 \, dt - \langle x_\theta^1, \varphi^1 \rangle + \langle x_\theta^0, \varphi(0) \rangle, \tag{3.2}$$

and φ is the solution to the adjoint system with datum at time T given by φ^1.

Furthermore, this minimizer can be equivalently determined as the solution to the system

$$\Lambda_\theta \varphi_\theta^1 = x_\theta^1 - e^{T A_\theta} x_\theta^0, \tag{3.3}$$

where Λ_θ is the controllability Gramian associated to (A_θ, B_θ). The last system is well posed due to the assumed controllability property which implies that Λ_θ is a regular matrix. In particular, it defines the mapping $\theta \in \Theta \mapsto \varphi_\theta^1 \in \mathbb{R}^N$, whose smoothness is transferred from the mappings of A_θ and B_θ at all levels. In the sequel by $\varphi^1(\Theta) \subset \mathbb{R}^N$ we denote the image of that mapping.

Having assumed these maps are Lipschitz continuous and the parameter $\theta \in \Theta$ varies on a compact set, both the set of all minimizers $\varphi^1(\Theta)$, as well as the set of corresponding optimal controls $u(\Theta)$ are compact sets. However, as the control manifold consists of time dependent functions, while the minimizers φ_θ^1 are vectors from $\mathbb{R}^N$, we prefer to work in the latter framework and to construct an approximating subspace for the set $\varphi^1(\Theta)$. By using the one-to-one correspondence between the minimizers φ_θ^1 and the associated controls u_θ, this will in turn provide approximation of the control manifold as well. For these reasons we formulate our problem as follows.

1. *Construct a reduced basis V_n that approximates the manifold $\varphi^1(\Theta)$ within the given precision δ.*
2. *For an arbitrary given parameter value $\theta \in \Theta$ construct an approximate control of the form $u_\theta^\star = B_\theta^* e^{(T-t)A_\theta^*} \varphi_\theta^{1,\star}$, with $\varphi_\theta^{1,\star} \in V_n$ such that it steers the system* (1.1) *to the prescribed target x_θ^1 as close as possible.*

The first part of the problem corresponds to the offline phase, while the second one should enable efficient computation of an approximate control during the online procedure.

Of course, the problem is feasible due to the compactness of the parameter set Θ and the continuous-dependence assumption of dynamics with respect to θ. However, the goal is to minimize the number of basis vectors n and to obtain the most efficient approximation. The type of the approximation error (uniform, mean square, ...) to be considered depends on the method selected for the construction of reduced basis, and will be discussed in the next subsection.

In the subsection devoted to the online phase we shall see how to construct approximation $\varphi_\theta^{1,\star}$ for a particular given parameter value, and discuss performance of the associated approximated controls.

3.1 Selection of a reduced basis

When dealing with a continuous range of parameter values, in order to do any kind of numerics, the first step requires discretization of the parameter set. From now on, by Θ_h we denote a finite set of cardinality $m = |\Theta_h|$ obtained by intersecting a regular lattice with Θ with the property

$$\forall \theta \in \Theta \quad \mathrm{dist}(\theta, \Theta_h) < h.$$

3.1.1 Proper Orthogonal Decomposition (POD)

Proper Orthogonal Decomposition is the reduced basis method that provides an optimal approximation space in the root-mean-square (RMS) error sense. Beside their numerous successful applications in different areas, here we just refer to those related to the control theory (Benner et al., 2015, Chapter 1 and the references therein).

In general, for a finite set of vectors $\{y_1, ..., y_m\}$ in $\mathbb{R}^N$, POD results in the $n < m$ dimensional space that minimizes the quantity

$$\sqrt{\frac{1}{m}\sum_i \text{dist}(y_i, V_n)^2}$$

over all n-dimensional subspaces V_n of the Span$\{y_1, \dots, y_m\}$.

To construct the POD space one introduces the correlation matrix $C = \frac{1}{m}Y^*Y \in \mathbb{R}^{N\times N}$, where Y is the matrix consisting of column vectors y_i, $i \in \{1, \dots, m\}$. Then the (orthogonal) POD basis V_n consists of eigenvectors of C corresponding to its largest n eigenvalues λ_i (sorted in a decreasing order), and it satisfies the following estimate

$$\frac{1}{m}\sum_i \text{dist}(y_i, V_n)^2 = \sum_{i=n+1}^{N} \lambda_i, \tag{3.4}$$

where we sum the remaining eigenvalues of the correlation matrix.

Thus, the construction of the POD basis corresponds to the singular value decomposition (SVD) of the matrix $\frac{1}{\sqrt{m}}Y$.

In the context of parameter dependent control problems, we formulate the corresponding algorithm as follows.

Algorithm 5 POD Control Algorithm

Initialize: Given precision $\delta > 0$.

1: Choose discretization constant $h \leqslant \delta/(\sqrt{2}L_\varphi)$, where L_φ is the Lipschitz constant of the mapping $\theta \mapsto \varphi^1_\theta$.

2: Construct Y, the matrix consisting of column vectors φ^1_i, $i \in \{1, \dots, m\}$, associated to the ordered values $\theta_1, \dots, \theta_m$ of Θ_h.

3: Perform SVD of $\frac{1}{\sqrt{m}}Y$. Denote by λ_i obtained singular values indexed in a decreasing order, and by ξ_i corresponding left-singular vectors.

4: Denote by n the smallest integer satisfying $\sum_{i=n+1}^{m} \lambda_i < \delta^2/2$

Return: The reduced POD subspace V_n is given by Span$\{\xi_1, \dots, \xi_n\}$.

The estimate (3.4) implies directly that the subspace V_n constructed by Algorithm 5 approximates the set $\varphi^1(\Theta_h) = \{\varphi^1_\theta \mid \theta \in \Theta_h\}$ with the RMS error less than $\delta/\sqrt{2}$. In addition, the above algorithm provides the mean square approximation over the whole range of parameter set Θ, i.e., one obtains an estimate on the expected value of dist$(\varphi^1_\theta, V_n)^2$, where we assume that θ obeys some probability distribution law.

For simplicity let us assume that θ is uniformly distributed over the set Θ. Let us denote by $(\Theta_i)_{i\in\{1,\dots,m\}}$ a disjoint union of Θ consisting of equal measure sets *centered* around $\theta_i \in \Theta_h$. More precisely we suppose $\Theta_i \subseteq B(\theta_i, h)$ (where

h is the discretization constant from Algorithm 5) and $|\Theta_i| = 1/m$ for every index i. Then

$$E(\mathrm{dist}(\varphi_\theta^1, V_n)^2) = \frac{1}{|\Theta|}\int_\Theta \mathrm{dist}(\varphi_\theta^1, V_n)^2 \, \mathrm{d}\theta$$
$$\leqslant \frac{1}{|\Theta|}\sum_i \int_{\Theta_i} |\varphi_\theta^1 - \varphi_i^1|^2 \, \mathrm{d}\theta + \frac{1}{m}\sum_i \mathrm{dist}(\varphi_i^1, V_n)^2. \quad (3.5)$$

By exploring the Lipschitz continuity of the mapping $\theta \mapsto \varphi_\theta^1$ and the choice of the discretization constant h it follows

$$|\varphi_\theta^1 - \varphi_{\theta_i}^1|^2 \leqslant L_\varphi |\theta - \theta_i|^2 \leqslant \delta^2/2.$$

Finally, combining the last estimate with the RMS error bound of the set $\varphi^1(\Theta_h)$ we obtain

$$\sqrt{E(\mathrm{dist}(\varphi_\theta^1, V_n)^2)} \leqslant \delta. \quad (3.6)$$

Remark 3.1. It should be noted that the Lipschitz constant L_φ for a general system might not be available in practice. However, as there exists an explicit expression for φ_θ (cf. (3.3)), L_φ can be bounded from above in terms of the Lipschitz constants of the mappings $\theta \to (A_\theta, B_\theta, \mathrm{x}_\theta^0, \mathrm{x}_\theta^1)$. Thus, for specific examples, where the dependence of the system dynamics and data on the parameter is given explicitly, one can calculate the latter constants, and use the upper bound for L_φ in order to estimate an optimal value of the discretization constant h.

In general, when we lack an estimate on L_φ, Algorithm 5 still provides the approximation over the discretized parameter set Θ_h, while the discretization constant h has to be chosen arbitrarily small.

3.1.2 Greedy algorithm

One of the most popular methods for constructing reduced basis relies on the so called greedy algorithms. These were successfully introduced and analyzed in works of A. Patera and his collaborators two decades ago (cf. Veroy et al., 2003) with the aim of solving parametric PDEs. Later they were applied to optimal control problems (Hernández-Santamaría et al., 2019; Kärcher and Grepl, 2014a,b), while the application to controllability of linear systems was introduced in Lazar and Zuazua (2016).

The method provides a linear subspace of a relatively small dimension that approximates a compact subset $\mathcal{K}$ in a Banach space X with a given precision δ. A general greedy algorithm is given in Algorithm 6 below.

The choice of the constant γ equal to one corresponds to the pure greedy algorithm. Reduction of its value simplify implementation of the algorithm as the optimization problem (3.7) does not have to be solved fully but only up to some fixed percentage extent. On the other side, as we shall see below, the

Algorithm 6 (Weak) Greedy Algorithm

Initialize: Fix a constant $\gamma \in (0, 1]$ and precision $\delta > 0$.

Choose $x_1 \in \mathcal{K}$ such that $\|x_1\|_X \geqslant \gamma \max_{x\in\mathcal{K}} \|x\|_X$.

$V_1 = \mathrm{Span}\{x_1\}$ and $\sigma_1(\mathcal{K}) = \max_{x\in\mathcal{K}} \mathrm{dist}(x, V_1)$.

$k \leftarrow 1$.

while $\sigma_k(\mathcal{K}) \geqslant \delta$ **do**

  Choose $x_{k+1} \in \mathcal{K}$ such that $\mathrm{dist}(x_{k+1}, V_k) \geqslant \gamma\, \sigma_k(\mathcal{K})$.

  $k \leftarrow k + 1$.

  Set $V_k = \mathrm{Span}\{x_1, \dots, x_k\}$ and compute

$$\sigma_k(\mathcal{K}) := \max_{x\in\mathcal{K}} \mathrm{dist}(x, V_k). \tag{3.7}$$

end while

Return: $V_n = \mathrm{Span}\{x_1, \dots, x_n\}$, where n is a number of chosen snapshots, that uniformly approximates set K with precision δ.

relaxation of the pure greedy algorithm to a weak greedy (with $\gamma \in (0, 1)$) does not significantly reduce the efficiency of the algorithm.

One of great advantages of greedy algorithms is their approximation efficiency, which is comparable with the optimal ones expressed through Kolmogorov widths. The latter is defined as

$$d_n(\mathcal{K}) := \inf_{\dim Y = n} \sup_{x\in\mathcal{K}} \inf_{y\in Y} \|x - y\|_X,$$

and they measure how well $\mathcal{K}$ can be approximated by a subspace in X of a fixed dimension n.

The following theorem (taken from DeVore et al., 2013, Corollary 3.3) gives a comparison of $\sigma_n(\mathcal{K})$ with $d_n(\mathcal{K})$.

Theorem 3.2. *For the weak greedy algorithm with constant $\gamma \in (0, 1]$ in a Hilbert space X, we have the following: If the compact set $\mathcal{K}$ is such that, for some $\alpha > 0$ and $C_0 > 0$*

$$d_n(\mathcal{K}) \leqslant C_0 n^{-\alpha} \qquad (n \in \mathbb{N}),$$

then

$$\sigma_n(\mathcal{K}) \leqslant C_1 n^{-\alpha} \qquad (n \in \mathbb{N}),$$

where $C_1 := \gamma^{-2} 2^{5\alpha+1} C_0$.

This theorem implies that in the case of a polynomial decay the (weak) greedy algorithms perform with the same rate as Kolmogorov widths. A similar estimate holds for subexponential decays as well (cf. DeVore et al., 2013).

Note that the weak greedy constant γ does not influence decay rates, as it enters the above estimate only through the multiplicative constant C_1.

However, in practical implementations, the compact set $\mathcal{K}$ to be approximated is often unknown (e.g. it consists of solutions to parameter dependent problems, as in our case) which poses a significant challenge when dealing with the maximization problem (3.7). Actually, it requires to evaluate the distance of a vector from the constructed reduced space without knowing that vector! Of course, computing all the elements of the set $\mathcal{K}$ is not a remedy, as it would require to solve a problem for each parameter value what is exactly we want to avoid. In order to solve this issue, instead of calculating the exact distance appearing in (3.7), one use some surrogate distance which is easier to compute.

The next theorem demonstrates that such replacement still provides a greedy approximation of the set $\mathcal{K}$ and it only affects value of the weak greedy constant γ.

Theorem 3.3. *Let $f : X \times X \mapsto \mathbb{R}$ be a function satisfying*

$$c\,\|x-y\|_X \leqslant f(x,y) \leqslant C\,\|x-y\|_X \qquad (x,y \in \mathcal{K}), \tag{3.8}$$

for some constants $C, c > 0$. Perform Algorithm 6 with the distance $\operatorname{dist}(x, V_k)$ *replaced by* $\inf_{y \in V_k} f(x,y)$. *Then the such modified algorithm provides a weak greedy search of the set $\mathcal{K}$ with the weak greedy constant $\gamma' = \gamma c/C$.*

Proof. In order to demonstrate the theorem it is enough to show that the vectors selected by the modified algorithm in each step satisfy $\operatorname{dist}(x_{k+1}, V_k) \geqslant \gamma' \max_{x \in \mathcal{K}} \operatorname{dist}(x, V_k)$.

To this end note that for any $x \in \mathcal{K}$ we have

$$c\operatorname{dist}(x, V_k) \leqslant \inf_{y \in V_k} f(x,y) \leqslant \frac{1}{\gamma} \inf_{y \in V_k} f(x_{k+1}, y) \leqslant \frac{C}{\gamma} \operatorname{dist}(x_{k+1}, V_k), \tag{3.9}$$

where the middle inequality results from the selection criteria of the modified algorithm for the next reduced basis vector x_{k+1}, while the remaining two from the equivalence relation (3.8). □

Therefore, for practical implementations of (weak) greedy algorithms it is essential to detect a surrogate distance which is easy to compute even without knowing the vectors itself.

In the context of control problems for parameter dependent systems, the set to be approximated is the manifold $\varphi^1(\Theta)$ that consists of all minimizers of a functional J_θ given by (3.2) as the parameter ranges over the compact set Θ. Of course, we do not have these minimizers at our disposal, thus we need to find a surrogate that will enable implementation of the greedy algorithm.

Let us suppose that φ_1^1 is a reduced basis vector selected through the weak greedy search of $\varphi^1(\Theta)$. In order to estimate distance of other vectors from the space spanned by φ_1^1, note that for any $\theta \in \Theta$ we have

$$\Lambda_- \|\varphi_\theta^1 - \varphi_1^1\| \leqslant \|\Lambda_\theta \varphi_\theta^1 - \Lambda_\theta \varphi_1^1\| \leqslant \Lambda_+ \|\varphi_1^1 - \varphi_1^1\|, \tag{3.10}$$

where the uniform coercive and boundedness constants on the Gramian operators (i.e. $\Lambda_- I \leqslant \Lambda_\theta \leqslant \Lambda_+ I$ for every $\theta \in \Theta$) follow from the compactness of the set Θ and the regularity of mappings $\theta \mapsto A_\theta, B_\theta$.

Although this note provides $\|\Lambda_\theta \varphi_\theta^1 - \Lambda_\theta \varphi_1^1\|$ as a surrogate distance of φ_θ^1 from φ_1^1, it will turn useful only if we can calculate it (much) easier than the exact one. However, by using characterization of the optimal adjoint datum (3.3) we know that $\Lambda_\theta \varphi_1^1$ can be expressed in terms of the corresponding initial and target data, which are prescribed in advance, i.e., we have them at our disposal. As for the other term entering the expression of the surrogate distance, it represents the final state of the system corresponding to the parameter value θ run by a control of the form $u = B_\theta^* \varphi$, where φ is the solution of the adjoint system with datum at time T equal to φ_1^1. Its calculation thus requires solving consecutively the adjoint plus primal system, which of course is much cheaper than calculating φ_θ^1.

Based on the above discussion we propose the greedy control algorithm (Algorithm 7) as follows, see Lazar and Zuazua (2016).

Obviously, the set $\{\varphi_1^1, \ldots, \varphi_n^1\}$ is linearly independent due to the selection criteria (3.11). Furthermore, as it consists of vectors in $\mathbb{R}^N$, the algorithm stops after, at most, $n \leqslant N$ iterations, and it fulfills the requirements of the weak greedy theory. More precisely Theorem 3.4 holds.

Theorem 3.4 (Theorem 4.1 of Lazar and Zuazua, 2016). *Algorithm 7 provides a week greedy approximation of the manifold $\varphi^1(\Theta)$ with the constant $\gamma = \Lambda_-/(2\Lambda_+)$ and the uniform approximation error less than δ.*

Comparing the last algorithm with the POD method, we note that the first requires solving the control problem only for n chosen parameter values, for which the corresponding minimizers of a functional J_θ given by (3.2) constitute the reduced basis. On the other hand, the latter method requires the optimization problem for (3.2) to be solved for every parameter value from the training set Θ_h, and eventually extracts a small number of most important directions out of collected solutions. Although the greedy selection, in addition to construction of the reduced basis vectors, requires computation of residual in each iteration for all parameter values from Θ_h, its total cost is still cheaper than the cost of calculating φ_θ^1 for all values of Θ_h (a precise estimate on its computational cost can be found in Lazar and Zuazua (2016)), which makes the greedy control strategy more efficient than the POD one.

Of course, existence of an affordable surrogate (given by (3.10)) plays an essential role in the above analysis. For general problems, that might lack an appropriate surrogate, the choice of an optimal method is not straightforward, and depends on particular characteristics of each problem of interest.

Algorithm 7 Greedy Control Algorithm

Initialize: Given precision $\delta > 0$.

1: Choose discretization constant $h \leqslant \delta/2L_\varphi$, where L_φ is the Lipschitz constant of the mapping $\theta \mapsto \varphi_\theta^1$.
2: **if** $\max_{\theta\in\Theta_h} |\mathrm{x}_\theta^1 - e^{TA_\theta}\mathrm{x}_\theta^0| < \frac{\delta}{2}\Lambda_-$ **then**
3: Stop the algorithm.
4: **end if**
5: Select the first snapshot φ_1^1 as the minimizer of functional J_θ corresponding to $\theta = \theta_1$, where $\theta_1 = \mathrm{argmax}_{\theta\in\Theta_h} |x_\theta^1 - e^{TA_\theta}x_\theta^0|$.
6: Compute $\Lambda_\theta\varphi_1^1$ and set $\Lambda_\theta\Phi_1^1 = \mathrm{Span}\{\Lambda_\theta\varphi_1^1\}$, and compute $\sigma_1 = \max_{\theta\in\Theta_h} \mathrm{dist}(x_\theta^1 - e^{TA_\theta}x_\theta^0, \Lambda_\theta\Phi_1^1)$.
7: $k \leftarrow 1$.
8: **while** $\sigma_k \geqslant \frac{\delta}{2}\Lambda_-$ **do**
9: Set $\theta_{k+1} := \mathrm{argmax}_{\theta\in\Theta_h} \mathrm{dist}(x_\theta^1 - e^{TA_\theta}x_\theta^0, \Lambda_\theta\Phi_k^1)$.
10: $k \leftarrow k+1$.
11: Select φ_k^1 as the minimizer of J_θ corresponding to $\theta = \theta_k$.
12: Compute $\Lambda_\theta\varphi_k^1$ and set $\Lambda_\theta\Phi_k^1 = \mathrm{Span}\{\Lambda_\theta\varphi_1^1, \ldots, \Lambda_\theta\varphi_k^1\}$, and compute

$$\sigma_k := \max_{\theta\in\Theta_h} \mathrm{dist}(x_\theta^1 - e^{TA_\theta}x_\theta^0, \Lambda_\theta\Phi_k^1) \tag{3.11}$$

13: **end while**

Return: The approximating space $\Phi_n^1 = \mathrm{Span}\{\varphi_1^1, \ldots, \varphi_n^1\}$, where n is a number of chosen snapshots (specially $\Phi_0^1 = \{0\}$ for $n = 0$).

3.2 The online procedure

In this subsection, given any parameter value $\theta \in \Theta$, we aim to build an approximation $\varphi_\theta^{1,\star}$ of an optimal adjoint datum by selecting an appropriate linear combination of reduced basis vectors. The selection criteria are prescribed by the Problem formulation, where the ultimate goal is to reach the prescribed target x_θ^1 as close as possible. Hereby we assume that we have at our disposal a reduced basis space $V_n = [\varphi_1^1, \ldots, \varphi_n^1]$ constructed in the offline phase by some of the methods explained above.

The state of the system at final time T driven by the approximate control determined by $\varphi_\theta^{1,\star}$ is given by

$$x_\theta^\star(T) = \Lambda_\theta\varphi_\theta^{1,\star} + e^{TA_\theta}\mathrm{x}_\theta^0.$$

By exploring relation (3.3) we see that its distance from the prescribed target equals $|\mathrm{x}_\theta^1 - x_\theta^\star(T)| = |\Lambda_\theta(\varphi_\theta^1 - \varphi_\theta^{1,\star})|$, and it reaches the minimal value if $\varphi_\theta^{1,\star}$

is selected as the solution to the problem

$$\varphi_\theta^{1,\star} = \underset{\varphi^1 \in V_n}{\operatorname{argmin}} \|\Lambda_\theta(\varphi_\theta^1 - \varphi^1)\| = \underset{\varphi^1 \in V_n}{\operatorname{argmin}} \|(\mathrm{x}_\theta^1 - e^{TA_\theta}\mathrm{x}_\theta^0) - \Lambda_\theta \varphi^1\|. \tag{3.12}$$

This choice of $\varphi_\theta^{1,\star}$ corresponds to the minimization of the functional J_θ, determined by (3.2), over the space V_n.

The minimum in (3.12) is obtained if $\Lambda_\theta \varphi_\theta^{1,\star}$ equals the projection of $\mathrm{x}_\theta^1 - e^{TA_\theta}\mathrm{x}_\theta^0$ to the space $\mathrm{Span}\{\Lambda_\theta\varphi_1^1, \ldots, \Lambda_\theta\varphi_n^1\}$. As Λ_θ is a linear operator, the coefficients of that projection in the latter basis will also provide the optimal choice of $\varphi_\theta^{1,\star}$.

For these reasons we formulate the online algorithm as follows (see Lazar and Zuazua, 2016).

Algorithm 8 Control algorithm-online part

Initialize: A parameter value $\theta \in \Theta$ is given.

1: Calculate $\Lambda_\theta \varphi_\theta^1 = \mathrm{x}_\theta^1 - e^{TA_\theta}\mathrm{x}_\theta^0$.

2: For $i \in \{1, \ldots, n\}$, calculate $\Lambda_\theta\varphi_i^1 = e^{(T-t)A_\theta} B u_i^\star$, where $u_i^\star = B_\theta^* e^{(T-t)A_\theta^*}\varphi_i^1$.

3: Project $\Lambda_\theta\varphi_\theta^1$ to $\mathrm{Span}\{\Lambda_\theta\varphi_1^1, \ldots, \Lambda_\theta\varphi_n^1\}$. Denote the projection by $P_{\Lambda_\theta,n}(\Lambda_\theta\varphi_\theta^1)$.

4: Solve the system $P_{\Lambda_\theta,n}(\Lambda_\theta\varphi_\theta^1) = \sum_i \alpha_i \Lambda_\theta \varphi_i^1$ for $\alpha_i, i \in \{1, \ldots, n\}$.

Return: The approximate control $u_\theta^\star$ given by $u_\theta^\star = \sum_i \alpha_i u_i^\star$, where $u_i^\star$ are determined within Step 2.

The algorithm provides an approximation of the optimal control u_θ. Its gain is the computational saving compared to the construction of u_θ by standard methods. The cost is typically reduced by factor n/N. We address the interested reader to Lazar and Zuazua (2016, Section 5), providing a detailed analysis of computational cost of greedy control algorithms, both in the offline and online phase.

Remark 3.5. Note that for a parameter value θ that belongs to the discretization set Θ_h used in the construction of the reduced basis space V_n some steps of the last algorithm can be skipped as the corresponding terms have already been calculated within the offline part of the algorithm. In particular, steps 1-3 in the case of a greedy selection of V_n, while for $\theta \in \Theta_h$ the whole algorithm turns superfluous if the POD method was used in the offline phase. As explained above, this is due to the specific characteristic of the latter approach that first explores the solution over a large number parameter, and subsequently compress relevant information out of it in order to construct the reduced basis.

Algorithm 8 differs from the standard online procedure used in reduced basis methods (cf. Hesthaven et al., 2016). The latter choose the approximate solution to the problem as its Galerkin projection to the selected reduced basis. In particular, in our setting this would determine an approximate adjoint datum as Galerkin projection to $\mathrm{Span}\{\varphi_1^1, \dots, \varphi_n^1\}$ considered as solution to $\Lambda_\theta \varphi_\theta^1 = \mathrm{x}_\theta^1 - e^{TA_\theta}\mathrm{x}_\theta^0$. More precisely it would result in a vector $\varphi_\theta^{1,\mathrm{GAL}} = \sum \alpha_i^{\mathrm{GAL}} \varphi_i^1$, where the coefficients α_i^{GAL} are chosen as solution to the system

$$\sum_i \langle \Lambda_\theta \varphi_i^1, \varphi_j^1 \rangle \alpha_i^{\mathrm{GAL}} = \langle \mathrm{x}_\theta^1 - e^{TA_\theta}\mathrm{x}_\theta^0, \varphi_j^1 \rangle.$$

On the other hand, the coefficients α_i obtained by Algorithm 8 are determined by orthogonal projection of $\mathrm{x}_\theta^1 - e^{TA_\theta}\mathrm{x}_\theta^0$ to the $\mathrm{Span}\{\Lambda_\theta\varphi_1^1, \dots, \Lambda_\theta\varphi_n^1\}$, and they satisfy the system

$$\sum_i \langle \Lambda_\theta \varphi_i^1, \Lambda_\theta \varphi_j^1 \rangle \alpha_i = \langle \mathrm{x}_\theta^1 - e^{TA_\theta}\mathrm{x}_\theta^0, \Lambda_\theta \varphi_j^1 \rangle.$$

The reason for specific design of the online control algorithm lies in the particular goal of our problem – that is to steer the system to the prescribed target as close as possible, and the proposed algorithm ensures the best control performance.

Finally, the precise estimates on the deviation of the system from the target follow from estimates on the reduced basis approximation developed through the online phase. Indeed, as $\varphi_\theta^{1,\star}$ is the solution to the minimization problem (3.12) it follows

$$|\mathrm{x}_\theta^1 - x_\theta^\star(T)| = |\Lambda_\theta(\varphi_\theta^1 - \varphi_\theta^{1,\star})| \leqslant |\Lambda_\theta(\varphi_\theta^1 - P_n\varphi_\theta^1)| \leqslant \Lambda_+|(\varphi_\theta^1 - P_n\varphi_\theta^1)|, \tag{3.13}$$

where P_n denotes the projection to the reduced basis space V_n, while Λ_+ is a positive constant satisfying $\Lambda_\theta \leqslant \Lambda_+ I$ for every $\theta \in \Theta$. Note that the boundedness of Λ_+ follows from the assumed Lipschitz dependence on the parameter and compactness of the set Θ.

For these reasons, the estimates obtained in the offline procedure on the approximation of the manifold $\varphi^1(\Theta)$ by the reduced basis transfer to the online phase. In particular, taking into account (3.6), for the POD selection we obtain the RMS bound

$$\sqrt{E(|\mathrm{x}_\theta^1 - x_\theta^\star(T)|^2)} \leqslant \Lambda_+\delta.$$

Similarly, the uniform estimate obtained by the greedy procedure provides

$$|\mathrm{x}_\theta^1 - x_\theta^\star(T)| \leqslant \Lambda_+\delta \qquad (\theta \in \Theta).$$

The last two estimates allow one to carefully design the whole process (including both the offline and online phase) in order to ensure a desired control performance. In particular, if we want to steer the system to the prescribed target within the precision ε, than the reduced basis error should satisfy $\delta \leqslant \varepsilon/\Lambda_+$.

We finish this subsection by a numerical example.

Example 3.6. We consider the control system (1.1) with the governing matrix of the form $A_\theta = \theta(N+1)^2\widetilde{A}$, where

$$\widetilde{A} = \begin{pmatrix} -2 & 1 & & 0 \\ 1 & -2 & \ddots & \\ & \ddots & \ddots & 1 \\ 0 & & 1 & -2 \end{pmatrix} \in \mathbb{R}^{N\times N}.$$

The control operator is assumed to be of the parameter-independent form

$$B = \left(0, \ldots, 0, (N+1)^2\right)^* \in \mathbb{R}^N.$$

Such system corresponds to the space-discretization of the heat equation problem with N internal grid points and the control on the right boundary:

$$\begin{cases} \partial_t v - \theta\partial_{xx}v = 0, & (t,x) \in (0,T)\times(0,1), \\ v(t,0) = 0, & t \in (0,T) \\ v(t,1) = u_\theta(t), & t \in (0,T) \\ v(0,x) = v_0(x), & x \in (0,1). \end{cases} \tag{3.14}$$

The parameter θ represents the diffusion coefficient and is supposed to range within the set $\Theta = [1, 10]$. We aim to control the system from the initial state $v_0(x) = \sin(\pi x)$ to zero in time $T = 1/10$.

The offline greedy algorithm has been applied for the system of dimension $N = 100$ with the stopping constant $\Lambda\_\delta/2 = 10^{-4}$, and the uniform discretization of Θ in $k = 100$ values. The algorithm stops after only three iterations, choosing parameter values 1.00, 1.18, and 1.45 (out of 100 eligibles).

The corresponding three minimizers φ_i^1 constitute the reduced basis V_3, and the approximate controls corresponding to all other parameter values are constructed as their suitable linear combination during the online phase. A rather small number of snapshots required to obtain approximation performance within the given accuracy is partially due to the strong dissipation effect of the heat equation, providing an exponential solution decay even in the absence of any control.

Here we present the results of the online algorithm applied for a specific parameter value $\theta = \pi$.

The corresponding approximate control is plotted on Fig. 3a. It exhibits rather strong oscillations when approaching the final time which is an intrinsic feature for the heat equation. The elapsed time of the online procedure was `3.58` seconds only, compared to `286` seconds required to construct the optimal exact control. This huge difference is due to a rather small dimension of the reduced basis, and it reconfirms computational efficiency of the greedy approach when compared to the standard one.

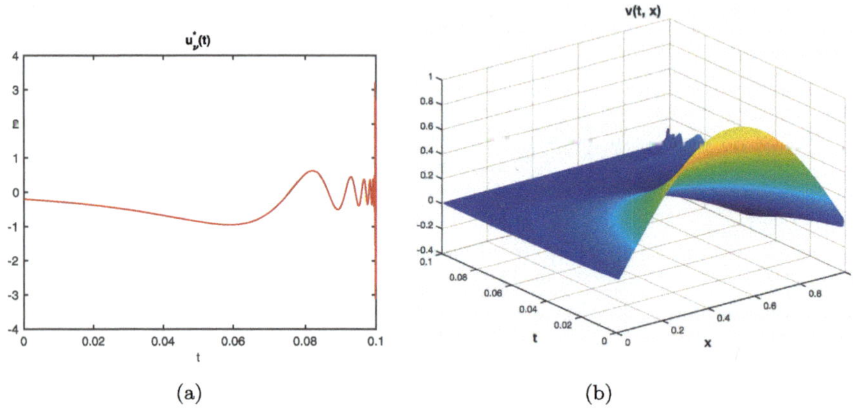

FIGURE 3 Evolution of the approximate control (a) and of the solution to the semi-discretized problem (3.14) governed by the approximate control $u_\theta^\star$ (b) for $\theta = \pi$.

Evolution of the solution, presented by a 3-D plot is given by Fig. 3b. The system is driven to the zero state, as required, within the error $|x(T)| = 1 \cdot 10^{-6}$.

3.3 Other approaches

Throughout this section we have dealt with a problem of constructing a reduced basis of the set $\varphi^1(\Theta)$ of all minimizers of the functional J_θ, given by (3.2), as θ ranges over the compact set Θ. By exploring the one-to-one correspondence between the minimizers φ_θ^1 and the associated optimal controls u_θ such approach enables rapid approximation of the latter during the online stage.

Of course, there are other methods at disposal by which one can construct (approximatively) a control function steering the system of interest under particular circumstances, determined by the exact realization of the parameter, to a prescribed target.

Construction of reduced basis for $u(\Theta)$**.** As an example, instead of dealing with the set $\varphi^1(\Theta)$, one can apply the reduced basis approach to the control manifold directly. A suggested method in that case would rely on a POD-Greedy strategy. This one is explored in case of time-dependent problems and it successfully combines two, above defined procedures (e.g. Haasdonk, 2013). In particular, each greedy step is accompanied by a corresponding POD process. The first selects a snapshot, which is a solution to time dependent problem for a specific parameter value, while the latter selects a single vector that approximates in the optimal manner the values of the snapshot evaluated at different time instants. It is important to mention that the method preserves convergences rates of greedy algorithms. More precisely, it maintains algebraic convergence rates of the Kolmogorov n-widths with the same exponent, and preserves their exponential decay with a slightly lower exponent (cf. Haasdonk, 2013, Section 4.2).

The POD-Greedy approach has been applied to evolutional equations and parabolic control problems (e.g. Hesthaven et al., 2016; Kärcher and Grepl, 2014a), but, as far as we know, it has not yet been developed in the context of controllability.

Construction of reduced basis for $\Lambda(\Theta)$. The analysis and methods discussed so far apply to control problems where the initial and target data are prescribed a priori, either depending on the parameters or not. In particular, the above presented algorithms provide a reduced basis for a particular data selection, and each change of it requires running the algorithm from scratch again.

From the point of view of applications it would be therefor interesting to develop efficient methods for control of parameter dependent systems that are robust with respect to these data. This requires construction of a reduced basis for the approximation of the Gramians Λ_θ, within the space of bounded linear operators.

A development of appropriate greedy procedure in that case, as explained above, requires identification of a feasible and easy executable surrogate. The key point in that direction is that Gramian operators satisfy a linear matrix relation: a differential Lyapunov equation in a case of a finite time Gramian, or an algebraic Lyapunov one for an infinite time horizon. In particular, the infinite time Gramian associated to the system (1.1) is the solution to

$$A_\theta \Lambda_\theta + \Lambda_\theta A_\theta^* = -B_\theta B_\theta^* \tag{3.15}$$

and is given by $\Lambda_\theta = \int_0^\infty e^{tA_\theta} B_\theta B_\theta^* e^{tA_\theta^*}\, dt$.

In order for the last integral to be convergent, one has to assume that A_θ is an exponentially stable operator for every $\theta \in \Theta$, thereby restricting the analysis to dissipative systems.

As for a surrogate distance of two Gramians associated to different parameters θ_1, θ_2, one attempts to follow the lines of Section 3.1.2 and estimate it by a residual obtained by plugging one operator into the equation satisfied by the other one:

$$c\|\Lambda_{\theta_1} - \Lambda_{\theta_2}\| \leqslant \|L_{A_{\theta_1}}(\Lambda_{\theta_1} - \Lambda_{\theta_2})\| \leqslant C\|\Lambda_{\theta_1} - \Lambda_{\theta_2}\|,$$

where L_A stands for the Lyapunov operator given by $L_A(P) = AP + PA^*$.

The above inequalities are easy to obtain in case of finite dimensional systems (cf. Son and Stykel, 2017), while appropriate result in case of an unbounded operator A_θ requires subtle analysis and development of necessary estimates in appropriate functional spaces (Lazar and Weston, 2021).

4 Conclusion

In this chapter we have presented three approaches for the computation of a control for parameter dependent system.

In the first approach, averaged controllability, we only build a control which steers the average of the system with respect to the parameter. This control is independent of the parameter and can be easily obtained. However, for a given realization of the parameter, there is no guarantee that the system solution will be close to the target.

In the second approach, ensemble controllability, we build a single, parameter independent control such that all the realizations of the parameter dependent system approximately reach the target. This is the ideal situation. However, not all systems are ensemble controllable, and the numerical computation of the control involves to solve a system of high dimension which might be ill conditioned.

Finally, the last approach is to compute a control for each realization of the parameter value. This approach is time-consuming if we have to repeat the control construction for every realization of the parameter. For this reason, we proposed a reduced basis approach, allowing a fast computation of an approximate control what ever the realization of the parameter is. This approach implies an offline part consisting in building an adequate reduced basis.

For all these approaches, we have provided corresponding theoretical background (including convergence results), numerical algorithms and illustrative examples.

Some specific comments and open question are listed below.

Robust control. Roughly speaking, the aim of robust control is to build a feedback controller K, independent of the parameter, such that the origin is a stable point for the system $\dot{x}_\theta = (A_\theta + B_\theta K)x_\theta$, for every realization of the parameter θ. This approach has been proposed in many text books and has not been developed in this chapter. We then refer the interested reader to books (Ackermann et al., 1993; Chang, 2014; Dorato et al., 1992; Yedavalli, 2014).

Averaged controllability. In this chapter, we illustrated averaged controllability with the mid-point quadrature rule. However, if the set of parameters is a compact in $\mathbb{R}^d$, and if we have enough regularity with respect to the parameter, then any numerical integration will be fine. Obviously, the choice of the numerical method will depend on the regularity of $\theta \mapsto (A_\theta, B_\theta)$ and on the dimension of the parameter space. When the parameter set is unbounded, the major difficulty is to find or develop a convergent method for the numerical integration on this parameter set. Some more efficient approaches have been proposed in Remarks 2.7 and 2.10.

Ensemble controllability. In this chapter, we only consider the uniform ensemble controllability. As far as we know, there do not exist numerical approaches for finding a control for L^q-ensemble controllability (for $1 \leqslant q < \infty$), without assuming that the system is L^∞-ensemble controllable.

Let us also point out that up to our best knowledge, there exist no ensemble controllability results for PDE systems (with an infinite set of parameters). In regard to the sufficient conditions given in Proposition 2.13, checking the assumptions on the spectral map is a real challenge due to the infinite number of eigenvalues for PDE operators.

Reduced basis for solutions to differential Lyapunov equations. In Section 3.3 we have described the procedure of obtaining a greedy approximation to solutions of parameter dependent, algebraic Lyapunov equations. It would be interesting to develop a similar algorithm for their differential counterparts. As finite time Gramians are solutions to the latter, this would in turn allow application to control problems in finite time horizon, and this for general systems, either of dissipative nature or not. However, it remains to check the efficiency of such approach, as potential residuals are a priory time-dependent which might enlarge the computational cost and the required memory capacities.

Model order reduction. For all the methods proposed in this paper, the control is obtained by solving the primal and dual systems (of potentially large dimension). In particular, when applying a reduced basis approach, we repeat this in the offline phase for each value from the discretized parameter set, which makes its computational cost very high.

In order to bypass that problem, it would be useful to implement model reduction techniques which enable one to substitute a high dimensional system by an approximate one of a smaller dimension. Recent overviews of model order reduction methods for parametric (time) dependent problems are given in Benner et al. (2015, 2017, Chapter 9) (cf. also the other chapters of the same book). These have been successfully implemented for elliptic and parabolic optimal control problems (Kärcher and Grepl, 2014b,a), but, as far as we know, they have not been developed in the context of controllability. Preliminary work in that direction is presented in Fabrini et al. (2018), but rigorous error bounds still remain open.

Neural networks. A data-driven approach based on combination of reduced basis (RB) methods and machine learning has been the object of intense investigation during the last years. Although there is a vast literature on application of neural networks in mathematics, the available references significantly reduce when it comes to the issue of solving parameter dependent problems. In Hesthaven and Ubbiali (2018); Santo et al. (2020) non-intrusive RB methods that rely on neural networks were successfully applied for computing solutions to parametric PDEs.

It would be interesting to develop a similar approach with the aim of efficiently treating parameter dependent control problems. The reduced basis algorithms were successfully developed in this context during the last decade, with the offline phase mainly exploring POD or a greedy sampling procedure. However, the cost of the corresponding online phase might still appear high, as the computation of projection (of solution to reduced basis) relies on the full-order model. An alternative approach would employ neural networks and train them to accurately predict the coefficients of solutions in reduced basis, with a computational cost independent of the dimension of the full-order model.

In addition, similar approach can be used in the context of simultaneous or ensemble controllability. Practical applications would include power and gas networks where recent attempts have been made at developing optimal control policies by using machine learning techniques in the offline phase (Ng et al., 2018).

PDEs. In this chapter we have restricted the exposition to finite dimensional linear systems. However, it is important to stress that a number of positive control results has been obtained in the context of parameter dependent PDEs. This is in particular the case for averaged controllability and reduced basis approach. Note that all the methods proposed in this chapter will work well for discretized PDEs. However, some finer numerical analysis ensuring appropriate convergence results in that case is still missing. More precisely, one has to design the discretization of the PDE in terms of the discretization of the parameter set (or inversely), so that convergence of the method is ensured when both discretization constants go to 0. This remains as a challenging open problem for future investigations.

Acknowledgments

This work was supported by the Croatian Science Foundation through the project Control of Dynamical Systems (ConDyS, IP-2016-06-2468)

Appendix A Proof of technical results related to Section 2

A.1 Proof of Lemma 2.6

Let us first observe that,

$$\int_{\Theta} f(\theta)\,\mathrm{d}\mu = \int_{\Theta} f(\theta)\,w(\theta)\mathrm{d}\theta = \int_0^1 f\left(W^{-1}(z)\right)\,\mathrm{d}z$$
$$= \sum_{k=1}^{K} \int_{z_{k-1}^K}^{z_k^K} f\left(W^{-1}(z)\right)\,\mathrm{d}z.$$

Note that $f \circ W^{-1}$ is of class C^2 on $(0, 1)$ but might not be defined at 0 or 1. Hence, for $k \in \{2, \ldots, K-1\}$, we classically have

$$\left|\int_{z_{k-1}^K}^{z_k^K} f \circ W^{-1}\mathrm{d}z - m_k^K f(\theta_k^K)\right| \leqslant \frac{(z_k^K - z_{k-1}^K)^3}{24} \sup_{(z_{k-1}^K, z_k^K)} \left|\frac{\mathrm{d}^2 f \circ W^{-1}}{\mathrm{d}z^2}\right|.$$

We observe that $\frac{\mathrm{d}^2 f\circ W^{-1}}{\mathrm{d}z^2} \circ W = \frac{1}{w^2}\left(f'' - f'\frac{w'}{w}\right)$ and hence, we have,

$$\left|\int_{z_{k-1}^K}^{z_k^K} f \circ W^{-1}\mathrm{d}z - m_k^K f(\theta_k^K)\right| \leqslant M_k(z_k^K - z_{k-1}^K)^3.$$

Note that this estimate is also valid for $k = 1$ (respectively $k = K$) if $f \circ W^{-1}$ is continuous (or admits a continuous extension) at 0 (respectively 1). When this does not hold, we can however use the following estimate, for $k = 1$ or $K = K$,

$$\left|\int_{z_{k-1}^K}^{z_k^K} f \circ W^{-1}\,\mathrm{d}z - m_k^K f(\theta_k^K)\right| \leqslant \int_{z_{k-1}^K}^{z_k^K} g \circ W^{-1}\,\mathrm{d}z + m_k^K g(\theta_k^K)$$
$$(k \in \{1, \dots, K\}).$$

A.2 Estimates related to Example 2.11

First note that we have,

$$w(\theta)^{-2}\left|f''(\theta) - f'(\theta)\frac{w'(\theta)}{w(\theta)}\right| = e^{2\theta}\left|f''(\theta) + f'(\theta)\right| \qquad (\theta \in \mathbb{R}_+).$$

In order to estimate f' and f'', one can use the following procedure.

• For $f(\theta) = e^{TA_\theta}\mathrm{x}_\theta^0$, let us define for every $t \in [0, T]$, $\xi_\theta(t) = e^{tA_\theta}\mathrm{x}_\theta^0$, so that $\xi_\theta(T) = f(\theta)$. Then, ξ_θ, ξ_θ' and ξ_θ'' (where "$'$" denotes the differentiation with respect to θ) are solution of

$$\begin{pmatrix} \dot\xi_\theta \\ \dot\xi_\theta' \\ \dot\xi_\theta'' \end{pmatrix} = \begin{pmatrix} A_\theta & 0 & 0 \\ A_\theta' & A_\theta & 0 \\ A_\theta'' & 2A_\theta' & A_\theta \end{pmatrix}\begin{pmatrix} \xi_\theta \\ \xi_\theta' \\ \xi_\theta'' \end{pmatrix}, \qquad \begin{pmatrix} \xi_\theta(0) \\ \xi_\theta'(0) \\ \xi_\theta''(0) \end{pmatrix} = \begin{pmatrix} \mathrm{x}_\theta^0 \\ x_\theta^{0'} \\ x_\theta^{0''} \end{pmatrix}.$$

In our particular case, we get,

$$\begin{pmatrix} \dot\xi_\theta \\ \dot\xi_\theta' \\ \dot\xi_\theta'' \end{pmatrix} = \begin{pmatrix} \begin{pmatrix} 0 & -\theta^2 \\ 1 & 0 \end{pmatrix} & 0 & 0 \\ \begin{pmatrix} 0 & -2\theta \\ 0 & 0 \end{pmatrix} & \begin{pmatrix} 0 & -\theta^2 \\ 1 & 0 \end{pmatrix} & 0 \\ \begin{pmatrix} 0 & -2 \\ 0 & 0 \end{pmatrix} & \begin{pmatrix} 0 & -4\theta \\ 0 & 0 \end{pmatrix} & \begin{pmatrix} 0 & -\theta^2 \\ 1 & 0 \end{pmatrix} \end{pmatrix}\begin{pmatrix} \xi_\theta \\ \xi_\theta' \\ \xi_\theta'' \end{pmatrix},$$

$$\begin{pmatrix} \xi_\theta(0) \\ \xi'_\theta(0) \\ \xi''_\theta(0) \end{pmatrix} = \begin{pmatrix} \theta \\ \sin\theta \\ 1 \\ \cos\theta \\ 0 \\ -\sin\theta \end{pmatrix}$$

and easily obtain

$$e^{2\theta}|f''(\theta)+f'(\theta)| = e^{2\theta}|\xi''_\theta(T)+\xi'_\theta(T)| \leqslant C_\varepsilon e^{(2+\varepsilon)\theta}, \tag{A.1a}$$

for every $\varepsilon > 0$ and for some constant $C_\varepsilon > 0$ (indeed $|\xi_\theta(T)|^2 + |\xi'_\theta(T)|^2 + |\xi''_\theta(T)|$ is polynomial in (T, θ) and hence can be bounded by an exponential). In addition, since the eigenvalues of A_θ are purely complex numbers, we have,

$$|f(\theta)| = |\mathrm{x}^0_\theta| \leqslant \theta + 1. \tag{A.1b}$$

• Similarly, for $f(\theta) = e^{tA_\theta}B_\theta$, we consider, $\varphi^1 \in \mathbb{R}^N$ and φ_θ solution of $-\dot\varphi_\theta = A^*_\theta\varphi_\theta$ with $\varphi_\theta(T) = \varphi^1$, we then have that,

$$-\begin{pmatrix} \dot\varphi_\theta \\ \dot\varphi'_\theta \\ \dot\varphi''_\theta \end{pmatrix} = \begin{pmatrix} A^*_\theta & 0 & 0 \\ A'^*_\theta & A^*_\theta & 0 \\ A''^*_\theta & 2A'^*_\theta & A^*_\theta \end{pmatrix}, \qquad \begin{pmatrix} \varphi_\theta(0) \\ \varphi'_\theta(0) \\ \varphi''_\theta(0) \end{pmatrix} = \begin{pmatrix} \varphi^1 \\ 0 \\ 0 \end{pmatrix}$$

and that,

$$\begin{aligned}\left(f''+f'\right)^*\varphi^1 &= B''^*_\theta\varphi_\theta(t) + B'^*_\theta(2\varphi'_\theta(t)+\varphi_\theta(t)) + B^*_\theta(\varphi''_\theta(t)+\varphi'_\theta(t)) \\ &= \mathrm{e}^*_2\left((\theta-1)\varphi''_\theta(t) + (\theta+1)\varphi'_\theta(t) + \varphi_\theta(t)\right),\end{aligned}$$

with $\mathrm{e}^*_2 = (0\ 1)$. Here also, we get that,

$$\begin{aligned} e^{2\theta}\left|\left(f''(\theta)+f'(\theta)\right)^*\varphi^1\right| &= e^{2\theta}|(\theta-1)\varphi''_\theta(t) + (\theta+1)\varphi'_\theta(t) + \varphi_\theta(t)| \\ &\leqslant C_\varepsilon e^{(2+\varepsilon)\theta}|\varphi^1|, \end{aligned} \tag{A.2a}$$

for every $\varepsilon > 0$, every $\varphi^1 \in \mathbb{R}^2$ and for some constant $C_\varepsilon > 0$ independent of φ^1 and t. In addition, we have,

$$|f(\theta)^*\varphi^1| = |\varphi_\theta(t)| = |\varphi^1| \leqslant (\theta+1)|\varphi^1| \qquad (\varphi^1 \in \mathbb{R}^2,\ t \in [0, T]). \tag{A.2b}$$

A.3 Proof of Lemma 2.15

The existence of a control $u^\varepsilon_K \in L^2(0, T)^M$ such that (2.17) holds directly follows from Definition 2. We then deduce the existence of a unique minimal L^2-norm control such that (2.17) holds. This control minimizes, $u \in L^2(0, T)^M \mapsto$

$f(u) + g(\Phi u) \in \mathbb{R}$, with

$$f(u) = \frac{1}{2}\|u\|^2_{L^2(0,T)^M}, \qquad \Phi u = \left(\int_0^T e^{(T-t)A_{\theta_k}} B_{\theta_k} u(t)\,\mathrm{d}t\right)_{k\in\{1,\dots,K\}}$$

$$(u \in L^2(0,T)^M),$$

$$g\left((X_k)_{k\in\{1,\dots,K\}}\right) = \begin{cases} +\infty & \text{if } \sup_{k\in\{1,\dots,K\}} \left| \mathrm{x}^1_{\theta_k} - e^{TA_{\theta_k}} \mathrm{x}^0_{\theta_k} - X_k \right| > \varepsilon, \\ 0 & \text{otherwise} \end{cases}$$

$$(X_1, \dots, X_K \in \mathbb{R}^N).$$

Using Fenchel-Rockafellar duality, see for instance Borwein and Zhu (2005); Ekeland and Témam (1999), we obtain that the minimizer u^ε_K of $u \mapsto f(u) + g(\Phi u)$ is given by $u^\varepsilon_K = \nabla f^*(-\Phi^* P^1)$, where $P^1 = (\varphi^1_k)_{k\in\{1,\dots,K\}} \in (\mathbb{R}^N)^K$ minimizes $P \in (\mathbb{R}^N)^K \mapsto f^*(\Phi^* P) + g^*(P)$. In these expressions, Φ^* is the adjoint of Φ and $f^\star$ (respectively $g^\star$) is the convex conjugate of f (respectively g). More precisely, we have,

$$\left(\Phi^*(\varphi^1_k)_{k\in\{1,\dots,K\}}\right)(t) = \sum_{k=1}^K B^*_{\theta_k} e^{(T-t)A^*_{\theta_k}} \varphi^1_k,$$

that is to say that $\Phi^*(\varphi^1_k)_{k\in\{1,\dots,K\}} = \sum_{k=1}^K B^*_{\theta_k}\varphi_k$, with φ_k defined by (2.19). We also have,

$$f^\star(\varphi) = \sup_{u\in L^2(0,T)^M} (\langle \varphi, u\rangle - f(u)) = \frac{1}{2}\|\varphi\|^2_{L^2(0,T)^M},$$

$$\begin{aligned} g^\star\left((\varphi^1_k)_{k\in\{1,\dots,K\}}\right) &= \sup_{X_1,\dots,X_K\in\mathbb{R}^N} \left(\sum_{k=1}^K \langle \varphi^1_k, X_k\rangle - g\left((X_k)_{k\in\{1,\dots,K\}}\right)\right) \\ &= \sup_{\substack{Y_1,\dots,Y_K\in\mathbb{R}^N \\ |Y_1|,\dots,|Y_K|\leqslant 1}} \sum_{k=1}^K \langle \varphi^1_k, \mathrm{x}^1_{\theta_k} - e^{TA_{\theta_k}}\mathrm{x}^0_{\theta_k} - \varepsilon Y_k\rangle \\ &= \sum_{k=1}^K \langle \varphi^1_k, \mathrm{x}^1_{\theta_k} - e^{TA_{\theta_k}}\mathrm{x}^0_{\theta_k}\rangle + \varepsilon \sum_{k=1}^K |\varphi^1_k|. \end{aligned}$$

All these expressions lead to the claim of Lemma 2.15.

References

Abdelli, M., Castro, C., 2020. Numerical approximation of the averaged controllability for the wave equation with unknown velocity of propagation.

Ackermann, J., Bartlett, A., Kaesbauer, D., Sienel, W., Steinhauser, R., 1993. Robust Control. Communications and Control Engineering Series. Springer-Verlag, London, Ltd., London.

Agrachev, A., Baryshnikov, Y., Sarychev, A., 2016. Ensemble controllability by Lie algebraic methods. ESAIM Control Optim. Calc. Var. 22 (4), 921–938.
Asier Bárcena-Petisco, J., Zuazua, E., 2020. Averaged dynamics and control for heat equations with random diffusion. Working paper or preprint.
Avetisyan, A.S., Khurshudyan, A.Z., 2019. Averaged controllability of Euler-Bernoulli beams with random material characteristics: the Green's function approach. Izv. Nat. Akad. Nauk Arm. Mekh. 72 (4), 7–18.
Beauchard, K., Coron, J.-M., Rouchon, P., 2010. Controllability issues for continuous-spectrum systems and ensemble controllability of Bloch equations. Commun. Math. Phys. 296 (2), 525–557.
Benner, P., Gugercin, S., Willcox, K., 2015. A survey of projection-based model reduction methods for parametric dynamical systems. SIAM Rev. 57 (4), 483–531.
Benner, P., Ohlberger, M., Cohen, A., Willcox, K.E. (Eds.), 2017. Model Reduction and Approximation: Theory and Algorithms. SIAM, Philadelphia.
Borwein, J.M., Zhu, Q.J., 2005. Techniques of Variational Analysis. CMS Books in Mathematics/Ouvrages de Mathématiques de la SMC, vol. 20. Springer-Verlag, New York.
Caflisch, R.E., 1998. Monte Carlo and quasi-Monte Carlo methods. In: Acta Numerica, 1998. In: Acta Numer., vol. 7. Cambridge Univ. Press, Cambridge, pp. 1–49.
Chang, X.-H., 2014. Robust Output Feedback H-Infinity Control and Filtering for Uncertain Linear Systems. Studies in Systems, Decision and Control, vol. 7. Springer, Heidelberg.
Chen, X., 2019. Structure theory for ensemble controllability, observability, and duality. Math. Control Signals Syst. 31 (2), 7.
Coulson, J., Gharesifard, B., Mansouri, A.-R., 2019. On average controllability of random heat equations with arbitrarily distributed diffusivity. Automatica J. IFAC 103, 46–52.
DeVore, R., Petrova, G., Wojtasczyk, P., 2013. Greedy algorithms for reduced bases in Banach spaces. Constr. Approx. 37 (1), 455–466.
Dick, J., Kuo, F.Y., Sloan, I.H., 2013. High-dimensional integration: the quasi-Monte Carlo way. Acta Numer. 22, 133–288.
Dirr, G., Schönlein, M., 2020. Uniform and L^q-ensemble reachability of parameter-dependent linear systems.
Dorato, P., Fortuna, L., Muscato, G., 1992. Robust Control for Unstructured Perturbations—an Introduction. Lecture Notes in Control and Information Sciences, vol. 168. Springer-Verlag, Berlin.
Ekeland, I., Témam, R., 1999. Convex Analysis and Variational Problems. Classics in Applied Mathematics, vol. 28. Society for Industrial and Applied Mathematics (SIAM), Philadelphia, PA. English edition.
Fabrini, G., Iapichino, L., Volkwein, S., 2018. Reduced-order greedy controllability of finite dimensional linear systems. IFAC-PapersOnLine 51 (2), 296–301.
Haasdonk, B., 2013. Convergence rates of the pod–greedy method. ESAIM: M2AN 47 (3), 859–873.
Helmke, U., Schönlein, M., 2014. Uniform ensemble controllability for one-parameter families of time-invariant linear systems. Syst. Control Lett. 71, 69–77.
Hernández-Santamaría, V., Lazar, M., Zuazua, E., 2019. Greedy optimal control for elliptic problems and its application to turnpike problems. Numer. Math. 141 (1), 455–493.
Hesthaven, J., Ubbiali, S., 2018. Non-intrusive reduced order modeling of nonlinear problems using neural network. J. Comput. Phys. 363, 55–78.
Hesthaven, J.S., Rozza, G., Stamm, B., 2016. Certified Reduced Basis Methods for Parametrized Partial Differential Equations. Springer International Publishing, Berlin.
Kärcher, M., Grepl, M.A., 2014a. A posteriori error estimation for reduced order solutions of parametrized parabolic optimal control problems. ESAIM: Math. Model. Numer. Anal. 48 (6), 1615–1638.
Kärcher, M., Grepl, M.A., 2014b. A certified reduced basis method for parametrized elliptic optimal control problems. ESAIM Control Optim. Calc. Var. 20 (2), 416–441.
Klamka, J., Khurshudyan, A.Z., 2019. Averaged controllability of heat equation in unbounded domains with random geometry and location of controls: the Green's function approach. Arch. Control Sci. 29(65) (4), 573–584.

Kreindler, E., Sarachik, P.E., 1964. On the concepts of controllability and observability of linear systems. IEEE Trans. Autom. Control AC-9, 129–136.
Kronrod, A.S., 1965. Nodes and Weights of Quadrature Formulas. Sixteen-Place Tables. Consultants Bureau, New York. Authorized translation from the Russian.
Lazar, M., 2018. Stability of observations of partial differential equations under uncertain perturbations. ESAIM Control Optim. Calc. Var. 24 (1), 45–61.
Lazar, M., Lohéac, J., 2020. Output controllability in a long-time horizon. Automatica J. IFAC 113, 108762.
Lazar, M., Weston, J., 2021. Greedy algorithm for parameter dependent operator Lyapunov equations. Syst. Control Lett. 154, 104968.
Lazar, M., Zuazua, E., 2014. Averaged control and observation of parameter-depending wave equations. C. R. Math. Acad. Sci. Paris 352 (6), 497–502.
Lazar, M., Zuazua, E., 2016. Greedy controllability of finite dimensional linear systems. Automatica 74, 327–340.
Li, J.-S., 2011. Ensemble control of finite-dimensional time-varying linear systems. IEEE Trans. Autom. Control 56 (2), 345–357.
Li, J.-S., Khaneja, N., 2009. Ensemble control of Bloch equations. IEEE Trans. Autom. Control 54 (3), 528–536.
Li, J.-S., Qi, J., 2016. Ensemble control of time-invariant linear systems with linear parameter variation. IEEE Trans. Autom. Control 61 (10), 2808–2820.
Li, J.-S., Zhang, W., Tie, L., 2020. On separating points for ensemble controllability. SIAM J. Control Optim. 58 (5), 2740–2764.
Lions, J.-L., 1988. Contrôlabilité exacte, perturbations et stabilisation de systèmes distribués. Tome 1. Recherches en Mathématiques Appliquées (Research in Applied Mathematics), vol. 8. Masson, Paris.
Lohéac, J., Zuazua, E., 2016. From averaged to simultaneous controllability. Ann. Fac. Sci. Toulouse Math. (6) 25 (4), 785–828.
Lohéac, J., Zuazua, E., 2017. Averaged controllability of parameter dependent conservative semigroups. J. Differ. Equ. 262 (3), 1540–1574.
Lü, Q., Zuazua, E., 2016. Averaged controllability for random evolution partial differential equations. J. Math. Pures Appl. (9) 105 (3), 367–414.
Maday, Y., 2006. Reduced basis method for the rapid and reliable solution of partial differential equations. In: International Congress of Mathematicians, 2006. European Mathematical Society, pp. 1255–1270.
Marín, F.J., Martínez-Frutos, J., Periago, F., 2017. Robust averaged control of vibrations for the Bernoulli-Euler beam equation. J. Optim. Theory Appl. 174 (2), 428–454.
Marín, F.J., Martínez-Frutos, J., Periago, F., 2018. A polynomial chaos-based approach to risk-averse piezoelectric control of random vibrations of beams. Int. J. Numer. Methods Eng. 115 (6), 738–755.
Martínez-Frutos, J., Periago Esparza, F., 2018. Optimal Control of PDEs Under Uncertainty. SpringerBriefs in Mathematics. Springer/BCAM Basque Center for Applied Mathematics, Cham/Bilbao.
Nakić, I., Tomljanović, Z., Truhar, N., 2019. Mixed control of vibrational systems. Z. Angew. Math. Mech. 99 (9), e201800328.
Ng, Y., Misra, S., Roald, L., Backhaus, S., 2018. Statistical learning for dc optimal power flow. In: Power Systems Computation Conference (PSCC). IEEE, pp. 1–7.
Quarteroni, A., Manzoni, A., Negri, F., 2016. Reduced Basis Methods for Partial Differential Equations. Unitext, vol. 92. Springer, Cham.
Robert, C.P., Casella, G., 2004. Monte Carlo Statistical Methods, second edition. Springer Texts in Statistics. Springer-Verlag, New York.
Russell, D.L., 1986. The Dirichlet-Neumann boundary control problem associated with Maxwell's equations in a cylindrical region. SIAM J. Control Optim. 24 (2), 199–229.

Santo, N.D., Deparis, S., Pegolotti, L., 2020. Data driven approximation of parametrized pdes by reduced basis and neural networks. J. Comput. Phys. 416, 109550.

Shao, T.S., Chen, T.C., Frank, R.M., 1964. Tables of zeros and Gaussian weights of certain associated Laguerre polynomials and the related generalized Hermite polynomials. Math. Comput. 18, 598–616.

Son, N., Stykel, T., 2017. Solving parameter-dependent Lyapunov equations using the reduced basis method with application to parametric model order reduction. SIAM J. Matrix Anal. Appl. 38 (2), 478–504.

Trefethen, L.N., 2000. Spectral Methods in MATLAB. Software, Environments, and Tools, vol. 10. Society for Industrial and Applied Mathematics (SIAM), Philadelphia, PA.

Trefethen, L.N., 2008. Is Gauss quadrature better than Clenshaw-Curtis? SIAM Rev. 50 (1), 67–87.

Triggiani, R., 1975. Controllability and observability in Banach space with bounded operators. SIAM J. Control 13, 462–491.

Tucsnak, M., Weiss, G., 2009. Observation and Control for Operator Semigroups. Birkhäuser-Verlag AG, Berlin.

Veroy, K., Prud'Homme, C., Rovas, D., Patera, A., 2003. A posteriori error bounds for reduced-basis approximation of parametrized noncoercive and nonlinear elliptic partial differential equations. In: 16th AIAA Computational Fluid Dynamics Conference. Orlando, United States.

Yedavalli, R.K., 2014. Robust Control of Uncertain Dynamic Systems. Springer, New York.

Zeng, S., Allgöwer, F., 2016. A moment-based approach to ensemble controllability of linear systems. Syst. Control Lett. 98, 49–56.

Zuazua, E., 2006. Controllability and observability of partial differential equations: some results and open problems. In: Dafermos, C.M., Feireisl, E. (Eds.), Handbook of Differential Equations: Evolutionary Equations, Vol. 3. Elsevier Science, Amsterdam, pp. 527–621.

Zuazua, E., 2014. Averaged control. Automatica J. IFAC 50 (12), 3077–3087.

Chapter 9

Space-time POD-Galerkin approach for parametric flow control

Francesco Ballarin[a], Gianluigi Rozza[b,*], and Maria Strazzullo[b]

[a]*Department of Mathematics and Physics, Catholic University of the Sacred Heart, Brescia, Italy,*
[b]*mathLab, Mathematics Area, SISSA, Trieste, Italy*
**Corresponding author: e-mail address: gianluigi.rozza@sissa.it*

Contents

Abstract

In this contribution we propose reduced order methods to fast and reliably solve parametrized optimal control problems governed by time dependent nonlinear partial differential equations. Our goal is to provide a tool to deal with the time evolution of several nonlinear optimality systems in many-query context, where a system must be analyzed for various physical and geometrical features. Optimal control can be used in order to fill the gap between collected data and mathematical model and it is usually related to very time consuming activities: inverse problems, statistics, etc. Standard discretization techniques may lead to unbearable simulations for real applications. We aim at showing how reduced order modeling can solve this issue. We rely on a space-time POD-Galerkin reduction in order to solve the optimal control problem in a low dimensional reduced space in a fast way for several parametric instances. The proposed algorithm is validated

Handbook of Numerical Analysis, Volume 23, ISSN 1570-8659, https://doi.org/10.1016/bs.hna.2021.12.009

with a numerical test based on environmental sciences: a reduced optimal control problem governed by viscous Shallow Waters Equations parametrized not only in the physics features, but also in the geometrical ones. We will show how the reduced model can be useful in order to recover desired velocity and height profiles more rapidly with respect to the standard simulation, not loosing in accuracy.

Keywords
Reduced order modeling, Optimal control problems, Time dependent nonlinear partial differential equations, Lagrangian approach

MSC Codes
35A35, 49J20, 65M60

1 Motivations and historical background

This contribution is rooted in control systems and controllability theory for partial differential equation. A control problem is a system on which you can act through suitable external variables said *controls* (Coron, 2007). The controllability theory answers to the need of steering a system towards a desired configuration. Is it always possible? Under which conditions can I reach an exact prescribed profile for my system?

The problem is quite fascinating and of utmost usefulness in many applications. Even if linear partial differential control systems have many complex aspects and features to be analyzed both theoretically and numerically (Coron, 2007; Glowinski, 1992; Lions, 1968, 1988; Lions and Zuazua, 2017), our main focus will concern nonlinearity in fluid dynamics. In this setting, the problem becomes more challenging and, besides the growing complexity, the need of a control tool increases. The case of nonlinear partial differential equations is much more complicated to handle. The control theory for fluid models, e.g. Navier-Stokes equations, prospered in the eighties thanks to the research of J. L. Lions. The main idea of his production relied on the role which nonlinearity plays as a control itself, giving the possibility or preventing the achievement of peculiar motion behaviors (Lions, 1990). This intuition paves the way to a wide range of literature which addresses the problem (Agrachev and Sarychev, 2005; Coron, 1996; Lions, 1969; Lions and Zuazua, 1997, 1998; Vázquez et al., 2008).

However, in many applied contexts, it is clear that not all the systems are controllable and, furthermore, it is not possible to prove the existence of controls which give the *exact* desired solution profile one wants to reach. This is the reason why the controllability theory expands towards optimization and optimal control theory. Namely, the new objective is to find a way to reach *the most similar configuration* with respect to the desired one, satisfying the underlying partial differential equation constraint. In the next section, we will introduce optimal control theory in a parametrized setting, addressing the great importance of a such mathematical model in scientific and engineering contexts.

2 Introduction

This chapter deals with parametrized optimal control problems (OCP($\boldsymbol{\mu}$)s) governed by time dependent nonlinear parametrized partial differential equations (PDE($\boldsymbol{\mu}$)s). Optimal control is a versatile mathematical tool which has been exploited in many fields of applications: shape optimization, see e.g. Delfour and Zolésio (2011); Haslinger and Mäkinen (2003); Mohammadi and Pironneau (2010), fluid dynamics, see e.g. Dedè (2007); Negri et al. (2015); Pošta and Roubíček (2007); de los Reyes and Tröltzsch (2007), heamodynamics (Ballarin et al., 2017; Lassila et al., 2013; Strazzullo et al., 2021b; Zainib et al., 2020), environmental predictions (Quarteroni et al., 2005, 2007; Strazzullo et al., 2018, 2021b) and more. In a parametrized setting, where a parameter $\boldsymbol{\mu} \in \mathcal{P} \subset \mathbb{R}^d$ could represent physical or geometrical features, OCP($\boldsymbol{\mu}$)s are exploited in order to study different configurations, which can be used to better understand the phenomenon one is dealing with.

The main motivation for using OCP($\boldsymbol{\mu}$)s relies in the need of simulations which have to be similar to a given observation or collected data. If on one hand, OCP($\boldsymbol{\mu}$)s can be useful in many research fields, on the other, they are very complex and demanding from a computational point of view and this issue limits their applicability, most of all in a time dependent framework. Time dependence in OCP($\boldsymbol{\mu}$)s has been widely described in literature, see e.g. Hinze et al. (2008a); Iapichino et al. (2016); Leugering et al. (2014); Seymen et al. (2014); Stoll and Wathen (2010, 2013). In this context, the required computational resources for OCP($\boldsymbol{\mu}$)s simulations drastically increase, and parametrized simulations can be unbearable to be performed with standard discretization techniques. To achieve a real-time OCP($\boldsymbol{\mu}$) model, able to reconstruct observable profiles varying with respect to a parameter, a rapid and suitable approximation technique is needed. To this end, we rely on reduced order methods (ROMs). This discretization approach builds a low dimensional framework which can be exploited in order to solve several parametric instances to give real-time information on the model at hand. Working in this reduced space allows us to solve the parametrized optimality system in a small amount of time, by reducing the involved computational costs: for an introduction to the methodology the interested reader may refer to Antoulas et al. (2020); Benner et al. (2017); Hesthaven et al. (2015); Prud'Homme et al. (2002); Rozza et al. (2013, 2008); Quarteroni et al. (2015), for example.

If we focus on OCP($\boldsymbol{\mu}$)s applications, there is a wide production concerning steady linear governing equations, see e.g. the following far-from-exhaustive list (Bader et al., 2016, 2015; Dedè, 2010; Gerner and Veroy, 2012; Iapichino et al., 2017; Kärcher and Grepl, 2014; Kärcher et al., 2018; Kunisch and Volkwein, 2008; Negri et al., 2015, 2013; Quarteroni et al., 2007). Then, the consolidated knowledge about ROM for steady OCP($\boldsymbol{\mu}$)s has been extended to time dependent OCP($\boldsymbol{\mu}$)s: moving from Hinze et al. (2008a); Stoll and Wathen (2010, 2013); Yilmaz and Karasözen (2014) as a starting point, in Iapichino et al. (2016); Strazzullo et al. (2020b, 2021a,b) the main effort is to enlarge the model in a parametrized setting, and generalize the standard algorithm used to build the reduced space framework to time dependency. From now on, we will only focus on POD-Galerkin approach. The motivation relies in its versatility, since it can be even applied to every kind of governing equations: see e.g. Pichi et al. (2020); Strazzullo et al. (2021b); Zainib et al. (2020) for POD employed in steady OCP($\boldsymbol{\mu}$)s for nonlinear governing equations, or Strazzullo et al. (2020a) for the application of such an algorithm to a space-time nonlinear OCP($\boldsymbol{\mu}$). In this contribution, the proposed techniques have been tested through parametrized numerical simulations given by distributed control for viscous Shallow Waters Equations (SWEs), a model capable to simulate coastal current behavior and used in environmental sciences for monitoring plans.

We aim at providing a space-time POD-Galerkin strategy for time dependent nonlinear OCP($\boldsymbol{\mu}$)s. This work is outlined as follows. In Section 3, we introduce the theoretical formulation for OCP($\boldsymbol{\mu}$)s following Hinze et al. (2008b) and Tröltzsch (2010), focusing on time dependent nonlinear problems. Furthermore, a brief introduction of the space-time algebraic formulation is presented. Section 4 deals with the main idea behind reduced order approximation (Hesthaven et al., 2015) and with its application to space-time nonlinear OCP($\boldsymbol{\mu}$)s. When dealing with the numerical approximation of time dependent nonlinear OCP($\boldsymbol{\mu}$)s, we will restrict ourselves to the case of semi-linear PDEs with quadratic nonlinearity in the state variable, to comply with the numerical test presented in Section 5: an application in coastal marine management through an OCP($\boldsymbol{\mu}$) governed by SWEs. The proposed numerical test is inspired by Strazzullo et al. (2020a), but in this case we consider not only physical, but also geometrical parametrization. Conclusions follow in Section 6.

3 Nonlinear time dependent parametrized optimal flow control problems

This section provides the continuous formulation for nonlinear time dependent OCP($\boldsymbol{\mu}$)s. We will introduce the Lagrangian approach technique (Gunzburger, 2003; Hinze et al., 2008b) to minimize a quadratic cost functional constrained to nonlinear time dependent PDE($\boldsymbol{\mu}$)s. We remark that the analysis is still valid in the Banach spaces setting, but for the sake of clarity, we will restrict ourselves to the simpler case of the real Hilbert spaces.

3.1 Problem formulation

Let us suppose to have a spatial domain Ω where a physical event described by a time dependent nonlinear PDE($\boldsymbol{\mu}$) occurs. In order to mathematically represent such an evolution in the time interval $[0, T]$, we define the spaces

$$\mathbb{Y} = \{y \in L^2(0, T; Y) \text{ such that } y_t \in L^2(0, T; Y^*)\} \text{ and } \mathbb{Q} = L^2(0, T; Y), \quad (1)$$

for Y Hilbert space. Indeed, we are provided by a nonlinear *state equation* $G : \mathbb{Y} \times \mathcal{P} \to \mathbb{Q}^*$ of the form

$$G(y; \boldsymbol{\mu}) = f, \tag{2}$$

where $y := y(\boldsymbol{\mu}) \in \mathbb{Y}$ is the *state variable*, namely the physical quantity we are interested in, while $f \in \mathbb{Q}^*$ is a forcing term and $\mathcal{P} \subset \mathbb{R}^P$ is a parameter space of dimension $P \geq 1$, which can describe physical and/or geometrical features of the system at hand. Moreover, we call $\mathcal{L}(\cdot, \cdot)$ the space of the continuous linear functions between two spaces. The considered PDE($\boldsymbol{\mu}$)s present the following form:

$$G(y; \boldsymbol{\mu}) = y_t + E_{n\ell}(y; \boldsymbol{\mu}) + E_{\ell}(y; \boldsymbol{\mu}), \tag{3}$$

where $E_{\ell} \in \mathcal{L}(\mathbb{Y}, \mathbb{Q}^*)$ and $E_{n\ell}$ represent the linear and nonlinear contributions to the equation, respectively. The term y_t describes the time evolution of the equation.

We now want to change the behavior of the state variable steering it to a desired profile, say $y_{\text{d}} := y_{\text{d}}(\boldsymbol{\mu}) \in \mathbb{Y}_{\text{obs}} \supseteq \mathbb{Y}$, thanks to the action of a *control variable* $u := u(\boldsymbol{\mu}) \in \mathbb{U}$, where $\mathbb{U} = L^2(0, T; U)$ with U another Hilbert space. Thanks to all the previous definitions, we are able to define the *controlled equation* $\mathcal{E}(y, u; \boldsymbol{\mu})$, where $\mathcal{E} : \mathbb{Y} \times \mathbb{U} \times \mathcal{P} \to \mathbb{Q}^*$ of the form:

$$\mathcal{E}(y, u; \boldsymbol{\mu}) := G(y; \boldsymbol{\mu}) - C(u) - f = 0, \tag{4}$$

where $C \in \mathcal{L}(\mathbb{U}, \mathbb{Q}^*)$ is related to the control variable action on the system in order to change the original state variable. The OCP($\boldsymbol{\mu}$) reads: given a $\boldsymbol{\mu} \in \mathcal{P}$, find the pair $(y, u) \in \mathbb{Y} \times \mathbb{U}$ which solves

$$\min_{y \in \mathbb{Y}_{\text{ad}} \subset \mathbb{Y}, u \in \mathbb{U}_{\text{ad}} \subset \mathbb{U}} J(y, u; y_{\text{d}}) \text{ subject to } \mathcal{E}(y, u; \boldsymbol{\mu}) = 0, \tag{5}$$

where $J : \mathbb{Y} \times \mathbb{U} \times \mathbb{Y}_{\text{obs}} \to \mathbb{R}$ is the *cost functional* defined by

$$J(y, u; y_{\text{d}}) := \frac{1}{2} \|y - y_{\text{d}}\|^2_{\mathbb{Y}_{\text{obs}}} + \frac{\alpha}{2} \|u\|^2_{\mathbb{U}}, \tag{6}$$

and $\alpha \in (0, 1]$ is a *penalization parameter*. It is clear that the smaller is the value of α, the more the control variable will influence the system. Problem (5) admits a solution if Hinze et al. (2008b, Section 1.5.2):

- **(i)** $\mathbb{U}_{\text{ad}}$ is convex, bounded and closed;
- **(ii)** $\mathbb{Y}_{\text{ad}}$ is convex and closed;
- **(iii)** for every $\boldsymbol{\mu} \in \mathcal{P}$, the controlled system $\mathcal{E}(y, u; \boldsymbol{\mu}) = 0$ has a bounded solution map $u \in \mathbb{U} \mapsto y(u) \in \mathbb{Y}$;
- **(iv)** for a given $\boldsymbol{\mu} \in \mathcal{P}$, the map $(y, u, \boldsymbol{\mu}) \in \mathbb{Y} \times \mathbb{U} \times \mathcal{P} \to \mathcal{E}(y, u; \boldsymbol{\mu}) \in \mathbb{Q}^*$ is weakly continuous with respect to (w.r.t.) the first two arguments;
- **(v)** for a given $y_{\text{d}} \in \mathbb{Y}_{\text{obs}}$, the cost functional $J(y, u; y_{\text{d}})$ is weakly lower semicontinuous w.r.t. the first two arguments.

To apply the Lagrangian theory, we define $z := z(\boldsymbol{\mu}) \in \mathbb{Y}^{**} = \mathbb{Y} \subset \mathbb{Q}$ be an arbitrary *adjoint variable*, which allows to translate the (5) in an unconstrained minimization problem. Indeed, calling with $X := X(\boldsymbol{\mu}) = (y(\boldsymbol{\mu}), u(\boldsymbol{\mu}), z(\boldsymbol{\mu})) \in \mathbb{X} := \mathbb{Y} \times \mathbb{U} \times \mathbb{Y}$ the global variable of the system at hand, we can build the *Lagrangian functional* $\mathscr{L} : \mathbb{X} \times \mathbb{Y}_{\text{obs}} \times \mathcal{P} \to \mathbb{R}$ as

$$\mathscr{L}(X; y_{\text{d}}, \boldsymbol{\mu}) := J(y, u; y_{\text{d}}) + \langle z, \mathcal{E}(y, u; \boldsymbol{\mu}) \rangle_{\mathbb{Q}\mathbb{Q}^*}, \tag{7}$$

where with $\langle \cdot, \cdot \rangle_{\mathbb{Q}\mathbb{Q}^*}$ we mean the duality pairing between $\mathbb{Q}$ and $\mathbb{Q}^*$. Let us assume that the following hold:

- **(vi)** $\mathbb{U}$ is nonempty;
- **(vii)** $J : \mathbb{Y} \times \mathbb{U} \times \mathbb{Y}_{\text{obs}} \to \mathbb{R}$ and $\mathcal{E} : \mathbb{Y} \times \mathbb{U} \times \mathcal{P} \to \mathbb{Q}^*$ are continuously Fréchet differentiable w.r.t. the first two arguments;
- **(viii)** given $\boldsymbol{\mu} \in \mathcal{P}$, the controlled system $\mathcal{E}(y, u; \boldsymbol{\mu}) = 0$ has a unique solution $y = y(u) \in \mathbb{Y}$ for all $u \in \mathbb{U}$;
- **(ix)** given $\boldsymbol{\mu} \in \mathcal{P}$, $D_y\mathcal{E}(y, u; \boldsymbol{\mu}) \in \mathcal{L}(\mathbb{Y}, \mathbb{Q}^*)$ has a bounded inverse for all control variables u.

From (ix), it is clear that $D_\star$ indicates the Fréchet derivative w.r.t. a variable $\star$ and we will exploit this notation from now on. Thanks to hypotheses (vi)–(ix), given a solution $(y, u) \in \mathbb{Y} \times \mathbb{U}$ of (5) for a given $\boldsymbol{\mu} \in \mathcal{P}$, there exists an adjoint variable $z \in \mathbb{Y}$ which satisfies the following *optimality system* (Hinze et al., 2008b):

$$\begin{cases} D_y\mathscr{L}(X; y_{\text{d}}, \boldsymbol{\mu})[\omega] = 0 & \forall \omega \in \mathbb{Q}, \\ D_u\mathscr{L}(X; y_{\text{d}}, \boldsymbol{\mu})[\kappa] = 0 & \forall \kappa \in \mathbb{U}, \\ D_z\mathscr{L}(X; y_{\text{d}}, \boldsymbol{\mu})[\zeta] = 0 & \forall \zeta \in \mathbb{Q}, \end{cases} \tag{8}$$

or equivalently, in strong form

$$\begin{cases} y + D_y\mathcal{E}(y, u; \boldsymbol{\mu})^*(z) = y_{\text{d}}, \\ \alpha u - C^*(z) = 0, \\ \mathcal{E}(y, u; \boldsymbol{\mu}) = 0, \end{cases} \tag{9}$$

where $D_y\mathcal{E}(y, u; \boldsymbol{\mu})^* \in \mathcal{L}(\mathbb{Q}, \mathbb{Y}^*)$ is the adjoint operator of the Fréchet linearization of $\mathcal{E}(y, u; \boldsymbol{\mu})$ w.r.t. the state variable, while $C^* \in \mathcal{L}(\mathbb{Y}, \mathbb{U}^*)$ is the adjoint of the control operator. Furthermore, the system (8) can be recast in compact form: given $\boldsymbol{\mu} \in \mathcal{P}$, find $X \in \mathbb{X}$ such that

$$\mathcal{G}(X; \boldsymbol{\mu}) = \mathcal{F}, \tag{10}$$

with

$$\mathcal{G}(X; \boldsymbol{\mu}) := \begin{bmatrix} y + D_y\mathcal{E}(y, u; \boldsymbol{\mu})^*(z) \\ \alpha u - C^*(z) \\ G(y, \boldsymbol{\mu}) - C(u) \end{bmatrix} \quad \text{and} \quad \mathcal{F} := \begin{bmatrix} y_{\text{d}} \\ 0 \\ f \end{bmatrix}.$$

We underline that the dual variable z is considered in $\mathbb{Y}$ in order to guarantee a proper definition of the optimality system due to the presence of a backward time evolution of the form $-z_t$ in the expression of $D_y\mathcal{E}(y, u; \boldsymbol{\mu})^*$.

In this context, we will always assume that the state equation is local invertible for every parametric instance, i.e. assumptions (viii) and (ix) always hold. We underline that in the nonlinear case it may happen to find multiple solutions for a given parameter. However, it will not be the case of this work, since we are restricting ourselves to a well-posed setting for the state equation. Indeed, we stress that in the nonlinear case it is only possible to recover existence results for the system (5). Uniqueness results are strongly related to the state equation one is dealing with and multiple solutions for the optimality system (10) can be found, see for example Pichi et al. (2020).

In the next section, we will show the space-time approximation of the optimality system at hand, generalizing the strategy already presented in Glas et al. (2017); Urban and Patera (2012); Yano (2014); Yano et al. (2014). We will focus on the case of semi-linear governing equations with quadratic dependence in the state variable, guided by the numerical results provided in Section 5.

3.2 The space-time approximation

The next step to be taken is the numerical approximation of the optimal solution X of (10) in order to investigate its features varying $\boldsymbol{\mu} \in \mathcal{P}$ in a space-time fashion. This approximation approach is a versatile tool already exploited in the discretization of parabolic equations and several OCP($\boldsymbol{\mu}$)s, see e.g. Glas et al. (2017); Hinze et al. (2008a,b, 2012); Langer et al. (2021); Strazzullo et al. (2020b, 2021a, 2020a); Yano (2014); Yano et al. (2014); Urban and Patera (2012). We first focus on the space discretization, which is performed through the Finite Element (FE) technique. Let us define the FE function space $Y^{N_{\text{FE}}^y} = Y \cap \mathbb{K}_{r_y}$ and $U^{N_{\text{FE}}^u} = U \cap \mathbb{K}_{r_u}$, where

$$\mathbb{K}_r = \{v \in C^0(\overline{\Omega}) \,:\, v|_K \in \mathbb{P}^r, \ \ \forall K \in \mathcal{T}\},$$

with $\mathbb{P}^r$ is the space of all the polynomials of degree at most equal to r and K is an element of a triangularization $\mathcal{T}$ of the spatial domain Ω. We can now consider the semi-discrete function spaces $\mathbb{Y}^{N_{\text{FE}}^y} = \left\{ y \in L^2(0, T; Y^{N_{\text{FE}}^y}) \text{ s.t. } y_t \in L^2(0, T; (Y^{N_{\text{FE}}^y})^*) \right\}$, $\mathbb{Q}^{N_{\text{FE}}^y} = L^2(0, T; Y^{N_{\text{FE}}^y})$ and $\mathbb{U}^{N_{\text{FE}}^u} = L^2(0, T; U^{N_{\text{FE}}})$. Once made the spatial discretization step, a time discretization over the interval $[0, T]$ must be taken into account resulting in the final space-time discrete spaces, which are denoted by $\mathbb{Y}_{N_t}^{N_{\text{FE}}^y}$, $\mathbb{Q}_{N_t}^{N_{\text{FE}}^y}$ and $\mathbb{U}_{N_t}^{N_{\text{FE}}^u}$, where N_t is the number of the considered timesteps. Following Langer et al. (2021), we decided to use the same space-time approximation for $\mathbb{Y}_{N_t}^{N_{\text{FE}}^y}$ and $\mathbb{Q}_{N_t}^{N_{\text{FE}}^y}$. For the sake of notation, we will refer to the space-time function spaces as $\mathbb{Y}^{\mathcal{N}_y} \equiv \mathbb{Q}^{\mathcal{N}_y}$ and $\mathbb{U}^{\mathcal{N}_u}$ and, as a consequence, $\mathbb{X}^{\mathcal{N}} := \mathbb{Q}^{\mathcal{N}_y} \times \mathbb{U}^{\mathcal{N}_u} \times \mathbb{Q}^{\mathcal{N}_y}$, i.e. $\mathcal{N} = 2\mathcal{N}_y + \mathcal{N}_u$, with $\mathcal{N}_y = N_{\text{FE}}^y \cdot N_t$

and $\mathcal{N}_u = N_{\text{FE}}^u \cdot N_t$. In this finite dimensional setting the problem to be solved reads: given $\boldsymbol{\mu} \in \mathcal{P}$ and observation $y_{\text{d}}^{\mathcal{N}} \in \mathbb{Q}^{\mathcal{N}_y}$ find $X^{\mathcal{N}} := X^{\mathcal{N}}(\boldsymbol{\mu}) \in \mathbb{X}^{\mathcal{N}}$ such that

$$\mathcal{G}(X^{\mathcal{N}}; \boldsymbol{\mu}) = \mathcal{F}. \tag{11}$$

In the following, we make clear the algebraic structure we exploited in the numerical experiment presented in Section 5.1. First of all, let us divide the time interval $[0, T]$ in N_t equispaced subintervals of length Δt and let us call $t_k = k\Delta t$ for $k = 0, \dots, N_t$ a generic time instance. The variables $y_k^{\mathcal{N}}$, $u_k^{\mathcal{N}}$ and $z_k^{\mathcal{N}}$ for a specific timestep, can be represented with FE basis $\{\phi^i\}_{i=1}^{N_{\text{FE}}^y}$ and $\{\psi^i\}_{i=1}^{N_{\text{FE}}^u}$ for $Y^{N_{\text{FE}}^y}$ and $U^{N_{\text{FE}}^u}$, respectively as follows

$$y_k^{\mathcal{N}} = \sum_{i=1}^{N_{\text{FE}}^y} y_k^i \phi^i, \qquad u_k^{\mathcal{N}} = \sum_{i=1}^{N_{\text{FE}}^u} u_k^i \psi^i, \qquad \text{and} \qquad z_k^{\mathcal{N}} = \sum_{i=1}^{N_{\text{FE}}^y} z_k^i \phi^i.$$

We now define the space-time state, control and adjoint vectors

$$\mathsf{y} = \begin{bmatrix} \bar{y}_1 \\ \vdots \\ \bar{y}_{N_t} \end{bmatrix}, \quad \mathsf{u} = \begin{bmatrix} \bar{u}_1 \\ \vdots \\ \bar{u}_{N_t} \end{bmatrix}, \quad \text{and} \quad \mathsf{z} = \begin{bmatrix} \bar{z}_1 \\ \vdots \\ \bar{z}_{N_t} \end{bmatrix}.$$

Namely, $\bar{y}_k$, $\bar{u}_k$ and $\bar{z}_k$ are the column vectors with FE coefficients of the variables at time instance t_k, with $k = 1, \dots, N_t$, i.e.

$$\bar{y}_k = \begin{bmatrix} y_k^1 \\ \vdots \\ y_k^{N_{FE}^y} \end{bmatrix}, \quad \bar{u}_k = \begin{bmatrix} u_k^1 \\ \vdots \\ u_k^{N_{FE}^u} \end{bmatrix}, \quad \text{and} \quad \bar{z}_k = \begin{bmatrix} z_k^1 \\ \vdots \\ z_k^{N_{FE}^y} \end{bmatrix}.$$

Applying the same strategy to the initial time condition, to the desired state and the forcing term, we can define

$$\mathsf{y}_0 = \begin{bmatrix} \bar{y}_0 \\ 0 \\ \vdots \\ 0 \end{bmatrix}, \quad \mathsf{y}_d = \begin{bmatrix} \bar{y}_{d1} \\ \bar{y}_{d2} \\ \vdots \\ \bar{y}_{dN_t} \end{bmatrix}, \quad \text{and} \quad \mathsf{f} = \begin{bmatrix} \bar{f}_1 \\ \bar{f}_2 \\ \vdots \\ \bar{f}_{N_t} \end{bmatrix},$$

respectively, where $\bar{y}_0$, $\bar{y}_{dk}$, and $\bar{f}_k$, in analogy with the aforementioned space-time variables, are the column vectors given by the FE coefficients in their respectively function spaces, with $k = 1, \dots, N_t$. We begin the discretization analysis from the state equation, for the sake of clarity. Concerning the space discretization level, once applied the controlled equation to the FE basis, we can derive the matrices $\mathsf{E}_{n\ell}(\mathsf{y};\boldsymbol{\mu}) + \mathsf{E}_{\ell}(\boldsymbol{\mu}) - \mathsf{C}$, omitting for the moment the time evolution. Furthermore, we need the mass matrices M_y and M_u for state/adjoint variables and control, respectively. At every time instance, after performing a backward Euler discretization in time, we have to solve the following system:

$$\mathsf{M}_y \bar{y}_k + \Delta t \underbrace{(\mathsf{E}_{n\ell}(\mathsf{y};\boldsymbol{\mu}) + \mathsf{E}_{\ell}(\boldsymbol{\mu}))}_{\mathsf{E}(\mathsf{y},\boldsymbol{\mu})} \bar{y}_k - \Delta t \mathsf{C} \bar{u}_k = \mathsf{M}_y \bar{y}_{k-1} + \bar{f}_k \Delta t \quad \text{for } 1 \le k \le N_t. \tag{12}$$

Thus, the whole space-time system reads

$$\underbrace{\begin{bmatrix} \mathsf{M}_y + \Delta t \mathsf{E}(\mathsf{y},\boldsymbol{\mu}) & & & & \\ -\mathsf{M}_y & \mathsf{M}_y + \Delta t \mathsf{E}(\mathsf{y},\boldsymbol{\mu}) & & & \\ & -\mathsf{M}_y & \mathsf{M}_y + \Delta t \mathsf{E}(\mathsf{y},\boldsymbol{\mu}) & & \\ & & \ddots & \ddots & \\ & & & -\mathsf{M}_y & \mathsf{M}_y + \Delta t \mathsf{E}(\mathsf{y},\boldsymbol{\mu}) \end{bmatrix}}_{\mathsf{K}(\mathsf{y};\boldsymbol{\mu})} \begin{bmatrix} \bar{y}_1 \\ \bar{y}_2 \\ \bar{y}_3 \\ \vdots \\ \bar{y}_{N_t} \end{bmatrix}$$

$$- \Delta t \begin{bmatrix} \mathsf{C} & & & \\ & \mathsf{C} & & \\ & & \mathsf{C} & \\ & & & \ddots & \\ & & & & \mathsf{C} \end{bmatrix} \begin{bmatrix} \bar{u}_1 \\ \bar{u}_2 \\ \bar{u}_3 \\ \vdots \\ \bar{u}_{N_t} \end{bmatrix} = \begin{bmatrix} \mathsf{M}_y \bar{y}_0 + \Delta t \bar{f}_1 \\ 0 + \Delta t \bar{f}_2 \\ 0 + \Delta t \bar{f}_3 \\ \vdots \\ 0 + \Delta t \bar{f}_{N_t} \end{bmatrix}. \tag{13}$$

We underline that the system presents a nonlinear dependence from the state in the term $\mathsf{E}_{n\ell}(\boldsymbol{\mu})$, related to the nonlinear contribution of the state operator $E_{n\ell}(y;\boldsymbol{\mu})$. The space-time state equation can be written in compact form as

$$\mathsf{K}(\mathsf{y};\boldsymbol{\mu})\mathsf{y} - \Delta t \mathsf{C}_{\mathrm{st}} \mathsf{u} = \mathsf{M}_{\mathsf{y}_{\mathrm{st}}} \mathsf{y}_0 + \Delta t \mathsf{f}, \tag{14}$$

where the subscript "st" indicates the all-at-once matrices, namely C_{st} is the block-diagonal matrix which entries are given by C of dimension $\mathbb{R}^{\mathcal{N}_y} \times \mathbb{R}^{\mathcal{N}_u}$ and $\mathsf{M}_{\mathsf{y}_{\mathrm{st}}}$ is the matrix made by M_y on the diagonal, of dimension $\mathbb{R}^{\mathcal{N}_y} \times \mathbb{R}^{\mathcal{N}_y}$.

We now take into account the optimality equation, which easily reads:

$$\alpha \Delta t \mathsf{M}_u \bar{u}_k - \Delta t \mathsf{C}^T \bar{z}_k = 0 \qquad \text{for } 1 \le k \le N_t, \tag{15}$$

or in compact form

$$\alpha \Delta t \mathsf{M}_{\mathsf{u}_{\mathrm{st}}} \mathsf{u} - \Delta t \mathsf{C}_{\mathrm{st}}^T \mathsf{z} = 0, \tag{16}$$

where $\mathsf{M}_{\mathsf{u}\,\mathrm{st}}$ is the block-diagonal matrix which entries are given by M_u of dimension $\mathbb{R}^{\mathcal{N}_u} \times \mathbb{R}^{\mathcal{N}_u}$. We can now take into account the adjoint equation. To this purpose, we have to explicit the algebraic structure of $D_y\mathcal{E}(y, u; \boldsymbol{\mu})$. The assumption of the quadratic nonlinearity in the state variable gives the following form to the Fréchet derivative of the controlled state equation with respect to the state y: $\mathsf{E}'_{n\ell}[\mathsf{y}](\boldsymbol{\mu}) + \mathsf{E}_{\ell}(\boldsymbol{\mu})$, where the linear state structure remains the same, while the nonlinear operator is linearized in $\mathsf{E}'_{n\ell}[\mathsf{y}]$. It is clear that the control operator C has to disappear since it does not depend on the state variable. Thus, performing a forward Euler method which is equivalent to an implicit scheme due the backward parabolic nature of the adjoint equation, at each time instance it reads:

$$\mathsf{M}_y \bar{z}_{k-1} = \mathsf{M}_y \bar{z}_k + \Delta t(-\mathsf{M}_{\mathrm{obs}} \bar{y}_{k-1} \underbrace{-\mathsf{E}'_{n\ell}[\mathsf{y}]^T(\boldsymbol{\mu}) - \mathsf{E}_{\ell}^T(\boldsymbol{\mu})}_{-\mathsf{E}^{\mathrm{adj}^T}(\boldsymbol{\mu})} \bar{p}_{k-1} + \mathsf{M}_{\mathrm{obs}} \bar{y}_{d_{k-1}})$$
$$\text{for } 1 < k \leq N_t, \tag{17}$$

where $\mathsf{M}_{\mathrm{obs}}$ is the state mass matrix restricted to the observation domain. We stress that $\mathsf{E}^{\mathrm{adj}}(\boldsymbol{\mu})$ hides the state variable dependence. Furthermore, also in this case, we can write the whole all-at-once system:

$$\underbrace{\begin{bmatrix} \mathsf{M}_y + \Delta t \mathsf{E}^{\mathrm{adj}^T}(\boldsymbol{\mu}) & -\mathsf{M}_y & & & \\ & \mathsf{M}_y + \Delta t \mathsf{E}^{\mathrm{adj}^T}(\boldsymbol{\mu}) & -\mathsf{M}_y & & \\ & & \ddots & \ddots & \\ & & & \mathsf{M}_y + \Delta t \mathsf{E}^{\mathrm{adj}^T}(\boldsymbol{\mu}) & -\mathsf{M}_y \\ & & & & \mathsf{M}_y + \Delta t \mathsf{E}^{\mathrm{adj}^T}(\boldsymbol{\mu}) \end{bmatrix}}_{\mathsf{K}'(\boldsymbol{\mu})} \begin{bmatrix} \bar{z}_1 \\ \bar{z}_2 \\ \bar{z}_3 \\ \vdots \\ \bar{z}_{N_t} \end{bmatrix}$$
$$+ \Delta t \begin{bmatrix} \mathsf{M}_{\mathrm{obs}} & & & & \\ & \mathsf{M}_{\mathrm{obs}} & & & \\ & & \mathsf{M}_{\mathrm{obs}} & & \\ & & & \ddots & \\ & & & & \mathsf{M}_{\mathrm{obs}} \end{bmatrix} \begin{bmatrix} \bar{y}_1 \\ \bar{y}_2 \\ \bar{y}_3 \\ \vdots \\ \bar{y}_{N_t} \end{bmatrix} = \begin{bmatrix} \Delta t \mathsf{M}_{\mathrm{obs}} \bar{y}_{d_1} \\ \Delta t \mathsf{M}_{\mathrm{obs}} \bar{y}_{d_2} \\ \Delta t \mathsf{M}_{\mathrm{obs}} \bar{y}_{d_3} \\ \vdots \\ \Delta t \mathsf{M}_{\mathrm{obs}} \bar{y}_{d_{N_t}} \end{bmatrix}.$$

Then, in a more compact notation, the adjoint equation reads:

$$\mathsf{K}'(\boldsymbol{\mu})\mathsf{z} + \Delta t \mathsf{M}_{\mathrm{obs\,st}} \mathsf{y} = \Delta t \mathsf{M}_{\mathrm{obs\,st}} \mathsf{y}_{\mathsf{d}}, \tag{18}$$

where $\mathsf{M}_{\mathrm{obs\,st}} \in \mathbb{R}^{\mathcal{N}_y} \times \mathbb{R}^{\mathcal{N}_y}$ is the block-diagonal matrix which entries are given by $\mathsf{M}_{\mathrm{obs}}$. Finally, we are able to collect all the information and build the global all-at-once system:

$$\overbrace{\begin{bmatrix} \Delta t \mathsf{M}_{\mathsf{obs\,st}} & 0 & \mathsf{K}'(\boldsymbol{\mu}) \\ 0 & \alpha \Delta t \mathsf{M}_{\mathsf{u\,st}} & -\Delta t \mathsf{C}_{\mathsf{st}}^T \\ \mathsf{K}(\mathsf{y};\boldsymbol{\mu}) & -\Delta t \mathsf{C}_{\mathsf{st}} & 0 \end{bmatrix}}^{\mathsf{G}(\mathsf{X};\boldsymbol{\mu})} \underbrace{\begin{bmatrix} \mathsf{y} \\ \mathsf{u} \\ \mathsf{z} \end{bmatrix}}_{\mathsf{X}} = \overbrace{\begin{bmatrix} \Delta t \mathsf{M}_{\mathsf{obs\,st}} \mathsf{y}_d \\ 0 \\ \mathsf{M}_{\mathsf{y}_{\mathsf{st}}} \mathsf{y}_0 + \Delta t \mathsf{f} \end{bmatrix}}^{\mathsf{F}}. \tag{19}$$

We remark that the nonlinearity of $\mathsf{G}(\mathsf{X};\boldsymbol{\mu})$ derives from $\mathsf{K}(\mathsf{y};\boldsymbol{\mu})$ and $\mathsf{K}'(\boldsymbol{\mu})$, due to the nonlinear terms $\mathsf{E}_{n\ell}(\mathsf{y};\boldsymbol{\mu})$ and $\mathsf{E}'_{n\ell}[\mathsf{y}]^T(\boldsymbol{\mu})$. However, for the sake of notation, we omitted the direct X-dependence from the matrices.

The nonlinear system, then, can be recast in residual formulation as

$$\mathsf{R}(\mathsf{X};\boldsymbol{\mu}) := \mathsf{G}(\mathsf{X};\boldsymbol{\mu}) - \mathsf{F} = 0, \tag{20}$$

where $\mathsf{R}(\mathsf{X};\boldsymbol{\mu})$ will be called the *global residual* of the optimality system. To solve system (20), we employed Netwon's method: namely we iteratively solve

$$\mathsf{X}^{j+1} = \mathsf{X}^j + \mathsf{Jac}(\mathsf{X}^j;\boldsymbol{\mu})^{-1}(\mathsf{F} - \mathsf{G}(\mathsf{X}^j;\boldsymbol{\mu})), \qquad j \in \mathbb{N}, \tag{21}$$

until a residual based convergence criterion is satisfied. We recall that the matrix $\mathsf{K}'(\boldsymbol{\mu})$ still depends on the state vector y in the term $\mathsf{E}'_{n\ell}[\mathsf{y}]^T$. Then, the linearization w.r.t. y of $\mathsf{K}'(\boldsymbol{\mu})$ leads to a new term in the formulation:

$$\mathsf{D}_{\mathsf{y}}(\mathsf{E}'_{n\ell}[\mathsf{y}^j]^T)[\mathsf{z}^j]. \tag{22}$$

Namely, the Jacobian matrix will have the following form:

$$\mathsf{Jac}(\mathsf{X}^j;\boldsymbol{\mu}) = \begin{bmatrix} \Delta t \mathsf{M}_{\mathsf{obs\,st}} + \mathsf{D}_{\mathsf{y}}(\mathsf{E}'_{n\ell}[\mathsf{y}^j]^T)[\mathsf{z}^j] & 0 & \mathsf{K}^{\mathsf{l}}(\mathsf{y}_j;\boldsymbol{\mu})^T \\ 0 & \alpha \Delta t \mathsf{M}_{u\,\mathsf{st}} & -\Delta t \mathsf{C}_{\mathsf{st}}^T \\ \mathsf{K}^{\mathsf{l}}(\mathsf{y}_j;\boldsymbol{\mu}) & -\Delta t \mathsf{C}_{\mathsf{st}} & 0 \end{bmatrix}, \tag{23}$$

where each matrix taken into consideration is now linear in the j-th value of one of the involved variables and $\mathsf{K}^{\mathsf{l}}(\mathsf{y}_j;\boldsymbol{\mu})$ is the linearized version of $\mathsf{K}(\mathsf{y},\boldsymbol{\mu})$. Thanks to this remark we are able to show the *saddle point* structure of the system at hand. Indeed, Eq. (23) can be written as

$$\mathsf{Jac}(\mathsf{X}^j;\boldsymbol{\mu}) = \begin{bmatrix} \mathsf{A} & \mathsf{B}^T \\ \mathsf{B} & 0 \end{bmatrix}, \tag{24}$$

where

$$\mathsf{A} = \begin{bmatrix} \Delta t \mathsf{M}_{\mathsf{obs\,st}} + \mathsf{D}_{\mathsf{y}}(\mathsf{E}'_{n\ell}[\mathsf{y}^j]^T)[\mathsf{z}^j] & 0 \\ 0 & \alpha \Delta t \mathsf{M}_{u\,\mathsf{st}} \end{bmatrix} \text{ and}$$

$$\mathsf{B} = \begin{bmatrix} \mathsf{K}^1(\mathsf{y}_j; \boldsymbol{\mu}) & -\Delta t \mathsf{C}_{\mathrm{st}} \end{bmatrix}. \tag{25}$$

The proposed framework is very common in many CFD applications, from Stokes equations to PDEs constrained optimization. In the saddle point setting, following the Brezzi theory (Brezzi, 1974), to guarantee the existence of a unique solution of (24), the matrix A should be invertible and the following *Brezzi inf-sup condition* should be verified (Benzi et al., 2005; Brezzi, 1974):

$$\beta^{\mathcal{N}}(\boldsymbol{\mu}) := \inf_{0 \neq \mathsf{z}} \sup_{0 \neq \mathsf{x}} \frac{\mathsf{z}^T \mathsf{B} \mathsf{x}}{\|\mathsf{x}\|_{\mathbb{Y} \times \mathbb{U}} \|\mathsf{z}\|_{\mathbb{Y}}} \geq \hat{\beta}(\boldsymbol{\mu})^{\mathcal{N}} > 0, \tag{26}$$

where $\mathsf{x} = [\mathsf{y}, \mathsf{u}]^T$. For linear state equations hypotheses (i)–(ix) assure the well-posedness of the saddle point structure (Hinze et al., 2008b; Negri et al., 2015, 2013; Strazzullo et al., 2020b). This is not the case for nonlinear settings, where the fulfillment of (i)–(ix) does not ensure uniqueness and the Brezzi theorem should be verified from case to case. In the FE context, the inequality (26) holds when the function spaces for state and adjoint coincide (Negri et al., 2015, 2013). The assumption $z \in \mathbb{Y}$, will guarantee the fulfillment of the inf-sup stability condition in the space-time approximation once provided at the continuous level. It is clear that in order to solve the all-at-once optimization problem in a parametrized setting, for a given $\boldsymbol{\mu} \in \mathcal{P}$, we have to deal with high-dimensional systems. The solution of many-query and/or real-time tasks require growing computational resources and computational time for simulations. In this context, the space-time approximation has some limitations most of all when one relies, as in our case, on the direct solution of the optimality system. To lighten this issue, besides the employment of more computational resources, proper preconditioners and multigrid approaches can be used, see e.g. Benzi et al. (2005); Schöberl and Zulehner (2007); Stoll and Wathen (2010, 2013) and the references therein. In the next section we will introduce ROMs for space-time nonlinear OCP($\boldsymbol{\mu}$)s, providing a general approximation strategy which can solve the issue of the huge amount of computational costs that are usually associated to a standard space-time solution process.

4 ROMs for nonlinear space-time OCP($\boldsymbol{\mu}$)s

This section introduces ROMs for nonlinear space-time OCP($\boldsymbol{\mu}$)s. We refer to Strazzullo et al. (2018, 2021b); Zainib et al. (2020) for previous contributions to ROM for nonlinear OCP($\boldsymbol{\mu}$)s and to Strazzullo et al. (2020a) for their extension to time dependent nonlinear governing equations. First, we introduce the ROMs ideas and we will briefly focus on the standard approaches to make the strategy efficient in terms of computational resources. Then, we will describe the POD-Galerkin basis construction algorithm, see Ballarin et al. (2015); Benner et al. (2017); Burkardt et al. (2006); Chapelle et al. (2013); Hesthaven et al. (2015) as general references. We will exploit the classical aggregated spaces technique, following the linear OCP($\boldsymbol{\mu}$)s fashion, as already presented in Bader et al. (2016, 2015); Dedè (2010); Gerner and Veroy (2012); Kärcher and Grepl (2014); Kärcher et al. (2018); Negri et al. (2015, 2013); Quarteroni et al. (2007). We underline that the proposed strategy is strictly related to the linear quadratic case, i.e. OCP($\boldsymbol{\mu}$)s governed by linear PDE($\boldsymbol{\mu}$)s with a quadratic cost functional. However, it represents a classic choice to deal with more general nonlinear frameworks, see e.g. Quarteroni et al. (2015).

4.1 General reduction strategy

In Section 2, we introduced the importance of parametric optimal control in several fields of applications. Parameters can represent physical features and/or geometrical ones and, in many-query and real-time contexts, there exists the need to study several parametric instances to better understand the properties of a system. For this task, the space-time formulation can be unbearable and limit the knowledge capability of OCP($\boldsymbol{\mu}$)s due to the huge amount of computational resources needed for their simulations. The ROM aim at building a *low-dimensional* surrogate function space, in order to decrease the needed time for a simulation, guaranteeing a better parametric analysis in a small amount of time. We now provide the main ROM ideas for OCP($\boldsymbol{\mu}$)s. Let us consider the global variable $X(\boldsymbol{\mu}) = (y(\boldsymbol{\mu}), u(\boldsymbol{\mu}), z(\boldsymbol{\mu}))$, parametric solution of (10). In this section we make the parameter dependence explicit, for the sake of clarity. Indeed, it will be useful to understand the basics and main features of this discretization approach. The first phase of the ROM relies in the construction of basis functions to represent the *high fidelity solution* $X^{\mathcal{N}}(\boldsymbol{\mu})$. This goal is reached through the employment of *snapshots*, i.e. properly chosen solutions of (11). The low-dimensional function space is a subset of $\mathbb{X}^{\mathcal{N}}$ and in order to solve the optimality system for a new value of $\boldsymbol{\mu} \in \mathcal{P}$ it is sufficient to perform a standard Galerkin projection over the reduced space. In the following, we assume to have already built the reduced space[1] for the global variable $X(\boldsymbol{\mu})$, say $\mathbb{X}_N \subset \mathbb{X}^{\mathcal{N}}$. We stress that the *reduced dimension* N verifies $N \ll \mathcal{N}$. The building procedure is part of the *offline phase* where not only the basis functions are computed, but also the $\boldsymbol{\mu}$-independent quantities are assembled and stored.

Once the offline phase is concluded, the *reduced optimality system* reads: given $\boldsymbol{\mu} \in \mathcal{P}$, find $X_N(\boldsymbol{\mu}) := (y_N(\boldsymbol{\mu}), u_N(\boldsymbol{\mu}), z_N(\boldsymbol{\mu})) \in \mathbb{X}_N := \mathbb{Y}_N \times \mathbb{U}_N \times \mathbb{Y}_N$ such that it holds:

$$\begin{cases} D_y\mathscr{L}(X_N; y_{\mathrm{d}}, \boldsymbol{\mu})[\omega] = 0 & \forall \omega \in \mathbb{Y}_N, \\ D_u\mathscr{L}(X_N; y_{\mathrm{d}}, \boldsymbol{\mu})[\kappa] = 0 & \forall \kappa \in \mathbb{U}_N, \\ D_z\mathscr{L}(X_N; y_{\mathrm{d}}, \boldsymbol{\mu})[\zeta] = 0 & \forall \zeta \in \mathbb{Y}_N. \end{cases} \tag{27}$$

The solution is given in an *online phase*, where for every new evaluation of $\boldsymbol{\mu} \in \mathcal{P}$, the optimality system (27) is assembled and solved. From the latest

1 The description of the algorithm used to build the spaces is postponed in Section 4.2.

arguments, it is natural to deduce that one of the main ingredients of this procedure is the efficient division between a possibly expensive offline phase, which is performed only once, and a fast projection phase to solve the system for several parameters. This may be possible if we assume an affine decomposition for (27), i.e. the involved equations have the following form:

$$\begin{aligned} D_y\mathscr{L}(X_N, y_d, \boldsymbol{\mu})[\omega] &= \sum_{q=1}^{Q_y} \Theta_y^q(\boldsymbol{\mu}) D_y\mathscr{L}^q(X_N; y_d)[\omega], \\ D_u\mathscr{L}(X_N; y_d, \boldsymbol{\mu})[\kappa] &= \sum_{q=1}^{Q_u} \Theta_u^q(\boldsymbol{\mu}) D_u\mathscr{L}^q(X_N; y_d)[\kappa], \\ D_z\mathscr{L}(X_N; y_d, \boldsymbol{\mu})[\zeta] &= \sum_{q=1}^{Q_z} \Theta_z^q(\boldsymbol{\mu}) D_z\mathscr{L}^q(X_N; y_d)[\zeta]. \end{aligned} \tag{28}$$

In other words, the system can be recast as the product of $\boldsymbol{\mu}$-dependent smooth functions $\Theta_y^q(\boldsymbol{\mu})$, $\Theta_u^q(\boldsymbol{\mu})$, $\Theta_z^q(\boldsymbol{\mu})$ and $\boldsymbol{\mu}$-independent forms $D_y\mathscr{L}^q(X_N, y_d)[\omega]$, $D_u\mathscr{L}^q(X_N; y_d)[\kappa]$, and $D_z\mathscr{L}^q(X_N; y_d)[\zeta]$.

When this is the case, the online phase does not depend on $\mathcal{N}$ and usually guarantees the solution of the system in a small amount of time.

Remark 4.1. For nonlinear systems, even if structure (28) is fulfilled, the involved nonlinear forms still depend on $X^{\mathcal{N}}(\boldsymbol{\mu})$ and this affects the computational advantage in using a POD-Galerkin approach, since it involves the assembly (and projection) of the high fidelity solution during the online stage. To overcome this issue, hyper-reduction techniques based on the Empirical Interpolation Method (EIM) may be employed, see e.g. Barrault et al. (2004) or Hesthaven et al. (2015, Chapter 5).

In the next section, we will describe the space-time POD-Galerkin strategy to build the reduced space $\mathbb{X}_N$.

4.2 Offline and online phase: from space-time POD algorithm for OCP(μ)s to Galerkin projection

The ROM building process has mainly been addressed through two techniques: the POD (Ballarin et al., 2015; Burkardt et al., 2006; Chapelle et al., 2013; Hesthaven et al., 2015) and the greedy algorithm (Gerner and Veroy, 2012; Hesthaven et al., 2015; Negri et al., 2015, 2013; Rozza and Veroy, 2007). In this work, we will focus on the first one since it can be applied to any state equation: indeed, greedy constructions are based on the employment of an error estimator, which is still not available for nonlinear time dependent OCP($\boldsymbol{\mu}$)s.

The POD-Galerkin algorithm samples $N_{\max}$ parameters in $\mathcal{P}$ and computes the related snapshots. After this *exploratory phase*, a compressing stage starts, where $N < N_{\max}$ basis functions are provided after snapshots manipulation, aiming to get rid of the redundant information in the parametrized system.

Let us define the subset $\mathcal{P}_{N_{\max}} \subset \mathcal{P}$ given by the sampled parameters, which has cardinality $N_{\max}$. The chosen snapshots will form the following *sampled* manifold

$$\mathcal{M}^{\mathcal{N}}_{N_{\max}} = \{X^{\mathcal{N}}(\boldsymbol{\mu}) \mid \boldsymbol{\mu} \in \mathcal{P}_{N_{\max}}\} \subset \mathcal{M}^{\mathcal{N}},$$

where we assume $N_{\max}$ large enough to let $\mathcal{M}^{\mathcal{N}}_{N_{\max}}$ be a reliable representation $\mathcal{M}^{\mathcal{N}}$. The POD algorithm is applied separately for each involved variable, in a *partitioned space-time approach*, where the procedure provides spaces of dimension N which minimizes the following quantities:

$$\sqrt{\frac{1}{N_{\max}} \sum_{\boldsymbol{\mu} \in \mathcal{P}_{N_{\max}}} \min_{\omega_N \in \mathbb{Y}_N} \left\| y^{\mathcal{N}_y}(\boldsymbol{\mu}) - \omega_N \right\|^2_{\mathbb{Y}}},$$

$$\sqrt{\frac{1}{N_{\max}} \sum_{\boldsymbol{\mu} \in \mathcal{P}_{N_{\max}}} \min_{\kappa_N \in \mathbb{U}_N} \left\| u^{\mathcal{N}_u}(\boldsymbol{\mu}) - \kappa_N \right\|^2_{\mathbb{U}}},$$

$$\sqrt{\frac{1}{N_{\max}} \sum_{\boldsymbol{\mu} \in \mathcal{P}_{N_{\max}}} \min_{\zeta_N \in \mathbb{Y}_N} \left\| z^{\mathcal{N}_y}(\boldsymbol{\mu}) - \zeta_N \right\|^2_{\mathbb{Y}}}.$$

In the following we describe the snapshots manipulation to build the reduce space only for one variable, say the state $y(\boldsymbol{\mu})$. The proposed arguments can be identically replied for the other variables as well. First of all, we consider the ordered set of parameters $\boldsymbol{\mu}_1, \ldots, \boldsymbol{\mu}_{N_{max}} \in \mathcal{P}_{N_{\max}}$ to which correspond a set of order snapshots $y^{\mathcal{N}_y}(\boldsymbol{\mu}_1), \ldots, y^{\mathcal{N}_y}(\boldsymbol{\mu}_{N_{max}})$. The correlation matrix $\mathbf{C}^y \in \mathbb{R}^{N_{max} \times N_{max}}$ of snapshots of the state variable, i.e.:

$$\mathbf{C}^y_{ml} = \frac{1}{N_{max}} (y^{\mathcal{N}_y}(\boldsymbol{\mu}_m), y^{\mathcal{N}_y}(\boldsymbol{\mu}_l))_{\mathbb{Y}}, \qquad 1 \le m, l \le N_{max}.$$

First, we solved the following eigenvalue problem:

$$\mathbf{C}^y x^y_n = \lambda^y_n x^y_n, \qquad 1 \le n \le N,$$

with $\|x^y_n\|_{\mathbb{Y}} = 1$. Let us assume to have sorted the eigenvalues $\lambda^y_1, \ldots, \lambda^y_{N_{\max}}$ in decreasing order and to consider only the first N ones, namely $\lambda^y_1, \ldots, \lambda^y_N$, and the corresponding eigenvectors $x^y_1, \ldots, x^y_N$. Let $(x^y_n)_m$ be *m-th* component of the state eigenvector $x^y_n \in \mathbb{R}^{N_{max}}$. Thus, the POD basis functions are given by the following relation:

$$\chi^y_n = \frac{1}{\sqrt{\lambda^y_n}} \sum_{m=1}^{N_{max}} (x^y_n)_m y^{\mathcal{N}_y}(\boldsymbol{\mu}_m), \qquad 1 \le n \le N. \tag{29}$$

The relation (29) is standard in data-compression algorithms such as POD, see e.g. Hesthaven et al. (2015); Quarteroni et al. (2015). It represents how the bases can be written in terms of the POD eigenvalues-eigenvectors pairs. We remark that the time instances are not separated in the POD procedure: i.e. the snapshots preserve the space-time structure. The values N_{max} and N can be guided by the analysis of the POD eigenvalues, since the following holds (Hesthaven et al., 2015; Quarteroni et al., 2015):

$$\sqrt{\frac{1}{N_{\max}} \sum_{m=1}^{N_{\max}} \| y^{\mathcal{N}_y}(\boldsymbol{\mu}_m) - P_N(y^{\mathcal{N}_y}(\boldsymbol{\mu}_m)) \|_{\mathbb{Y}}^2} = \sqrt{\sum_{m=N+1}^{N_{\max}} \lambda_m^y}, \tag{30}$$

where $P_N : \mathbb{Y} \to \mathbb{Y}_N$ is the projection for functions in $\mathbb{Y}$ onto the reduced space $\mathbb{Y}_N$. Another aspect to take care of is the sampling of the N_{max} parameters for the POD, which can be related to some previous knowledge one has of the system at hand.

As already specified in Section 3.2, the linearized OCP($\boldsymbol{\mu}$) leads to the solution of a saddle point system at each iteration of the Newton's method. To prove the existence and uniqueness of the solution, the matrix B of system (24) must verify the inf-sup stability condition (26) for every $\boldsymbol{\mu} \in \mathcal{P}$. The relation, at the space-time level, holds thanks to the same discretization technique used for state and adjoint variable. However, this assumption does not guarantee the fulfillment of the inf-sup stability at the reduced level. Indeed, the basis has to be manipulated in order to achieve this goal since the standard space-time POD process may lead to different reduced spaces for state and adjoint, even if the high fidelity discretization is the same for both the variables. To overcome this issue, we exploit *aggregated spaces* technique. This strategy is very common and well known in ROM literature for OCP($\boldsymbol{\mu}$)s, see Bader et al. (2016, 2015); Dedè (2010); Gerner and Veroy (2012); Kärcher and Grepl (2014); Kärcher et al. (2018); Negri et al. (2015, 2013); Quarteroni et al. (2007) as references. The main idea is to build a common function space which represents both state and adjoint variables

$$\mathbb{Y}_N = \text{span}\,\{\chi_n^y, \chi_n^z, \; n = 1, \ldots, N\}, \tag{31}$$

while for the control variable we rely on the standard space

$$\mathbb{U}_N = \text{span}\{\chi_n^u, \; n = 1, \ldots, N\}. \tag{32}$$

We can now define the *basis* matrix

$$\mathsf{Z} = \begin{bmatrix} \mathsf{Z}_\mathsf{x} \\ \mathsf{Z}_\mathsf{z} \end{bmatrix}, \qquad \text{and} \qquad \mathsf{Z}_\mathsf{x} = \begin{bmatrix} \mathsf{Z}_\mathsf{y} \\ \mathsf{Z}_\mathsf{u} \end{bmatrix}$$

where $\mathsf{Z}_\mathsf{y} \equiv \mathsf{Z}_\mathsf{z} = [\chi_1^y | \cdots | \chi_N^y | \chi_1^z | \cdots | \chi_N^z] \in \mathbb{R}^{\mathcal{N}_y \times 2N}$ and $\mathsf{Z}_\mathsf{u} = [\chi_1^u | \cdots | \chi_N^u] \in \mathbb{R}^{\mathcal{N}_u \times N}$. The Z spans the reduced space $\mathbb{X}_N$. In this framework we can solve the optimality system in a small amount of time for every parametric instance through a Galerkin projection into the reduced spaces. Thus, the final system reads:

$$\mathsf{G}_N(\mathsf{X}_N;\boldsymbol{\mu})\mathsf{X}_N = \mathsf{F}_N, \tag{33}$$

with

$$\mathsf{G}_N(\mathsf{X}_N;\boldsymbol{\mu}) := \mathsf{Z}^T\mathsf{G}(\mathsf{Z}\mathsf{X}_N;\boldsymbol{\mu}), \qquad \text{and} \qquad \mathsf{F}_N := \mathsf{Z}^T\mathsf{F}.$$

The system (33) inherits the nonlinearity feature from the high fidelity one and the Newton's method can be employed also in this case

$$\mathsf{X}_N^{j+1} := \mathsf{X}_N^j + \mathsf{Jac}_N(\mathsf{X}_N^j;\boldsymbol{\mu})^{-1}(\mathsf{F}_N - \mathsf{G}(\mathsf{X}_N^j;\boldsymbol{\mu})\mathsf{X}_N^j), \qquad j \in \mathbb{N}, \tag{34}$$

presenting the saddle point structure for the Fréchet derivative, i.e.

$$\mathsf{Jac}_N(\mathsf{X}_N;\boldsymbol{\mu})\mathsf{X}_N = \begin{bmatrix} \mathsf{A}_N & \mathsf{B}_N^T \\ \mathsf{B}_N & 0 \end{bmatrix} \begin{bmatrix} \mathsf{x}_N \\ \mathsf{z}_N \end{bmatrix}, \tag{35}$$

with $\mathsf{Jac}_N(\mathsf{X}_N;\boldsymbol{\mu}) = \mathsf{Z}^T\mathsf{Jac}(\mathsf{Z}\mathsf{X}_N;\boldsymbol{\mu})\mathsf{Z}$, $\mathsf{A}_N = \mathsf{Z}_\mathsf{x}^T\mathsf{A}\mathsf{Z}_\mathsf{x}$ and $\mathsf{B}_N = \mathsf{Z}_\mathsf{z}^T\mathsf{B}\mathsf{Z}_\mathsf{x}$.

In the reduced framework, the *reduced inf-sup condition* reads

$$\beta_N(\boldsymbol{\mu}) := \inf_{0\neq \mathsf{z}_N} \sup_{0\neq \mathsf{x}_N} \frac{\mathsf{z}_N^T\mathsf{B}_N\mathsf{x}_N}{\|\mathsf{x}_N\|_{\mathbb{Y}\times\mathbb{U}}\|\mathsf{z}_N\|_{\mathbb{Y}}} \geq \hat{\beta}_N(\boldsymbol{\mu}) > 0, \tag{36}$$

which is verified due to the aggregated spaces definition. It is known that using the state and adjoint function spaces built on their respective snapshots without any kind of manipulation might be not sufficient to guarantee the well-posedness of a linear saddle point system (35). Indeed, by means of standard POD function spaces, the fulfillment of (36) is not assured. To avoid this issue, we follow the strategy already employed in Negri et al. (2013) and previously in Dedè (2010): we use the same space for state and adjoint, made by the union of the two different bases obtained by the POD applied to state and adjoint, respectively, as defined in (31). We remark that, even if the reduced system increases its final dimension through this approach, it is usually much smaller than $\mathcal{N}$. Moreover, it is clear that to have a good reduced approximation, the high fidelity solution must be a good representation of the continuous framework. However, as already specified in Section 3, it has to be verified case by case. This kind of analysis goes beyond the goal of this contribution and we will always assume that the space-time approximation is a valid discrete representation of the continuous problem. In the next section we will show with some numerical examples on how convenient ROMs can be in the framework of space-time nonlinear OCP($\boldsymbol{\mu}$)s.

5 Application to shallow waters equations

This section shows how the previously proposed methodologies can be applied to the viscous SWEs model. We stress that the described strategies are general and can be applied to several state equations. Furthermore, they easily adapt to simpler settings like steady and/or linear problems, which are still useful in many engineering and scientific fields, as already specified in Section 2. However, in the following, we will focus on environmental sciences and, more specifically, in coastal management: a growing impact field of research that needs the support of accurate real-time simulations. To this end, first, we will introduce the viscous SWEs state equations and some motivations for the primary importance of such a model in coastal fluid dynamics. Then, we will briefly describe the optimality system formulation at the continuous level before proposing some numerical results to test our methodology in a physical and geometrical parametrized setting.

5.1 Main motivations and problem formulation

This section is motivated by the growing demand of fast and reliable simulations in the framework of coastal management. Indeed, the marine environment is related to social and economic growth, biodiversity and ecosystem preservation, monitoring plans for possibly dangerous events related to weather factors or anthropic behavior.

The viscous SWEs result in a very versatile model in coastal sciences which can represent marine impact on shores and coasts, planetary currents, tsunamis waves... (Cavallini and Crisciani, 2013; Vreugdenhil, 2013). The state equation and the control problem have been studied analytically and numerically in several works, see for example Agoshkov et al. (1994a, 1993, 1994b, 2007); Ferrari and Saleri (2004); Miglio et al. (1999, 2005b,a); Ricchiuto et al. (2007); Ricchiuto and Bollermann (2009); Takase et al. (2010). In a parametrized setting, they have been explored as state equation in Ştefănescu and Navon (2013); Ştefănescu et al. (2014) and in Strazzullo et al. (2020a) in an optimal control framework.

First of all we define our parameter, $\boldsymbol{\mu} = (\mu_1, \mu_2, \mu_3, \mu_4) \in \mathcal{P} \subset \mathbb{R}^4$. The first three parameters are related to the physics of the problem, while the latter describes the geometry of the considered spatial domain: the specific description of the parametrized setting is postponed later in the section. For this peculiar state equation, we define the function spaces $Y_v = H^1_{\Gamma_{D_v}(\mu_4)}(\Omega(\mu_4))$, $Y_h = L^2(\Omega(\mu_4))$ and the space $U = L^2(\Omega(\mu_4))$, where $\Gamma_{D_v}(\mu_4)$ is a portion of the boundary domain $\partial\Omega(\mu_4)$ where Dirichlet boundary conditions have been imposed. The involved variables are the vertically averaged velocity profile of the wave $\boldsymbol{v}$ and the surface elevation variable h, respectively considered in Y_v and Y_h. We will focus on the simpler setting, where the bottom bathymetry z_b is defined as a constant function: it can be generalized to more realistic bathymetry, see e.g. Ferrari and Saleri (2004); Ricchiuto and Bollermann (2009); Ricchiuto et al. (2007). In Fig. 1 we provide a description of the physical phenomenon we are dealing with, together with the notations we will exploit. We used the standard 2D-model presented in Agoshkov et al. (1993); Miglio et al. (1999).

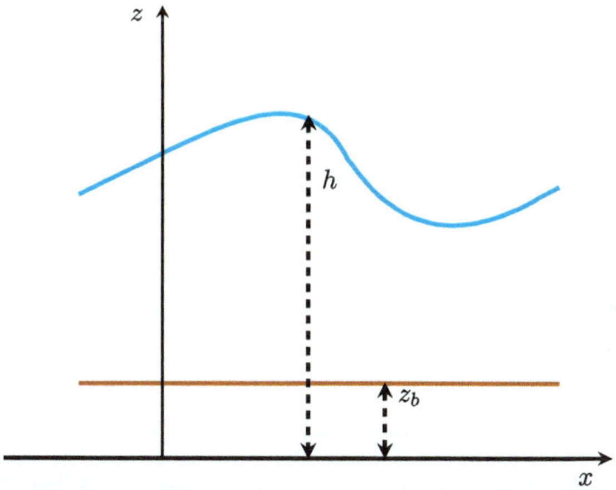

FIGURE 1 Notations: schematic representation.

The state solution velocity and height are defined, respectively, in

$$\mathbb{Y}_v = \left\{ \boldsymbol{v} \in L^2(0,T;[Y_v]^2) \text{ such that } \boldsymbol{v}_t \in L^2(0,T;[Y_v^*]^2) \right\},$$

and

$$\mathbb{Y}_h = \left\{ h \in L^2(0,T;Y_h) \text{ such that } h_t \in L^2(0,T;Y_h^*) \right\}.$$

Thus, we can define the global state function space given by $\mathbb{Y} = \mathbb{Y}_v \times \mathbb{Y}_h$, where the state variable $(\boldsymbol{v}, h)$ will be sought. Furthermore, let $\boldsymbol{u} \in \mathbb{U} := L^2(0,T;[U]^2)$ be the control variable of the system. For the numerical results, we will deal with a distributed optimal control, where the control variable acts as an external forcing term such as atmospheric pressure, bottom friction, and wind action. Namely, you are not actually controlling the system but one can interpret the optimal control framework as an inverse problem capable to guess what are the physical conditions that can represent a desired velocity-height profile $(\boldsymbol{v}_d, h_d) \in \mathbb{Y}_{\text{obs}} := \mathbb{Q}_v \times \mathbb{Q}_h$, where $\mathbb{Q}_v := L^2(0,T;[L^2(\Omega(\mu_4))]^2)$ and $\mathbb{Q}_h := L^2(0,T;L^2(\Omega(\mu_4)))$. The problem at hand reads: given $\boldsymbol{\mu} \in \mathcal{P}$ find $(\boldsymbol{v}, h) \in \mathbb{Y}$ which minimizes $J((\boldsymbol{v}, h), \boldsymbol{u}, (\boldsymbol{v}_{\text{d}}, h_{\text{d}}))$, where

$$\begin{aligned} J((\boldsymbol{v}, h), \boldsymbol{u}; (\boldsymbol{v}_{\text{d}}, h_{\text{d}})) = & \frac{1}{2}\int_0^T \int_{\Omega(\mu_4)} (h - h_{\text{d}}(\mu_3))^2 d\Omega(\mu_4) dt \\ & + \frac{1}{2}\int_0^T \int_{\Omega(\mu_4)} (\boldsymbol{v} - \boldsymbol{v}_{\text{d}}(\mu_3))^2 d\Omega(\mu_4) dt \end{aligned} \tag{37}$$

$$+\frac{\alpha}{2}\int_0^T\int_{\Omega(\mu_4)} \boldsymbol{u}^2 d\Omega(\mu_4)dt,$$

constrained to the following equation

$$\begin{cases} \boldsymbol{v}_t - \mu_1\Delta\boldsymbol{v} + \mu_2(\boldsymbol{v}\cdot\nabla)\boldsymbol{v} + g\nabla h - \boldsymbol{u} = 0 & \text{in } \Omega(\mu_4)\times(0,T), \\ h_t + \text{div}(h\boldsymbol{v}) = 0 & \text{in } \Omega(\mu_4)\times(0,T), \\ \boldsymbol{v} = \boldsymbol{v}_0 & \text{on } \Omega(\mu_4)\times\{0\}, \\ h = h_0 & \text{on } \Omega(\mu_4)\times\{0\}, \\ \boldsymbol{v} = \boldsymbol{0} & \text{on } \partial\Omega(\mu_4)\times(0,T). \end{cases} \tag{38}$$

We recall that $\boldsymbol{v} \in L^2(0,T;[Y_v]^2)$, $\boldsymbol{v}_t \in L^2(0,T;[Y_v{}^*]^2)$, $h \in L^2(0,T;Y_h)$, $h_t \in L^2(0,T;Y_h{}^*)$, and $\boldsymbol{u} \in L^2(0,T;[U]^2)$, where $Y_h = U = L^2(\Omega(\mu_4))$ and $Y_v = H^1_{\Gamma_D}(\Omega(\mu_4))$. Now, noticing that

$$Y_v \hookrightarrow U \hookrightarrow Y_v^*,$$

we can interpret the first equation as a sum of terms in $L^2(0,T;[Y_v{}^*]^2)$. We now focus on the second equation. Using standard regularity results for Navier-Stokes equations (Constantin and Foias, 1988), h is in $L^2(0,T;Y_v)$. Exploiting Sobolev embedding theorems in dimension three (considering space and time), we have $h\boldsymbol{v} \in L^2(0,T;[Y_h]^2)$ and, consequently, $\text{div}(h\boldsymbol{v}) \in L^2(0,T;Y_v^*)$. Thus, the second equation is a sum of terms in $L^2(0,T;Y_v{}^*)$, since $Y_h^* \hookrightarrow Y_v^*$. Even if the setting presented in Section 3 includes a broad class of state equations, to the best of our knowledge no results about OCP($\boldsymbol{\mu}$)s governed by the two-dimensional SWEs are known in the Lagrangian context. However, we believe that one could apply the same techniques already used in Hinze et al. (2008b, Section 1.8.1) for Navier-Stokes equations to state the well-posedness of the problem and to recover hypothesis (i)–(ix). However, this topic goes beyond the goal of this contribution and we will restrict ourselves to the case of well-posedness for this specific test case. As already specified in Section 3, the value of $\alpha \in (0,1]$ influences the control action: the smaller is α, the larger the action of the control variable is. The state equation (38) models free surface incompressible flows in hydrostatic pressure: this assumption is verified when shallow depths are considered, namely when the water height is much lower than the wavelength, which is classical for coastal phenomena. We deal with two physical parameters μ_1 and μ_2 representing diffusive and convective action of the system considered, respectively. Moreover, μ_3 will define the desired solution profile, while μ_4, as already specified, will affect the spatial geometry we will deal with. Let us suppose to have recast the state equation (38) in weak formulation as $\mathcal{E}((\boldsymbol{v},h),\boldsymbol{u};\boldsymbol{\mu})$, thus we can define the following Lagrangian functional

$$\begin{aligned} &\mathscr{L}((\boldsymbol{v},h),\boldsymbol{u},(\boldsymbol{z},q),(\boldsymbol{v}_\text{d},h_\text{d});\boldsymbol{\mu}) \\ &\quad = J((\boldsymbol{v},h),\boldsymbol{u};(\boldsymbol{v}_\text{d},h_\text{d})) + \langle(\boldsymbol{z},q),\mathcal{E}((\boldsymbol{v},h),\boldsymbol{u};\boldsymbol{\mu})\rangle_{\mathbb{Q}\mathbb{Q}^*}, \end{aligned} \tag{39}$$

where the variable $(z, q) \in \mathbb{Y}$ is the adjoint variable. Here we do not specify all the bilinear forms involved, the interested reader may refer to Strazzullo et al. (2020a). The optimality system is, then, obtained through the differentiation with respect to the all involved variables. The minimization problem thus reads: given $\boldsymbol{\mu} \in \mathcal{P}$, find $((\boldsymbol{v}, h), \boldsymbol{u}, (z, q)) \in \mathbb{Y} \times \mathbb{U} \times \mathbb{Y}$ such that

$$\begin{cases} D_{\boldsymbol{v}}\mathscr{L}((\boldsymbol{v}, h), \boldsymbol{u}, (z, q))[\boldsymbol{\zeta}] = 0 & \forall \boldsymbol{\zeta} \in \mathbb{Q}_v, \\ D_h\mathscr{L}((\boldsymbol{v}, h), \boldsymbol{u}, (z, q))[\theta] = 0 & \forall \theta \in \mathbb{Q}_h, \\ D_{\boldsymbol{u}}\mathscr{L}((\boldsymbol{v}, h), \boldsymbol{u}, (z, q))[\boldsymbol{\tau}] = 0 & \forall \boldsymbol{\tau} \in \mathbb{U}, \\ D_z\mathscr{L}((\boldsymbol{v}, h), \boldsymbol{u}, (z, q))[\boldsymbol{\kappa}] = 0 & \forall \boldsymbol{\kappa} \in \mathbb{Q}_v, \\ D_q\mathscr{L}((\boldsymbol{v}, h), \boldsymbol{u}, (z, q))[\xi] = 0 & \forall \xi \in \mathbb{Q}_h. \end{cases} \tag{40}$$

For the sake of completeness, we report the system (40) in strong formulation:

$$\begin{cases} \boldsymbol{v} - z_t - \mu_1 \Delta z - \mu_2(\boldsymbol{v} \cdot \nabla) z & \\ \qquad + \mu_2 (\nabla \boldsymbol{v})^T z - h \nabla q = \boldsymbol{v}_d(\mu_3) & \text{in } \Omega(\mu_4) \times (0, T), \\ h - q_t - \boldsymbol{v} \cdot \nabla q - g\,\mathrm{div}(z) = h_d(\mu_3) & \text{in } \Omega(\mu_4) \times (0, T), \\ z = \boldsymbol{0} & \text{on } \partial\Omega(\mu_4) \times (0, T) \\ z = \boldsymbol{0} & \text{on } \Omega(\mu_4) \times \{T\}, \\ q = 0 & \text{on } \Omega(\mu_4) \times \{T\}, \\ \alpha \boldsymbol{u} = z & \text{in } \Omega(\mu_4) \times (0, T) \\ \boldsymbol{v}_t - \mu_1 \Delta \boldsymbol{v} + \mu_2 (\boldsymbol{v} \cdot \nabla) \boldsymbol{v} + g \nabla h = \boldsymbol{u} & \text{in } \Omega(\mu_4) \times (0, T), \\ h_t + \mathrm{div}(h \boldsymbol{v}) = 0 & \text{in } \Omega(\mu_4) \times (0, T), \\ \boldsymbol{v} = \boldsymbol{0} & \text{on } \partial\Omega(\mu_4) \times (0, T), \\ \boldsymbol{v} = \boldsymbol{v}_0 & \text{on } \Omega(\mu_4) \times \{0\}, \\ h = h_0 & \text{on } \Omega(\mu_4) \times \{0\}. \end{cases} \tag{41}$$

In the next section we will present the numerical results for a distributed OCP($\boldsymbol{\mu}$)s with physical and geometrical parameters. They are an extension of Strazzullo et al. (2020a) to parametrized spatial domain. Furthermore, we will briefly discuss the high fidelity space-time approximation and how we perform a POD-Galerkin in this specific framework in order to build the reduced space approximation for this specific OCP($\boldsymbol{\mu}$)s.

5.2 Numerical results

This section aims at validating the performances of the POD-Galerkin projection reduced approach proposed in Section 4. We followed the test case presented in Ferrari and Saleri (2004), recasting it in the parametrized setting briefly introduced in Section 5.1, where not only physical parameter is considered, but also the geometry of the spatial domain changes with respect to the choice of $\boldsymbol{\mu} \in \mathcal{P} = (0.00001, 1.0) \times (0.01, 0.5) \times (0.1, 1.0) \times (0.8, 1.5)$. Indeed, the spatial domain is given by $\Omega(\mu_4) = [0, 10\mu_4] \times [0, 10]$. In an inverse problem fashion, the optimal control setting gives information about the forcing term needed to achieve a given desired solution profile. To perform simulations and consequent reduction, we took into account a pull back of the optimality system (40) in the *reference domain* given by $\Omega = [0, 10] \times [0, 10]$, i.e. the reference parameter is $\mu_4 = 1$. In the considered framework, we exploited $z_b = 0$ for the flat bathymetry and we studied the wave evolution in the time interval $[0, T]$ where the final time is $T = 0.8$ s. Let x_1 and x_2 be the spatial coordinates. The problem at hand aims at reducing the impact of the spreading of a mass of water with an initial Gaussian distributed elevation under a null initial velocity: i.e.

$$\boldsymbol{v}_0 = \boldsymbol{0}, \qquad \text{and} \qquad h_0 = 0.2(1 + 5e^{(-(\frac{x_1}{\mu_4}-5)^2-(x_2-5)^2+1))}).$$

Namely we want to study under which physical conditions, say wind action and bottom friction, the system can reach the desired state $(\mu_3\boldsymbol{v}_\mathrm{d}, \mu_3 h_\mathrm{d})$, where $(\boldsymbol{v}_\mathrm{d}, h_\mathrm{d})$ is the solution at time T of the uncontrolled state equation (38), with

$$\boldsymbol{v}_{\mathrm{d}0} = \boldsymbol{0}, \qquad \text{and} \qquad h_{\mathrm{d}0} = 2e^{(-(\frac{x_1}{\mu_4}-5)^2-(x_2-5)^2+1)},$$

and no forcing term, i.e. $\boldsymbol{u} = \boldsymbol{0}$.

Let us briefly analyze the high fidelity approximation. We exploited FE discretization, and the optimal solution fields have been discretized through linear polynomial, namely $N_{\mathrm{FE}}^y = N_{\mathrm{FE}}^v + N_{\mathrm{FE}}^h$, where N_{FE}^v and N_{FE}^h are the FE dimensions for state and adjoint velocity and height profile and control variable, respectively, obtained using $r_v = r_h = 1$, as proposed in Saleri and Miglio (2007). The same polynomial degree has been exploited for the control variable, giving a FE discretization of dimension N_{FE}^u. Concerning the time approximation, we perform Euler methods, dividing the time interval $N_t = 8$ timesteps, with a $\Delta t = 0.1$ s. The number of timesteps can be increased following the iterative techniques presented in Hinze et al. (2008a, 2012); Stoll and Wathen (2010, 2013). Although, in this work, for the sake of simplicity and clearness, we exploited a direct solver for the optimality system (41) and this affects the resolution of the high fidelity approximation. In the end, at the truth approximation level, we deal with a system of a total dimension $\mathcal{N} = 76'352$, although, we underline that it is sufficient to validate the reduced approach proposed in Section 4. Let us focus our attention on the construction phase of the low dimensional function spaces. As specified in Section 4.2, we applied a partitioned approach, where several POD compressions, over the five different correlation matrices of $N_{\max} = 100$ snapshots have been carried out for all the involved variables. The number of snapshots $N_{\max}$ was heuristically chosen observing that the POD eigenvalue decay is comparable to the one obtained by $N_{\max} = 150$, see Fig. 2. Besides, we could not explore further the parametric space since the space-time approximation drastically affects the computational time needed for

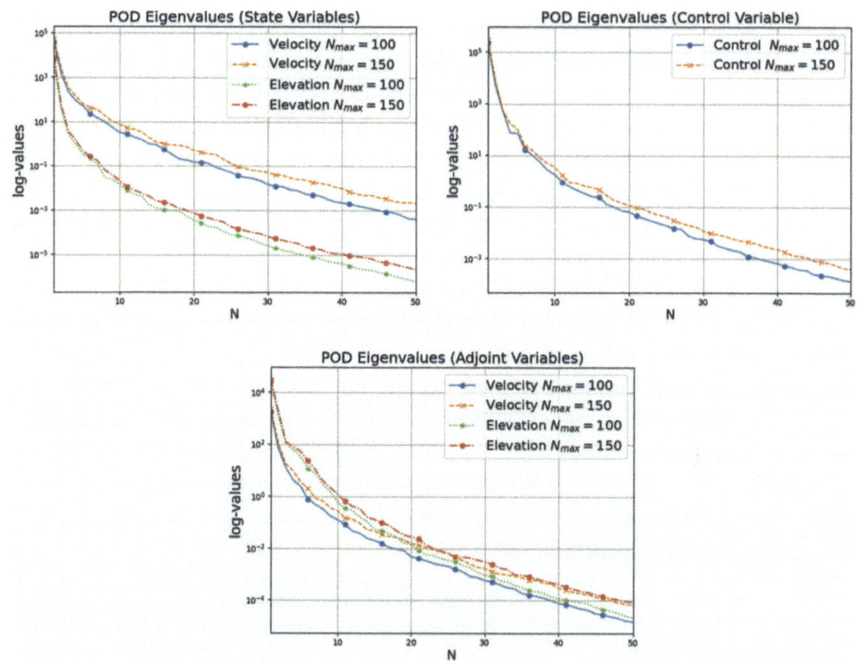

FIGURE 2 Comparison of POD eigenvalues for $N_{\text{max}} = 100$ and $N_{\text{max}} = 150$. *Top Left*: state variables. *Top Right*: control variable. *Bottom*: adjoint variables.

the offline phase of the POD algorithm, i.e. the snapshots collection. Then, the application of the strategy described in Section 4.2 will lead to the following spaces:

$$\begin{aligned}
\mathbb{Y}_N^v &= \text{span}\{\chi_n^v,\ n = 1, \dots, N\},\\
\mathbb{Y}_N^h &= \text{span}\{\chi_n^h,\ n = 1, \dots, N\},\\
\mathbb{U}_N &= \text{span}\{\chi_n^u,\ n = 1, \dots, N\},\\
\mathbb{Y}_N^z &= \text{span}\{\chi_n^z,\ n = 1, \dots, N\},\\
\mathbb{Y}_N^q &= \text{span}\{\chi_n^q,\ n = 1, \dots, N\},
\end{aligned}$$

where we retained $N = 30$ basis functions,[2] given a global reduced space dimension of $9N = 270$. The value N has been heuristically chosen as a trade-off between relative errors values and computational time saving in the reduced framework. Thus, we define

$$\mathbb{Z}_N^{vz} = \text{span}\ \{\chi_n^v, \chi_n^z,\ n = 1, \dots, N\} \ \text{ and } \ \mathcal{Z}_N^{hq} = \text{span}\ \{\chi_n^h, \chi_n^q,\ n = 1, \dots, N\}.$$

2 For the sake of simplicity, we exploit the same value of N for all the spaces.

For state and adjoint velocity-height variable we will use $\mathbb{Z}_N = \mathbb{Z}_N^{vz} \times \mathbb{Z}_N^{hq}$ so that the inf-sup condition (36) holds. Indeed, we remark that, this strategy guarantees the reduced inf-sup stability condition (36) to be verified for the linearized problem at each step of the Newton's Method. Despite the increasing of the dimensionality, we will see that the performances of the reduced projection are still convenient with respect to the high fidelity simulations: comments on the computational advantages are postponed later in the section. We stress that the used POD-Galerkin approach might be sub-optimal for the state equation at hand as underlined in several works, see e.g. Grimberg et al. (2020); Taddei and Zhang (2021). However, the proposed optimal control numerical setting does not suffer much the reduced representation: indeed we are dealing with a viscous model and, furthermore, the controlled framework seems to act itself as a sort of stabilization, where the forcing term changes in order to achieve a less convection-dominated solution. This is confirmed both by the comparison between projection-based ROM and best-fit projection averaged relative log-errors for a testing set of 20 uniformly distributed parameters of Table 1 and by the eigenvalue decay represented in Fig. 2.

TABLE 1 Projection-based ROM and best-fit projection averaged relative log-errors comparison.

N	Averaged relative errors proj-ROM					Averaged relative errors best-fit projection				
	v	h	u	w	q	v	h	u	w	q
6	4.68e–2	5.42e–3	2.30e–2	5.26e–2	2.39e–2	2.52e–2	4.05e–3	1.17e–2	2.11e–2	2.21e–2
10	1.48e–2	2.28e–3	4.53e–3	1.28e–2	7.34e–3	1.22e–2	1.58e–3	4.47e–3	1.04e–2	5.89e–3
16	6.82e–3	9.01e–4	1.97e–3	5.72e–3	3.27e–3	5.65e–3	5.88e–4	1.97e–3	4.73e–3	2.50e–3
20	4.14e–3	5.83e–4	1.34e–3	3.96e–3	2.07e–3	3.44e–3	3.89e–4	1.34e–3	3.40e–3	1.52e–3
26	2.43e–3	3.17e–4	7.94e–4	4.47e–3	1.16e–3	2.07e–3	2.04e–4	7.88e–4	2.13e–3	8.39e–4
30	1.81e–3	2.25e–4	5.48e–4	1.75e–3	7.73e–4	1.15e–3	1.44e–4	5.48e–4	1.52e–3	5.54e–4

All the features of the offline and online phase of the experiment are reported in Table 2.

The basis function considered allowed us to well describe the full order approximated system in the reduced framework, as the reader can notice from the average relative errors represented in Fig. 3 with the following norms:

$$\int_0^T \left\| \boldsymbol{v}^{N_{\mathrm{FE}}^{v}} - \boldsymbol{v}_N \right\|_{H^1}^2 dt, \quad \int_0^T \left\| h^{N_{\mathrm{FE}}^{h}} - h_N \right\|_{L^2}^2 dt,$$

$$\int_0^T \left\| \boldsymbol{u}^{N_{\mathrm{FE}}^{u}} - \boldsymbol{u}_N \right\|_{L^2}^2 dt, \quad \int_0^T \left\| \boldsymbol{z}^{N_{\mathrm{FE}}^{v}} - \boldsymbol{z}_N \right\|_{H^1}^2 dt, \quad \text{and}$$

$$\int_0^T \left\| q^{N_{\mathrm{FE}}^{h}} - q_N \right\|_{L^2}^2 dt.$$

TABLE 2 Data for the distributed OCP(μ) governed by SWEs.

Data	Values
$\mathcal{P}$	$(0.00001, 1) \times (0.01, 0.5) \times (0.1, 1) \times (0.8, 1.5)$
$[0, T]$	$[0\text{ s}, 0.8\text{ s}]$
values of $(\mu_1, \mu_2, \mu_3, \mu_4, \alpha)$	$(0.1, 0.01, 0.1, 1.5, 0.1)$
$N_{\max}$	100
N	30
Sampling Distribution	Uniform
$\mathcal{N}$	76′352
ROM System Dimension	270

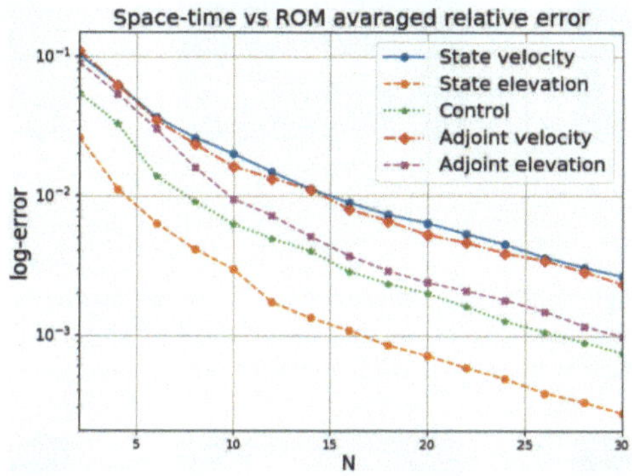

FIGURE 3 Averaged relative error between the space-time and ROM approximation for all the involved variables.

The errors are averaged over a testing set of 20 parameters uniformly distributed: from the plots in Fig. 3, one can observe how the POD-Galerkin approach leads to a good approximation of all the involved quantities. Indeed, using $N = 30$, state and adjoint velocity reach values around $2 \cdot 10^{-3}$, while the relative error for control, state and adjoint elevation is below 10^{-3}. Furthermore, the accuracy of the reduced model can be understood also from the comparison between the space-time solutions and the ROM solutions presented in Figs. 4, 5, and 6 for state elevation, state velocity and control at $t = 0.1$ s, 0.4 s, 0.8 s, respectively. Let us comment on the computational time performances between space-time approximation and ROM simulations. Indeed, not only the reduced bases are able to reproduce several time instances in an accurate way, but there is a gain in the computational time needed for parametrized simulations.

We call *speedup index* the number of reduced simulations which can be performed in the time of the solution of one high fidelity optimality system. For this specific test case, the speedup is of the order of $O(20)$ for $N = 1, \ldots, 30$, namely, it is lightly influenced by the reduced spaces dimension. The speedup index tells us that performing a Galerkin projection in the aggregated spaces is

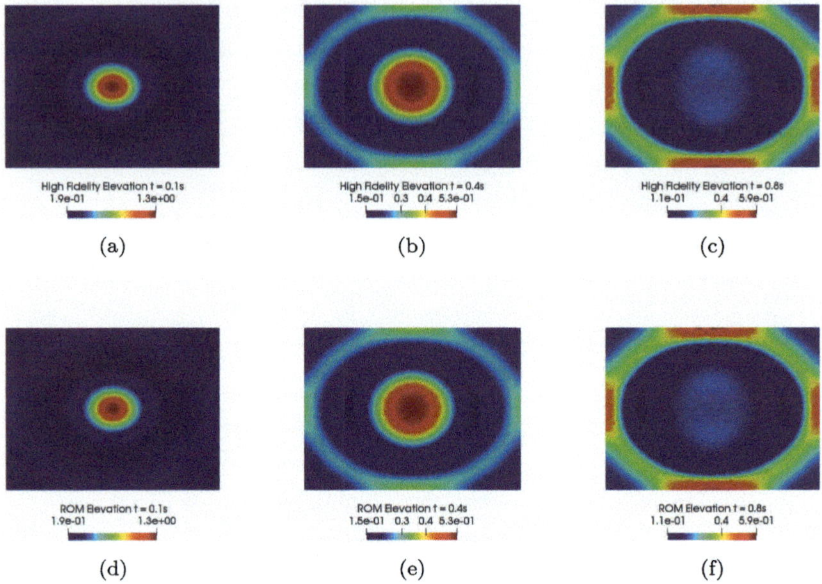

FIGURE 4 Optimal high fidelity and reduced state elevation variable with $\alpha = 0.1$ and $\boldsymbol{\mu} = (0.1, 0.01, 0.1, 1.5)$. High fidelity solutions for $t = 0.1$ s, 0.4 s, 0.8 s in (a), (b), (c), respectively, and reduced solutions for $t = 0.1$ s, 0.4 s, 0.8 s in (d), (e), (f).

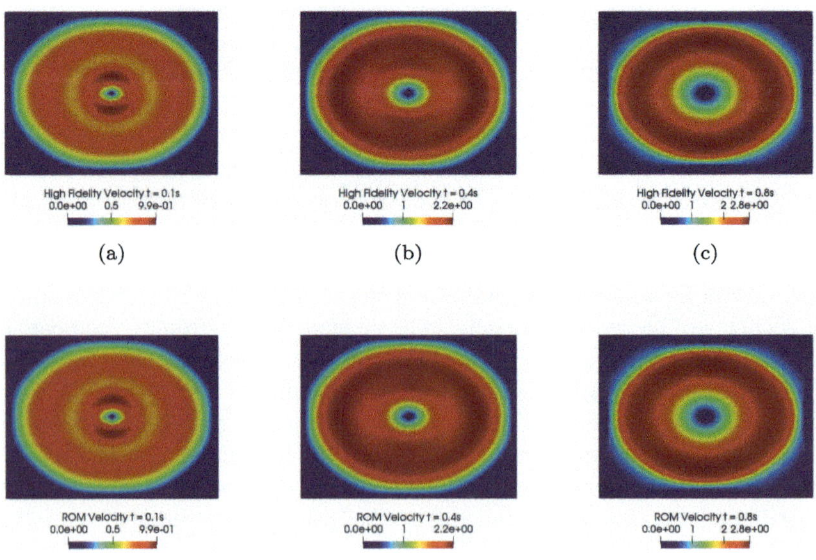

FIGURE 5 Optimal high fidelity and reduced state velocity variable with $\alpha = 0.1$ and $\boldsymbol{\mu} = (0.1, 0.01, 0.1, 1.5)$. High fidelity solutions for $t = 0.1$ s, 0.4 s, 0.8 s in (a), (b), (c), respectively, and reduced solutions for $t = 0.1$ s, 0.4 s, 0.8 s in (d), (e), (f).

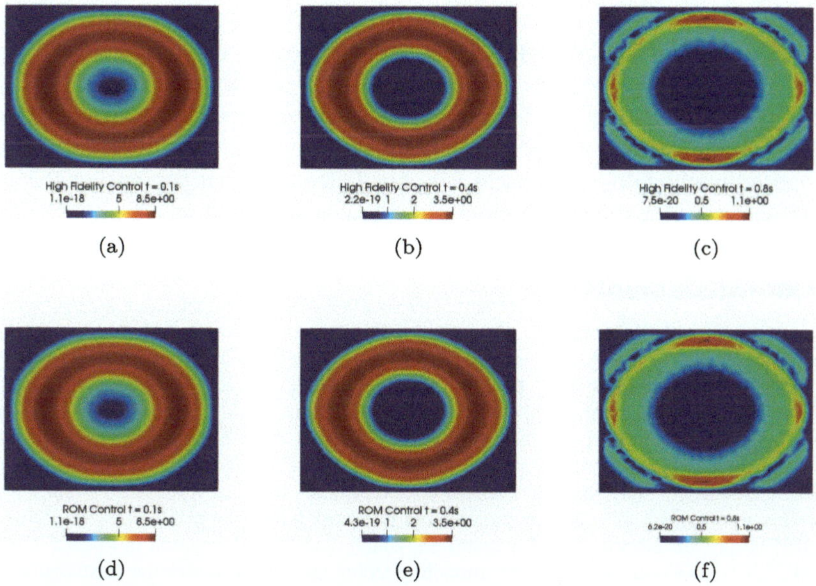

FIGURE 6 Optimal high fidelity and reduced control variable with $\alpha = 0.1$ and $\boldsymbol{\mu} = (0.1, 0.01, 0.1, 1.5)$. High fidelity solutions for $t = 0.1$ s, 0.4 s, 0.8 s in (a), (b), (c), respectively, and reduced solutions for $t = 0.1$ s, 0.4 s, 0.8 s in (d), (e), (f).

still convenient with respect the solution of the whole space-time OCP($\boldsymbol{\mu}$). The saved computational time can be used to study and analyze several parametric configurations in a real-time or many-query context.

In the next section some comments and conclusions follow.

6 Conclusions

In this contribution, we propose ROMs as a reliable and fast strategy to deal with parametrized nonlinear time dependent OCP($\boldsymbol{\mu}$). The analysis we propose is general and can be applied to several PDE($\boldsymbol{\mu}$)s constrained optimization processes. We described the optimality system used to reach a desired profile both at the continuous and discrete level, in a space-time fashion, presenting it not only from a theoretical point of view, but also in its algebraic form, underlining the hidden saddle point structure of the linearized system. Thus, we moved towards model order reduction in order to alleviate the issue of the complexity of the optimality system, which results in a high dimensional problem. We propose a space-time POD-Galerkin approach: the choice was led by the need of an algorithm which could be applied also to very complicated equations, such as the nonlinear time dependent optimality systems. This work relies in building a reduced framework for time dependent nonlinear OCP($\boldsymbol{\mu}$)s, capable of filling the gap between data and physical model. We validate the method in environmental applications such as marine ecosystem management and coastal engineering, though a solution tracking optimization problem governed by viscous SWEs under physical and geometrical parametrization. Our aim was to show how ROMs could be a suitable tool to rapidly simulate marine environment, deeply characterized by a growing demanding computational effort. Indeed, the general proposed methodology results in fast simulations without paying in accuracy with respect to the time consuming space-time approximation.

Acknowledgments

We acknowledge the support by European Union Funding for Research and Innovation – Horizon 2020 Program – in the framework of European Research Council Executive Agency: Consolidator Grant H2020 ERC CoG 2015 AROMA-CFD project 681447 "Advanced Reduced Order Methods with Applications in Computational Fluid Dynamics". We also acknowledge the PRIN 2017 "Numerical Analysis for Full and Reduced Order Methods for the efficient and accurate solution of complex systems governed by Partial Differential Equations" (NA-FROM-PDEs) and the INDAM-GNCS project "Tecniche Numeriche Avanzate per Applicazioni Industriali". The computations in this work have been performed with RBniCS (2015) library, developed at SISSA mathLab, which is an implementation in FEniCS (Logg et al., 2012) of several reduced order modeling techniques; we acknowledge developers and contributors to both libraries.

References

Agoshkov, V.I., Ambrosi, D., Pennati, V., Quarteroni, A., Saleri, F., 1993. Mathematical and numerical modelling of shallow water flow. Computational Mechanics 11 (5), 280–299.

Agoshkov, V.I., Ovchinnikov, E., Quarteroni, A., Saleri, F., 1994a. Recent developments in the numerical simulation of shallow water equations II: temporal discretization. Mathematical Models and Methods in Applied Sciences 4 (04), 533–556.

Agoshkov, V.I., Quarteroni, A., Saleri, F., 1994b. Recent developments in the numerical simulation of shallow water equations I: boundary conditions. Applied Numerical Mathematics 15 (2), 175–200.

Agoshkov, V.I., Saleri, F., Miglio, E., 2007. An optimal control approach for 1D-2D shallow water equations coupling. In: Communications to SIMAI Congress, Vol. 1.

Agrachev, A., Sarychev, A., 2005. Navier–Stokes equations: controllability by means of low modes forcing. Journal of Mathematical Fluid Mechanics 7 (1), 108–152.

Antoulas, A.C., Beattie, C.A., Güğercin, S., 2020. Interpolatory Methods for Model Reduction. SIAM.

Bader, E., Kärcher, M., Grepl, M.A., Veroy, K., 2016. Certified reduced basis methods for parametrized distributed elliptic optimal control problems with control constraints. SIAM Journal on Scientific Computing 38 (6), A3921–A3946.

Bader, E., Kärcher, M., Grepl, M.A., Veroy-Grepl, K., 2015. A certified reduced basis approach for parametrized linear-quadratic optimal control problems with control constraints. IFAC-PapersOnLine 48 (1), 719–720.

Ballarin, F., Faggiano, E., Manzoni, A., Quarteroni, A., Rozza, G., Ippolito, S., Antona, C., Scrofani, R., 2017. Numerical modeling of hemodynamics scenarios of patient-specific coronary artery bypass grafts. Biomechanics and Modeling in Mechanobiology 16 (4), 1373–1399.

Ballarin, F., Manzoni, A., Quarteroni, A., Rozza, G., 2015. Supremizer stabilization of POD–Galerkin approximation of parametrized steady incompressible Navier–Stokes equations. International Journal for Numerical Methods in Engineering 102 (5), 1136–1161.

Barrault, M., Maday, Y., Nguyen, N.C., Patera, A.T., 2004. An empirical interpolation method: application to efficient reduced-basis discretization of partial differential equations. Comptes Rendus. Mathématique 339 (9), 667–672.

Benner, P., Ohlberger, M., Cohen, A., Willcox, K., 2017. Model Reduction and Approximation: Theory and Algorithms. SIAM.

Benzi, M., Golub, G.H., Liesen, J., 2005. Numerical solution of saddle point problems. Acta Numerica 14, 1–137.

Brezzi, F., 1974. On the existence, uniqueness and approximation of saddle-point problems arising from Lagrangian multipliers. Revue Française d'Automatique, Informatique, Recherche Opérationnelle. Analyse Numérique 8 (2), 129–151.

Burkardt, J., Gunzburger, M., Lee, H., 2006. POD and CVT-based reduced-order modeling of Navier–Stokes flows. Computer Methods in Applied Mechanics and Engineering 196 (1–3), 337–355.

Cavallini, F., Crisciani, F., 2013. Quasi-Geostrophic Theory of Oceans and Atmosphere: Topics in the Dynamics and Thermodynamics of the Fluid Earth, Vol. 45. Springer Science & Business Media, New York.

Chapelle, D., Gariah, A., Moireau, P., Sainte-Marie, J., 2013. A Galerkin strategy with proper orthogonal decomposition for parameter-dependent problems: analysis, assessments and applications to parameter estimation. ESAIM: Mathematical Modelling and Numerical Analysis 47 (6), 1821–1843.

Constantin, P., Foias, C., 1988. Navier-Stokes Equations. University of Chicago Press.

Coron, J.M., 1996. On the controllability of the 2-D incompressible Navier-Stokes equations with the Navier slip boundary conditions. ESAIM. Control, Optimisation and Calculus of Variations 1, 35–75.

Coron, J.M., 2007. Control and Nonlinearity, Number 136. American Mathematical Soc.

de los Reyes, J.C., Tröltzsch, F., 2007. Optimal control of the stationary Navier-Stokes equations with mixed control-state constraints. SIAM Journal on Control and Optimization 46 (2), 604–629.

Dedè, L., 2007. Optimal flow control for Navier-Stokes equations: drag minimization. International Journal for Numerical Methods in Fluids 55 (4), 347–366.

Dedè, L., 2010. Reduced basis method and a posteriori error estimation for parametrized linear-quadratic optimal control problems. SIAM Journal on Scientific Computing 32 (2), 997–1019.

Delfour, M.C., Zolésio, J., 2011. Shapes and Geometries: Metrics, Analysis, Differential Calculus, and Optimization, Vol. 22. SIAM, Philadelphia.

Ferrari, S., Saleri, F., 2004. A new two-dimensional shallow water model including pressure effects and slow varying bottom topography. ESAIM: Mathematical Modelling and Numerical Analysis 38 (2), 211–234.

Gerner, A.L., Veroy, K., 2012. Certified reduced basis methods for parametrized saddle point problems. SIAM Journal on Scientific Computing 34 (5), A2812–A2836.

Glas, S., Mayerhofer, A., Urban, K., 2017. Two Ways to Treat Time in Reduced Basis Methods. Springer International Publishing, Cham, pp. 1–16.

Glowinski, R., 1992. Ensuring well-posedness by analogy; Stokes problem and boundary control for the wave equation. Journal of Computational Physics 103 (2), 189–221.

Grimberg, S., Farhat, C., Youkilis, N., 2020. On the stability of projection-based model order reduction for convection-dominated laminar and turbulent flows. Journal of Computational Physics 419, 109681.

Gunzburger, M.D., 2003. Perspectives in Flow Control and Optimization, Vol. 5. SIAM, Philadelphia.

Haslinger, J., Mäkinen, R.A.E., 2003. Introduction to Shape Optimization: Theory, Approximation, and Computation. SIAM, Philadelphia.

Hesthaven, J.S., Rozza, G., Stamm, B., 2015. Certified Reduced Basis Methods for Parametrized Partial Differential Equations. SpringerBriefs in Mathematics. Springer, Milano.

Hinze, M., Köster, M., Turek, S., 2008a. A hierarchical space-time solver for distributed control of the Stokes equation. Technical Report, SPP1253-16-01.

Hinze, M., Köster, M., Turek, S., 2012. A space-time multigrid method for optimal flow control. In: Constrained Optimization and Optimal Control for Partial Differential Equations. Springer, pp. 147–170.

Hinze, M., Pinnau, R., Ulbrich, M., Ulbrich, S., 2008b. Optimization with PDE Constraints, Vol. 23. Springer Science & Business Media, Antwerp.

Iapichino, L., Trenz, S., Volkwein, S., 2016. Reduced-order multiobjective optimal control of semilinear parabolic problems. In: Karasözen, B., Manguoğlu, M., Tezer-Sezgin, M., Göktepe, S., Uğur, Ö. (Eds.), Numerical Mathematics and Advanced Applications ENUMATH 2015. Springer International Publishing, Cham, pp. 389–397.

Iapichino, L., Ulbrich, S., Volkwein, S., 2017. Multiobjective pde-constrained optimization using the reduced-basis method. Advances in Computational Mathematics 43 (5), 945–972.

Kärcher, M., Grepl, M.A., 2014. A certified reduced basis method for parametrized elliptic optimal control problems. ESAIM. Control, Optimisation and Calculus of Variations 20 (2), 416–441.

Kärcher, M., Tokoutsi, Z., Grepl, M.A., Veroy, K., 2018. Certified reduced basis methods for parametrized elliptic optimal control problems with distributed controls. Journal of Scientific Computing 75 (1), 276–307.

Kunisch, K., Volkwein, S., 2008. Proper orthogonal decomposition for optimality systems. ESAIM: Mathematical Modelling and Numerical Analysis 42 (1), 1–23.

Langer, U., Steinbach, O., Tröltzsch, F., Yang, H., 2021. Unstructured space-time finite element methods for optimal control of parabolic equations. SIAM Journal on Scientific Computing 43 (2), A744–A771.

Lassila, T., Manzoni, A., Quarteroni, A., Rozza, G., 2013. A reduced computational and geometrical framework for inverse problems in hemodynamics. International Journal for Numerical Methods in Biomedical Engineering 29 (7), 741–776.

Leugering, G., Benner, P., Engell, S., Griewank, A., Harbrecht, H., Hinze, M., Rannacher, R., Ulbrich, S., 2014. Trends in PDE Constrained Optimization. Springer, New York.

Lions, J.L., 1968. Contrôle optimal de systèmes gouvernés par des équations aux dérivées partielles. Dunod.

Lions, J.L., 1969. Quelques méthodes de résolution des problemes aux limites non linéaires. Dunod.

Lions, J.L., 1988. Exact controllability, stabilization and perturbations for distributed systems. SIAM Review 30 (1), 1–68.

Lions, J.L., 1990. Are there connections between turbulence and controllability? In: Analyse et Optimisation de Systemes. In: Lecture Notes in Control and Information Sciences, vol. 144. Springer-Verlag, Berlin-Heidelberg-New York.

Lions, J.L., Zuazua, E., 1997. Contrôlabilité exacte des approximations de Galerkin des équations de Navier-Stokes. Comptes Rendus de L'Académie des Sciences. Series 1, Mathematics 324 (9), 1015–1021.

Lions, J.L., Zuazua, E., 1998. Exact boundary controllability of Galerkin's approximations of Navier-Stokes equations. Annali Della Scuola Normale Superiore Di Pisa. Classe Di Scienze 26 (4), 605–621.

Lions, J.L., Zuazua, E., 2017. A generic uniqueness result for the Stokes system and its control theoretical consequences. In: Partial Differential Equations and Applications: Collected Papers in Honor of Carlo Pucci, p. 221.

Logg, A., Mardal, K., Wells, G., 2012. Automated Solution of Differential Equations by the Finite Element Method. Springer-Verlag, Berlin.

Miglio, E., Perotto, S., Saleri, F., 2005a. Model coupling techniques for free-surface flow problems: Part I. Nonlinear Analysis 63 (5–7), e1885–e1896.

Miglio, E., Perotto, S., Saleri, F., 2005b. Model coupling techniques for free-surface flow problems: Part II. Nonlinear Analysis 63 (5–7), e1897–e1908.

Miglio, E., Quarteroni, A., Saleri, F., 1999. Finite element approximation of quasi-3D shallow water equations. Computer Methods in Applied Mechanics and Engineering 174 (3–4), 355–369.

Mohammadi, B., Pironneau, O., 2010. Applied Shape Optimization for Fluids. Oxford University Press, New York.

Negri, F., Manzoni, A., Rozza, G., 2015. Reduced basis approximation of parametrized optimal flow control problems for the Stokes equations. Computers & Mathematics with Applications 69 (4), 319–336.

Negri, F., Rozza, G., Manzoni, A., Quarteroni, A., 2013. Reduced basis method for parametrized elliptic optimal control problems. SIAM Journal on Scientific Computing 35 (5), A2316–A2340.

Pichi, F., Strazzullo, M., Ballarin, F., Rozza, G., 2020. Driving bifurcating parametrized nonlinear pdes by optimal control strategies: application to Navier-Stokes equations and model reduction. Submitted for publication. arXiv:2010.13506.

Pošta, M., Roubíček, T., 2007. Optimal control of Navier–Stokes equations by Oseen approximation. Computers & Mathematics With Applications 53 (3), 569–581.

Prud'Homme, C., Rovas, D.V., Veroy, K., Machiels, L., Maday, Y., Patera, A., Turinici, G., 2002. Reliable real-time solution of parametrized partial differential equations: reduced-basis output bound methods. Journal of Fluids Engineering 124 (1), 70–80.

Quarteroni, A., Manzoni, A., Negri, F., 2015. Reduced Basis Methods for Partial Differential Equations: an Introduction, Vol. 92. Springer.

Quarteroni, A., Rozza, G., Dedè, L., Quaini, A., 2005. Numerical approximation of a control problem for advection-diffusion processes. In: Ceragioli, F., Dontchev, A., Futura, H., Marti, K., Pandolfi, L. (Eds.), System Modeling and Optimization. International Federation for Information Processing, CSMO Conference on System Modeling and Optimization, Vol. 199. Springer, Boston, pp. 261–273.

Quarteroni, A., Rozza, G., Quaini, A., 2007. Reduced basis methods for optimal control of advection-diffusion problems. In: Advances in Numerical Mathematics, Number CMCS-CONF-2006-003. RAS and University of Houston, pp. 193–216.

RBniCS, 2015. RBniCS - reduced order modelling in FEniCS. http://mathlab.sissa.it/rbnics.

Ricchiuto, M., Abgrall, R., Deconinck, H., 2007. Application of conservative residual distribution schemes to the solution of the shallow water equations on unstructured meshes. Journal of Computational Physics 222 (1), 287–331.

Ricchiuto, M., Bollermann, A., 2009. Stabilized residual distribution for shallow water simulations. Journal of Computational Physics 228 (4), 1071–1115.

Rozza, G., Huynh, D., Manzoni, A., 2013. Reduced basis approximation and a posteriori error estimation for Stokes flows in parametrized geometries: roles of the inf-sup stability constants. Numerische Mathematik 125 (1), 115–152.

Rozza, G., Huynh, D., Patera, A., 2008. Reduced basis approximation and a posteriori error estimation for affinely parametrized elliptic coercive partial differential equations: application to transport and continuum mechanics. Archives of Computational Methods in Engineering 15 (3), 229–275.

Rozza, G., Veroy, K., 2007. On the stability of the reduced basis method for Stokes equations in parametrized domains. Computer Methods in Applied Mechanics and Engineering 196 (7), 1244–1260.

Saleri, F., Miglio, E., 2007. Geometric multiscale approach by optimal control for shallow water equations. In: Applied and Industrial Mathematics in Italy II. World Scientific, pp. 537–548.

Schöberl, J., Zulehner, W., 2007. Symmetric indefinite preconditioners for saddle point problems with applications to PDE-constrained optimisation problems. SIAM Journal on Matrix Analysis and Applications 29 (3), 752–773.

Seymen, Z.K., Yücel, H., Karasözen, B., 2014. Distributed optimal control of time-dependent diffusion–convection–reaction equations using space–time discretization. Journal of Computational and Applied Mathematics 261, 146–157.

Ştefănescu, R., Navon, I.M., 2013. POD/DEIM nonlinear model order reduction of an ADI implicit shallow water equations model. Journal of Computational Physics 237, 95–114.

Ştefănescu, R., Sandu, A., Navon, I.M., 2014. Comparison of POD reduced order strategies for the nonlinear 2D shallow water equations. International Journal for Numerical Methods in Fluids 76 (8), 497–521.

Stoll, M., Wathen, A., 2010. All-at-once solution of time-dependent PDE-constrained optimization problems.

Stoll, M., Wathen, A., 2013. All-at-once solution of time-dependent Stokes control. Journal of Computational Physics 232 (1), 498–515.

Strazzullo, M., Ballarin, F., Mosetti, R., Rozza, G., 2018. Model reduction for parametrized optimal control problems in environmental marine sciences and engineering. SIAM Journal on Scientific Computing 40 (4), B1055–B1079.

Strazzullo, M., Ballarin, F., Rozza, G., 2020a. POD-Galerkin model order reduction for parametrized nonlinear time dependent optimal flow control: an application to Shallow Water Equations. Journal of Numerical Mathematics. In press. arXiv:2003.09695.

Strazzullo, M., Ballarin, F., Rozza, G., 2020b. POD-Galerkin model order reduction for parametrized time dependent linear quadratic optimal control problems in saddle point formulation. Journal of Scientific Computing 83 (55).

Strazzullo, M., Ballarin, F., Rozza, G., 2021a. Certified reduced basis for linear parametrized parabolic optimal control problems in space-time formulation. Submitted for publication. arXiv:2103.00460.

Strazzullo, M., Zainib, Z., Ballarin, F., Rozza, G., 2021b. Reduced order methods for parametrized nonlinear and time dependent optimal flow control problems: towards applications in biomedical and environmental sciences. In: Numerical Mathematics and Advanced Applications ENUMATH 2019.

Taddei, T., Zhang, L., 2021. Space-time registration-based model reduction of parameterized one-dimensional hyperbolic PDEs. ESAIM: Mathematical Modelling and Numerical Analysis 55 (1), 99–130.

Takase, S., Kashiyama, K., Tanaka, S., Tezduyar, T.E., 2010. Space–time SUPG formulation of the shallow-water equations. International Journal for Numerical Methods in Fluids 64 (10–12), 1379–1394.

Tröltzsch, F., 2010. Optimal Control of Partial Differential Equations. Graduate Studies in Mathematics, vol. 112. Verlag, Wiesbad.

Urban, K., Patera, A.T., 2012. A new error bound for reduced basis approximation of parabolic partial differential equations. Comptes Rendus. Mathématique 350 (3–4), 203–207.

Vázquez, R., Trélat, E., Coron, J.M., 2008. Control for fast and stable laminar-to-high-Reynolds-numbers transfer in a 2D Navier-Stokes channel flow. Discrete and Continuous Dynamical Systems. Series B 10 (4), 925–956.

Vreugdenhil, C.B., 2013. Numerical Methods for Shallow-Water Flow, Vol. 13. Springer Science & Business Media.

Yano, M., 2014. A space-time Petrov–Galerkin certified reduced basis method: application to the Boussinesq equations. SIAM Journal on Scientific Computing 36 (1), A232–A266.

Yano, M., Patera, A.T., Urban, K., 2014. A space-time hp-interpolation-based certified reduced basis method for Burgers' equation. Mathematical Models and Methods in Applied Sciences 24 (09), 1903–1935.

Yilmaz, F., Karasözen, B., 2014. An all-at-once approach for the optimal control of the unsteady Burgers equation. In: Recent Advances in Applied and Computational Mathematics: ICACM-IAM-METU. Journal of Computational and Applied Mathematics 259, 771–779.

Zainib, Z., Ballarin, F., Fremes, S., Triverio, P., Jiménez-Juan, L., Rozza, G., 2020. Reduced order methods for parametric optimal flow control in coronary bypass grafts, toward patient-specific data assimilation. International Journal for Numerical Methods in Biomedical Engineering, e3367.

Chapter 10

Moments and convex optimization for analysis and control of nonlinear PDEs

Milan Korda[a,b], Didier Henrion[a,b], and Jean Bernard Lasserre[a,c,*]
[a]*CNRS-LAAS, Toulouse, France,* [b]*Faculty of Electrical Engineering, Czech Technical University in Prague, Prague, Czech Republic,* [c]*Institute of Mathematics of Toulouse, Toulouse, France*
**Corresponding author: e-mail address: lasserre@laas.fr*

Contents

Abstract

This work presents a convex-optimization-based framework for analysis and control of nonlinear partial differential equations. A measure-valued approach uses a particular weak embedding of the nonlinear PDE and results in *linear* equations in appropriate space of Borel measures. These equations are then used as linear constraints of an infinite-dimensional linear programming problem (LP). This LP is then approximated by a hierarchy of convex, finite-dimensional, semidefinite programming problems (SDPs). For analysis of uncontrolled PDEs, optimal solutions to these SDPs provide bounds on any specified polynomial functional integral of the solutions to the PDE. For PDE control, optimal solutions to these SDPs provide bounds on the optimal value of a given optimal control problem as well as suboptimal feedback controllers. The entire approach is based purely on convex optimization and is mesh-free, i.e., does not rely on spatio-temporal gridding, even though the PDE addressed can be fully nonlinear. It is applicable

Handbook of Numerical Analysis, Volume 23, ISSN 1570-8659, https://doi.org/10.1016/bs.hna.2021.12.010

to a very broad class of nonlinear PDEs with polynomial data. Computational complexity is analyzed and several complexity reduction procedures are described. Numerical examples demonstrate the approach.

Keywords
Partial differential equations, Occupation measure, Optimal control, Semidefinite programming, Convex optimization

MSC Codes
90C23, 90C22, 65M99, 35C99, 35D30

1 Introduction

This paper considers the problem of analysis and optimal control of nonlinear partial differential equations (PDEs) with polynomial data. The proposed approach proceeds from a measure-valued weak formulation which embeds the nonlinear PDE into infinite-dimensional spaces of Borel measures. The result is a system of *linear* equations involving the so called *occupation* and *boundary* measures as variables. These equations, along with a user-specified objective functional, give rise to an infinite-dimensional linear programming problem (LP) in appropriate spaces of Borel measures. In the uncontrolled case, any solution to this infinite-dimensional LP provides a bound on a given functional (e.g., energy) of the solution to the PDE while in the controlled case it provides a bound on the optimal value of the optimal control problem considered.

This infinite-dimensional LP is then approximated using the classical Lasserre's hierarchy of finite-dimensional *convex* semidefinite programming (SDP) relaxations. In the uncontrolled case, this hierarchy provides a monotone sequence of upper or lower bounds on the optimum of the infinite dimensional LP and hence also upper and lower bounds on the given functional of the solution to the PDE. In the controlled case, the solutions to the SDPs provide bounds on the optimal value of the optimal control problem considered as well as a sequence of suboptimal polynomial feedback controllers. The sequence of SDP relaxations is proven to converge to the optimal value of the infinite dimensional LP as the relaxation degree tends to infinity, under standard assumptions. The SDP relaxations can be readily solved by standard semidefinite programming solvers such as MOSEK or SeDuMi (Pólik et al., 2007).

One major advantage of the presented approach is its overall convexity, thereby avoiding the inherent issues associated to non-convex optimization such as local minima. This comes at the expense of having to solve a sequence of convex semidefinite programming problems of increasing size, governed by the number of unknown functions in the PDE plus the number of spatio-temporal dimensions of the PDE and the number of derivatives appearing nonlinearly. In order for the approach to be applicable to large-scale problems such as those coming from computational fluid dynamics (e.g., the Navier-Stokes equations in three spatial dimensions), structure/sparsity exploitation (e.g. using Waki et al., 2006) or tailored moment-sum-of-squares solvers (e.g., Papp and Yildiz, 2019) may be required. In this paper, we briefly describe several of the most obvious ways for structure exploitation and complexity reduction, although a more detailed analysis of the problem structure is likely to uncover more.

At this point we would like to stress that the goal of this work is not to present a new PDE *solver* but rather a numerical procedure, based on convex optimization, for obtaining *bounds* on functionals of the solutions to the PDE and for *control* design. Besides being of independent theoretical interest, such bounds are important in robustness analysis (e.g., to verify that the maximum force on a given construction does not exceed the breakage point of the material used) as well as for validation of numerical solvers. The use of the control design portion of our work is broad, spanning virtually all applications where feedback control of PDEs is required as well as the field of PDE-constrained optimization (e.g., Hinze et al., 2008) where the "control" may not necessarily represent a feedback control in the classical sense but, for example, a parametrization of the shape of a material to be designed. Also such bounds can be used to evaluate the suboptimality of existing methods for optimal control of PDEs by comparing the value achieved by these methods with the guaranteed lower bounds provided by our approach.

Another distinguishing and important feature of the proposed approach is its mesh-free character. In contrast to other numerical approaches, it does *not* rely on spatio-temporal gridding but rather works with moment sequences of measures supported on the graphs of the solution to the PDE. In fact the bounds or the polynomial feedback controller are obtained without ever numerically solving the PDE. This is especially useful in the case of complex dynamical behavior (e.g., a high Reynolds number fluid) where fine gridding is necessary to ensure convergence of the numerical scheme for solving the PDE, thereby making the solution very costly. Moreover, once the number of moments is fixed, the *size* of the resulting SDP to solve is not affected by the complexity of the dynamical behavior studied. Of course, the *quality* of bounds or controller obtained with a given number of moments may depend on the complexity of the dynamical behavior. Nevertheless the bounds are always valid, irrespective of the number of moments considered (up to the numerical precision of the SDP solution).

An open theoretical question remains: Under what conditions is the infinite-dimensional LP equivalent to the original problem? The question has been answered for optimal control problems with ordinary differential equations in Vinter (1993) and later for problems of region of attraction and invariant set computation in Henrion and Korda (2014) and Korda et al. (2014b). For nonlinear PDEs the situation appears to be far more complex; in the uncontrolled case, one possible pathway for establishing this equivalence is to invoke weak-strong uniqueness results (e.g., from Brenier et al., 2011) which suggest expanding the set of constraints of the infinite-dimensional LP with appropriate additional inequalities satisfied by any strong solution.

The presented approach can be seen as a generalization to nonlinear partial differential equations of the approach initiated in Lasserre et al. (2008) for nonlinear ordinary differential equations.

To the best of the authors' knowledge, this is the first time that nonlinear PDEs are addressed at this level of generality, using convex optimization and semidefinite programming. Besides the concurrent work (Marx et al., 2020) focusing on hyperbolic conservation laws with the emphasis on convergence aspects, we are aware of only a few other attempts in the technical literature. For example, the early work (Rubio, 1995) exploited the linear measure-valued formulation of non-linear (semi-linear elliptic) PDE control problems, although without convergence proofs and using a computationally inefficient linear programming discretization. The moment-SOS hierarchy (Henrion et al., 2020; Lasserre, 2019) was used systematically in Mevissen et al. (2008) for solving approximately non-linear PDEs, formulated after time and domain discretization as large-scale sparse non-convex polynomial optimization problems. Bounds on functionals of solutions were obtained with SOS polynomials for linear elliptic PDEs in Kawai (2015), for non-linear PDEs arising in fluid dynamics in Chernyshenko et al. (2014) and for the non-linear Kuramoto-Sivashinsky PDE in Goluskin and Fantuzzi (2019). These works, however, provide bounds with no convergence guarantees and do not exploit the "primal" formulation of the problem on measures, which we believe to be crucial for convergence analysis. In the context of classical linear matrix inequality (LMI) Lyapunov approaches, there have been recently several works that developed LMI conditions for PDE analysis and controller design, but only in the linear case; see e.g. Fridman and Orlov (2009); Barreau et al. (2018); Lamare et al. (2016); Valmorbida et al. (2016); Gahlawat and Peet (2017) and references therein. Finally, occupation measures and the moment-SOS hierarchy were used in Magron and Prieur (2020) for Riesz-spectral linear PDE control by state-mode discretization and in Bertsimas and Caramanis (2006) for obtaining bounds on functionals of solutions to linear PDEs. Our contribution and novelty with respect to these references are as follows:

- we consider non-linear PDE analysis and control;
- we introduce the notion of occupation and boundary measures in the PDE context;
- we do not use time, space, or spectral discretization[1] of the PDE. We rather use the full potential of the measure-valued formulation and approximate *moments* of optimal measure-solutions;

[1] The approach used in this paper bears some similarity to the classical Galerkin discretization since the PDE is enforced (in the weak form) on a certain subspace of smooth test functions (in our case on all polynomials up to a given degree). However, contrary to the Galerkin method, the solution to the PDE is *not* sought as a linear combination of these test functions but rather represented by the moments of the associated occupation measure whose positivity and support localization are enforced using semidefinite programming.

- we provide rigorous convergence proofs for the moment-SOS hierarchy in the PDE context.

The paper is organized as follows. Section 2 describes the problem setting for the analysis of PDEs. Section 3 introduces the occupation and boundary measures and uses them to derive a linear embedding of the nonlinear PDE, resulting in an infinite-dimensional LP. Section 4 presents the hierarchy of SDP relaxations providing the upper and lower bounds and describes several generalizations of the approach. Section 5 describes the problem setup for PDE control. Section 6 describes the linear embedding for the controlled case, leading an infinite-dimensional LP. Section 7 presents the SDP relaxations for the controlled case and a way to extract a feedback controller from its solutions. Section 8 describes a generalization to higher-order PDEs. Section 9 presents numerical examples and we conclude and given an outlook in Section 10.

2 Problem statement (analysis)

We consider the system of nonlinear partial differential equations[2] (PDEs)

$$F(x, y(x), \mathcal{D}y(x)) + \sum_{i,j} B_{i,j}(x, y(x)) \frac{\partial^2 y}{\partial x_i \partial x_j} = 0 \text{ in } \Omega^{\mathrm{o}} \tag{1a}$$

$$G(x, y(x), \mathcal{D}y(x)) = 0 \text{ on } \partial\Omega, \tag{1b}$$

where $y : \mathbb{R}^n \to \mathbb{R}^{n_y}$ is the unknown, possibly vector-valued, function defined on a domain $\Omega \subset \mathbb{R}^n$ and $F : \mathbb{R}^n \times \mathbb{R}^{n_y} \times \mathbb{R}^{n_y \times n} \to \mathbb{R}^{n_F}$ and $G : \mathbb{R}^n \times \mathbb{R}^{n_y} \times \mathbb{R}^{n_y \times n} \to \mathbb{R}^{n_G}$ are given, possibly vector-valued, functions. The function F is assumed to be a multivariate polynomial in its arguments whereas G is allowed to be piecewise polynomial.[3] The symbol $\mathcal{D}y$ denotes the Jacobian matrix of y, i.e., $(\mathcal{D}y)_{ij} = \frac{\partial y_i}{\partial x_j}$, where $\frac{\partial y_i}{\partial x_j}$ denotes the weak derivative of y_i with respect to x_j (in Sobolev space sense). The coefficients $B_{i,j}(x, y(x)) : \mathbb{R}^n \times \mathbb{R}^{n_y} \to \mathbb{R}^{n_F \times n_y}$ are fixed matrix functions with each entry being a polynomial in (x, y) and

$$\frac{\partial^2 y}{\partial x_i \partial x_j} = \left[\frac{\partial^2 y_1}{\partial x_i \partial x_j}, \ldots, \frac{\partial^2 y_{n_y}}{\partial x_i \partial x_j} \right]^\top,$$

[2] The summation indices in (1a) run through $i \leq n,\ j \leq n,\ i \leq j$ (due to the symmetry of the Hessian matrix); we omit the summation bounds for conciseness.

[3] Allowing G to be piecewise polynomial provides a significant modeling freedom for the boundary condition (1b). For example, setting $G = 0$ on a certain piece of the boundary results in the trivial constraint $0 = 0$, thereby not assigning any boundary condition on this piece (which is commonly encountered in PDEs involving space and time where the initial condition is typically assigned whereas the terminal condition is free or vice versa).

where $\cdot^\top$ denotes the transpose. The domain Ω is assumed to be a compact basic semialgebraic[4] set of the form

$$\Omega = \{x \in \mathbb{R}^n \mid g_i^\Omega(x) \geq 0,\ i = 1, \ldots, n^\Omega\}, \tag{2}$$

where g_i^Ω are polynomials and $n^\Omega \in \mathbb{N}$. The set Ω is assumed to be equal to the closure of its interior Ω^o and its boundary $\partial\Omega$ is assumed to be (locally) Lipschitz (in the sense of Leoni, 2017, Definition 12.9).

Ω being a basic semialgebraic set, its boundary is piecewise smooth and hence admits a surface measure σ satisfying the integration by parts formula (a special case of Stokes' formula)

$$\int_{\partial\Omega} h(x)\eta_m(x)\, d\sigma(x) = \int_\Omega \frac{\partial h}{\partial x_m}(x)\, dx,\ \ m = 1, \ldots, n \tag{3}$$

for any $h \in W^{1,\infty}(\Omega^\mathrm{o})$, where $\eta(x) = (\eta_1(x), \ldots, \eta_n(x))$ is the outward-pointing normal unit vector to $\partial\Omega$. The boundary $\partial\Omega$ can be decomposed as

$$\partial\Omega = \bigcup_{i=1}^{n_\mathrm{b}} \Omega_{\partial,i} \tag{4}$$

with each $\Omega_{\partial,i}$ being basic semialgebraic and smooth, i.e.,

$$\Omega_{\partial,i} = \{x \in \mathbb{R}^n \mid g_j^{\partial,i}(x) \geq 0,\ j = 1, \ldots, n_{\partial,i}\}, \tag{5}$$

where $g_j^{\partial,i}$ are polynomials and $n_{\partial,i} \in \mathbb{N}$. The partition (4) can be taken disjoint up to a set of zero surface measure σ (e.g., corner points). The piecewise polynomial function G is assumed to be polynomial on each element $\Omega_{\partial,i}$ of the partition.

Remark 1 (Weak solutions, regularity, and uniqueness). Throughout this paper we consider solutions $y(\cdot)$ to the PDE (1) to lie in the Sobolev space $W^{k,\infty}$ with $k = 1$ if $B_{i,j} = 0$ and $k = 2$ otherwise. Eqs. (1a) and (1b) are assumed to be satisfied almost everywhere with respect to the Lebesgue measure on Ω^o respectively the surface measure σ on $\partial\Omega$. In some cases, this notion of solution is too wide to be of practical interest; if this the case, Eqs. (1a) and (1b) must be supplemented with additional conditions that single out the solutions of interest. These conditions may, for example, be the semiconcavity for Hamilton-Jacobi type PDEs or the entropy inequalities for conservation laws (see, e.g., Sections 3.3.3 and 3.4.3 of Evans (2010)). We note that the $W^{k,\infty}$ regularity assumption is made in order to avoid technicalities; all results presented in this paper hold also with $y(\cdot) \in W^{k,p}$ with sufficiently large p depending on the degree of the polynomials F and G.

[4] A basic semialgebraic set is the intersection of super or sub-level sets of finitely many multivariate polynomials. This is a broad class of sets covering most of domains considered in practice such as boxes, spheres, ellipsoids, and intersections thereof.

This is a very general setting encompassing a large number of equations of interest, including the compressible as well as incompressible Navier-Stokes equations.

Remark 2 (Higher-order PDEs). Higher order nonlinear PDEs can be handled using the proposed approach by introducing additional variables. This is described in Section 8.

For the time being, we consider the problem of analysis of such a PDE (control is treated in Section 5); in particular we are interested in establishing bounds on a functional of the form

$$J\big(y(\cdot)\big) := \int_{\Omega} L\left(x, y(x), \mathcal{D}y(x)\right) dx + \int_{\partial\Omega} L_{\partial}\left(x, y(x), \mathcal{D}y(x)\right) d\sigma(x), \quad (6)$$

where $L : \mathbb{R}^n \times \mathbb{R}^{n_y} \times \mathbb{R}^{n_y \times n} \to \mathbb{R}$ is a given multivariate polynomial and $L_{\partial} : \mathbb{R}^n \times \mathbb{R}^{n_y} \times \mathbb{R}^{n_y \times n} \to \mathbb{R}$ is a piecewise multivariate polynomial such that each restriction of L_{∂} to $\Omega_{\partial,i}$ is a multivariate polynomial. This is a fairly general form of a functional; for example, if y represents the velocity field of a fluid, then (6) with $L(x, y, z) = \sum_i y_i^2$ and $L_{\partial} = 0$ is proportional to the average kinetic energy of the fluid.

Now we are ready to state the first goal of this paper:

Problem 1. *Find upper and lower bounds on (6) evaluated along the solutions of (1).*

Formally speaking, this task can be expressed as finding upper and lower bounds on the optimal value of the optimization problem

$$\begin{array}{ll} \inf/\sup\limits_{y(\cdot)} & J\big(y(\cdot)\big) \\ \text{subject to} & \text{(1a), (1b).} \end{array} \quad (7)$$

If minimization is considered in (7), then lower bounds on the optimal value of this optimization problem yield lower bounds on (6) evaluated along the solutions of (1a) and (1b). Similarly, if maximization is considered in (7), then upper bounds on the optimal value of (7) yield upper bounds on (6).

We note that since the PDE (1a) and (1b) is nonlinear, problem (7) is an infinite-dimensional *nonconvex* optimization problem. One of the main contributions of this work is a method of finding bounds on the optimal value of this infinite-dimensional nonconvex problem in terms of the solution to a finite-dimensional *convex* optimization problem.

Remark 3 (Uniqueness). We make no a priori assumptions on the uniqueness of the solutions to (1). If (1) has a unique solution, then the feasible set of the optimization problem (7) is a singleton in which case the infimum and supremum of (7) coincide; otherwise, if the solution to (1) is not unique, the infimum and supremum may differ. In either case, the approach presented provides upper bounds on the supremum and lower bounds on the infimum. See also Remark 1.

Remark 4 (Non-polynomial Dirichlet boundary conditions). The presented approach can be extended to non-polynomial Dirichlet boundary conditions of the form $y(x) = h_i(x)$, $x \in \partial\Omega_i$, $i \in \mathcal{I}$, $\mathcal{I} \subset \{1, \ldots, n_b\}$, where h_i is an arbitrary integrable function such that the integrals $\int_{\partial\Omega_i} x^\alpha h_i^\beta(x)\, d\sigma(x)$ are computable (either analytically or numerically). This is treated in the extended version of the chapter (Korda et al., 2018, Section 4.1). Note that this generalization can also be used with polynomial h_i, leading to a simpler representation than when using the general form (1b).

Remark 5 (Periodic boundary conditions). The presented approach can also be extended to handle periodic boundary conditions. This is treated in Korda et al. (2018, Section 4.2).

3 Occupation measures for nonlinear PDEs

The goal of this section is to transform the nonconvex optimization problem (7) to an infinite-dimensional *linear* (and hence convex) programming problem. The key ingredient for doing so is the so-called occupation and boundary measures associated to the solutions of the PDE (1). For notational simplicity we assume that the solutions to (1) satisfy

$$y(x) \in \mathbf{Y} \subset \mathbb{R}^{n_y}, \quad \mathcal{D}y(x) \in \mathbf{Z} \subset \mathbb{R}^{n_y \times n}.$$

Given any solution to (1), the *occupation measure* associated to $y(\cdot)$ is defined by

$$\mu(A \times B \times C) = \int_\Omega \mathbb{I}_{A\times B\times C}\,(x, y(x), \mathcal{D}y(x))\, dx \tag{8}$$

for all Borel sets $A \subset \Omega$, $B \subset \mathbf{Y}$, $C \subset \mathbf{Z}$. Similarly, the *boundary measure* is defined by

$$\mu_\partial(A \times B \times C) = \int_{\partial\Omega} \mathbb{I}_{A\times B\times C}\,(x, y(x), \mathcal{D}y(x))\, d\sigma(x), \tag{9}$$

for all Borel sets $A \subset \partial\Omega$, $B \subset \mathbf{Y}$, $C \subset \mathbf{Z}$. The occupation measure captures the behavior of the solution $y(\cdot)$ and its first derivative in Ω whereas μ_∂ captures the behavior on the boundary. The measures μ and μ_∂ are nonnegative Borel measures on $\Omega \times \mathbf{Y} \times \mathbf{Z}$ respectively $\partial\Omega \times \mathbf{Y} \times \mathbf{Z}$. It follows immediately from the definition of μ and μ_∂ that for any bounded Borel measurable function $h : \Omega \times \mathbf{Y} \times \mathbf{Z} \to \mathbb{R}$ we have

$$\int_\Omega h\,(x, y(x), \mathcal{D}y(x))\, dx = \int_{\Omega\times\mathbf{Y}\times\mathbf{Z}} h\,(x, y, z)\, d\mu(x, y, z) \tag{10a}$$

and

$$\int_{\partial\Omega} h\,(x, y(x), \mathcal{D}y(x))\, d\sigma(x) = \int_{\partial\Omega\times\mathbf{Y}\times\mathbf{Z}} h\,(x, y, z)\, d\mu_\partial(x, y, z). \tag{10b}$$

We note that on the right-hand side of (10a) and (10b), y and z are not functions of x anymore but rather independent variables; their dependence is now encoded in the occupation and boundary measures μ and μ_∂. Therefore, the measures μ and μ_∂ allow us to evaluate integral functionals of the solutions to the PDE (1) using integration with respect to ordinary nonnegative Borel measures defined on the Euclidean space subsets $\Omega \times \mathbf{Y} \times \mathbf{Z}$ respectively $\partial\Omega \times \mathbf{Y} \times \mathbf{Z}$. This simple observation will turn out to be crucial in constructing computable bounds on the optimal value of (7).

3.1 Linear representation

Now we will use the occupation and boundary measures to derive a *linear* equation in the space of Borel measures which is satisfied by any pair of occupation and boundary measures arising from a solution to (1).

3.1.1 Scalar case

To begin, we will demonstrate the derivation in the scalar case ($n = n_y = 1$) and with $B_{i,j} = 0$. To this end, let $y(\cdot)$ be a solution to (1) and denote $z(x) := \frac{\partial y}{\partial x}(x)$. Given any test function $\phi \in C^\infty(\Omega \times \mathbf{Y})$ we have using (3)

$$\begin{aligned}\int_{\partial\Omega} \phi(x, y(x))\eta(x) d\sigma(x) &= \int_\Omega \frac{d}{dx}\phi(x, y(x))\, dx = \int_\Omega \frac{\partial\phi}{\partial x} + \frac{\partial\phi}{\partial y}\frac{\partial y}{\partial x}\, dx \\ &= \int_\Omega \frac{\partial\phi}{\partial x} + \frac{\partial\phi}{\partial y} z(x)\, dx,\end{aligned}$$

where $\eta(x)$ is the outward-pointing normal unit vector to $\partial\Omega$. Using (10a) and (10b), we get

$$\int_{\partial\Omega\times\mathbf{Y}\times\mathbf{Z}} \phi(x, y)\eta(x)\, d\mu_\partial(x, y, z) - \int_{\Omega\times\mathbf{Y}\times\mathbf{Z}} \frac{\partial\phi}{\partial x} + \frac{\partial\phi}{\partial y} z\, d\mu(x, y, z) = 0. \tag{11}$$

Similarly, we also have

$$\int_\Omega \phi(x, y(x))F(x, y(x), z(x))\, dx = 0,$$

$$\int_{\partial\Omega} \phi(x, y(x))G(x, y(x), z(x))\, d\sigma(x) = 0.$$

Using (10a) and (10b) these equations translate to

$$\int_{\Omega\times\mathbf{Y}\times\mathbf{Z}} \phi(x, y)F(x, y, z)\, d\mu(x, y, z) = 0 \tag{12}$$

$$\int_{\partial\Omega\times\mathbf{Y}\times\mathbf{Z}} \phi(x, y)G(x, y, z)\, d\mu_\partial(x, y, z) = 0. \tag{13}$$

Again notice that in (11), (12), (13), y and z are no longer functions of x but rather independent variables; their dependence is encoded in the occupation and boundary measures μ and μ_∂. In addition, crucially, we observe that these equations depend *linearly* on (μ, μ_∂). Therefore, we have shown that any solution to the nonlinear PDE (1) satisfies the system of *linear* equations (11), (12), (13) in variables (μ, μ_∂) for all test functions $\phi \in C^\infty(\Omega \times \mathbf{Y})$.

3.1.2 General case

We have the following theorem:

Theorem 1. *Let $y(\cdot) \in W^{k,\infty}(\Omega^o; \mathbf{Y}) \cap C(\Omega; \mathbf{Y})$ ($k = 1$ if $B_{i,j} = 0$, $k = 2$ otherwise) satisfy (1a) and (1b) almost everywhere with respect to the Lebesgue measure on Ω^o respectively the surface measure σ on $\partial\Omega$. If μ and μ_∂ are defined by (8) and (9), then for all $\phi \in C^\infty(\Omega \times \mathbf{Y})$*

$$\int_{\Omega\times\mathbf{Y}\times\mathbf{Z}} \left[\frac{\partial\phi}{\partial x_m} + \sum_{k=1}^{n_y} \frac{\partial\phi}{\partial y_k} z_{k,m}\right] d\mu(x,y,z) - \int_{\partial\Omega\times\mathbf{Y}\times\mathbf{Z}} \phi(x,y)\eta_m(x)\, d\mu_\partial(x,y,z) = 0 \tag{14a}$$

for all $m \in \{1, \ldots, n\}$ and

$$\begin{aligned}
&\int_{\Omega\times\mathbf{Y}\times\mathbf{Z}} \phi(x,y)F(x,y,z)\, d\mu(x,y,z) \\
&- \sum_{i,j} \int_{\Omega\times\mathbf{Y}\times\mathbf{Z}} \phi(x,y) \left[\frac{\partial B_{i,j}}{\partial x_j} + \sum_{k=1}^{n_y} \frac{\partial B_{i,j}}{\partial y_k} z_{k,j}\right] z_{\bullet,i}\, d\mu(x,y,z) \\
&- \sum_{i,j} \int_{\Omega\times\mathbf{X}\times\mathbf{Y}} \left[\frac{\partial\phi}{\partial x_j} + \sum_{k=1}^{n_y} \frac{\partial\phi}{\partial y_k} z_{k,j}\right] B_{i,j}(x,y) z_{\bullet,i}\, d\mu(x,y,z) \\
&+ \sum_{i,j} \int_{\partial\Omega\times\mathbf{Y}\times\mathbf{Z}} B_{i,j}(x,y)\phi(x,y) z_{\bullet,i}\eta_j(x)\, d\mu_\partial(x,y,z) = 0
\end{aligned} \tag{14b}$$

and

$$\int_{\partial\Omega\times\mathbf{X}\times\mathbf{Z}} \phi(x,y)G(x,y,z)\, d\mu_\partial(x,y,z) = 0, \tag{14c}$$

where $z_{\bullet,i}$ denotes the ith column of the matrix variable $z \in \mathbb{R}^{n_y \times n}$.

Proof. To prove (14a), fix $m \in \{1, \ldots, n\}$ and $\phi \in C^\infty(\Omega \times \mathbf{Y})$. Since y is continuous and $y \in W^{k,\infty}$ with $k \in \{1, 2\}$ and $\partial\Omega$ is Lipschitz, y is in fact Lipschitz on Ω (this follows from Evans (2010, Theorem 4, p. 294) and the extension theorem in Leoni (2017, Theorem 12.15) used in the last step of the proof).

Consequently, $\phi(x, y(x))$ is also Lipschitz on Ω and therefore the integration by parts formula (3) can be applied to obtain

$$\int_{\partial\Omega} \phi(x, y(x))\eta_m(x)\, d\sigma(x) = \int_{\Omega} \left[\frac{\partial \phi}{\partial x_m} + \sum_{k=1}^{n_y} \frac{\partial \phi}{\partial y_k} \frac{\partial y_k}{\partial x_m} \right] dx.$$

Using (10a) and (10b), this is equivalent to

$$\int_{\partial\Omega\times\mathbf{X}\times\mathbf{Z}} \phi(x, y)\eta_m(x)\, d\mu_\partial(x, y, z)$$
$$= \int_{\Omega\times\mathbf{X}\times\mathbf{Z}} \left[\frac{\partial \phi}{\partial x_m} + \sum_{k=1}^{n_y} \frac{\partial \phi}{\partial y_k} z_{k,m} \right] d\mu(x, y, z),$$

which is (14a).

To prove (14b), fix $\phi \in C^\infty(\Omega \times \mathbf{Y})$ and assume $y \in W^{2,\infty}(\Omega^\circ; \mathbf{Y})$ satisfies (1a) for almost all $x \in \Omega$. Then the function $\phi(\cdot, y(\cdot))$ is in $W^{2,\infty}(\Omega; \mathbb{R})$. In addition, since F and $B_{i,j}$ are polynomials in all its arguments, each component of the vector-valued function $F(\cdot, y(\cdot), \mathcal{D}y(\cdot))$ is in $L_\infty(\Omega; \mathbb{R})$ and each component of the matrix-valued function $B_{i,j}(\cdot, y(\cdot))$ is in $W^{2,\infty}(\Omega; \mathbb{R})$. Therefore, each term on the left-hand side of (1a) is in $L_\infty(\Omega)$ and since (1a) holds for almost all $x \in \Omega^\circ$ and Ω is compact, we have

$$\int_{\Omega} \phi(x, y(x)) F(x, y(x), \mathcal{D}y(x))\, dx$$
$$+ \sum_{i,j} \int_{\Omega} \phi(x, y(x)) B_{i,j}(x, y(x)) \frac{\partial^2 y}{\partial x_i \partial x_j}\, dx = 0.$$

Eq. (14b) follows by using (10a) on the first term of the above expression and (10a) and (10b) on the second term after using the integration by parts formula (3) and the chain rule, which are justified under the regularity conditions stated above since both $\phi(\cdot, y(\cdot))B_{i,j}(\cdot, y(\cdot))$ and $\frac{\partial y}{\partial x_j}$ are Lipschitz.

Finally, Eq. (14c) follows immediately by integrating (1b) against a test function $\phi \in C^\infty(\Omega \times \mathbf{Y})$ and using (10b). □

We note that the assumption $y(\cdot) \in W^{k,\infty}$ is made in order to avoid technicalities and can be relaxed; see Remark 1.

More importantly, we note that as in the scalar case, Eqs. (14a), (14b), (14c) depend *linearly* on (μ, μ_∂). We also note that x, y, z are all integrated variables; in particular y and z are *not* a function of x in these equations. Notice also that if F, G, and $B_{i,j}$ are polynomial, all integrands in (14) are polynomials in (x, y, z) provided that the test function $\phi(x, y)$ is polynomial and provided that the normal vector $\eta(x)$ depends polynomially[5] on x. This will be crucial in developing

[5] The assumption of η being polynomial is made just for simplicity of exposition. If $\partial\Omega_i \subset \{x : g_i(x) = 0\}$ with $g_i \in \mathbb{R}[x]$, then $\eta(x) = \nabla g_i(x)/\|\nabla g(x)\|$. Therefore on $\partial\Omega_i$ one replaces the

a tractable finite-dimensional approximation of these infinite-dimensional equations.

Note also that Eqs. (14a)-(14c) are in general *relaxations* in the sense that the set of all measures (μ, μ_∂) satisfying these equations may be larger than the set of all occupation and boundary measure pairs corresponding to the solutions to the PDE (1). Note in particular that the set of all measures satisfying (14a)-(14c) is convex, whereas the set of all occupation and boundary measure pairs satisfying (1) may be nonconvex (unless the solution is unique, which we do not assume). Therefore, in the very least, the set of all solutions to (14a)-(14c) contains the entire closed *convex hull* of the set of all boundary and occupation measure pairs. It remains an open question whether the solution set of (14a)-(14c) in fact coincides with this convex hull. This question is not only of independent theoretical interest but also a prerequisite for tightness of the bounds developed in the following sections; see Remark 7 for a further discussion.

Remark 6 (Relation to Marx et al., 2020). The philosophy of this work is the same as that of Marx et al. (2020) and historically the first drafts of both works were produced at around the same time. The main differences between the two works are as follows:

- The work (Marx et al., 2020) focuses on scalar hyperbolic conservation laws whereas our work is applicable to much broader class of systems of multivariate fully nonlinear PDEs of the form (1).
- The work (Marx et al., 2020) focuses on analysis only, whereas this work addresses both analysis and optimal control of PDEs.
- The work (Marx et al., 2020) utilizes test functions that depend only on (t, x) whereas our work uses test functions depending on (t, x, y), thereby generalizing previous works on ODEs (e.g., Vinter, 1993; Lasserre et al., 2008). However, in Marx et al. (2020), additional entropy inequalities are added that guarantee the concentration of the measure-valued solutions. These entropy inequalities can be seen as a subset of the constraints imposed in our work by taking $\phi(t, x, y) = \psi(t, x)\eta'(y)$ where ψ is a test function in $C^\infty(\Omega)$ and η comes from an entropy pair (η, q) (see Definition 2 in Marx et al., 2020 or DiPerna, 1985). Since in our work we impose these entropy constraints in the form of equalities, we require stronger regularity assumptions on the solution to the PDE ($W^{1,\infty}$ versus L^∞).

3.2 Infinite-dimensional linear program

Now we are ready to write down an infinite-dimensional LP whose optimal value will provide a lower or upper bound on the optimal value of (7). We denote

---

boundary measure σ with $\sigma' := \sigma/\|\nabla g_i(x)\|$ and then $\eta(x)d\mu_{\partial_i}$ is replaced everywhere with $\nabla g_i(x)d\mu_{\partial_i}$; see Section 4.3 of the extended version (Korda et al., 2018). Moments of σ_i' can be obtained from those of σ_i as described in Lasserre and Magron (2020).

by $\mathcal{M}(A)_+$ the set of all nonnegative Borel measures with supports included in the set A. We shall decompose the boundary measure μ_∂ according to the partition of the boundary of Ω (4), i.e., we write

$$\mu_\partial = \sum_{i=1}^{n_b} \mu_{\partial_i}, \quad \mu_{\partial_i} \in \mathcal{M}(\partial\Omega_i \times \mathbf{Y} \times \mathbf{Z})_+. \tag{15}$$

We note that the decomposition (15) is unique since $\mu_{\partial_i}(\partial\Omega_j \times \mathbf{Y} \times \mathbf{Z}) = 0$ whenever $i \neq j$; this follows from the fact that, by (9), the x-marginal of μ_∂ is equal to the surface measure σ and $\sigma(\partial\Omega_i \cap \partial\Omega_j) = 0$ whenever $i \neq j$.

With this notation, the infinite-dimensional LP reads

$$\begin{array}{ll} \inf/\sup\limits_{(\mu,\mu_{\partial_1},\ldots,\mu_{\partial_{n_b}})} & \displaystyle\int_{\Omega\times\mathbf{Y}\times\mathbf{Z}} L(x,y,z)\,d\mu + \sum_{i=1}^{n_b} \int_{\partial\Omega\times\mathbf{Y}\times\mathbf{Z}} L_\partial(x,y,z)\,d\mu_{\partial_i} \\ \text{subject to} & \text{(14a), (14b), (14c) [with } \mu_\partial = \textstyle\sum_{i=1}^{n_b} \mu_{\partial_i}] \\ & \forall \phi \in C^\infty(\Omega \times \mathbf{Y}) \\ & \displaystyle\int_{\partial\Omega_i\times\mathbf{X}\times\mathbf{Z}} \psi(x)\,d\mu_{\partial_i}(x,y,z) = \int_{\partial\Omega_i} \psi(x)\,\sigma(x) \\ & \forall \psi \in C^\infty(\Omega),\ i \in \{1,\ldots,n_b\} \\ & (\mu, \mu_{\partial_1}, \ldots, \mu_{\partial_{n_b}}) \in \mathcal{K}, \end{array} \tag{16}$$

where

$$\mathcal{K} = \mathcal{M}(\Omega \times \mathbf{Y} \times \mathbf{Z})_+ \times \mathcal{M}(\partial\Omega_1 \times \mathbf{Y} \times \mathbf{Z})_+ \times \ldots \times \mathcal{M}(\partial\Omega_{n_b} \times \mathbf{Y} \times \mathbf{Z})_+.$$

Since $\mathcal{M}(\cdot)_+$ is a *convex cone*, so is $\mathcal{K}$. As a result, since all equality constraints of (16) are affine in $(\mu, \mu_{\partial_1}, \ldots, \mu_{\partial_{n_b}})$, the optimization problem (16) is indeed an infinite dimensional LP. The last equality constraint of (16) is a normalization constraint prescribing that the x-marginal of μ_{∂_i} is equal to the surface measure restricted to Ω_i, which trivially holds for any boundary measure defined by (9). Using the integration by parts formula (3) and (14a) with $\phi(x, y) = x^\alpha$ for all $\alpha \in \mathbb{N}^n$, this also implies that the x-marginal of μ is equal to the Lebesgue measure on Ω, which holds for any occupation measure defined by (8).

Importantly, the following result states that whenever the pair of measures (μ, μ_∂) feasible in (16) is supported on a graph of a sufficiently regular function $x \mapsto (y(x), z(x))$, then necessarily $\frac{\partial y_k}{\partial x_m} = z_{k,m}(x)$.

Lemma 1. *Let (μ, μ_∂) be a pair of measures feasible in (16) and assume that μ and μ_∂ are supported on a graph $\{(x, y(x), z(x)) \mid x \in \Omega\} \subset \Omega \times \mathbf{Y} \times \mathbf{Z}$ for some mappings $y(\cdot) \in W^{1,\infty}$ and $z(\cdot) \in L_\infty$. Then for all $m \in \{1, \ldots, n\}$ and all $k \in \{1, \ldots, n_y\}$*

$$\frac{\partial y_k}{\partial x_m}(x) = z_{k,m}(x) \tag{17}$$

holds almost everywhere in Ω. In particular, if $y(\cdot) \in C^1(\Omega)$ and $z(\cdot) \in C(\Omega)$, then (17) holds for all $x \in \Omega$.

Proof. Fix k and m and select $\phi(x, y) = \psi(x) y_k$ with $\psi(x) \in C^\infty(\Omega)$. Using the fact that (μ, μ_∂) are feasible in (16) and hence satisfy (14a) and by the assumption that μ and μ_∂ are supported on the graphs, we obtain

$$\int_\Omega \left[y_k(x) \frac{\partial \psi}{\partial x_m} + \psi(x) z_{k,m}(x) \right] dx = \int_{\partial\Omega} \eta_m(x) y_k(x) \psi(x) \, d\sigma(x),$$

where we disintegrated the measures μ and μ_∂ and used the facts that their respective x-marginals are the Lebesgue measure on Ω and the surface measure σ. Applying integration by parts (3) to the right hand side of this equation yields

$$\int_\Omega \psi(x) \left[z_{k,m}(x) - \frac{\partial y_k}{\partial x_m}(x) \right] dx = 0.$$

Since Ω is compact and ψ an arbitrary smooth function, this implies that (17) holds for almost all $x \in \Omega$. If $y_k \in C^1$ and $z_k \in C$, this implies that in fact (17) holds for all $x \in \Omega$. □

Now we will prove a result establishing that the LP (16) provides bounds on the optimal value of the nonconvex optimization problem (7) and therefore on the functional (6) evaluated along the solutions to the PDE (1).

Theorem 2. *Let p_{inf} and p_{sup} denote the optimal values of (7) with infimum respectively supremum and similarly let $\hat{p}_{\text{inf}}$ and $\hat{p}_{\text{sup}}$ denote the respective optimal values for (16). Then we have*

$$\hat{p}_{\text{inf}} \le p_{\text{inf}}, \tag{18a}$$

$$\hat{p}_{\text{sup}} \ge p_{\text{sup}}. \tag{18b}$$

Proof. The result follows immediately from Theorem 1. Indeed, given any $y(\cdot)$ satisfying the constraints of (7) (i.e., solving the PDE (1)), Theorem 1 guarantees that (14a)-(14c) are satisfied by the corresponding occupation and boundary measures (μ, μ_∂) defined by (8) and (9), which are nonnegative measures with supports in $(\Omega \times \mathbf{Y} \times \mathbf{Z})$ respectively $(\partial\Omega \times \mathbf{Y} \times \mathbf{Z})$ and therefore feasible in (16), after performing the decomposition (15). The value of the objective functional in (16) associated to this occupation and boundary measures coincides with $J(y(\cdot))$ by virtue of (10a) and (10b). Therefore, the feasible set of (7) is a subset of the feasible set of (16) and the objective functionals coincide on the intersection of the two. As a result, (18a) and (18b) necessarily hold. □

Remark 7. It is an interesting and challenging question to investigate whether the inequalities in (18a) and (18b) are strict. In other words, to investigate whether the infinite-dimensional LP (16) is a relaxation or an equivalent reformulation of (7). If (18a) and (18b) turn out to be relaxations (i.e., the inequalities are strict), one could try to add additional linear or convex constraints to (16) that are necessarily satisfied by the occupation and boundary measures of any solution to (1), thereby tightening the constraint set of (16), with the aim to reduce the gap between the optimal values. One type of inequalities, valid for a class of conservation laws, are the so-called entropy inequalities originally introduced in DiPerna (1985) in a scalar setting and subsequently generalized in, e.g., Brenier et al. (2011). The situation, however, appears to be more complex for the general formulation treated in this work, especially in the controlled setting treated in Section 5. Let us mention that the absence of such a gap was proven in Vinter (1993) for the problem of optimal control of ordinary differential equations and in Henrion and Korda (2014); Korda et al. (2014b) for the problems of region of attraction and invariant set computation.

4 Computable bounds using SDP relaxations

In order to turn the previous developments into a computational method, we rely on the so-called Moment-SOS hierarchy of semidefinite relaxations of infinite-dimensional LPs in the space of measures. This approach is now well established and therefore we only state the main convergence result of the resulting numerical scheme. For a more detailed exposition the interested reader is referred to the extended version of this work available online (Korda et al., 2018) and to Lasserre (2009) and Henrion et al. (2020) for some of its various applications.

Such relaxations can be constructed using the high-level software package Gloptipoly 3 (Henrion et al., 2009). The resulting SDP relaxations are indexed by a degree d which controls the quality of the approximation and appears in the following theorem whose proof can be found in Korda et al. (2018).

Theorem 3. *Let $\hat{p}^{\mathrm{SDP}}_{\mathrm{inf},d}$ resp. $\hat{p}^{\mathrm{SDP}}_{\mathrm{sup},d}$ denote the optimal values of the degree-d SDP relaxations of (16) with infimum resp. supremum. Then:*

$$\hat{p}^{\mathrm{SDP}}_{\mathrm{inf},d} \le \hat{p}^{\mathrm{SDP}}_{\mathrm{inf},d+1} \le \hat{p}_{\mathrm{inf}} \le p_{\mathrm{inf}}\,; \quad \hat{p}^{\mathrm{SDP}}_{\mathrm{sup},d} \ge \hat{p}^{\mathrm{SDP}}_{\mathrm{sup},d+1} \ge \hat{p}_{\mathrm{sup}} \ge p_{\mathrm{sup}} \tag{19}$$

for all $d \in \mathbb{N}$. In addition,

$$\lim_{d\to\infty} \hat{p}^{\mathrm{SDP}}_{\mathrm{inf},d} = \hat{p}_{\mathrm{inf}} \quad \textit{and} \quad \lim_{d\to\infty} \hat{p}^{\mathrm{SDP}}_{\mathrm{sup},d} = \hat{p}_{\mathrm{sup}}, \tag{20}$$

provided that Ω, $\mathbf{Y}$ and $\mathbf{Z}$ are compact and the polynomials defining these sets satisfy the Archimedean condition.[6]

[6] The Archimedean condition (e.g., Laurent, 2009, Definition 3.18) is a nonrestrictive algebraic condition implying compactness. This condition can always be satisfied by adding redundant ball constraints of the form $N^2 - x^\top x \ge 0$ to the description of the *compact* basic semialgebraic sets Ω, $\mathbf{Y}$, $\mathbf{Z}$ for a sufficiently large N. Adding such ball constraints also ensures that the infimum or supremum in the SDP relaxations is always attained. However, Theorem 3 is valid whenever the Archimedean condition holds, irrespective of whether redundant ball constraints are present.

Generalizations

Generalization of the approach to the case of non-polynomial boundary conditions, periodic boundary conditions and to the case of non-polynomial surface normal vector can be found in the extended version (Korda et al., 2018), Sections 4.1–4.3.

Computational complexity and complexity reduction

Computational complexity and complexity reduction are discussed in detail in the extended version (Korda et al., 2018, Section 9).

5 Problem statement (control)

In this section, we describe the problem setting for control of nonlinear PDEs. We consider a controlled nonlinear PDE of the form

$$F(x, y(x), \mathcal{D}y(x)) + \sum_{i,j} B_{i,j}(x, y(x))\frac{\partial^2 y}{\partial x_i \partial x_j} = C(x, y(x))u(x), \qquad x \in \Omega, \tag{21a}$$

$$\forall i \in \{1, \ldots, n_b\}\ G_i(x, y(x), \mathcal{D}y(x)) = C_{\partial_i}(x, y(x))u_{\partial_i}(x), \ \ x \in \partial\Omega_i, \tag{21b}$$

where $u(\cdot)$ is the distributed control and $u_{\partial_i}(\cdot)$ are the boundary controls defined on each element of the partition of the boundary (4). The functions $C : \Omega \to \mathbb{R}^{n_F \times n_u}$ and $C_{\partial_i} : \partial\Omega_i \to \mathbb{R}^{n_{G_i} \times n_{u_i}}$ are given matrix functions with each entry being a multivariate polynomial in (x, y). Similarly, the vector functions G_i : $\Omega \times \mathbb{R}^{n_y} \times \mathbb{R}^{n_y \times n} \to \mathbb{R}^{n_{G_i}}$ are assumed to be multivariate polynomials in all variables. Otherwise, the setup is the same as in Section 2.

The control goal is to minimize the functional

$$\begin{aligned} J\big(y(\cdot), u(\cdot)\big) := &\int_\Omega L\,(x, y(x), \mathcal{D}y(x))\, dx \\ &+ \sum_{i=1}^{n_b} \int_{\Omega_{\partial,i}} L_{\partial_i}\,(x, y(x), \mathcal{D}y(x))\, d\sigma(x) \\ &+ \int_\Omega L_u\,(x, y(x))^\top u(x)\, dx \\ &+ \sum_{i=1}^{n_b} \int_{\Omega_{\partial,i}} L_{u_i}\,(x, y(x))^\top u_{\partial_i}(x)\, d\sigma(x), \end{aligned} \tag{22}$$

subject to constraints on y, $\mathcal{D}y$ and the control inputs. Here L and L_{∂_i} are polynomials and L_u and L_{u_i} are vectors of polynomials. This leads to the following optimal control problem:

$$
\begin{array}{lll}
\inf/\sup\limits_{y(\cdot),u(\cdot)} & J\big(y(\cdot),u(\cdot)\big) & \\
\text{subject to} & \text{(21a), (21b)} & \\
& y(x)\in \mathbf{Y}\;\forall x\in\Omega, & y(x)\in \mathbf{Y}_{\partial_i}\;\;\forall x\in\partial\Omega_i,\;\; i\in\{1,\ldots,n_{\mathrm{b}}\} \\
& \mathcal{D}y(x)\in \mathbf{Z}\;\forall x\in\Omega, & \mathcal{D}y(x)\in \mathbf{Z}_{\partial_i}\;\;\forall x\in\partial\Omega_i,\;\; i\in\{1,\ldots,n_{\mathrm{b}}\} \\
& u(x)\in \mathbf{U}\;\forall x\in\Omega, & u_{\partial_i}(x)\in \mathbf{U}_{\partial_i}\;\;\forall x\in\partial\Omega_i,\;\; i\in\{1,\ldots,n_{\mathrm{b}}\}
\end{array}
\tag{23}
$$

where $\mathbf{Y}$ and $\mathbf{Z}$ as well as $\mathbf{Y}_{\partial_i}$ and $\mathbf{Z}_{\partial_i}$, $i=1,\ldots,n_{\mathrm{b}}$, are basic semialgebraic sets. The sets $\mathbf{U}$ and $\mathbf{U}_{\partial_i}$ are assumed to be the unit boxes

$$\mathbf{U}=[0,1]^{n_u},\quad \mathbf{U}_{\partial_i}=[0,1]^{n_{u_i}}\,. \tag{24}$$

We note that any box constraint of the form $[-u_1^{\min},u_1^{\max}]\times\ldots\times[-u_{n_u}^{\min},u_{n_u}^{\max}]$ can be transformed to the unit boxes by a simple affine transformation of the functions C, F resp. C_{∂_i}, G_i. The input constraints (24) therefore cover all box constraints which are the most commonly encountered type of constraint in practice.

6 Linear representation (control)

In this section we derive an infinite-dimensional linear program in the space of Borel measures whose optimal value provides a lower bound on (23); we also show how an optimal solution to this LP can be used to extract a controller in a feedback form. This infinite-dimension LP is then approximated in Section 7 via finite-dimensional SDP relaxations, the solutions of which provide controllers for the PDE (21) in a feedback form. The approach closely follows the method of Korda et al. (2014a); Majumdar et al. (2014) developed for controlled ordinary differential equations.

As in Section 3.1 we consider the occupation and boundary measures μ and μ_∂ defined in (8) and (9), with the decomposition of the boundary measure (15). In addition, we define the vectors of *control measures* $\nu\in\mathcal{M}(\Omega\times\mathbf{Y})_+^{n_u}$ and $\nu_{\partial_i}\in\mathcal{M}(\partial\Omega_i\times\mathbf{Y})_+^{n_{u_i}}$ by

$$d\nu=u\,d\bar{\mu} \tag{25a}$$

$$d\nu_{\partial_i}=u_{\partial_i}d\bar{\mu}_{\partial_i}, \tag{25b}$$

where $\bar{\mu}$ respectively $\bar{\mu}_{\partial_i}$ are the (x,y) marginals[7] of μ respectively μ_{∂_i}. In other words, we have $\nu\ll\bar{\mu}$ and $\nu_{\partial_i}\ll\bar{\mu}_{\partial_i}$ with density (i.e., the Radon-Nikodym derivative) equal to $u(\cdot)$, respectively $u_{\partial_i}(\cdot)$. Here $\nu\ll\bar{\mu}$ signifies that the measure ν is absolutely continuous w.r.t. the measure $\bar{\mu}$. Therefore, for any bounded

7 The (x,y) marginal of a measure $\mu\in\mathcal{M}(\Omega\times\mathbf{Y}\times\mathbf{Z})_+$ is defined by $\bar{\mu}(A\times B)=\mu(A\times B\times\mathbf{Z})$ for all Borel sets $A\subset\Omega$, $B\subset\mathbf{Y}$.

Borel measurable function $h : \Omega \times \mathbf{Y} \to \mathbb{R}$:

$$\int_{\Omega} h(x, y(x))\, u(x)\, dx = \int_{\Omega\times\mathbf{Y}} h(x, y)\, d\nu(x, y), \tag{26a}$$

$$\int_{\partial\Omega_i} h(x, y(x))\, u_{\partial_i}(x)\, \sigma(x) = \int_{\partial\Omega_i\times\mathbf{Y}} h(x, y)\, d\nu_{\partial_i}(x, y). \tag{26b}$$

The following theorem is an immediate generalization of Theorem 1.

Theorem 4. *Let $u(\cdot) \in L_\infty(\Omega, dx)$, $u_{\partial_i}(\cdot) \in L_\infty(\Omega, d\sigma)$, $i = 1, \ldots, n_{\mathrm{b}}$, and let $y(\cdot) \in W^{k,\infty}(\Omega^{\mathrm{o}}; \mathbf{Y}) \cap C(\Omega; \mathbf{Y})$ ($k = 1$ if $B_{i,j} = 0$, $k = 2$ otherwise) satisfy (21a) and (21b) almost everywhere with respect to the Lebesgue measure on Ω^{o} respectively the surface measure σ on $\partial\Omega$. If μ and μ_∂ are defined by (8) and (9) and ν and ν_{∂_i} by (25a) and (25b), then for all $\phi \in C^\infty(\Omega \times \mathbf{Y})$*

$$\int_{\Omega\times\mathbf{Y}\times\mathbf{Z}} \left[\frac{\partial\phi}{\partial x_m} + \sum_{k=1}^{n_y} \frac{\partial\phi}{\partial y_k} z_{k,m}\right] d\mu(x, y, z) - \int_{\partial\Omega\times\mathbf{Y}\times\mathbf{Z}} \phi(x, y)\eta_m(x)\, d\mu_\partial(x, y, z) = 0 \tag{27a}$$

for all $m \in \{1, \ldots, n\}$ and

$$\begin{aligned}
&\int_{\Omega\times\mathbf{Y}\times\mathbf{Z}} \phi(x, y) F(x, y, z)\, d\mu(x, y, z)\\
&\quad - \sum_{i,j} \int_{\Omega\times\mathbf{Y}\times\mathbf{Z}} \phi(x, y) \left[\frac{\partial B_{i,j}}{\partial x_j} + \sum_{k=1}^{n_y} \frac{\partial B_{i,j}}{\partial y_k} z_{k,j}\right] z_{\bullet,i}\, d\mu(x, y, z)\\
&\quad - \sum_{i,j} \int_{\Omega\times\mathbf{X}\times\mathbf{Y}} \left[\frac{\partial\phi}{\partial x_j} + \sum_{k=1}^{n_y} \frac{\partial\phi}{\partial y_k} z_{k,j}\right] B_{i,j}(x, y) z_{\bullet,i}\, d\mu(x, y, z)\\
&\quad + \sum_{i,j} \int_{\partial\Omega\times\mathbf{Y}\times\mathbf{Z}} B_{i,j}(x, y)\phi(x, y) z_{\bullet,i}\eta_j(x)\, d\mu_\partial(x, y, z)\\
&= \sum_{k=1}^{n_u} \int_{\Omega\times\mathbf{Y}} C_{\bullet,k}(x, y)\phi(x, y) d\nu^k(x, y)
\end{aligned} \tag{27b}$$

and

$$\int_{\partial\Omega\times\mathbf{X}\times\mathbf{Z}} \phi(x, y) G(x, y, z)\, d\mu_\partial(x, y, z) = \sum_{i=1}^{n_{\mathrm{b}}} \sum_{k=1}^{n_{u_i}} \int_{\partial\Omega_i\times\mathbf{Y}} C^i_{\bullet,k}(x, y)\phi(x, y) d\nu^k_{\partial_i}(x, y), \tag{27c}$$

where $z_{\bullet,i}$ denotes the ith column of the matrix variable $z \in \mathbb{R}^{n_y \times n}$ and $C_{\bullet,k}$ respectively $C^i_{\bullet,k}$ the kth columns of C respectively C_{∂_i}. The symbols ν^k and $\nu^k_{\partial_i}$ denote the kth components of the vector measures ν and ν_{∂_i}.

Proof. The proof is analogous to the proof of Theorem 1. □

Note that again Eqs. (27a)-(27c) are *linear* in the variables $(\mu, \mu_\partial, \nu, \nu_{\partial_1}, \ldots, \nu_{\partial_{n_b}})$. This will allow us to write down an infinite-dimensional LP relaxation of the optimal control problem (23).

Before doing so we describe how to handle the requirement that $\nu \ll \bar{\mu}$ and $\nu_{\partial_i} \ll \bar{\mu}_{\partial_i}$ with density (= the control input) in [0, 1]. This is equivalent to requiring that $\nu \leq \bar{\mu}$ and $\nu_{\partial_i} \leq \bar{\mu}_{\partial_i}$, where $\leq$ is the ordinary inequality sign (i.e. $\nu(A) \leq \mu(A)$ for all Borel sets A). This in turn is equivalent to the existence of nonnegative slack measures $\hat{\nu} \in \mathcal{M}(\Omega \times \mathbf{Y})_+$, $\hat{\nu}_{\partial_i} \in \mathcal{M}(\partial\Omega \times \mathbf{Y})_+$, $i = 1, \ldots, n_b$, such that

$$\nu + \hat{\nu} = \bar{\mu}$$
$$\nu_{\partial_i} + \hat{\nu}_{\partial_i} = \bar{\mu}_{\partial_i},\ i \in \{1, \ldots, n_b\}.$$

Finally, these equalities are equivalent to

$$\int_{\Omega\times\mathbf{Y}} \phi(x,y)\,d\nu(x,y) + \int_{\Omega\times\mathbf{Y}} \phi(x,y)\,d\hat{\nu}(x,y) = \int_{\Omega\times\mathbf{Y}\times\mathbf{Z}} \phi(x,y)\,d\mu(x,y,z) \tag{28a}$$

$$\int_{\partial\Omega_i\times\mathbf{Y}} \phi(x,y)\,d\nu_{\partial_i}(x,y) + \int_{\partial\Omega_i\times\mathbf{Y}} \phi(x,y)\,d\hat{\nu}_{\partial_i}(x,y) = \int_{\partial\Omega_i\times\mathbf{Y}\times\mathbf{Z}} \phi(x,y)\,d\mu_{\partial_i}(x,y,z),\ i \in \{1, \ldots, n_b\} \tag{28b}$$

holding for all test functions $\phi \in C^\infty(\Omega \times \mathbf{Y})$.

6.1 Infinite dimensional LP (control)

Now we are ready to state the infinite-dimensional linear programming relaxation of (23). The relaxation reads

$$\begin{array}{ll} \inf\limits_{\substack{\mu,\mu_{\partial_1},\ldots,\mu_{\partial_{n_b}},\\ \nu,\nu_{\partial_1},\ldots,\nu_{\partial_{n_b}},\\ \hat{\nu},\hat{\nu}_{\partial_1},\ldots,\hat{\nu}_{\partial_{n_b}}}} & J_{\mathrm{LP}}(\mu, \mu_1, \ldots, \mu_{n_b}, \nu, \nu_1, \ldots, \nu_{n_b}) \\ \text{subject to} & \text{(27a), (27b), (27c), (28a), (28b)} \\ & [\text{with } \mu_\partial = \sum_{i=1}^{n_b} \mu_{\partial_i}] \quad \forall \phi \in C^\infty(\Omega \times \mathbf{Y}) \end{array}$$

$$\int_{\partial\Omega_i \times \mathbf{X} \times \mathbf{Z}} \psi(x)\, d\mu_{\partial_i}(x, y, z) = \int_{\partial\Omega_i} \psi(x)\sigma(x)$$
$$\forall \psi \in C^\infty(\Omega),\ i \in \{1, \ldots, n_b\}$$
$$(\mu, \mu_{\partial_1}, \ldots, \mu_{\partial_{n_b}}, \nu, \nu_{\partial_1}, \ldots, \nu_{\partial_{n_b}}, \hat{\nu}, \hat{\nu}_{\partial_1}, \ldots, \hat{\nu}_{\partial_{n_b}}) \in \mathcal{K}, \quad (29)$$

where

$$J_{\mathrm{LP}} = \int_{\Omega \times \mathbf{Y} \times \mathbf{Z}} L(x, y, z)\, d\mu(x, y, z) + \sum_{i=1}^{n_b} \int_{\partial\Omega \times \mathbf{Y} \times \mathbf{Z}} L_{\partial_i}(x, y, z)\, d\mu_{\partial_i}(x, y, z)$$
$$+ \sum_{k=1}^{n_u} \int_{\Omega} L_u^k(x, y) d\nu^k(x, y) + \sum_{i=1}^{n_b} \sum_{k=1}^{n_{u_i}} \int_{\partial\Omega} L_{u_i}^k(x, y) d\nu_{\partial_i}(x, y), \quad (30)$$

with L_u^k and $L_{u_i}^k$ denoting the kth components of the vector polynomials L_u and L_{u_i}, and $\mathcal{K}$ the convex cone:

$$\mathcal{K} = \mathcal{M}(\Omega \times \mathbf{Y} \times \mathbf{Z})_+ \times \mathcal{M}(\partial\Omega_1 \times \mathbf{Y} \times \mathbf{Z})_+ \times \ldots \times \mathcal{M}(\partial\Omega_{n_b} \times \mathbf{Y} \times \mathbf{Z})_+$$
$$\times\ \mathcal{M}(\Omega \times \mathbf{Y})_+ \times \mathcal{M}(\partial\Omega_1 \times \mathbf{Y})_+ \times \ldots \times \mathcal{M}(\partial\Omega_{n_b} \times \mathbf{Y})_+$$
$$\times\ \mathcal{M}(\Omega \times \mathbf{Y})_+ \times \mathcal{M}(\partial\Omega_1 \times \mathbf{Y})_+ \times \ldots \times \mathcal{M}(\partial\Omega_{n_b} \times \mathbf{Y})_+.$$

7 Control design using SDP relaxations

Since all integrands in the infinite-dimensional LP (29) are polynomials and all measures are defined on basic semialgebraic sets, a finite-dimensional SDP relaxation can be constructed following exactly the same procedure as described in Section 4. The result is the following finite-dimensional SDP:

$$\inf_{s} \{ c_d^\top s \ :\ A_d\, s = b_d\,; \quad s \in \mathcal{K}_d \}, \quad (31)$$

where

$$\mathcal{K}_d = \mathcal{M}_d(\Omega \times \mathbf{Y} \times \mathbf{Z})_+$$
$$\times\, \mathcal{M}_d(\partial\Omega_1 \times \mathbf{Y} \times \mathbf{Z})_+ \times \ldots \times \mathcal{M}_d(\partial\Omega_{n_b} \times \mathbf{Y} \times \mathbf{Z})_+$$
$$\times\, \mathcal{M}_d(\Omega \times \mathbf{Y})_+ \times \mathcal{M}_d(\partial\Omega_1 \times \mathbf{Y})_+ \times \ldots \times \mathcal{M}_d(\partial\Omega_{n_b} \times \mathbf{Y})_+$$
$$\times\, \mathcal{M}_d(\Omega \times \mathbf{Y})_+ \times \mathcal{M}_d(\partial\Omega_1 \times \mathbf{Y})_+ \times \ldots \times \mathcal{M}_d(\partial\Omega_{n_b} \times \mathbf{Y})_+,$$

and where the decision s is partitioned as

$$s := [s_\mu^\top,\ s_{\mu_1}^\top,\ \ldots,\ s_{\mu_{n_b}}^\top,\ s_\nu^\top,\ s_{\nu_1}^\top,\ \ldots,\ s_{\nu_{n_b}}^\top,\ s_{\hat{\nu}}^\top,\ s_{\hat{\nu}_1}^\top,\ \ldots,\ s_{\hat{\nu}_{n_b}}^\top]^\top, \quad (32)$$

corresponding to truncated moment sequences of the respective measures, up to degree d. The optimization problem (31) is a convex, finite-dimensional, semidefinite program that can be solved by any of the available public software such as MOSEK or SeDuMi (Pólik et al., 2007). In addition, the matrices and vectors A_d, b, c_d as well as the cone $\mathcal{K}_d$ can be assembled automatically using Gloptipoly 3 (Henrion et al., 2009), starting from a high level description closely resembling the form of the infinite-dimensional LP (29).

The following theorem is an exact analogue of Theorem 3 in the setting with control.

Theorem 5. *Let p_d denote the optimal value of the SDP (31), let $\hat{p}$ denote the optimal value of the infinite-dimensional LP (29) and let $p^\star$ denote the optimal value of the optimal control problem (23). Then*

$$p_d \leq p_{d+1} \leq \hat{p} \leq p^\star \tag{33}$$

for all $d \in \mathbb{N}$. In addition,

$$\lim_{d\to\infty} p_d = \hat{p}, \tag{34}$$

provided that Ω, $\mathbf{Y}$ and $\mathbf{Z}$ are compact and the polynomials defining these sets satisfy the Archimedean condition (see Footnote 6).

Proof. The proof is a verbatim copy of the proof of Theorem 3, once boundedness of the first components of the moments sequences of ν, $\hat{\nu}$, ν_{∂_i}, $\hat{\nu}_{\partial_i}$ has been established. But this follows immediately from boundedness of the first component of the moment sequence of μ and μ_{∂_i} and from (28a) and (28b) with $\phi = 1$. □

7.1 Controller extraction

In this section we describe one way to extract a feedback controller from an optimal solution of the SDP (31). The idea is to use the fact that the measures ν, ν_{∂_i} are absolutely continuous with respect to the measures μ, μ_{∂_i}, with density being the control input. The information at our disposal are the truncated approximate moment sequences of these measures obtained from an optimal solution of the SDP (31). Let s^d denote such a solution for a given even value of d, partitioned as in (32). In general, there are a number of ways to obtain an approximation of a density, given truncated moment sequences. In our case, we use a simple moment matching technique with a polynomial density. That is, we seek polynomials $\kappa^d(x, y)$ and $\kappa^d_{\partial_i}(x, y)$ of degree $d/2$ such that

$$\ell_{s^d_\nu}(\beta_{d/2}(x, y)) = \ell_{s_\mu}(\beta_{d/2}(x, y)\kappa^d(x, y)) \tag{35a}$$

$$\ell_{s^d_{\nu_i}}(\beta_{d/2}(x, y)) = \ell_{s_{\mu_i}}(\beta_{d/2}(x, y)\kappa^d_{\partial_i}(x, y)), \tag{35b}$$

where $\ell_s : \mathbb{R}[x, y] \to \mathbb{R}$ is the Riesz functional:

$$p\,(= \sum_{\beta\in\mathbb{N}^{n_x+n_y}} p_\beta (x, y)^\beta\,) \mapsto \ell_s(p) = \sum_\beta p_\beta s_\beta, \quad p \in \mathbb{R}[x, y],$$

and $\beta_{d/2}$ is the vector of monomials in (x, y) up to the total degree $d/2$. Denoting c_κ and c_{κ_i} the coefficients of κ and $\kappa^d_{\partial_i}$, we can write $k^d = \beta^\top_{d/2} c_\kappa$ and $k^d_{\partial_i} = \beta^\top_{d/2} c_{\kappa_i}$. Inserting these into (35a) and (35b) leads to

$$\tilde{s}^d_\nu = M(s^d_\mu) c_\kappa \tag{36a}$$

$$\tilde{s}^d_{\nu_i} = M(s^d_{\mu_i}) c_{\kappa_i}, \tag{36b}$$

where $\tilde{s}^d_\nu$ respectively $\tilde{s}^d_{\nu_i}$ denote the first $\binom{n+d/2}{n}$ components of s^d_ν respectively $s^d_{\nu_i}$ (i.e., all moments up to degree $d/2$) and $M(s^d_\mu)$ and $M(s^d_{\mu_i})$ denote the moment matrices associated to s^d_μ and $s^d_{\mu_i}$ defined in Korda et al. (2018, Eq. (20)). Eqs. (36a) and (36b) are a linear system of equations that can be solved for c_κ and c_{κ_i} (exactly if the moment matrices are invertible or in a least-squares sense otherwise), thereby obtaining the polynomial feedback controllers κ^d and $\kappa^d_{\partial_i}$. The control inputs $u(x)$ and $u_{\partial_i}(x)$ for the PDE (21) are then given in a feedback form:

$$u(x) = \kappa^d(x, y(x)), \tag{37a}$$

$$u_{\partial_i}(x) = \kappa^d_{\partial_i}(x, y(x)). \tag{37b}$$

8 Higher-order PDEs

In this section we briefly describe how the proposed approach extends to higher-order PDEs. We do not present a general theory but rather illustrate the approach on an example. The idea is to introduce additional variables for higher order derivatives. For instance consider the Dym equation (Strauss, 1990)

$$\frac{\partial y_1}{\partial x_1} - y_1^3 \frac{\partial^3 y_1}{\partial x_2^3} = 0,$$

where typically x_1 is time, x_2 the spatial variable, and $y_1(x_1, x_2)$ is the unknown function. Introduce the variables

$$y_2 = \frac{\partial y_1}{\partial x_2}, \quad y_3 = \frac{\partial y_2}{\partial x_2}$$

with which we obtain

$$\frac{\partial y_1}{\partial x_1} - y^3 \frac{\partial y_2}{\partial x} = 0, \quad y_2 - \frac{\partial y_1}{\partial x_2} = 0, \quad y_3 - \frac{\partial y_2}{\partial x_2} = 0.$$

This is a system of PDEs of the form (1a) with

$$F(x, y, \mathcal{D}y) = \begin{bmatrix} \frac{\partial y_1}{\partial x_1} - y^3 \frac{\partial y_2}{\partial x_2} \\ y_2 - \frac{\partial y_1}{\partial x_2} \\ y_3 - \frac{\partial y_2}{\partial x_2} \end{bmatrix}, \quad B_{i,j} = 0,$$

where $y = (y_1, y_2, y_3)$ and $x = (x_1, x_2)$. From here, the approach proceeds as described in the previous sections. However note that in this case most derivatives of y appear *linearly* in F and therefore the computational complexity of the approach can be significantly reduced by using one of the complexity reduction techniques described in Korda et al. (2018).

Convergence results of Theorem 3 are preserved as long as the so-called *Running Intersection Property* (Lasserre, 2006) is satisfied. See also Nie and Demmel (2008) for a discussion of computational complexity of different sparse relaxation techniques.

9 Numerical examples

9.1 Burgers' equation – analysis

Consider Burgers' equation (e.g., Evans, 2010)

$$\frac{\partial y}{\partial x_1} + y\frac{\partial y}{\partial x_2} = 0, \tag{38}$$

where typically, x_1 is time and x_2 the spatial variable. The goal is to find bounds on the functional (6) evaluated along the solutions of the PDE (38), with domain $\Omega = [0, T] \times [0, L]$, and $T = 5$, $L = 1$. For numerical stability reasons, we scaled the x_1 variable to $x_1 \in [0, 1]$ (when working with the monomial basis, we always advise to scale all variables to unit boxes or balls). Consider a periodic boundary condition on the spatial variable x_2, i.e., $y(x_1, 0) = y(x_1, L)$ for all $x_1 \in [0, T]$; this boundary condition is imposed as described in the extended version (Korda et al., 2018, Section 4.2). We also impose the initial condition $y(0, x_2) = y_0(x_2) = 10(x_2(1 - x_2))^2$; this Dirichlet-type initial condition is imposed as described in Korda et al. (2018, Section 4.1). For the integral cost functional we first choose $L(x, y) = y^2$ and $L_\partial(x, y) = 0$ in (6), i.e., we are seeking bounds on $\int_\Omega y(x)^2\,dx$, which can be interpreted as the average kinetic energy of the solution. It is straightforward to prove that in this case the moments $a_k(x_1) = \int_0^L y^k(x_1, x_2)\,dx_2$ are preserved (i.e., $\frac{d}{d_{x_1}} a_k(x_1) = 0$) for the solution of the PDE (38). Therefore

$$\begin{aligned}\int_{[0,T]\times[0,L]} y(x)^2\,dx &= T a_2(0) = \int_0^L y^2(0, x_2)\,dx_2 \\ &= 5\int_0^1 [10(x_2(1 - x_2))^2]^2\,dx_2 = \frac{50}{63} \approx 0.79365079365.\end{aligned}$$

Solving the SDP relaxation at step $d = 4$ of the hierarchy (19), yields:

$$\hat{p}^{\text{SDP}}_{\text{inf},d} \approx 0.79365079357, \quad \hat{p}^{\text{SDP}}_{\text{sup},d} \approx 0.79365080188,$$

i.e., achieving precision of nine respectively six significant digits for the lower and upper bounds. However, this extremely high accuracy is due to the special choice of the objective functional along with the invariance of $a_2(x_1)$ and cannot be expected in general. Indeed, the invariance condition implies that $s_{(0,2)} = T\int_0^L y_0(x_2)\,dx_2 (= 50/63)$ is among the constraints of the SDP relaxation and hence the objective functional $c^\top s = s_{(0,2)}$ is fully specified by equality constraints only.

Next, we consider an objective functional where the value of the objective functional is not fully determined by the constraints. By choosing $L(x, y) := x_2^2 y^2$ and $L_\partial := 0$, we want to find bounds on $\int_{[0,T]\times[0,L]} x_2^2 y^2(x_1, x_2)\,dx_1 dx_2$. The results are summarized in Table 1. We have set $\mathbf{Y} = \mathbb{R}$ and $\mathbf{Z} = \mathbb{R}$, since we assume to have no *à priori* bound on the supremum of $|y|$ and since $\frac{\partial y}{\partial x_2}$ is known to be discontinuous in this case; therefore the convergence result in Theorem 3 is not guaranteed in this case. Nevertheless, one still obtains bounds of increasing accuracy, with a guaranteed interval $[0.276, 0.283]$ at step $d = 8$ of the hierarchy.

TABLE 1 Uncontrolled Burgers equation: Upper and lower bounds on the cost functional (6) with $L = x_2^2 y^2$ and $L_\partial = 0$.

d	4	6	8
Lower bound (SeDuMi)	0.206	0.263	0.276
Upper bound (SeDuMi)	0.380	0.297	0.283
Parse time (Gloptipoly 3)	2.91 s	3.41 s	6.23 s
SDP solve time (SeDuMi / MOSEK)	2.62 / 1.63 s	2.61 / 1.32 s	20.67 / 7.05[a] s

[a] *For $d = 8$ the MOSEK solver reported an ill-posed dual problem even though the objective value converged to a valid bound (up to a numerical roundoff error). This may be caused by the absence of Slater's condition. Investigation of the dual SDP, its theoretical interpretation, practical use and numerical properties, are left for future work.*

9.2 Burgers' equation – control

We now consider controlling Burgers' equation with distributed control, i.e.,

$$\frac{\partial y}{\partial x_1} + y\frac{\partial y}{\partial x_2} = u(x_1, x_2), \tag{39}$$

where $u(\cdot,\cdot)$ is the control input to be designed subject to the constraint $u(x_1, x_2) \in [-1, 1]$. The domain is given by $\Omega = [0, T] \times [0, L]$ with $T = 3$ and $L = 1$ and we consider a periodic boundary constraint on the x_2 variable, enforced as described Section 9.1. The control goal is to minimize the energy $\int_{[0,T]\times[0,L]} y(x_1, x_2)^2\,dx$ of the solution, starting from the initial condition $y(0, x_2) = y_0(x_2) = 10(x_2(1 - x_2))^2$. This is enforced by selecting $L(x, y) = y^2$ and $L_\partial = L_u^k = L_{u_i}^k = 0$ in the objective functional (22); the initial condition is imposed as described in the extended version (Korda et al., 2018, Section 4.1).

We solve (31) with $d = 6$, where we used the complexity reduction method described in Korda et al. (2018, Section 9.1.1) to derive the SDP. From an optimal solution of the SDP, we extracted a polynomial controller of degree three of the form (37a), by using the moment-matching equation (36a). For a closed-loop simulation, the PDE was discretized using a finite-difference scheme with time-step (i.e., for x_1) of 0.01 s, and spatial discretization of the x_2 variable with 100 grid points. One observes a stable behavior with the solution converging to zero as x_1 (= time) increases. See Fig. 1.

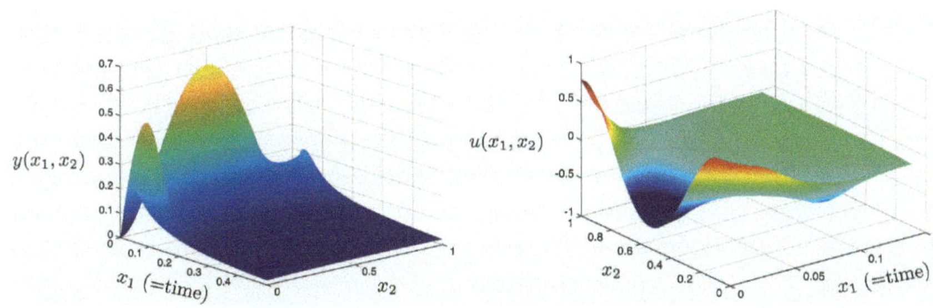

FIGURE 1 Control of Burgers' equation (39). The control input $u(x_1, x_2)$ is given by $u(x_1, x_2) = \kappa^d(x_1, x_2, y(x_1, x_2))$, where κ^d is a polynomial of degree three in (x_1, x_2, y). CPU-time to solve the SDP (31) was ≈ 1.2 s, using Matlab® + MOSEK 8, 2 GHz intel i7. Evaluation of the polynomial feedback control law for each time step on the entire spatial grid of 100 points takes approximately 1.5 ms.

10 Conclusion

We have presented a convex-optimization-based method for analysis and control of nonlinear partial differential equations (PDEs). The method proceeds by embedding the problem at hand into spaces of Borel measures, with the so called occupation and boundary measures as variables. This measure-valued weak formulation leads to an infinite-dimensional linear program (LP), which is then approximated by a sequence of finite-dimensional convex semidefinite programming problems (SDPs), with proven convergence to the optimal value of the LP. Optimal solutions of these SDPs provide either bounds of increasing accuracy on (possibly nonlinear) integral functionals of solutions to the PDE in the uncontrolled case, or polynomial feedback controllers in the controlled case. Two distinguishing features of the approach are (i) its overall convexity, and (ii) its mesh-free character, i.e., no spatio-temporal gridding is needed. We have discussed computational complexity of the approach and outlined several ways of complexity reduction, exploiting the structure of the problem.

There are several open questions left for future research. The first one is the equivalence of the infinite-dimensional LP to the original problem. A starting point here could be the theory of weak-strong solutions (Brenier et al., 2011) initiated in Lions (1996) where additional convex constraints are introduced to ensure uniqueness of the weak (i.e., measure-valued) solution whenever the strong solution is unique. These so-called entropy constraints could be added to the infinite-dimensional LP considered here, thereby proving the equivalence (at least in the uncontrolled case and for the class of Euler PDEs considered in Brenier et al., 2011). In the more general setting that we have considered, the situation is more complex and this equivalence remains an open question, especially in the controlled case.

Another issue is the computational complexity reduction. We have outlined several of the most obvious ways of doing so in the extended version (Korda et al., 2018) Nevertheless, we believe that a more detailed case by case inspection of the problem structure would lead to further reduction. Numerical conditioning of the resulting SDPs as a function of the problem data is also an important and not well understood question. Finally, a problem-adapted basis is very likely to be superior to the generic monomial basis used here in terms of numerical conditioning, thereby allowing the SDP relaxations to be accurately computed for higher values of d, possibly with first-order optimization methods that scale much more favorably with the problem size but struggle with numerical conditioning.

Acknowledgments

The first author would like to thank Hassan Arbabi for kindly providing the numerical solver for the Burgers' equation as well as for an insightful discussion and comments on the paper. The authors would also like to thank Swann Marx for pointing out the work (DiPerna, 1985) and a stimulating discussion on the topic. This work also benefited from interesting discussions with Martin Kružík and Josef Málek. The research of M. Korda was supported by the Swiss National Science Foundation under grant P2ELP2_165166. The research of D. Henrion and J.B. Lasserre was funded by the European Research Council (ERC) under the European's Union Horizon 2020 research and innovation program (grant agreement 666981 TAMING).

References

Barreau, M., Seuret, A., Gouaisbaut, F., Baudouin, L., 2018. Lyapunov stability analysis of a string equation coupled with an ordinary differential system. IEEE Transactions on Automatic Control 63 (11), 3850–3857.

Bertsimas, D., Caramanis, C., 2006. Bounds on linear PDEs via semidefinite optimization. Mathematical Programming 108 (1), 135–158.

Brenier, Y., De Lellis, C., Székelyhidi, L., 2011. Weak-strong uniqueness for measure-valued solutions. Communications in Mathematical Physics 305 (2), 351.

Chernyshenko, S.I., Goulart, P., Huang, D., Papachristodoulou, A., 2014. Polynomial sum of squares in fluid dynamics: a review with a look ahead. Philosophical Transactions of the Royal Society. Series A 372 (2020), 20130350.

DiPerna, R.J., 1985. Measure-valued solutions to conservation laws. Archive for Rational Mechanics and Analysis 88 (3), 223–270.

Evans, L.C., 2010. Partial Differential Equations, second edition. American Mathematical Society.

Fridman, E., Orlov, Y., 2009. An LMI approach to H_∞ boundary control of semilinear parabolic and hyperbolic systems. Automatica 45 (9), 2060–2066.

Gahlawat, A., Peet, M.M., 2017. A convex sum-of-squares approach to analysis, state feedback and output feedback control of parabolic PDEs. IEEE Transactions on Automatic Control 62 (4), 1636–1651.

Goluskin, D., Fantuzzi, G., 2019. Bounds on mean energy in the Kuramoto-Sivashinsky equation computed using semidefinite programming. Nonlinearity 32 (5), 1705.

Henrion, D., Korda, M., 2014. Convex computation of the region of attraction of polynomial control systems. IEEE Transactions on Automatic Control 59 (2), 297–312.

Henrion, D., Korda, M., Lasserre, J.B., 2020. The Moment-SOS Hierarchy: Lectures in Probability, Statistics, Computational Geometry, Control and Nonlinear PDEs. World Scientific, Singapore.

Henrion, D., Lasserre, J.B., Löfberg, J., 2009. Gloptipoly 3: moments, optimization and semidefinite programming. Optimization Methods & Software 24, 761–779.

Hinze, M., Pinnau, R., Ulbrich, M., Ulbrich, S., 2008. Optimization with PDE Constraints, Vol. 23. Springer Science & Business Media.

Kawai, R., 2015. Explicit hard bounding functions for boundary value problems for elliptic partial differential equations. Computers & Mathematics with Applications 70 (12), 2822–2837.

Korda, M., Henrion, D., Jones, C.N., 2014a. Controller design and region of attraction estimation for nonlinear dynamical systems. In: Proceedings of the 19th World Congress of the International Federation of Automatic Control, Vol. 47(3), pp. 2310–2316.

Korda, M., Henrion, D., Jones, C.N., 2014b. Convex computation of the maximum controlled invariant set for polynomial control systems. SIAM Journal on Control and Optimization 52 (5), 2944–2969.

Korda, M., Henrion, D., Lasserre, J.B., 2018. Moments and convex optimization for analysis and control of nonlinear partial differential equations. arXiv preprint. arXiv:1804.07565.

Lamare, P.-O., Girard, A., Prieur, C., 2016. An optimisation approach for stability analysis and controller synthesis of linear hyperbolic systems. ESAIM. Control, Optimisation and Calculus of Variations 22 (4), 1236–1263.

Lasserre, J.B., 2006. Convergent SDP-relaxations in polynomial optimization with sparsity. SIAM Journal on Optimization 17 (3), 822–843.

Lasserre, J.B., 2009. Moments, Positive Polynomials and Their Applications, first edition. Imperial College Press.

Lasserre, J.B., 2019. The moment-SOS hierarchy. In: Sirakov, B., de Souza, P.N., Viana, M. (Eds.), Proceedings of the International Congress of Mathematicians (ICM 2018), Vol. 4. World Scientific, pp. 3773–3794.

Lasserre, J.B., Henrion, D., Prieur, C., Trélat, E., 2008. Nonlinear optimal control via occupation measures and LMI-relaxations. SIAM Journal on Control and Optimization 47 (4), 1643–1666.

Lasserre, J.B., Magron, V., 2020. Computing the Hausdorff boundary measure of semi-algebraic sets. SIAM Journal on Applied Algebra and Geometry 4 (3), 441–469.

Laurent, M., 2009. Sums of squares, moment matrices and optimization over polynomials. In: Emerging Applications of Algebraic Geometry. Springer, pp. 157–270.

Leoni, G., 2017. A First Course in Sobolev Spaces. American Mathematical Soc.

Lions, P.L., 1996. Mathematical Topics in Fluid Mechanics: Volume 2: Compressible Models, Vol. 2. Oxford University Press on Demand.

Magron, V., Prieur, C., 2020. Optimal control of linear pdes using occupation measures and sdp relaxations. IMA Journal of Mathematical Control and Information 37 (1), 159–174.

Majumdar, A., Vasudevan, R., Tobenkin, M.M., Tedrake, R., 2014. Convex optimization of nonlinear feedback controllers via occupation measures. The International Journal of Robotics Research 33 (9), 1209–1230.

Marx, S., Weisser, T., Henrion, D., Lasserre, J.B., 2020. A moment approach for entropy solutions to nonlinear hyperbolic pdes. Mathematical Control and Related Fields 10 (1), 113–140.

Mevissen, M., Kojima, M., Nie, J., Takayama, N., 2008. Solving partial differential equations via sparse SDP relaxations. Pacific Journal of Optimization 4 (2), 213–241.

Nie, J., Demmel, J., 2008. Sparse SOS relaxations for minimizing functions that are summations of small polynomials. SIAM Journal on Optimization 19 (4), 1534–1558.

Papp, D., Yildiz, S., 2019. Sum-of-squares optimization without semidefinite programming. SIAM Journal on Optimization 29 (1), 822–851.

Pólik, I., Terlaky, T., Zinchenko, Y., 2007. SeDuMi: a package for conic optimization. In: IMA Workshop on Optimization and Control.

Rubio, J., 1995. The global control of nonlinear diffusion equations. SIAM Journal on Control and Optimization 33 (1), 308–322.

Strauss, W.A., 1990. Nonlinear Wave Equations. Number 73. American Mathematical Society.

Valmorbida, G., Ahmadi, M., Papachristodoulou, A., 2016. Stability analysis for a class of partial differential equations via semidefinite programming. IEEE Transactions on Automatic Control 61 (6), 1649–1654.

Vinter, R., 1993. Convex duality and nonlinear optimal control. SIAM Journal on Control and Optimization 31 (2), 518–538.

Waki, H., Kim, S., Kojima, M., Muramatsu, M., 2006. Sums of squares and semidefinite program relaxations for polynomial optimization problems with structured sparsity. SIAM Journal on Optimization 17 (1), 218–242.

Chapter 11

Turnpike properties in optimal control

An overview of discrete-time and continuous-time results

Timm Faulwasser[a] **and Lars Grüne**[b,*]
[a]*Institute for Energy Systems, Energy Efficiency and Energy Economics, Department of Electrical Engineering and Information Technology, TU Dortmund University, Dortmund, Germany,* [b]*Chair of Applied Mathematics, Mathematical Institute, University of Bayreuth, Bayreuth, Germany*
**Corresponding author: e-mail address: lars.gruene@uni-bayreuth.de*

Contents

Abstract

The turnpike property refers to the phenomenon that in many optimal control problems, the solutions for different initial conditions and varying horizons approach a neighborhood of a specific steady state, then stay in this neighborhood for the major part of the time horizon, until they may finally depart. While early observations of the phenomenon can be traced back to works of Ramsey and von Neumann on problems in economics in 1928 and 1938, the turnpike property received continuous interest in economics since the 1960s and recent interest in systems and control. The present chapter provides an introductory overview of discrete-time and continuous-time results in finite and infinite-dimensions. We comment on dissipativity-based approaches and infinite-horizon results, which enable the exploitation of turnpike properties for the numerical solution of problems with long and infinite horizons. After drawing upon numerical examples, the chapter concludes with an outlook on time-varying, discounted, and open problems.

Handbook of Numerical Analysis, Volume 23, ISSN 1570-8659, https://doi.org/10.1016/bs.hna.2021.12.011

Keywords
Turnpike properties, Optimal control, Dissipativity, Numerical solution

MSC Codes
49-02, 49J15, 93D20, 49K40

1 Introduction and historical origins

Optimal control has a rich history, which can be traced back at least to the 1600s (Sussmann and Willems, 1997). Moreover, around the mid of the 20th century two complementary breakthroughs—the Pontryagin Minimum Principle (PMP) (Boltyanskii et al., 1960) and Dynamic Programming (DP) (Bellman, 1954)—have been achieved for which the investigations into the Calculus of Variations have paved the road, see (Pesch, 2013). The number of papers, books, and PhD theses triggered by the PMP and DP is countless. This evident scientific appeal is likely due to optimal control being a unique combination of powerful mathematical concepts applicable to the entire spectrum of dynamics systems from finite to infinite dimensions—including differential algebraic settings, hybrid formulations, and inclusions. At the same time, optimal control is nowadays readily applicable in a variety of contexts ranging from analysis of phenomena in nature and society towards the design and automation of technical systems. Hence, it comes without surprise that the list of successful applications of optimal control is ever growing since and includes highlights such as: the moon landing in 1969 and numerous space missions; insights into brain evolution in humans (González-Forero and Gardner, 2018); understanding the complex interplay between economic growth, climate change, and green house gas emissions (Nordhaus, 1992); design and understanding of methods for machine learning (E et al., 2017; Esteve et al., 2020a; Li et al., 2017); as well as numerous mass-consumer products, which entail feedback control loops designed and implemented via optimal control theory. In the view of automatic control applications, a tremendously successful off-spring of the family of methods subsumed as optimal control is its receding horizon variant, which is also denoted as model predictive control (Rawlings et al., 2017).

From Ramsey and von Neumann to aircraft design

This chapter is concerned with providing an overview on the turnpike phenomenon and its numerical application, which occurs frequently in real-world Optimal Control Problems (OCPs). It refers to qualitative and quantitative similarity of optimal solutions to OCPs for different initial conditions and varying horizons. Specifically, the solutions approach a neighborhood of the best steady state, stay within this neighborhood for some time, and might leave this neighborhood towards the end of the optimization horizon. The notion of turnpike properties in OCPs has been coined by (Dorfman et al., 1958, p. 331) with the following lines:

It is exactly like a turnpike paralleled by a network of minor roads. [...], if origin and destination are far enough part, it will always pay to get on to the turnpike and cover distance at the best rate of travel, even if this means adding a little mileage at either end.

Prior to this foundational description, the phenomenon has been reported by von Neumann (1938) and by Ramsey (1928). Indeed, without using the term *turnpike*, Samuelson (1949) implicitly conjectured corresponding theorems. Given that these early works on the subject investigated specific problems arising in economics, it is not surprising that turnpike theory was driven by research on applying optimal control in economics and econometrics (McKenzie, 1976; Carlson et al., 1991). Interestingly, also in systems and control the phenomenon was observed at least as early as (Wilde and Kokotovic, 1972)—though Kokotovic and Wilde used a different name, referring to it as a *dichotomy in optimal control.*[1] Historically, the majority of turnpike results look at specific problems and exploit structure of the optimality conditions. Recent papers have proposed to leverage turnpikes in applications as diverse as shape optimization in aircraft design (Lance et al., 2020), membrane-filtration systems (Kalboussi et al., 2018), or control of chemical reactors with uncertain models (Faulwasser and Pannocchia, 2019).

Remarkably, in the systems and control community, turnpikes have received little attention. One of the earliest comments on turnpike properties in receding-horizon optimal control appears to be (Rawlings et al., 2008), who later coined the phrase economic MPC (Rawlings and Amrit, 2009) for receding horizon optimal control with generic objectives.[2] Eventually, the interest in economic MPC has led to a new line of research towards turnpike properties centered around dissipativity notions, which have been introduced by Jan Willems (Willems, 1971, 1972a,b). Early works in this direction include (Grüne, 2013; Faulwasser et al., 2014), while Grüne and Müller (2016) and Faulwasser et al. (2017) investigate equivalence of turnpike and dissipativity properties of OCPs in discrete-time, respectively, continuous-time settings. Given the close relation between dissipativity and stability of dynamic systems, it is far from surprise, that the recent elaboration of the dissipativity route to turnpikes establishes a close relation between turnpikes and stability in infinite-horizon optimal control (Grüne et al., 2017; Faulwasser and Kellett, 2021). All these works refer to steady-state turnpikes. We will comment on the progress with respect to time-varying and periodic turnpikes in the outlook in Section 5.

The remainder of this chapter is structured as follows: In Section 1 we conclude with the problem setup and a review of the necessary optimality conditions. Section 2 attempts a taxonomy of turnpike properties, this way providing

[1] More recently, the name *hyper-sensitive* optimal control problems was also used (Rao and Mease, 1999).

[2] *Generic* refers to objectives that do not merely penalize the distance to some reference setpoint.

guidance in the considerable amount of literature on this subject. Section 3 details the generating mechanisms as well as necessary and sufficient conditions for the occurrence of the phenomenon. Finally, Section 4 comments on the exploitation of turnpikes in numerics and receding-horizon optimal control. This chapter concludes with an outlook on advanced turnpike settings and open problems of interest.

Notation

The set $\mathbb{N}_{[a,b]} \doteq \{a, \ldots, b\} \subset \mathbb{N}$ denotes an interval of natural numbers. Functions $u : [0, T] \to \mathcal{U}$ and sequences $u : \{a, a+1, \ldots, b-1, b\} \to \mathcal{U}$ are written as $u(\cdot)$. Partial derivatives are indicated by subscripts, i.e. ℓ_x is the partial derivative of ℓ w.r.t. x. Throughout this chapter, optimal solutions to optimization and optimal control problems are indicated by superscript $(\cdot)^\star$. Variables at steady state are marked by superscript $\bar{(\cdot)}$; optimal steady-state variables are indicated by $\bar{(\cdot)}^\star$.

1.1 Discrete-time and continuous-time setting

We consider discrete-time OCPs of the following form

$$V_T(x_0) \doteq \min_{u(\cdot)} \sum_{t=0}^{T-1} \ell(x(t), u(t)) + \varphi(x(T)) \tag{1a}$$

subject to

$$x(t+1) = f(x(t), u(t)),\ x(0) = x_0 \qquad t \in \mathbb{N}_{[0,T-1]}, \tag{1b}$$

$$0 \geq g(x(t), u(t)), \qquad t \in \mathbb{N}_{[0,T-1]}, \tag{1c}$$

$$0 \geq \psi(x(T)). \tag{1d}$$

The constraint set for $z \doteq (x, u) \in \mathcal{X} \times \mathcal{U}$, where $\mathcal{X}$ and $\mathcal{U}$ are normed metric spaces, is defined as

$$\mathbb{Z} = \{z = (x, u) \in \mathcal{X} \times \mathcal{U} \mid g_j(z) \leq 0,\ j \in \mathbb{N}_{[1,n_g]}\}, \tag{2}$$

where the constraints (1c) are given by $g(x, u) = \begin{bmatrix} g_1(x, u) & \ldots & g_{n_g}(x, u) \end{bmatrix}^\top$ with $g_j : \mathcal{X} \times \mathcal{U} \to \mathbb{R}$, $j \in \mathbb{N}_{[1,n_g]}$.

The projection of $\mathbb{Z}$ onto $\mathcal{X}$ is denoted as $\mathbb{X}$ and the projection of $\mathbb{Z}$ onto $\mathcal{U}$ is written as $\mathbb{U}$.[3] The continuous functions $\ell : \mathcal{X} \times \mathcal{U} \to \mathbb{R}$ and $\varphi : \mathcal{X} \to \mathbb{R}$ are the Lagrange function and, respectively, the Mayer term. Moreover, $T \in \mathbb{N}$ is the horizon.

We remark that, throughout this chapter, unless otherwise stated, we do not make specific assumptions on ℓ and φ such as quadratic structures or convexity.

3 In case this projection depends on the state x, we write $\mathbb{U}(x)$.

Instead we will work with a rather generic dissipativity assumption to be introduced later. Eq. (1d) defines the terminal constraint; the more general case of coupled constraints on initial and terminal states is left out due to space limitations.

The continuous-time counterpart to OCP (1) reads

$$V_T(x_0) \doteq \min_{u(\cdot)\in\mathcal{L}([0,T],\mathcal{U})} \int_0^T \ell(x(t),u(t))\mathrm{d}t + \varphi(x(T)) \tag{3a}$$

subject to

$$\dot{x} = f(x,u),\ x(0) = x_0 \qquad t\in[0,T], \tag{3b}$$

$$0 \geq g(x(t),u(t)), \qquad t\in[0,T], \tag{3c}$$

$$0 \geq \psi(x(T)). \tag{3d}$$

The projected constraint sets are similarly defined as in case of OCP (1). The horizon is $T \in \mathbb{R}^+$. However, the underlying space is considered to be a real-valued vector space of appropriate finite dimension, i.e., (3) refers to a finite-dimensional OCP. In order guarantee existence of optimal solutions for continuous-time OCPs, one will typically invoke further differentiability properties of the problem data and feasibility assumptions. For the sake of brevity, we skip the details and refer to classical texts such as (Lee and Markus, 1967; Berkovitz, 1974) or more recent treatments (Hartl et al., 1995; Vinter, 2010; Gerdts, 2011).

Comparing the discrete-time and the continuous-time variants—OCP (1), respectively, OCP (3)—shows that as OCP (1) is defined in normed metric spaces it admits the interpretation as an infinite-dimensional system. This implies that a large variety of discrete-time results to be touched upon later can be—or already have been—formulated in this general setting. In contrast, it is more difficult to formalize infinite-dimensional continuous-time optimal control problems in a unified way. Hence we have limited (3) to finite-dimensional problems. Nevertheless, we will comment on infinite-dimensional continuous-time problems at several places throughout this chapter.

1.2 Optimality conditions

Dual variables are key to the understanding of any optimization problem. Hence, we now formulate the first-order Necessary Conditions of Optimality (NCOs) for OCP (1), respectively, OCP (3).

The NCOs for OCP (1) are built using the point-wise Lagrangian function

$$\mathcal{L}(0) \doteq \lambda_0^\top(x(0) - x_0), \tag{4a}$$

$$\mathcal{L}(t) \doteq \ell(x(t),u(t)) + \lambda(t+1)^\top(f(x(t),u(t)) - x(t+1)) + \mu(t)^\top g(x(t),u(t)), \qquad t\in\mathbb{N}_{[0,T-1]}, \tag{4b}$$

$$\mathcal{L}(T) \doteq \varphi(x(T)) + \mu(T)^\top\psi(x(T)), \tag{4c}$$

which directly leads to the overall Lagrangian

$$L(x(\cdot), u(\cdot), \lambda(\cdot), \mu(\cdot)) \doteq \sum_{t=0}^{T} \mathcal{L}(t),$$

with $\mathcal{L}(t)$ from (4). Here, x and u are called the primal variables while λ and μ are called the dual variables or adjoints/costates (in case of λ).

Let the objective functional in OCP (1) be written as

$$J_T(u(\cdot), x_0) = \sum_{t=0}^{T-1} \ell(x(t), u(t)) + \varphi(x(T))$$

and the one in OCP (3) as

$$J_T(u(\cdot), x_0) = \int_0^T \ell(x(t), u(t))\mathrm{d}t + \varphi(x(T)).$$

Then, for finite horizons $T < \infty$, optimality of primal solutions $(x(t), u(t))$ is characterized by

$$J_T(u^\star(\cdot), x_0) \leq J_T(u(\cdot), x_0) \quad \text{for all admissible } u(\cdot)$$

A dual solution is called optimal if it satisfies the NCOs defined below together with the primal optimal solution. Optimal primal and dual solutions, provided they exist, are denoted by superscript $\cdot^\star$, i.e. we write $\chi^\star(\cdot; x_0)$, $\chi \in \{x, u, \lambda, \mu\}$. The argument x_0 will be dropped whenever the considered initial condition can be inferred from the context. The first-order NCOs are given by $\nabla L = 0$, which entails

$$x^\star(t+1) = f(x^\star(t), u^\star(t)), \qquad x^\star(0) = x_0, \tag{5a}$$

$$\lambda^\star(t) = f_x^\top \lambda^\star(t+1) + \ell_x + g_x^\top \mu^\star(t), \; \lambda^\star(T) = \left(\varphi_x + \psi_x^\top \mu^\star\right)\Big|_{t=T}, \tag{5b}$$

$$0 = f_u^\top \lambda^\star(t) + \ell_u + g_u^\top \mu^\star(t). \tag{5c}$$

Observe that the NCOs as listed above leave out parts of standard KKT conditions such as primal feasibility, complementary slackness, or higher-order conditions. However, for the purposes of the subsequent exposition, the conditions above suffice.

In correspondence to the discrete-time OCP we consider the steady-state optimization problem

$$\min_{\bar{u}, \bar{x}} \; \ell(\bar{x}, \bar{u}) \tag{6a}$$

subject to

$$\bar{x} = f(\bar{x}, \bar{u}), \tag{6b}$$
$$0 \geq g(\bar{x}, \bar{u}) \tag{6c}$$

and its Lagrangian $\bar{L} = \ell(\bar{x}, \bar{u}) + \bar{\lambda}^\top (f(\bar{x}, \bar{u}) - \bar{x}) + \bar{\mu}^\top g(\bar{x}, \bar{u})$. Superscript $\bar{\cdot}$ denotes variables at steady state and hence optimal steady-state variables are indicated by $\bar{\cdot}^\star$.[4] The optimality conditions of this problem entail

$$\bar{x}^\star = f(\bar{x}^\star, \bar{u}^\star), \tag{7a}$$
$$\bar{\lambda}^\star = f_x^\top \bar{\lambda}^\star + \ell_x + g_x^\top \bar{\mu}^\star, \tag{7b}$$
$$0 = f_u^\top \bar{\lambda}^\star + \ell_u + g_u^\top \bar{\mu}^\star. \tag{7c}$$

Notice that the optimality conditions (7) of the stationary problem (6) coincide with the full NCOs (5) of OCP (1) if only constant solutions are considered.

The continuous-time counterpart to (5) can be obtained via a classical direct adjoining approach, i.e. one considers the Hamiltonian

$$H(x, u, \lambda, \mu) \doteq \lambda^0 \ell(x, u) + \lambda^\top f(x, u) + \mu^\top g(x, u). \tag{8}$$

Excluding abnormal problems we set $\lambda^0 \equiv 1$ and then the Pontryagin Minimum Principle reads

$$\dot{x}^\star = f(x^\star, u^\star), \qquad x^\star(0) = x_0, \tag{9a}$$
$$\dot{\lambda}^\star = -f_x^\top \lambda^\star - \ell_x - g_x^\top \mu^\star, \ \lambda^\star(T) = \left(\varphi_x + \psi_x^\top \mu^\star\right)\Big|_{t=T}, \tag{9b}$$
$$0 = f_u^\top \lambda^\star + \ell_u + g_u^\top \mu^\star, \tag{9c}$$
$$H(x^\star, u^\star, \lambda^\star, \mu^\star) \leq \min_{u \in \mathbb{U}(x^\star)} H(x^\star, u, \lambda^\star, \mu^\star). \tag{9d}$$

Similar to the discrete-time case, we have left out some technical conditions which do not affect the further developments. Specifically, in the time-invariant setting considered here the Hamiltonian is constant along optimal solutions $H(x^\star, u^\star, \lambda^\star, \mu^\star) = const$ and the adjoints satisfy $(\lambda_0, \lambda^\star(t)) \neq 0$ for all $t \in [0, T]$. Moreover, in analyzing the continuous-time OCPs state path constraints (3c) can lead to difficulties. Hence, one usually either restricts the setting to mixed-input-state constraints combined with a linear independence condition, or one limits the degree of pure state path constraints, or one considers optimality conditions formulated in terms of bounded variation. We refer, e.g., to Hartl et al. (1995) for comments and insights. Here, however, we will be able to avoid those technicalities by assuming for OCP (1), respectively, OCP (3) that the turnpike properties to be analyzed occur in the interior of the constraint set $\mathbb{Z}$.

[4] A feasible primal pair $(\bar{x}, \bar{u})$ satisfying (6b) in discrete time or (10b) in continuous time is called a *controlled equilibrium*.

The steady-state optimization problem corresponding to OCP (3) reads

$$\min_{\bar{u},\bar{x}} \ \ell(\bar{x},\bar{u}) \tag{10a}$$

subject to

$$0 = f(\bar{x},\bar{u}), \tag{10b}$$

$$0 \geq g(\bar{x},\bar{u}). \tag{10c}$$

The reader will face no difficulties in verifying that also in continuous time the optimality conditions of the stationary problem (10) correspond to the full NCOs (9) of OCP (3) if constant solutions are considered.

2 Definition and taxonomy of turnpike properties

In many applications it is of interest to consider the solutions of OCPs for varying problem data such as parameters of the underlying dynamics—(1b) respectively (3b)—, different initial conditions x_0 of said model, or varying length of the optimization horizon T in (1), respectively, in (3). Classical sensitivity theory (Fiacco, 1976) provides a handle for locally analyzing the former, i.e. for small parametric variations. In contrast, turnpike theory aims at a *non-local* analysis of the structure of solutions for varying initial conditions x_0 and horizons T over a large range of values. Hence, we consider OCP (1) / (3) for a set of initial conditions $\mathbb{X}_0 \subseteq \mathbb{X} \subseteq \mathcal{X}$ and horizons $T \in \mathbb{R}^+/\mathbb{N}$. Notice that, depending on the considered input constraints and the properties of (1b) / (3b), the set $\mathbb{X}_0$ may be open or closed and bounded or unbounded. However, in most relevant cases it will be uncountable. As we will see later, the size of the maximal set $\mathbb{X}_0$ depends on the reachability properties of (1b) / (3b).

Recall the notational convention for optimal primal and dual solutions $\chi^\star(\cdot; x_0)$, $\chi \in \{x, u, \lambda, \mu\}$. Let $\bar{\chi}$ be some controlled equilibrium satisfying (6b) / (10b) (see Footnote 4 for an informal definition), and let $\mathcal{B}_\varepsilon(\bar{\chi})$ be a ball of radius $\varepsilon > 0$ centered at $\bar{\chi}$. Turnpike theory tries to establish properties of the following type:

$$\forall x_0 \in \mathbb{X}_0, \forall T > 0, \forall \varepsilon > 0, \mid \left[\text{time } \chi^\star(\cdot; x_0) \notin \mathcal{B}_\varepsilon(\bar{\chi})\right] \leq \nu(\varepsilon) < \infty \tag{11}$$

where $\nu : [0, \infty) \to \mathbb{R} \cap \infty$ is independent of the actual horizon length T and of the initial condition x_0. Naturally, the conceptual property above calls for precisions and further comments: With respect to which variables $\chi \in \{x, u, \lambda, \mu\}$ shall the closeness be established? How to measure/express the time spent outside of $\mathcal{B}_\varepsilon(\bar{\chi})$? Besides these technical aspects it is imperative to notice two issues: (a) The conceptual statement above becomes trivial, if the considered horizons T are bounded from above by a finite number. (b) The property is to be established for all $x_0 \in \mathbb{X}_0$. Thus, given $\mathbb{X}_0$ is more complex than a simple ε-ball, one aims at establishing a non-local (similarity) property of optimal solutions for varying initial conditions and varying horizon length.

2.1 Motivating example

To illustrate the turnpike phenomenon, we consider a simple example dating back to Cliff and Vincent (1973). The OCP at hand reads

$$\min_{u(\cdot)\in\mathcal{L}([0,T],\mathcal{U})} \int_0^T \left[ax(t) + bu(t) - cx(t)u(t)\right] \mathrm{d}t, \tag{12a}$$

subject to

$$\dot{x} = x(x_s - x - u), \quad x(0) = x_0 > 0 \tag{12b}$$

$$\forall t \in [0, T| : u(t) \in [0, u_{max}] \tag{12c}$$

$$\forall t \in [0, T| : x(t) \geq x_{min} \tag{12d}$$

The state x is the fish density in a certain habitat, the control u is the fishing rate, and the parameter x_s describes the highest sustainable fish density. In slightly modified form—i.e., with an additional terminal constraint and the state constraint $x(\tau) > 0$—this OCP is analyzed by Cliff and Vincent (1973) and in (Carlson et al., 1991, Chap. 3.3).

The parameter values are $a = 1$, $b = c = 2$, $x_s = u_{max} = 5$, $x_{min} = 0.1$. In case the state constraint $x \geq x_{min}$ is active on some interval $[\tau_{act}, \tau_{deact}]$, it is easily verified that $u(t) = x_s - x_{min} > 0$, $\forall t \in [\tau_{act}, \tau_{deact}]$. However, for sufficiently small but positive x_{min}, if $x_0 = x_{min}$, the optimal solutions depart from the state constraint immediately.

The considered OCP has exactly one singular arc of order one along which

$$\bar{x} = \frac{1}{2}x_s + \frac{b-a}{2c} > x_{min}, \quad \bar{u} = x_s - \bar{x}, \quad \bar{\lambda} = c - \frac{b}{\bar{x}}, \quad \bar{\mu} = 0, \tag{13}$$

i.e. primal and dual variables are at steady state. We refer to (Bryson and Ho, 1969; Zelikin and Borisov, 1994) for details on the analysis of singular arcs and to (Carlson et al., 1991, Chap. 3.3) for details on the derivation of the singular arc in this example. As long as the state constraint does not become active, the optimal input signal for this OCP takes only three values $u^\star(t; x_0) \in \{0, \bar{u}, u_{max}\}$ for all $t \in [0, T]$ and all $x_0 \geq x_{min}$.

Fig. 1 shows numerical results for initial conditions $x_0 \in \mathbb{X}_0 = \{0.5,\ 2.5,\ 4.5\}$ and horizons $T \in \{1.0,\ 1.1,\ 1.2\}$. As one can see, the optimal solutions all comprise three distinct arcs: the turnpike entry arc, the turnpike arc, and the leaving arc. Put differently, all solutions enter the steady state at $(\bar{x}, \bar{u})$ exactly.[5] Moreover, while the primal variables enter $(\bar{x}, \bar{u})$, the adjoint λ also reaches its steady value $\bar{\lambda}$. Notice that, despite different initial conditions and different horizons, the optimal solutions are qualitatively similar, i.e. they have the middle arc along which the optimal solutions are stationary in common. As demanded

[5] We remark that for turnpike analysis it is in general *not crucial* that the optimal solutions enter the turnpike at $(\bar{x}, \bar{u})$ *exactly*. Indeed, as we comment subsequently, in many cases the solutions will only get arbitrarily close, see also Remark 1 and the specific turnpike definitions below.

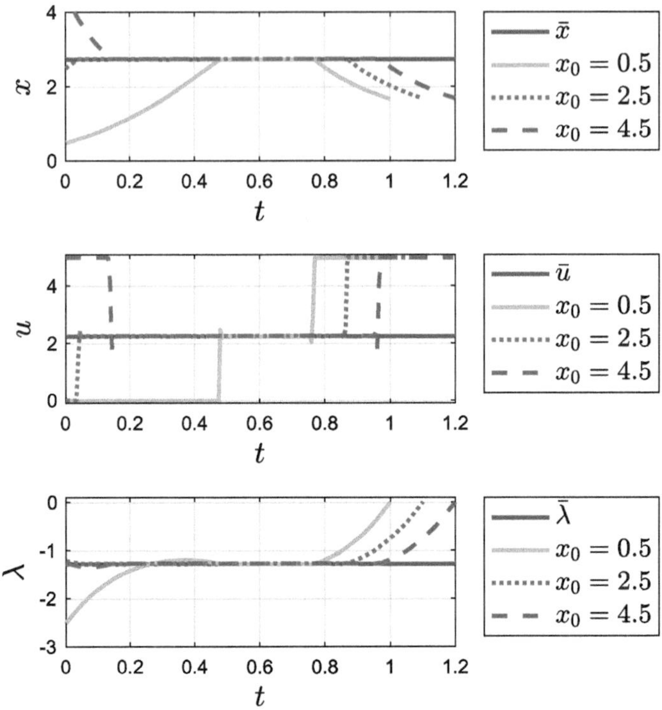

FIGURE 1 Fish harvest example with bilinear objective with pairs of x_0 and T: $x_0 = 0.5$, $T = 1.0$; $x_0 = 2.5$, $T = 1.1$; and $x_0 = 4.5$, $T = 1.2$. Depicted are optimal state, input, and adjoint trajectories.

by (11), the time the optimal solutions spent outside of the steady state triplet $(\bar{x}, \bar{u}, \bar{\lambda})$ is bounded independent of the horizon length.

2.2 Turnpike definitions

In order to be more precise, we turn towards a discrete-time definition of the turnpike phenomenon which was introduced by Grüne (2013), see also (Zaslavski, 2006, first paragraph of Chapter 8) for an earlier though different definition. To this end, consider the placeholder variable $\xi \in \{x, (x, u), (x, u, \lambda, \mu)\}$, which refers to time-dependent functions of the state, input, and dual variables and its steady-state counterpart $\bar{\xi}$.

Definition 1 (Discrete-time (cardinality) turnpike). OCP (1) is said to have a turnpike at $\bar{\xi}$, $\xi \in \{x, (x, u), (x, u, \lambda, \mu)\}$ if there exists a function $\nu_\xi : [0, \infty) \to [0, \infty]$ such that, for all $x_0 \in \mathbb{X}_0$ and all $T \in \mathbb{N}$,

$$\#\mathbb{Q}_{\xi,T}(\varepsilon) \leq \nu_\xi(\varepsilon) < \infty \quad \forall \varepsilon > 0, \tag{14}$$

where

$$\mathbb{Q}_{\xi,T}(\varepsilon) \doteq \{t \in \{0, \ldots, T-1\} \mid \|\xi^\star(t; x_0) - \bar{\xi}\| > \varepsilon\} \tag{15}$$

and $\#\mathbb{Q}_{\xi,T}(\varepsilon)$ is the cardinality of $\mathbb{Q}_{\xi,T}(\varepsilon)$.

- If $\xi = x$, then OCP (1) is said to have a *state turnpike* at $\bar{x}$.
- If $\xi = (x, u)$, then OCP (1) is said to have an *input-state turnpike* at $(\bar{x}, \bar{u})$.
- If $\xi = (x, u, \lambda, \mu)$, then OCP (1) is said to have a *primal-dual turnpike* at $(\bar{x}, \bar{u}, \bar{\lambda}, \bar{\mu})$.

Moreover, if

$$\lim_{\varepsilon \searrow 0} \nu_\xi(\varepsilon) < \infty$$

then the turnpike property is said to be *exact*.

Usually, one refers to the steady-state variables $\{\bar{x}, (\bar{x}, \bar{u}), (\bar{x}, \bar{u}, \bar{\lambda}, \bar{\mu})\}$ simply as the *turnpike*. Historically, many results have addressed the case of state turnpikes, while the input-state and the primal-dual case have received attention more recently. One important aspect in these developments has been the research on converse turnpike results, i.e. analyzing the implications if an OCP is assumed to exhibit a specific turnpike property. We will come back to this point later in Section 3.1.

The continuous-time counterpart to Definition 1 was proposed in (Faulwasser et al., 2014; Faulwasser and Bonvin, 2015). However, the conceptual idea has already been used in earlier results (Carlson et al., 1991, Thm. 4.2). Using again the placeholder variable ξ, we define the set

$$\Theta_{\xi,T}(\varepsilon) \doteq \left\{t \in [0, T] \mid \left\|\xi^\star(t; x_0) - \bar{\xi}\right\| > \varepsilon\right\}. \tag{16}$$

Definition 2 (Continuous-time (measure) turnpike). OCP (3) is said to have a turnpike at $\bar{\xi}$, $\xi \in \{x, (x, u), (x, u, \lambda, \mu)\}$ if there exists a function $\nu_\xi : [0, \infty) \to [0, \infty]$ such that, for all $x_0 \in \mathbb{X}_0$ and all $T \in \mathbb{R}^+$, we have

$$\mu\left[\Theta_{\xi,T}(\varepsilon)\right] \leq \nu_\xi(\varepsilon) < \infty \quad \forall \varepsilon > 0, \tag{17}$$

where $\mu[\cdot]$ is the Lebesgue measure on $\mathbb{R}$.

- If $\xi = x$, then OCP (3) is said to have a *state turnpike* at $\bar{x}$.
- If $\xi = (x, u)$, then OCP (3) is said to have an *input-state turnpike* at $(\bar{x}, \bar{u})$.
- If $\xi = (x, u, \lambda, \mu)$, then OCP (3) is said to have a *primal-dual turnpike* at $(\bar{x}, \bar{u}, \bar{\lambda}, \bar{\mu})$.

Moreover, if

$$\lim_{\varepsilon \searrow 0} \nu_\xi(\varepsilon) < \infty$$

then the turnpike property is said to be *exact*.

The two previous definitions (in their non-exact variants) state the most general turnpike properties we will encounter in this chapter. The reason being that, in principle, the measure/cardinality based properties allow for cases where optimal solutions would temporarily depart and return from the turnpike neighborhood. For the case of exact state turnpikes it is easy to verify that this cannot be the case (Faulwasser and Bonvin, 2017, Lem. 1).[6] For the non-exact case, which was called approximate in (Faulwasser and Bonvin, 2015), similar statements can be made using dissipativity conditions (Faulwasser et al., 2020).

To continue our exploration of different turnpike concepts, we turn toward a stronger notion where instead of measures and cardinalities the deviation of optimal solutions from their turnpike values is bounded in time by an exponential function. This kind of turnpike property was first observed in the economics literature, see, e.g., (Bewley, 1982, Theorem (4.5)) or (McKenzie, 1986, Theorem 10.1). Here we give a definition that appears in similar form, e.g., in (Porretta and Zuazua, 2013) or (Damm et al., 2014).

Definition 3 (Exponential turnpike). The discrete-time OCP (1) is said to have an exponential turnpike at $\bar{\xi}$, $\xi \in \{x, (x,u), (x,u,\lambda,\mu)\}$ if there exist $C > 0$, $\rho \in [0, 1)$ such that, for all $x_0 \in \mathbb{X}_0$ and all $T \in \mathbb{N}$,

$$\|\xi^\star(t; x_0) - \bar{\xi}\| \leq C\left(\rho^t + \rho^{T-t}\right). \tag{18}$$

Similarly, the continuous-time OCP (3) is said to have an exponential turnpike at $\bar{\xi}, \xi \in \{x, (x,u), (x,u,\lambda,\mu)\}$ if there exist $C > 0$, $\gamma > 0$ such that, for all $x_0 \in \mathbb{X}_0$ and all $T \in \mathbb{R}$, if (18) holds with $\rho \doteq e^{-\gamma}$.

One can show that the exponential turnpike property implies the cardinality/measure variant as one computes from (18) that

$$\mu\left[\Theta_{\xi,T}(\varepsilon)\right] = \max\{0, 2t_1(\varepsilon, T)\},$$

$$t_1(\varepsilon, T) = -\frac{1}{\gamma}\ln\left(\frac{\varepsilon}{2C} + \sqrt{\frac{\varepsilon^2}{4C^2} - e^{-\gamma T}}\right).$$

Here $t_1(\varepsilon, T)$ can be interpreted as the worst-case time the optimal solutions enter the ε-ball around the turnpike. This derivation uses that the symmetry of the right hand side in (18) implies $T - t_1(\varepsilon, T) = t_2(\varepsilon, T)$, where $t_2(\varepsilon, T)$ is the worst-case time the optimal solutions leave the ε-ball. Hence, one has that

$$\mu\left[\Theta_{\xi,T}(\varepsilon)\right] \leq \max\left\{0, -\frac{2}{\gamma}\ln\frac{\varepsilon}{2C}\right\}.$$

6 Occasionally, exact turnpike is also denoted as *finite* turnpikes (Gugat et al., 2020). Here, however, we stick to the wording coined in (Carlson et al., 1991, Chap. 2).

Similarly, in the discrete-time case (18) implies

$$\#\mathbb{Q}_{\xi,T}(\varepsilon) \leq \min\left\{k \in \mathbb{N} \,\middle|\, -\frac{2}{\gamma}\log_\rho \frac{\varepsilon}{2C} \leq k\right\}.$$

Summing up, the definitions of the cardinality/measure turnpike and the exponential variants provide the blueprint for a number of specific variants such as, e.g., exponential input-state turnpikes etc. That is, one can create taxonomy of turnpike properties with respect to two different dimensions:

- characterization by the asymptotics of *how* the closeness of optimal solutions to the turnpike is measured—i.e. the difference of Definition 1/2 and Definition 3—, and
- characterization by the arguments/variables of *what* is close to its turnpike value—i.e., the difference between subvariants in Definition 1/2.

This taxonomy is sketched in Fig. 2 and we notice that it is not entirely complete. For example Gugat and Hante (2019) investigate a stronger-than-measure-turnpike weaker-than-exponential-turnpike result for OCPs comprising infinite-dimensional hyperbolic systems and strict convexity assumptions for the objective. Specifically, it is shown that the optimal control profiles are close to their stationary values in a time-averaged sense. In other words, instead of an exponential bound—see (18) for the discrete-time version—a weaker integrable bound is established. Theorem 1 and Theorem 2 in (Gugat and Hante, 2019) establish input turnpike properties. The latter under integrality constraints for the controls. For problems arising in OCPs comprising finite-dimensional mechanical systems, (Faulwasser et al., 2019, 2021) recently proposed the consideration of hyperbolic bounds.

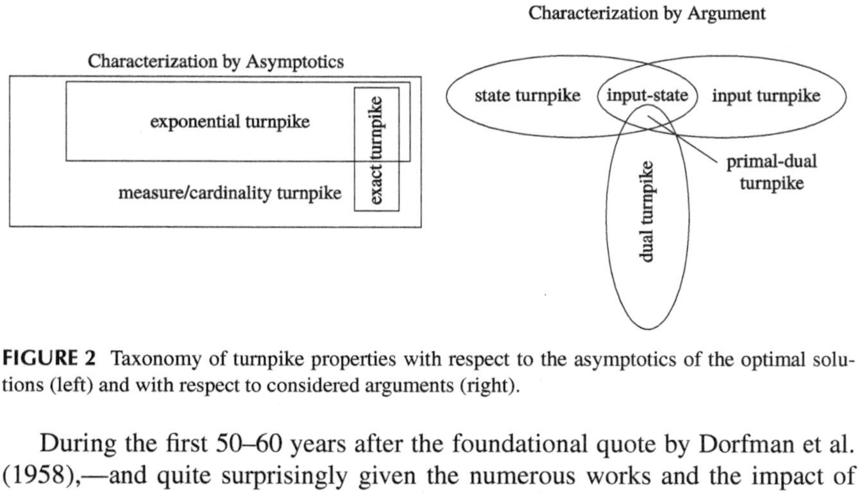

FIGURE 2 Taxonomy of turnpike properties with respect to the asymptotics of the optimal solutions (left) and with respect to considered arguments (right).

During the first 50–60 years after the foundational quote by Dorfman et al. (1958),—and quite surprisingly given the numerous works and the impact of turnpike results—precise definitions of turnpike properties have been, to the best of our knowledge, not been common in the literature. Actually, the book of

Zaslavski (2006) appears to be the first to present an explicit turnpike definition. This implies that the vast majority of classical turnpike results does not adhere to any kind of taxonomy and defines the property under consideration in an ad-hoc manner. Rather most of those results are denoted simply as *turnpike results*.

Putting the taxonomy proposed above in use, we note that originally state turnpike properties have been in the focus of many works in economics (McKenzie, 1976; Carlson et al., 1991). The need to extend to input-state turnpikes arose in receding-horizon and predictive control (Faulwasser and Bonvin, 2015) and in the investigation of converse turnpike results (Grüne and Müller, 2016; Faulwasser et al., 2017). We remark that especially converse turnpike analysis relies on precise definitions of the phenomenon. In parallel to the developments driven by receding-horizon and predictive control, the move towards primal-dual (exponential) turnpikes was done by Porretta and Zuazua (2013); Trélat and Zuazua (2015). We will comment in Section 3.1 on the differences in the technical analysis of the varying approaches. Moreover, in Section 3.2 we will comment on the extension towards infinite horizons.

3 Generating mechanisms

Before entering the technical analysis we informally illustrate the mechanisms that generate turnpike phenomena. To this end, we consider the three distinct arcs of turnpike solutions—the entry arc, the turnpike arc, and the leaving arc—separately. Notice that the characteristic leaving arc is not required by Definitions 1–3.

For the sake of illustration, we revisit the fish harvest problem from Section 2.1. Fig. 3 illustrates the geometry of OCP (12) in the $x - u$ plane (right) while also recalling the optimal trajectories (left). The dashed black line in the right hand side figure depicts the set of steady states given by $0 = x(x_s - x - u)$, and the colored lines are the contour lines of $\ell(x, u)$. The red cross marks the optimal steady state at which the set of steady states is tangent to the contour lines of the objective. This point coincides with the primal variables of the singular arc (13).

Fig. 3 provides an intuitive explanation why the problem at hand exhibits the turnpike phenomenon: It is easily verified that the singular arc (13) corresponds to the unique KKT point of the steady-state optimization problem (10) for OCP (12). Moreover, one can directly infer that above the steady-state set (= dashed black line), the dynamics of (12b) points in the direction of lower x values, while below the steady-state set it points in the direction of increasing x values.

In other words, the fact that the optimal solutions approach the turnpike clearly requires (at least asymptotic) controllability properties of the dynamics. Moreover, reaching the turnpike implicitly expresses that there does not exist any other set or state than the turnpike $(\bar{x}, \bar{u})$ where the system can

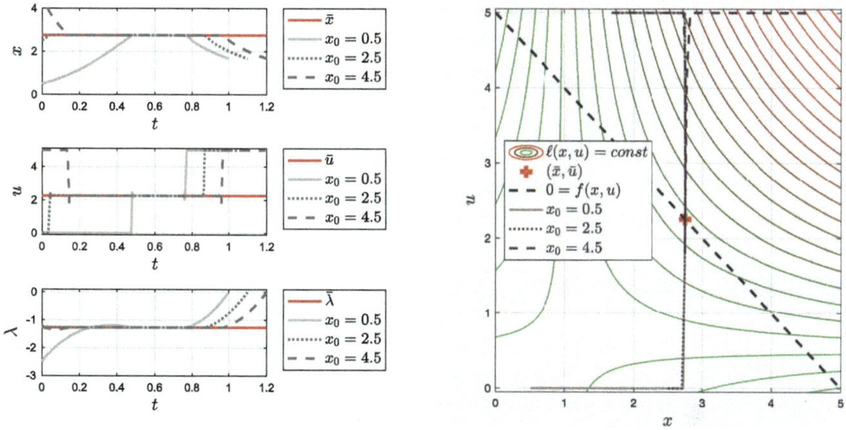

FIGURE 3 Fish harvest example with bilinear objective with pairs of x_0 and T: $x_0 = 0.5$, $T = 1.0$; $x_0 = 2.5$, $T = 1.1$; and $x_0 = 4.5$, $T = 1.2$. Left: optimal state, input and adjoint trajectories. Right: $x - u$ plane with optimal trajectories and contour lines of the Lagrange function $\ell(x, u)$ (colored), and the set of admissible steady states (dashed black).

stay infinitely long and which has better time-averaged performance than $\ell(\bar{x}, \bar{u})$.[7]

Finally, comparing the left and right hand side plots in Fig. 3, one may observe that the characteristic leaving arc steers the system to the transient domain above the steady-state set, where for the final part of the time horizon the optimal solutions are at a domain of lower values of ℓ. Note that, however, the system can only stay a limited amount of time in this domain due to the system dynamics and the constraints.

At first glance, one may conjecture that the characteristic turnpike phenomenon is induced by the fact that for sufficiently long horizons OCP (12) exhibits the unique and stationary singular arc (13). However, as we show next, this is not the case. To this end, we modify OCP (12) as follows

$$\ell(x,u) = \frac{1}{2}\left((x - x_c)^2 q + (u - u_c)^2\right), \quad \varphi(x) = -\frac{1}{2}x^2, \quad \psi(x) = 0 \quad (19)$$

with $x_c = 4$, $u_c = 5$ and $q = 10$. Due to the quadratic input penalization, OCP (12) with (19) is not singular. The left hand side of Fig. 4 shows its numerical solutions. As one can see, despite the changes to the objective, the optimal

7 This observation indicates that the turnpike phenomenon is closely linked to a property which was coined at optimal operation at steady state, which in turn is closely linked to overtaking optimality (Carlson et al., 1991). This property requires that in a time-averaged sense a dynamic system is optimally operated at a specific steady state. For details and the link to dissipativity of OCPs see (Müller et al., 2015) for the discrete-time case and (Faulwasser et al., 2017) for continuous-time results.

solutions also show the turnpike phenomenon. The right hand side of Fig. 4 illustrates again the geometry in the $x-u$ plane. Recall that without a Mayer term (and with bilinear Lagrange term ℓ) the optimal solutions of OCP (12) exhibited a leaving arc approaching the domain with lower values of ℓ, cf. Fig. 3. Observe that in OCP (12) with (19) the Mayer term $\varphi(x) = -\frac{1}{2}x^2$ is dominant compared to the Lagrange cost ℓ. Hence, the turnpike leaving arcs approach a part of the $x-u$ plane with higher running cost ℓ but lower terminal cost φ, see Fig. 4.

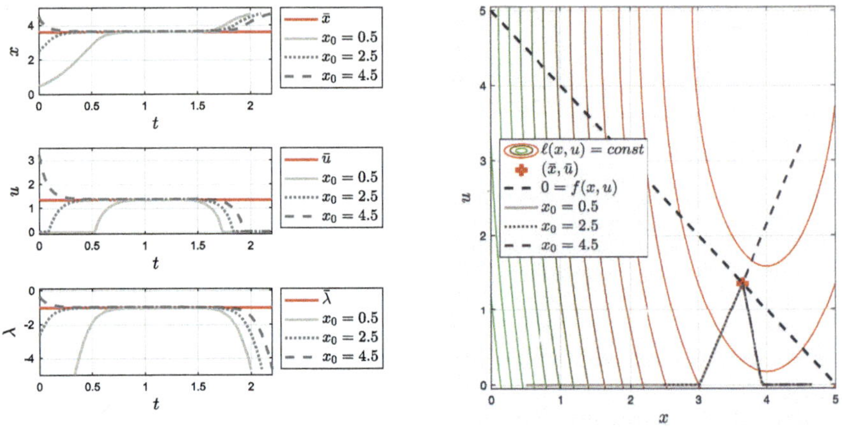

FIGURE 4 Fish harvest example with quadratic objective and Mayer term (19) with pairs of x_0 and T: $x_0 = 0.5$, $T = 2.0$; $x_0 = 2.5$, $T = 2.1$; and $x_0 = 4.5$, $T = 2.2$. Left: optimal state, input and adjoint trajectories. Right: $x-u$ plane with optimal trajectories and contour lines of the Lagrange function $\ell(x,u)$ (colored), and the set of admissible steady states (dashed black).

Finally, we consider a third variation of OCP (12) by changing (19) to

$$\ell(x,u) = \frac{1}{2}\left((x-x_c)^2 q + (u-u_c)^2\right), \quad \varphi(x) = 0, \quad \psi(x) = x - x_f, \tag{20}$$

with $x_c = 3.64$, $u_c = 1.37$, $q = 10$, and $x_f = 3$. The left hand side of Fig. 5 shows the numerical solutions for the arising problem with the resulting terminal equality constraint $\psi(x) = x - 3 = 0$.

Observe that, despite the fact that the center point of the convex quadratic objective $x_c = 3.64$, $u_c = 1.37$ being an unconstrained minimizer of ℓ and a steady state, optimal solutions still show the turnpike leaving arc. This is because the problem data (20) implies that the optimal solutions have to depart from $(\bar{x}, \bar{u})$ to meet the terminal equality constraint $\psi(x(T)) = 0$.

Based on the three considered variants of the optimal fish harvest problem, one can identify three different mechanisms which induce the characteristic turnpike leaving arc, i.e. the departure from the neighborhood of the optimal steady state towards the end of the horizon. The middle column of Table 1 concisely recalls the primal perspective, which we already discussed. Therein,

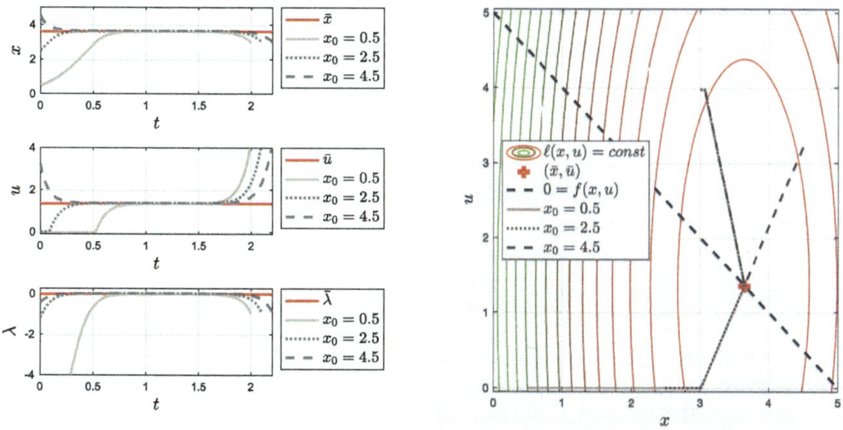

FIGURE 5 Fish harvest example with quadratic objective and terminal constraint (20) with pairs of x_0 and T: $x_0 = 0.5$, $T = 2.0$; $x_0 = 2.5$, $T = 2.1$; and $x_0 = 4.5$, $T = 2.2$. Left: optimal state, input and adjoint trajectories. Right: $x - u$ plane with optimal trajectories and contour lines of the Lagrange function $\ell(x, u)$ (colored), and the set of admissible steady states (dashed black).

$\tau < T$ denotes a time after which the running cost along the leaving arc is smaller than at the turnpike.

While the middle column describes the turnpike phenomenon that we see in the primal variables, the right hand side column of Table 1 shows the dual perspective, which provides additional valuable insights. Indeed, besides the different Lagrange functions ℓ, the three considered OCPs differ in terms of the transversality conditions, i.e. the terminal constraint for the adjoint variable λ.

TABLE 1 Comparison of the three variants of the optimal fish harvest problem.

OCP	Primal Perspective	Adjoint Perspective
(12)	$\exists \tau < T \, \forall t \in [\tau, T] : \ell(x^\star(t), u^\star(t)) < \ell(\bar{x}, \bar{u})$	$\bar{\lambda} \neq \lambda^\star(T) = 0$
(12) & (19)	$\int_0^T \ell(x^\star(t), u^\star(t)) \mathrm{d}t + \varphi(x^\star(T))$	$\bar{\lambda} \neq \lambda^\star(T) = \varphi_x(x^\star(T))$
(12) & (20)	$\int_0^T \ell(x^\star(t), u^\star(t)) \mathrm{d}t$ with $\psi(x^\star(T)) = 0$	$\bar{\lambda} \neq \lambda^\star(T)$

In case of the original OCP (12), the absence of a Mayer term implies $\lambda^\star(T) = 0$, which differs from the steady value of $\bar{\lambda}$ given in (13). For (12) with (19), the Mayer term implies $\lambda^\star(T) = \varphi_x(x^\star(T))$. Here we have chosen the Mayer term such that $\bar{\lambda} \neq \lambda^\star(T) = \varphi_x(x^\star(T))$. In other words, for the first two variants of the fish harvest problem, it is the difference between the turnpike value of the adjoint and the corresponding boundary conditions—$\bar{\lambda} \neq \lambda^\star(T)$—which creates the leaving arc. The third variant—i.e., (12) with (20)—includes a terminal equality constraint $\psi(x^\star(T)) = 0$. As we consider $\psi(x) = x - 3$, and $\bar{x} = 3.64 \neq x^\star(T)$, in this case it is the primal terminal constraint, which forces the optimal solutions to leave the optimal steady state $(\bar{x}, \bar{u})$.

In summary, we arrive at the following observations:

- The turnpike entry requires asymptotic reachability properties. It is triggered by the fact for sufficiently long horizons it is optimal in a time averaged sense to approach the turnpike.
- The turnpike as such is the globally optimal steady state of the dynamics under the considered Lagrange function ℓ.
- The characteristic leaving arc is induced whenever the terminal constraints (either for the primal or the dual variables) present in the optimality system (9) do not correspond to the turnpike variables. Specifically, this implies that either a primal mismatch between the turnpike state and the terminal equality constraints $\bar{x} \notin \{x \in \mathbb{R}^{n_x} \mid \psi(x) = 0\}$ or a dual mismatch between $\bar{\lambda} \neq \lambda^\star(T)$ occurs. In the latter case, it is well-known that the transversality conditions for $\lambda^\star(T)$ (5b) / (9b) are governed by the interplay of the Mayer term φ and the terminal inequality constraints $\psi(x) \leq 0$, see (Bryson and Ho, 1969).

Remark 1 (Exactness of turnpikes). The careful reader will have noticed that in case of OCP (12) the turnpike property is indeed exact as the optimal solutions enter the optimal steady state, which coincides with the singular arc (13). Intuitively, it is clear that whenever a turnpike property is exact this induces additional structure to the optimality system (9). Indeed under fairly mild technical assumptions it has been shown in (Faulwasser and Bonvin, 2017, Thm. 3) that whenever the only singular arc is a steady state of (9), then, if they exist, turnpikes are exact. More generally, whenever the set of control values taken is finite, the turnpike will be exact with respect to the inputs. This occurs frequently in case of mixed-integer OCPs, see Faulwasser and Murray (2020) for first steps in this direction. We remark that recently it has been shown L_1 tracking terms can also induce exact turnpikes phenomena in finite dimensional OCPs (Gugat et al., 2020) and that certain objectives induce exactness of turnpikes also in Hilbert spaces.

Finally, in (Zanon and Faulwasser, 2018, Remark 3 and Lemma 3) we argued that turnpikes in the interior of $\mathbb{Z}$ are never exact provided that the Jacobian linearization at $(\bar{x}, \bar{u})$ is controllable and the steady state problem satisfies differentiability assumptions and LICQ, see also Faulwasser and Zanon (2018) for the discrete-time analysis.

3.1 Necessary and sufficient conditions for turnpikes

As the examples in the previous section showed, the turnpike property is easily observed numerically by looking at simulated optimal trajectories. However, the fact that closed-form expressions of optimal trajectories are hardly ever available indicates that this is not a feasible way to obtain *rigorous* statements about the presence of the turnpike property.

To this end, sufficient and necessary conditions have been developed in the literature, which we survey in this section. One obvious necessary condition is approximate asymptotic controllability to $\bar{\xi}$. This property demands that for all neighborhoods $\mathcal{N}$ of $\bar{\xi}$ there exists $\tau > 0$ such that for all initial conditions $x_0 \in \mathbb{X}_0$ and all $T > 0$ there is an admissible control $u(\cdot)$ with $\xi(t; x_0) \in \mathcal{N}$ for all $t \in [\tau, T]$. For affine linear systems $\dot{x} = Ax + Bu + c$ and $(\bar{x}, \bar{u}) \in \operatorname{int} \mathbb{Z}$, one easily sees that this property is equivalent to stabilizability of the pair (A, B).

The existing sufficient conditions can be divided into two classes: (i) conditions based on necessary optimality conditions and (ii) conditions based on dissipativity. Regarding (i), Trélat and Zuazua (2015) provide a prototype example for finite dimensional systems. Essentially, suitable conditions on the Hamiltonian (8) of the OCP are formulated, which allow to conclude the continuous-time primal-dual exponential turnpike property. This approach can be extended to larger classes of systems, for instance, to optimal control problems governed by PDEs, as in, e.g., (Porretta and Zuazua, 2013, 2016; Zuazua, 2017; Grüne et al., 2020c; Pighin, 2020) or to shape optimization problems (Lance et al., 2020). It should be mentioned that several of these references consider linear-quadratic problems as a special case or as a local approximation of more general nonlinear problems. This allows to use the Riccati equation as an alternative way to formulate necessary conditions for optimality, and thus also to derive sufficient conditions for turnpike behavior, typically in form of detectability conditions.

Regarding (ii), dissipativity-based approaches rely on the systems-theoretic notion of dissipativity introduced by Willems (1972a,b), typically in a strict form. This property requires the existence of so-called storage and supply functions $S : \mathbb{X} \to \mathbb{R}$ and $w : \mathbb{X} \times \mathbb{U} \to \mathbb{R}$, which satisfy the inequality

$$\nabla S^{\top}(x) f(x, u) \leq w(x, u) - \alpha(\|x - \bar{x}\| + \|u - \bar{u}\|) \tag{21}$$

for continuous-time problems and

$$S(f(x, u)) - S(x) \leq w(x, u) - \alpha(\|x - \bar{x}\| + \|u - \bar{u}\|) \tag{22}$$

for discrete-time problems, where[8] $\alpha \in \mathcal{K}_\infty$ and S is bounded from below. For some purposes it is sufficient that only $\|x - \bar{x}\|$ appears in the argument of α in (21) or (22). The link to the OCPs introduced in the last section is then established by specifying $w(x, u) \doteq \ell(x, u) - \ell(\bar{x}, \bar{u})$.

The earliest result we are aware of that proves that strict dissipativity (plus an appropriate controllability property) implies the turnpike property can be found in (Carlson et al., 1991, Theorem 4.2). The respective Assumption 4.2 in this reference requires the existence of a linear storage functions $S(x) = \bar{p}^{\top} x$. The proof, however, is easily extended to general S, see (Grüne, 2013, Theorem 5.6) in discrete time and (Faulwasser et al., 2014, Theorem 1) or (Faulwasser et al.,

8 As usual, $\mathcal{K}_\infty$ denotes the space of functions $\alpha : [0, \infty) \to [0, \infty)$ that are continuous, strictly increasing and unbounded with $\alpha(0) = 0$.

2017, Theorem 2) in continuous time. The proof idea is actually very easy; we sketch the idea in discrete time for state turnpike in a problem with terminal cost $\varphi \equiv 0$: assume that the optimal trajectory $x^\star$ stays outside an ε-ball around $\bar{x}$ for an amount of time instances $K(T)$ that grows unboundedly as T tends to infinity. Then summing up (22) along the optimal trajectory yields

$$\begin{aligned} V_T(x_0) &= \sum_{t=0}^{T-1} \ell(x^\star(t), u^\star(t)) \geq T\ell(\bar{x}, \bar{u}) - S(x_0) + S(x^\star(T)) + K(T)\alpha(\varepsilon) \\ &\geq C + T\ell(\bar{x}, \bar{u}) + K(T)\alpha(\varepsilon), \end{aligned}$$

where C is a constant that is independent of T (here the lower bound on S is important because $x^\star(T)$ may vary with T). Now, if in addition the inequality

$$V_T(x_0) \leq \widetilde{C} + T\ell(\bar{x}, \bar{u}) \tag{23}$$

holds for a constant $\widetilde{C}$ independent of T, then we arrive at a contradiction, implying that $K(T)$ must be bounded as $T \to \infty$, from which the turnpike property can be concluded. Inequality (23) can be ensured by controllability or stabilizability properties of the dynamics, see, e.g., Theorem 6.4 in (Grüne, 2013). An assumption of this kind is imposed in all dissipativity based proofs of the turnpike property. Notice that the proof as sketched above requires the dissipation inequalities and (21) and (22) to hold only along optimal solutions. Moreover, in the continuous-time case, the integral counterpart to (21) suffices.[9]

As this sketch already indicates, dissipativity-based proofs of the turnpike property typically only deliver measure or cardinality turnpike. Yet, under additional conditions on the problem data one can also obtain exponential turnpike (Damm et al., 2014), which is particularly interesting because it provides a direct way to establish exponential turnpike for nonlinear systems without having to resort to linearization arguments. Strict dissipativity inequalities as in (21) or (22) yield the input-state turnpike property, while if one uses $\alpha(\|x - \bar{x}\|)$ instead of $\alpha(\|x - \bar{x}\| + \|u - \bar{u}\|)$ in these inequalities one obtains the state turnpike property. As there is no direct connection between strict dissipativity and necessary optimality conditions, there is no direct access to the dual variables in this approach and hence—to the best of our knowledge—no direct way to

9 The integral counterpart to (21) reads

$$S(x(T)) - S(x_0) \leq \int_0^T \ell(x(t), u(t))\mathrm{d}t - \int_0^T \alpha(\|x(t) - \bar{x}\| + \|u(t) - \bar{u}\|)\mathrm{d}t.$$

This can also be written as

$$S(x(T)) - S(x_0) \leq V_T(x_0) - \int_0^T \alpha(\|x(t) - \bar{x}\| + \|u(t) - \bar{u}\|)\mathrm{d}t,$$

which highlights that the dissipation inequality—specifically the storage function S—provides a lower bound on V_T, see also (Willems, 1971; Faulwasser and Kellett, 2021).

obtain primal-dual turnpike properties from strict dissipativity. However, under suitable assumptions on the optimal control problem, the primal-dual turnpike property follows from the input-state turnpike property (Faulwasser et al., 2020).

As we will discuss in Section 4, below, strict dissipativity is a very useful property in the context of model predictive control, which allows for deriving qualitative properties of the approximately optimal trajectory generated by this method. It is thus natural to ask how much stronger strict dissipativity is compared to the mere occurrence of the turnpike property. In (Grüne and Müller, 2016) it was shown for discrete-time controllable systems, that strict dissipativity (with $\alpha(\|x-\bar{x}\|)$ instead of $\alpha(\|x-\bar{x}\|+\|u-\bar{u}\|)$) is equivalent to a robust variant of the state-turnpike property. This robust variant demands that not only optimal but also near-optimal trajectories exhibit turnpike behavior, i.e. the turnpike property is robust with respect to perturbations of optimal trajectories. These results can be refined for linear-quadratic problems in both discrete and continuous time, see (Grüne and Guglielmi, 2018, 2021).

3.2 Finite and infinite-horizon cases

So far we have considered turnpike properties on finite time horizons T. While it is important to allow for arbitrarily long horizons T, until now each T under consideration was finite. It is, however, also possible (and not uncommon in the literature) to define the turnpike property for optimal control problems on an infinite time horizon. In the discrete-time case, the optimization criterion (1a) then becomes

$$V_\infty(x_0) \doteq \inf_{u(\cdot)} \sum_{k=0}^{\infty} \ell(x(k), u(k)), \tag{24}$$

the conditions (1b) and (1c) are required for all $k \in \mathbb{N}_0$, and the terminal constraint (1d) is dropped. In continuous time, the changes are analogous.

Definition (24) must be interpreted with care, because in general the infinite sum need not converge. In case an optimal equilibrium exists, the usual remedy is to replace $\ell(x,u)$ by $\ell(x,u)-\ell(\bar{x},\bar{u})$. This implies that $V_\infty(\bar{x})<\infty$ and then additional regularity assumptions are imposed (e.g., bounds and continuity assumptions on the storage function S and on V_∞) in order to ensure that the problem is well posed. This procedure is adopted, e.g., in (Grüne and Pannek, 2017, Chapter 7). Alternatively, optimality can be defined using the notion of overtaking optimality (Carlson et al., 1991; Blot and Hayek, 2014), which can in particular be exploited for analyzing turnpike properties for time-varying systems (Grüne et al., 2018).

Note that the optimal trajectories are now defined for all $k \in \mathbb{N}_0$. The translation of the turnpike properties from finite to infinite time horizon can then be performed almost formally: it suffices to replace "T" by "∞" in the definitions. For instance, the definition of cardinality turnpike in the discrete-time case becomes:

Definition 4 (Discrete-time (cardinality) turnpike for infinite horizon problem). OCP (24) is said to have a turnpike at $\bar{\xi}$, $\xi \in \{x, (x, u), (x, u, \lambda, \mu)\}$ if there exist a function $\nu_\xi : [0, \infty) \to [0, \infty]$ such that, for all $x_0 \in \mathbb{X}_0$,

$$\#\mathbb{Q}_{\xi,\infty}(\varepsilon) \leq \nu_\xi(\varepsilon) < \infty \quad \forall\, \varepsilon > 0, \tag{25}$$

where

$$\mathbb{Q}_{\xi,\infty}(\varepsilon) \doteq \{t \in \mathbb{N} \mid \|\xi^\star(t; x_0) - \bar{\xi}\| > \varepsilon\} \tag{26}$$

and $\#\mathbb{Q}_{\xi,\infty}(\varepsilon)$ is the cardinality of $\mathbb{Q}_{\xi,\infty}(\varepsilon)$.

- If $\xi = x$, then OCP (24) is said to have a *state turnpike* at $\bar{x}$.
- If $\xi = (x, u)$, then OCP (24) is said to have an *input-state turnpike* at $(\bar{x}, \bar{u})$.
- If $\xi = (x, u, \lambda, \mu)$, then OCP (24) is said to have a *primal-dual turnpike* at $(\bar{x}, \bar{u}, \bar{\lambda}, \bar{\mu})$.

It is easy to check that this definition implies that for each $x_0 \in \mathbb{X}_0$ and each $\varepsilon > 0$ there is $T_\varepsilon > 0$ such that

$$\|\xi^\star(t; x_0) - \bar{\xi}\| \leq \varepsilon$$

for all $t \geq T_\varepsilon$—which is nothing else than saying that $\xi^\star(t; x_0)$ converges to $\bar{\xi}$ as $t \to \infty$. The infinite-horizon cardinality turnpike property hence demands that all optimal trajectories (possibly including the corresponding inputs and costates) converge towards the same steady state.

Interestingly, the analogous infinite-horizon variant of the continuous-time measure turnpike property (which is obtained replacing T by ∞ in (16) and (17)) does not imply this convergence. This is because the mere fact that the measure $\nu_\xi(\varepsilon)$ is finite does not exclude the existence of arbitrary large times $t \in \Theta_{\xi,\infty}$. However, for instance for the state trajectory, convergence can be recovered if the state trajectory is uniformly continuous in t, because the Lebesgue measure of $\Theta_{\xi,\infty} \cap [t_1, \infty)$ must tend to 0 as t_1 tends to ∞. This means that the length of time intervals contained in $\Theta_{\xi,\infty} \cap [t_1, \infty)$ tends to zero, and uniform continuity then implies that the deviation of the trajectories from the turnpike during these time intervals also becomes arbitrarily small as t_1 tends to ∞. Alternatively, (Carlson et al., 1991, Chap. 4.6) discuss an approach to guarantee primal-dual stability of infinite-horizon continuous-time optimal solutions via the Lyapunov function candidate $\lambda^\top x$, certain convexity assumptions and LaSalle's invariance principle. Moreover, (Faulwasser and Kellett, 2021, Thm. 2 and Thm. 4) show convergence and local exponential stability of primal solutions via a strict dissipativity assumption.

In the case of the exponential turnpike, the infinite-horizon version is obtained replacing (18) by

$$\|\xi^\star(t; x_0) - \bar{\xi}\| \leq C\rho^t,$$

which is required for all $t \geq 0$. This implies exponential convergence of the optimal trajectories towards the turnpike $\bar{\xi}$ in both discrete and continuous time. In case the stronger estimate $\|\xi^\star(t; x_0) - \bar{\xi}\| \leq C\|x_0 - \bar{x}\|\rho^t$ holds, we even obtain exponential stability of the optimally controlled system. Conversely, for infinite-horizon OCPs, one can show as a local result that exponential stability of the infinite-horizon dual variables implies strict input-state dissipativity (Faulwasser and Kellett, 2021).

The generating mechanisms for infinite-horizon turnpike behavior are very similar to those in the finite-horizon case. Under suitable regularity conditions on the optimal control problem, it can even be rigorously shown that finite-horizon turnpike behavior occurs if and only if infinite-horizon turnpike behavior occurs. Conditions under which this is true are given, e.g., in (Zaslavski, 2006, Theorem 3.1.4) or in (Zaslavski, 2008, Theorem 3) in continuous time, and in (Grüne et al., 2017) in discrete time (see also (Zaslavski, 2014, Theorem 2.2) for conditions under which the infinite-horizon turnpike property implies the finite-horizon property in discrete time). To give an impression about the flavor of such conditions, we briefly explain the two conditions needed in (Grüne et al., 2017, Theorem 3.1): The first condition is that the optimal value functions for finite and infinite horizon are bounded on bounded sets by bounds that are independent of the horizon length. The second condition demands that trajectories with bounded objective remain in bounded sets, again uniformly, i.e. the size of the bounded set is determined by the bound on the objective and does not depend on the time horizon. Here it should be mentioned that the robust definition of the turnpike property discussed in the last paragraph of Section 3.1 is needed for this result.

Remark 2 (Relation between $\bar{\lambda}$ and V_∞). In case of a primal-dual turnpike as per Definition 2 one immediately has that for sufficiently large horizons T, there exists $\tau \in [0, T] \setminus \Theta_{(x,u,\lambda,\mu),T}(\varepsilon)$ such that

$$\nabla V_{T-\tau}(x(\tau)) = \lambda^\star(\tau) \approx \bar{\lambda},$$

i.e. close to the turnpike $\bar{\lambda}$ is an approximation of $\nabla V_{T-\tau}$. Under appropriate regularity assumptions we can also make statements about the dual variables in the infinite-horizon case. For instance, under suitable differentiability assumptions it can be shown that the gradient of the optimal value function of the infinite-horizon OCP, V_∞, satisfies

$$\nabla V_\infty(\bar{x}) = \bar{\lambda} = -\nabla S(\bar{x}),$$

cf. (Faulwasser and Kellett, 2021, Thm. 10).

In view of Table 1 it also appears natural to conjecture that for $T = \infty$ the optimal adjoint trajectory $\lambda^\star$ approaches its turnpike value. Indeed, a recent analysis in (Faulwasser and Kellett, 2021) shows that assuming dissipativity one can provide a positive answer to this classical problem of adjoint boundary conditions for infinite-horizon OCPs posed by Halkin (1974).

4 Exploitation of turnpikes in numerics and receding-horizon control

There are at least three different ways how the turnpike property can be exploited for the numerical computation of optimal trajectories: (i) splitting the optimization horizon at the turnpike, (ii) receding-horizon approximation, and (iii) exploiting the turnpike phenomenon in the numerical discretization appearing in (ii). All of these approaches are particularly efficient when optimal trajectories on long, possible infinitely long time intervals $[0, T]$, $T \in \mathbb{R} \cup \{\infty\}$, are sought. We note that when the length of the time horizon T increases, the problem size—e.g., the number of decision variables in direct numerical solution methods—grows while the conditioning of the problem often deteriorates. At the same time, the effort to solve underlying linear systems of equations grows super-linearly with their dimension. Put differently, we have the following empirical insight for OCPs more complex than a linear-quadratic regulator: the longer the optimization horizon the more difficult it is to solve an OCP numerically. Moreover, the direct computation of infinite-horizon optimal trajectories is challenging and often only possible using dynamic programming techniques, which become numerically infeasible already for moderate state dimension, see, e.g., (Bertsekas, 1995). This motivates techniques that circumvent the direct computation of optimal trajectories on long and infinite horizons.

The *first approach* is the most straightforward one: Given the turnpike equilibrium $\bar{x}$, the optimization horizon $T > 0$ (possibly $T = \infty$), an initial condition x_0 and, in case $T < \infty$, a terminal condition $x(T) \in \mathbb{X}_f \doteq \{x \in \mathbb{X} \mid \psi(x) \leq 0\}$, we compute

- an optimal trajectory $x_1(\cdot)$ with finite horizon $T_1 < T$ and initial and terminal conditions $x_1(0) = x_0$, $x_1(T_1) = \bar{x}$
- in case $T < \infty$, an optimal trajectory $x_2(\cdot)$ with horizon $T_2 < T - T_1$ and initial and terminal conditions $x_2(0) = \bar{x}$, $x_2(T_2) \in \mathbb{X}_f$.

An approximation of the optimal trajectory is then obtained by

$$\tilde{x}(t) = \begin{cases} x_1(t), & t \in [0, T_1] \\ x^e, & t \in [T_1, T - T_2] \\ x_2(t - T + T_2), & t \in [T - T_2, T] \end{cases}$$

in case $T < \infty$ and

$$\tilde{x}(t) = \begin{cases} x_1(t), & t \in [0, T_1] \\ x^e, & t \in [T_1, \infty) \end{cases}$$

in case $T = \infty$. The idea for this construction goes back at least as far as (Wilde and Kokotovic, 1972), see also (Anderson and Kokotović, 1987), in which also the error of the method is analyzed for particular classes of optimal control problems. This construction has also been suggested for PDEs, see (Hernández-Santamaría et al., 2019, Rem. 7). If the turnpike is not exact, the method causes an error because the approaching arc—which is approximated by $x_1(\cdot)$—does not reach the turnpike equilibrium $\bar{x}$ exactly but only a neighborhood of $\bar{x}$. Likewise, the middle part of the optimal solution is only near $\bar{x}$ but does not exactly coincide with $\bar{x}$. Finally, the leaving arc, which is approximated by $x_2(\cdot)$, does in general not start exactly at $\bar{x}$ but only near $\bar{x}$. The resulting error can be estimated if the speed of convergence towards the turnpike is known (as in the case of exponential turnpike) and if the sensitivity of the optimal value with respect to the initial condition $x_0 \approx \bar{x}$ is independent of T, i.e. if the optimal value for initial conditions x_0 satisfying $\|x_0 - \bar{x}\| \leq \delta$ satisfies $|V_T(x_0) - V_T(\bar{x})| \leq \varepsilon$ with $\varepsilon > 0$ independent of T and $\varepsilon \to 0$ as $\delta \to 0$. In this case the error of each of the trajectory pieces constituting $\tilde{x}(\cdot)$ and thus the overall error can be estimated.

The *second approach* uses Model Predictive Control (MPC), also known as Receding Horizon Control. In this approach, we fix a finite horizon $T_{opt} < T$ and a control horizon $\delta < T_{opt}$, set $t := 0$, $x_{MPC}(0) := x_0$, and proceed as follows:

1. Solve the optimal control problem with horizon T_{opt} (possibly with additional or modified terminal conditions) with initial condition $x_0 = x_{MPC}(t)$ and denote the numerical approximation of the optimal solution by $x_{opt}(\cdot)$
2. set $x_{MPC}(t + \tau) := x_{opt}(\tau)$ for $\tau \in [0, \delta]$, set $t := t + \delta$
3. if $T < \infty$, set $T_{opt} := \min\{T_{opt}, T - t\}$, $\delta := \min\{\delta, T - t\}$
4. if $T_{opt} > 0$, go to **1.**

Compared to the first approach, MPC has the advantage that the optimization is carried out repeatedly. Efficient modern algorithms can perform one of the optimizations in Step **1.** of the MPC algorithm in the range of milliseconds or even below (Houska et al., 2011). This means that for many applications the approach is real-time capable and thus provides the control in form of a feedback law that is able to react to perturbations or modeling errors. Yet, even if for a specific OCP the MPC methodology is not real-time capable, e.g., due to the dimensionality of the problem or due to the complexity of the dynamics, MPC is a powerful approach to reduce the computational complexity of the problem and allows to approximate problems that cannot be solved directly. As such, MPC can be seen as a complexity reduction technique in time.

Under essentially the same conditions as for the first approach it can be shown that this method yields approximately optimal trajectories (Grüne, 2016). If one additionally assumes strict dissipativity (21) or (22), it can also be shown that x_{MPC} has the turnpike property, at least approximately (Grüne and Stieler, 2014). Suitable terminal conditions on the optimal control problem to be solved in Step **1.** can improve both the performance properties as well as the qualitative behavior of x_{MPC} (Zanon and Faulwasser, 2018; Faulwasser et al., 2018; Grüne and Pannek, 2017, Chapter 7). Interestingly, so far this analysis appears to be limited to the case $T = \infty$ as we are not aware of a rigorous analysis for the case of a finite T. Yet, we expect that similar arguments work for finite T.

The *third approach* is intertwined with the second approach, as it concerns efficient numerical discretization for computing the solutions of the optimal control problem in Step **1.** of the MPC algorithm, particularly for problems governed by partial differential equations (PDEs). It relies on the obvious fact that only $x_{opt}|_{[0,\delta]}$ is needed for generating x_{MPC} in Step **2.** of the MPC algorithm, hence there is no need to compute $x_{opt}(\tau)$ accurately for $\tau \in (\delta, T_{opt}]$.

To this end, the following interpretation of the turnpike property is useful: it implies that optimal trajectories are essentially independent of the initial condition and—if present—of the terminal condition, for times that are not close to the initial or terminal time, respectively. In the particular case of the exponential turnpike property, one can even quantify that the influence of initial and terminal condition is damped exponentially as t increases or decreases, respectively. For several classes of PDE governed optimal control problems it turns out (Grüne et al., 2019, 2020c) that the same mechanism that generates the exponential turnpike property also implies that perturbations acting on the system at a certain point t_{pert} in time are damped out exponentially, i.e. their influence on the solution decreases exponentially in $|t - t_{pert}|$. This property can be called an *exponential sensitivity* property and was also established recently for other types of optimal control problems, see, e.g., Shin et al. (2019). Since discretization errors can be interpreted as perturbations, this implies that increasingly coarse discretizations towards the end of the horizon do only marginally affect the accuracy of $x_{opt}|_{[0,\delta]}$. This observation forms the basis for the design of efficient adaptive discretizations in time and space, which can be generated using appropriate goal oriented error estimators (Grüne et al., 2020b). Moreover, the third approach has also been considered for solving OCPs concerning chemical processes wherein in the middle of the horizon a less dense discretization scheme was used while no error bounds are given, see (Sahlodin and Barton, 2015).

We illustrate the third approach and its use in MPC by means of a numerical example. We consider the linear heat equation with distributed control

$$\begin{aligned} \dot{x}(t,\omega) &= 0.1\Delta x(t,\omega) + u && \text{for } (t,\omega) \in (0,T) \times \Omega \\ x(t,\omega) &= 0 && \text{for } (t,\omega) \in (0,T) \times \partial\Omega \\ x(0,\omega) &= 0 && \text{for } \omega \in \Omega, \end{aligned} \tag{27}$$

with $\Omega = [0,3] \times [0,1]$ and $T = 10$. We consider the cost function

$$\ell(x(t,\cdot), u(t,\cdot)) = \frac{1}{2}\|x(t,\cdot) - x_{ref}(t,\cdot)\|^2_{L_2(\Omega)} + \alpha\|u\|^2_U,$$

with $\alpha = 10^{-3}$, where x_{ref} may either be the constant reference

$$x_{ref}(t,\omega) = x_c(\omega) := g\left(\left\|\omega - \begin{pmatrix}1.5\\0.5\end{pmatrix}\right\|\right)$$

depicted in Fig. 6, or the time-varying reference

$$x_{ref}(t,\omega) = x_{tv}d(t,\omega) := g\left(\|\omega - \omega(t)\|\right),$$

with

$$\omega(t) = \left(1.5 - \cos\left(\pi\left(\frac{t}{10}\right)\right), \left|\cos\left(\pi\left(\frac{t}{10}\right)\right)\right|\right)^{\top}.$$

In both cases we use $g(r) = 10\exp(1 - 1/(1 - 100r^2/9))$ for $s < 1$ and $g(r) = 0$ else. It was shown in (Grüne et al., 2019) that this problem has the exponential turnpike and the exponential sensitivity property. Fig. 7 shows the adaptive grid over time for the numerical solution of one optimal control problem in Step **1.** of the MPC algorithm for the static reference. On the right, the adaptivity criterion is obtained from a usual error estimator on $[0, T]$, resulting in a grid that is refined in almost all time steps. On the left, a goal oriented error estimator is used, that only aims at providing an accurate solution on the interval $[0, \delta]$ that is actually implemented in the MPC algorithm, here with $\delta = 0.5$. The fineness of the grid is chosen such that the total number of space grid points is approximately equal on both sides. Just as the turnpike and sensitivity theory predicts, the accuracy of the solution for $t \in [0, \delta]$ is only marginally affected by errors for large t, hence the error estimator allows for coarse grids.

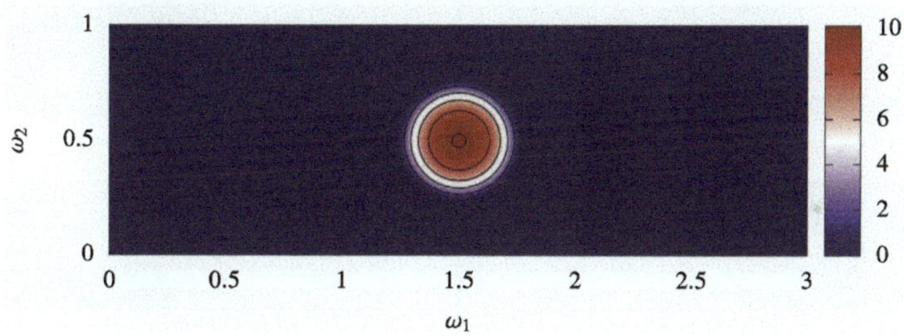

FIGURE 6 Constant reference trajectory $x_c(\omega)$.

The advantage of using a goal oriented error estimator that exploits the turnpike structure also shows up when looking at the value of the MPC closed-loop trajectory in dependence of the number of space grid points. It is particularly pronounced for the time varying reference, the respective values can be seen in Fig. 8 (for other settings see (Grüne et al., 2020b)). As this figure shows, with the goal oriented error estimator the closed-loop cost is significantly lower for the same number of grid points.

For further numerical examples, including adaptive time discretization as well as unstable and semilinear PDEs we refer to (Grüne et al., 2019, 2020b).

5 Topics not discussed and open problems

In this section we briefly sketch parts of the theory that we did not present in this chapter and discuss some open questions. A class of optimal control problems that we did not discuss here are discounted problems, in which the cost is multiplied with a time varying, decreasing factor, typically an exponential of the form

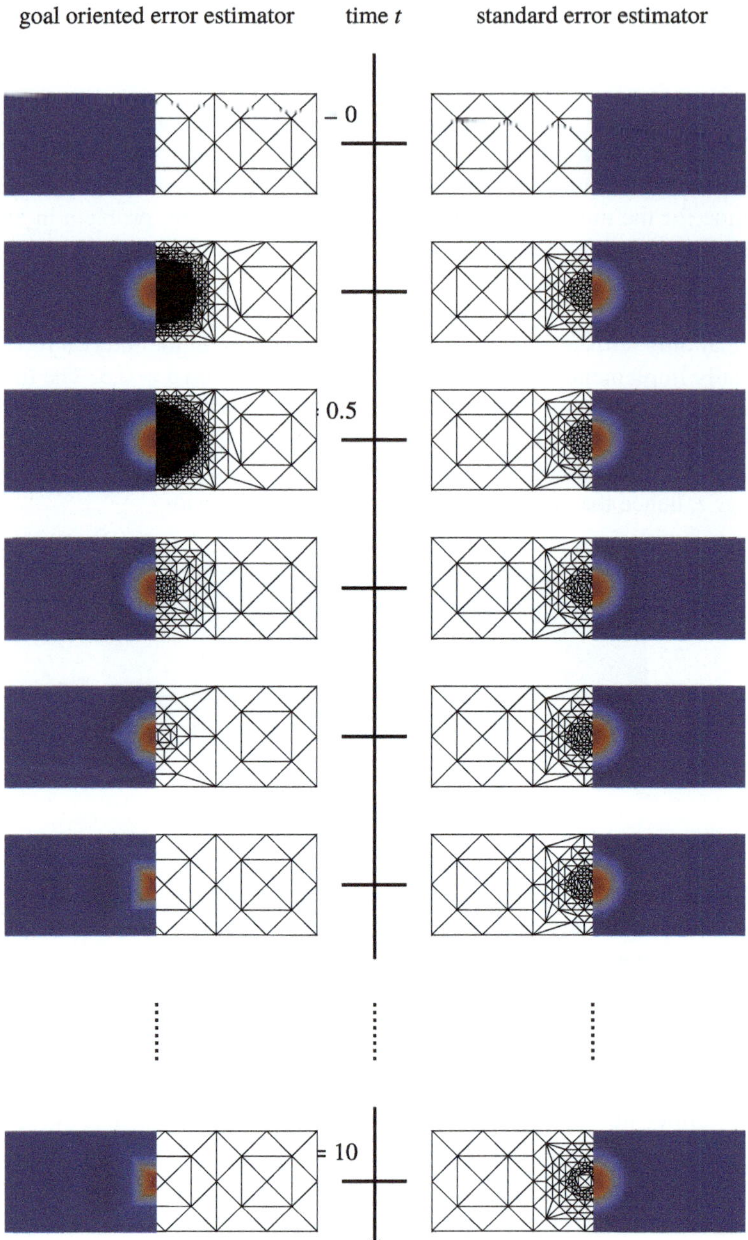

FIGURE 7 Evolution of adaptively refined space grids for an MPC adapted goal oriented error estimator (left) and a standard error estimator (right) with 5825 and 5478 total space grid points, respectively.

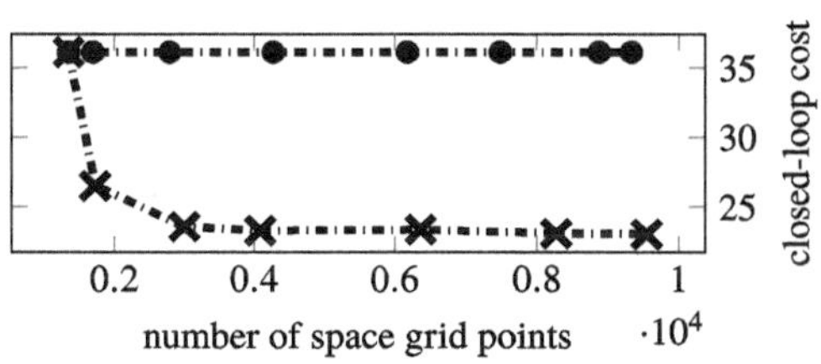

FIGURE 8 Comparison of the cost of the MPC closed-loop trajectories for different error estimators used for spatial refinement for the time-varying reference x_{tv}. • = standard error estimator, **x** = MPC adapted goal oriented error estimator.

$e^{-\delta t}$, $\delta > 0$. In discrete time, many of the results discussed in this paper carry over to discounted problems, see (Gaitsgory et al., 2018; Grüne et al., 2020a). In continuous time, however, this still seems to be largely open.

This situation is similar for time-varying problems. Here, one has to distinguish on the one hand between problems with time-invariant data that exhibit a time-varying turnpike, e.g., a periodic orbit instead of an equilibrium. On the other hand, time-varying turnpikes appear naturally in problems depending on time-varying data. Motivated by MPC, both problems have been studied recently, see, e.g., (Grüne et al., 2018; Grüne and Pirkelmann, 2019; Müller and Grüne, 2016; Zanon et al., 2017). However, these investigations are again mostly limited to discrete time.

Both in discrete-time and continuous-time the existing turnpike results focus mostly on the turnpike being in the interior of the constraints, i.e., a comprehensive treatment with active constraints at the turnpike appears to be open. Moreover, problems with mixed-integer inputs and/or hybrid dynamics have only been discussed to a very limited extent, see (Faulwasser and Murray, 2020) for discrete-time dissipativity-based results and (Gugat and Hante, 2019) for results on infinite-dimensional hyperbolic systems with integrality constraints on the inputs under strict convexity assumptions for the objective.

In recent years the analysis of turnpike properties for infinite-dimensional continuous-time optimal control problems—typically problems governed by PDEs—has received great attention, see, e.g., (Faulwasser et al., 2020; Gugat and Hante, 2019; Grüne et al., 2020c; Lance et al., 2020; Porretta and Zuazua, 2013; Zuazua, 2017; Pighin, 2020; Porretta and Zuazua, 2016). Interestingly, this line of research is usually based on necessary optimality conditions rather than on dissipativity notions with Trélat and Zhang (2018) being one of the few exceptions. A comprehensive theory linking dissipativity and turnpike phenomena for this problem class (as it is available, e.g., for finite-dimensional linear quadratic problems (Grüne and Guglielmi, 2021)) is still missing, even for the relatively simply linear-quadratic case.

More generally, the connection between the optimality-condition-based and the dissipativity-based approach to turnpike analysis is still not completely understood. Both approaches are usually used in parallel and as an alternative to each other, but many aspects of the relation between these concepts—for instance the question whether there are problem classes for which one of the approaches works while the other fails—have not yet been clarified. Moreover, the implication of turnpike properties on the Hamilton-Jacobi-Bellman-Equation and its solution is not widely explored. First steps have been taken in Esteve et al. (2020b); Faulwasser and Kellett (2021).

Finally, in mathematical economy there is a huge body of literature on turnpike behavior in stochastic optimal control problems, see (Marimon, 1989; Kolokoltsov and Yang, 2012; Sethi et al., 1992) for only a small selection of representative papers. While this property has been established for many specific optimal control problems, a thorough mathematical treatment based, e.g., on stochastic notions of dissipativity, as well as an exploitation of this property for numerical purposes are missing so far.

Acknowledgment

We thank Manuel Schaller for providing the figures of the numerical examples in Section 4.

References

Anderson, B.D.O., Kokotović, P.V., 1987. Optimal control problems over large time intervals. Automatica 23 (3), 355–363.

Bellman, R., 1954. The theory of dynamic programming. Bulletin of the American Mathematical Society 60 (6), 503–515.

Berkovitz, L., 1974. Optimal Control Theory. Applied Mathematical Sciences, vol. 12. Springer.

Bertsekas, D.P., 1995. Dynamic Programming and Optimal Control. Vol. 1 and 2. Athena Scientific, Belmont, MA.

Bewley, T., 1982. An integration of equilibrium theory and turnpike theory. Journal of Mathematical Economics 10 (2–3), 233–267.

Blot, J., Hayek, N., 2014. Infinite-Horizon Optimal Control in the Discrete-Time Framework. SpringerBriefs in Optimization. Springer, New York.

Boltyanskii, V., Gamkrelidze, R., Pontryagin, L., 1960. Theory of optimal processes. I. The maximum principle. Izvestiya Rossiiskoi Akademii Nauk. Seriya Matematicheskaya 24 (1), 3–42.

Bryson, A., Ho, Y.-C., 1969. Applied Optimal Control. Ginn and Company, Waltham, Massachusetts.

Carlson, D.A., Haurie, A.B., Leizarowitz, A., 1991. Infinite Horizon Optimal Control. Deterministic and Stochastic Systems, 2nd edn. Springer, Berlin.

Cliff, E., Vincent, T., 1973. An optimal policy for a fish harvest. Journal of Optimization Theory and Applications 12 (5), 485–496.

Damm, T., Grüne, L., Stieler, M., Worthmann, K., 2014. An exponential turnpike theorem for dissipative discrete time optimal control problems. SIAM Journal on Control and Optimization 52 (3), 1935–1957.

Dorfman, R., Samuelson, P., Solow, R., 1958. Linear Programming and Economic Analysis. McGraw-Hill, New York.

E, W., Han, J., Jentzen, A., 2017. Deep learning-based numerical methods for high-dimensional parabolic partial differential equations and backward stochastic differential equations. Communications in Mathematics and Statistics 5 (4), 349–380.

Esteve, C., Geshkovski, B., Pighin, D., Zuazua, E., 2020a. Large-time asymptotics in deep learning. Preprint. arXiv:2008.02491.

Esteve, C., Kouhkouh, H., Pighin, D., Zuazua, E., 2020b. The turnpike property and the long-time behavior of the Hamilton-Jacobi equation. arXiv preprint. arXiv:2006.10430.

Faulwasser, T., Bonvin, D., 2015. On the design of economic NMPC based on approximate turnpike properties. In: Proc. of 54th IEEE Conference on Decision and Control. Osaka, Japan, pp. 4964–4970.

Faulwasser, T., Bonvin, D., 2017. Exact turnpike properties and economic NMPC. European Journal of Control 35, 34–41.

Faulwasser, T., Flaßkamp, K., Ober-Blöbaum, S., Worthmann, K., 2019. Towards velocity turnpikes in optimal control of mechanical systems. IFAC-PapersOnLine 52 (16), 490–495.

Faulwasser, T., Flaßkamp, K., Ober-Blöbaum, S., Worthmann, K., 2021. A dissipativity characterization of velocity turnpikes in optimal control problems for mechanical systems. IFAC-PapersOnLine 54 (9), 624–629.

Faulwasser, T., Grüne, L., Humaloja, J.-P., Schaller, M., 2020. The interval turnpike property for adjoints. Pure and Applied Functional Analysis. In press. Preprint available from arXiv:2005.12120.

Faulwasser, T., Grüne, L., Müller, M.A., 2018. Economic nonlinear model predictive control. Foundations and Trends® in Systems and Control 5 (1), 1–98.

Faulwasser, T., Kellett, C., 2021. On continuous-time infinite horizon optimal control – dissipativity, stability and transversality. Automatica 134, 109907.

Faulwasser, T., Korda, M., Jones, C., Bonvin, D., 2014. Turnpike and dissipativity properties in dynamic real-time optimization and economic MPC. In: Proc. of the 53rd IEEE Conference on Decision and Control. Los Angeles, California, USA, pp. 2734–2739.

Faulwasser, T., Korda, M., Jones, C.N., Bonvin, D., 2017. On turnpike and dissipativity properties of continuous-time optimal control problems. Automatica 81, 297–304.

Faulwasser, T., Murray, A., 2020. Turnpike properties in discrete-time mixed integer optimal control. IEEE Control Systems Letters 4, 704–709. arXiv:2002.02049.

Faulwasser, T., Pannocchia, G., 2019. Towards a unifying framework blending real-time optimization and economic model predictive control. Industrial and Chemical Engineering Research 58 (30), 13583–13598.

Faulwasser, T., Zanon, M., 2018. Asymptotic stability of economic NMPC: the importance of adjoints. IFAC-PapersOnLine 51 (20), 157–168.

Fiacco, A., 1976. Sensitivity analysis for nonlinear programming using penalty methods. Mathematical Programming 10 (1), 287–311.

Gaitsgory, V., Grüne, L., Höger, M., Kellett, C.M., Weller, S.R., 2018. Stabilization of strictly dissipative discrete time systems with discounted optimal control. Automatica 93, 311–320.

Gerdts, M., 2011. Optimal Control of ODEs and DAEs. Walter de Gruyter.

González-Forero, M., Gardner, A., 2018. Inference of ecological and social drivers of human brain-size evolution. Nature 557 (7706), 554–557.

Grüne, L., 2013. Economic receding horizon control without terminal constraints. Automatica 49 (3), 725–734.

Grüne, L., 2016. Approximation properties of receding horizon optimal control. Jahresbericht der Deutschen Mathematiker-Vereinigung 118 (1), 3–37.

Grüne, L., Guglielmi, R., 2018. Turnpike properties and strict dissipativity for discrete time linear quadratic optimal control problems. SIAM Journal on Control and Optimization 56 (2), 1282–1302.

Grüne, L., Guglielmi, R., 2021. On the relation between turnpike properties and dissipativity for continuous time linear quadratic optimal control problems. Mathematical Control and Related Fields 11, 169–188. https://doi.org/10.3934/mcrf.2020032.

Grüne, L., Kellett, C.M., Weller, S.R., 2017. On the relation between turnpike properties for finite and infinite horizon optimal control problems. Journal of Optimization Theory and Applications 173 (3), 727–745.

Grüne, L., Müller, M.A., 2016. On the relation between strict dissipativity and the turnpike property. Systems & Control Letters 90, 45–53.

Grüne, L., Müller, M.A., Kellett, C.M., Weller, S.R., 2020a. Strict dissipativity for discrete time discounted optimal control problems. Mathematical Control and Related Fields. In press.

Grüne, L., Pannek, J., 2017. Nonlinear Model Predictive Control. Theory and Algorithms, 2nd edn. Springer-Verlag, London.

Grüne, L., Pirkelmann, S., 2019. Economic model predictive control for time-varying system: performance and stability results. In: Special Issue on MPC for Energy Systems: Economic and Distributed Approaches. Optimal Control Applications & Methods 41 (1), 42–64.

Grüne, L., Pirkelmann, S., Stieler, M., 2018. Strict dissipativity implies turnpike behavior for time-varying discrete time optimal control problems. In: Feichtinger, G., Kovacevic, R.M., Tragler, G. (Eds.), Control Systems and Mathematical Methods in Economics: Essays in Honor of Vladimir M. Veliov. In: Lecture Notes in Econom. and Math. Systems, vol. 687. Springer, Cham, pp. 195–218.

Grüne, L., Schaller, M., Schiela, A., 2019. Sensitivity analysis of optimal control for a class of parabolic PDEs motivated by model predictive control. SIAM Journal on Control and Optimization 57 (4), 2753–2774.

Grüne, L., Schaller, M., Schiela, A., 2020b. Efficient MPC for parabolic PDEs with goal oriented error estimation. SIAM Journal on Scientific Computing. In press. Preprint available from arXiv:2007.14446.

Grüne, L., Schaller, M., Schiela, A., 2020c. Exponential sensitivity and turnpike analysis for linear quadratic optimal control of general evolution equations. Journal of Differential Equations 268 (12), 7311–7341.

Grüne, L., Stieler, M., 2014. Asymptotic stability and transient optimality of economic MPC without terminal conditions. Journal of Process Control 24 (8), 1187–1196.

Gugat, M., Hante, F., 2019. On the turnpike phenomenon for optimal boundary control problems with hyperbolic systems. SIAM Journal on Control and Optimization 57 (1), 264–289.

Gugat, M., Schuster, M., Zuazua, E., 2020. The finite-time turnpike phenomenon for optimal control problems: stabilization by non-smooth tracking terms. Preprint. arXiv:2006.07051.

Halkin, H., 1974. Necessary conditions for optimal control problems with infinite horizons. Econometrica, 267–272.

Hartl, R., Sethi, S., Vickson, R., 1995. A survey of the maximum principles for optimal control problems with state constraints. SIAM Review 37 (2), 181–218.

Hernández-Santamaría, V., Lazar, M., Zuazua, E., 2019. Greedy optimal control for elliptic problems and its application to turnpike problems. Numerische Mathematik 141 (2), 455–493.

Houska, B., Ferreau, H.J., Diehl, M., 2011. An auto-generated real-time iteration algorithm for nonlinear MPC in the microsecond range. Automatica 47 (10), 2279–2285.

Kalboussi, N., Rapaport, A., Bayen, T., Amar, N., Ellouze, F., Harmand, J., 2018. Optimal control of membrane-filtration systems. IEEE Transactions on Automatic Control 64 (5), 2128–2134.

Kolokoltsov, V., Yang, W., 2012. Turnpike theorems for Markov games. Dynamic Games and Applications 2 (3), 294–312.

Lance, G., Trélat, E., Zuazua, E., 2020. Shape turnpike for linear parabolic PDE models. Systems & Control Letters 142, 104733.

Lee, E., Markus, L., 1967. Foundations of Optimal Control Theory. The SIAM Series in Applied Mathematics. John Wiley & Sons, New York, London, Sydney.

Li, Q., Chen, L., Tai, C., Weinan, E., 2017. Maximum principle based algorithms for deep learning. Journal of Machine Learning Research 18 (1), 5998–6026.

Marimon, R., 1989. Stochastic turnpike property and stationary equilibrium. Journal of Economic Theory 47 (2), 282–306.

McKenzie, L., 1976. Turnpike theory. Econometrica 44 (5), 841–865.

McKenzie, L.W., 1986. Optimal economic growth, turnpike theorems and comparative dynamics. In: Handbook of Mathematical Economics, Vol. III. In: Handbooks in Economics, vol. 1. North-Holland, Amsterdam, pp. 1281–1355.

Müller, M., Angeli, D., Allgöwer, F., 2015. On necessity and robustness of dissipativity in economic model predictive control. IEEE Transactions on Automatic Control 60 (6), 1671–1676.

Müller, M.A., Grüne, L., 2016. Economic model predictive control without terminal constraints for optimal periodic behavior. Automatica 70, 128–139.

Nordhaus, W.D., 1992. An optimal transition path for controlling greenhouse gases. Science 258, 1315–1319.

Pesch, H., 2013. Carathéodory's royal road of the calculus of variations: missed exits to the maximum principle of optimal control theory. Numerical Algebra, Control and Optimization 3 (1), 161–173.

Pighin, D., 2020. The turnpike property in semilinear control. arXiv preprint. arXiv:2004.03269.

Porretta, A., Zuazua, E., 2013. Long time versus steady state optimal control. SIAM Journal on Control and Optimization 51 (6), 4242–4273.

Porretta, A., Zuazua, E., 2016. Remarks on long time versus steady state optimal control. In: Mathematical Paradigms of Climate Science. Springer, pp. 67–89.

Ramsey, F.P., 1928. A mathematical theory of saving. The Economic Journal 38 (152), 543–559.

Rao, A., Mease, K., 1999. Dichotomic basis approach to solving hyper-sensitive optimal control problems. Automatica 35 (4), 633–642.

Rawlings, J., Amrit, R., 2009. Optimizing process economic performance using model predictive control. In: Magni, L., Raimondo, D., Allgöwer, F. (Eds.), Nonlinear Model Predictive Control - Towards New Challenging Applications. In: Lecture Notes in Control and Information Sciences, vol. 384. Springer, Berlin, pp. 119–138.

Rawlings, J., Bonné, D., Jorgensen, J., Venkat, A., Jorgensen, S., 2008. Unreachable setpoints in model predictive control. IEEE Transactions on Automatic Control 53 (9), 2209–2215.

Rawlings, J., Mayne, D., Diehl, M., 2017. Model Predictive Control: Theory, Computation, and Design. Nob Hill Publishing, Madison, WI.

Sahlodin, A., Barton, P., 2015. Optimal campaign continuous manufacturing. Industrial & Engineering Chemistry Research 54 (45), 11344–11359.

Samuelson, P.A., 1949. Market mechanisms and maximization. Technical report. Rand Corporation. https://www.rand.org/content/dam/rand/pubs/papers/2008/P69.pdf.

Sethi, S., Soner, H.M., Zhang, Q., Jiang, J., 1992. Turnpike sets and their analysis in stochastic production planning problems. Mathematics of Operations Research 17 (4), 932–950.

Shin, S., Faulwasser, T., Zanon, M., Zavala, V.M., 2019. A parallel decomposition scheme for solving long-horizon optimal control problems. In: Proceedings of the 58th IEEE Conference on Decision and Control—CDC 2019. Nice, France, pp. 5264–5271.

Sussmann, H., Willems, J., 1997. 300 years of optimal control: from the brachystochrone to the maximum principle. IEEE Control Systems 17 (3), 32–44.

Trélat, E., Zhang, C., 2018. Integral and measure-turnpike properties for infinite-dimensional optimal control systems. MCSS. Mathematics of Control, Signals and Systems 30 (1), 3.

Trélat, E., Zuazua, E., 2015. The turnpike property in finite-dimensional nonlinear optimal control. Journal of Differential Equations 258 (1), 81–114.

Vinter, R., 2010. Optimal Control. Birkhäuser, Basel, Boston, Berlin.

von Neumann, J., 1938. Über ein ökonomisches Gleichungssystem und eine Verallgemeinerung des Brouwerschen Fixpunktsatzes. In: Menger, K. (Ed.), Ergebnisse eines Mathematischen Seminars.

Wilde, R., Kokotovic, P., 1972. A dichotomy in linear control theory. IEEE Transactions on Automatic Control 17 (3), 382–383.

Willems, J., 1971. Least squares stationary optimal control and the algebraic Riccati equation. IEEE Transactions on Automatic Control 16 (6), 621–634.

Willems, J.C., 1972a. Dissipative dynamical systems. I. General theory. Archive for Rational Mechanics and Analysis 45, 321–351.

Willems, J.C., 1972b. Dissipative dynamical systems. II. Linear systems with quadratic supply rates. Archive for Rational Mechanics and Analysis 45, 352–393.

Zanon, M., Faulwasser, T., 2018. Economic MPC without terminal constraints: gradient-correcting end penalties enforce stability. Journal of Process Control 63, 1–14.

Zanon, M., Grüne, L., Diehl, M., 2017. Periodic optimal control, dissipativity and MPC. IEEE Transactions on Automatic Control 62 (6), 2943–2949.
Zaslavski, A.J., 2006. Turnpike Properties in the Calculus of Variations and Optimal Control. Springer, New York.
Zaslavski, A.J., 2008. Turnpike properties of approximate solutions of autonomous variational problems. Control and Cybernetics 37 (2), 491–512.
Zaslavski, A.J., 2014. Turnpike Phenomenon and Infinite Horizon Optimal Control. Springer International Publishing.
Zelikin, M., Borisov, V., 1994. Theory of Chattering Control. Springer.
Zuazua, E., 2017. Large time control and turnpike properties for wave equations. Annual Reviews in Control 44, 199–210.

Chapter 12

Some challenging optimization problems for logistic diffusive equations and their numerical modeling

Idriss Mazari[a], Grégoire Nadin[b], and Yannick Privat[c,d,*]
[a]*CEREMADE, UMR CNRS 7534, Université Paris-Dauphine, Université PSL, Place du Maréchal De Lattre De Tassigny, Paris cedex 16, France,* [b]*CNRS, Sorbonne Université, Laboratoire Jacques-Louis Lions, Paris, France,* [c]*IRMA, Université de Strasbourg, CNRS UMR 7501, Strasbourg, France,* [d]*Institut Universitaire de France (IUF), Paris cedex 05, France*
**Corresponding author: e-mail address: yannick.privat@unistra.fr*

Contents

Abstract

This chapter is dedicated to the study of several shape optimization problems arising in population dynamics. We provide some answers to the generic question: what is the optimal way to arrange resources inside a domain? Of course, the wording "optimal" refers to different criteria, which may be the survival ability of a given species or the total population size. We consider here a very simple spatial ecology model in which the evolution of the population is governed by the logistic diffusive equation parametrized by the resources distribution inside the domain, denoted $m(\cdot)$, which is the main optimization variable and models the favorable and unfavorable parts of the habitat. We investigate here two optimal design problems: the first one is related to the species persistence for large times. It boils down to the optimization of the principal eigenvalue associated with

Handbook of Numerical Analysis, Volume 23, ISSN 1570-8659, https://doi.org/10.1016/bs.hna.2021.12.012

an elliptic operator with respect to the resources distributions $m(\cdot)$. The second one deals with steady-states of the aforementioned reaction-diffusion equation and aims at maximizing the total size of the population with respect to resources distributions. In our analysis, we mainly focus on qualitative properties of maximizers, and illustrate it with the help of numerical illustrations. We also highlight related open problems and interesting numerical issues that remain to be investigated.

Keywords
Extremal eigenvalue problem, Shape optimization, Symmetrization techniques

MSC Codes
49J15, 49K20, 49R05, 49M05

1 Introduction and bio-mathematical background

1.1 General introduction

The study of population dynamics is now well-established as a field of paramount importance, both in life sciences and in mathematics. The first models for population dynamics, which originated in the seminal (Fisher, 1937; Kolmogorov et al., 1937) and were then studied at length, relied on a spatial homogeneity assumption, and the influence of spatial heterogeneity and of the shape of the domain remained elusive for some time. It was successfully used for the first time in an ecological modeling framework in Skellam (1951). In this paper, Skellam proved, among other things, that, assuming the boundary of the environment is lethal then no population can survive if this environment is too small. The next mathematical works addressing in a crude way the influence of the environment on population dynamics was the series of articles by Cantrell and Cosner (1989, 1991b,a, 1998), and this opened the way for a finer understanding of these mathematical and biological queries. Since then, this line of research has been extremely active. In this Chapter, we present some recent works that are part of the endeavor to grasp a fuller understanding of spatially heterogeneous equations. Here, spatial heterogeneity will be taken into account via the **resources distribution**, and the main focus is the following informal question:

What is the optimal way to spread resources in a domain?

More precisely, we investigate two main issues:

1. the optimal survival ability: how can we spread resources so as to optimize the survival ability of a population? The relevant results are presented in Section 2.
2. the maximal total population size: how should we design the resources distribution in order to maximize the total population size? We present the main results related to this query in Section 3.

It should be noted that we will thus be investigating **optimization problems**; since, for such problems, obtaining a complete description of the solutions is in general impossible, we focus here on **qualitative properties** of these solutions (e.g. symmetry properties or bang-bang property-see Section 1.1.3), and the related numerical issues.

Structure of the chapter. In Section 1.1.1, we present the biological model that will be the central focus of this Chapter; the two optimization problems informally stated above are formalized in Section 1.1.2, and the properties under investigation are described in Section 1.1.3. Results on the optimal survival ability are gathered in Section 2, while the ones devoted to the population size are presented in Section 3. We conclude this chapter by discussing, in Section 4, possible generalizations of the works presented here and several open problems, both theoretical and numerical, that remain to be investigated.

1.1.1 The main biological model

We use the classical Fisher-KPP model (Fisher, 1937; Kolmogorov et al., 1937) which, as has been acknowledged (DeAngelis et al., 2020; Skellam, 1951; Zhang et al., 2017), captures several of the essential features of population dynamics. In this model, the population is assumed to live an habitat Ω, which we assume satisfies the following regularity assumption:

$$\Omega \text{ is a bounded connected domain in } \mathbb{R}^d,\ d \geq 1 \text{ with a Lipschitz boundary.} \tag{$\mathcal{H}_{\text{reg}}$}$$

Additional smoothness assumptions on Ω will be specified for each result.

At the modeling level, we make the following assumptions:

1. The number of individuals is large enough that the population can be modeled as a density $u : \mathbb{R}_+ \times \Omega \to \mathbb{R}_+$.
2. There is a non-linear crowding effect $-u^2$ which amounts for intra-specific competition.
3. The population can access resources. These resources are modeled through a function $m \in L^\infty(\Omega)$, taken into account *via* a reaction term mu. The subset $\{m > 0\}$ corresponds to a favorable zone, while the zone $\{m \leq 0\}$ corresponds to a lethal region.
4. The population disperses at random in the domain with a characteristic dispersal rate $\sqrt{\mu}$. This leads to the diffusion term $\mu\Delta u$. We mention results related to other types of dispersal (i.e. that take into account the spatial heterogeneity) in Section 4.
5. Boundary conditions are either of Dirichlet or Neumann type, i.e.

$$u = 0 \text{ on } \partial\Omega \qquad \text{or} \qquad \partial_n u = 0 \text{ on } \partial\Omega.$$

We discuss some results available for Robin boundary conditions in Section 4.

6. We consider a non-trivial initial condition, i.e. a nonnegative and bounded function u_0 in Ω, $u_0 \neq 0$.

Finally, let us introduce the boundary conditions operator B defined by

$$Bu = u \text{ in the Dirichlet case}, \qquad Bu = \partial_n u \text{ in the Neumann case}.$$

These considerations lead to the *logistic diffusive equation*, two versions of which are stated here: the parabolic and the stationary one.

We first state its evolutionary (parabolic) version, reading

$$\begin{cases} u_t(t,x) = \mu\Delta u(t,x) + u(t,x)[m(x) - u(t,x)] & (t,x) \in \mathbb{R}_+ \times \Omega, \\ Bu(t,x) = 0 & (t,x) \in \mathbb{R}_+ \times \partial\Omega, \\ u(0,x) = u_0(x) & x \in \overline{\Omega}. \end{cases} \tag{1}$$

Of particular importance to understand the asymptotic behavior of the solution u of (1), as $t \to +\infty$, is the existence of **non-trivial steady states**, that is, of solutions $\theta_{m,\mu}$ of the stationary logistic diffusive equation

$$\begin{cases} \mu\Delta\theta_{m,\mu}(x) + \theta_{m,\mu}(x)(m(x) - \theta_{m,\mu}(x)) = 0 & x \in \Omega, \\ B\theta_{m,\mu}(x) = 0 & x \in \partial\Omega, \\ \theta_{m,\mu}(\cdot) \geq 0, \theta_{m,\mu}(\cdot) \not\equiv 0. \end{cases} \tag{2}$$

Indeed, we expect that, as $t \to \infty$, the solution $u(t,\cdot)$ of (1) will converge to such steady-states. As it turns out, both the questions of existence of solutions $\theta_{m,\mu}$ of (2) and of convergence of $u(t,\cdot)$ to such steady states can be solved by the investigation of a simple **spectral criterion**. Considering the linearization of (2) around $z \equiv 0$ leads to considering the linear differential operator

$$\mathcal{L}^B_{m,\mu} : u \in \mathcal{D}(\mathcal{L}^B_{m,\mu}) \mapsto -\mu\Delta u - mu, \qquad B \in \{D, N\},$$

where the superscript B stands for the boundary conditions: we will use $B = N$ for Neumann boundary conditions, and $B = D$ for Dirichlet boundary conditions, so that

$$\mathcal{D}(\mathcal{L}^D_{m,\mu}) = W^{2,2}(\Omega) \cap W^{1,2}_0(\Omega)$$
$$\mathcal{D}(\mathcal{L}^N_{m,\mu}) = \{u \in W^{2,2}(\Omega) \mid \partial_n u = 0 \text{ on } \partial\Omega\} \text{ (if } m \text{ does not vanish identically).}$$

Consider the first eigenvalue $\lambda^B(m,\mu)$ of $\mathcal{L}^B_{m,\mu}$. We recall that it can be defined through the minimization of a Rayleigh quotient:

$$\lambda^D(m,\mu) := \inf_{u \in W^{1,2}_0(\Omega)\setminus\{0\}} \mathfrak{R}_{(m,\mu)}[u] \quad \text{and}$$
$$\lambda^N(m,\mu) := \inf_{u \in W^{1,2}(\Omega)\setminus\{0\}} \mathfrak{R}_{(m,\mu)}[u], \tag{3}$$

where

$$\mathfrak{R}_{(m,\mu)}[u] = \frac{\mu \int_\Omega |\nabla u|^2 - \int_\Omega m u^2}{\int_\Omega u^2}.$$

This eigenvalue is well defined whenever $m \in L^\infty(\Omega)$. It is often called *principal eigenvalue*, meaning that its associated eigenfunction, often dubbed *principal eigenfunction*, does not change sign. The link with the existence of solutions of (2) and the convergence of solutions (1) was first observed in Shigesada and Kawasaki (1997) and formalized in the following form in Berestycki et al. (2005): let $B \in \{D, N\}$.

1. First case: $\lambda^B(m, \mu) < 0$. There exists a unique solution $\theta_{m,\mu}$ of (2), called *persistence steady-state*. Moreover, any solution $u = u(t, x)$ of (1) associated with a non-zero initial condition u_0 converges in $C^1(\overline{\Omega})$ to $\theta_{m,\mu}$ In other words, **the population survives**. Moreover, if $B = N$ (Neumann case), by taking $u(\cdot) = 1$ in the Rayleigh quotient formulation, we note that this condition is fulfilled provided $\int_\Omega m > 0$.
2. Second case: $\lambda^B(m, \mu) \geq 0$. In that case, no solution of (2) exists. Any solution $u = u(t, x)$ of (1) associated with a non-zero initial condition u_0 converges in $C^1(\overline{\Omega})$ to 0, called *extinction steady-state*. In other words, **the population goes extinct**.

A possible interpretation of these results is that the eigenvalue $\lambda^B(m, \mu)$ with $B \in \{D, N\}$ quantifies the **survival ability** of the population.

Regarding the total population size, we note that if $\lambda^N(m, \mu) < 0$, the uniqueness of the solution of (2) allows us to define the **total population size** associated with a resources distribution

$$F(m, \mu) = \int_\Omega \theta_{m,\mu}. \tag{4}$$

In what follows, we will not consider the Dirichlet case when dealing with the total population size (4). Indeed, the proofs of the results presented here rely on an asymptotic analysis of the solutions as $\mu \to +\infty$. In the case of Dirichlet boundary conditions, if $m(\cdot) \in L^\infty(\Omega)$ is given, then, for large values of μ (namely $\mu \geq \frac{\|m\|_{L^\infty}}{\lambda_1^D(\Omega)}$ where $\lambda_1^D(\Omega)$ denotes the first Dirichlet eigenvalue on Ω), one has $\lambda^D(m, \mu) \geq 0$, meaning that the stationary steady-state $\theta_{m,\mu}$ is trivial. Hence, for Dirichlet boundary conditions, such asymptotic analysis can no longer hold.

1.1.2 The optimization problems

The previous considerations led several authors (Berestycki et al., 2005; Kao et al., 2008; Lamboley et al., 2016; Lou, 2008b; Mazari et al., 2020a; Nagahara and Yanagida, 2018; Mazari and Ruiz-Balet, 2021) to consider optimization problems related to the survival ability and the total population size. Such problems are set under natural constraints, the first of which deals with the total amount of available resources: we introduce a fixed parameter $m_0 \in (0, \kappa)$ and require that the resources distributions m considered satisfy

$$\int_\Omega m = m_0|\Omega|.$$

This accounts for the fact that, in a given domain, only a limited amount of resources is available. The second constraint is a pointwise one, and accounts for natural limitations of the environment, i.e. the fact that, in a single spot, only a maximum amount of resources may be available. This is modeled by introducing a real parameter $\kappa > 0$ and requiring that admissible resources distributions m satisfy

$$\|m\|_{L^\infty(\Omega)} \leq \kappa.$$

In this chapter, we will further simplify this constraint and require the following stronger assumption:

$$0 \leq m \leq \kappa \quad \text{a.e. in } \Omega.$$

For optimization problems without such a constraint, see for instance Bai et al. (2016). In other words, the admissible class is

$$\mathcal{M}(\Omega) := \left\{ m \in L^\infty(\Omega), \int_\Omega m = m_0|\Omega|, \ 0 \leq m \leq \kappa \right\},$$

and we assume $\kappa > m_0$. As was noted, the condition $m \in \mathcal{M}(\Omega)$ ensures that $\lambda(m, \mu) < 0$, and so $\theta_{m,\mu}$ is indeed well-defined.

The first optimization problem deals with the optimal survival ability. It writes

$$\boxed{\inf_{m \in \mathcal{M}(\Omega)} \lambda^B(m, \mu).} \qquad (B \in \{D, N\}) \tag{I}$$

The relevant results for this problem are gathered in Section 2, but let us underline that this optimization problem ultimately deals with an evolution problem, as it governs the long-time behavior of the evolution equation (1).

Note that most of the results available in the Neumann case extend to periodic boundary conditions, using the arguments developed in Roques and Hamel (2007).

The second optimization problem deals with the total population size, and reads

$$\boxed{\sup_{m \in \mathcal{M}(\Omega)} \int_\Omega \theta_{m,\mu},} \tag{II}$$

where $\theta_{m,\mu}$ solves (2) with Neumann boundary conditions.

We present the available results in Section 3. These optimization problems involve the unique solution of a stationary elliptic equation.

Remark 1. It is also relevant from the biological point of view to consider sign-changing resources distribution, which amounts to replacing $\mathcal{M}(\Omega)$ with

$$\widehat{\mathcal{M}}(\Omega) := \left\{ m \in L^\infty(\Omega)\,,\, \int_\Omega m = -m_0|\Omega|,\ -m_- \leq m \leq m_+ \right\},$$

where (m_0, m_-, m_+) denote a triple of positive numbers. Indeed, since the weight m represents the resources distribution, its positive part $\{m > 0\}$ represents the favorable part of habitat for the species whereas its negative one $\{m < 0\}$ stands for the unfavorable one.

This requires a little more technicality to apply the usual reasoning, but at the end, we obtain similar results. Regarding Problem (**II**), we can also extend with a little effort all the results presented in this chapter. We refer to Lamboley et al. (2016) for a discussion of this issue for Problem (**I**) and to Mazari et al. (2020a, Section 3.2) for a discussion of this issue for Problem (**II**).

1.1.3 Type of properties investigated

Let us now state which properties will be investigated. Whether it be spectral optimization or optimal control problems for non-linear equations, obtaining a complete description of maximizers is in general impossible. In this Chapter, we focus on *qualitative properties* of maximizers: following the seminal works (Berestycki et al., 2005; Shigesada and Kawasaki, 1997), we are looking for general paradigms of the form: *maximizers should satisfy the bang-bang property* or *Concentrating resources is favorable for survival of species* and, in general, for (non)-symmetry of optimizers.

Regarding the survival ability of species, we mainly follow the lines of Lamboley et al. (2016) in order to address such intricate questions. Regarding the total population size, we will mostly follow, for our presentation, the articles (Mazari et al., 2020a; Mazari and Ruiz-Balet, 2021; Nagahara and Yanagida, 2018).

Let us now present these two properties.

The bang-bang property: the first relevant property is the so-called **bang-bang property**. Namely, since we are working with L^∞ constraints, are the solutions of (**I**) and (**II**) of the form $m^* = \kappa \mathbb{1}_E$ for some subset $E \subset \Omega$? When such a property is satisfied, which is always the case for (**I**), it proves to be very convenient for numerical applications. We note that, for the spectral optimization problem (**I**), this property is an easy consequence of the concavity of the functional under consideration and of the convexity of the set of admissible resources distributions, see Theorem 1.

Fragmentation and concentration of resources: other particularly relevant features of (**I**) and (**II**) are *concentration* and, conversely, *fragmentation* of resources. To explain these features, let us assume that some maximizers m_{I} and

m_{II} of, respectively, (**I**) and (**II**) satisfy the bang-bang properties. In other words, there exist two subsets E_{I} and E_{II} such that

$$m_{\text{I}} = \kappa \mathbb{1}_{E_{\text{I}}}, \quad m_{\text{II}} = \kappa \mathbb{1}_{E_{\text{II}}}.$$

What do E_{I} and E_{II} look like? How many connected components do they have? If the set Ω enjoys some symmetry properties, do E_{I} and E_{II} enjoy the same symmetries?

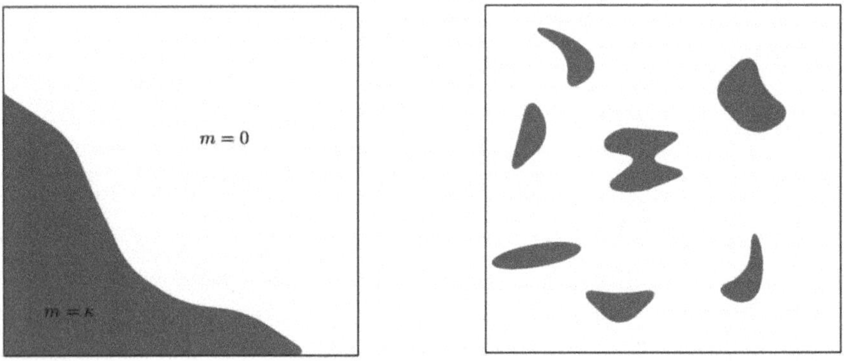

FIGURE 1 $\Omega = (0, 1)^2$. The resources distribution on the left is "more concentrated" than the one on the right.

In the case of spectral optimization (**I**) this property is known to hold in certain sets, for instance in orthotopes $\Omega = (0, 1)^n$, and corresponds to the paradigm that *concentrating resources favors survival* (Berestycki et al., 2005; Shigesada and Kawasaki, 1997). This is essentially due to the *energetic nature of spectral functionals*, which allows for the use of rearrangement inequalities. For the total population size however, no such results were available until (Mazari et al., 2020a; Mazari and Ruiz-Balet, 2021), and we will show that the answer to this question is, for this problem, highly dependent on the rate of the characteristic dispersal rate: concentration does hold for large diffusivities μ, while fragmentation does for small diffusivities.

Let us mention that effects of movement and spatial heterogeneity on population dynamics via reaction-diffusion-advection models, focusing on the persistence, competition, and evolution of organisms in spatially heterogeneous environments are much discussed in the recent survey article (Lam et al., 2020).

2 Optimal eigenvalue problem

This section is devoted to the results available for Problem (**I**).

2.1 Qualitative analysis

The first natural issue, when dealing with Problem (**I**) is related to the existence and *bang-bang* property of minimizers. It has been investigated in particular in Hintermüller et al. (2012); Jha and Porru (2011); Lou and Yanagida (2006); Roques and Hamel (2007) for several boundary conditions and in several dimensions. More recently, an elegant argument has been proposed in Derlet et al. (2010), yielding at the same time the existence of a *bang-bang minimizer* m^* and that the minimizing set $\{m^* = \kappa\}$ is an upper-level set of the principal eigenfunction. We reproduce this argument hereafter in the case of Neumann boundary conditions.

Theorem 1 (Derlet et al., 2010; Hintermüller et al., 2012; Jha and Porru, 2011; Lou and Yanagida, 2006; Roques and Hamel, 2007). *Let $B \in \{D, N\}$. Problem* (**I**) *has a solution m^*, and moreover there exists a measurable subset $E^* \subset \Omega$ such that, up to a set of zero Lebesgue measure, there holds*

$$m^* = m_{E^*} \quad \textit{where} \quad m_{E^*} = \kappa \mathbb{1}_{E^*} \textit{ a.e. in } \Omega.$$

Furthermore, let u denote an associated nonnegative eigenfunction (that is, associated to the eigenvalue problem (3) *with $m = m^*$), which is unique up to a multiplicative constant. There exists $\eta > 0$ such that $E^* = \{u \geq \eta\}$ up to a negligible set.*

From Derlet et al., 2010. Let us assume without loss of generality that $B = N$, the proof in the Dirichlet case being an immediate adaptation of what follows. Observe first that the existence is standard, see for instance Henrot (2006, Thm 8.1.2) where it is shown that $m \mapsto \lambda^N(m)$ is continuous for the L^∞ weak-$\star$ topology and the set of admissible weights $\mathcal{M}(\Omega)$ is compact for this topology. Notice furthermore that, as the infimum of linear functionals, $m \mapsto \lambda^N(m)$ is concave.

Let m^* be a minimizer for Problem (**I**) and denote by u its associated eigenfunction. Direct computations show that

$$\lambda^N(m^*) = \frac{\mu \int_\Omega |\nabla u|^2 - \int_\Omega m^* u^2}{\int_\Omega u^2} \geq \frac{\mu \int_\Omega |\nabla u|^2 - \sup_{\tilde{m} \in \mathcal{M}(\Omega)} \int_\Omega \tilde{m} u^2}{\int_\Omega u^2}.$$

According to the so-called *bathtub principle* (see for instance Lieb and Loss, 2001, Theorem 1.14), there exists a measurable subset $E^* \subset \Omega$ such that

$$\sup_{\tilde{m} \in \mathcal{M}(\Omega)} \int_\Omega \tilde{m} u^2 = \kappa \int_\Omega \mathbb{1}_{E^*} u^2$$

and

$$\{u > t\} \subset E^* \subset \{u \geq t\} \quad \text{and} \quad |E^*| = \frac{m_0 |\Omega|}{\kappa},$$

for a given $t > 0$. Since, from Lou (2008a, Corollary 1.1), the level sets of u have zero measure, E^* is uniquely defined up to a zero-measure set. Thus,

$$\lambda^N(m^*) \geq \frac{\mu \int_\Omega |\nabla u|^2 - \kappa \int_\Omega \mathbb{1}_{E^*} u^2}{\int_\Omega u^2} \geq \lambda^N(\kappa \mathbb{1}_{E^*}).$$

By minimality of $\lambda^N(m^*)$, it follows that all inequalities above are in fact equalities. Furthermore, since all the level sets of the eigenfunction u have zero Lebesgue measure, one easily shows that if m^* is not *bang-bang*, the first inequality above is strict, leading to a contradiction. The expected conclusion follows. □

Let us now comment on the regularity of the optimal set E^* introduced in the statement of Theorem 1.

Remark 2 (On the regularity of the free boundary). Proving the regularity of the free boundary $\Gamma := \partial E^* \backslash \partial\Omega$ is a very difficult question in general. It follows from classical elliptic regularity that the principal eigenfunction u is $\mathcal{C}^{1,a}(\Omega)$ for every $a \in [0, 1)$. Hence, as $E^* = \{u > \alpha\}$ up to a set of Lebesgue measure zero, the boundary Γ is $C^{1,a}$-smooth at any point where $\nabla u \neq 0$ and therefore, using a bootstrap argument, one infers the local analytic regularity of Γ in this case, see Chanillo et al. (2000b). The regularity problem is thus reduced to the one of the degeneracy of the eigenfunction u on its level line Γ.

When Dirichlet conditions are imposed on the boundary $\partial\Omega$, then it has been proved in Chanillo et al. (2008), when $d = 2$, that $u \in \mathcal{C}^{1,1}(\Omega)$, that ∂E does not hit the boundary and consists of finitely many disjoint, simple and closed real-analytic curves. In higher dimensions, it is only known that Γ is smooth up to a closed set of Hausdorff dimension $d - 1$ (Chanillo and Kenig, 2008). However, the situation is much more complicated since one could expect, as for some other free boundary problems, the emergence of stable singularities.

Before focusing on symmetry properties of optimizers, let us highlight that all solutions are known in the simple one-dimensional case, namely $d = 1$.

Proposition 1 (Chanillo et al., 2000a; Lou and Yanagida, 2006). *Let us assume that* $\Omega = (0, 1)$, $\kappa > 0$ *and* $m_0 \in (0, \kappa)$.

- *Dirichlet case (Chanillo et al., 2000a). Problem* (**I**) *has a unique solution* m^* *corresponding to a concentrated and centered resources distributions, namely:*

$$m^* = \kappa \mathbb{1}_{((1-m_0)/2, (1+m_0)/2)}.$$

- *Neumann case. (Lou and Yanagida, 2006) Problem* (**I**) *has exactly two solutions* m_1^* *and* m_2^* *corresponding to concentrated resources distributions meeting the boundary of* Ω*, namely*

$$m_1^* = \kappa \mathbb{1}_{(0, m_0)} \quad \text{and} \quad m_2^* = m_1^*(1 - \cdot).$$

These results, the proofs of which rely on symmetrization arguments, are illustrated on Fig. 2.

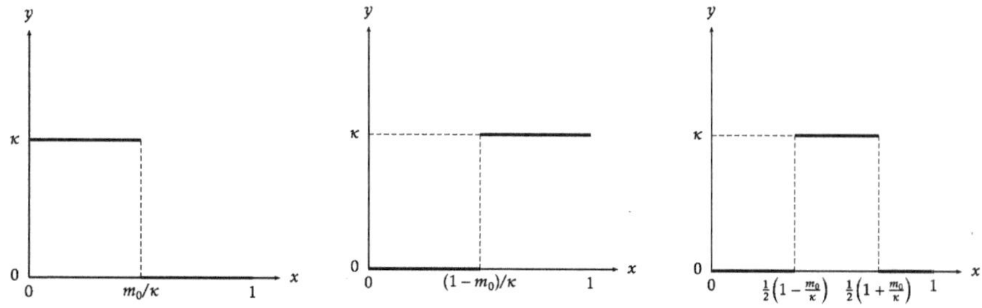

FIGURE 2 $\Omega = (0, 1)$. Left and middle: optimal resources distribution in the Neumann case. Right: optimal resource distribution in the Dirichlet case.

The following theorem is dedicated to the analysis of optimal domains for Problem (**I**), when dealing with Dirichlet boundary conditions. It follows from an easy adaptation of the results from Chanillo et al. (2000a) and Lamboley et al. (2016).

Theorem 2 (Dirichlet case, Chanillo et al., 2000a; Lamboley et al., 2016). *Let us assume that $B = D$, $d \geq 2$ and $\kappa > 0$. Let E^* be a subset of Ω such that $m^* = \kappa \mathbb{1}_{E^*}$ solves Problem* (**I**). *Then,*

- *The complement set of E^* in Ω contains a tubular neighborhood of the boundary $\partial\Omega$.*
- *There exists $\mu_0(\Omega, m_0, \kappa) > 0$ such that if $\mu \geq \mu_0(\Omega, m_0, \kappa)$, then every connected component of the complement set of E^* hits the boundary $\partial\Omega$.*
- *Assume that the domain Ω is symmetric and convex with respect to the hyperplane $\mathscr{H} := \{x_1 = 0\}$. Then, E^* is symmetric and convex with respect to $\mathscr{H}$. Furthermore, the associated nonnegative eigenfunction u is decreasing in x_1 for $x_1 \geq 0$.*
- *If one assumes that Ω is convex and has a $\mathscr{C}^2$ boundary, there exists $\mu_0(\Omega, m_0, \kappa) > 0$ such that if $\mu \geq \mu_0(\Omega, m_0, \kappa)$, then E^* is convex and $\partial E^* \cap \Omega$ is real analytic.*
- *If E^* or $\Omega \backslash E^*$ is rotationally symmetric (i.e. a union of concentric rings, whose center is denoted O) and has a finite number of connected components, then E^* and Ω are concentric balls.*
- *If Ω is a ball centered at O, then so is E^*.*

The next result is dedicated to the analysis of optimal domains for Problem (**I**), when dealing with Neumann boundary conditions. A common conjecture in dimension 2 in this framework is that the minimizing set has constant curvature, that is, it would be a quarter of ball, a stripe, or the complementary of a quarter of ball depending on the parameters (see e.g. Kao et al., 2008; Roques and Hamel, 2007 and Fig. 4 hereafter). The results below, taken from Lamboley et al. (2016, Section 5), disprove this conjecture.

Theorem 3 (Neumann case, Lamboley et al., 2016). *Let us assume that $B = N$, $d \geq 2$ and $\kappa > 0$. Let E^* be a subset of Ω such that $m^* = \kappa \mathbb{1}_{E^*}$ solves Problem* (**I**).

- *If $\Omega = \Pi_{k=1}^{N}(0, L_k)$, then*
  1. *(Steiner symmetry) m_{E^*} is monotonic with respect to $x_k \in (0, L_k)$ for all k.*
  2. *If $\partial E^* \cap \Omega$ is analytic, then $\partial E^* \cap \Omega$ does not contain any piece of sphere.*
- *If $\Omega = B(0, 1)$, then*
  1. *(Circular Symmetry) there exists $\theta_0 \in [0, 2\pi)$ such that E^* is symmetric with respect to the half straight line $\{\theta = \theta_0\}$ in the radial coordinates (r, θ). Moreover, for all $r \in (0, 1)$, $\{\theta \in [0, 2\pi), (r, \theta) \in E^*\}$ is an interval.*
  2. *E^* is not a ball.*

2.2 Numerical investigations

Let us first introduce the numerical algorithm we used for computing optimal sets. This approach has been first introduced in Chanillo et al. (2000a) and used in Hintermüller et al. (2012); Lamboley et al. (2016). It strongly rests upon the characterization of the optimal set E^* provided in Theorem 1, as a level set of the principal eigenfunction u. Let us explain the method's principle before summarizing the complete algorithm. The aim is to implement a kind of fixed point procedure resting upon the formula

$$E^* = \{u_{E^*} \geq \eta_{E^*}\}$$

where E^* denotes an optimizer for Problem (**I**), u_{E^*} its associated eigenfunction (uniquely chosen to be nonnegative and normalized in $L^2(\Omega)$) and $\eta_{E^*} > 0$ the corresponding Lagrange multiplier associated to the volume constraint of E^*, see Theorem 1.

Consider an arbitrary set E_k with measure $m_0|\Omega|/\kappa$, and its associated eigenfunction u_{E_k} (uniquely defined as above). Then, since the Lagrange multiplier can be seen as a parameter adjusted to preserve the volume constraint, we update E_k by setting

$$E_{k+1} = \{u_k \geq \eta_k\},$$

where $\eta_k > 0$ denotes the unique number such that $|\{u_k \geq \eta_k\}| = m_0|\Omega|/\kappa$.

It is notable that this leads to a descent algorithm, as observed in Hintermüller et al. (2012), although, to the best of our knowledge, there is no available result regarding the complete convergence analysis.

This method has been used to compute optimal sets for Dirichlet boundary conditions (Chanillo et al., 2000a), Neumann boundary conditions in squares and ellipses (Kao et al., 2008), Robin boundary conditions in squares (Hintermüller et al., 2012). In general these solutions look like strips, balls, or complements of balls, depending on the parameters. For example, the optimal sets in

Algorithm 1: Fixed point procedure to compute local optimizer E^*_{num}

Initialization:

Let E_0 be an arbitrary subset of Ω, maxiter $\in \mathbb{N}^*$ and $\varepsilon_{\text{tol}} > 0$ be given;
Set $k = 0$ and $\delta = \varepsilon_{\text{tol}} + 1$;
Compute (λ_0^B, u_0), the principal eigenpair associated to E_0, with u_0 chosen to be nonnegative and $L^2(\Omega)$ normalized;

While $k \leq$ *maxiter and* $\delta > \varepsilon_{\text{tol}}$ **do**

1: compute $\eta_k > 0$ so that $|\{u_k \geq \eta_k\}| = m_0|\Omega|/\kappa$ by implementing a standard bisection method;
2: compute the level set $E_{k+1} = \{u_k \geq \eta_k\}$;
3: compute $(\lambda_{k+1}^B, u_{k+1})$, the principal eigenpair associated to E_{k+1}, with u_{k+1} chosen to be nonnegative and $L^2(\Omega)$ normalized;
4: update $\delta \leftarrow |\lambda_{k+1} - \lambda_k|$;
5: $k \leftarrow k + 1$;

End:

Set $E^*_{\text{num}} = E_k$.

the Dirichlet case for different values of the area constraint on E in a square are shown in Fig. 3.

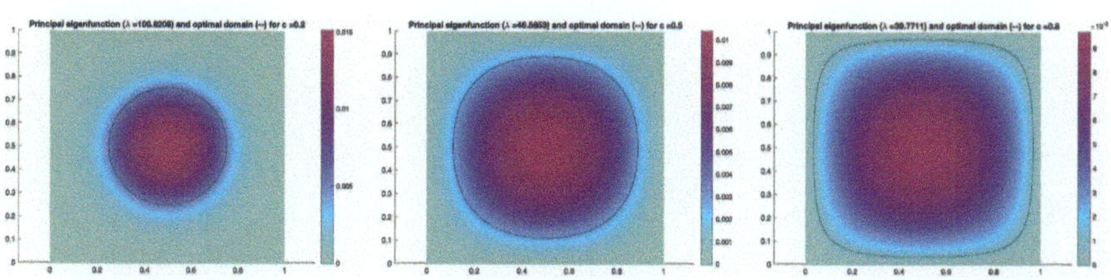

(a) $|E| = 0.2$ (b) $|E| = 0.5$ (c) $|E| = 0.8$

FIGURE 3 $\Omega = (0, 1)^2$. Optimal domains in the Dirichlet case with $\kappa = 0.5$ and several volume constraints on E. Colors correspond to the intensity levels of the eigenfunction and the domain E encloses the zone corresponding to the warmest colors. Note that we know by Theorem 2 that the boundaries of the patterns observed in (a), (b), and (c) do not contain any part with constant curvature.

Numerical results for the square and the ball in the Neumann case are gathered in Figs. 4 and 6. Convergence curves illustrating the efficiency of the method are drawn on Figs. 5 and 7. On Fig. 8, we provide additional examples in the case of an ellipse with semi-axis 1 and $1/\sqrt{\pi}$. In particular, as an illustration, we provide examples of optimal domains in a case that has not been studied theoretically: that of mixed Neumann and Dirichlet conditions.

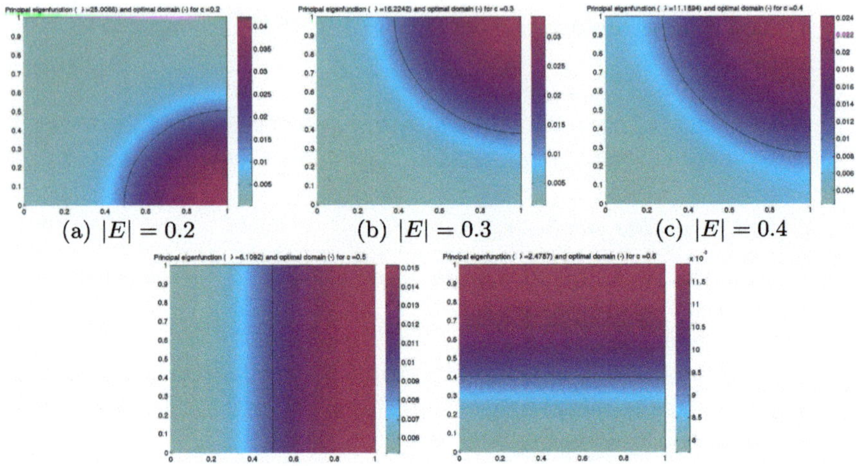

(a) $|E| = 0.2$ (b) $|E| = 0.3$ (c) $|E| = 0.4$

(d) $|E| = 0.5$ (e) $|E| = 0.6$

FIGURE 4 $\Omega = (0, 1)^2$. Optimal domains in the Neumann case with $\kappa = 0.5$ and several volume constraints on E. Colors correspond to the intensity levels of the eigenfunction and the domain E encloses the zone corresponding to the warmest colors. Note that we know by Theorem 3 that the boundaries of the patterns observed in (a), (b), and (c) do not contain any part with constant curvature.

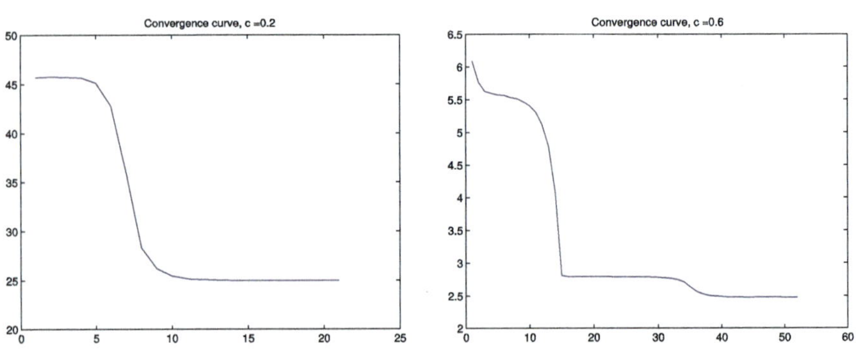

(a) $|E| = 0.2$ - convergence curve (b) $|E| = 0.6$ - convergence curve

FIGURE 5 $\Omega = (0, 1)^2$. Two examples of convergence curves in the Neumann case ($\beta = 0$) with $\kappa = 0.5$, $c = 0.2$ (left) and $c = 0.6$ (right).

3 Maximizing the total population size

This section is devoted to the analysis of Problem (**II**).

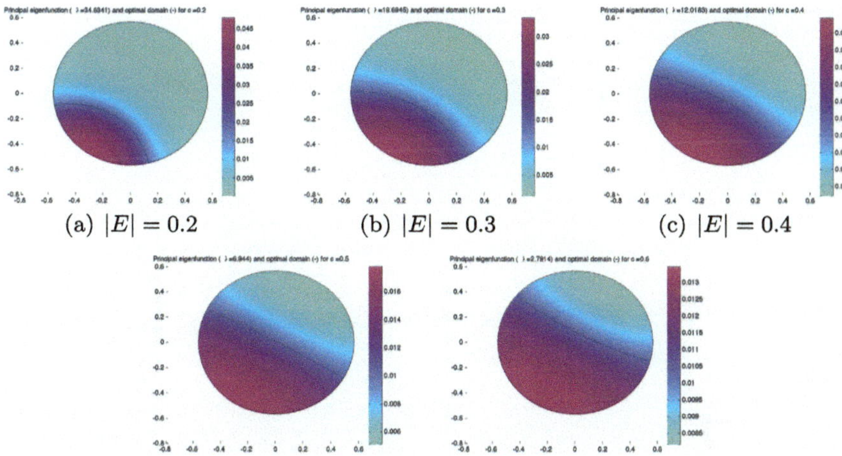

(a) $|E| = 0.2$ (b) $|E| = 0.3$ (c) $|E| = 0.4$

(d) $|E| = 0.5$ (e) $|E| = 0.6$

FIGURE 6 $\Omega = B(0, 1/\sqrt{\pi})$. Optimal domains in the Neumann case with $\kappa = 0.5$ and several volume constraints on E. Colors correspond to the intensity levels of the eigenfunction and the domain E encloses the zone corresponding to the warmest colors.

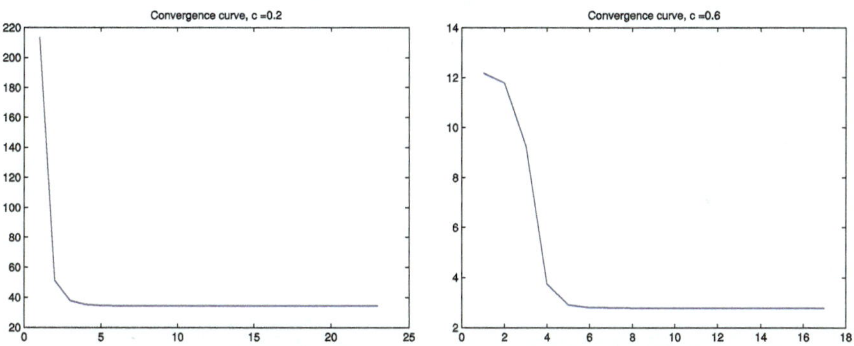

(a) $c = 0.2$ - convergence curve (b) $c = 0.6$ - convergence curve

FIGURE 7 $\Omega = B(0, 1/\sqrt{\pi})$. Two examples of convergence curves in the Neumann case with $\kappa = 0.5$ and $|E| = 0.2$ (left) or $|E| = 0.6$ (right).

3.1 Qualitative analysis

We state here two important properties of solutions of **(II)**, whose proof can be found in Mazari et al. (2020a); Mazari and Ruiz-Balet (2021).

In what follows, we assume that $\Omega \subset \mathbb{R}^d$, with $d \in \mathbb{N}^*$ is a connected open set with a $\mathscr{C}^2$ boundary, or that $\Omega = (0, 1)^d$.

Recall that our aim is to solve Problem **(II)**

$$\sup_{m \in \mathcal{M}(\Omega)} F_\mu(m)$$

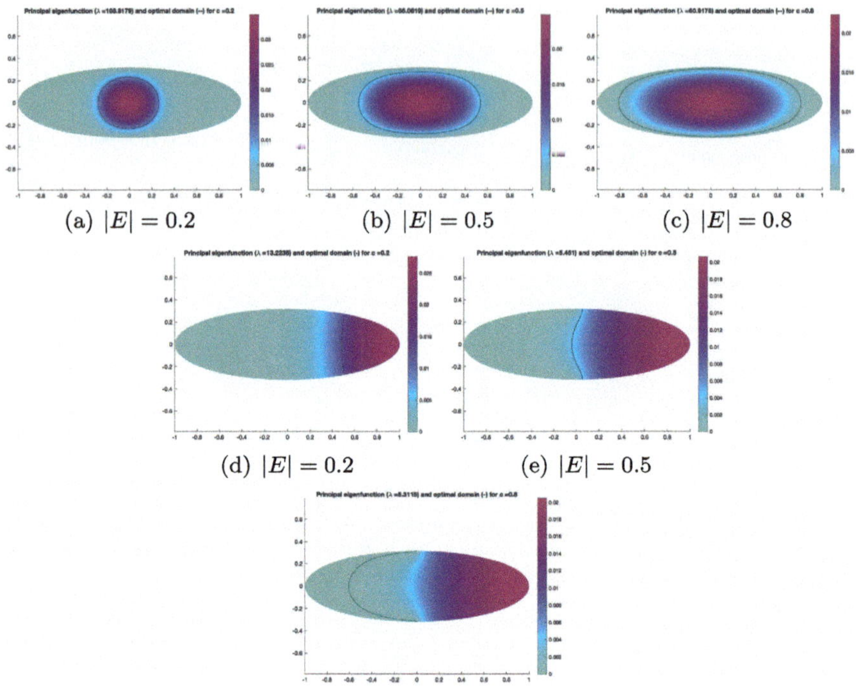

(a) $|E| = 0.2$ (b) $|E| = 0.5$ (c) $|E| = 0.8$

(d) $|E| = 0.2$ (e) $|E| = 0.5$

(f) $|E| = 0.8$

FIGURE 8 $\Omega = \{(x, y) \in \mathbb{R}^2 \mid x^2 + y^2/\pi = 1\}$. Top: optimal domains in the Dirichlet case with $\kappa = 0.5$ and several volume constraints on E. Bottom: optimal domains in the case of mixed Dirichlet-Neumann boundary conditions (Dirichlet on the left part of the ellipse and Neumann in the right one) with $\kappa = 0.5$ and several volume constraints on E. Colors correspond to the intensity levels of the eigenfunction and the domain E encloses the zone corresponding to the warmest colors curvature.

where the functional F_μ is defined by

$$F_\mu(m) = \int_\Omega \theta_{m,\mu},$$

and $\theta_{m,\mu}$ solves (2).

This problem has been stated by Lou in his survey article (Lou, 2008b). The main difficulty in studying this problem stems from the facts that $\theta_{m,\mu}$ solves a nonlinear PDE and that the criterion F_μ does not derive from an energy. An important property that needs to be tackled is that of the *bang-bang property* for optimal resources distributions. It has been partially addressed in Ding et al. (2010), in which the authors apply the so-called Pontryagin principle, show the Gâteaux-differentiability of the functional and carry out several numerical simulations backing up the conjecture that its maximizers are of bang-bang type, in other words equal to 0 or κ a.e. in Ω.

However, proving this bang-bang property is challenging. The analysis of optimality conditions appears rather intricate. Indeed, the sensitivity of the total population size functional with respect to the variations of $m(\cdot)$ is directly related to the solution of the adjoint state, defined as the solution of:

$$\begin{cases} \mu\Delta p_{m,\mu} + p_{m,\mu}(m - 2\theta_{m,\mu}) = 1 & \text{in } \Omega, \\ \frac{\partial p_{m,\mu}}{\partial \nu} = 0 & \text{on } \partial\Omega. \end{cases} \tag{5}$$

Note that $p_{m,\mu}$ belongs to $W^{1,2}(\Omega)$ and is unique, according to the Fredholm alternative. Then, one can prove that, for all $\mu > 0$, the application F_μ is Gâteaux-differentiable with respect to m in direction h and its Gâteaux derivative writes

$$dF_\mu(m)[h] = -\int_\Omega h\theta_{m,\mu}p_{m,\mu}.$$

Now consider a maximizer m^*. To derive optimality conditions, we introduce, for a given $m \in \mathcal{M}(\Omega)$, the cone $\mathcal{T}_{m,\mathcal{M}(\Omega)}$ of admissible perturbations at m, namely the set of functions $h \in L^\infty(\Omega)$ such that, for any sequence of positive real numbers ε_n decreasing to 0, there exists a sequence of functions $h_n \in L^\infty(\Omega)$ converging to h as $n \to +\infty$, and $m + \varepsilon_n h_n \in \mathcal{M}(\Omega)$ for every $n \in \mathbb{N}$.

If m^* is a maximizer, then for every perturbation h in the cone of admissible perturbations $\mathcal{T}_{m^*,\mathcal{M}(\Omega)}$, there holds $dF_\mu(m^*)[h] \geq 0$. The analysis of these optimality condition leads to the following result.

Proposition 2. *Let us define $\varphi_{m,\mu} = \theta_{m,\mu}p_{m,\mu}$, where $\theta_{m,\mu}$ and $p_{m,\mu}$ solve respectively Eqs.* (2) *and* (5). *There exists $c \in \mathbb{R}$ such that*

$$\{\varphi_{m,\mu} < c\} = \{m = \kappa\}, \quad \{\varphi_{m,\mu} = c\} = \{0 < m < \kappa\}, \quad \{\varphi_{m,\mu} > c\} = \{m = 0\}.$$

This property shows that exploiting properties of optimal configurations needs hence a deep understanding of the behavior of $\theta_{m,\mu}$ as well as that of the adjoint state. Using an analogous property, Nagahara and Yanagida (2018) proved that if the optimal resources distribution is Riemann Integrable then it is of bang-bang type. Their proof is valid for all $\mu > 0$, while the result we will present below in Theorem 4 is only valid for large μ's. But their regularity hypothesis might be restrictive. Indeed, we will display in Section 3.2 some numerical simulations showing that the maximizer might oscillate a lot when μ becomes small, thus leading to potential difficulties from the regularity point of view.

Property no. 1: pointwise constraints, bang-bang property. In order to overcome these difficulties, we introduced in Mazari et al. (2020a) a new method based on series expansions in powers of the diffusivity μ asymptotic in order to work out optimality conditions.

Theorem 4 (Mazari et al., 2020a). *Let $\mu > 0$, $\kappa > 0$, $m_0 \in (0, \kappa)$. There exists a positive number $\mu^* = \mu^*(\Omega, \kappa, m_0)$ such that, for every $\mu \geq \mu^*$, the functional F_μ is strictly convex. As a consequence, for $\mu \geq \mu^*$, every maximizer of Problem* (**II**) *is bang-bang, that is, $m^* = \kappa \mathbb{1}_E$, where $\mathbb{1}_E$ is the characteristic function of a (measurable) resources set E.*

The proof rests upon a tricky expansion of the solution $\theta_{m,\mu}$ of (2), as a series involving the solutions of a sequence of cascade systems.

Property no. 2: concentration-fragmentation of maximizers. It is well-known (see e.g. Berestycki et al., 2005; Lamboley et al., 2016) that concentrating resources, meaning that the resources distribution m is decreasing in each direction,[1] favors the survival of species. On the contrary, we will say that a resources set is fragmented whenever it is disconnected (see Fig. 1).

Theorem 5 (Mazari et al., 2020a). *Consider $\Omega = (0, 1)^d$. Any family of maximizers $\{m^*_\mu\}_{\mu>0}$ for Problem* (**II**) *converges as $\mu \to +\infty$ in $L^1(\Omega)$ to the characteristic function of a set E which is concentrated (meaning that its characteristic function $\mathbb{1}_E$ is monotone with respect to each space variable).*

In the one-dimensional case, we also prove that if the diffusivity is large enough, there are only two maximizers, that are simple crenels meeting either the left or the right boundary (see Fig. 9).

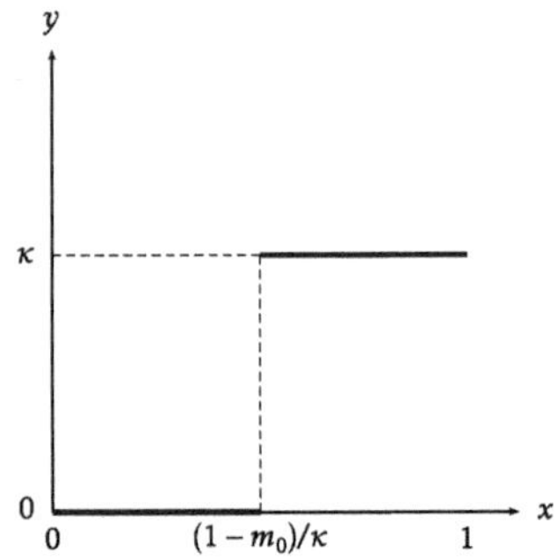

FIGURE 9 A solution of (**II**) in 1D, for large diffusivities μ.

In order to prove these results, we show that the maximizers converge, when μ tends to $+\infty$, to the maximizers of a limit problem. Indeed, let us introduce the function space

$$X := \left\{ u \in W^{1,2}(\Omega),\ \int_\Omega u = 0 \right\}$$

[1] A given function $m \in L^1(\Omega)$ is said *concentrated* in $\Omega = (0; 1)^d$ whenever the functions $x \mapsto m(x, x_2, \dots, x_d)$, $x \mapsto m(x_1, \dots, x_{d-1}, x)$, and $x \mapsto m(x_1, \dots, x_{k-1}, x, x_{k+1}, \dots, x_d)$ all index $k \in [\![1, d-1]\!]$ are non-increasing in $(0; 1)$.

and the energy functional

$$\mathcal{E}_m : X \ni u \mapsto \frac{1}{2}\int_\Omega |\nabla u|^2 - m_0 \int_\Omega mu.$$

Theorem 6. *(Mazari et al., 2020a, Γ-convergence property). For any $\mu > 0$, let m_μ^* be a solution of Problem* (**II**)*. Any L^1 closure point of $\{m_\mu^*\}_{\mu>0}$ as $\mu \to \infty$ is a solution of the optimal design problem*

$$\min_{m\in\mathcal{M}(\Omega)} \min_{u\in X} \mathcal{E}_m(u).$$

The above results are derived from an investigation of the minimization of $\mathcal{E}_m$.

We now present a more surprising property, first observed in Mazari et al. (2020a): **fragmentation may be better than concentration for small diffusivities.**

Theorem 7 (Mazari et al., 2020a). *Let $\Omega = (0,1)$. The function $\tilde{m} = \kappa\chi_{(1-\ell,1)}$ (and $\tilde{m}(1-\cdot) = \kappa\chi_{(0,\ell)}$) does not solve Problem* (**II**) *for small values of μ. More precisely, if we extend $\tilde{m}$ outside of $(0,1)$ by periodicity, there exists $\mu > 0$ such that*

$$\int_\Omega \theta_{\tilde{m},\mu} < \int_\Omega \theta_{\tilde{m}(2\,\cdot),\mu}.$$

This property has been investigated more precisely in Mazari and Ruiz-Balet (2021). Define for $M > 0$ the class

$$\mathcal{M}_M(\Omega) := \left\{m \in \mathcal{M}(\Omega),\, \|m\|_{BV(\Omega)} \le M\right\}.$$

Theorem 8 (Mazari and Ruiz-Balet, 2021). *Consider a family of maximizers $\{m_\mu^*\}_{\mu>0}$ of Problem* (**II**)*. There holds*

$$\left\|m_\mu^*\right\|_{BV(\Omega)} \xrightarrow[\mu\to 0^+]{} +\infty. \tag{6}$$

More precisely:

$$\forall M > 0, \quad \exists \mu_M > 0 \text{ s.t. } \forall 0 < \mu \le \mu_M \quad \sup_{m\in\mathcal{M}_M(\Omega)} \int_\Omega \theta_{m,\mu} < \sup_{m\in\mathcal{M}(\Omega)} \int_\Omega \theta_{m,\mu}. \tag{7}$$

Let us finally mention that a discretized version of this problem has been studied in Lou et al. (2020), and has led to a very precise identification of discrete minimizers, highlighting that the number of related components increases as μ goes to 0.

3.2 Numerical investigations

We provide hereafter several numerical simulations of optimizers of the total population size, solving Problem (**II**). These simulations are taken from Mazari and Ruiz-Balet (2021) and have been obtained in the following way:

- In the optimization procedure described below, one encodes the control $m(\cdot)$ through its Fourier coefficients. To avoid the emergence of local maximizers, one picks several random initial guesses (by randomizing the first Fourier coefficients and applying an affine transformation to them to guarantee that $m(\cdot)$ satisfies the constraints).
- One works with a uniform space discretization.
- For each initial guess of $m(\cdot)$, one computes the solution θ of the logistic equation (2), and the solution to the adjoint-state (5) by using a finite differences method.
- We then implement a constrained adapted gradient type descent method, where the perturbation h of the control m is obtained by minimizing $h \mapsto dF_\mu(m)[h]$ over the set of all admissible perturbations $h(\cdot)$, namely the set of all elements $h \in L^\infty(\Omega)$ such that $\int_\Omega h = 0$ and $m + h$ belongs almost everywhere to $[0, \kappa]$.
- This allows to select numerical optimizers and to perform a last gradient descent to ensure the robustness of the found solution.

We refer to Mazari and Ruiz-Balet (2021) for further explanations on the employed method. See Figs. 10 and 11.

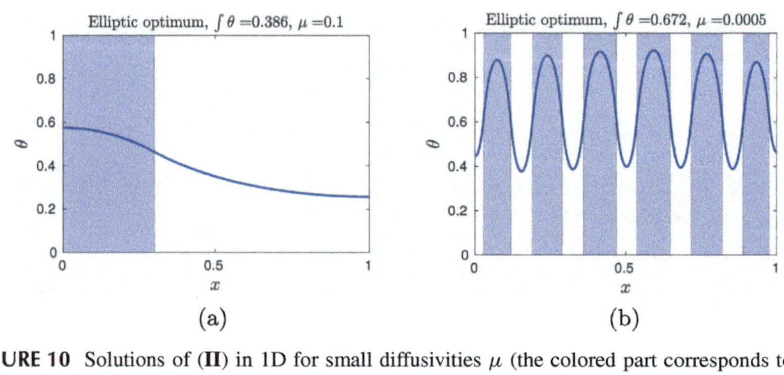

FIGURE 10 Solutions of (**II**) in 1D for small diffusivities μ (the colored part corresponds to the set $\{m = \kappa\}$ and the white ones to $\{m = 0\}$), and the associated eigenfunction in dark blue.

4 Generalization and perspectives

In this chapter, we have summarized the known results related to the problem of maximizing population survival when controlled by a resource term. The problems considered in this chapter can be extended/generalized in various directions in a natural way. We discuss in the next section the generalization of

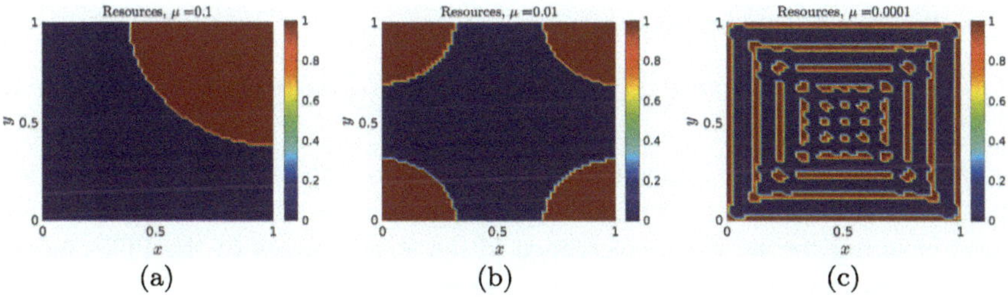

FIGURE 11 $\Omega = (0,1) \times (0,1)$. Optimal domains (in red) in the Neumann case with $\kappa = 1$, $m_0 = 0.3$ and various dispersal rates μ (according to Mazari and Ruiz-Balet, 2021).

the previous models to a drift operator and conclude this chapter by mentioning various problems that are still open or natural generalizations of the issues discussed here.

We mention nevertheless that we choose to not tackle the generalization to Problems (**I**) and (**II**) to other boundary conditions in the present article. Nevertheless, the analysis of such conditions when dealing with Problems (**I**) can be found in Lamboley et al. (2016).

4.1 Spectral optimization for the biased movement of species

For $\varepsilon \geq 0$, let us consider a population density with a resource term $m(\cdot)$ temporarily assumed to be differentiable and whose flux is $-\nabla u + \varepsilon u \nabla m$. The term $u\nabla m$ stands for a bias in the population movement, modeling a tendency of the population to disperse along the gradient of resources and hence move to favorable regions. The parameter ε quantifies the influence of the resources distribution on the movement of the species. The complete associated reaction diffusion equation, called "logistic diffusive equation", reads

$$\frac{\partial u}{\partial t} = \nabla \cdot \left(\nabla u - \varepsilon u \nabla m \right) + mu - u^2 \quad \text{in } \Omega,$$

completed with suitable boundary conditions. Plugging the change of variable $v = e^{-\varepsilon m} u$ in this equation leads to

$$\frac{\partial v}{\partial t} = \Delta v + \varepsilon \nabla m \cdot \nabla v + mv - e^{\varepsilon m} v^2 \quad \text{in } \Omega.$$

It is known (see e.g. Belgacem and Cosner, 1995; Belgacem, 1997; Murray, 1993) that the asymptotic behavior of this equation is driven by the principal eigenvalue of the operator $\mathcal{L} : \psi \mapsto -\Delta \psi - \varepsilon \nabla m \cdot \nabla \psi - m\psi$. At this point, we are deliberately imprecise about the definition of the operator $\mathcal{L}$. We should specify its domain of definition, but of course this depends very strongly on the boundary conditions, which we have not yet specified. The associated principal eigenfunction ψ satisfies in particular

$$-\nabla\cdot\left(e^{\varepsilon m}\nabla\psi\right)-me^{\varepsilon m}\psi=\lambda_\varepsilon(m)\psi e^{\varepsilon m}\quad\text{in }\Omega,$$

completed by adequate boundary conditions. Following the approach developed in Lamboley et al. (2016), optimal configurations of resources correspond to the ones ensuring the fastest convergence to the steady-states of the PDE above, which comes to minimizing $\lambda_\varepsilon(m)$ with respect to m.

As previously, we will distinguish between two standard boundary conditions on the eigenfunction ψ: Dirichlet and Neumann ones, for which the corresponding eigenvalues are respectively denoted with a D or N superscript. The resulting optimization problems read

$$\inf_{m\in\mathcal{M}(\Omega)}\lambda_\varepsilon^D(m)\quad\text{and}\quad\inf_{m\in\mathcal{M}(\Omega)}\lambda_\varepsilon^N(m)\tag{8}$$

where

$$\lambda_\varepsilon^D(m)=\inf_{u\in W_0^{1,2}(\Omega)}\frac{\frac{1}{2}\int_\Omega e^{\varepsilon m}|\nabla u|^2-\int_\Omega me^{\varepsilon m}u^2}{\int_\Omega e^{\varepsilon m}u^2}\quad\text{(Dirichlet case)}$$

$$\lambda_\varepsilon^N(m)=\inf_{u\in W^{1,2}(\Omega)}\frac{\frac{1}{2}\int_\Omega e^{\varepsilon m}|\nabla u|^2-\int_\Omega me^{\varepsilon m}u^2}{\int_\Omega e^{\varepsilon m}u^2}\quad\text{(Neumann case).}\tag{9}$$

It is notable that the results strongly differ depending on if one considers the one-dimensional case ($\Omega=(0,1)$) or the multi-dimensional one.

Theorem 9 (1D case, Caubet et al., 2017). *Let $\Omega=(0,1)$, $\kappa>0$, $m_0\in(0,\kappa)$, and $\varepsilon\geq 0$. The optimal design problem* (8) *has a solution m^*. Moreover, there exists $\varepsilon_0>0$ depending on the problem parameters such that for all $\varepsilon<\varepsilon_0$:*

- *in the Dirichlet case, there is a unique minimizer given by $m^*=\kappa\mathbb{1}_{((1-m_0)/2,(1+m_0)/2)}$;*
- *in the Neumann case, there are exactly two minimizers given by $m^*=\kappa\mathbb{1}_{(0,m_0)}$ and $m^*=\kappa\mathbb{1}_{(1-m_0,1)}$.*

In the 1D case, one recovers similar results to the ones obtained in Proposition 1 corresponding to the case $\varepsilon=0$, whereas we do not expect existence in the higher dimensional cases, except if Ω is a ball.

Theorem 10 (Multi-dimensional case, Dirichlet conditions, Mazari et al., 2020b). *Let Ω be a bounded connected subset of $\mathbb{R}^n$ with a connected Lipschitz boundary, let $\varepsilon>0$ and $n\geq 2$. Let us consider the Dirichlet case. If the optimization problem* (8) *has a solution m^*, then it necessarily writes $m^*=\kappa\mathbb{1}_{E^*}$, where E^* is a measurable subset of Ω. Moreover, if ∂E^* is a C^2 hypersurface and if Ω is connected, then Ω is a ball.*

Theorem 10 can be interpreted as follows: assuming that Ω is not a ball and that the population density moves along the gradient of the resources, it is not possible to lay the resources in an optimal way: because of the homogenization phenomenon, if we are given a distribution of resources through a density m, there will always exist another distribution decreasing the eigenvalue we aim at optimizing. In the 1D case, optimal configurations for more general boundary conditions of Robin type (including the case of homogeneous Dirichlet and Neumann boundary conditions) have been obtained, by using a new rearrangement technique. Finally, let us mention the related result (Hamel et al., 2011, Theorem 2.1), dealing with Faber-Krahn type inequalities for general elliptic operators involving a drift term.

4.2 Open problems

We conclude this section by formulating open problems, interesting from the theoretical or numerical point of view, that remain to be investigated or developed.

In what follows, let $\kappa > 0$, $m_0 \in (0, \kappa)$, and E^* be an optimizer for Problem (**I**).

Open problem 1. *In the Neumann case, investigate the validity of the property: "let Ω satisfy ($\mathcal{H}_{\text{reg}}$) (see Section 1.1.1), then one has $E^* \cap \partial\Omega \neq \emptyset$."*

This conjecture is supported by the analysis of the particular case where Ω is a square (Theorem 3) and numerical computations (see for instance Figs. 4, 6, and 8).

Open problem 2. *In the Neumann case, if Ω is a ball, can $\partial E^* \cap \Omega$ be a piece of sphere? Note that if Ω is a square in dimension 2, we already know that $\partial E^* \cap \Omega$ is not a piece of sphere according to Theorem 3, in spite of the numerical results on Fig. 4 highlighting that the curvature of several optimizers seems to be almost constant.*

Open problem 3. *Investigate convergence properties of Algorithm 1. In particular:*

- *Can one ensure that, starting from any initial configuration E_0 in Ω, Algorithm 1 converges to a local minimizer for Problem* (**I**)*?*
- *In that case, can the convergence speed be identified?*

The following open problems are related to Problem (**II**).

Open problem 4. *For general, possibly smooth, open connected domains Ω, obtain a sharp estimate of the nonnegative number μ^* introduced in Theorem 4.*

According to the main result of Nagahara and Yanagida (2018), we conjecture that maximizers are bang-bang functions for any $\mu > 0$, in other words that $\mu^* = 0$.

Open problem 5. *For general, possibly nonsmooth, and non-connected domains Ω, can the main results/approaches of the present article be generalized? In any case, can one numerically look for maximizers with the help of a dedicated shape optimization algorithm?*

Open problem 6. *Let $\Omega = (0, 1)$. Given that, for $\mu > 0$ small enough, the optimal configurations for $\lambda_1(\cdot, \mu)$ and F_μ do not coincide according to Proposition 1 and Theorem 7, it would be natural and biologically relevant to investigate the maximization of a convex combination of F_μ and $\lambda_1(\cdot, \mu)$ over $\mathcal{M}(\Omega)$.*

Open problem 7. *For general, possibly smooth, open connected domains Ω, investigate the asymptotic behavior of maximizers for Problem* (**II**) *as the parameter μ decreases to 0. Such an issue appears intricate since it requires a refine study of singular limits for the involved operators.*

In the spirit of the problems studied in this chapter, it would be very interesting to study the case of similar optimal control problems for (non-scalar) reaction-diffusion systems. Few results exist in this field. The study of such problems generally poses an obvious difficulty: the usual tools of the "principle of comparison" type generally fail. Moreover, it is not always easy to establish usable persistence criteria.

Let us state hereafter a generic kind of such problems.

Open problem 8. *Let Ω be a bounded connected open set. Formally, let $n_1(t)$ denote a density of wild individuals (typically predators whose evolution needs to be controlled) and $n_2(t)$ the density at time t of a controlled population that we introduced in the environment Ω where the first population lives. For $\mu_1 > 0$ and $\mu_2 > 0$, we consider as model of population density dynamics the competitive compartmental system*

$$\begin{cases} \dfrac{\partial n_1}{\partial t}(t,x) - \mu_1 \Delta n_1(t,x) = f_1(n_1(t,x), n_2(t,x)) & (t,x) \in \mathbb{R}_+ \times \Omega \\ \dfrac{\partial n_2}{\partial t}(t,x) - \mu_2 \Delta n_2(t,x) = f_2(n_1(t,x), n_2(t,x)) + u(t,x) & (t,x) \in \mathbb{R}_+ \times \Omega \\ n_1(0,\cdot) = n_1^0(\cdot), \quad n_2(0,\cdot) = n_2^0(\cdot), & \text{in } \Omega, \end{cases} \tag{10}$$

where u is a non-negative function standing for a control. Hence, a typical issue consists in acting on the second population to steer (n_1, n_2) as close as possible to a stable steady-state (n_1^*, n_2^*) of the system without control. If the horizon of time $T > 0$ is given, we are then led to minimize the distance of (n_1, n_2) to either (n_1^*, n_2^*) or its attraction basin, by considering admissible controls $u \in \mathcal{M}_0(\Omega)$. A standard important issue to derive optimal control strategies is to make the control structure precise. Typically, in what case can one ensure that optimal controls (whenever they exist) are bang-bang?

It is notable that such a problem is a simplified control model of population replacement strategies, where one aims at controlling a population of wild *Aedes* mosquitoes by means of *Wolbachia* infected ones. For this problem, the functions f_i, $i = 1, 2$, are defined by

$$f_1(n_1, n_2) = b_1 n_1 \left(1 - s_h \frac{n_2}{n_1 + n_2}\right)\left(1 - \frac{n_1 + n_2}{K}\right) - d_1 n_1, \tag{11}$$

$$f_2(n_1, n_2) = b_2 n_2 \left(1 - \frac{n_1 + n_2}{K}\right) - d_2 n_2, \tag{12}$$

where s_h, K, b_i, and d_i, $i = 1, 2$ denote positive constants. In Almeida et al. (2019), a similar problem, where the criterion is a least square functional and the steady-state corresponds to the extinction of the first population, has been investigated.

References

Almeida, L., Privat, Y., Strugarek, M., Vauchelet, N., 2019. Optimal releases for population replacement strategies: application to wolbachia. SIAM J. Math. Anal. 51 (4), 3170–3194.

Bai, X., He, X., Li, F., 2016. An optimization problem and its application in population dynamics. Proc. Am. Math. Soc. 144 (5), 2161–2170.

Belgacem, F., 1997. Elliptic Boundary Value Problems with Indefinite Weights, Variational Formulations of the Principal Eigenvalue, and Applications. Chapman & Hall/CRC Research Notes in Mathematics Series. Taylor & Francis.

Belgacem, F., Cosner, C., 1995. The effect of dispersal along environmental gradients on the dynamics of populations in heterogeneous environment. Can. Appl. Math. Q. 3 (379–397), 01.

Berestycki, H., Hamel, F., Roques, L., 2005. Analysis of the periodically fragmented environment model: I – species persistence. J. Math. Biol. 51 (1), 75–113.

Cantrell, R.S., Cosner, C., 1989. Diffusive logistic equations with indefinite weights: population models in disrupted environments. Proc. R. Soc. Edinb., Sect. A 112 (3–4), 293–318.

Cantrell, R.S., Cosner, C., 1991a. Diffusive logistic equations with indefinite weights: population models in disrupted environments. II. SIAM J. Math. Anal. 22 (4), 1043–1064.

Cantrell, R.S., Cosner, C., 1991b. The effects of spatial heterogeneity in population dynamics. J. Math. Biol. 29 (4), 315–338.

Cantrell, R.S., Cosner, C., 1998. On the effects of spatial heterogeneity on the persistence of interacting species. J. Math. Biol. 37 (2), 103–145.

Caubet, F., Deheuvels, T., Privat, Y., 2017. Optimal location of resources for biased movement of species: the 1D case. SIAM J. Appl. Math. 77 (6), 1876–1903.

Chanillo, S., Grieser, D., Imai, M., Kurata, K., Ohnishi, I., 2000a. Symmetry breaking and other phenomena in the optimization of eigenvalues for composite membranes. Commun. Math. Phys. 214 (2), 315–337.

Chanillo, S., Grieser, D., Kurata, K., 2000b. The free boundary problem in the optimization of composite membranes. In: Differential Geometric Methods in the Control of Partial Differential Equations. Boulder, CO, 1999. In: Contemp. Math., vol. 268. Amer. Math. Soc., Providence, RI, pp. 61–81.

Chanillo, S., Kenig, C.E., 2008. Weak uniqueness and partial regularity for the composite membrane problem. J. Eur. Math. Soc. 10 (3), 705–737.

Chanillo, S., Kenig, C.E., To, T., 2008. Regularity of the minimizers in the composite membrane problem in $\mathbb{R}^2$. J. Funct. Anal. 255 (9), 2299–2320.

DeAngelis, D., Zhang, B., Ni, W.-M., Wang, Y., 2020. Carrying capacity of a population diffusing in a heterogeneous environment. Mathematics 8 (1), 49.

Derlet, A., Gossez, J.-P., Takáč, P., 2010. Minimization of eigenvalues for a quasilinear elliptic Neumann problem with indefinite weight. J. Math. Anal. Appl. 371 (1), 69–79.

Ding, W., Finotti, H., Lenhart, S., Lou, Y., Ye, Q., 2010. Optimal control of growth coefficient on a steady-state population model. Nonlinear Anal., Real World Appl. 11 (2), 688–704.

Fisher, R.A., 1937. The wave of advances of advantageous genes. Annu. Eugen. 7 (4), 355–369.

Hamel, F., Nadirashvili, N., Russ, E., 2011. Rearrangement inequalities and applications to isoperimetric problems for eigenvalues. Ann. Math. 174 (2), 647–755.

Henrot, A., 2006. Extremum Problems for Eigenvalues of Elliptic Operators. Frontiers in Mathematics. Birkhäuser Verlag, Basel.

Hintermüller, M., Kao, C.-Y., Laurain, A., 2012. Principal eigenvalue minimization for an elliptic problem with indefinite weight and Robin boundary conditions. Appl. Math. Optim. 65 (1), 111–146.

Jha, K., Porru, G., 2011. Minimization of the principal eigenvalue under Neumann boundary conditions. Numer. Funct. Anal. Optim. 32 (11), 1146–1165.

Kao, C.-Y., Lou, Y., Yanagida, E., 2008. Principal eigenvalue for an elliptic problem with indefinite weight on cylindrical domains. Math. Biosci. Eng. 5 (2), 315–335.

Kolmogorov, A., Pretrovski, I., Piskounov, N., 1937. Étude de l'équation de la diffusion avec croissance de la quantité de matière et son application à un problème biologique. Mosc. Univ. Bull. Math. 1, 1–25.

Lam, K.-Y., Liu, S., Lou, Y., 2020. Selected topics on reaction-diffusion-advection models from spatial ecology. Math. Appl. Sci. Eng. 1 (2), 91–206.

Lamboley, J., Laurain, A., Nadin, G., Privat, Y., 2016. Properties of optimizers of the principal eigenvalue with indefinite weight and Robin conditions. Calc. Var. Partial Differ. Equ. 55 (6).

Lieb, E., Loss, M., 2001. Analysis. American Mathematical Society.

Lou, H., 2008a. On singular sets of local solutions to p-Laplace equations. Chin. Ann. Math., Ser. B 29 (5), 521–530.

Lou, Y., 2008b. Some Challenging Mathematical Problems in Evolution of Dispersal and Population Dynamics. Springer Berlin Heidelberg, Berlin, Heidelberg, pp. 171–205.

Lou, Y., Nagahara, K., Yanagida, E., 2020. Maximizing the total population with logistic growth in a patchy environment. Preprint.

Lou, Y., Yanagida, E., 2006. Minimization of the principal eigenvalue for an elliptic boundary value problem with indefinite weight, and applications to population dynamics. Jpn. J. Ind. Appl. Math. 23 (3), 275–292.

Mazari, I., Nadin, G., Privat, Y., 2020a. Optimal location of resources maximizing the total population size in logistic models. J. Math. Pures Appl. 9 (134), 1–35.

Mazari, I., Nadin, G., Privat, Y., 2020b. Shape optimization of a weighted two-phase Dirichlet eigenvalue. Preprint.

Mazari, I., Ruiz-Balet, D., 2021. A fragmentation phenomenon for a non-energetic optimal control problem: optimisation of the total population size in logistic diffusive models. SIAM J. Appl. Math. 81 (1), 153–172.

Murray, J.D., 1993. Mathematical Biology. Springer Berlin Heidelberg.

Nagahara, K., Yanagida, E., 2018. Maximization of the total population in a reaction–diffusion model with logistic growth. Calc. Var. Partial Differ. Equ. 57 (3), 80.

Roques, L., Hamel, F., 2007. Mathematical analysis of the optimal habitat configurations for species persistence. Math. Biosci. 210 (1), 34–59.

Shigesada, N., Kawasaki, K., 1997. Biological Invasions: Theory and Practice. Oxford University Press.

Skellam, J.G., 1951. Random dispersal in theoretical populations. Biometrika 38 (1–2), 196–218.

Zhang, B., Kula, A., Mack, K.M.L., Zhai, L., Ryce, A.L., Ni, W.-M., DeAngelis, D.L., Dyken, J.D.V., 2017. Carrying capacity in a heterogeneous environment with habitat connectivity. Ecol. Lett. 20 (9), 1118–1128.

Chapter 13

Gradient flows and nonlinear power methods for the computation of nonlinear eigenfunctions

Leon Bungert[a] and Martin Burger[b,*]
[a]*Hausdorff Center for Mathematics, University of Bonn, Bonn, Germany,* [b]*Department of Mathematics, Friedrich–Alexander University, Erlangen, Germany*
**Corresponding author: e-mail address: martin.burger@fau.de*

Contents

Abstract
This chapter describes how gradient flows and nonlinear power methods in Banach spaces can be used to solve nonlinear eigenvector-dependent eigenvalue problems, and how these can be approximated using Γ-convergence. We review several flows from literature, which were proposed to compute nonlinear eigenfunctions, and show that they all relate to normalized gradient flows. Furthermore, we show that the implicit Euler

Handbook of Numerical Analysis, Volume 23, ISSN 1570-8659, https://doi.org/10.1016/bs.hna.2021.12.013

discretization of gradient flows gives rise to a nonlinear power method of the proximal operator and we demonstrate their convergence to nonlinear eigenfunctions. Finally, we prove that Γ-convergence of functionals implies convergence of their ground states.

Keywords
Gradient flows, Nonlinear eigenvalue problems, Nonlinear power methods, Ground states, Gamma-convergence

MSC Codes
35B40, 35P30, 47J10

1 Introduction

Nonlinear eigenvalue problems appear in different applications in physics, (Cohen and Gilboa, 2018; Weinstein, 1982), mathematics (Amann, 1976; Cancès et al., 2010; Rabinowitz, 1971), and also in modern disciplines like data science (Bühler and Hein, 2009; Bungert et al., 2019; Gilboa, 2018; Hein and Bühler, 2010). While linear eigenvalue problems have the form $\lambda u = Au$, where A is a linear operator or a matrix, with *nonlinear eigenvalue problems* we refer to equations of the form $\lambda B(u) = A(u)$ which depend non-linearly on the eigenvector. For instance, solutions of such nonlinear eigenvalue problems can be used to describe non-Newtonian fluids or wave propagation in a nonlinear medium. Only in the last years it was noticed that nonlinear eigenvalue problems can also be used for solving tasks arising in data science.

Note that eigenfunctions of linear operators have been used for decomposing and filtering data like audio data for decades. Their most prominent occurrence is the Fourier transform which decomposes signals into trigonometric functions, in other words eigenfunctions of the Laplacian operator. While this is the most popular technique for audio processing, for the filtering of images the Fourier transform is only of limited use since it has a hard time resolving the discontinuities which naturally arise as edges in images. Instead, nonlinear operators like the 1-Laplacian have turned out to be suitable tools for defining nonlinear spectral decompositions and filtering of data (Benning et al., 2017; Burger et al., 2015, 2016; Fumero et al., 2020; Gilboa, 2013, 2014, 2018; Gilboa et al., 2016)). In these works the total variation flow $\partial_t u(t) \in -\partial \operatorname{TV}(u(t))$ with $u(0) = f$ (Andreu et al., 2002) was utilized to define a nonlinear spectral representation of f as $f = \int_0^T \phi(t)\mathrm{d}t$, where the functions $\phi(t)$ for $t > 0$ are computed from the gradient flow and are (at least close to) eigenfunctions of the 1-Laplacian operator $\Delta_1 u = \operatorname{div}(\nabla u/|\nabla u|)$, see Gilboa (2014). Bungert et al. (2021a) investigated these decompositions theoretically and proved conditions for the functions $\phi(t)$ to be nonlinear eigenfunctions. The representation above can be used for multiple image processing tasks. For example, in Fig. 1 we show an application to face fusion (Benning et al., 2017), where the spectral representations of two images are combined.

FIGURE 1 Nonlinear spectral face fusion using method by Benning et al. (2017). Left to right: Saint Nicholas, Fusion "Saint Nichoclaus", Fusion "Santolas", Santa Claus.

Besides spectral decompositions, also the computation of ground states, i.e., eigenfunctions with minimal eigenvalue has important applications. A prototypical example in data science is graph clustering, where eigenfunctions of the graph p-Laplacian operators (Elmoataz et al., 2015) are used for partitioning graphs. As observed by Bühler and Hein (2009); Bungert et al. (2019); Hein and Bühler (2010), eigenfunctions of the nonlinear 1-Laplacian are more suitable for clustering than Laplacian eigenfunctions since they allow for sharp discontinuities, see Fig. 2.

(a) Two-moon graph (b) 2-Laplacian eigenfunction (c) 1-Laplacian eigenfunction

FIGURE 2 Spectral graph clustering.

In this chapter we focus on the following class of variational nonlinear eigenvalue problems on a Banach space $\mathcal{X}$, where the nonlinear operators are subdifferentials. For a convex function $f : \mathcal{X} \to (-\infty, \infty]$ its subdifferential is defined as

$$\partial f(u) := \{\zeta \in \mathcal{X}^* : f(u) + \langle \zeta, v - u \rangle \leq f(v),\ \forall v \in \mathcal{X}\}. \tag{1.1}$$

Let us also define the so-called duality mapping of $\mathcal{X}$ with respect to a number $p \in [1, \infty)$ as $\Phi^p_{\mathcal{X}} = \partial \frac{1}{p} \|\cdot\|^p$ (see Section 2 for details).

Definition 1.1 (Nonlinear eigenvectors). Let $\mathcal{J} : \mathcal{X} \to (-\infty, \infty]$ be proper and convex, and $p \in [1, \infty)$. A tuple $(u, \hat{u}) \in \mathcal{X} \times \arg\min \mathcal{J}$ is called p-eigenvector of $\mathcal{J}$ with eigenvalue $\lambda \in \mathbb{R}$ if

$$\lambda \Phi^p_{\mathcal{X}}(\hat{u} - u) + \partial \mathcal{J}(u) \ni 0. \tag{1.2}$$

Note that one could replace the duality map with the subdifferential of a different functional, however, for most nonlinear eigenvalue problems which are interesting in applications this is not necessary. The reason why we define eigenvectors as tuples instead of single vectors will become clear in a second.

Of particular interest in many applications ranging from physics to data science are eigenvectors with minimal eigenvalue which are called *ground states* and defined as follows:

Definition 1.2 (Ground states). Let $\mathcal{J} : \mathcal{X} \to (-\infty, \infty]$ be proper and $p \in [1, \infty)$. A tuple $(u_*, \hat{u}_*) \in \mathcal{X} \times \arg\min \mathcal{J}$ is called p-ground state of $\mathcal{J}$ if

$$(u_*, \hat{u}_*) \in \underset{\hat{u} \in \arg\min \mathcal{J}}{\arg\max} \ \underset{u \in \mathcal{X}}{\arg\min} \frac{\mathcal{J}(u) - \mathcal{J}(\hat{u})}{\left\| u - \hat{u} \right\|^p}. \tag{1.3}$$

Note that throughout the chapter arg min and arg max are subsets of $\mathcal{X}$. Proposition 2.2 below states that ground states are eigenvectors in the sense of Definition 1.1 with minimal eigenvalue (Benning and Burger, 2013). Defining eigenvectors and ground states as tuples prevents minimizers of $\mathcal{J}$ from being trivial ground states. We call the quotient appearing in (1.3) Rayleigh quotient since it is a generalization of the standard Rayleigh quotient $\langle u, Au \rangle / \|u\|^2$ from linear algebra.

While Definition 1.2 and (1.2) are quite abstract, they cover most interesting nonlinear eigenvalue problems from different fields. To give an example from physics, for the energy $\mathcal{J}(u) = \int_{\mathbb{R}^d} \frac{1}{2} |\nabla u|^2 - \frac{1}{p+2} |u|^{p+2} \, \mathrm{d}x$ on $\mathcal{X} = L^2(\mathbb{R}^d)$ the eigenvalue problem (1.2) is characterized by $\hat{u}_* = 0$ and u_* solving the nonlinear Schrödinger equation (Weinstein, 1982) $\lambda u = -\Delta u - |u|^p u$.

In this chapter we explain how nonlinear eigenvectors and ground states can be computed. Note that these problems are hard to solve for multiple reasons. Firstly, the eigenvalues in (1.2) are unknown, secondly, the saddle point problem (1.3) is neither convex in u nor concave in $\hat{u}$, and lastly the functional $\mathcal{J}$ and the Banach space norm $\|\cdot\|$ are non-smooth in interesting applications. However, thanks to the variational structure of Definitions 1.1 and 1.2 it turns out that gradient flows and their time discretizations are a useful tool.

Gradient flows are ubiquitous in mathematics and the natural sciences and in a Hilbert space setting (Aubin and Cellina, 2012; Brezis, 1973) they take the form

$$u'(t) + \partial \mathcal{J}(u(t)) \ni 0. \tag{1.4}$$

In the context of the nonlinear eigenvalue problem (1.2), we have to work with more gradient flows in a Banach space. To do this, one observes that the subdifferential (1.1) naturally lives in the dual space and hence one has to "push" the time derivative to the dual space. This can be achieved by the duality mapping $\Phi^p_{\mathcal{X}} = \partial \frac{1}{p} \|\cdot\|^p$ and leads to the equation

$$\Phi^p_{\mathcal{X}}(u'(t)) + \partial \mathcal{J}(u(t)) \ni 0, \tag{1.5}$$

which is referred to as p-gradient flow or short p-flow. More generally, one could replace $\Phi^p_{\mathcal{X}}(u')$ by a functional of the form $\partial_{u'} \mathcal{D}(u, u')$ (Mielke, 2016).

Eq. (1.5) is a special case of a metric gradient flow, for which a comprehensive theory is laid out in the monograph by Ambrosio et al. (2008).

The relation between the gradient flow (1.5) and the nonlinear eigenvalue problem (1.2) is not completely obvious although they share structural similarities. The simplest example for relations between gradient flows and ground states is the linear heat equation $u'(t) = \Delta u(t)$, posed on a bounded domain with Dirichlet boundary conditions. Expanding the solution in terms of an eigenbasis of the Laplacian operator one can show that if the initial datum satisfies $u(0) \geq 0$, then the normalized heat flow converges to a Laplacian eigenfunction:

$$\lim_{t\to\infty} \left\| \frac{u(t)}{\|u(t)\|_{L^2}} - w \right\|_{L^2} = 0, \quad \text{where} \quad \lambda w = -\Delta w.$$

The limit w is referred to as *asymptotic profile* or *extinction profile* and describes the shape of the solution just before it converges. The existence of asymptotic profiles which are eigenvectors of the spatial differential operator has been proved for numerous (non)linear PDEs, including the total variation flow and p-Laplacian evolution equations (Andreu et al., 2002; Andreu-Vaillo et al., 2004; Bonforte and Figalli, 2012; Ghidaglia and Marzocchi, 1991; Kamin and Vázquez, 1988; Portilheiro and Vázquez, 2013; Vázquez, 2018), porous medium (Stan et al., 2018; Vázquez, 2007) and fast diffusion equations (Bonforte and Figalli, 2020; Bonforte et al., 2012).

These results were generalized to the abstract equations (1.4) and (1.5) by Bungert (2020); Bungert and Burger (2020); Bungert et al. (2021a); Hynd and Lindgren (2017); Varvaruca (2004) where asymptotic profiles were shown to satisfy (1.2) or its Hilbertian version $\lambda u \in \partial\mathcal{J}(u)$. A different branch of research (Aujol et al., 2018; Cohen and Gilboa, 2018; Feld et al., 2019; Nossek and Gilboa, 2018) studied Rayleigh quotient minimizing flows arising as gradient flows of the Rayleigh quotient $R(u) = \mathcal{J}(u)/\|u\|^p$ for $u \neq 0$. The flows work well in practice and solve the same nonlinear eigenvalue problems at convergence, however, their rigorous mathematical analysis is more challenging. A comprehensive overview of these methods is given in the chapter by Gilboa (2021). From a numerical perspective two major questions arise:

- Do also the time discretizations of all these flows, in particular of (1.5), converge to solutions of the eigenvalue problems (1.2)?
- Can the eigenvalue problem (1.2) be approximated through discretization?

The first question was partially answered by Bungert (2020); Bungert et al. (2021b), where it was observed that a time-discretization of normalized gradient flows yields a power method for the proximal operator of the energy functional, whose convergence was proved for absolutely one-homogeneous functionals on Hilbert spaces. The second question was addressed for a special case by Roith and Bungert (2022), where it was proved that Γ-convergence of a sequence of L^∞-type functionals on weighted graphs implies convergence of their ground states, i.e., eigenvectors with minimal eigenvalue.

In this chapter we provide a comprehensive theory for the two questions, raised above. In Section 3 we first show that both gradient flows of general convex functionals and their implicit time discretization possess the astonishing property that they decrease the nonlinear Rayleigh quotient in (1.3). Section 4 then reviews several flows from the literature, which were proposed to solve nonlinear eigenvalue problems associated to homogeneous functionals. In particular, we show that all these flows are equivalent to a renormalization of the gradient flow. In Section 5 we then thoroughly analyze a nonlinear power method based on the proximal operator, naturally arising as time-discretization of the flows from Section 4. We prove convergence of the power method to a nonlinear eigenvector, show angular convergence, and prove that the power method preserves positivity. In Section 6 we then answer the second question by proving that Γ-convergence of general convex functionals implies convergence of the corresponding ground states. We illustrate our theoretical findings in Section 7 where we show numerical results of proximal power iterations, study their numerical convergence, and visualize the convergence of eigenvectors under Γ-convergence. Our numerical experiments are carried out both on regular grids and on general weighted graphs.

The appendix contains some of our proofs and provides further details on the phenomena of exact reconstruction and finite extinction which can occur in the evaluation of proximal operators and have to be taken into account for well-defined proximal power methods.

2 Convex analysis and nonlinear eigenvalue problems

If $(\mathcal{X}, \|\cdot\|)$ is a Banach space we denote the dual space of $\mathcal{X}$ by $(\mathcal{X}^*, \|\cdot\|_*)$ where

$$\|\zeta\|_* = \sup_{\substack{u\in\mathcal{X}\\ \|u\|=1}} \langle \zeta, u\rangle, \quad \zeta \in \mathcal{X}^* \tag{2.1}$$

is the dual norm. For $p \in [1, \infty)$ the so-called p-proximal operator of a convex functional $\mathcal{J} : \mathcal{X} \to (-\infty, \infty]$ is defined as the set-valued function

$$\operatorname{prox}^p_{\tau\mathcal{J}}(f) = \operatorname*{arg\,min}_{u\in\mathcal{X}} \frac{1}{p}\|u - f\|^p + \tau\mathcal{J}(u), \quad \tau \geq 0, \tag{2.2}$$

where the minimization problem has a solution under suitable conditions on the functional $\mathcal{J}$. For $\mathcal{X} = \mathcal{H}$ being a Hilbert space and $\mathcal{J}$ being convex, the p-proximal coincides with the standard proximal operator which is single-valued. The p-proximal operator has strong relations with the duality map of $\mathcal{X}$, given by

$$\Phi^p_{\mathcal{X}}(u) := \begin{cases} \{\zeta \in \mathcal{X}^* : \langle\zeta, u\rangle = \|\zeta\|_* \|u\|, \ \|\zeta\|_* = \|u\|^{p-1}\}, & p > 1, \\ \{\zeta \in \mathcal{X}^* : \langle\zeta, u\rangle = \|u\|, \ \|\zeta\|_* \leq 1\}, & p = 1. \end{cases} \tag{2.3}$$

Note that the two definitions differ only in the case $u = 0$ where the duality map coincides with $\{0\}$ for $p > 1$ and with $\{\zeta \in \mathcal{X}^* : \|\zeta\| \leq 1\}$ for $p = 1$. The following proposition collects some important properties of the duality map and the p-proximal operator. This operator is in fact a special case of a larger class of proximity operators on Banach spaces studied by Combettes and Reyes (2013); Penot (1998); Wexler (1973).

Proposition 2.1 (Properties of duality map and p-proximal). *It holds that*

- $\Phi^p_{\mathcal{X}}(u) \neq \emptyset$ *for all* $u \in \mathcal{X}$,
- $\Phi^p_{\mathcal{X}}$ *is a* $p-1$*-homogeneous map in the sense that*

$$\Phi^p_{\mathcal{X}}(cu) = c|c|^{p-2}\Phi^p_{\mathcal{X}}(u), \qquad \forall u \in \mathcal{X},\ c \in \mathbb{R}, \tag{2.4}$$

- $\Phi^p_{\mathcal{X}} = \partial\left(\frac{1}{p}\|\cdot\|^p\right)$, *where* ∂ *denotes the subdifferential,*
- $u \in \operatorname{prox}^p_{\tau\mathcal{J}}(f)$ *if and only if*

$$\exists\, \eta \in \Phi^p_{\mathcal{X}}(u-f),\ \zeta \in \partial\mathcal{J}(u) : 0 = \eta + \tau\zeta. \tag{2.5}$$

Many statements in this chapter concern ground states of a functional $\mathcal{J}$, see Definition 1.2. As already mentioned in the introduction, ground states of convex functionals are solutions to a doubly nonlinear eigenvalue problem involving the duality map and the subdifferential of $\mathcal{J}$, as the following proposition states, the proof of which can be found in the appendix.

Proposition 2.2. *Let* $\mathcal{J} : \mathcal{X} \to (-\infty, \infty]$ *be convex and proper. Let* $(u_*, \hat{u}_*) \in \mathcal{X} \times \arg\min \mathcal{J}$ *be a* p*-ground state and define* $\lambda_p := p\frac{\mathcal{J}(u_*)-\mathcal{J}(\hat{u}_*)}{\|u_*-\hat{u}_*\|^p}$. *Then it holds*

$$\lambda_p \Phi^p_{\mathcal{X}}(\hat{u}_* - u_*) + \partial\mathcal{J}(u_*) \ni 0. \tag{2.6}$$

Note that our definition of ground states and the relation to the nonlinear eigenvalue problem (1.2) is valid for general convex functionals and does not require any homogeneity, which is typically assumed for nonlinear eigenvalue problems. Still, some of our results are only valid for *absolutely α-homogeneous* functionals for some degree $\alpha \in [1, \infty)$, which by definition satisfy

$$\mathcal{J}(cu) = |c|^\alpha \mathcal{J}(u), \quad \forall u \in \operatorname{dom}\mathcal{J}. \tag{2.7}$$

Important examples for such functionals are

$$\mathcal{J}(u) = \sup_{\|\phi\|_\infty = 1} \int_\Omega u \operatorname{div}\phi \,\mathrm{d}x, \quad \text{the total variation,} \qquad \alpha = 1,$$

$$\mathcal{J}(u) = \int_\Omega |\nabla u|^p \mathrm{d}x, \qquad \text{the } p\text{-Dirichlet energy,} \quad \alpha = p.$$

For instance, for the p-Dirichlet energy with homogeneous Dirichlet boundary conditions the abstract eigenvalue problem (1.2) becomes the p-Laplacian eigenvalue problem

$$\lambda_p |u|^{p-2} u = -\Delta_p u.$$

For convex α-homogeneous functionals we define their nullspace as

$$\mathcal{N}(\mathcal{J}) := \{u \in \mathcal{X} : \mathcal{J}(u) = 0\}. \tag{2.8}$$

Because of the convexity and the homogeneity this is indeed a vector space, and for lower semicontinuous functionals it is additionally closed.

In the following we state the assumptions which we use in different parts of this chapter. For a compact presentation we make the following fundamental assumption which we assume without further notice and reference.

Assumption. The functional $\mathcal{J} : \mathcal{X} \to (-\infty, \infty]$ is proper, convex, and lower semicontinuous.

For studying convergence of rescaled gradient flows and proximal power methods to nonlinear eigenvectors in Section 4 and Section 5 we need the following assumption.

Assumption 1. The sub-level sets of $\|\cdot\| + \mathcal{J}(\cdot)$ are relatively compact, in the sense that every sequence $(u_n) \subset \mathcal{X}$ such that $\sup_{n\in\mathbb{N}} \|u_n\| + \mathcal{J}(u_n) < \infty$ admits a subsequence (which we do not relabel) and an element $u \in \mathcal{X}$ such that $\|u_n - u\| \to 0$ as $n \to \infty$.

In Section 5 we pose an assumption concerning absolutely α-homogeneous functionals which demands that their nullspace be trivial. We will see that this is no restriction of generality and makes the presentation more concise.

Assumption 2. The functional $\mathcal{J} : \mathcal{X} \to (-\infty, \infty]$ is absolutely α-homogeneous for some $\alpha \in [1, \infty)$ and satisfies $\mathcal{N}(\mathcal{J}) = \{0\}$.

3 Gradient flows and decrease of Rayleigh quotients

This section is dedicated to proving that under general conditions the p-flow (1.5) and its implicit Euler discretization in time, often referred to as *minimizing movement scheme*, both admit the astonishing property that they do not only decrease the energy $\mathcal{J}$ but also the nonlinear Rayleigh quotient in (1.3). This property was already observed for the continuous-time gradient flows of homogeneous functionals, e.g., by Bungert and Burger (2020); Ghidaglia and Marzocchi (1991). However, for general convex functionals and for the time-discrete flows this statement appears to be novel. Before we turn to the proofs we recap some basic facts about the p-flow and the associated minimizing movement scheme.

3.1 The p-gradient flow and minimizing movements

The equation for a large part of this chapter is the p-gradient flow (1.5) of a functional $\mathcal{J}$ on a Banach space $\mathcal{X}$ with initial datum $f \in \mathcal{X}$, which we repeat here for convenience:

$$\Phi^p_{\mathcal{X}}(u'(t)) + \partial\mathcal{J}(u(t)) \ni 0, \qquad u(0) = f.$$

It is a doubly-nonlinear generalization of Hilbertian gradient flows (1.4), which are recovered by choosing $\mathcal{X}$ as a Hilbert space and $p = 2$. The right concept of solutions for the p-flow is the one of p-curves of maximal slope:

Definition 3.1. Let $1 < p < \infty$. A p-curve of maximal slope for $\mathcal{J}$ is an absolutely continuous map $u : [0, \infty) \to \mathcal{X}$ that satisfies

$$\frac{\mathrm{d}}{\mathrm{d}t}\mathcal{J}(u(t)) \le -\frac{1}{p}\left|u'\right|^p(t) - \frac{1}{q}\left|\partial\mathcal{J}\right|^q(u(t)), \quad a.e.\, t > 0, \tag{3.1}$$

where $\left|u'\right|(t) := \lim_{h\to 0} \frac{\|u(t+h)-u(t)\|}{|h|}$ denotes the *metric derivative* of u in $t > 0$ and $|\partial\mathcal{J}|(u) := \inf\left\{\|\zeta\|_* : \zeta \in \partial\mathcal{J}(u)\right\}$ denotes the *local slope* of $\mathcal{J}$ in $u \in \mathcal{X}$.

Note that the opposite inequality to (3.1) is always satisfied thanks to the Cauchy-Schwarz and Young inequalities. This enables one to show that p-curves of maximal slope are in one-to-one correspondence to almost everywhere differentiable solutions of (1.5), see Hynd and Lindgren (2017). Existence of solutions is discussed, e.g., by Ambrosio et al. (2008); Santambrogio (2017) (see also Stefanelli (2021) for a new approach) but in this chapter we rather focus on qualitative properties of the flows.

For proving existence, most approaches use the concept of minimizing movements (Ambrosio, 1995; Ambrosio et al., 2008; De Giorgi, 1993), which is an implicit Euler discretization in time of the p-flow (1.5). Besides being a theoretical tool, it also serves as a numerical scheme for approximating solutions. The scheme arises by discretizing the values of $u(t)$ one discrete time steps $k\tau$ where $k \in \mathbb{N}_0$ and $\tau > 0$ is a time step size. Hence, the implicit discretization of (1.5) takes the form

$$\Phi^p_{\mathcal{X}}\left(\frac{u^{k+1} - u^k}{\tau}\right) + \partial\mathcal{J}(u^{k+1}) \ni 0, \tag{3.2a}$$

$$\iff u^{k+1} \in \operatorname*{arg\,min}_{u\in\mathcal{X}} \frac{1}{p}\left\|u - u^k\right\|^p + \tau^{p-1}\mathcal{J}(u), \tag{3.2b}$$

$$\iff u^{k+1} \in \operatorname{prox}^p_{\tau^{p-1}\mathcal{J}}(u^k), \tag{3.2c}$$

where we set $u^0 := f$. As the time step size τ tends to zero, under generic conditions a time interpolation of the sequence generated by (3.2) converges to a curve of maximal slope which is a solution of (1.5) (see, e.g., Ambrosio et al. (2008); Santambrogio (2017)). This yields both existence of solutions and a numerical method to compute them.

3.2 Decrease of the Rayleigh quotients

First, we show that the p-flow (1.5) of a convex functional decreases the generalized Rayleigh quotient from (1.3).

Proposition 3.1. *Let $u : [0, \infty) \to \mathcal{X}$ be an absolutely continuous solution of* (1.5) *and let $\mathcal{J}$ be convex. Then it holds for all $\hat{u} \in \arg\min \mathcal{J}$*

$$\frac{\mathrm{d}}{\mathrm{d}t} \frac{\mathcal{J}(u(t)) - \mathcal{J}(\hat{u})}{\|u(t) - \hat{u}\|} \le 0, \quad \forall 0 < t < \infty. \tag{3.3}$$

If $\mathcal{J}$ is absolutely α-homogeneous for $\alpha \in [1, \infty)$, it even holds

$$\frac{\mathrm{d}}{\mathrm{d}t} \frac{\alpha \mathcal{J}(u(t))}{\|u(t) - \hat{u}\|^\alpha} \le 0, \quad \forall 0 < t < \infty. \tag{3.4}$$

Proof. Using the triangle inequality it can be easily seen (Hynd and Lindgren, 2017) that

$$\left| \frac{\mathrm{d}}{\mathrm{d}t} \|u(t) - \hat{u}\| \right| \le \|u'(t)\| .$$

Furthermore, one has by properties of p-flows (Hynd and Lindgren, 2017)

$$\frac{\mathrm{d}}{\mathrm{d}t} \mathcal{J}(u(t)) = - \|u'(t)\|^p .$$

Furthermore, for all $\zeta \in \partial \mathcal{J}(u(t))$ one has

$$\mathcal{J}(u(t)) - \mathcal{J}(\hat{u}) \le \|\zeta\|_* \|u(t) - \hat{u}\| .$$

By definition of the p-flow, for every $t > 0$ there exists $\eta(t) \in \Phi^p_{\mathcal{X}}(u'(t))$ and $\zeta(t) \in \partial \mathcal{J}(u(t))$ such that $\eta(t) + \zeta(t) = 0$. Putting these four statements together, one obtains

$$\begin{aligned}
\frac{\mathrm{d}}{\mathrm{d}t} \frac{\mathcal{J}(u(t)) - \mathcal{J}(\hat{u})}{\|u(t) - \hat{u}\|} &= - \frac{\left(\mathcal{J}(u(t)) - \mathcal{J}(\hat{u})\right) \frac{\mathrm{d}}{\mathrm{d}t} \|u(t) - \hat{u}\|}{\|u(t) - \hat{u}\|^2} + \frac{\frac{\mathrm{d}}{\mathrm{d}t} \mathcal{J}(u(t))}{\|u(t) - \hat{u}\|} \\
&\le \frac{1}{\|u(t) - \hat{u}\|} \left[\|\zeta(t)\|_* \|u'(t)\| - \|u'(t)\|^p \right] \\
&= \frac{1}{\|u(t) - \hat{u}\|} \left[\|\eta(t)\|_* \|u'(t)\| - \|u'(t)\|^p \right] \\
&= \frac{1}{\|u(t) - \hat{u}\|} \left[\|u'(t)\|^p - \|u'(t)\|^p \right] = 0,
\end{aligned}$$

where we used the properties of the duality mapping $\Phi^p_{\mathcal{X}}$.

In the case that $\mathcal{J}$ is absolutely α-homogeneous we can use (Bungert and Burger, 2020)

$$\alpha \mathcal{J}(u(t)) \leq \|\zeta\|_* \left\|u(t) - \hat{u}\right\|, \quad \forall \zeta \in \partial \mathcal{J}(u(t)),$$

to compute

$$\begin{aligned}
\frac{\mathrm{d}}{\mathrm{d}t} \frac{\alpha \mathcal{J}(u(t))}{\left\|u(t) - \hat{u}\right\|^\alpha} &= -\frac{\alpha \mathcal{J}(u(t)) \frac{\mathrm{d}}{\mathrm{d}t} \left\|u(t) - \hat{u}\right\|^\alpha}{\left\|u(t) - \hat{u}\right\|^{2\alpha}} + \frac{\alpha \frac{\mathrm{d}}{\mathrm{d}t} \mathcal{J}(u(t))}{\left\|u(t) - \hat{u}\right\|^\alpha} \\
&= \frac{\alpha \mathcal{J}(u(t))\alpha \left\|u(t) - \hat{u}\right\|^{\alpha-1} \left\|u'(t)\right\|}{\left\|u(t) - \hat{u}\right\|^{2\alpha}} - \frac{\alpha \left\|u'(t)\right\|^p}{\left\|u(t) - \hat{u}\right\|^\alpha} \\
&\leq \frac{\alpha}{\left\|u(t) - \hat{u}\right\|^\alpha} \left[\frac{\alpha \mathcal{J}(u(t)) \left\|u'(t)\right\|}{\left\|u(t) - \hat{u}\right\|} - \left\|u'(t)\right\|^p \right] \\
&\leq \frac{\alpha}{\left\|u(t) - \hat{u}\right\|^\alpha} \left[\|\zeta(t)\|_* \left\|u'(t)\right\| - \left\|u'(t)\right\|^p \right] \leq 0,
\end{aligned}$$

where the last inequality follows again from $-\zeta(t) = \eta(t) \in \Phi^p_{\mathcal{X}}(u'(t))$. □

Next, we show that also the sequence $(u^k)_{k \in \mathbb{N}} \subset \mathcal{X}$ generated by the minimizing movement scheme (3.2) decreases the Rayleigh quotient in the precise same way, which is an important consistency property for this scheme.

Proposition 3.2. *Let $\tau > 0$, $u^{k+1} \in \mathrm{prox}^p_{\tau \mathcal{J}}(u^k)$, and $\hat{u} \in \arg\min \mathcal{J}$. Then it holds that*

$$\frac{\mathcal{J}(u^{k+1}) - \mathcal{J}(\hat{u})}{\left\|u^{k+1} - \hat{u}\right\|} \leq \frac{\mathcal{J}(u^k) - \mathcal{J}(\hat{u})}{\left\|u^k - \hat{u}\right\|} \tag{3.5}$$

with the convention $\frac{0}{0} := 0$. If $\mathcal{J}$ is absolutely α-homogeneous for $\alpha \in [1, \infty)$, it even holds

$$\frac{\alpha \mathcal{J}(u^{k+1})}{\left\|u^{k+1} - \hat{u}\right\|^\alpha} \leq \frac{\alpha \mathcal{J}(u^k)}{\left\|u^k - \hat{u}\right\|^\alpha}, \quad \forall \hat{u} \in \mathcal{N}(J). \tag{3.6}$$

Proof. Let us fix $\hat{u} \in \arg\min \mathcal{J}$. From the minimality of u^{k+1} we infer

$$\frac{1}{p} \left\|u^{k+1} - u^k\right\|^p + \tau \mathcal{J}(u^{k+1}) \leq \frac{1}{p} \left\|v - u^k\right\|^p + \tau \mathcal{J}(v), \quad \forall v \in \mathcal{X}. \tag{3.7}$$

We now distinguish two cases, starting with the easier one.

Case 1: Here, we assume that $\left\|u^{k+1} - \hat{u}\right\| \geq \left\|u^k - \hat{u}\right\|$. Choosing $v = u^k$ in (3.7) one obtains $\mathcal{J}(u^{k+1}) \leq \mathcal{J}(u^k)$. We can combine these two estimates to

$$\frac{\mathcal{J}(u^{k+1}) - \mathcal{J}(\hat{u})}{\left\|u^{k+1} - \hat{u}\right\|} \leq \frac{\mathcal{J}(u^k) - \mathcal{J}(\hat{u})}{\left\|u^k - \hat{u}\right\|}.$$

Case 2: Now, we assume that $\left\|u^{k+1}-\hat{u}\right\| \leq \left\|u^k-\hat{u}\right\|$. Choosing

$$v := \frac{\left\|u^{k+1}-\hat{u}\right\|}{\left\|u^k-\hat{u}\right\|}(u^k-\hat{u})+\hat{u},$$

which satisfies

$$\begin{aligned}\left\|v-u^k\right\| &= \left\|\frac{\left\|u^{k+1}-\hat{u}\right\|}{\left\|u^k-\hat{u}\right\|}(u^k-\hat{u})-(u^k-\hat{u})\right\| = \left|\frac{\left\|u^{k+1}-\hat{u}\right\|}{\left\|u^k-\hat{u}\right\|}-1\right|\left\|u^k-\hat{u}\right\| \\ &= \left|\left\|u^{k+1}-\hat{u}\right\|-\left\|u^k-\hat{u}\right\|\right| \leq \left\|u^{k+1}-u^k\right\|,\end{aligned}$$

we obtain

$$\begin{aligned}\mathcal{J}(u^{k+1}) \leq \mathcal{J}(v) &= \mathcal{J}\left(\frac{\left\|u^{k+1}-\hat{u}\right\|}{\left\|u^k-\hat{u}\right\|}(u^k-\hat{u})+\hat{u}\right) \\ &= \mathcal{J}\left(\frac{\left\|u^{k+1}-\hat{u}\right\|}{\left\|u^k-\hat{u}\right\|}u^k+\left(1-\frac{\left\|u^{k+1}-\hat{u}\right\|}{\left\|u^k-\hat{u}\right\|}\right)\hat{u}\right).\end{aligned}$$

Since $\left\|u^{k+1}-\hat{u}\right\| \leq \left\|u^k-\hat{u}\right\|$, we obtain using the convexity of $\mathcal{J}$

$$\mathcal{J}(u^{k+1}) \leq \frac{\left\|u^{k+1}-\hat{u}\right\|}{\left\|u^k-\hat{u}\right\|}\mathcal{J}(u^k)+\left(1-\frac{\left\|u^{k+1}-\hat{u}\right\|}{\left\|u^k-\hat{u}\right\|}\right)\mathcal{J}(\hat{u}).$$

This inequality can be reordered to (3.5).

If $\mathcal{J}$ is absolutely α-homogeneous, we can use that $\mathcal{J}(u^{k+1}+\hat{u}) = \mathcal{J}(u^{k+1})$ for all $\hat{u} \in \arg\min \mathcal{J} = \mathcal{N}(J)$ (see Bungert and Burger (2020)) and obtain

$$\mathcal{J}(u^{k+1}) \leq \left(\frac{\left\|u^{k+1}-\hat{u}\right\|}{\left\|u^k-\hat{u}\right\|}\right)^{\alpha}\mathcal{J}(u^k),$$

which can be reordered to (3.6). □

4 Flows for solving nonlinear eigenproblems

In this section we review several flows from the literature whose large time behavior has been brought into correspondence with ground states according to Definition 1.2 or—more generally—nonlinear eigenfunctions solving (1.2). The relation between gradient flows and eigenvectors of general subdifferential operators was studied by Bungert and Burger (2020); Bungert et al. (2021a, 2019); Hynd and Lindgren (2017); Varvaruca (2004), see also Bungert (2020). Rayleigh quotient minimizing flows were investigated by Aujol et al. (2018); Cohen and Gilboa (2018); Feld et al. (2019); Nossek and Gilboa (2018), see also

Gilboa (2018, 2021). In the following we review these flows and discover that they have strong relations and are connected through time reparametrizations and normalizations.

The first three approaches below are all based on the p-flow (1.5) or gradient flows on Hilbert spaces (1.4). The fourth class of flows we study here are Rayleigh quotient minimizing flows on Hilbert spaces which are seemingly different from gradient flows. However, we show that they all arise as time rescalings of normalized gradient flows. Note that all approaches utilize absolutely α-homogeneous functionals (see (2.7)).

4.1 Rescaled p-flows

The first approach was investigated by Hynd and Lindgren (2017) and considers the p-flow (1.5) of absolutely p-homogeneous functionals on Banach spaces, i.e., the homogeneity of the functional coincides with p. The authors showed that in this case solutions of the p-flow decay exponentially and that their exponential rescalings converge to a limit, which was characterized to be zero or a nonlinear eigenfunction in the sense of (1.2).

To set the scene, let $\mathcal{J} : \mathcal{X} \to (-\infty, \infty]$ be absolutely p-homogeneous and define

$$\lambda_p := \inf_{u \neq 0} \frac{p\mathcal{J}(u)}{\|u\|^p}. \tag{4.1}$$

The value λ_p is called simple if any two minimizers are linearly dependent. Furthermore, from Proposition 2.2 we see that λ_p is the smallest eigenvalue of the eigenvalue problem (1.2).

Theorem 4.1 (Theorem 1.3 by Hynd and Lindgren (2017)). *Let $p > 1$ and $u : [0, \infty) \to \mathcal{X}$ be a solution of* (1.5) *with $\mathcal{J}$ being absolutely p-homogeneous. Assume that $\lambda_p > 0$ and simple and denote $\mu_p := \lambda_p^{\frac{1}{p-1}}$. Then it holds*

- *The limit $w := \lim_{t\to\infty} e^{\mu_p t} u(t)$ exists.*
- *If $w \neq 0$, then w is a p-ground state and it holds $\lambda_p = \lim_{t\to\infty} \frac{p\mathcal{J}(u(t))}{\|u(t)\|^p}$.*

Example 4.1. In this example we let $\mathcal{X} = L^p(\Omega)$ for $p > 1$ and consider the functional $\mathcal{J}(u) = \int_\Omega |\nabla u|^p \mathrm{d}x$ if $u \in W_0^{1,p}(\Omega)$ and $\mathcal{J}(u) = \infty$ else. If the initial datum of (1.5) is positive, i.e., $f(x) \geq 0$ for almost all $x \in \Omega$, then w is non-zero and is a solution of the p-Laplacian eigenvalue problem $\lambda_p |u|^{p-2} u = -\Delta_p u$.

4.2 Rescaled gradient flows

The second approach we review here was investigated by Bungert (2020); Bungert and Burger (2020); Bungert et al. (2021a) and deals with Hilbertian gradient flows of absolutely α-homogeneous functionals for $\alpha \in [1, \infty)$, meaning that the homogeneity of the functional not necessarily coincides $p = 2$, which is the appropriate parameter for the duality map in (1.5) in the case of Hilbert space gradient flows. It was shown that, depending the homogeneity α, solutions decay either in finite time, exponentially, or algebraically (see also Hauer and Mazón (2019) for analogous statements for metric gradient flows) and that suitable rescalings converge. For all $\alpha \neq 2$ the limit was shown to be a non-zero solution of the eigenvalue problem $\lambda u \in \partial \mathcal{J}(u)$ which is the Hilbert space version of the eigenvalue problem (1.2). For $\alpha = 2$ which coincides with $p = 2$ the limit can be zero, just as for the previous approach where the homogeneity of the functional coincides with the duality parameter p.

We now adopt the following setting: Let $\mathcal{X} = \mathcal{H}$ be a Hilbert space and $\mathcal{J} : \mathcal{H} \to \mathbb{R} \cup \{\infty\}$ be absolutely α-homogeneous. Setting $p = 2$ in (1.5) yields the gradient flow

$$u'(t) + \partial \mathcal{J}(u(t)) \ni 0, \qquad u(0) = f. \tag{4.2}$$

In order to characterize the decay of solution to this flow one considers the following initial value problem, depending on $\lambda > 0$

$$a'(t) = -\lambda a(t)^{\alpha-1}, \qquad a(0) = 1, \tag{4.3}$$

which arises as special case of (4.2) by setting $\mathcal{H} = \mathbb{R}$, $\mathcal{J}(x) = \frac{\lambda}{\alpha}|x|^\alpha$, and $f = 1$.

Slightly more general as above we define the minimal eigenvalue as

$$\lambda_\alpha := \inf_{u \in \mathcal{N}(\mathcal{J})^\perp} \frac{\alpha \mathcal{J}(u)}{\|u\|^\alpha} = \inf_{u \notin \mathcal{N}(\mathcal{J})} \frac{\alpha \mathcal{J}(u)}{\|u - \overline{u}\|^\alpha}, \tag{4.4}$$

where the unique orthogonal projection of $u \in \mathcal{H}$ onto the nullspace $\mathcal{N}(\mathcal{J})$ is defined by

$$\overline{u} := \arg\min_{v \in \mathcal{N}(\mathcal{J})} \|u - v\|. \tag{4.5}$$

Since for absolutely homogeneous functionals it holds $\arg\min \mathcal{J} = \mathcal{N}(\mathcal{J})$ and $\min \mathcal{J} = 0$, problem (4.4) coincides precisely the eigenvalue of ground states in the sense of Definition 1.2.

As shown by Bungert and Burger (2020) the dynamics of (4.3) completely describe the asymptotic behavior of the much more general gradient flow (4.2). Furthermore, rescaling the gradient flow with a solution of (4.3), yields convergence to an eigenfunction.

Theorem 4.2 (Theorems 2.3, 2.4 by Bungert and Burger (2020)). *Let $u : [0, \infty) \to \mathcal{H}$ be a solution of* (4.2) *with $\mathcal{J}$ being absolutely α-homogeneous. Let furthermore $T_{\text{ex}} := \inf\{t > 0 : u(t) = \overline{f}\}$ denote the extinction time of* (4.2) *and*

$$\lambda := \begin{cases} \frac{1}{(2-\alpha)T_{\mathrm{ex}}}, & 1 \leq \alpha < 2, \\ \lambda_\alpha \, \|f\|^{\alpha-2}, & 2 \leq \alpha < \infty. \end{cases}$$

Then there exists an increasing sequence $(t_k)_{k\in\mathbb{N}}$ with $t_k \nearrow T_{\mathrm{ex}}$, and an element $w \in \mathcal{H}$ with

$$\lim_{k\to\infty} \frac{u(t_k) - \overline{f}}{a_\lambda(t_k)} = w, \qquad \lambda w \in \partial\mathcal{J}(w),$$

where $a_\lambda : [0,\infty) \to [0,1]$ denotes the solution of (4.3). *Furthermore, it holds:*

- *If $\alpha \neq 2$ then $w \neq 0$.*
- *If $\alpha = 2$ and $w \neq 0$, then w is a 2-ground state and it holds $\lambda_2 = \lim_{t\to\infty} \frac{2\mathcal{J}(u(t))}{\|u(t)-\overline{f}\|^2}$.*

Example 4.2. Here we let $\mathcal{H} = L^2(\Omega)$ and define the absolutely one-homogeneous functional $\mathcal{J}(u) = \operatorname{ess\,sup}_{x\in\Omega} |\nabla u(x)|$ if $u \in W_0^{1,\infty}(\Omega)$ and $\mathcal{J}(u) = \infty$ else. If the initial datum of (4.2) is positive, i.e., $f(x) \geq 0$ for almost all $x \in \Omega$, then w is non-zero and a solution of the eigenvalue problem $\lambda u \in \partial\mathcal{J}(u)$, which was shown by Bungert et al. (2020b) to equal a multiple of the Euclidean distance function $x \mapsto \operatorname{dist}(x, \partial\Omega)$ (see Roith and Bungert (2022) for the analogous statement for geodesic distance functions).

4.3 Normalized gradient flows

The previous rescaling factors $e^{\mu_p t}$ and $1/a_\lambda(t)$, respectively, allow one to investigate whether the rescaled solutions converge to ground states, i.e., eigenvectors with minimal eigenvalue. However, both rescalings depend on numbers which are not available in applications (the eigenvalue λ_p and the extinction time T_{ex}). Therefore, one can study rescalings with the norm of the solution itself, i.e.,

$$\begin{cases} u'(t) + \partial\mathcal{J}(u(t)) \ni 0, & u(0) = f, \\ w(t) = \frac{u(t)-\overline{f}}{\|u(t)-\overline{f}\|}, \end{cases} \tag{4.6}$$

where $\mathcal{J}$ is as before. Since the proofs of the previous theorems largely depend on the statement of Proposition 3.1, it does not matter too much which rescaling one chooses as long as one can make sure that the rescaled gradient flow remains uniformly bounded.

The following result was proven by Varvaruca (2004) for functionals that are locally subhomogeneous, however, for brevity we state it for homogeneous functionals. It asserts that the rescalings from (4.6) converge to eigenvectors, however, does not give conditions for this eigenvector being a ground state.

Theorem 4.3 (Theorem 4.1 by Varvaruca (2004)). *Let $\alpha \geq 2$ and $u : [0, \infty) \to \mathcal{H}$ be a solution of* (4.2) *with $\mathcal{J}$ being absolutely α-homogeneous. Then there exists an increasing sequence $(t_k)_{k\in\mathbb{N}}$ with $t_k \to \infty$, and an element $w \in \mathcal{H}$ such that*

$$\lim_{k\to\infty} \frac{u(t_k) - \overline{f}}{\left\|u(t_k) - \overline{f}\right\|^{\alpha}} = w, \qquad \lim_{k\to\infty} \frac{\alpha\mathcal{J}(u(t_k))}{\left\|u(t_k) - \overline{f}\right\|^{\alpha}} = \lambda, \qquad \lambda w \in \partial\mathcal{J}(w).$$

Remark 4.1. Using the techniques from Bungert and Burger (2020) and introducing extinction times, it is easy to prove that this theorem also holds true for all $\alpha \geq 1$. One only has to replace the sequences $t_k \to \infty$ with sequences $t_k \to T_{\mathrm{ex}}$.

We conclude this section by an interesting observation. One can show that time rescalings $w(t)$ in (4.6) also solve a suitable flow, which is the statement of the following proposition.

Proposition 4.1. *Let $w(t)$ be given by* (4.6) *for $t \in [0, T_{\mathrm{ex}})$ and let $v(t) := w\left(\|u(t)\|^{2-\alpha} t\right)$. Then v solves*

$$v' \in \alpha\mathcal{J}(v)v - \partial\mathcal{J}(v), \quad v(0) = \frac{f - \overline{f}}{\left\|f - \overline{f}\right\|}. \tag{4.7}$$

Proof. Since $t \mapsto u(t)$ is Lipschitz continuous and right differentiable for $t > 0$ and $u(t) \neq 0$ for all $t \in [0, T_{\mathrm{ex}})$, the same holds true for $t \mapsto \|u(t)\|$ and $t \mapsto w(t)$. Using the quotient rule and letting $\zeta(t) := -u'(t) \in \partial\mathcal{J}(u(t))$, one gets

$$\begin{aligned} w'(t) &= \frac{\|u(t)\|\, u'(t) - \langle u'(t), u(t)\rangle \frac{u(t)}{\|u(t)\|}}{\|u(t)\|^2} = \frac{\alpha\mathcal{J}(u(t))}{\|u(t)\|^2} w(t) - \frac{\zeta(t)}{\|u(t)\|} \\ &= \|u(t)\|^{\alpha-2}\alpha\mathcal{J}(w(t))w(t) - \frac{\zeta(t)}{\|u(t)\|} \\ &= \|u(t)\|^{\alpha-2}\left[\alpha\mathcal{J}(w(t))w(t) - \frac{\zeta(t)}{\|u(t)\|^{\alpha-1}}\right]. \end{aligned}$$

Now we use that $\partial\mathcal{J}$ is $(\alpha - 1)$-homogeneous to obtain that $\zeta(t)/\|u(t)\|^{\alpha-1} \in \partial\mathcal{J}(w(t))$. Hence, the time rescaling $v(t) := w(\|u(t)\|^{2-\alpha} t)$ solves the flow (4.7). □

4.4 Rayleigh quotient minimizing flows

Flows similar to (4.7) have been proposed in the literature a lot, however, without explicitly relating to gradient flows. A comprehensive overview is given by Gilboa (2021). For instance, Feld et al. (2019) studied a Rayleigh quotient minimizing flow which takes the form

$$\begin{cases} u(0) = f, \\ u' = R(u)q - \zeta, \quad q \in \partial H(u),\ \zeta \in \partial \mathcal{J}(u). \end{cases} \tag{4.8}$$

The flow was derived in order to minimize the Rayleigh quotient

$$R(u) = \frac{\mathcal{J}(u)}{H(u)}, \tag{4.9}$$

for absolutely one-homogeneous functionals $\mathcal{J}$ and H defined on a Hilbert space $\mathcal{H}$ and extends previous results from Aujol et al. (2018). In applications, often $H(u) = \|u\|_{\mathcal{X}}$ where $\mathcal{X}$ is a Banach space larger than $\mathcal{H}$, e.g., $\mathcal{X} = L^1(\Omega)$ and $\mathcal{H} = L^2(\Omega)$.

The flow (4.8) has the property that $\frac{\mathrm{d}}{\mathrm{d}t}\frac{1}{2}\|u(t)\|^2 = 0$, hence preserving the norm of the initial condition. Consequently, for the choice $H(u) = \|u\|$ and $\|f\| = 1$ the flow (4.8) reduces to (4.7), which is equivalent to the normalized gradient flow (4.6), as we showed in the previous section.

A similar evolution was initially studied by Nossek and Gilboa (2018) for the minimization of $\mathcal{J}(u)/\|u\|$ for one-homogeneous $\mathcal{J}$. For the minimization of the general Rayleigh quotient (4.9) we can generalize it as follows

$$\begin{cases} v(0) = f, \\ v' = r - \frac{\eta}{H_*(\eta)}, \quad r \in \partial H(v),\ \eta \in \partial \mathcal{J}(v). \end{cases} \tag{4.10}$$

Here H_* is the dual seminorm (see, e.g., Bungert (2020); Bungert et al. (2021a) for properties) to the absolutely 1-homogeneous functional H, defined as

$$H_*(\zeta) = \sup_{\substack{u \in \mathcal{H} \\ H(u)=1}} \langle \zeta, u \rangle, \quad \zeta \in \mathcal{N}(H)^\perp. \tag{4.11}$$

Interestingly, as we show in the following proposition, this flow is asymptotically equivalent to (4.8). In particular, for $H(u) = \|u\|$ the original method from Nossek and Gilboa (2018) is asymptotically equivalent to the normalized gradient flow (4.6).

Proposition 4.2. *Let v be a solution of* (4.10) *and let $\phi : [0, \infty) \to [0, \infty)$ solve the initial value problem*

$$\phi'(t) = \frac{\mathcal{J}(v(\phi(t)))}{H(v(\phi(t)))}, \quad \phi(0) = 0. \tag{4.12}$$

Then $u(t) := v(\phi(t))$ solves

$$u' = R(u)q - \frac{\mathcal{J}(u)}{H(u)H_*(\zeta)}\zeta, \quad q \in \partial H(u),\ \zeta \in \partial \mathcal{J}(u). \tag{4.13}$$

Proof. Defining $\zeta(t) := \eta(\phi(t)) \in \partial\mathcal{J}(u(t))$ and $q(t) := r(\phi(t)) \in \partial H(u(t))$, one computes

$$\begin{aligned} u'(t) &= \phi'(t)v'(\phi(t)) = \phi'(t)\left[q(t) - \frac{\zeta(t)}{H_*(\zeta(t))}\right] \\ &= \frac{\mathcal{J}(u(t))}{H(u(t))}q(t) - \frac{\mathcal{J}(u(t))}{H(u(t))H_*(\zeta(t))}\zeta(t) \\ &= R(u(t))q(t) - \frac{\mathcal{J}(u(t))}{H(u(t))H_*(\zeta(t))}\zeta(t). \end{aligned}$$

□

Remark 4.2. By the definition of the dual functional H_* (cf. (4.11)) it holds

$$\frac{\mathcal{J}(u(t))}{H(u(t))H_*(\zeta(t))} = \frac{\langle \zeta(t), u(t) \rangle}{H(u(t))H_*(\zeta(t))} \leq 1, \quad \forall t > 0.$$

Since, however, (4.13) converges to an eigenvector as $t \to \infty$, one can even show

$$\lim_{t\to\infty} \frac{\mathcal{J}(u(t))}{H(u(t))H_*(\zeta(t))} = 1.$$

This makes (4.13) asymptotically equivalent to (4.8).

Remark 4.3. Note that the initial value problem (4.12) admits a unique solution, e.g., if $\mathcal{J}(v(0))/H(v(0)) < \infty$. As Nossek and Gilboa (2018) one can show that $\frac{\mathrm{d}}{\mathrm{d}t}\frac{\mathcal{J}(v(t))}{H(v(t))} \leq 0$, such that the right hand side of (4.12) is a continuous and non-increasing function, for which existence of the initial value problem is guaranteed by the classical Peano theorem and uniqueness follows from the monotonicity of the right hand side.

5 Nonlinear power methods for homogeneous functionals

In this section we first show that the time discretization of the normalized gradient flow (4.6) yields a nonlinear power method of the proximal operator. Afterwards, we analyze general nonlinear power methods of p-proximal operators in Banach spaces.

A nonlinear power method for the computation of matrix norms has already been investigated in the early work by Boyd (1974). Bühler and Hein (2009); Hein and Bühler (2010) applied nonlinear power methods for 1-Laplacian and p-Laplacian eigenvectors to graph clustering. The convergence of p-Laplacian inverse power methods for ground states and second eigenfunctions was investigated by Bozorgnia (2016, 2020). Nonlinear power methods and Perron-Frobenius theory for order-preserving multihomogeneous maps were analyzed by Gautier et al. (2020, 2019a,b). Finally, Bungert et al. (2021b) first investigated proximal power methods, focusing on absolutely one-homogeneous functionals on Hilbert spaces. In the following, we extend these results to our general setting.

5.1 Normalized gradient flows and power methods

The normalized gradient flow (4.6), analyzed above, already bears a lot of similarity with a power method. Indeed, discretizing it in time using the minimizing movement scheme (3.2) yields a power method for the proximal operator as the following proposition shows.

Proposition 5.1. *Let $\mathcal{J} : \mathcal{H} \to (-\infty, \infty]$ be an absolutely α-homogeneous functional on a Hilbert space $\mathcal{H}$, and let the sequences u^k and w^k be generated by the iterative scheme*

$$\begin{cases} u^0 &= f, \\ u^{k+1} &= \operatorname{prox}_{\tau^k \mathcal{J}}(u^k), \quad k \in \mathbb{N}_0, \\ w^{k+1} &= \frac{u^{k+1}-\overline{f}}{\|u^{k+1}-\overline{f}\|}, \end{cases} \tag{5.1}$$

where $(\tau^k)_{k\in\mathbb{N}} \subset \mathbb{R}_+$ is a sequence of step sizes. Then it holds that

$$w^{k+1} = \frac{\operatorname{prox}_{\sigma^k}(w^k)}{\|\operatorname{prox}_{\sigma^k}(w^k)\|}, \tag{5.2}$$

where the step sizes σ^k are given by

$$\sigma^k := \tau^k \left\| u^k - \overline{f} \right\|^{\alpha-2}. \tag{5.3}$$

Proof. Without loss of generality we can assume $\overline{f} = 0$. Using the homogeneity of $\mathcal{J}$, we can compute

$$\begin{aligned} u^{k+1} &= \operatorname{prox}_{\tau^k \mathcal{J}}(u^k) = \operatorname{prox}_{\tau^k \mathcal{J}}\left(\left\|u^k\right\| w^k\right) \\ &= \operatorname*{arg\,min}_{u\in\mathcal{H}} \frac{1}{2}\left\|u - \left\|u^k\right\| w^k\right\|^2 + \tau^k \mathcal{J}(u) \\ &= \operatorname*{arg\,min}_{u\in\mathcal{H}} \frac{1}{2}\left\|\frac{u}{\|u^k\|} - w^k\right\|^2 + \tau^k \left\|u^k\right\|^{-2} \mathcal{J}(u) \\ &= \operatorname*{arg\,min}_{u\in\mathcal{H}} \frac{1}{2}\left\|\frac{u}{\|u^k\|} - w^k\right\|^2 + \tau^k \left\|u^k\right\|^{\alpha-2} \mathcal{J}\left(\frac{u}{\|u^k\|}\right) \\ &= \left\|u^k\right\| \operatorname{prox}_{\sigma^k}(w^k). \end{aligned}$$

This readily implies

$$w^{k+1} = \frac{u^{k+1}}{\|u^{k+1}\|} = \frac{\operatorname{prox}_{\sigma^k \mathcal{J}}(w^k)}{\|\operatorname{prox}_{\sigma^k \mathcal{J}}(w^k)\|}.$$

□

5.2 Analysis of nonlinear power methods

As we have seen above, basically all flows for the computation of nonlinear eigenfunctions are equivalent to the normalized gradient flow (4.6). Since the latter gives rise to a nonlinear power method, as shown in Proposition 5.1, we thoroughly analyze such power methods in larger generality in the following. Here, we generalize the proof strategy, which was developed by Bungert et al. (2021b) for proximal power methods of absolutely one-homogeneous functionals in Hilbert spaces. In particular, we utilize a Banach space framework and investigate the power method of the p-proximal operator of an absolutely α-homogeneous functional.

To avoid heavy notation we pose Assumption 2 for the rest of this section, stating that $\mathcal{J}$ is absolutely α-homogeneous with trivial nullspace. This is a common assumption for eigenvalue problems of homogeneous functionals, see, e.g., Bungert and Burger (2020); Hynd and Lindgren (2017), and can always be achieved by replacing the space $\mathcal{X}$ with the quotient space $\mathcal{X}/\mathcal{N}(\mathcal{J})$.

Example 5.1 (The p-Dirichlet energies). Let us study the case that $\mathcal{J}(u) = \int_\Omega |\nabla u|^p \mathrm{d}x$ equals the p-Dirichlet energy for $p > 1$ or $\mathcal{J}(u) = \mathrm{TV}(u)$ equals the total variation for $p = 1$. By extending to ∞ these functionals can be defined for $u \in L^p(\Omega)$, however, the appropriate space to make sure $\mathcal{N}(\mathcal{J}) = \{0\}$ is given by $\mathcal{X} := L^p(\Omega)/\{u \equiv const.\}$, equipped with the norm $\|u\|_{\mathcal{X}} := \inf_{c\in\mathbb{R}} \|u - c\|_{L^p(\Omega)}$. Here, Poincaré's inequality makes sure that

$$\lambda_p := \inf_{u\in L^p(\Omega)} \frac{\mathcal{J}(u)}{\|u\|_{\mathcal{X}}^p} = \inf_{u\in L^p(\Omega)} \frac{\mathcal{J}(u)}{\inf_{c\in\mathbb{R}} \|u - c\|_{L^p(\Omega)}^p} > 0.$$

We now analyze the nonlinear power method

$$\begin{cases} u^{k+1/2} \in \operatorname{prox}^p_{\sigma^k \mathcal{J}}(u^k), \\ u^{k+1} \quad = \frac{u^{k+1/2}}{\|u^{k+1/2}\|}, \end{cases} \tag{5.4}$$

where $\sigma^k > 0$ are regularization parameters that can depend on u^k, and u^0 is chosen such that $\|u^0\| = 1$. Note that the iteration (5.4) chooses an arbitrary point in the (possibly multivalued) p-proximal operator which is reminiscent of subgradient methods. If the Banach space $\mathcal{X}$ is strictly convex and $p > 1$, this choice is unique (Schuster et al., 2012).

For proving convergence, we use similar arguments as Bungert et al. (2021b) where proximal power iterations on Hilbert spaces were investigated. To this end we first introduce suitable parameter choice rules for the parameter σ^k in (5.4), which make sure that the power method is well-defined and we can control their convergence.

Definition 5.1 (Parameter rules). Let $0 < c < 1$ be a constant.

- The *constant* parameter rule is given by $\sigma^k = \frac{c}{\mathcal{J}(u^0)}$ for all $k \in \mathbb{N}$.
- The *variable* parameter rule is given by $\sigma^k = \frac{c}{\mathcal{J}(u^k)}$ for all $k \in \mathbb{N}$.

Our first statement collects two important properties of the proximal power method (5.4), namely that its iterates are normalized and their energy decreases. The proof is based on Proposition 3.2 and works as the one by Bungert et al. (2021b).

Proposition 5.2. *Let Assumption 2 hold and σ^k be given by a rule in Definition 5.1. Then for all $k \in \mathbb{N}_0$ the iterative scheme* (5.4) *is well-defined and satisfies*

1. $\|u^k\| = 1$,
2. $\mathcal{J}(u^{k+1}) \leq \mathcal{J}(u^k)$.

Our first statement is concerns the angle between $u^{k+1/2}$ and u^k, which necessarily should converge to zero as the proximal power method (5.4) converges. The proof can be found in the appendix.

Proposition 5.3 (Angular convergence). *The iterates of the proximal power method* (5.4) *satisfy*

$$\lim_{k\to\infty} \left\| u^{k+1/2} - u^k \right\|^p - \left| \left\| u^{k+1/2} \right\| - 1 \right|^p = 0. \tag{5.5}$$

If $\|\cdot\|$ is a Hilbert norm and $p = 2$ this can be simplified to

$$\lim_{k\to\infty} \frac{\langle u^{k+1/2}, u^k \rangle}{\left\| u^{k+1/2} \right\| \left\| u^k \right\|} = 1, \tag{5.6}$$

which shows that the cosine of the angle converges to 1 and hence the angle to zero.

Now we can show subsequential convergence of the proximal power method (5.4) to an eigenvector of the p-proximal operator. The proof which is very similar to Bungert et al. (2021b) can be found in the appendix.

Theorem 5.1 (Convergence of the proximal power method). *Let Assumption 1 and Assumption 2 be fulfilled and assume that $\mathcal{J}$ satisfies the coercivity condition*

$$\inf_{u \in \mathcal{X}} \frac{\mathcal{J}(u)}{\|u\|^\alpha} > 0. \tag{5.7}$$

Let the sequence $(u^k)_{k\in\mathbb{N}}$ *be generated by* (5.4) *with a step size rule from Definition 5.1. Then it holds*

- *the step sizes* $(\sigma^k)_{k\in\mathbb{N}}$ *are non-decreasing and converge to* $\sigma^* \in (0,\infty)$,
- *the sequence* $(u^k)_{k\in\mathbb{N}}$ *admits a subsequence converging to some* $u^* \in \mathcal{X}\setminus\{0\}$,
- *there exists* $\mu \in (0,1]$ *such that*

$$\mu u^* \in \operatorname{prox}^p_{\sigma^*\mathcal{J}}(u^*). \tag{5.8}$$

Furthermore, if $p > 1$ *it holds* $\mu < 1$.

Remark 5.1. For $p = 1$ it can happen that $u^* \in \operatorname{prox}^p_{\sigma^*\mathcal{J}}(u^*)$ although $u^* \neq 0$ is not a minimizer of $\mathcal{J}$, which is known as exact penalization (Bungert and Burger, 2019; Bungert et al., 2020a). It occurs for positive $\sigma^* > 0$ smaller than the value $\sigma_*(u^*)$ (cf. Proposition A.1). The underlying reason that in this case the proximal power iteration may converge to eigenvectors with eigenvalue one is that for $p = 1$ the upper bound of the exact reconstruction time (A.2) coincides with the lower bound for the extinction time (B.2) and both become sharp for eigenvectors.

For $p > 1$ eigenvectors of the p-proximal are in one-to-one correspondence to eigenvectors in the sense of (1.2). Hence, in this case the limit of the proximal power method is a solution to this nonlinear eigenvalue problem.

Theorem 5.2 (Convergence to subdifferential eigenvector). *Assume that* $p > 1$. *Then under the conditions of Theorem 5.1 there exists* $\lambda > 0$ *and* $u^* \in \mathcal{X}$ *such that, up to a subsequence, the sequence* $(u^k)_{k\in\mathbb{N}}$ *generated by* (5.4) *converges to* u^* *which satisfies*

$$-\lambda\Phi^p_{\mathcal{X}}(u^*) + \partial\mathcal{J}(u^*) \ni 0.$$

We conclude this section by proving that under generic conditions the proximal power method (5.4) preserves positivity. In many scenarios ground states in the sense of Definition 1.2 can be characterized as unique non-negative eigenvectors (Boyd, 1974; Bungert and Burger, 2020; Bungert et al., 2020b; Roith and Bungert, 2022), in which case Theorem 5.2 implies convergence to the unique ground state.

Proposition 5.4. *Assume that* $\mathcal{X}$ *is a strictly convex Banach lattice and* $\mathcal{J}(|u|) \leq \mathcal{J}(u)$ *for all* $u \in \mathcal{X}$. *If the proximal power method* (5.4) *is initialized with* $u^0 \geq 0$, *then all iterates of* (5.4) *are non-negative and hence also the eigenvector* u^* *from Theorem 5.1 is non-negative.*

Proof. The proof works inductively. Assuming that $u^k \geq 0$ we would like to show that $u^{k+1/2} \geq 0$ which implies the same for u^{k+1}. To this end, we define a competitor for $u^{k+1/2}$ by setting $\tilde{u} := (u^{k+1/2} + |u^{k+1/2}|)/2 \geq 0$. Using the assumptions, we can calculate

$$\frac{1}{p}\left\|\tilde{u} - u^k\right\|^p + \sigma^k\mathcal{J}(\tilde{u})$$

$$
\begin{aligned}
&= \frac{1}{p}\left\| \frac{u^{k+1/2}+|u^{k+1/2}|}{2} - \frac{u^k+|u^k|}{2} \right\|^p + \sigma^k \mathcal{J}\left(\frac{u^{k+1/2}+|u^{k+1/2}|}{2}\right) \\
&\le \frac{1}{2p}\left\| u^{k+1/2} - u^k \right\|^p + \frac{\sigma^k}{2}\mathcal{J}(u^{k+1/2}) + \frac{1}{2p}\left\| |u^{k+1/2}| - |u^k| \right\|^p \\
&\quad + \frac{\sigma^k}{2}\mathcal{J}(|u^{k+1/2}|) \\
&\le \frac{1}{p}\left\| u^{k+1/2} - u^k \right\|^p + \sigma^k \mathcal{J}(u^{k+1/2}) \le \frac{1}{p}\left\| u - u^k \right\|^p + \sigma^k \mathcal{J}(u), \quad \forall u \in \mathcal{X}.
\end{aligned}
$$

Hence, $\tilde{u} \in \operatorname{prox}^p_{\sigma^k \mathcal{J}}(u^k)$ and using the strict convexity of $\mathcal{X}$ we infer $u^{k+1/2} = \tilde{u} \ge 0$. □

6 Γ-convergence implies convergence of ground states

Having studied the both flows and nonlinear power methods which converge to solutions of nonlinear eigenvalue problems, we now study how eigenfunctions behave under approximation. More precisely, we consider the convergence of ground states of Γ-converging functionals, as defined in Definition 1.2.

An important field of application for our results is continuum limits, which study Γ-convergence of functionals defined on grids or weighted graphs towards their continuum versions as the grids / graphs become denser. Because of their applicability in graph clustering and other data science tasks, energies depending on the gradient of a function have been investigated a lot in the last years. E.g., these studies proved continuum limits for the graph and grid total variation (Chambolle and Pock, 2021; García Trillos and Slepčev, 2015), the p-Dirichlet energies (Slepčev and Thorpe, 2019), and a functional related to the Lipschitz constant (Roith and Bungert, 2022).

While most of these works applied their results to minimization problems featuring the respective functionals, Roith and Bungert (2022) observed that Γ-convergence of the investigated absolutely one-homogeneous functional implies also convergence of its ground states. Apart from this paper, there is not much literature on the stability of ground states or more general eigenvector problems under Gamma convergence; the only other reference we found is Alvarez-Caudevilla et al. (2010), which studies linear eigenvalue problems of a specific form and Brasco et al. (2016) which deal with the convergence of fractional to local p-Laplacian eigenfunctions.

In this section we generalize the result from Roith and Bungert (2022) to general convex functionals. We start with recapping the definition of Γ-convergence (Braides et al., 2002) and then prove the convergence result.

Definition 6.1 (Γ-convergence). A sequence of functionals $(\mathcal{J}_k)_{k\in\mathbb{N}}$ on a metric space $\mathcal{X}$ is said to Γ-converge to a functional $\mathcal{J}$ on $\mathcal{X}$ (written as $\mathcal{J}_k \xrightarrow{\Gamma} \mathcal{J}$) if the following two conditions hold.

- **liminf inequality:** For all sequences $(u_k)_{k\in\mathbb{N}} \subset \mathcal{X}$ which converge to $u \in \mathcal{X}$ it holds

$$\mathcal{J}(u) \leq \liminf_{k\to\infty} \mathcal{J}_k(u_k).$$

- **limsup inequality:** For all $u \in \mathcal{X}$ there exists a sequence $(u_k)_{k\in\mathbb{N}} \subset \mathcal{X}$ (called recovery sequence) which converges to $u \in \mathcal{X}$ and satisfies

$$\mathcal{J}(u) \geq \limsup_{k\to\infty} \mathcal{J}_k(u_k).$$

We can now prove the main theorem of this section, stating that a convergent sequence of ground states converges to a ground state of the limiting functional. For this we need to assume that the minimizers of the limiting functional can be approximated by the ones of the approximating functionals in the following distance

$$\mathrm{d}(A, B) = \sup_{x\in A} \inf_{y\in B} \|x - y\|, \quad A, B \subset \mathcal{X},$$

which constitutes a part of the Hausdorff distance. Note that this is quite a weak condition and in typical applications, as the ones described above, the set of minimizers of the approximating and limiting functionals even coincide.

Theorem 6.1. *Assume that*

$$\mathcal{J}_k \xrightarrow{\Gamma} \mathcal{J}, \quad k \to \infty, \tag{6.1a}$$

$$\mathrm{d}(\arg\min \mathcal{J}, \arg\min \mathcal{J}_k) \to 0, \quad k \to \infty. \tag{6.1b}$$

Let $(u_k, \hat{u}_k) \subset \mathcal{X} \times \arg\min \mathcal{J}_k$ *be a sequence of* p*-ground states of* $\mathcal{J}_k$ *and assume that* $u_k \to u_*$ *and* $\hat{u}_k \to \hat{u}_*$. *Then* $(u_*, \hat{u}_*) \subset \mathcal{X} \times \arg\min \mathcal{J}$ *is a* p*-ground state of* $\mathcal{J}$ *and it holds*

$$\frac{\mathcal{J}(u_*) - \mathcal{J}(\hat{u}_*)}{\|u_* - \hat{u}_*\|^p} = \lim_{k\to\infty} \frac{\mathcal{J}_k(u_k) - \mathcal{J}_k(\hat{u}_k)}{\|u_k - \hat{u}_k\|^p}. \tag{6.2}$$

Remark 6.1. A few remarks regarding assumption (6.1b) are in order.

- It is satisfied for all the examples above, where $\arg\min \mathcal{J}$ coincides with the set of constant functions and $\mathcal{J}_k$ is a discrete p-Dirichlet energy on a graph (Roith and Bungert, 2022; Slepčev and Thorpe, 2019). Its set of minimizers consists of functions which are constant on each connected component of the graph. Hence, one even has the inclusion $\arg\min \mathcal{J} \subset \arg\min \mathcal{J}_k$ which implies $\mathrm{d}(\arg\min \mathcal{J}, \arg\min \mathcal{J}_k) = 0$.
- For general Γ-converging functionals (6.1b) is wrong, e.g., for $f_k(x) = |x|^k$ with $\arg\min \mathcal{J}_k = \{0\}$ and $f(x) = \chi_{[-1,1]}(x)$ with $\arg\min \mathcal{J} = [-1, 1]$.

- If $\mathcal{J}$ has a unique minimizer, meaning $\arg\min \mathcal{J}$ is a singleton, and $\mathcal{J}_k$ is a sequence of equi-coercive functionals (see Braides et al. (2002) for definitions) condition (6.1b) is also satisfied. In this case any sequence of minimizers (u_k) of $\mathcal{J}_k$ converges to a minimizer of $\mathcal{J}$. If this minimizer is uniquely determined it can therefore be approximated with elements $u_k \in \arg\min \mathcal{J}_k$.

Proof. By (6.1a) it holds $\hat{u}_* \in \arg\min \mathcal{J}$. Also, $(\hat{u}_k)_{k\in\mathbb{N}}$ is a recovery sequence of $\hat{u}_*$. To see this, take another recovery sequence $(\tilde{v}_k)_{k\in\mathbb{N}}$ of $\hat{u}_*$. Since $\hat{u}_k \in \arg\min \mathcal{J}_k$, it holds

$$\limsup_{k\to\infty} \mathcal{J}_k(\hat{u}_k) \le \limsup_{k\to\infty} \mathcal{J}_k(\tilde{v}_k) \le \mathcal{J}(\hat{u}_*),$$

which means that $(\hat{u}_k)_{k\in\mathbb{N}}$ is a recovery sequence of $\hat{u}_*$. Using the liminf inequality we get

$$\lim_{k\to\infty} \mathcal{J}_k(\hat{u}_k) = \mathcal{J}(\hat{u}_*).$$

Let now $w \in \mathcal{X}$ be arbitrary and $(w_k)_{k\in\mathbb{N}}$ be a recovery sequence for w. Using the assumptions we can compute

$$\begin{aligned}\frac{\mathcal{J}(u_*) - \mathcal{J}(\hat{u}_*)}{\left\|u_* - \hat{u}_*\right\|^p} &\overset{(6.1a)}{\le} \liminf_{k\to\infty} \frac{\mathcal{J}_k(u_k) - \mathcal{J}_k(\hat{u}_k)}{\left\|u_k - \hat{u}_k\right\|^p} \\ &\overset{(1.3)}{\le} \liminf_{k\to\infty} \frac{\mathcal{J}_k(w_k) - \mathcal{J}_k(\hat{u}_k)}{\left\|w_k - \hat{u}_k\right\|^p} \overset{(6.1a)}{\le} \frac{\mathcal{J}(w) - \mathcal{J}(\hat{u}_*)}{\left\|w - \hat{u}_*\right\|^p},\end{aligned}$$

hence, u_* solves the minimization problem in (1.3). Choosing $w = u_*$ shows (6.2).

Similarly, let $\hat{w} \in \arg\min \mathcal{J}$ be arbitrary and let $(w_k)_{k\in\mathbb{N}}$ be a recovery sequence for $\hat{w}$. By assumption (6.1b) there exists a sequence $(\hat{w}_k)_{k\in\mathbb{N}} \subset \mathcal{X}$ such that $\hat{w}_k \in \arg\min \mathcal{J}_k$ for all $k \in \mathbb{N}$ and $\left\|\hat{w} - \hat{w}_k\right\| \to 0$ as $k \to \infty$. We claim that $(\hat{w}_k)_{k\in\mathbb{N}}$ is also a recovery sequence for $\hat{w}$. To see this we first we observe that—since $\hat{w}_k \in \arg\min \mathcal{J}_k$—it holds

$$\limsup_{k\to\infty} \mathcal{J}_k(\hat{w}_k) \le \limsup_{k\to\infty} \mathcal{J}_k(w_k) \le \mathcal{J}(\hat{w}).$$

Hence, $(\hat{w}_k)_{k\in\mathbb{N}}$ is also a recovery sequence for $\hat{w}$ and we can compute

$$\begin{aligned}\frac{\mathcal{J}(u_*) - \mathcal{J}(\hat{u}_*)}{\left\|u_* - \hat{u}_*\right\|^p} &\overset{(6.2)}{=} \lim_{k\to\infty} \frac{\mathcal{J}_k(u_k) - \mathcal{J}_k(\hat{u}_k)}{\left\|u_k - \hat{u}_k\right\|^p} \\ &\overset{(1.3)}{\ge} \limsup_{k\to\infty} \frac{\mathcal{J}_k(u_k) - \mathcal{J}_k(\hat{w}_k)}{\left\|u_k - \hat{w}_k\right\|^p} \overset{(6.1a)}{\ge} \frac{\mathcal{J}(u_*) - \mathcal{J}(\hat{w})}{\left\|u_* - \hat{w}\right\|^p},\end{aligned}$$

hence $\hat{u}_*$ solves the maximization problem in (1.3) and $(u_*, \hat{u}_*)$ is a p-ground state. □

Remark 6.2 (Compactness and Normalization). To be able to apply Theorem 6.1 one needs some additional compactness and suitable normalization which ensure that ground states of $\mathcal{J}_k$ converge. For instance, if $\mathcal{J}_k$ and $\mathcal{J}$ are absolutely p-homogeneous, arbitrary multiples of p-ground states are again p-ground states, which means that a priori a sequence of ground states does not need to converge. However, in this case one can restrict oneself to normalized ground states solving

$$u_* \in \arg\min \{\mathcal{J}(u) : u \in \mathcal{X}, \|u\|_\sim = 1\},$$

where $\|\cdot\|_\sim$ is a quotient norm, defined as $\|u\|_\sim := \inf_{\hat{u} \in \mathcal{N}(\mathcal{J})} \|u - \hat{u}\|$. Using the limsup-inequality one can then show as Roith and Bungert (2022) that $\limsup_{k\to\infty} \mathcal{J}_k(u_k) \le \mathcal{J}(u_*)$, where u_k and u_* denote p-ground states of $\mathcal{J}_k$ and $\mathcal{J}$, respectively. Hence, in this case a compactness assumption of the kind

$$\limsup_{k\in\mathbb{N}} \{\|u_k\| + \mathcal{J}_k(u_k)\} < \infty \implies (u_k)_{k\in\mathbb{N}} \text{ is relatively compact}$$

implies the convergence of normalized p-ground states using Theorem 6.1.

Besides the applications on grids and weighted graphs, where the conditions of Theorem 6.1 have been verified in the papers referenced at the beginning of this section, one can also apply the theorem for Galerkin approximations using Finite Elements.

Example 6.1 (Galerkin discretizations). Let $\mathcal{J} : \mathcal{X} \to (-\infty, \infty]$ be a continuous functional. Assume that $(\mathcal{X}_k)_{k\in\mathbb{N}} \subset \mathcal{X}$ is a sequence of embedded approximation spaces such that for all $u \in \mathcal{X}$ there exists a sequence $(u_k)_{k\in\mathbb{N}} \subset \mathcal{X}$ such that $u_k \in \mathcal{X}_k$ for all $k \in \mathbb{N}$ and $u_k \to u$ as $k \to \infty$, in other words it holds $\mathrm{d}(\mathcal{X}, \mathcal{X}_k) \to 0$ as $k \to \infty$. One can define the functionals

$$\mathcal{J}_k : \mathcal{X} \to (-\infty, \infty], \qquad u \mapsto \begin{cases} \mathcal{J}(u) & \text{if } u \in \mathcal{X}_k, \\ \infty & \text{else,} \end{cases} \tag{6.3}$$

which clearly Γ-converge to $\mathcal{J}$ and hence satisfy (6.1a).

The prototypical example for this is a Finite Element Galerkin approximation of the base space $\mathcal{X} = W_0^{1,p}(\Omega)$ with approximating spaces $\mathcal{X}_k$ containing piecewise polynomials. Interesting functionals in this case are integral functional of the form

$$\mathcal{J}(u) = \int_\Omega \Phi(x, u(x), \nabla u(x))\,\mathrm{d}x, \qquad u \in W_0^{1,p}(\Omega). \tag{6.4}$$

If the spaces $\mathcal{X}_k$ contain piecewise polynomials of high degree or if Φ is strongly non-linear, the functional $\mathcal{J}(u)$ for $u \in \mathcal{X}_k$ cannot be evaluated accurately. In this case, one can also study Γ-convergence of quadrature approximations of (6.4) as done by Ortner (2004).

7 Applications

In this section we present numerical results of the proximal power method (5.4) applied to ground state problems on grids and weighted graphs. A weighted graph is a tuple $G = (\widehat{\Omega}, \omega)$, where $\widehat{\Omega}$ is a finite set of vertices and $\omega : \widehat{\Omega} \times \widehat{\Omega} \to [0, \infty)$ is a function which assigns edge weights to pairs of vertices. Weighted graphs are very handy objects for approximating variational problems since they allow for natural definitions of differential operators (Elmoataz et al., 2015) and facilitate discrete to continuum analysis using Γ-convergence or PDE techniques (see, e.g., Bozorgnia et al. (2020); Calder (2018); García Trillos and Slepčev (2015); Roith and Bungert (2022); Slepčev and Thorpe (2019)). We define the set of vertex and edge functions on a graph G as

$$\mathcal{H}(\widehat{\Omega}) = \{u : \widehat{\Omega} \to \mathbb{R}\}, \qquad \mathcal{H}(\widehat{\Omega} \times \widehat{\Omega}) = \{h : \widehat{\Omega} \times \widehat{\Omega} \to \mathbb{R}\}, \tag{7.1}$$

together with p-norms

$$\|u\|_p^p := \sum_{x \in \widehat{\Omega}} |u(x)|^p, \qquad \|u\|_\infty := \max_{x \in \widehat{\Omega}} |u(x)|, \qquad u \in \mathcal{H}(\widehat{\Omega}), \tag{7.2}$$

$$\|h\|_p^p := \sum_{x,y \in \widehat{\Omega}} |h(x,y)|^p, \quad \|h\|_\infty := \max_{x,y \in \widehat{\Omega}} |h(x,y)|, \quad h \in \mathcal{H}(\widehat{\Omega} \times \widehat{\Omega}). \tag{7.3}$$

Furthermore, one can define a gradient operator $\nabla_G : \mathcal{H}(\widehat{\Omega}) \to \mathcal{H}(\widehat{\Omega} \times \widehat{\Omega})$ and a divergence operator $\mathrm{div}_G : \mathcal{H}(\widehat{\Omega} \times \widehat{\Omega}) \to \mathcal{H}(\widehat{\Omega})$ as

$$\nabla_G u(x,y) := \sqrt{\omega(x,y)}(u(y) - u(x)), \tag{7.4}$$

$$\mathrm{div}_G h(x) := \sum_{y \in \widehat{\Omega}} \sqrt{\omega(x,y)}(h(y,x) - h(x,y)). \tag{7.5}$$

For illustrating the convergence of the proximal power method for the special case of absolutely one-homogeneous functionals $\mathcal{J}$ on a Hilbert space, we define an eigenvector affinity as

$$\mathrm{aff}_k := \frac{\|\zeta^k\|^2}{\mathcal{J}(\zeta^k)}, \qquad \zeta^k := \frac{u^k - u^{k+1/2}}{\sigma^k}. \tag{7.6}$$

Note that by the optimality conditions of the 2-proximal operator on a Hilbert space one can easily show that $\mathrm{aff}_k \in [0, 1]$ and $\mathrm{aff}_k = 1$ if and only if ζ^k is an eigenvector of the subdifferential $\partial\mathcal{J}$ (and hence also of the proximal operator). We accurately evaluate all proximal operators using a primal-dual algorithm (Chambolle and Pock, 2011).

7.1 Calibrable sets

In this first example the considered graph $G = (\widehat{\Omega}, \omega)$ is just a standard rectangular grid, where the differential operators defined above simply coincide with finite difference approximations. To be precise, $\widehat{\Omega} = \{(ih, ih) : i = 0, \dots, N\}$ where $N \in \mathbb{N}$ and $h = 1/N$. We consider the discrete total variation functional with central differences

$$\mathcal{J}_G(u) := \frac{1}{2h} \sum_{x \in \widehat{\Omega}} \sqrt{[u(x + he_1) - u(x - he_1)]^2 + [u(x + he_2) - u(x - he_2)]^2}, \tag{7.7}$$

where e_i is the i-th unit vector (see Chambolle and Pock (2021) for a plethora of other, in particular, better discretizations). We compute the p-proximal power iteration (5.4) of $\mathcal{J}_G$ for $p \in \{1, 2\}$, using the norms $\|\cdot\|_1$ and $\|\cdot\|_2$, respectively. The first five iterates of the method are depicted in Fig. 3. As predicted by Theorem 5.1, the power methods converge to eigenvectors of the proximal operators. However, only in the case $p = 2$, where we used the Hilbert norm for the proximal operator, the limit is a calibrable set (Alter et al., 2005), which is an eigenvector of the subdifferential operator in the sense of (1.2). Hence, Theorem 5.2, which states that only for $p > 1$ proximal eigenvectors are also subdifferential eigenvectors, is sharp, in general.

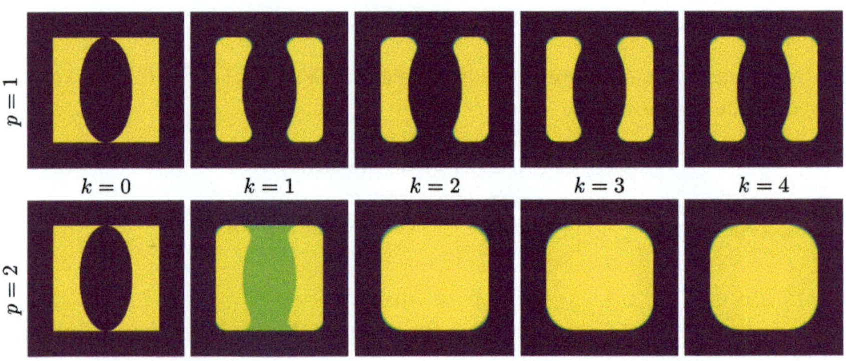

FIGURE 3 **Top row:** Five iterates for $p = 1$. Limit is no calibrable set. **Bottom row:** Five iterates for $p = 2$. Limit is a calibrable set.

7.2 Graph clustering

Here, we consider graph clustering using ground states of the graph p-Dirichlet energy

$$\mathcal{J}_{G,p}(u) := \|\nabla_G u\|_p^p = \sum_{x,y \in \widehat{\Omega}} \omega(x, y)^{\frac{p}{2}} |u(y) - u(x)|^p, \tag{7.8}$$

where the case $p = 1$ is the graph total variation. If G is connected, the nullspace of these functionals coincides with the constant graph functions. Ground states are solutions to

$$\min_{u \in \mathcal{N}(\mathcal{J}_{G,p})^{\perp}} \frac{\mathcal{J}_{G,p}(u)}{\|u\|_2}, \tag{7.9}$$

and correspond to eigenfunctions of the graph p-Laplacian operator

$$\Delta_{G,p} u(x) := \sum_{y \in \widehat{\Omega}} \omega(x,y)^{\frac{p}{2}} |u(y) - u(x)|^{p-2}(u(y) - u(x)). \tag{7.10}$$

On the left side of Fig. 4 we show a graph representing the so-called "two moons" data set together with its ground states for $p = 2$ and $p = 1$. As observed by Aujol et al. (2018); Bühler and Hein (2009); Bungert et al. (2019) the ground state for $p = 1$ yield a sharp clustering of the graph into two subgraphs whereas the standard graph Laplacian eigenfunction for $p = 2$ yields a very diffuse interface.

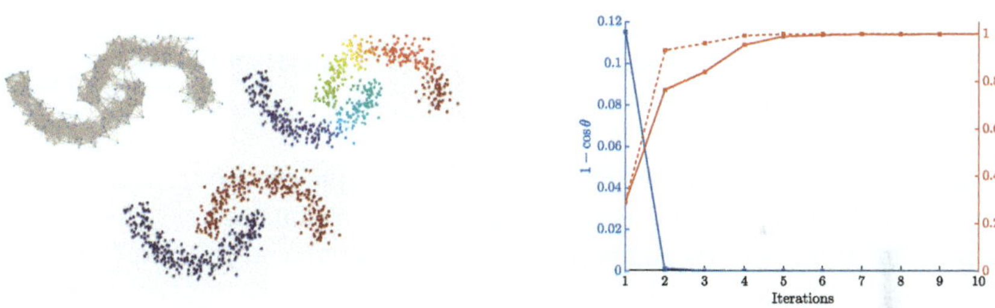

FIGURE 4 Spectral graph clustering. **Left:** Two-moon graph, Laplacian eigenfunction, 1-Laplacian eigenfunction. **Right:** Angle (blue) and the eigenvector affinity (red) plotted over the iterations. Solid lines correspond to constant parameter rule, dashed lines to variable.

On the right side of Fig. 4 we plot the convergence of the proximal power method (5.4) for constant (solid lines) and variable parameter rule (dashed lines), cf. Definition 5.1. The blue lines represent the quantity $1 - \cos\theta$ where θ is the angle between $u^{k+1/2}$ and u^k, see (5.6). Proposition 5.3 states that this quantity converges to zero which is reflected in our numerical experiments. The red lines depict the eigenvector affinity (7.6) which converges to one, as desired. Notably, here the variable parameter rule leads to a quicker convergence than the constant one. Fig. 4 makes clear that after only six iterations the proximal power method has converged to the 1-Laplacian ground state, depicted on the bottom left.

7.3 Geodesic distance functions on graphs

In this example we present an application to distance functions on graphs. To this end, we choose a subset $\widehat{\mathcal{O}} \subset \widehat{\Omega}$ of the vertex set which play the role of a constraint set. We define the functional $\mathcal{J}_{G,\infty} : \mathcal{H}(\widehat{\Omega}) \to \mathbb{R} \cup \{\infty\}$ as

$$\mathcal{J}_{G,\infty}(u) := \|\nabla_G u\|_\infty = \max_{x,y\in\widehat{\Omega}} \sqrt{\omega(x,y)}|u(y) - u(x)| \tag{7.11}$$

if $u \in \mathcal{H}(\widehat{\Omega})$ satisfies $u = 0$ on $\widehat{\mathcal{O}}$ and $\mathcal{J}_{G,\infty}(u) = \infty$ else. The value $\mathcal{J}_{G,\infty}(u)$ can be interpreted as largest local Lipschitz constant of u which satisfies the constraints on $\widehat{\mathcal{O}}$. Since the nullspace of $\mathcal{J}_{G,\infty}$ is trivial when the graph is connected, ground states solve

$$\min_{u\in\mathcal{H}(\widehat{\Omega})} \frac{\mathcal{J}_{G,\infty}(u)}{\|u\|_2}. \tag{7.12}$$

Bungert et al. (2020b) characterized solutions of this problem as multiples of the geodesic graph distance function to the set $\widehat{\mathcal{O}}$. Furthermore, Roith and Bungert (2022) proved that, if the graph vertices $\widehat{\Omega}$ and $\widehat{\mathcal{O}}$ converge to continuum domain Ω and a closed constraint set $\mathcal{O} \subset \overline{\Omega}$ in the Hausdorff distance and the weights ω are scaled appropriately, the Γ-limit of the functionals $\mathcal{J}_{G,\infty}$ is given by

$$\mathcal{J}_\infty(u) = \begin{cases} \|\nabla u\|_{L^\infty(\Omega)}, & \text{if } u \in W^{1,\infty}(\Omega),\ u = 0 \text{ on } \mathcal{O}, \\ \infty, & \text{else.} \end{cases} \tag{7.13}$$

It was proved that the corresponding ground states converge (cf. Theorem 6.1), as well. The limiting ground states of $\mathcal{J}_\infty$ where also characterized as geodesic distance functions to $\mathcal{O}$ by Bungert et al. (2020b); Roith and Bungert (2022).

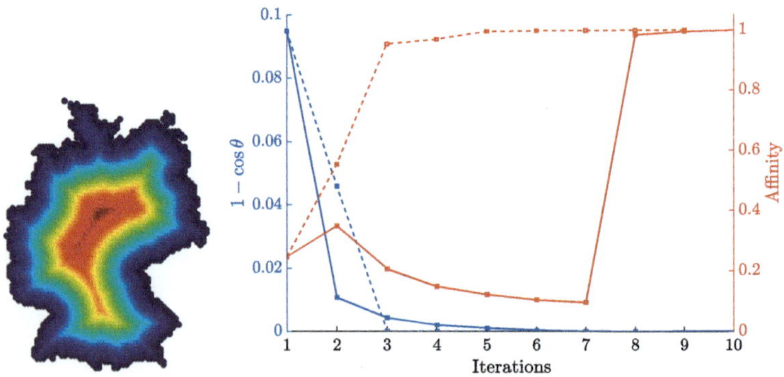

FIGURE 5 **Left:** Distance function of the German border. **Right:** Angle (blue) and the eigenvector affinity (red) plotted over the iterations. Solid lines correspond to constant parameter rule, dashed lines to variable.

Figs. 5 and 6 show ground states of $\mathcal{J}_{G,\infty}$ on two types of weighted graphs. The first one in Fig. 5 is a grid graph representing a map of Germany. The constraint set $\widehat{\mathcal{O}}$ is chosen as the boundary and hence the computed ground state coincides with the distance function to the border of Germany. Similarly, in

Fig. 6 we show the computed ground states for different resolutions of a graph which consists of random samples from a two-dimensional dumbbell-shaped manifold, embedded in $\mathbb{R}^3$. The constraint set $\widehat{\mathcal{O}}$ is chosen to be a fixed vertex in the top left of the manifold and hence the computed ground states coincide with the geodesic distance to this point.

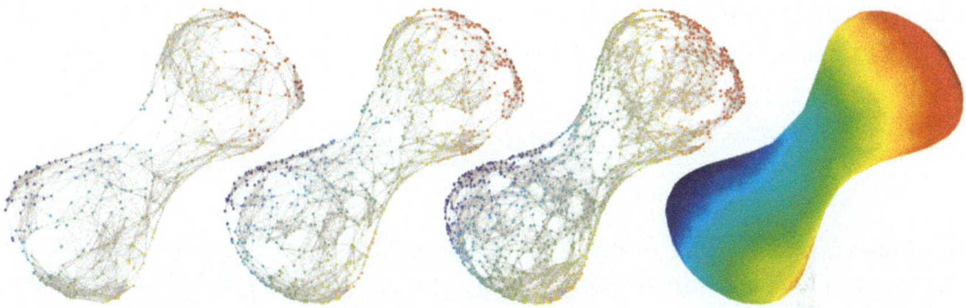

FIGURE 6 Geodesic distance functions to a point at the left of a discretized manifold with increasing resolution. Last image is a rendered version of the penultimate one.

On the right side of Fig. 5 we again plotted the quantitative convergence behavior of the proximal power method (5.4) in terms of convergence of the angle (5.6) and the eigenvector affinity (7.6). Here we make similar observations as in Fig. 4, in particular the variable parameter rule leads to a quicker convergence than the constant one and both the angle and the eigenvector affinity converge after approximately six iterations of the method. Interestingly, for the constant parameter rule the eigenvector affinity first decreases before it finally reaches a value close to one, whereas the angle decreases monotonously. This is no numerical artifact but is due to the fact that the affinity (7.6) is a much stronger convergence criterion than the angle since the former depends on the functional at hand (here $\mathcal{J}_{G,\infty}$) whereas the latter does not.

Appendices

The statements regarding exact reconstruction and finite extinction proved below are generalizations of Bungert and Burger (2019).

Appendix A Exact reconstruction time

Proposition A.1. *It holds that*

$$f \in \operatorname*{arg\,min}_{u \in \mathcal{X}} \|u - f\| + \sigma \mathcal{J}(u)$$

if and only if $\sigma \le \sigma_*(f)$, *where*

$$\sigma_*(f) := \frac{1}{\inf\{\|\zeta\|_* : \zeta \in \partial\mathcal{J}(f)\}}. \tag{A.1}$$

Furthermore, it holds

$$\sigma_*(f) \le \inf_{\hat{u}\in\arg\min\mathcal{J}} \frac{\|f-\hat{u}\|}{\mathcal{J}(f)-\mathcal{J}(\hat{u})}. \tag{A.2}$$

Proof. For the first implication we assume that f solves the minimization problem. Then the optimality condition (2.5) implies that there is $\eta \in \Phi^1_{\mathcal{X}}(0)$ and $\zeta \in \partial\mathcal{J}(f)$ such that $0 = \eta + \sigma\zeta$. Using that $\Phi^1_{\mathcal{X}}(0) = \{\eta \in \mathcal{X}^* : \|\eta\|_* \le 1\}$, we obtain

$$\sigma = \frac{\|\eta\|_*}{\|\zeta\|_*} \le \frac{1}{\inf\{\|\zeta\|_* : \zeta \in \partial\mathcal{J}(f)\}} = \sigma_*(f).$$

Conversely, assume that $\sigma \le \sigma_*(f)$. Because of weak-* closedness of $\partial\mathcal{J}(f)$ and weak-* lower semicontinuity of $\|\cdot\|_*$, we can choose $\zeta \in \partial\mathcal{J}(f)$ such that $\sigma_*(f) = 1/\|\zeta\|_*$. We define $\eta := -\sigma\zeta$ and compute

$$\langle \eta, v\rangle = -\sigma\langle\zeta, v\rangle \le \sigma_* \|\zeta\|_* \|v\| = \|v\|, \qquad \forall v \in \mathcal{X}.$$

This proves that $\eta \in \Phi^1_{\mathcal{X}}(0)$ and hence the optimality condition (2.5) is satisfied.

To show the upper bound (A.2), we use convexity of $\mathcal{J}$ to derive for all $\zeta \in \partial\mathcal{J}(f)$

$$\begin{aligned}
&\mathcal{J}(f) + \langle\zeta, \hat{u} - f\rangle \le \mathcal{J}(\hat{u}) \\
\Longrightarrow\ &\mathcal{J}(f) - \mathcal{J}(\hat{u}) \le \|\zeta\|_* \|f - \hat{u}\| \\
\Longleftrightarrow\ &\|\zeta\|_* \ge \frac{\mathcal{J}(f)-\mathcal{J}(\hat{u})}{\|f-\hat{u}\|}, \qquad \forall v \in \arg\min\mathcal{J}, \\
\Longleftrightarrow\ &\|\zeta\|_* \ge \sup_{\hat{u}\in\arg\min\mathcal{J}} \frac{\mathcal{J}(f)-\mathcal{J}(\hat{u})}{\|f-\hat{u}\|}.
\end{aligned}$$

Taking the infimum in $\zeta \in \partial\mathcal{J}(f)$ this implies the desired estimate. □

Appendix B Extinction time

Proposition B.1. *Assume that* $\arg\min\mathcal{J} \ne \emptyset$ *and that there exists* $\lambda_1 > 0$ *such that* $\lambda_1 \|u - \hat{u}\| \le \mathcal{J}(u) - \mathcal{J}(\hat{u})$ *for all* $\hat{u} \in \arg\min\mathcal{J}$ *and* $u \in \mathcal{X}$ *Then it holds that*

$$(\arg\min\mathcal{J}) \cap \left(\arg\min_{u\in\mathcal{X}} \frac{1}{p}\|u-f\|^p + \sigma\mathcal{J}(u)\right) \ne \emptyset$$

if and only if $\sigma \ge \sigma_{**}(f)$*, where*

$$\sigma_{**}(f) := \inf_{\hat{u}\in\arg\min\mathcal{J}} \inf_{\eta\in\Phi^p_{\mathcal{X}}(\hat{u}-f)} \sup_{v\in\mathcal{X}} \frac{\langle -\eta, v-\hat{u}\rangle}{\mathcal{J}(v)-\mathcal{J}(\hat{u})}. \tag{B.1}$$

Furthermore, one has the estimates

$$\sigma_{**}(f) \geq \inf_{\hat{u} \in \arg\min \mathcal{J}} \frac{\|f - \hat{u}\|^p}{\mathcal{J}(f) - \mathcal{J}(\hat{u})} \tag{B.2}$$

$$\sigma_{**}(f) \leq \frac{1}{\lambda_1} \inf_{\hat{u} \in \arg\min \mathcal{J}} \|f - \hat{u}\|^{p-1} < \infty. \tag{B.3}$$

Proof. If $\hat{u} \in \arg\min \mathcal{J}$ solves (2.2), from (2.5) we get $\eta \in \Phi^p_{\mathcal{X}}(\hat{u} - f)$ and $\zeta \in \partial \mathcal{J}(\hat{u})$ such that $0 = \eta + \sigma \zeta$. This implies

$$\sigma_{**}(f) \leq \sup_{v \in \mathcal{X}} \frac{\langle -\eta, v - \hat{u} \rangle}{\mathcal{J}(v) - \mathcal{J}(\hat{u})} = \sigma \sup_{v \in \mathcal{X}} \frac{\langle \zeta, v - \hat{u} \rangle}{\mathcal{J}(v) - \mathcal{J}(\hat{u})} \leq \sigma.$$

Conversely, let $\sigma \geq \sigma_{**}(f)$. Using the assumed coercivity together with closedness of $\arg\min \mathcal{J}$ and the duality map, one can show that the infima in (B.1) are attained and we denote the minimizers by $\hat{u} \in \arg\min \mathcal{J}$ and $\eta \in \Phi^p_{\mathcal{X}}(\hat{u} - f)$. Defining $\zeta := -\eta/\sigma$, it holds

$$\begin{aligned}
\mathcal{J}(\hat{u}) + \langle \zeta, u - \hat{u} \rangle &= \mathcal{J}(\hat{u}) + \frac{1}{\sigma} \langle -\eta, u - \hat{u} \rangle \\
&= \mathcal{J}(\hat{u}) + \frac{\mathcal{J}(u) - \mathcal{J}(\hat{u})}{\sigma} \frac{\langle -\eta, u - \hat{u} \rangle}{\mathcal{J}(u) - \mathcal{J}(\hat{u})} \\
&\leq \mathcal{J}(\hat{u}) + \frac{\mathcal{J}(u) - \mathcal{J}(\hat{u})}{\sigma_{**}(f)} \sup_{v \in \mathcal{X}} \frac{\langle -\eta, v - \hat{u} \rangle}{\mathcal{J}(v) - \mathcal{J}(\hat{u})} \\
&\leq \mathcal{J}(u), \qquad \forall u \in \mathrm{dom}(\mathcal{J}).
\end{aligned}$$

This shows $\zeta \in \partial \mathcal{J}(\hat{u})$ which means that $\hat{u}$ satisfies the optimality condition (2.5).

Next we prove the bounds on $\sigma_{**}(f)$. By choosing $u = f$ in the supremum it holds

$$\sigma_{**}(f) \geq \inf_{\hat{u} \in \arg\min \mathcal{J}} \inf_{\eta \in \Phi^p_{\mathcal{X}}(\hat{u} - f)} \frac{\langle -\eta, f - \hat{u} \rangle}{\mathcal{J}(f) - \mathcal{J}(\hat{u})} = \inf_{\hat{u} \in \arg\min \mathcal{J}} \frac{\|f - \hat{u}\|^p}{\mathcal{J}(f) - \mathcal{J}(\hat{u})}.$$

Using the coercivity of $\mathcal{J}$ one obtains

$$\begin{aligned}
\sigma_{**}(f) &\leq \frac{1}{\lambda_1} \inf_{\hat{u} \in \arg\min \mathcal{J}} \inf_{\eta \in \Phi^p_{\mathcal{X}}(\hat{u} - f)} \sup_{v \in \mathcal{X}} \frac{\|\eta\|_* \|v - \hat{u}\|}{\|v - \hat{u}\|} \\
&= \frac{1}{\lambda_1} \inf_{\hat{u} \in \arg\min \mathcal{J}} \|f - \hat{u}\|^{p-1} < \infty.
\end{aligned}$$

□

Appendix C Remaining proofs

Proof of Proposition 2.2. Since the duality map is never empty, we can choose $\eta \in \Phi^p_{\mathcal{X}}(\hat{u}_* - u_*)$ and define $\zeta := -\lambda_p \eta$. We claim that $\zeta \in \partial\mathcal{J}(u_*)$, which would conclude the proof. To see this we compute

$$\begin{aligned}\mathcal{J}(u_*) + \langle\zeta, u - u_*\rangle &= \mathcal{J}(u_*) + \lambda_p\langle\eta, u_* - u\rangle \\ &\leq \mathcal{J}(u_*) + \lambda_p\left[\frac{1}{p}\left\|\hat{u}_* - u\right\|^p - \frac{1}{p}\left\|\hat{u}_* - u_*\right\|^p\right] \\ &\leq \mathcal{J}(u_*) + (\mathcal{J}(u) - \mathcal{J}(\hat{u}_*)) - (\mathcal{J}(u_*) - \mathcal{J}(\hat{u}_*)) \\ &= \mathcal{J}(u), \qquad \forall u \in \mathcal{X},\end{aligned}$$

where we used Proposition 2.1 and (1.3). □

Proof of Proposition 5.3. By the definition of $u^{k+1/2}$ it holds

$$\frac{1}{p}\left\|u^{k+1/2} - u^k\right\|^p + \sigma^k\mathcal{J}(u^{k+1/2}) \leq \frac{1}{p}\left\|u - u^k\right\|^p + \sigma^k\mathcal{J}(u), \quad \forall u \in \mathcal{X}.$$

Choosing $u = \left\|u^{k+1/2}\right\| u^k$ and reordering the resulting inequality yields

$$\begin{aligned}&\frac{1}{p}\frac{1}{\sigma^k}\left(\left\|u^{k+1/2} - u^k\right\|^p - \left|\left\|u^{k+1/2}\right\| - 1\right|^p\right) \\ &\leq \mathcal{J}\left(\left\|u^{k+1/2}\right\| u^k\right) - \mathcal{J}(u^{k+1/2}) \\ &\leq \left\|u^{k+1/2}\right\|^\alpha\left(\mathcal{J}(u^k) - \mathcal{J}(u^{k+1})\right) \\ &\leq C\left(\mathcal{J}(u^k) - \mathcal{J}(u^{k+1})\right),\end{aligned}$$

where we used that $u^{k+1/2}$ converges and is therefore bounded. Summing both sides and using that $k \mapsto \mathcal{J}(u^k)$ is non-increasing, one obtains that the left hand side is summable. Since by the triangle inequality it is non-negative, it therefore converges to zero. The reformulation for Hilbert norms and $p = 2$ is straightforward by expanding the square. □

Before we can prove the convergence statement of the proximal power method, we need a continuity property of the p-proximal operator, which relies on the compactness Assumption 1. The proof works precisely as the one by Bungert et al. (2021b).

Lemma C.1. *Let Assumption 1 be fulfilled. Let $v_n, v \in \mathcal{X}$ such that $\|v_n - v\| \to 0$ as $n \to \infty$ and let $u_n \in \operatorname{prox}^p_{\sigma(v_n)\mathcal{J}}(v_n)$ for $n \in \mathbb{N}$, where the regularization parameters are chosen in such a way that $\sigma(v_n) \to \sigma$ as $n \to \infty$. Then there exists $u \in \mathcal{X}$ such that (up to a subsequence) it holds $\|u_n - u\| \to 0$ and $u \in \operatorname{prox}^p_{\sigma\mathcal{J}}(v)$.*

Proof of Theorem 5.1. First we note that the step sizes σ^k converge. For the constant parameter rule this is trivial. For the variable rule, we observe that according to Proposition 5.2

$$\sigma^k = \frac{c}{\mathcal{J}(u^k)} \leq \frac{c}{\mathcal{J}(u^{k+1})} = \sigma^{k+1},$$

hence the step sizes are an increasing sequence. Furthermore, by (5.7) the step sizes are bounded from above and hence converge to some $\sigma^* > 0$.

From Proposition 5.2 we infer that the sequence $(u^k)_{k\in\mathbb{N}}$ is relatively compact, meaning that up to a subsequence $\|u^k - u^*\| \to 0$ for some element $u^* \in \mathcal{X}$ with $\|u^*\| = 1$. After another subsequence refinement Lemma C.1 implies that also the sequence $u^{k+1/2}$ can be assumed to converge to an element $\tilde{u}$ which satisfies $\tilde{u} \in \mathrm{prox}^p_{\sigma^*}(u^*)$.

For the constant parameter rule it holds $\sigma^* = \sigma(u^*)$ and for the variable rule because of lower semicontinuity:

$$\sigma^* = \lim_{k\to\infty} \sigma^k \leq \frac{c}{\liminf_{k\to\infty} \mathcal{J}(u^k)} \leq \frac{c}{\mathcal{J}(u^*)} = \sigma(u^*).$$

Since furthermore $\sigma(u^*) < \sigma_{**}(u^*)$ by definition (cf. (B.2)), we infer from Proposition B.1 that $\tilde{u} \neq 0$.

From the scheme (5.4) and the convergence of u^k and $u^{k+1/2}$ it now follows that

$$u^* = \frac{\tilde{u}}{\|\tilde{u}\|}, \qquad \tilde{u} \in \mathrm{prox}^p_{\sigma^*}(u^*).$$

If we define $\mu := \|\mathrm{prox}^p_{\sigma^*}(u^*)\| > 0$ this can be reordered to (5.8).

It remains to show that $\mu \leq 1$ for $p = 1$ and $\mu < 1$ for $p > 1$. We start with the case $p = 1$. In this case (5.8) is equivalent to

$$\|\mu u^* - u^*\| + \sigma^* \mathcal{J}(\mu u^*) \leq \|v - u^*\| + \sigma^* \mathcal{J}(v), \quad \forall v \in \mathcal{X}.$$

Choosing $v = u^*$ then yields

$$|\mu - 1| \, \|u^*\| + \sigma \mu \mathcal{J}(u^*) \leq \sigma^* \mathcal{J}(u^*).$$

If we now assume that $\mu > 1$ we obtain

$$(\mu - 1)\,\|u^*\| \leq 0$$

which is a contradiction to $u^* \neq 0$. Hence, we have shown that $\mu \leq 1$ if $p = 1$.

In the case $p > 1$ we note that the optimality conditions for (5.8) read

$$\begin{aligned} 0 &\in \Phi^p_{\mathcal{X}}((\mu - 1)u^*) + \sigma^* \partial \mathcal{J}(\mu u^*) \\ &= |\mu - 1|^{p-2}(\mu - 1)\Phi^p_{\mathcal{X}}(u^*) + \sigma^* \mu^{\alpha-1} \partial \mathcal{J}(u^*), \end{aligned}$$

where we used that $\Phi^p_{\mathcal{X}}$ and $\partial\mathcal{J}$ are homogeneous with degree $p-1$ and $\alpha-1$, respectively (cf. Proposition 2.1 and Bungert and Burger (2020)). Using the properties of the duality map and the subdifferential of absolutely p-homogeneous functionals this implies

$$0 = |\mu - 1|^{p-2}(\mu - 1)\left\|u^*\right\|^p + \sigma^*\mu^{\alpha-1}\mathcal{J}(u^*).$$

If we assume that $\mu \geq 1$, this equality implies $u^* = 0$ which is a contradiction. □

Proof of Theorem 5.2. The statement follows from setting $\lambda = (1-\mu)|1-\mu|^{p-2}/(\sigma^*\mu^{p-1}) \geq 0$ where μ is as in the previous proof. □

References

Alter, François, Caselles, Vicent, Chambolle, Antonin, 2005. A characterization of convex calibrable sets in $\mathbb{R}^N$. Mathematische Annalen 332 (2), 329–366.

Alvarez-Caudevilla, Pablo, Lemenant, Antoine, et al., 2010. Asymptotic analysis for some linear eigenvalue problems via Gamma-Convergence. Advances in Differential Equations 15 (7/8), 649–688.

Amann, Herbert, 1976. Fixed point equations and nonlinear eigenvalue problems in ordered Banach spaces. SIAM Review 18 (4), 620–709.

Ambrosio, Luigi, 1995. Minimizing movements. Rendiconti Della Accademia Nazionale Delle Scienze Detta Accademia Dei XL. Parte I. Memorie Di Matematica 19, 191–246.

Ambrosio, Luigi, Gigli, Nicola, Savaré, Giuseppe, 2008. Gradient Flows: in Metric Spaces and in the Space of Probability Measures. Springer Science & Business Media.

Andreu, Fuensanta, Caselles, Vicent, Díaz, Jesus Ildefonso, Mazón, José M., 2002. Some qualitative properties for the total variation flow. Journal of Functional Analysis 188 (2), 516–547.

Andreu-Vaillo, Fuensanta, Caselles, Vicent, Mazón, José M., 2004. Parabolic Quasilinear Equations Minimizing Linear Growth Functionals, Vol. 223. Springer Science & Business Media.

Aubin, J-P., Cellina, Arrigo, 2012. Differential Inclusions: Set-Valued Maps and Viability Theory, Vol. 264. Springer Science & Business Media.

Aujol, Jean-François, Gilboa, Guy, Papadakis, Nicolas, 2018. Theoretical analysis of flows estimating eigenfunctions of one-homogeneous functionals. SIAM Journal on Imaging Sciences 11 (2), 1416–1440.

Benning, Martin, Burger, Martin, 2013. Ground states and singular vectors of convex variational regularization methods. Methods and Applications of Analysis 20 (4), 295–334.

Benning, Martin, Möller, Michael, Nossek, Raz Z., Burger, Martin, Cremers, Daniel, Gilboa, Guy, Schönlieb, Carola-Bibiane, 2017. Nonlinear spectral image fusion. In: International Conference on Scale Space and Variational Methods in Computer Vision. Springer, pp. 41–53.

Bonforte, Matteo, Figalli, Alessio, 2012. Total variation flow and sign fast diffusion in one dimension. Journal of Differential Equations 252 (8), 4455–4480.

Bonforte, Matteo, Figalli, Alessio, 2020. Sharp extinction rates for fast diffusion equations on generic bounded domains. Communications on Pure and Applied Mathematics.

Bonforte, Matteo, Grillo, Gabriele, Vazquez, Juan Luis, 2012. Behaviour near extinction for the Fast Diffusion Equation on bounded domains. Journal de Mathématiques Pures et Appliquées 97 (1), 1–38.

Boyd, David W., 1974. The power method for ℓ^p norms. Linear Algebra and Its Applications 9, 95–101.

Bozorgnia, Farid, 2016. Convergence of inverse power method for first eigenvalue of p-Laplace operator. Numerical Functional Analysis and Optimization 37 (11), 1378–1384.

Bozorgnia, Farid, 2020. Approximation of the second eigenvalue of the p-Laplace operator in symmetric domains. arXiv:1907.13390 [math.SP].

Bozorgnia, Farid, Bungert, Leon, Tenbrinck, Daniel, 2020. The infinity Laplacian eigenvalue problem: reformulation and a numerical scheme. arXiv:2004.08127 [math.NA].

Braides, Andrea, et al., 2002. Gamma-Convergence for Beginners, Vol. 22. Clarendon Press.

Brasco, Lorenzo, Parini, Enea, Squassina, Marco, 2016. Stability of variational eigenvalues for the fractional p-Laplacian. Discrete and Continuous Dynamical Systems. Series A 36, 1813–1845.

Brezis, Haim, 1973. Operateurs maximaux monotones et semi-groupes de contractions dans les espaces de Hilbert. Elsevier.

Bühler, Thomas, Hein, Matthias, 2009. Spectral clustering based on the graph p-Laplacian. In: Proceedings of the 26th Annual International Conference on Machine Learning. ACM, pp. 81–88.

Bungert, Leon, 2020. Nonlinear Spectral Analysis with Variational Methods. PhD thesis. Friedrich-Alexander-Universität Erlangen-Nürnberg (FAU).

Bungert, Leon, Burger, Martin, 2019. Solution paths of variational regularization methods for inverse problems. Inverse Problems 35 (10), 105012.

Bungert, Leon, Burger, Martin, 2020. Asymptotic profiles of nonlinear homogeneous evolution equations of gradient flow type. Journal of Evolution Equations 20 (3), 1061–1092.

Bungert, Leon, Burger, Martin, Chambolle, Antonin, Novaga, Matteo, 2021a. Nonlinear spectral decompositions by gradient flows of one-homogeneous functionals. Analysis & PDE 14 (3), 823–860.

Bungert, Leon, Burger, Martin, Korolev, Yury, Schönlieb, Carola-Bibiane, 2020a. Variational regularisation for inverse problems with imperfect forward operators and general noise models. Inverse Problems 36 (12), 125014.

Bungert, Leon, Burger, Martin, Tenbrinck, Daniel, 2019. Computing nonlinear eigenfunctions via gradient flow extinction. In: International Conference on Scale Space and Variational Methods in Computer Vision. Springer, pp. 291–302.

Bungert, Leon, Hait-Fraenkel, Ester, Papadakis, Nicolas, Gilboa, Guy, 2021b. Nonlinear power method for computing eigenvectors of proximal operators and neural networks. SIAM Journal on Imaging Sciences 14 (3), 1114–1148.

Bungert, Leon, Korolev, Yury, Burger, Martin, 2020b. Structural analysis of an L-infinity variational problem and relations to distance functions. Pure and Applied Analysis 2 (3), 703–738.

Burger, Martin, Eckardt, Lina, Gilboa, Guy, Moeller, Michael, 2015. Spectral representations of one-homogeneous functionals. In: International Conference on Scale Space and Variational Methods in Computer Vision. Springer, pp. 16–27.

Burger, Martin, Gilboa, Guy, Moeller, Michael, Eckardt, Lina, Cremers, Daniel, 2016. Spectral decompositions using one-homogeneous functionals. SIAM Journal on Imaging Sciences 9 (3), 1374–1408.

Calder, Jeff, 2018. The game theoretic p-Laplacian and semi-supervised learning with few labels. Nonlinearity 32 (1), 301.

Cancès, Eric, Chakir, Rachida, Maday, Yvon, 2010. Numerical analysis of nonlinear eigenvalue problems. Journal of Scientific Computing 45 (1), 90–117.

Chambolle, Antonin, Pock, Thomas, 2011. A first-order primal-dual algorithm for convex problems with applications to imaging. Journal of Mathematical Imaging and Vision 40 (1), 120–145.

Chambolle, Antonin, Pock, Thomas, 2021. Chapter 6 - approximating the total variation with finite differences or finite elements. In: Geometric Partial Differential Equations - Part II. In: Handbook of Numerical Analysis, vol. 22. Elsevier, pp. 383–417.

Cohen, Ido, Gilboa, Guy, 2018. Energy dissipating flows for solving nonlinear eigenpair problems. Journal of Computational Physics 375, 1138–1158.

Combettes, Patrick L., Reyes, Noli N., 2013. Moreau's decomposition in Banach spaces. Mathematical Programming 139 (1), 103–114.

De Giorgi, Ennio, 1993. New problems on minimizing movements. In: Ennio de Giorgi: Selected Papers, pp. 699–713.

Elmoataz, Abderrahim, Toutain, Matthieu, Tenbrinck, Daniel, 2015. On the p-Laplacian and ∞-Laplacian on graphs with applications in image and data processing. SIAM Journal on Imaging Sciences 8 (4), 2412–2451.

Feld, Tal, Aujol, Jean-François, Gilboa, Guy, Papadakis, Nicolas, 2019. Rayleigh quotient minimization for absolutely one-homogeneous functionals. Inverse Problems 35 (6), 064003.

Fumero, Marco, Möller, Michael, Rodolà, Emanuele, 2020. Nonlinear spectral geometry processing via the tv transform. ACM Transactions on Graphics 39 (6), 1–16.

García Trillos, Nicolás, Slepčev, Dejan, 2015. Continuum limit of total variation on point clouds. Archive for Rational Mechanics and Analysis (ISSN 1432-0673) 220 (1), 193–241.

Gautier, Antoine, Hein, Matthias, Tudisco, Francesco, 2020. Computing the norm of nonnegative matrices and the log-Sobolev constant of Markov chains. arXiv:2002.02447 [math.NA].

Gautier, Antoine, Tudisco, Francesco, Hein, Matthias, 2019a. A unifying Perron–Frobenius theorem for nonnegative tensors via multihomogeneous maps. SIAM Journal on Matrix Analysis and Applications 40 (3), 1206–1231.

Gautier, Antoine, Tudisco, Francesco, Hein, Matthias, 2019b. The Perron–Frobenius theorem for multihomogeneous mappings. SIAM Journal on Matrix Analysis and Applications 40 (3), 1179–1205.

Ghidaglia, J.M., Marzocchi, A., 1991. Exact decay estimates for solutions to semilinear parabolic equations. Applicable Analysis 42 (1–4), 69–81.

Gilboa, Guy, 2013. A spectral approach to total variation. In: International Conference on Scale Space and Variational Methods in Computer Vision. Springer, pp. 36–47.

Gilboa, Guy, 2014. A total variation spectral framework for scale and texture analysis. SIAM Journal on Imaging Sciences 7 (4), 1937–1961.

Gilboa, Guy, 2018. Nonlinear Eigenproblems in Image Processing and Computer Vision. Springer.

Gilboa, Guy, 2021. Iterative methods for computing eigenvectors of nonlinear operators. In: Handbook of Mathematical Models and Algorithms in Computer Vision and Imaging: Mathematical Imaging and Vision, pp. 1–28.

Gilboa, Guy, Moeller, Michael, Burger, Martin, 2016. Nonlinear spectral analysis via one-homogeneous functionals: overview and future prospects. Journal of Mathematical Imaging and Vision 56 (2), 300–319.

Hauer, Daniel, Mazón, José, 2019. Kurdyka–Łojasiewicz–Simon inequality for gradient flows in metric spaces. Transactions of the American Mathematical Society 372 (7), 4917–4976.

Hein, Matthias, Bühler, Thomas, 2010. An inverse power method for nonlinear eigenproblems with applications in 1-spectral clustering and sparse PCA. In: Lafferty, J., Williams, C., Shawe-Taylor, J., Zemel, R., Culotta, A. (Eds.), Advances in Neural Information Processing Systems, Vol. 23. Curran Associates, Inc.

Hynd, Ryan, Lindgren, Erik, 2017. Approximation of the least Rayleigh quotient for degree p homogeneous functionals. Journal of Functional Analysis 272 (12), 4873–4918.

Kamin, Shoshana, Vázquez, Juan Luis, 1988. Fundamental solutions and asymptotic behaviour for the p-Laplacian equation. Revista Matemática Iberoamericana 4 (2), 339–354.

Mielke, Alexander, 2016. On evolutionary Γ-convergence for gradient systems. In: Macroscopic and Large Scale Phenomena: Coarse Graining, Mean Field Limits and Ergodicity. Springer, pp. 187–249.

Nossek, Raz Z., Gilboa, Guy, 2018. Flows generating nonlinear eigenfunctions. Journal of Scientific Computing 75 (2), 859–888.

Ortner, Christoph, 2004. Γ-Limits of Galerkin Discretizations with Quadrature. Tech. Rep.

Penot, Jean-Paul, 1998. Proximal mappings. Journal of Approximation Theory 94 (2), 203–221.

Portilheiro, Manuel, Vázquez, Juan Luis, 2013. Degenerate homogeneous parabolic equations associated with the infinity-Laplacian. Calculus of Variations and Partial Differential Equations 46 (3–4), 705–724.

Rabinowitz, Paul H., 1971. Some global results for nonlinear eigenvalue problems. Journal of Functional Analysis 7 (3), 487–513.

Roith, Tim, Bungert, Leon, 2022. Continuum limit of Lipschitz learning on graphs. Foundations of Computational Mathematics. Forthcoming. arXiv:2012.03772 [cs.LG], 2020.

Santambrogio, Filippo, 2017. {Euclidean, metric, and Wasserstein} gradient flows: an overview. Bulletin of Mathematical Sciences 7 (1), 87–154.

Schuster, Thomas, Kaltenbacher, Barbara, Hofmann, Bernd, Kazimierski, Kamil S., 2012. Regularization Methods in Banach Spaces, Vol. 10. Walter de Gruyter.

Slepčev, Dejan, Thorpe, Matthew, 2019. Analysis of p-Laplacian regularization in semisupervised learning. SIAM Journal on Mathematical Analysis 51 (3), 2085–2120.

Stan, Diana, del Teso, Félix, Vázquez, Juan Luis, 2018. Porous medium equation with nonlocal pressure. In: Current Research in Nonlinear Analysis. Springer, pp. 277–308.

Stefanelli, Ulisse, 2021. A new minimizing-movements scheme for curves of maximal slope. arXiv: 2103.00846 [math.AP].

Varvaruca, Eugen, 2004. Exact rates of convergence as $t \to \infty$ for solutions of nonlinear evolution equations. Journal of Evolution Equations 4 (4), 543–565.

Vázquez, Juan Luis, 2007. The Porous Medium Equation: Mathematical Theory. Oxford University Press.

Vázquez, Juan Luis, 2018. Asymptotic behaviour for the fractional heat equation in the Euclidean space. Complex Variables and Elliptic Equations 63 (7–8), 1216–1231.

Weinstein, Michael I., 1982. Nonlinear Schrödinger equations and sharp interpolation estimates. Communications in Mathematical Physics 87 (4), 567–576.

Wexler, D., 1973. Prox-mappings associated with a pair of Legendre conjugate functions. Modélisation Mathématique et Analyse Numérique 7 (R2), 39–65.

Chapter 14

Dynamic Programming versus supervised learning

Optimal fishing quotas

Gilles Pagès[a] and Olivier Pironneau[b,*]

[a]*Sorbonne Université, Laboratoire de Probabilités, Statistique et Modélisation, Paris, France,*
[b]*Sorbonne Université, Laboratoire Jacques-Louis Lions, Paris, France*
**Corresponding author: e-mail address: olivier.pironneau@sorbonne-universite.fr*

Contents

Abstract

This article deals with the design of optimal fishing quotas within a model used in Senegal for the long term evolution of the fish biomass.

Handbook of Numerical Analysis, Volume 23, ISSN 1570-8659, https://doi.org/10.1016/bs.hna.2021.12.014

The numerical tests validate supervised learning for stochastic control in this case. However the control is not Markovian, but for fishing quotas this does not seem to be a problem.
For comparison the same problem is solved with Dynamic Programming. Three implementations are tested using either a probabilistic method or the Hamilton-Jacobi partial differential equations.

Keywords
Stochastic control, Dynamic Programming, Partial differential equations, Supervised learning

MSC Codes
49J20, 60H10, 60H35, 68T20

1 Introduction

Stochastic control of differential systems with random data or modeling errors, is usually solved by Dynamic Programming, a technique introduced in the 1950 by Richard Bellman (1957). Numerically one has to solve a backward in time differential equation possibly coupled with the forward stochastic differential equation (*SDE*) of the problem. The method has various discrete implementations (see Pagès et al., 2004, Lauriere and Pironneau, 2016, etc). Calculus of variations and optimal control is another approach, which, in principle is more expensive because the backward adjoint Partial Differential Equation (*PDE*) and the forward Kolmogorov *PDE* for the PDF of the process are coupled and must be solved iteratively by a gradient method.

A third method has been proposed several years ago (see Balakrishnan and Biega, 1996, Kamalapurkar et al., 2018, Bertsekas, 2019, Auger and Pironneau, 2020, and others): Supervised Learning with Neural Networks: like off-line/on-line reduction methods, one is ready to pay a high CPU price in a "training" phase so as to solve quickly the problem with new data with the "trained network". The novelty is that a trained neural network can be ported on a small unit like a smart phone and yet be fast and accurate.

In this chapter the three classes of methods will be tested on an easy yet practical problem studied in Brochier et al. (2018): the control of a fishing site in Senegal where the fish biomass is threatened by intensive fishing. In Auger and Pironneau (2020) it was shown that parameter identification from a few measurements near time zero could be obtained accurately and fast by supervised learning. Here the parameters are known but with some uncertainty and the objective is to apply optimal quotas to stabilize the fish biomass at a desired level in this stochastic setting.

The model has two ordinary differential equations (*ODE*), one for the fish biomass X_t at time t, and one for the fishing effort which, for us, will be the number of boats at sea E_t.

Quotas are meant to keep the biomass at a constant desired level X_d. Hence the total number of fish caught must be lower than the quota Q_t. By construction each boat can't catch more than qX_t, where q is a constant reflecting the "structural limit" of the fisherman's catch yet proportional to the biomass. Nevertheless each boat is told to keep their catch below Q_t/E_t.

The control problem is solved by minimizing a criterion, the integral over time of $\|X_t - X_d\|^2$. But quotas, imposed by the regulator, are unpopular. To avoid too low (constraining) quotas, we add to the criteria the integral over time of $-\alpha Q_t$ where α is constant and positive; consequently, a low quota will not be optimal because it gives a larger value to the criteria than a big quota.

Along the way we found that the solution of the problem is very often "bang-bang", i.e.; switching from the structural limit to the quota often. For the regulator, rapidly changing quotas may be also dangerously unpopular; hence in an attempt to avoid such frequent drastic changes we add to the criteria the integral over time of $\beta\|\dot{Q}_t\|^2$ where β is a positive constant.

More details on the modeling of the problem can be found in Appendix A.

The plan of the paper is as follows:

In Section 1, the control problem is stated. It is a stochastic control problem because the data are not known accurately. Existence of solutions and discretization methods are discussed. We identify 3 formulations:

1. Deterministic control where the state is the solution of an *ODE*.
2. Non-dynamic control, i.e. an *SDE* for the state equation but a deterministic control function of time only.
3. A fully stochastic state and control, usual to Dynamic Programming.

In Section 2, a Monte-Carlo method is proposed for the non-dynamic case. It is a brute force Monte-Carlo approach, much more expensive than the Dynamic Programming approach of Section 4, but it has an advantage: it can be ported to a Neural Network for training and testing. We have used the standard multilayered Recurrent Neural Networks of Chollet (2017) (RNN, see Fig. 3); LSTM is also available in Keras, but as there are no repeating patterns in the solution of a *SDE* with brownian noise, it is not clear that a Long-Short-Term-Memory feedback architecture would improve the (already good) performance of RNN.

Finally, in the last section *PDE* methods are investigated, based either on Ito Calculus or Kolmogorov equations for the probability density function (PDF) of the process. It will be seen that these methods do not handle very well the penalty term $\beta\|\dot{Q}_t\|^2$.

2 A model problem

To study the evolution of the biomass of a fishing site we consider the following stochastic optimal control problem

$$\min_{u\in\mathcal{U}} J(u) := \int_0^T \mathbb{E}\left[(X_t - X_d)^2 - \alpha u(X_t,t) + \beta|\partial_t u(X_t,t)|^2\right] \mathrm{d}t, \qquad (1)$$

$$\text{with } \mathrm{d}X_t = X_t(r - \kappa X_t - u_t)\mathrm{d}t + \sigma X_t \mathrm{d}W_t, \; X_0 = X^0 \geq 0, \tag{2}$$

where W_t is a standard Wiener process on a probability space $(\Omega, \mathcal{A}, \mathbb{P})$, $X_d, r, \kappa, \sigma, u_m, u_M$ are real constants which depend on the considered fishing site and

$$\mathcal{U} = \left\{u \in H^1\left(]0, T[, W^{1,1}\right) : u_m \leq u(t, X_t) \leq u_M \; a.s.\right\}.$$

This problem arises for the determination of fishing quotas (see Appendix A for details): X_t is the fish biomass at time t and $u_t X_t$ is the fishing quota at time t. Occasionally we will also display a variable proportional to the number of boats at sea, the "fishing effort" $E(t)$, modeled by

$$\mathrm{d}E_t = (a - u_t X_t - cE_t)\,\mathrm{d}t, \quad E_0 = E^0 \text{ given},$$

where a is related to the price of fish and c is the operating cost of the fisherman. As explained in Appendix A, $a - u_t X_t$ is the profit of the fisherman and it decreases when too many fish are caught because the price of fish collapse.

The control aims at keeping a biomass at a desired level X_d at all times, but, quotas being unpopular, there is a term to prevent low quotas, $-\alpha u_t$ and another to prevent quotas from changing too quickly: $\beta|\partial_t u|^2$.

It is important to notice that one may also consider the case where u is a deterministic function of t only:

$$\min_{u \in \mathcal{U}_d} J(u) := \int_0^T \mathbb{E}\left[(X_t - X_d)^2\right] - \alpha u(t) + \beta|\partial_t u(t)|^2 \Big] \mathrm{d}t \text{ with (2)}, \\ \text{with } \mathcal{U}_d = \left\{u \in H^1(]0, T[) : u_m \leq u(t) \leq u_M\right\}. \tag{3}$$

We shall refer to it as the non-dynamic problem by opposition to Dynamic Programming where $u(X_t, t)$ is stochastic because of X_t but Markovian. In the non-dynamic case, the optimal control is the same for all trajectories X_t and in the dynamic case u adapts to the trajectories. To our surprise the non-dynamic case turns out to be more precise and more suitable because the day-to-day noise may not be very meaningful.

Occasionally we shall also refer to the deterministic case, i.e. when $\sigma = 0$.

2.1 Existence of solution: the deterministic case

In the deterministic case the problem reads

$$\min_{u \in \mathcal{U}_d} \left\{ J(u) = \int_0^T |X - X_d|^2 - \alpha u + \beta|\dot{u}|^2 : \; \frac{\dot{X}}{X} = r - \kappa X - u, \; X(0) = X^0 \right\}. \tag{4}$$

As u is positive $0 \leq X(t) \leq X_M := X^0 e^{rT}$. Then obviously $F(X) := rX - \kappa X^2 - Xu$ is uniformly Lipschitz in X: let $\alpha = r + 2X_M + 2u_M$.

$$\forall t \in [0, T], \quad |F(X_1(t)) - F(X_2(t))| \leq |X_1(t) - X_2(t)|\big(r + u_m + 2\kappa X_M\big),$$

therefore the solution of the *ODE* exists and is unique. It is also uniformly Lipschitz continuous in u:

$$\dot{X}_i = rX_i - \kappa X_i^2 - u_i X_i,\ i = 1, 2, \Rightarrow$$
$$\dot{X}_1 - \dot{X}_2 = (X_1 - X_2)\big(r - \kappa(X_1 + X_2) - (u_1 + u_2)\big) - (u_1 - u_2)(X_1 + X_2) \Rightarrow$$
$$|X_1(t) - \dot{X}_2(t)| \le \alpha \int_0^t |X_1(s) - X_2(s)| \mathrm{d}s + 2X_M \int_0^t |u_1(s) - u_2(s)| \mathrm{d}s$$

so that by a Grönwall argument, one derives that, for every $t \in [0, T]$

$$|X_1(t) - X_2(t)| \le 2X_M e^{\alpha T} \|u_1 - u_2\|_{L^2(]0,T[)}.$$

This also shows that $u \mapsto \int_0^T |X(t) - X_d|^2 \mathrm{d}t$ is continuous on $L^2(]0, T[)$.

Proposition 1. *Problem (4) has a solution.*

Proof. Consider a minimizing sequence: $u^n \in \mathcal{U}_d$, $J(u^n) \to \inf_{u \in \mathcal{U}_d} J(u)$. $\mathcal{U}_d$ being non-empty, bounded and closed and $\|\dot{u}\|_0^2$ being part of the criteria, there is a subsequence with $u^n \to u^* \in \mathcal{U}_d$ weakly in $H^1(]0, T[)$ and strongly in $L^2(]0, T[)$. By the above established continuity, $X^n \to X^*$ in $L^2(]0, T[)$. Obviously $\int_0^T \alpha u^n \to \int_0^T \alpha u^*$ and by the lower semicontinuity of $u \to \|\dot{u}\|_0^2$,

$$J(u^*) \le \liminf J(u^n) = \inf_{u \in \mathcal{U}_d} J(u) \Rightarrow u^* \text{ is a solution.}$$

□

2.2 Existence of solution: the stochastic non-dynamic case

Proposition 2. *The solution of the SDE (2) exists and is positive.*

Proof. $X \mapsto (r - \kappa X - u_t)X$ is locally Lipschitz, uniformly in t since $u_t \in [u_m, u_M]$; hence for every realization $W_t(\omega)$ there is a unique strong solution until a blow-up time τ which is a stopping time for the filtration $\mathcal{F}_t^w = \sigma(\mathcal{N}_s, W_s, s \le t)$ where $\mathcal{N}_s$ denotes the $\mathbb{P}$-negligible sets of $\mathcal{A}$. On $[0, \tau[$ we may work with $Y_t = \log X_t$, solution (by Itô calculus) of

$$\begin{aligned} \mathrm{d}Y_t &= \big(r - \tfrac{\sigma^2}{2} - \kappa X_t - u_t\big)\mathrm{d}t + \sigma \mathrm{d}W_t, \quad \Rightarrow \forall t \in (0, \tau]: \\ X_t &= X^0 \exp\left((r - \tfrac{\sigma^2}{2})t - \int_0^t (\kappa X_s + u_s)\mathrm{d}s + \sigma W_t\right) \le X^0 e^{S_t(u_m)}, \end{aligned} \tag{5}$$

where $S_t(v) := (r - \frac{\sigma^2}{2})t - vt + \sigma W_t$. Hence $\int_0^\tau X_s \mathrm{d}s = +\infty$ is impossible unless $\tau = +\infty$, $\mathbb{P}$-a.s. Therefore, $X_t \in \mathbb{R}^+$ for every $t \ge 0$ and

$$X^0 e^{S_t(u_M)} \exp\left(-\kappa \int_0^t e^{S_s(u_M)} \mathrm{d}s\right) \le X_t \le X^0 e^{S_t(u_m)}. \tag{6}$$

□

Theorem 1. *Problem (1)(2) has a solution.*

Proof. Consider two trajectories driven by the same W_t but with different u, denote δX_t the difference and $\bar{X}_t$ their average; we have

$$\mathrm{d}\delta X_t = \big(\delta X_t(r - 2\kappa\bar{X}_t - \bar{u}) - \bar{X}_t\delta u\big)\mathrm{d}t + \sigma\delta X_t dW_t.$$

By Itô calculus:

$$(\delta X_t)^2 = 2\int_0^t \big((\delta X_s)^2(r + \tfrac{\sigma^2}{2} - 2\kappa\bar{X}_t - \bar{u}) - \bar{X}_s\delta u(s)\delta X_s\big)\mathrm{d}s + 2\int_0^t \sigma(\delta X_s)^2\mathrm{d}W_s$$

so that $\mathbb{E}[(\delta X_t)^2] \le 2(r + \frac{\sigma^2}{2} - u_m)\int_0^t \mathbb{E}[(\delta X_s)^2]\mathrm{d}s - \int_0^t \delta u(s)\mathbb{E}[\bar{X}_s\delta X_s]\mathrm{d}s$

$$\le 2(r + \tfrac{\sigma^2}{2} - u_m)\int_0^t \mathbb{E}[(\delta X_s)^2] + \int_0^t \delta u(s)\mathbb{E}[(\bar{X}_s)^2]\mathrm{d}s.$$

We have seen earlier that $0 \le X_t \le X_0 e^{(r-\frac{1}{2}\sigma^2)t+\sigma W_t}$, therefore $\mathbb{E}[(\bar{X}_s)^2] \le 4e^{(2r+\sigma^2)s}\mathbb{E}[X_0^2]$, so much so that by Grönwall's lemma,

$$\mathbb{E}[(\delta X_t)^2] \le 2(r + \tfrac{\sigma^2}{2} - u_m)\int_0^t \mathbb{E}[(\delta X_s)^2]\mathrm{d}s + 4e^{(2r+\sigma^2)T}\mathbb{E}[X_0^2]\int_0^t (\delta u(s))^2 ds$$

and by Grönwall's lemma again, $\mathbb{E}[(\delta X_t)^2] \le 4e^{2r+\sigma^2-2u_m)t}e^{(2r+\sigma^2)T} \times \mathbb{E}[X_0^2]\int_0^t |\delta u(s)|$. Hence, there exists $C > 0$, $\sup_{t\in[0,t]}\mathbb{E}[(\delta X_t)^2] \le C\,\mathbb{E}[X_0^2]\|\delta u\|_{L^1(]0,T[)}$. Continuity of $u \mapsto \int_0^T \mathbb{E}[(X_t - X_d)^2]\mathrm{d}t$ follows and existence is then shown as in (2.1). □

2.3 Existence of solution: the dynamic case

To establish existence by probabilistic methods proved to be rather hard. On the other hand, the problem can be converted into an equivalent distributed control problem by using the Kolmogorov equation of the PDF of X_t.

We will show in Section 7 that there is a solution when the $\mathbb{E}$ in the criteria and κ are multiplied by a cutoff for large X and, with $Q^+ := \mathbb{R}^+\times]0,T[$, and a modified $\mathcal{U}$:

$$\mathcal{U} = \big\{u \in L^2(W^{1,1}(\mathbb{R}^+)) : u \in [u_m, u_M],\ \partial_t u \in L^2(Q^+),\ \lim_{X\to+\infty} |\log X| u(X,t) < +\infty\big\}.$$

2.4 Discretization

Let $h = \frac{T}{M}$ and $t^m = mh$. For any f, f^m denotes an approximation of $f(mh)$.

Define the Euler – or Euler-Maruyama – scheme of the diffusion by

$$X^{m+1} = X^m \left(1 + h(r - \kappa X^m - u^m) + \sigma \delta W^m\right) \tag{7}$$

where $\delta W^m = W_{(m+1)h} - W_{mh}$. Note that positivity of X may not be preserved by this scheme, but we may choose to discretize (5). Alternatively, to avoid the computing cost of exponentials, we may use the Milstein scheme:

$$\tilde{X}^{m+1} = \tilde{X}^m \left(1 + h(r - \kappa \tilde{X}^m - u^m - \frac{\sigma^2}{2}) + \sigma \delta W^m + \frac{\sigma^2}{2}(\delta W^m)^2\right). \tag{8}$$

Proposition 3. *Let $\tilde{X}$ denote the solution of the continuous Milstein scheme given for every $t \in [mh, (m+1)h)$ by*

$$\tilde{X}_t = \tilde{X}^m \Big(1 + (t - mh)\big(r - \kappa \tilde{X}^m - u^m - \tfrac{\sigma^2}{2}\big) + \sigma(W_t - W^m) + \tfrac{\sigma^2}{2}(W_t - W^m)^2\Big).$$

Assume $\tilde{X}_0 = X^0 > 0$. Then, on event $A_h = \left\{\tilde{X}^m < \frac{1}{2\kappa h} + \frac{r - \sigma^2/2 - u_M}{\kappa}\right\}_{m=0}^{M-1}$, the continuous Milstein $(\tilde{X}_t)_{t \in [0,T]}$ almost surely has positive paths.

Proof. We proceed by induction. Starting from (8), we want the discriminant of the second degree polynomial equation in $W_t - W^m$ to be negative, namely after obvious simplification by $\sigma > 0$

$$1 - 2\Big(1 + h\big(r - \kappa \tilde{X}^m - u^m - \tfrac{\sigma^2}{2}\big)\Big) < 0 \Longleftrightarrow \tilde{X}^m < \frac{1}{2\kappa h} + \frac{r - \sigma^2/2 - u_M}{\kappa}.$$

□

Remark 1. It is clear that $\mathbb{P}(A_h) \uparrow 1$ as $h \to 0$. For more details on the positivity preserving property of the Milstein scheme, see Pagès (2018, Chapter 7.5).

3 Brute force solution of the non-dynamic control problem by Monte-Carlo

We wish to solve (a discretized version of) Problem (3) by a Monte Carlo simulation. Recall that $\delta W^m = \sqrt{h}\mathcal{N}_{0,1}^m$ where $(\mathcal{N}_{0,1}^m)_{m=1,\dots,M}$ denotes the numerical realization of an i.i.d. sequence of $\mathcal{N}(0,1)$-distributed random variables. For each of the above three schemes there is a function ζ_h such that

$$X^{m+1} = \zeta_h(X^m, u^m, \mathcal{N}_{0,1}^{m+1}),\ 0 \le m \le M-1,\ \ X^0 \text{ given.} \tag{9}$$

Let $(\mathcal{N}^m_{0,1,k})^{m\in[0,M)}_{k\in[1,K]}$, be $K\times M$ independent realizations of $\mathcal{N}_{0,1}$ and, for $\{u^m \in [u_m, u_M]\}_{m=1}^M$ given, independent of k, let $\{X_k^m\}_{k=1}^K$ be the K (independent) results of (9). By the law of large numbers:

$$\mathbb{E}[(X_t - X_d)^2]_{|t=mh} \approx \mathbb{E}_K[(X^m - X_d)^2] := \frac{1}{K}\sum_{k=1}^{K}(X_k^m - X_d)^2.$$

Then (3) can be approximated by

$$\min_{\{u^m\in[u_m,u_M]\}_{m=0}^{M-1}} J_K := h\sum_{m=0}^{M-2}\left[\mathbb{E}_K[(X^{m+1}-X_d)^2] - \alpha u^m + \frac{\beta}{h^2}|u^{m+1}-u^m|^2\right]. \tag{10}$$

3.1 Solution with a gradient method

Gradient methods update $\{u^m\}_{m=1}^{M-1}$ by $u^m \leftarrow u^m - \mu G^m$ where μ, the step size, is an appropriate scalar, and $G^m = \partial_{u^m} J_K$. Let X_k^m be computed by (7); let us differentiate this equation and multiply it by an adjoint variable P_k^m and sum:

$$\sum_{m=0}^{M-2}\Big[- P_k^m \delta X_k^{m+1} + P_k^m\Big(\delta X_k^m\Big(1 + h(r - 2\kappa X_k^m - u^m) + \sigma\delta W_k^m\Big)$$
$$- X_k^m h\delta u^m\Big)\Big] = 0.$$

As $\delta X^0 = 0$, if we set $P^{M-1} = P^{M-2} = 0$, $v_k^m = 1 + h(r - 2\kappa X_k^m - u^m) + \sigma\delta W_k^m$,

$$-\sum_{0}^{M-2} hP_k^m X_k^m \delta u^m = \sum_{1}^{M-2}(P_k^{m-1}\delta X_k^m - P_k^m\delta X^m v_k^m)$$
$$= \sum_{1}^{M-2}\delta X_k^m(P_k^{m-1} - P_k^m v_k^m).$$

Define P_k^m by $P_k^{m-1} = P_k^m v_k^m - 2(X_k^m - X_d)$, $m = M-2, M-3, \ldots$ Then

$$\sum_{m=1}^{M-1}\delta E_K[(X^m - X_d)^2] = \frac{2}{K}\sum_{k=1}^{K}\sum_{m=1}^{M-1}(X_k^m - X_d)\delta X_k^m$$
$$= \frac{1}{K}\sum_{k=1}^{K}\sum_{m=0}^{M-2} P_k^m X_k^m \delta u_k^m.$$

So the gradient is $G^m = \frac{1}{K}\sum_{k=1}^K P_k^m X_k^m - \alpha - 2\frac{\beta}{h^2}(u^{m+1} - 2u^m + u^{m-1})$. Note that the successive $(u^m)_m = 0, \ldots, M-1$ obtained by this gradient descent are not adapted and depend on all the $M \times K$ simulated $\mathcal{N}_{0,1,k}^m$.

3.2 Implementation and examples

We use the `L-BFGS-B` function of the Python library `minimize`. It is an accelerated projected gradient method for problems with box constraints and an automatic adjustment of the step size μ. We took 50 time steps and chose to solve a deterministic case with

$$\sigma = 0, K = 1, X_d = 1, \kappa = 1.2, X_0 = 0.9, c = 1, a = 1, u_m = 0.5,\ u_M = 1.$$

The computer program is in file `fishingQuotaBruteforce.py`. The results with $\alpha = \beta = 0$ are shown on the left in Fig. 1 with a saturation of the constraints $u = u_m$ when $m < 7$; but the solution is not bang-bang everywhere and $u = 0.8$ gives $X^{m+1} = X^m = X_d$, $P^m = 0$. On the right in Fig. 1 the results are shown for $\alpha = 0.01$, $\beta = 0.001$. On Fig. 2 a stochastic case of (10) is solved with the same parameters except $\sigma = 0.3$. The results are shown for $K = 1, 10, 100, 10000$ Monte-Carlo samples.

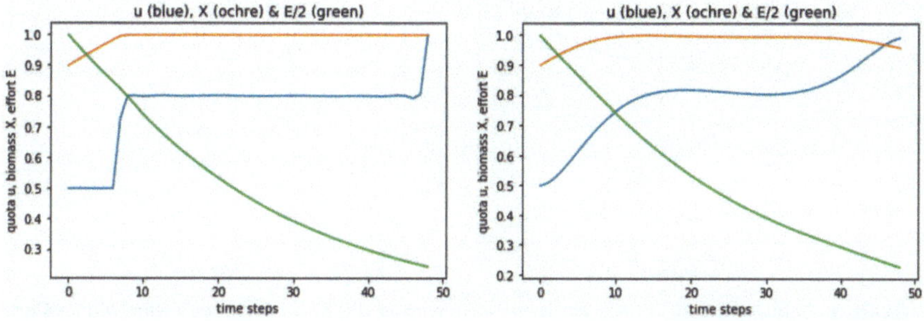

FIGURE 1 Solution of (1) discretized by (10), versus m, in the deterministic case ($\sigma = 0$, $K = 1$) with $\beta = 0$ (left), and $\beta = 0.001$ (right). The blue curve is the control u; the biomass X_t is in ochre and the fishing effort in green.

4 Solution of the non-dynamic control problem by supervised learning

We still aim at to solving (3) using Monte Carlo simulations (written here in continuous time for genericity)

$$\min_{u \in \mathcal{U}} J(u) := \int_0^T \left[\frac{1}{K} \sum_1^K (X_k(t) - X_d)^2 - \alpha u(t) + \beta(\partial_t u(t))^2 \right] dt \qquad (11)$$

$$\text{with}\ \ dX_k = X_k(t)\big(r - \kappa X_k(t) - u(t)\big)dt + \sigma X_k(t) dW_k(t),\ \ X(0) = X^0, \qquad (12)$$

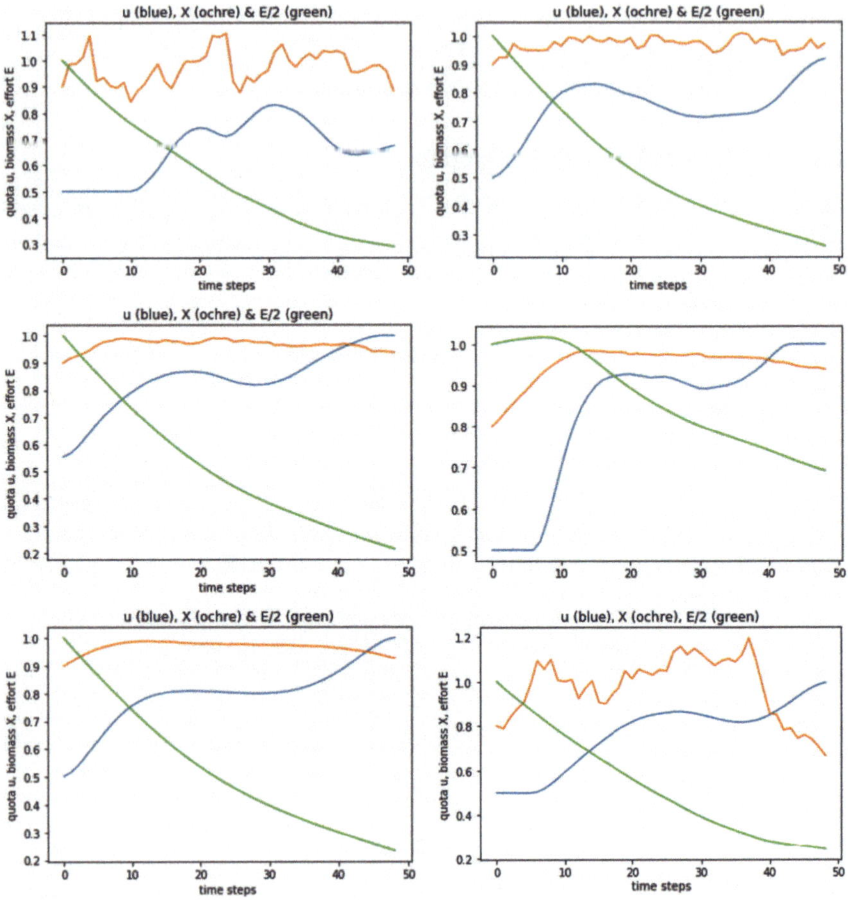

FIGURE 2 Solution of (10) in the stochastic case with $\sigma = 0.3$: convergence of the Monte-Carlo approximation. Left to right, up first: K=1, K=10, K=100,K=1000, K=10000. Already the results are fairly converged at K=100. In the last figure the optimal control is computed with K=100 and used to solve (2) with a new random noise, thereby showing that the solution is still acceptable. Note that the brute force solution for $K \gg 1$ is quite close to the noiseless solution of Fig. 1-b.

where W^k are K independent standard Brownian motions X_k is replaced by a numerical scheme. In practice To reduce CPU cost, we shall train a neural network with input parameters $[X^0, r, \kappa]$ and output an optimal $\{u^m\}_{m=1}^{M-1}$ for these parameters, computed by the brute force method above. There is no need to vary the parameter X_d because we can always return to the case $X_d = 1$ by scaling.

To build a training set we use a large number of synthetic solutions obtained with the brute force method of the previous section. We have used a small $K = 100$ to generate 10000 solutions (it takes about 90 minutes on an intel core i7 machine). It works as follows:

- Choose $[X^0, r, \kappa]$ uniformly random in $[X^0_m, X^0_M] \times [r_m, r_M] \times [\kappa_m, \kappa_M]$.
- Generate $K \cdot M$ normal gaussian random values for dW_t.
- Compute a discrete optimal solution $\{u^m\}_{m=1}^{M-1}$ of (11) with K=100 by the L-BFGS-B Python module.
- Then train a neural network $[X^0, r, \kappa] \to \{u^m\}_{m=1}^{M-1}$ with these 10000 synthetic solutions.
- Finally, check the precision by using the network's answer $\{u^m\}_{m=1}^{M-1}$ to a new set of values for $[X^0, r, \kappa]$ and compare with the brute force solution corresponding to these parameters.

For an introduction to Statistical Learning with Neural Networks see Goodfellow et al. (2016). We have used `Keras` (see Chollet, 2017) to generate a 4-layer Neural Network (see Fig. 3). The input layer has 3 inputs; the two hidden layers have 50 neurons each and the output layer has M nodes. All layers use the ReLU activation. The two hidden layers are separated by a batch-normalization.

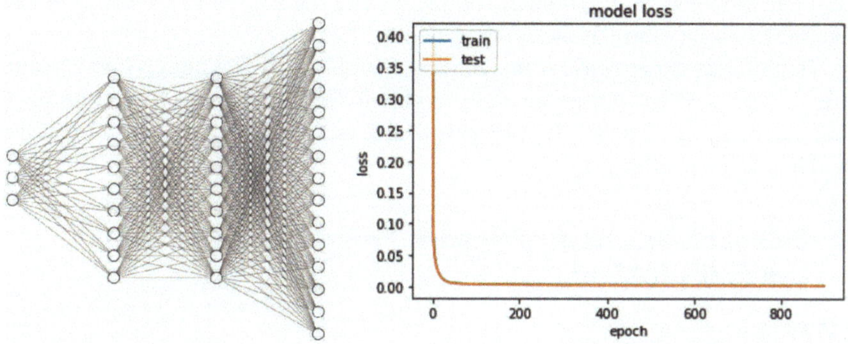

FIGURE 3 The Neural Network has 3 inputs, two hidden layers each with 50 neurons (10 shown here) and separated by batch-normalization and an output layer with M output (M=15 shown here). Right: typical convergence curve of the loss function during the training phase.

4.1 A numerical test

We have taken 50 time steps and K=100 for the brute force part. The following parameters are used:

$$T = 2,\ \alpha = 0.01,\ \beta = 0.001,\ \sigma = 0.3,\ X_d = 1,\ u_m = 0.5,\ u_M = 1.$$

For the Neural Network training we have used 10 000 samples in batches of 32 and a maximum of 2 000 epochs. The input parameters are

```
r = np.random.uniform(1.9,2.2),\;
kappa = np.random.uniform(1,1.3),\;
X0=np.random.uniform(0.7,1.3)
```

Finally for the fishing effort equation, $a = c = 1$. The computer program is written in Python and is in the file `fishingQuotaIA4.py`. The learning phase produced the following precision on each of the time steps:

```
Test loss: 0.03751661651
Average relative error |1-u_NN/u| for the 49 time steps =
[0.031 0.03954 0.04612 0.03440 0.03665 0.0432 0.03549 0.04364
 0.04247 0.03737 0.03448 0.03950 0.04441 0.04149 0.03366 0.04054
 0.03978 0.04160 0.0404  0.03319 0.03800 0.03229 0.04076 0.03475
 0.03333 0.03761 0.04427 0.03837 0.03909 0.03384 0.03707 0.03838
 0.03682 0.03278 0.03412 0.02993 0.03138 0.02646 0.02687 0.02550
 0.02197 0.01949 0.02680 0.02290 0.01801 0.01872 0.01909 0.014978 0.01907]
```

Notice that the error is about the same at all time steps.

In Fig. 4 the performance of the trained Neural Network on 6 new cases are shown. Each case corresponds to a set of parameters $[r, \kappa, X_0]$ and a new realization of the random noise. The "exact" control u is computed by the brute force method with $K = 100$ and compared with the solution u_{NN} given by the neural network (values at the 50 time steps).

In all cases except case 4 the control u_{NN} given by the network is close to the "exact" one; but even then, X_{NN} computed by (7) with u_{NN} is close to X computed with u "exact". This reveals that X is not that sensitive to u: that the problem is somewhat stiff.

5 Bellman's Stochastic Dynamic Programming for the dynamic problem

Consider the *value function*

$$V(t,X) = \min_{u\in\mathcal{U}} \left\{ \int_t^T \mathbb{E}\left[(X_\tau - X_d)^2 - \alpha u + \beta|\partial_t u|^2\right] \mathrm{d}\tau : \right.$$
$$\left. \mathrm{d}X_\tau = X_\tau \mathrm{d}\tau (r - \kappa X_\tau - u_\tau) + \sigma X_\tau \mathrm{d}W_\tau, \quad X(t) = X \right\}. \tag{13}$$

Let $[0, T] = \cup_1^M [(m-1)h, mh]$. Let $\zeta_h(X, u^m, z)$ denote one iterate of a numerical scheme (7)(8) for the *SDE* starting at X, for instance with (7)

$$\zeta_h(X, u, z) = X + X(r - \kappa X - u)h + X\sigma z\sqrt{h},$$
$$z \text{ being the realization of a } \mathcal{N}_{0,1} \text{ r.v.}$$

Let $v^M(X) = 0$ and let, for $m \in \{0, \ldots, M-1\}$, $v^m(X) =$

$$\min_{u\in\mathcal{U}} \int_{mh}^{(m+1)h} \mathbb{E}\left[|\zeta_h(X,u,z) - X_d|^2 - \alpha u + \beta|\partial_t u|^2\right] \mathrm{d}\tau$$
$$+ \mathbb{E}[v^{m+1}(\zeta_h(X,u,z))] \tag{14}$$

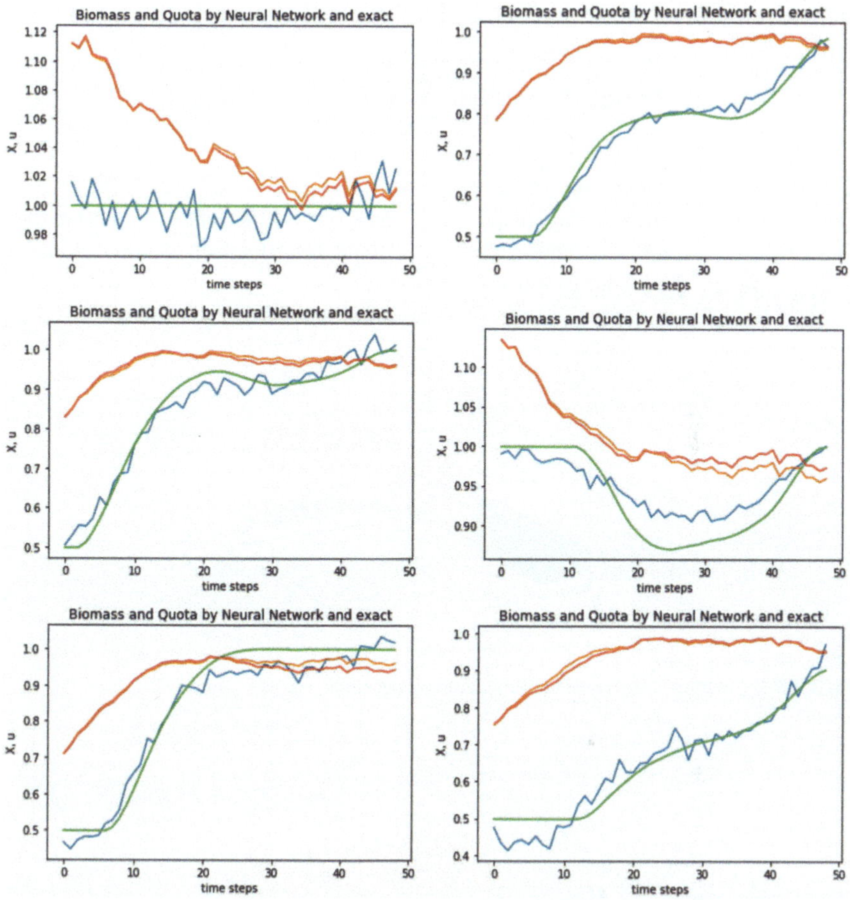

FIGURE 4 On each of the 6 plots X and u are shown: True discrete solution in green for X and red for u. NN solution in blue for X_{NN} and ochre for u_{NN}. The plots correspond to the following values of the parameters: $r = 2.02285$, $\kappa = 1.0203$, $X_0 = 1.1122$, $r = 2.0688$, $\kappa = 1.2641$, $X_0 = 0.7852$, $r = 1.9151$, $\kappa = 1.0141$, $X_0 = 0.8299$, $r = 2.0738$, $\kappa = 1.1876$, $X_0 = 1.1344$, $r = 2.0463$, $\kappa = 1.1224$, $X_0 = 0.7550$.

$$\approx \min_{u \in \mathcal{U}} \left\{ h\mathbb{E}\left[|\zeta_h(X,u,z) - X_d|^2\right] - \alpha h u + \frac{\beta}{h}|u^{m+1}(X) - u|^2 \right.$$
$$\left. + \mathbb{E}\left[v^{m+1}\left(\zeta_h(X,u,z)\right)\right] \right\}.$$

Evidently $\mathbb{E}\left[|\zeta_h(X,u,z) - X_d|^2\right] = X - X_d + hX(r - \kappa X - u)|^2 + hX^2\sigma^2$. To compute $\mathbb{E}[v^{m+1}(\zeta_h(X;u;z))]$, we use a quadrature formula with Q points $\{z_q\}_{q=1}^{Q}$ and weights $\{w_q\}_{q=1}^{Q}$ based on optimal quantization of the normal distribution $\mathcal{N}(0,1)$ (see Pagès and Printems, 2003 or Pagès, 2018, Chapter 5 and the website quantize.maths-fi.com for download of weighted grids) so that

$$\mathbb{E}[v^{m+1}(\zeta_h(X;u;z))] \approx \sum_{q=1}^{Q} w_q v^{m+1}\big(\zeta_h(X,u,z_q)\big).$$

Finally at every time step and every $X_j = jL/J$, $j = 0, ..., J$, with $L \gg 1$, the result is minimized with respect to $u \in \mathcal{U}$ by a dichotomy. In this fashion $\{u^m(X_j)\}_{j=1}^J$ is obtained and a piecewise linear interpolation is constructed to prepare for the next time step u^{m-1}.

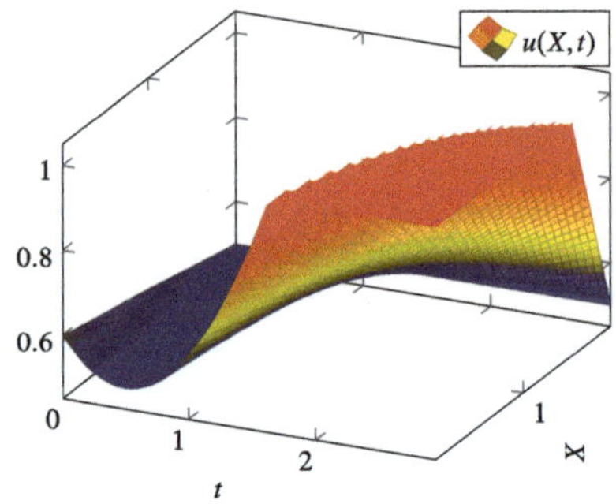

FIGURE 5 Solution of the Dynamic Programming equation: $u(x,t)$.

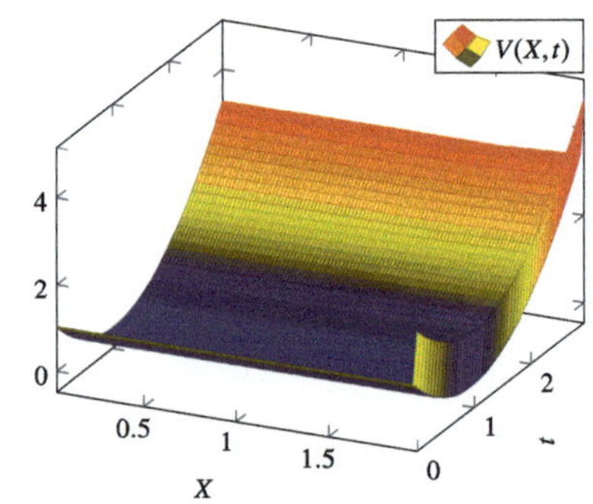

FIGURE 6 Solution of the Dynamic Programming equation: $V(x,t)$.

With the same parameters as above except $\beta = 0.1$, $L = 3$, $J = 50$, and $M = 50$, the results of Figs. 5, 6 have been obtained. Once $\{u^m(X)\}_0^{M-1}$ is known, for any realization of the fishing model we use it as follows:

$$X^{m+1} = X^m + hX^m\big(r - \kappa X^m - u^m(X^m)\big) + X^m\sigma\sqrt{h}\mathcal{N}_{0,1}^m, \quad X^0 = X_0.$$

Two such simulations are shown on Fig. 7, 8. The results are not as good as before, meaning that the dynamic setting gives a poor solution to the control of the fishing site. These results should be compared with those of Appendix A which deals with a very similar problem for which there is a polynomial in log X solution when $\alpha = 0$.

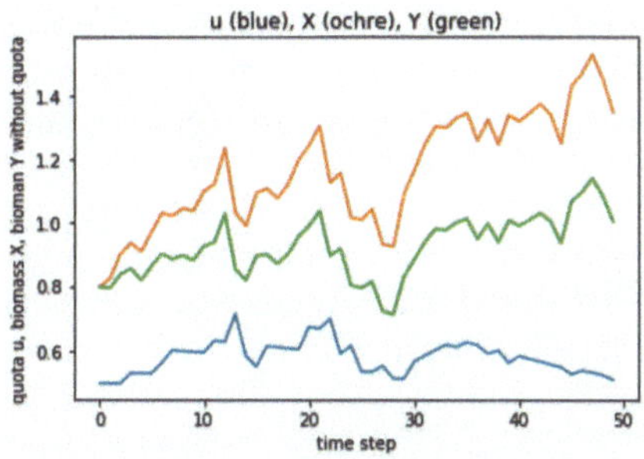

FIGURE 7 Simulation of the fishing model with a quota function computed by Dynamic Programming and $X_0 = 0.8$.

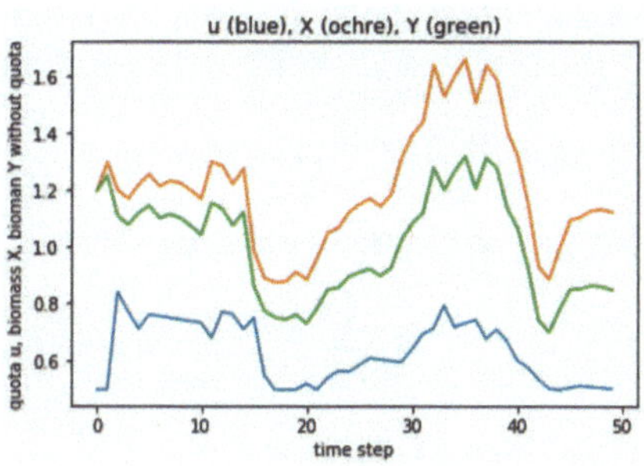

FIGURE 8 Simulation of the fishing model with a quota function computed by Dynamic Programming and $X_0 = 1.2$.

6 Solution with the Hamilton-Jacobi-Bellman partial differential equations

The Dynamic Programming equations can be written as a deterministic set of *PDE*: the Hamilton-Jacobi-Bellman (*HJB*) equations.

To this end we consider the same value function defined in (13).

Proposition 4. *The value function verifies* $V(\cdot, T) = 0$ *and*

$$0 = \inf_{u \in \mathcal{U}} \{(X - X_d)^2 - \alpha u + \beta |\partial_t u|^2 + \partial_t V + (r - \kappa X - u) X \partial_X V \\ + \frac{\sigma^2 X^2}{2} \partial_{XX} V\}. \tag{15}$$

Proof. The proof given below is heuristic and non standard but it gives a link with the previous section. For the standard argument see for instance (Yong and

Zhou, 1991). We return to (14) above. As u^M plays no role in the discretization we assume that $u^M = u^{M-1}$. Hence when $m = M-1$ the minimization reduces to

$$v^{M-1} = h \min_{u\in\mathcal{U}} \left\{\mathbb{E}\left[|X - X_d|^2\right] - \alpha u\right\} = h(|X - X_d|^2 - \alpha u_m)$$

and the minimizer is $u^{M-1} = u_m$. When $m < M-1$ a similar argument gives,

$$v^m(X) \approx \min_{u\in\mathcal{U}} \left\{\left(|X - X_d|^2 - \alpha u + \frac{\beta}{h^2}|u^{m+1}(X) - u|^2\right) h \right. \\ \left. + \mathbb{E}\left[v^{m+1}(\zeta_h(X,u,z))\right]\right\}. \tag{16}$$

With a Taylor expansion, let us approximate $v^{m+1}(X+h) \approx \sum_{q=0}^{q_M} v^{(q)m+1}(X)\frac{h^q}{q!}$. For clarity denote $\lambda = r - \kappa X - u$, so

$$\mathbb{E}\left[v^{m+1}(\zeta_h(X,u,z))\right] = v^{m+1}(X) + \mathbb{E}\left[\sum_0^{q_M} v^{(q)m+1} X^q h^{\frac{q}{2}} (\sigma z + \lambda\sqrt{h})^q\right]$$

$$\approx v^{m+1}(X) + X v'^{m+1}\lambda h + \frac{X^2}{2} v''^{m+1}(\lambda^2 h + \sigma^2)h + \frac{X^3}{6}\lambda\sigma^2 v^{(3)m+1} h^2 + \ldots$$

$$\mathbb{E}\left[v^{m+1}(\zeta_h(X,u,z))\right] \approx v^{m+1}(X) + X v'^{m+1}(r - \kappa X - u)h \\ + \frac{X^2}{2} v''^{m+1}\sigma^2 h + o(h).$$

Let us use it in (16). Then u must solve

$$\min\left\{-\alpha u + \frac{\beta}{h^2}\left(u^2 - 2uu^{m+1}\right) - uXv'^{m+1}\right\} \quad \Rightarrow$$

$$u = u^{m+1} + \frac{h^2}{2\beta}(\alpha + Xv'^{m+1})$$

but comply also with the box constraints:

$$u^m = \min\left\{u_M, \max\left\{u_m, u^{m+1} + \frac{h^2}{2\beta}\left(\alpha + Xv'^{m+1}\right)\right\}\right\} \text{ and}$$

$$v^m \approx v^{m+1} + h\left(Xv'^{m+1}(r - \kappa X - u^m) + \frac{X^2\sigma^2}{2}v''^{m+1} + |X - X_d|^2 - \alpha u^m \right. \\ \left. + \frac{\beta}{h^2}|u^{m+1} - u^m|^2\right).$$

As this corresponds to an explicit discretization of a parabolic *PDE*, h will have to be extremely small: $h = O(\delta X^2)$, where δX is the mesh size. □

6.1 Numerical results

We have implemented the method using `FreeFEM++` (Hecht, 2012) with a slight change: the term $\frac{X^2\sigma^2}{2}v''^{m+1}$ is changed to $\frac{X^2\sigma^2}{2}v''^{m}$ so as to increase numerical stability.

Numerical results shown on Fig. 9, 10 are obtained with the program in file `pagesbellmanstoch.edp` and the following parameters:

$$r=2,\ \kappa=1.2,\ \sigma=0.3,\ X_d=1,\ \alpha=0.01,\ \beta=0.1,\ u_m=0.5,\ u_M=1.$$

The computational domain is $X\in(0,3)$ divided in 400 intervals, $t\in(0,2)$, discretized with 40 time steps.

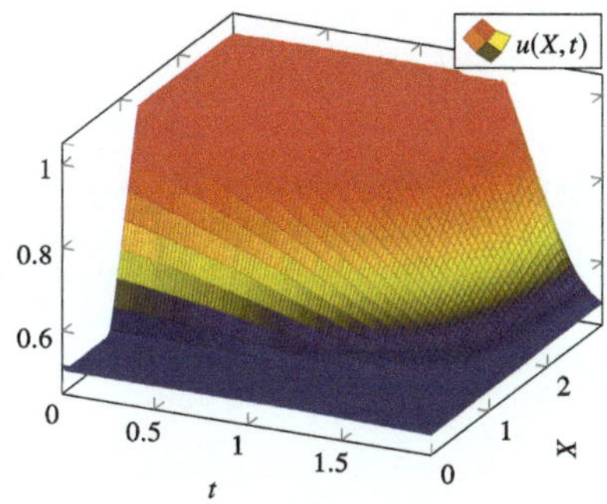

FIGURE 9 Solution of the Dynamic Programming equation: side view.

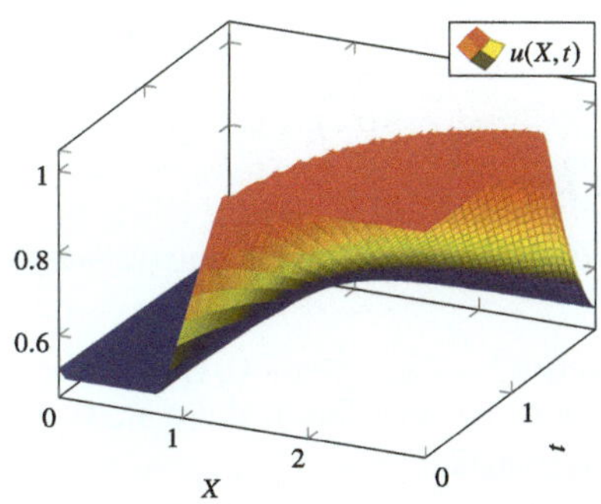

FIGURE 10 Solution of the Dynamic Programming equation: front view.

Once $\{u^m(X)\}_0^{M-1}$ is known, for any realization of the fishing model we use it as follows:

$$X^{m+1}=X^m+hX^m(r-\kappa X^m-u^m(X^m))+X^m\sigma\sqrt{h}\mathcal{N}_{0,1}^m,\quad X^0=X_0.$$

Two such simulations are shown on Fig. 11, 12.

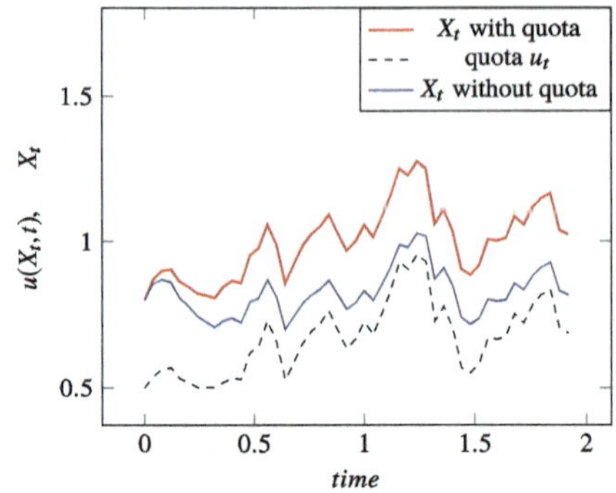

FIGURE 11 Simulation of the fishing model with a quota function computed by Dynamic Programming and $X_0 = 0.8$.

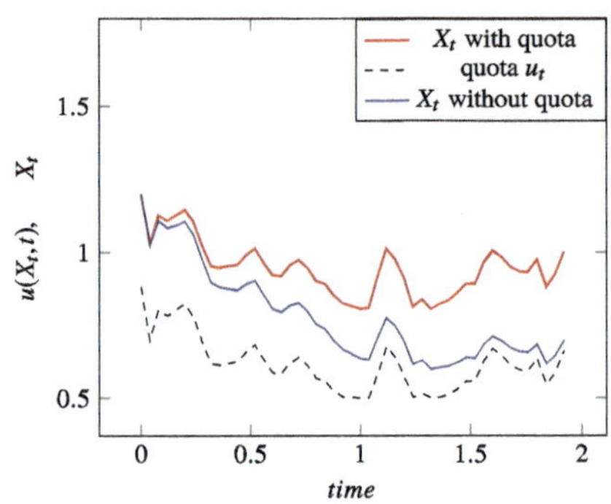

FIGURE 12 Simulation of the fishing model with a quota function computed by Dynamic Programming and $X_0 = 1.2$.

6.2 Analysis when $\beta = 0$

Proposition 5. *When $\beta = 0$, the solution is always bang-bang.*

Proof. With $\mathcal{U} = \{u = (u_m)_n,\ u_m \in [u_m, u_M]\}$ and $u_m > 0$, X being positive, the optimal control u^* must be such that if there are $0 < t_1, t_2 < T$ such that for all $t \in (t_1, t_2)$, $u(X, t)$ is constant then

1. $u^* = u_m$ when $X\partial_X V + \alpha < 0$, leading to $\partial_t u^* = 0$ and

$$\partial_t V + (r - \kappa X - u_m)X\partial_X V + \frac{\sigma^2 X^2}{2}\partial_{XX} V = \alpha u_m - (X - X_d)^2$$

2. $u^* = u_M$ when $X\partial_X V + \alpha > 0$, leading to $\partial_t u^* = 0$ and

$$\partial_t V + (r - \kappa X - u_M)X\partial_X V + \frac{\sigma^2 X^2}{2}\partial_{XX} V = \alpha u_M - (X - X_d)^2$$

3. Otherwise when $u_m < u^*(X,t) < u_M$, we must have $V(T)=0$ and

$$0=(X-X_d)^2-\alpha u^*+\partial_t V+(r-\kappa X)X\partial_X V+\frac{\sigma^2 X^2}{2}\partial_{XX}V,$$
$$X\partial_X V+\alpha=0$$

The third option being impossible, the solution of the problem is bang-bang. □

Remark 2. Note that this *HJB* dynamic solution is different from the non-dynamic solution of Fig. 1.

6.3 Discretization

One way to find u^* when it is neither u_m nor u_M is to discretize in time and solve

$$\inf_{u^m} g(u^m) := \left\{-\alpha u^m+\frac{\beta}{h^2}|u^m-u^{m+1}|^2-u^m X\partial_X V\right\}$$

The stationary point is at

$$-\alpha+\frac{2\beta}{h^2}(u^{m-1}-u^m)-X\partial_X V^{m-1}=0,\ \text{i.e. } \bar{u}^{m-1}=u^m+\frac{h^2}{2\beta}(\alpha+X\partial_X V^{m-1}).$$

So we have to compare $g(u_m)$, $g(\bar{u}^m)$ and $g(u_M)$ and choose the smallest. The end result is very similar to the previous section.

Two simulations are done with $\beta=0$ and shown on Fig. 13, 14. The control is indeed bang-bang.

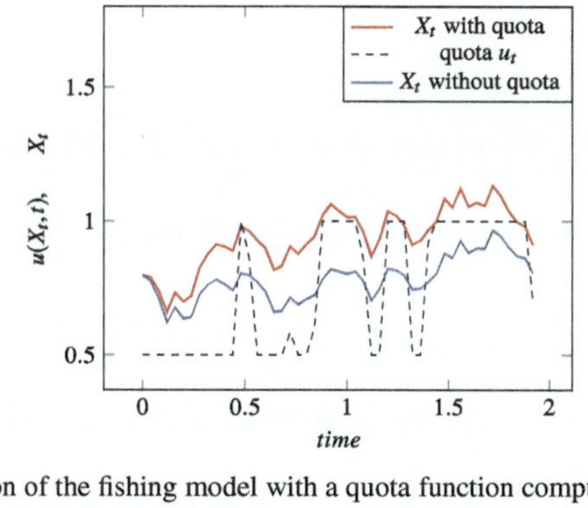

FIGURE 13 Simulation of the fishing model with a quota function computed by *HJB* when $\beta=0$ and $X_0=0.8$.

7 Solution with the Kolmogorov equation

The Kolmogorov equation for the probability density $\rho(X,t)$ of $\{X_t\}_0^T$ is:

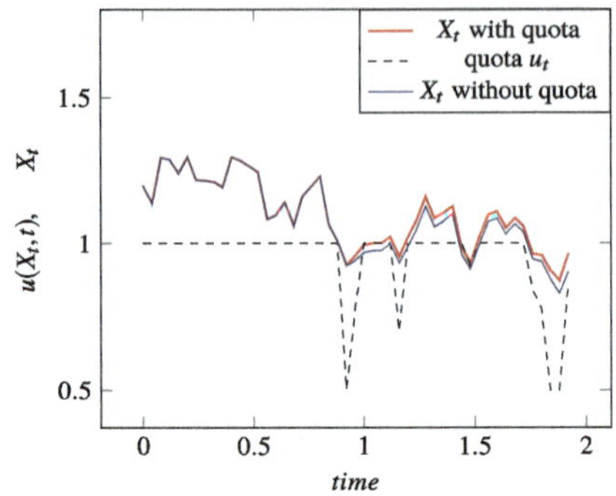

FIGURE 14 Simulation of the fishing model with a quota function computed by *HJB* when $\beta = 0$ and $X_0 = 1.2$.

$$\partial_t \rho + \partial_X[(r - \kappa X - u)X\rho] - \partial_{XX}[\tfrac{X^2\sigma^2}{2}\rho] = 0,\ \rho(X,0) = \rho^0(X),\ \forall X \in \mathbb{R}^+. \tag{17}$$

The optimal control problem (1) is: with (17)

$$\min_{u \in \mathcal{U}} J(u) := \int_{Q^+} \left[(X - X_d)^2 - \alpha u(X,t) + \beta|\partial_t u(X,t)|^2\right] \rho(X,t)\mathrm{d}X\mathrm{d}t. \tag{18}$$

The conditions for having equivalence between the stochastic control written as (1) and (18) are detailed in Le Bris and Lions (2008). To avoid using weighted Sobolev spaces we make a change variable $y = \log X$, set $v = r - \frac{\sigma^2}{2} - \kappa e^y - u(e^y, t)$ so that $\mathrm{d}Y_t = v\mathrm{d}t + \sigma \mathrm{d}W_t$ and then study the existence of a *PDE* for Y_t given by

$$\partial_t \rho + \partial_Y(v\rho) - \partial_{YY}[\tfrac{\sigma^2}{2}\rho] = 0,\ \rho(Y,0) = \rho^0(Y),\ \forall Y \in \mathbb{R},\ t \in]0,T[. \tag{19}$$

The *PDE* is understood in the sense that for all $\hat{\rho} \in H^1(\mathbb{R})$,

$$\int_{\mathbb{R}} \left[\hat{\rho}\partial_t \rho - v\rho\partial_Y\hat{\rho} + \frac{\sigma^2}{2}\partial_Y\rho\partial_Y\hat{\rho}\right] = 0,$$
$$a.e.\ t \in]0,T[;\ \rho(Y,0) = \rho^0(Y),\ \forall Y \in \mathbb{R}.$$

Note that v can be changed to $v - L$, $L \in \mathbb{R}$ by changing ρ to ρe^{-Lt}. Hence when $\lim_{Y\to\infty} u = u_\infty \neq 0$ we can take $L = r - \frac{\sigma^2}{2} - u_\infty$ and assume that $\lim_{Y\to\infty} v = 0$.

Proposition 6. *If* $v \in W^{1,1}_{loc}(\mathbb{R})$, $v/(1+|y|) \in L^1(\mathbb{R}) \cap L^\infty(\mathbb{R})$, $\partial_y v \in L^\infty(\mathbb{R})$ *then* Y_t *has a PDF,* $\rho \in L^\infty\left(L^2(\mathbb{R}) \cap L^\infty(\mathbb{R})\right) \cap L^2\left(H^1(\mathbb{R})\right)$, *given by (19).*

Consequently, for the problem to make sense, $(r - \kappa X - u)$ needs to be multiplied by a regular cutoff function, tending to zero fast enough when $X \to +\infty$. We shall henceforth assume that κ is a function of X which is zero when $X > X_M$, for some X_M.

Theorem 2. *Given X_d, σ, α, β, K, T, R in $\mathbb{R}^+$, v_m, v_M, K in $\mathbb{R}$ and $\rho^0 \in [0,1]$ with $\int_{\mathbb{R}} \rho^0(u)\mathrm{d}u = 1$, let $\mathcal{Q} = \mathbb{R}\times]0,T[$ and $\mathcal{Q}_R =]-R, R[\times]0,T[$ and,*

$$\mathcal{V} = \left\{v:\ v_m \le v \le v_M,\ \|\partial_t v\|_{L^2(\mathcal{Q})} \le K\right\} \text{ and consider}$$
$$\min_{v\in\mathcal{V}} J(v) := \int_{\mathcal{Q}_R} \left[(e^Y - X_d)^2 + \alpha v(Y,t) + \beta|\partial_t v(Y,t)|^2\right]\rho(Y,t)\mathrm{d}Y\mathrm{d}t$$
$$\text{with } \partial_t\rho + \partial_Y(v\rho) - \partial_{YY}[\tfrac{\sigma^2}{2}\rho] = 0,\ \rho(Y,0) = \rho^0(Y),\ \forall Y\in\mathbb{R},\ t\in]0,T[. \tag{20}$$

Problem (20) has a solution.

Proof. Note first that $J \ge \alpha v_m$. Note also that when u_M is large enough so that $v = 0 \in \mathcal{V}$, the solution of the *PDE* is $\tilde\rho(Y,t) = (4\pi)^{-\frac{1}{2}} \int_{\mathbb{R}} e^{-\frac{X^2}{4t}} \rho^0(X-Y)\mathrm{d}X$. As $\int_{-R}^R \tilde\rho(Y,t)e^{2Y}$ is finite and $\mathcal{V}$ is non empty the infimum is finite. Multiplying the *PDE* by ρ and integrating over $\mathbb{R}$ leads to

$$\int_{\mathbb{R}} [\partial_t|\rho^2| + \sigma^2|\partial_Y u|^2] \le 2\int_R |v|\rho\partial_Y\rho \le 2\|v\|_\infty \|\rho\|_{L^2(\mathbb{R})}\|\partial_Y\rho\|_{L^2(\mathbb{R})},$$
$$a.e.\ t\in]0,T[.$$

Consequently, and because $\|v\|_\infty \le v_M$, ρ is bounded in $L^2\left(H^1(\mathbb{R})\right) \cap L^\infty\left(L^2(\mathbb{R})\right) \cap L^2(\mathcal{Q})$ and $\partial_t\rho$ is bounded in $L^2\left(H^{-1}(\mathbb{R})\right)$. Consequently ρ is compact in $L^2(\mathcal{Q})$ when v varies in $\mathcal{V}$. From the maximum principle we also know that $0 \le \rho_n \le 1$ in $\mathcal{Q}$ and that $\int_{\mathbb{R}} \rho(Y,t)\mathrm{d}Y = 1$ a.e.

Now let us take a minimizing sequence $\{v_n\}_{n\in\mathbb{N}}$ for (20). Then ρ_n is bounded in $L^2\left(H^2(\mathbb{R})\right) \cap L^\infty\left(L^2(\mathbb{R})\right) \cap W^{1,2}\left(H^{-1}(\mathbb{R})\right)$ and compact in $L^2(\mathcal{Q}_R)$. It follows from above that there is a v such that, for a subsequence, $v_n \to v$ in L^∞-weak-star and weakly in $L^2(\mathcal{Q})$ and a ρ such that $\rho_n \to \rho$ strongly in $L^2(\mathcal{Q})$. Hence $\int_{\mathbb{R}} v_n\rho_n\partial_Y\hat\rho \to \int_{\mathbb{R}} v\rho\partial_Y\hat\rho$ for any $\hat\rho \in L^2\left(H^1(\mathbb{R})\right)$ and consequently ρ solves the *PDE* with v.

As $\rho_n \in [0,1]$ and by Fatou's lemma, $\int_{\mathcal{Q}_R}(e^Y - X_d)^2 \ge \liminf \int_{\mathcal{Q}_R}(e^Y - X_d)^2\rho_n \ge \int_{\mathcal{Q}}(e^Y - X_d)^2\rho$. By the weak (resp. strong) convergence of v_n (resp. ρ_n), $\int_{\mathcal{Q}_R} \alpha v_n\rho_n \to \int_{\mathcal{Q}_R} \alpha v\rho$.

To analyze the last term in the criteria we write it as

$$\beta\int_{\mathcal{Q}_R} |\partial_t v_n|^2\rho_n = \beta\int_{\mathcal{Q}_R} |\partial_t v_n|^2\rho + \beta\int_{\mathcal{Q}_R} |\partial_t v_n|^2(\rho_n - \rho).$$

By the lower semi-continuity of semi-norms, $\lim_{n\to\infty}\int_{\mathcal{Q}_R} |\partial_t v_n|^2\rho \ge \int_{\mathcal{Q}_R} |\partial_t v|^2\rho$. Finally the last term above is bounded by $\beta\|(\partial_t v_n)^2\|_{L^2(\mathcal{Q}_R)}\|\rho_n -$

$\rho\|_{L^2(Q_R)} \leq \beta K \|\rho_n - \rho\|_{L^2(Q_R)} \to 0$. Consequently, the criteria in (20) are l.s.c. □

Remark 3. If $\int_{\mathbb{R}^+} X^2 \rho^0(X) \mathrm{d}X < \infty$, it may be possible to let $R \to \infty$.

7.1 Computation of gradients

Consider the variational form of the Kolmogorov equation: find $\rho \in L^2(0, T, V)$ such that, for all $\hat{\rho} \in V$,

$$\int_{\mathbb{R}^+} \left(\hat{\rho}\partial_t \rho - (r - \kappa X - u) X \rho \partial_X \hat{\rho} + \frac{\sigma^2}{2} \partial_X (X^2 \rho) \partial_X \hat{\rho} \right) = 0, \quad \rho(0) \text{ given.}$$

Calculus of variations (see Lauriere and Pironneau, 2016 for details) leads to $\delta\rho(0) = 0$ and

$$\int_{\mathbb{R}^+} \left(\hat{\rho}\partial_t \delta\rho - (r - \kappa X - u) X \delta\rho \partial_X \hat{\rho} + \frac{\sigma^2}{2} \partial_X (X^2 \delta\rho) \partial_X \hat{\rho} \right)$$
$$= - \int_{\mathbb{R}^+} \rho X \partial_X \hat{\rho} \delta u.$$

Define the adjoint p' by $p'(T) = 0$ and, for all $\hat{p} \in V$:

$$\int_{\mathbb{R}^+} \Big(\hat{p}\partial_t p' + \hat{p}(r - \kappa X - u) X \partial_X p' - \frac{\sigma^2}{2} \partial_X (X^2 \hat{p}) \partial_X p'$$
$$+ \hat{p} \mathbf{1}_\Omega \big[(X - X_d)^2 - \alpha u + \beta |\partial_t u|^2 \big] \Big) = 0.$$

Add the two equations above with $\hat{\rho} = p'$ and $\hat{p} = \delta\rho$. It gives

$$\int_{\mathbb{R}^+} \partial_t (p' \delta\rho) + \int_\Omega [(X - X_d)^2 - \alpha u + \beta |\partial_t u|^2] \delta\rho = \int_{\mathbb{R}^+} \rho X \partial_X p' \delta u.$$

As $p'(T) = 0$ and $\delta\rho(0) = 0$, an integration in time gives

$$\int_{\Omega \times]0,T[} [(X - X_d)^2 - \alpha u + \beta |\partial_t u|^2] \delta\rho = - \int_{Q^+} \rho X \partial_X p' \delta u.$$

Finally, from (18),

$$\begin{aligned}
\delta J &= \langle \text{grad}_u J, \delta u \rangle \\
&= \int_{\Omega \times [0,T]} \Big[\left((X - X_d)^2 - \alpha u + \beta |\partial_t u|^2 \right) \delta\rho - \alpha \rho \delta u + 2 \rho \partial_t u \partial_t \delta u \Big] \\
&= - \int_{Q^+} \rho X \partial_X p' \delta u + \int_{\Omega \times]0,T[} \rho [-\alpha \delta u + 2 \partial_t u \partial_t \delta u].
\end{aligned}$$

7.2 Results

A steepest descent algorithm with fixed step size and projected gradient on the box constraints is

$$v = u^k - \mu \text{grad}_u J^k, \quad u^{k+1} = \max\{u_m, \min\{u_M, v\}\}. \tag{21}$$

Note that we have constrained u to satisfy $\partial_t u(0) = \partial_t u(T) = 0$ so as to integrate by parts the term $\partial_t u \partial_t \delta u$. This is only a slight loss of generality. The same parameters as above have been used except that the computational domain is $(0, 5)$ instead of $(0, 3)$. The PDF of X_0 is a Gaussian curve with $\sigma_0 = 0.2$. The computer program is in the file `GPOP2dxt.edp`. After 100 iterations J is a fifth of its initial value. Fig. 15 shows the surfaces u, P, and ρ versus X, t. Note that $(X, t) \mapsto u(X, t)$ is either 0.5 or 1, implying that all controls $u(X_t, t)$ will be bang-bang. Figs. 16 and 17 show two applications, one with $X_0 = 0.8$ and the other with $X_0 = 1.2$. The method achieves a reasonable stabilization of the fish biomass, but the quotas vary rapidly, as if the β-term was not playing its regularizing role.

FIGURE 15 From left to right: optimal ρ, p', u. Horizontal axis is $X \in]0, X_M[$, vertical is $t \in [0, T]$ with $X_M = 5$, $T = 2$, $X_d = 1$, $u_m = 0.5$, $u_M = 1$. Note that $(X, t) \mapsto u(X, t)$ is either 0.5 or 1, implying that all controls $u(X_t, t)$ will be bang-bang. There is a vertical scale factor of 0.5 applied.

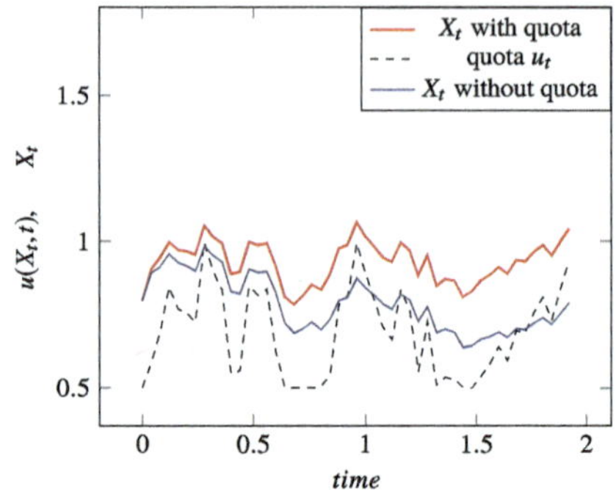

FIGURE 16 Optimal biomass and quota function computed by Optimal Control and compared with the solution with $u = u_M$ when $X_0 = 0.8$.

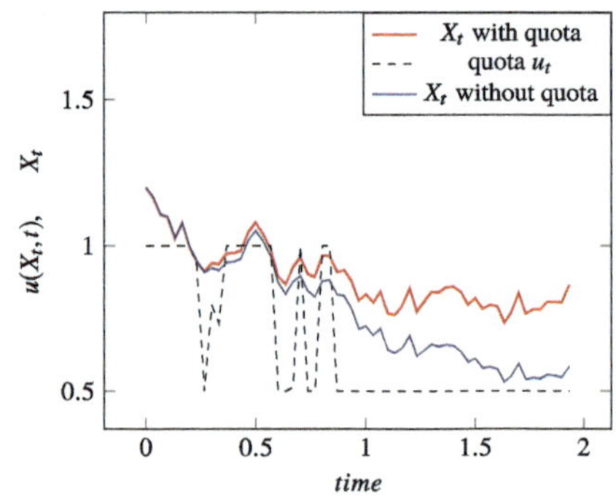

FIGURE 17 Optimal biomass and quota function computed by Optimal Control and compared with the solution with $u = u_M$ when $X_0 = 1.2$.

8 Solution by Itô calculus

In computational finance Itô calculus is preferred over the Kolmogorov setting, primarily because the initial condition for the PDF is singular.

Consider again Problem (1). Itô calculus applied to

$$j_t = \int_t^T \mathbb{E}\Big[(X_\tau - X_d)^2 + \beta(\partial_t u)^2 - \alpha u\Big]\mathrm{d}\tau,$$

yields[1]

[1] As a function of X_t random, j being deterministic has a *PDE* given by Itô calculus, roughly speaking as follows:

$$\mathrm{d}_t j = \partial_t j \mathrm{d}t + \mathbb{E}[\partial_X j \mathrm{d}X] + \tfrac{1}{2}\mathbb{E}[\partial_{XX} j \mathrm{d}\langle X\rangle]$$
$$= \mathrm{d}t\left(\partial_t j + \partial_X j(r - \kappa X - u)X + \frac{X^2\sigma^2}{2}\partial_{XX} j\right) = -\mathrm{d}t[(X - X_d)^2 + \beta(\partial_t u)^2 - \alpha u].$$

$$\partial_t j + (r - \kappa X - u)X\partial_X j + \frac{X^2\sigma^2}{2}\partial_{XX} j = -[(X - X_d)^2 + \beta(\partial_t u)^2 - \alpha u]. \tag{22}$$

Let us compute j as the solution at $\tau = 0$ and $X = X^0$ of the *PDE*, backward in time and initiated by $j(X, T) = 0$, $\forall X \in \mathbb{R}^+$, $\tau \in]0, T[$:

The optimal control problem (1) is now

$$\min_{u \in \mathcal{U}} \int_{\mathbb{R}} j(X, 0)\rho^0(X)\mathrm{d}X : \text{ subject to (22)} \tag{23}$$

where $\rho^0(X)$ is the PDF of X^0. We have to restrict $X < X_M$ for the right hand side of the *PDE* to be in $L^2(\mathbb{R})$. On the open interval $I := (0, X_M)$ the variational formulation of the *PDE* consists in finding j in a weighted Sobolev space:

$$j \in V := \{g \in L^2(\Omega) : X\partial_X g \in L^2(I)\},\ j(\cdot, T) = 0 \text{ and } \forall \hat{j} \in V, \tag{24}$$

$$\int_I (\partial_t j + (r - \kappa X - u)X\partial_X j)\hat{j} - \int_I \frac{\sigma^2}{2}\partial_X(X^2\hat{j})\partial_X j + \frac{\sigma^2}{2}X_M j(X_M)\hat{j}(X_M)$$
$$= -\int_I [(X - X_d)^2 + \beta(\partial_t u)^2 - \alpha u]\hat{j}.$$

8.1 Gradient computation

The gradient of j with respect to u results from the following calculus:

$$\int_{I\times]0,T[} \text{grad}_u j\delta u := \int_I \rho^0 \delta j(X^0, 0) \text{ subject to}$$
$$\int_I (\partial_t \delta j + (r - \kappa X - u)X\partial_X \delta j)\hat{j} - \int_I \frac{\sigma^2}{2}\partial_X(X^2\hat{j})\partial_X\delta j$$
$$+ \frac{\sigma^2}{2}X_M\delta j(X_M)\hat{j}(X_M) = \int_I \hat{j}\big[(X\partial_X j + \alpha)\delta u - 2\beta\partial_t u\partial_t\delta u\big],\ \forall \hat{j} \in V.$$

Define p by $p(0) = \rho^0$ and

$$\int_I \big(\hat{p}\partial_t p - (p(r - \kappa X - u)X\partial_X\hat{p}\big) + \int_I \frac{\sigma^2}{2}\partial_X(X^2 p)\partial_X\hat{p}$$
$$- \frac{\sigma^2}{2}X_M\hat{p}(X_M)p(X_M) = 0,$$

Then adding both equations with $\hat{p} = \delta j$ and $\hat{j} = p$ and remembering that $\delta j(X, T) = 0$, leads to

$$\int_I (p\partial_t\delta j + \delta j\partial_t p) = \int_{I\times]0,T[} [p(X\partial_X j + \alpha)\delta u - 2\beta\partial_t u\partial_t\delta u].$$

On the other hand,

$$-\int_I \rho^0\delta j(X, 0) = \int_I \delta j(X, T)p(T) - \int_I \delta j(X, 0)p(0) = \int_{I\times]0,T[} \partial_t(p\delta j).$$

Consequently

$$\int_{I\times]0,T[} \mathrm{grad}_u j\delta u = \int_I \rho^0 \delta j(X^0,0)$$
$$= -\int_{I\times]0,T[} [p(X\partial_X j + \alpha)\delta u - 2\beta\partial_t u \partial_t \delta u]\,.$$

Hence this method is a minor variation of the method with the Kolmogorov equation: ρ and p have similar equations and p' and j too.

9 Limit with vanishing volatility

We have seen earlier that the optimal solution of the non-dynamic control is very close to the solution of the deterministic control problem.

We explore here the convergence of the *PDE* method to the deterministic case by letting σ tend to zero. For clarity we take $\alpha = \beta = 0$.

Consider again (23). Let $Y(X,\tau)$ be the solution of

$$\begin{aligned} &\dot{Y}(\tau) = (r - \kappa Y(\tau) - u(Y(\tau),\tau))\,Y(\tau), \quad Y(t) = X; \\ &\text{then } \partial_t j + (r - \kappa X - u)X\partial_X j = \frac{\mathrm{d}}{\mathrm{d}\tau}\rho(Y(\tau,X),\tau)|_{\tau=t}. \end{aligned} \tag{25}$$

Hence (22) is rewritten as

$$\frac{\mathrm{d}}{\mathrm{d}\tau} j(Y(\tau,X),\tau)|_{\tau=t} = -\tfrac{\sigma^2}{2}\partial_{XX}(X^2\rho) - (X - X_d)^2.$$

Consequently, when $\sigma \to 0$, j becomes equal to $\tilde{j}$, the integral of $-(X - X_d)^2$ on the streamline (25) which passes through X^0 at $t = 0$:

$$\begin{aligned} &\tilde{j}(Y(T),T) - \tilde{j}(Y(0),0) = -j(X^0,0) = -\int_{Y(0)}^{Y(T)} (X(s) - X_d)^2 \\ &i.e.\ \tilde{j}(0) = \int_0^T (X(t) - X_d)^2 \mathrm{d}t : \ \dot{X} = (r - \kappa X - u)X, \quad X(0) = X^0. \end{aligned} \tag{26}$$

This analysis shows that *HJB* becomes numerically singular when σ is too small.

10 Conclusion

Let us compare the 3 classes of methods:

1. Non-dynamic control (NDC) solved by Monte-Carlo.
2. Dynamic Programming in a stochastic setting (DP).
3. Dynamic Programming solved with Partial Differential equations (*HJB*).

NDC is a mathematical oddity because it does not take into account the necessity of a Markovian solution forbidding an attempt to use the current noise level to predict the next step (feedback); but from the practical point of view it is certainly the best solution for the fishing site regulator, especially when it is coupled with Supervised Learning. After all, provided that the model fits the reality, a Markovian solution is not needed. By reinitializing the data every so often at $\{T_k\}_0^J$ and solving J control problem on smaller time intervals $[T_{j-1}, T_j]$, one could borrow from Model Predictive Control (see Boyd and Barratt, 1991). Yet, it can be seen on Fig. 4 that the precision is hardly better at early times than at later times.

On this fishing site control problem DP and *HJB* are different implementations of the same approach. All four implementations gave mathematically sound results, but hardly usable because bang-bang, because the Dirac ∂u_t when u is bang-bang, is not handled by the numerical algorithms. Note that a bang-bang saturation of constraints gives a solution which is not far from the elementary but intuitive strategy which consists in taking $u_t = u_m$ when $X_t > X_d$ and u_M otherwise; but we don't want such discontinuous policies. In other words the penalization of $|\partial_t u|$ works for NDC but not for DP and *HJB* as implemented here. A penalization of $|\partial_X u|$ is probably needed as well.

HJB with *PDE*s is the most taxing numerically and DP is much faster and also capable of generating approximate solution which do not saturate the constraints as much. NDC requires brute force Monte-Carlo, which is also very costly, yet with supervised learning NDC outperform *HJB* and DP methods.

This problem belongs to a class for which the user cannot measure precisely the random errors in the data and the model; the number of boats at sea and the quantity of fish caught everyday may just reflect the roughness of the model rather than the stochasticity of the measurements; nevertheless this point needs to be investigated further and in the mean time the non-dynamic control problem makes sense and gives better results. Furthermore, when coupled with a neural network the optimization process is very fast, once the learning phase is done. Finally, porting the trained network to a hand held device is also a major advantage.

Acknowledgments

We are thankful to François Murat for his help to minimize the hypotheses necessary for the existence of a solution to the Kolmogorov control problem.

The computer programs in Python and FreeFEM++ can be obtained by email from `olivier.pironneau@sorbonne-universite.fr`.

Appendix A A model with fishing quota

Let $X(t)$ be the biomass at time t, $E(t)$ the fishing effort – interpreted as the number of boats – and $Q(t)$ is the fishing capacity per boat. In Brochier et al. (2018) $Q(t) = qX(t)$, with the catchability q constant, meaning that the more fish there is the more fishermen will catch them. In this study a quota $Q(t)$, given to each fisherman, is imposed on the maximum weight of fish caught on a day

t; hence the total amount of fish caught on a day is $\min(qX(t), Q(t))E(t)$. The modified model is an equation for $X(t)$ which says that the biomass changes due to the natural growth/decay rate r, the long time limit κ' of X and the depletion due to fishing:

$$\dot{X}(t) = X(t)(r - \kappa' X(t)) - \min(qX(t), Q(t))E(t). \quad \text{(Biomass dynamics.)}$$

Let $F(t) = \min(qX(t), Q(t))$. The fishing effort is driven by the profit pF minus the operating cost of a boat c, where p being the price of fish.

$$\dot{E} = E(pF - c) \quad \text{(Fishing effort driven by profit,)}$$

The price is driven by the demand $D(p)$ and the resource FE:

$$\Phi\dot{p} = D(p) - FE \text{ with the demand } D(p) = \frac{a'}{1+\gamma p}$$

where Φ is the inverse time scale at which the fish market price adjusts. When $\Phi << 1$, it may be approximated by $D(p) = FE$ i.e. $pFE = \frac{a'}{\gamma} - \frac{FE}{\gamma}$. With $a = a'/\gamma$ and $\kappa = \kappa'\gamma$, the whole system is:

$$\dot{X} = X(r - \frac{\kappa}{\gamma}X) - \min(qX(t), Q(t))E, \quad \dot{E} = a - \min(qX(t), Q(t))\frac{E}{\gamma} - cE.$$

Now notice that by changing $X_0 \to X_0/\gamma$ we may take $\gamma = 1$.

Remark 4. Finally with $\tilde{t} = tq$, $\tilde{Q} = Q/q$, and $(\tilde{r}, \tilde{\kappa}'', \tilde{a}, \tilde{c}) = (r, \kappa, a, c)/q$ the above system is identical but now $q = 1$.

Remark 5. The *ODE* system is of the form $\dot{Y}(t) = \psi(Y(t))$, $t \in]0, T[$, $\psi : \mathbb{R}^2 \to \mathbb{R}^2$. Global existence of solution on $[0, T]$ is not known because ψ is only locally Lipschitz and not sublinear for large E, X.

A numerical simulation with and without optimal quota is shown on Fig. 18 with the following parameters (see file `fishingQuota0.py` and for the definition of α and β, see below):

$$r = 2, \kappa = 1.2, a = 1, c = 1, T = 2, M = 50, X_d = 1, q = 1.5$$
$$X_0 = 0.9, E_0 = 2, \alpha = 0.01, \beta = 0.001, u_m = 0.5, u_M = 1.5.$$

Notice that q is big enough so that $\min(qX(t), Q(t)) = Q(t)$, $\forall t$. Notice also that without quota the biomass tends to zero, in this case.

Appendix B Reformulation

To prevent extinction, a constraint is set on the total catch $\min(qX(t), Q(t))E(t))$. Optimization will then define $Q(t)$. It is expected that $Q(t) <$

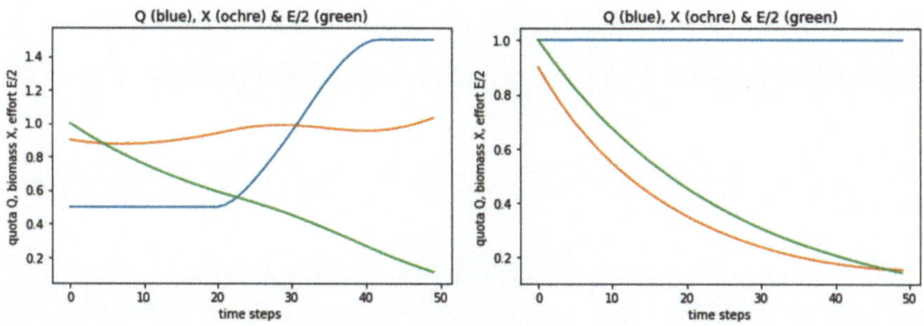

FIGURE 18 Left: Quota (blue) and fishing effort (ochre). No quota means that $Q(t) = 1$ for all t. Right: biomass with (blue) and without quota (ochre).

$qX(t)$, otherwise the quotas are irrelevant in the sense that the fisherman is given a maximum allowed catch which is greater than what he could possibly catch.

So let us add a constraint $Q(t) \leq Q_M$ so that $\min(qX(t), Q(t)) = Q(t)$. Denote $u(t) = E(t)Q(t)/X(t)$. Then we need $u(t) \leq qE(t)$ and the problem becomes

$$\min_{Q' \in \mathcal{Q}'} \left\{ \bar{J} := \int_0^T (X(t) - X_d(t))^2 \mathrm{d}t : \dot{X} = X\left(r - \kappa' X - u(t)\right), \; X(0) = X^0 \right\}$$

In the paper, the optimization is performed without computing E; then E is computed with the optimal u by solving $\dot{E} = a - cE - uX$, $E(0) = E^0$. As said before it is not feasible to impose too small a quota, so finally $\mathcal{U} = \left\{u_m \leq u(t) \leq u_M\right\}$ where u_M is less than $qE(t)$ at all time.

B.1 More constraints on quotas

Low quotas are unpopular. To avoid them we may use penalty and add to the criteria $-\int_0^T \alpha u(t)\mathrm{d}t$.

More over, it will be unpopular if $u(t) = u_m$ one day and $u(t) = u_M$ the next day, so to avoid too many changes we may add to the criteria a term like $\beta|\frac{du}{dt}|^2$.

An optimal policy is a solution of

$$\min_{Q \in \mathcal{Q}} J(u) := \int_0^T \left[|X - X_d|^2 - \alpha u + \beta |\frac{du}{dt}|^2 \right] : \text{ subject to}$$
$$\dot{X} = X\left(r - \kappa' X - u(t)\right), \quad X(0) = X^0.$$

Fig. 1 shows the solution with the same parameters as for Fig. 18. There are differences because it is a different set of constraints. The policy given to each fisherman is $Q(t)$ for the full system and $u(t)X(t)/E(t)$ for the reduced one.

Appendix C An analytical solution for a similar problem

It was said earlier that $Y_t = \log X_t$ simplifies the problem. If the goal $X_t \sim X_d$, is changed to $Y_t \sim Y_d$, then with $v_t = r - \frac{\sigma^2}{2} - \kappa e^{Y_t} - u$ the problem is

$$\min_{u\in\mathcal{U}}\left\{\int_0^T \mathbb{E}[(Y_t - Y_d)^2 - \alpha u + \beta|\partial_t u|^2 \text{ with } dY_t = v_t(u)dt + \sigma dW_t,\right\}$$
$$\Rightarrow 0 = \inf_{u\in\mathcal{U}}\left\{\partial_t V + v\partial_Y V + \tfrac{\sigma^2}{2}\partial_{YY}V + (Y - Y_d)^2 - \alpha u + \beta|\partial_t u|^2\right\}.$$

If the constraints are not active, the equation for the optimality of u is

$$\partial_Y V + \alpha + 2\beta\partial_{tt}u = 0.$$

When $\alpha = 0$ there is a polynomial solution to this system: $V = aY^2 + bY + c$, $v = KY + k$ where a, b, c, K, k are functions of time with $a(T) = b(T) = c(T) = \ddot{K}(T) = \dot{K}(T) = \ddot{k}(T) = \dot{k}(T) = 0$. By identification for all Y it is found that:

$$\dot{a} + 2aK + \beta\dot{K}^2 + 1 = 0, \quad \dot{b} + Kb + 2ak + 2\beta\dot{K}\dot{k} = 0,$$
$$\dot{c} + kb + \sigma^2 a + \beta\dot{k}^2 = 0, \quad \beta\ddot{K} + a = 0, \quad 2\beta\ddot{k} + b = 0.$$

One must also check that $u_t = r - \frac{\sigma^2}{2} - \kappa X_t - K(t)\log X_t - k(t)$ is in $\mathcal{U}$.

A solution procedure is to use the next to last equation in the first one and the last one into the second one:

$$-\dddot{K} - 2K\ddot{K} + \dot{K}^2 + \frac{1}{\beta} = 0,\ \ddot{K}(T) = \dot{K}(T) = K(T) = 0$$
$$-\dddot{k} - K\ddot{k} + \dot{K}\dot{k} - \ddot{K}k = 0,\ \ddot{k}(T) = \dot{k}(T) = k(T) = 0 \Rightarrow k(t) = 0\ \forall t.$$

Results, obtained with finite differences, are shown on Figs. 19 & 20.

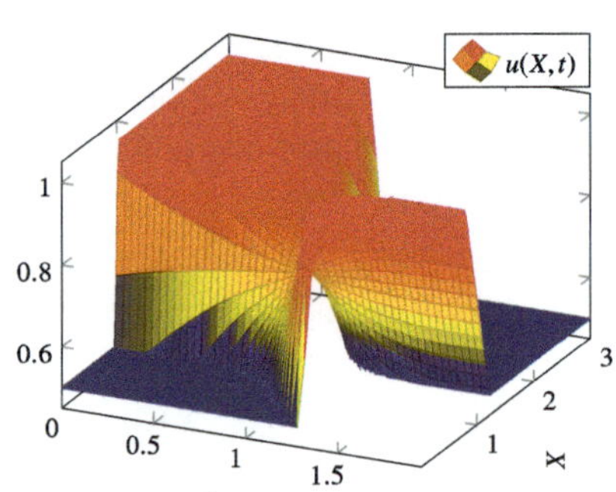

FIGURE 19 Solution of the Dynamic Programming equation: side view.

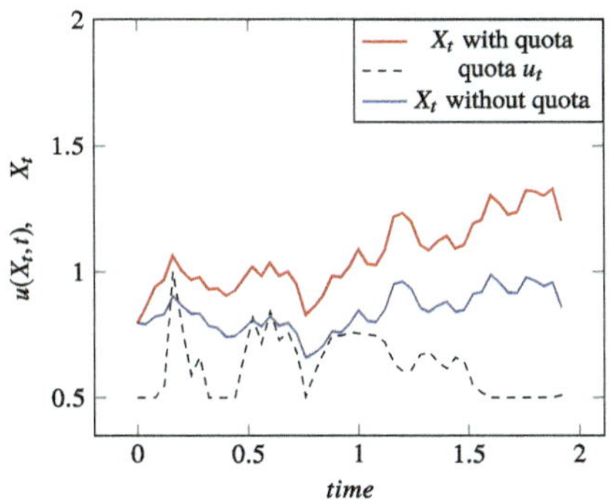

FIGURE 20 Optimal biomass and quota when $X_0 = 0.8$ and comparison with the solution with $u = u_M$.

References

Auger, P., Pironneau, O., 2020. Parameter identification by statistical learning of a stochastic dynamical system modelling a fishery with price variation. Comptes Rendus de L'Académie Des Sciences.

Balakrishnan, S., Biega, V., 1996. Adaptive-critic-based neural networks for aircraft optimal control. Journal of Guidance, Control, and Dynamics 19 (4), 893–898.

Bellman, R., 1957. Dynamic Programming. Princeton University Press, Princeton, NJ.

Bertsekas, D., 2019. Reinforced Learning & Optimal Control. Athena Scientific, Belmont Mass.

Boyd, S., Barratt, C., 1991. Linear Controller Design. Prentice-Hall.

Brochier, T., Auger, P., Thiao, D., Bah, A., Ly, S., Nguyen Huu, T., Brehmer, P., 2018. Can overexploited fisheries recover by self-organization? Reallocation of the fishing effort as an emergent form of governance. Marine Policy 95, 46–56.

Chollet, F., 2017. Deep Learning with Python. Manning Publications.

Goodfellow, I., Bengio, Y., Courville, A., 2016. Deep Learning. MIT-Bradford.

Hecht, F., 2012. New development in FreeFem++. Journal of Numerical Mathematics 20, 251–265. See also http://www.freefem.org.

Kamalapurkar, R., Walters, P., Rosenfeld, J., Dixon, W., 2018. Reinforcement Learning for Optimal Feedback Control. Springer.

Lauriere, M., Pironneau, O., 2016. Dynamic Programming for mean-field type control. Journal of Optimization Theory and Applications 169 (3), 902–924.

Le Bris, C., Lions, P.L., 2008. Existence and uniqueness of solutions to Fokker-Planck type eqs. with irregular coefficients. Communications in Partial Differential Equations 33, 1272–1317.

Pagès, G., 2018. Numerical Probability: An Introduction with Applications to Finance. Springer, Berlin, p. 574p.

Pagès, G., Pham, H., Printems, J., 2004. An optimal Markovian quantization algorithm for multidim. stochastic control problems. Stochastics and Dynamics 4 (4), 501–545.

Pagès, G., Printems, J., 2003. Optimal quadratic quantization for numerics: the Gaussian case. Monte Carlo Methods and Applications 9 (2), 135–165.

Yong, Jiongmin, Zhou, Xun Yu, 1991. Stochastic Controls Hamiltonian Systems and HJB Equations. Application of Mathematics Series, vol. 43. Springer.

Chapter 15

Data-driven modeling and control of large-scale dynamical systems in the Loewner framework

Methodology and applications

Ion Victor Gosea[a,*], Charles Poussot-Vassal[b], and Athanasios C. Antoulas[c,d,e]

[a]*Data-Driven System Reduction and Identification (DRI) Group, Max Planck Institute (MPI), Magdeburg, Germany,* [b]*Information Processing and Systems Department, ONERA, Toulouse, France,* [c]*Electrical and Computer Engineering (ECE) Department, Rice University, Houston, TX, United States,* [d]*Max Planck Institute, Magdeburg, Germany,* [e]*Baylor College of Medicine, Houston, TX, United States*

Corresponding author: e-mail address: gosea@mpi-magdeburg.mpg.de

Contents

Handbook of Numerical Analysis, Volume 23, ISSN 1570-8659, https://doi.org/10.1016/bs.hna.2021.12.015

Abstract
In this contribution we discuss the modeling and model reduction framework known as the Loewner framework. This is a data-driven approach, applicable to large-scale systems, which was originally developed for applications to linear time-invariant systems. In recent years, this method has been extended to a number of additional more complex scenarios, including linear parametric or nonlinear dynamical systems. We will provide here an overview of the latter two, together with time-domain extensions. Additionally, the application of the Loewner framework is illustrated by a collection of practical test cases. Firstly, for data-driven complexity reduction of the underlying model, and secondly, for dealing with control applications of complex systems (in particular, with feedback controller design).

Keywords
Data-driven modeling, Data-driven control, Loewner matrix, Rational approximation, Interpolation-based methods, Complex systems, Feedback controller design, Linear systems, Parametrized systems, Bilinear systems, Time-domain data, Time-delay systems

MSC Codes
93A15, 93A30, 93B11, 93B15, 93B52, 93C05, 93C10, 93C15

1 Introduction: data-driven modeling and control

The physical complexity of dynamical systems describing time-dependent processes stems from underlying non-linearities, the coupling dynamics, or the large amount of degrees of freedom (variables or parameters). The latter aspect is also related to enforcing specific accuracy requirements, that yield models of large dimension which are hence challenging to use for control purposes or for numerical simulations.

Simulating such complex dynamical systems is currently a common feature of many numerical software toolboxes, and is widely used both in industry and in academia. As numerical simulations become more involved, processing of increased amounts of data is required. Consequently, the number of variables under analysis is limited to the physical ones (even in the era of machine learning), while the rest are discarded. Computing simplified, easy to use dynamical models is one purpose of the model approximation and reduction discipline. Such models may then be used in a many query optimization and simulation process. That is why it is of critical importance to construct reliable surrogate models. Model reduction typically refers to a class of methodologies used for reducing the computational complexity of large-scale models of dynamical systems. The goal generally is to approximate the original system with a smaller and simpler system with the same structure and similar response characteristics as the original. For an overview of model reduction methods, we refer the reader to Antoulas (2005); Benner et al. (2015, 2017); Antoulas et al. (2020a), and to the references therein. In many practical scenarios, a complete mathematical description of the dynamical system under study is not always available or not fully known. Instead of depending only on physical laws (describing partial or ordinary differential equations), one can infer important properties of the system directly from measured or computed data.

With the increasing emergence of data-driven applications in many fields of the applied sciences, the need for incorporating measurements in the modeling and controlling stage has steadily grown over the last decades. The main challenge consists in using the available data in order to effectively construct surrogate models or controllers. In this latter case, the controller has to be designed based on experimental measurements, instead of a model description. Model-based methods can hence be replaced by data-driven strategies that construct the controller, directly from experimental data. Such techniques are also known as direct methods and can be useful when control-oriented models are either too complex or too costly to obtain.

The Loewner Framework (LF) is a data-driven model identification and reduction technique that was originally introduced in Mayo and Antoulas (2007). It is based on the Loewner-matrix interpolation method elaborated by the third author of the current paper, more than 20 years earlier, in the seminal contribution (Antoulas and Anderson, 1986). Using only measured data, the LF constructs surrogate models directly and with low computational effort. For recent tutorial papers on LF applied to linear dynamical systems, we refer the reader to Antoulas et al. (2017); Karachalios et al. (2020a). Extensions that use time-domain data were provided in Ionita and Antoulas (2012); Antoulas et al. (2018) (for a Hankel matrix approach) as well as in Peherstorfer et al. (2017) (for inferring transfer function measurements from time series). The Loewner framework has been recently extended to certain classes of nonlinear systems, such as bilinear systems in Antoulas et al. (2016a), and quadratic-bilinear systems in Gosea and Antoulas (2015, 2018). An adaptive extension of the original Loewner-based method in Antoulas and Anderson (1986), named the AAA (Adaptive-Antoulas-Anderson) algorithm, was recently proposed in Nakatsukasa et al. (2018); it is a data-driven rational approximation method that combines interpolation and least-squares (LS) fitting.

In the first part of this contribution, the Loewner framework is mainly used as a model identification and reduction tool. In the second part, the same framework is used for feedback controller design. In the proposed setup, the reference controller is not computed by means of a given model, but using input-output data of the system. Consequently, the Loewner framework is used for synthesizing a controller directly from measured data, which shows that it is also a data-driven control tool. Data-driven control strategies consist in recasting the control design problem as an identification one. By doing so, the model simplification process is shifted directly to the controller design step. The Loewner-based data-driven control methodology was extensively studied in recent years, starting with the original contribution (Kergus et al., 2017) and subsequently with Vuillemin et al. (2020); Gosea et al. (2021b); Poussot-Vassal et al. (2021a,b).

The main philosophy of the Loewner framework is as follows: starting with frequency response measurements (or, alternatively, with time-domain sequences of measured inputs and outputs), the data are arranged in a specific matrix format. Then, the dominant characteristics of the model are extracted by means of an appropriate projection (by means of **SVD**'s). Thus, simplified / reduced surrogate models can be computed without access to system's description. Such models typically need to preserve some of the original system's properties, such as stability or passivity. In what follows we will also cover practical issues such as stability preservation/enforcement, as presented in Gosea and Antoulas (2016).

We denote by $\mathbb{R}$ (resp. $\mathbb{C}$) the set of real (resp. complex) numbers, $\mathbb{C}_+$ ($\mathbb{C}_-$) the open right (left) half plane and the complex variable by $\imath = \sqrt{-1}$. Let $\mathbf{\Lambda}_{(\mathbf{A},\mathbf{E})} = \{z \in \mathbb{C} | \det(z\mathbf{E} - \mathbf{A}) = 0\}$ denote the set of singularities of the pencil $(\mathbf{A}, \mathbf{E})$ and, more generally, let $\mathbf{\Lambda}_{\mathbf{H}}$ denote the set of isolated singularities of the complex function $\mathbf{H}$.

The paper is organized as follows: after the introduction on data-driven modeling and control in Section 1, we continue with an overview on the Loewner framework for data-driven modeling in Section 2 with various subsections that cover specific extensions of the framework. Section 3 contains three model reduction examples in the Loewner framework, while Section 4 deals with the Loewner data-driven control rationale. This illustrates how the Loewner tool can be effective for both model-based or data-driven control approaches. Finally, 5 contains the concluding remarks together with a short summary of the paper.

2 The Loewner framework for data-driven modeling: an overview

2.1 Generalities on the Loewner framework and model structures

The Loewner framework is a data-driven method aimed at building a time invariant differential algebraic equation model / realization, with associated transfer function $\mathbf{H}_{\mathcal{I}}$ or $\mathbf{H}^{(\mathcal{J})}$ (defined later). This model interpolates data obtained from experimental measurements or the evaluation of a (collection of) transfer function(s). As made clearer later in this section, according to the mathematical structure and nature of the underlying system, $\mathbf{H}_{\mathcal{I}}$ has some specific properties.

Although in this section we mainly focus on the Loewner framework as a system identification tool, *i.e.*, by achieving (rational) interpolation properties, the main application of the method is indeed for model reduction purposes (when frequency response data are available). This will be illustrated in the third part of the manuscript, *i.e.*, in Section 3, with several examples of applying model order reduction. The model approximation and reduction provided in the Loewner framework is granted by means of numerical compression of the interpolation data (using a singular value decomposition). In this way, the compressed model attains similar interpolation properties as the uncompressed raw data model, but at the same time, it is of reduced order. Choosing a truncation value is often dictated by the application, and it is typically done as a compromise between enforcing high approximation quality at the cost of increasing the dimension of the compressed data models.

In its original form presented in Mayo and Antoulas (2007), $\mathbf{H}^{(N)}$ ($\mathcal{J} = \{N\}$) is a descriptor linear time invariant (**LTI**) dynamical model with transfer function $\mathbf{H}^{(N)} : \mathbb{C}\backslash\mathbf{\Lambda}_{(\mathbf{A},\mathbf{E})} \to \mathbb{C}^{p\times m}$, where $N \in \mathbb{N}$ denotes the number of collected data. We also denote with $\mathbf{H}_n$ the transfer function with McMillan degree n ($\mathcal{I} = \{n\}$). A complete description of this case is available in the recent surveys (Antoulas et al., 2017; Karachalios et al., 2020a). Extension to parametric **LTI** (**pLTI**) model structure also exists (Ionita and Antoulas, 2014). In this case, one obtains a multi-valued rational and polynomial transfer function $\mathbf{H}^{(N,M)} : (\mathbb{C} \times \mathbb{R})\backslash\mathbf{\Lambda}_{\mathbf{H}^{(N,M)}} \to \mathbb{C}^{p\times m}$ where $\mathcal{J} = \{N, M\}$ data are used (or $\mathbf{H}_{r,q}$, where $\mathcal{I} = \{r, q\}$), where $M \in \mathbb{N}$ is the number of data along the parameter variable (and $q \in \mathbb{N}$ is the rational and polynomial order along the parameter). The resulting function both interpolates the complex and real parametric variables. From a different perspective, extensions to nonlinear model structures have also been investigated. Among them, one can mention the bilinear and/or quadratic forms, explored in a series of papers (Gosea and Antoulas, 2015; Antoulas et al., 2016a; Gosea and Antoulas, 2018; Antoulas et al., 2019; Gosea et al., 2021a; Antoulas et al., 2020b). In these cases, the associated transfer function is a collection of multivariate coupled infinite cascade of linear systems reading as $\mathbf{H}^{(N_1)} : \mathbb{C}\backslash\mathbf{\Lambda}_{\mathbf{H}^{(N_1)}} \to \mathbb{C}^{p\times m}$, $\mathbf{H}^{(N_1,N_2)} : (\mathbb{C} \times \mathbb{C})\backslash\mathbf{\Lambda}_{\mathbf{H}^{(N_1,N_2)}} \to \mathbb{C}^{p\times m}$ and $\mathbf{H}^{(N_1,N_2,\ldots)} : (\mathbb{C} \times \mathbb{C} \times \ldots)\backslash\mathbf{\Lambda}_{\mathbf{H}^{(N_1,N_2,\ldots)}} \to \mathbb{C}^{p\times m}$ ($\mathcal{J} = \{N_1, N_2, \ldots\}$). The Loewner interpolation framework seeks a function interpolating the $N_1, N_2, \ldots$ data along each related multi-valued transfer functions $\mathbf{H}^{(N_1)}$, $\mathbf{H}^{(N_1,N_2,\ldots)}$ ($N_1, N_2, \cdots \in \mathbb{N}$). Similarly, we denote with $\mathbf{H}_{r_1,r_2,\ldots}$ the associated transfer function of order $r_1, r_2, \cdots \in \mathbb{N}$.

In all the cases mentioned here, the transfer function (or the set of transfer functions) is rational and polynomial, and it interpolates the data. In comparison to realization-driven model reduction, data-driven methods based on interpolation construct models that match the original transfer function(s) at well chosen points in the complex plane (also denoted as support points for barycentric representations (Berrut and Trefethen, 2006; Nakatsukasa et al., 2018)). As such, it provides a generalization of the Padé method to an arbitrary (set of) point(s). Data-driven methods based on rational and polynomial interpolation benefit also from the fact that it only requires the transfer function evaluation, whereas projection methods require the internal model (system matrices or operators). The latter are thus referred to as *intrusive* methods, while the former are *non-intrusive* or *data-driven* ones. In this section, a review of the Loewner framework is provided, together with some of its extensions. More specifically, Section 2.2 presents the Loewner framework in its original form, leading to a linear time invariant model. Extensions to linear parametric and to bilinear systems are sequentially illustrated in Sections 2.3 and 2.5. As a direct extension, the time-domain Loewner, dealing with sampled time-domain data instead of frequency domain data, is covered in Section 2.4.

2.2 The Loewner framework in the LTI case

The main ingredient of the Loewner framework is summarized next in the multi-input multi-output (**MIMO**) **LTI** case (Antoulas, 2005, chap. 4). Let us consider this system to be a m inputs p outputs dynamical one described by a n-th order differential algebraic equation (**DAE**) model $\mathcal{S} : (\mathbf{E}, \mathbf{A}, \mathbf{B}, \mathbf{C}, \mathbf{0})$[1] which explicitly reads as $\mathbf{E}\dot{\mathbf{x}}(t) = \mathbf{A}\mathbf{x}(t) + \mathbf{B}\mathbf{u}(t)$, $\mathbf{y}(t) = \mathbf{C}\mathbf{x}(t)$, where $\mathbf{E}, \mathbf{A} \in \mathbb{R}^{n\times n}$, $\mathbf{B} \in \mathbb{R}^{n\times m}$, $\mathbf{C} \in \mathbb{R}^{p\times n}$. Its associated transfer function $\mathbf{H} : \mathbb{C}\backslash\mathbf{\Lambda}_{(\mathbf{A},\mathbf{E})} \rightarrow \mathbb{C}^{p\times m}$ is

$$\mathbf{H}(\xi) = \mathbf{C}\mathbf{\Phi}(\xi)\mathbf{B} \text{ where } \mathbf{\Phi}(\xi) = (\xi\mathbf{E} - \mathbf{A})^{-1} \in \mathbb{C}^{n\times n}. \tag{1}$$

Importantly, as any rational and polynomial function, relation (1) can be characterized in its Lagrangian basis with distinct Lagrange nodes (or support points) $\lambda_i \in \mathbb{C}\backslash\mathbf{\Lambda}_{(\mathbf{A},\mathbf{E})}$. Then one can rewrite it in its barycentric formula form as follows (for $\alpha_i \neq 0$), $\mathbf{H}(\xi) = \sum_{i=1}^{n+1} \beta_i \mathbf{q}_i(\xi) / \sum_{i=1}^{n+1} \alpha_i \mathbf{q}_i(\xi)$ where $\mathbf{q}_i(\xi) = 1/(\xi - \lambda_i)$. Let this system generate right (column) data together with left (row) data, as

$$(\lambda_i, \mathbf{r}_i, \mathbf{w}_i) \text{ for } i = 1, \ldots, \overline{n} \text{ and } (\mu_j, \mathbf{l}_j^T, \mathbf{v}_j^T) \text{ for } j = 1, \ldots, \underline{n}, \tag{2}$$

where $\mathbf{w}_i = \mathbf{H}(\lambda_i)\mathbf{r}_i$ and $\mathbf{v}_j^T = \mathbf{l}_j^T\mathbf{H}(\mu_j)$, with $\mathbf{r}_i \in \mathbb{C}^{m\times 1}$, $\mathbf{l}_j \in \mathbb{C}^{p\times 1}$, $\mathbf{w}_i \in \mathbb{C}^{p\times 1}$, and $\mathbf{v}_j \in \mathbb{C}^{m\times 1}$ $(m, p \geq 1)$. In addition, we define the set of distinct interpolation points $\{z_k\}_{k=1}^N \subset \mathbb{C}$, leading to responses $\{\Phi_k\}_{k=1}^N \in \mathbb{C}^{p\times m}$, rearranged as follows ($N = \overline{n} + \underline{n}$),

$$\{z_k\}_{k=1}^N = \{\lambda_i\}_{i=1}^{\overline{n}} \cup \{\mu_j\}_{j=1}^{\underline{n}} \text{ and } \{\Phi_k\}_{k=1}^N = \{\Phi_i\}_{i=1}^{\overline{n}} \cup \{\Phi_j\}_{j=1}^{\underline{n}}. \tag{3}$$

The method then consists in building the *Loewner* matrix $\mathbb{L} \in \mathbb{C}^{\underline{n}\times\overline{n}}$ and *shifted Loewner* matrix $\mathbb{M} \in \mathbb{C}^{\underline{n}\times\overline{n}}$ defined as follows, for $i = 1, \ldots, \overline{n}$ and $j = 1, \ldots, \underline{n}$:

$$\begin{aligned}
\mathbb{L}_{(j,i)} &= \frac{\mathbf{v}_j^T\mathbf{r}_i - \mathbf{l}_j^T\mathbf{w}_i}{\mu_j - \lambda_i} = \frac{\mathbf{l}_j^T\big(\mathbf{H}(\mu_j) - \mathbf{H}(\lambda_i)\big)\mathbf{r}_i}{\mu_j - \lambda_i},\\
\mathbb{M}_{(j,i)} &= \frac{\mu_j\mathbf{v}_j^T\mathbf{r}_i - \lambda_i\mathbf{l}_j^T\mathbf{w}_i}{\mu_j - \lambda_i} = \frac{\mathbf{l}_j^T\big(\mu_j\mathbf{H}(\mu_j) - \lambda_i\mathbf{H}(\lambda_i)\big)\mathbf{r}_i}{\mu_j - \lambda_i}.
\end{aligned} \tag{4}$$

Additionally, let $\mathbb{W} = [\mathbf{w}_1, \cdots, \mathbf{w}_{\overline{n}}]$ and $\mathbb{V} = [\mathbf{v}_1, \cdots, \mathbf{v}_{\underline{n}}]^T$. Finally, let $\mathbf{\Lambda} = \mathrm{diag}(\lambda_1, \cdots, \lambda_{\overline{n}})$, $\mathbf{M} = \mathrm{diag}\big(\mu_1, \cdots, \mu_{\underline{n}}\big)$, $\mathbf{R} = [\mathbf{r}_1, \cdots, \mathbf{r}_{\overline{n}}]$, and $\mathbf{L} =$

[1] Fixing the last component to zero is not restrictive: see *e.g.* Karachalios et al. (2020a) where the direct feed-through term is equivalently incorporated in the $\mathbf{E}$ matrix.

$[\mathbf{l}_1, \cdots, \mathbf{l}_{\underline{n}}]$. The following Sylvester equations (Antoulas, 2005, chap. 6) are hence satisfied by the Loewner $\mathbb{L}$ and shifted Loewner $\mathbb{M}$ matrices:

$$\mathbf{M}\mathbb{L} - \mathbb{L}\mathbf{\Lambda} = \mathbb{V}\mathbf{R} - \mathbf{L}\mathbb{W} \text{ and } \mathbf{M}\mathbb{M} - \mathbb{M}\mathbf{\Lambda} = \mathbf{M}\mathbb{V}\mathbf{R} - \mathbf{L}\mathbb{W}\mathbf{\Lambda}. \quad (5)$$

Then, the descriptor realization $\mathcal{S}^{(N)}$ is given as,[2]

$$\mathbf{E}^{(N)}\dot{\mathbf{x}}(t) = \mathbf{A}^{(N)}\mathbf{x}(t) + \mathbf{B}^{(N)}\mathbf{u}(t) \text{ , } \mathbf{y}(t) = \mathbf{C}^{(N)}\mathbf{x}(t), \quad (6)$$

where $\mathbf{E}^{(N)} = -\mathbb{L} \in \mathbb{C}^{\underline{n}\times\overline{n}}$, $\mathbf{A}^{(N)} = -\mathbb{M} \in \mathbb{C}^{\underline{n}\times\overline{n}}$, $\mathbf{B}^{(N)} = \mathbb{V} \in \mathbb{C}^{\underline{n}\times m}$, and $\mathbf{C}^{(N)} = \mathbb{W} \in \mathbb{C}^{p\times\overline{n}}$ and which associated transfer function $\mathbf{H}^{(N)} : \mathbb{C}\backslash\mathbf{\Lambda}_{(\mathbb{L},\mathbb{M})} \rightarrow \mathbb{C}^{p\times m}$ (A^+ denotes the Moore–Penrose pseudo-inverse of A)

$$\mathbf{H}^{(N)}(\xi) = \mathbf{C}^{(N)}\mathbf{\Phi}^{(N)}(\xi)\mathbf{B}^{(N)} \text{ where } \mathbf{\Phi}^{(N)}(\xi) = (\xi\mathbf{E}^{(N)} - \mathbf{A}^{(N)})^{+} \in \mathbb{C}^{\underline{n}\times\overline{n}} \quad (7)$$

tangentially interpolates $\mathbf{H}$ at the given support points and directions defined in (2), *i.e.* satisfies the conditions

$$\mathbf{H}^{(N)}(\lambda_i)\mathbf{r}_i = \mathbf{H}(\lambda_i)\mathbf{r}_i \text{ and } \mathbf{l}_j^T\mathbf{H}^{(N)}(\mu_j) = \mathbf{l}_j^T\mathbf{H}(\mu_j). \quad (8)$$

Note that $\mathbf{H}^{(N)}$ is an interpolant of the data without any reduction. It refers to the realization constructed using the N available data.

From now on, let us assume that $\underline{n} = \overline{n}$, also referred to as the square case.[3] Moreover, assuming that the number $N = \underline{n} + \overline{n}$ of available data is large enough, then it was shown in Mayo and Antoulas (2007) that a minimal model $\mathbf{H}_r$ of dimension $r < \overline{n} = \underline{n}$ still satisfying the interpolatory conditions (8) can be computed by projecting the realization (6), provided that the following holds (for $k = 1, \ldots, N$)[4]

$$\mathbf{rank}(z_k\mathbb{L} - \mathbb{M}) = \mathbf{rank}([\mathbb{L}, \mathbb{M}]) = \mathbf{rank}([\mathbb{L}^H, \mathbb{M}^H]^H) = r, \quad (9)$$

where z_k are as in (3). Let $\mathbf{Y} \in \mathbb{C}^{\underline{n}\times r}$ (resp. $\mathbf{X} \in \mathbb{C}^{\overline{n}\times r}$) be the matrix containing the first r left (resp. right) singular vectors of $[\mathbb{L}, \mathbb{M}]$ (resp. $[\mathbb{L}^H, \mathbb{M}^H]^H$). Then, $\mathcal{S}_r : (\mathbf{E}_r, \mathbf{A}_r, \mathbf{B}_r, \mathbf{C}_r, \mathbf{0})$ where

$$\mathbf{E}_r = \mathbf{Y}^H\mathbf{E}^{(N)}\mathbf{X} \text{ , } \mathbf{A}_r = \mathbf{Y}^H\mathbf{A}^{(N)}\mathbf{X} \text{ , } \mathbf{B}_r = \mathbf{Y}^H\mathbf{B}^{(N)} \text{ and } \mathbf{C}_r = \mathbf{C}^{(N)}\mathbf{X}, \quad (10)$$

is a descriptor realization of $\mathbf{H}_r$, given as $\mathbf{H}_r(\xi) = \mathbf{C}_r\mathbf{\Phi}_r(\xi)\mathbf{B}_r$ where $\mathbf{\Phi}_r(\xi) = (\xi\mathbf{E}_r - \mathbf{A}_r)^{-1} \in \mathbb{C}^{r\times r}$ encoding a *minimal McMillan degree* equal to $\nu = \mathbf{rank}(\mathbb{L})$. Note that if r in (9) is greater than $\mathbf{rank}(\mathbb{L})$, then $\mathbf{H}_r$ may either have

2 The upper case letter N as superscript denotes here the number of considered data used to construct the realization.

3 The term square refers to the square shape of the dynamic matrices $\mathbf{A}$ and $\mathbf{E}$.

4 The lower case letter r as subscript denotes here the dimension of the realization instead of the number of data measurements (denoted with the upper case letter N).

a direct feed-through term or a polynomial part. Finally, the number r of singular vectors composing $\mathbf{Y}$ and $\mathbf{X}$ used to project the system $\mathbf{H}_r$ in (10) may be decreased at the cost of imposing an approximate interpolation of data, leading to the reduced order rational and polynomial model. This allows a trade-off between complexity of the resulting model and accuracy of the interpolation. The Loewner framework thus is a landmark appropriate for identification, approximationm and order reduction. Let us close this first part with two linear differential algebraic equations examples where the Loewner framework is applied. Both continuous and sampled-time cases are considered, highlighting how versatile this landmark is. More detailed, didactic examples are found in surveys (Antoulas et al., 2017; Karachalios et al., 2020a).

Remark 1. Because of its data-driven nature, the Loewner framework produces reduced-order models for which the approximation quality strongly depends on the data used in the process. One of the main attributes of the method is that it compresses the available information (interpolation conditions) and extracts relevant information. Moreover, it has been recently shown in Antoulas et al. (2017), Section 2.1.10, that the projected model indeed satisfies particular interpolation properties. There, the left and right interpolation points, values, and directions are explicitly provided for the projected Loewner model.

Remark 2. Other data-driven approaches such as the AAA algorithm (Nakatsukasa et al., 2018) enforce interpolation at particular data points, based on a greedy selection scheme. In this way, an indication of where to interpolate is provided.

Example 1 (Continuous-time rational and polynomial model interpolation). We consider the rational and polynomial model $\mathbf{H}(s) = s + 1/(s+1) = (s^2+s+1)/(s+1)$ with a realization $\mathcal{S} : (\mathbf{E}, \mathbf{A}, \mathbf{B}, \mathbf{C}, \mathbf{0})$ given by: $\mathbf{E} = [0\,1\,0;\ 0\,0\,1;\ 0\,0\,1]$, $\mathbf{A} = [1\,0\,0;\ 0\,1\,0;\ 0\,0\ -1]$, $\mathbf{B}^T = [0\,0\,1]$, and $\mathbf{C} = [1\,1\,1]$. By sampling $\mathbf{H}$ along $\lambda_i = \{1, 3, 5, 7\}$ and $\mu_j = \{2, 4, 6, 8\}$ with tangential directions $\mathbf{r}_i = \mathbf{l}_j = 1$ for $i, j = 1, \ldots, 4 = \overline{n} = \underline{n}$ ($N = 8$) leads to the measurements $\mathbf{w}_i = \{3/2, 13/4, 31/6, 57/8\}$ and $\mathbf{v}_j = \{7/3, 21/5, 43/7, 73/9\}$. Constructing the Loewner matrices as in (4), one obtains a 4-th order realization $\mathcal{S}_N : (-\mathbb{L}, -\mathbb{M}, \mathbb{V}, \mathbb{W})$. Following (9), the rank of the $[\mathbb{L}, \mathbb{M}]$ matrix is equal to $r = 3$. Computing the **SVD** of the $[\mathbb{L}, \mathbb{M}]$ matrix leads to the following normalized singular values $\sigma = \{1, 5.59 \cdot 10^{-2}, 6.8804 \cdot 10^{-4}, 5.8311 \cdot 10^{-17}\}$ and thus suggests to preserve the $r = 3$ first columns of $\mathbf{Y}$ and $\mathbf{X}$, as in (10). After projection, this leads to a minimal order realization which related transfer function exactly recovers the original $\mathbf{H}$, with McMillan degree of $\nu = 2$ and associated realization $r = 3$. In addition, computing the singularities of the associated pencil $(\mathbb{M}, \mathbb{L})$ gives $\{-1, \infty, \infty\}$, being exactly the one of the original model $\mathbf{H}$. The singularity in -1 is related to the rational part of $\mathbf{H}$, $1/(s+1)$ (finite dynamic mode). The two singularities in ∞ are related to the impulsive (polynomial part) and non-dynamic (direct feed-through term) modes.

Example 2 (Interpolation in the sampled-time). Let us consider the discrete-time model $\mathbf{H}(z) = z/(z - 1/2)$, with sampling period $h = 1$ second. We evaluate the function on the unit circle centered in zero. Then, by choosing $\lambda_i = \{e^{-\imath 0.1h}, e^{\imath 0.1h}, e^{-\imath 2h}, e^{\imath 2h}\}$, $\mu_j = \{e^{-\imath h}, e^{\imath h}, e^{-\imath 3h}, e^{\imath 3h}\}$ and tangential directions $\mathbf{r}_i = \mathbf{l}_j = 1$ for $i, j = 1, \ldots, 4 = \overline{n} = \underline{n}$ $(N = 8)$, one obtains $\mathbf{w}_i$ and $\mathbf{v}_j$. By construction, the Loewner matrices of dimension 4×4 contain complex entries. As data are provided in complex conjugate form, one may work with real arithmetic by projecting (see §2.5.4 Karachalios et al., 2020a for details).

By then solving (5), one readily obtains $\mathbb{L}$ and $\mathbb{M}$ and the associated 4-th order realization $\mathcal{S}_N : (-\mathbb{L}, -\mathbb{M}, \mathbf{V}, \mathbf{W}, \mathbf{0})$. Applying the rank revealing factorization (9) and (10), one obtains the McMillan degree $\nu = \mathbf{rank}(\mathbb{L}) = 1$ and $r = 2$. This suggests a constant term. By applying the procedure in Gosea et al. (2020), the direct term is reconstructed by the infinite eigenvalue computation of $(\mathbb{M}, \mathbb{L})$ pencil. In this case one finds $D = 1$. By removing it from the data and re-computing the Loewner procedure, one finds $\nu = r = 1$ and the sampled realization $(\mathbf{E}_1, \mathbf{A}_1, \mathbf{B}_1, \mathbf{C}_1, \mathbf{D}_1) = (2.897, 1.448, -0.9632, -1.504, 1)$, with transfer function $\mathbf{H}_1 = (z - 1.665 \times 10^{-16})/(z - 0.5)$, recovering almost perfectly $\mathbf{H}$.

2.3 Generalizations to parametric linear systems

The Loewner framework has been extended to parametric **LTI** (**pLTI**) systems, first in Antoulas et al. (2012) and in a more detailed manner in Ionita and Antoulas (2014).[5] In parametric model approximation and reduction, the aim is to construct reduced-order models that match the response of the original model / data, along the dynamical parameter ξ (usually complex) and along parameter ρ (real-valued). In what follows we will only show how the two variable case works, *i.e.* with one single parameter $\rho \in \mathbb{R}$ (for further extensions, see Ionita and Antoulas, 2014). We construct models which are reduced both with respect to the complex variable (frequency) and to the real one (parameter). In this configuration, let us consider such a m input p output ρ-parametrized dynamical system described by a n-th order differential algebraic equation (**DAE**) model denoted $\mathcal{S}(\rho) : (\mathbf{E}(\rho), \mathbf{A}(\rho), \mathbf{B}(\rho), \mathbf{C}(\rho), \mathbf{0})$ given as $\mathbf{E}(\rho)\dot{\mathbf{x}}(t) = \mathbf{A}(\rho)\mathbf{x}(t) + \mathbf{B}(\rho)\mathbf{u}(t)$, $\mathbf{y}(t) = \mathbf{C}(\rho)\mathbf{x}(t)$ where $\mathbf{E}(\rho), \mathbf{A}(\rho) \in \mathbb{R}^{n \times n}$, $\mathbf{B}(\rho) \in \mathbb{R}^{n \times m}$, $\mathbf{C}(\rho) \in \mathbb{R}^{p \times n}$, $\rho \in \mathbb{R}$, with associated transfer function $\mathbf{H} : (\mathbb{C} \times \mathbb{R}) \backslash \mathbf{\Lambda}_{(\mathbf{A}(\rho), \mathbf{E}(\rho))} \rightarrow \mathbb{C}^{p \times m}$ given for $\mathbf{\Phi}(\xi, \rho) = (\xi \mathbf{E}(\rho) - \mathbf{A}(\rho))^{-1} \in \mathbb{C}^{n \times n}$, as:

$$\mathbf{H}(\xi, \rho) = \mathbf{C}(\rho)\mathbf{\Phi}(\xi, \rho)\mathbf{B}(\rho). \tag{11}$$

[5] The approach developed in Ionita and Antoulas (2014) interpolates more frequencies and parameter combinations than the approach derived in Antoulas et al. (2012). This latter interpolates an extended Loewner matrix instead and leads to the coefficients of a rational and polynomial function given in the Barycentric form.

As for the Loewner case, let us assume that function (11) can be expressed using the distinct Lagrange support points λ_i and π_j, as (for $\alpha_{ij} \neq 0$)

$$\mathbf{H}(\xi,\rho) = \frac{\sum_{i=1}^{n+1}\sum_{j}^{m+1} \beta_{ij}\mathbf{q}_{ij}(\xi,\rho)}{\sum_{i=1}^{n+1}\sum_{j=1}^{m+1} \alpha_{ij}\mathbf{q}_{ij}(\xi,\rho)} \text{ where } \mathbf{q}_{ij}(\xi,\rho) = \frac{1}{(\xi-\lambda_i)(\rho-\pi_j)}. \tag{12}$$

Note that the above structure is not restrictive at all since it provides a parametrization as a function of support points (see Nakatsukasa et al., 2018; Berrut and Trefethen, 2006). Computation of the approximant is done in a similar way as for the non-parametric rational case: one seeks the β_{ij} and α_{ij} of the rational barycentric formula (12). Assuming that $\mathbf{H}(\xi,\rho)$ is sampled along the dynamical parameter ξ and the parametric one ρ as

$$\{z_k\}_{k=1}^{N} = \{\lambda_i\}_{i=1}^{\overline{n}} \cup \{\mu_j\}_{j=1}^{\underline{n}} \text{ and } \{p_l\}_{l=1}^{M} = \{\pi_i\}_{i=1}^{\overline{m}} \cup \{\nu_j\}_{j=1}^{\underline{m}}, \tag{13}$$

leading to $\mathbf{H}(z_k, p_l) = \Phi_{k,l}$. Thus the measurement matrix reads

$$\Phi = \begin{bmatrix} \Phi_{(11)} & \Phi_{(12)} \\ \Phi_{(21)} & \Phi_{(22)} \end{bmatrix} \in \mathbb{C}^{N\times M}, \tag{14}$$

where $\Phi_{(11)} = \Phi_{1,\ldots,\overline{n}/1,\ldots,\overline{m}} \in \mathbb{C}^{\overline{n}\times\overline{m}}$, $\Phi_{(12)} = \Phi_{1,\ldots,\overline{n}/1,\ldots,\underline{m}} \in \mathbb{C}^{\overline{n}\times\underline{m}}$, $\Phi_{(21)} = \Phi_{1,\ldots,\underline{n}/1,\ldots,\overline{m}} \in \mathbb{C}^{\underline{n}\times\overline{m}}$, and $\Phi_{(22)} = \Phi_{1,\ldots,\underline{n}/1,\ldots,\underline{m}} \in \mathbb{C}^{\underline{n}\times\underline{m}}$. The rows correspond to frozen values of z_k related to the dynamical (complex) ξ parameter. The columns correspond to frozen p_l values related to the (real) ρ parameter. Similarly to the non-parametric case mentioned in Section 2.2, one may construct the following one variable Loewner matrices: *(i)* $\mathbb{L}_2 \in \mathbb{C}^{\underline{nm}\times\overline{nm}}$ associated to $\Phi_{(11)}$ along $\lambda_i \bigcup \pi_j$, *(ii)* $\mathbb{L}_{\lambda_i} \in \mathbb{C}^{\underline{m}\times\overline{m}}$ associated to the i-th row of $[\Phi_{(11)}, \Phi_{(12)}]$ along p_l and *(iii)* $\mathbb{L}_{\pi_j} \in \mathbb{C}^{\underline{n}\times\overline{n}}$ associated to the j-th column of $[\Phi_{(11)}^H, \Phi_{(21)}^H]^H$ along z_k.

Then the global two dimensional Loewner matrix $\widehat{\mathbb{L}_2} = [\mathbb{L}_\lambda^H \, \mathbb{L}_\pi^H \, \mathbb{L}_2^H]^H \in \mathbb{C}^{(\overline{n}\underline{m}+\underline{n}\overline{m}+\underline{nm})\times(\overline{nm})}$, where $\mathbb{L}_\lambda = [(\mathbf{e}_1^T \otimes \mathbb{L}_{\lambda_1})^H \ldots (\mathbf{e}_{\overline{n}}^T \otimes \mathbb{L}_{\lambda_{\overline{n}}})^H]^H$ and $\mathbb{L}_\pi = [(\mathbb{L}_{\pi_1} \otimes \mathbf{e}_1^T)^H \ldots (\mathbb{L}_{\pi_{\overline{n}}} \otimes \mathbf{e}_{\overline{m}}^T)^H]^H$.

As in the non-parametric case, one important step is the determination of the minimal rational orders n and m in (12) hidden in the data collection. This is computed by evaluating the null-space of the single variable Loewner matrices combinations

$$r = \max_l \textbf{rank}\, \mathbb{L}_{p_l} \text{ and } q = \max_k \textbf{rank}\, \mathbb{L}_{z_k}, \tag{15}$$

where $\mathbb{L}_{p_l}$ and $\mathbb{L}_{z_k}$ are the one dimensional Loewner matrices associated to the k-th row and l-th column of Φ, respectively. Then, one can simply set $(\overline{n},\overline{m}) = (r+1, q+1)$, partition the data (13)-(14), and reconstruct $\widehat{\mathbb{L}_2}$. The two dimensional Lowner matrices ensure $\textbf{rank}\,\widehat{\mathbb{L}_2} = \textbf{rank}\,\mathbb{L}_2 = \overline{nm} - (\overline{n}-r)(\overline{m}-q) = \overline{nm} - 1$. The coefficients α_{ij} and β_{ij} of the two variables barycentric transfer function interpolating the data, are obtained by computing the null-space of $\widehat{\mathbb{L}_2}$ as $\mathbf{c} = \mathbf{ker}\,\widehat{\mathbb{L}_2}$ where $\mathbf{c} \in \mathbb{C}^{(r+1)\times(q+1)}$. Note that it is usually preferred to work with real arithmetic, *e.g.* for model time domain simulation or control design and analysis. In that case z_k are compiled in a closed conjugate form and support points are doubled (refer to §A.2 of Ionita and Antoulas, 2014 for detailed exposition). Note also that a trade-off between accuracy and complexity with both the frequency and the parameter variables can be obtained by decreasing the order r and q below the one given by (15).

Following the barycentric formulae, and Ionita and Antoulas (2014); Antoulas et al. (2012), one may reconstruct the associated multi-valued transfer function $\mathbf{H}_{r,q} : (\mathbb{C}\times\mathbb{R})\backslash\mathbf{\Lambda}_{\mathbf{H}_{r,q}} \to \mathbb{C}^{p\times m}$ as follows (where $\mathbf{\Phi}(\xi,\rho) \in \mathbb{C}^{N_{r,q}\times N_{r,q}}$, $N_{r,q} = r + 2q + 2$)

$$\mathbf{H}_{r,q}(\xi,\rho) = \mathbf{C}\mathbf{\Phi}^{-1}(\xi,\rho)\mathbf{B}$$
$$\text{where } \mathbf{\Phi}(\xi,\rho) = \begin{bmatrix} \mathbf{J}_{\lambda,r}(\xi) & \mathbf{0} & \mathbf{0} \\ \mathbb{A} & \mathbf{J}^T_{\pi,q}(\rho) & \mathbf{0} \\ \mathbb{B} & \mathbf{0} & [\mathbf{J}_{\pi,q}(\rho), \boldsymbol{\tau}] \end{bmatrix}, \tag{16}$$

with $\mathbf{B} = [\mathbf{0}, \boldsymbol{\tau}, \mathbf{0}]^T \in \mathbb{R}^{N_{r,q}}$ and $\mathbf{C} = [0, \ldots, 0, -1] \in \mathbb{R}^{N_{r,q}}$. Moreover, the following holds for $k = 1, \ldots, r$, $l = 1, \ldots, q$ and $\mathbf{w} = \text{vect}(\Phi_{(11)})$: $\mathbb{A}_{:,k} = [\mathbf{c}^H_{k,1} \cdots \mathbf{c}^H_{k,q+1}]^H$, $\mathbb{B}_{:,k} = [(\mathbf{c}_{k,1}\mathbf{w}_{k,1})^H \ldots (\mathbf{c}_{k,q+1}\mathbf{w}_{k,q+1})^H]^H$ and $\boldsymbol{\tau}_k^{-1} = \prod_{l=1,l\neq k}^{q+1}(\pi_k - \pi_l)$ and with

$$\mathbf{J}_{\eta,t}(x) = \begin{bmatrix} x-\eta_1 & \eta_2 - x & & \\ \vdots & & \ddots & \\ x-\eta_1 & & & \eta_{t+1} - x \end{bmatrix} \in \mathbb{C}^{t\times(t+1)}. \tag{17}$$

Notice that (16) depends only on the extended Loewner matrix null-space $\mathbf{c}$, support points $\{\lambda_i\}_{i=1}^{r+1}$, $\{\pi_j\}_{j=1}^{q+1}$ and the response data matrix $\{\Phi_{(11)}\}_{i,j=1}^{r+1,q+1}$.

Remark 3 (Minimal realization in the multi-parametric case). Realization (16) is no longer identical to the one in the single variable case as the resolvant $\mathbf{\Phi}(\xi,\rho)$ includes both the dynamic and parametric variables, leading instead to an order $N_{r,q}$ of r. Finding a minimal order realization is still an open problem.

Remark 4 (**SIMO**, **MISO**, and **MIMO** cases). Both **SIMO** and **MISO** cases can be addressed by tangential interpolating the data instead of the scalar (see §A.1 of Ionita and Antoulas, 2014 for details). Extension to the **MIMO** case is not solved yet. The tangential approach used in the non-parametric case and in most of multi-port interpolation frameworks (Gallivan et al., 2004) is not applicable as is. An alternative approach is presented in Lefteriu et al. (2011) but which "only" interpolates $\Phi_{(11)}$, forgetting $\Phi_{(12)}$ and $\Phi_{(21)}$.

Example 3 (Reynolds parameter dependent linearized Navier-Stokes model). Let us consider a two-dimensional open square cavity fluid-flow problem where air flows from left to right for three different Reynolds numbers $Re = 4000$, 5250, or 6000. Such a configuration, illustrated on Fig. 1 (top right), is described in Barbagallo et al. (2008) and in Poussot-Vassal and Sipp (2015). For simulation, Navier-Stokes equations are used along a mesh composed of 193,874 triangles, corresponding to $n = 680,974$ degrees of freedom for the velocity variables along the x and y axis. After linearization around three fixed points for Reynolds numbers, and discretization along the flow axis, $\{\mathbf{H}_l\}_{l=1}^{3}$ can be described as a **DAE** realization of order $n = 680,974$ where the input $\mathbf{u}(t)$ is the vertical pressure actuator located upstream of the cavity and the output $\mathbf{y}(t)$ is a shear stress sensor, located downstream of the cavity. A set of continuous-time n-th order realization $\mathcal{S}_l : (\mathbf{E}, \mathbf{A}_l, \mathbf{B}, \mathbf{C}, \mathbf{0})$ are obtained. In Poussot-Vassal and Sipp (2015), the **IRKA** approach (Gugercin et al., 2008) is used to sequentially approximate each realization with a low dimensional one. The interpolation along the parameter is done in a second step by interpolating each coefficients in the canonical basis of the obtained reduced models. Here instead, the parametric Loewner framework is applied. The frequency response of each configuration along $\{z_k\}_{k=1}^{N} = z_0 \bigcup \{\imath\omega_k, -\imath\omega_k\}_{k=1}^{100}$, where $z_0 \in \mathbb{R}_+$ and ω_k logarithmically-spaced frequencies. Then, twenty intermediate configurations between each Reynolds numbers $Re = 4000, 5250, 6000$ are constructed by linear interpolation. We obtain $\{z_k\}_{k=1}^{N=201}$, $\{p_l\}_{l=1}^{N=41}$ and thus $\Phi \in \mathbb{C}^{201\times 41}$. Our objective is to come up with a parametrized linear model that is able to faithfully reproduce the original transfer function data on a particular range of frequencies as well as on a parameter range.[6]

Fig. 1 (top left) depicts two types of singular values. The singular values drop indicates reduction orders $r = 30$ and $q = 20$ for building the two dimensional Loewner matrix. As a real valued rational function is preferred rather than a complex one, the twice more support points are considered and realization size is increased. The reduced linear parametric model is sampled over the same frequency and parameter range as before. When comparing to the original samples on Fig. 1 (bottom), the overall result is satisfactory, with a model of complexity $r = 60$ (instead of $n = 680,974$) and $q = 20$ (instead of a collection).

2.4 Generalization to modeling from time-domain data

For a linear, time-invariant **SISO** system, let the impulse response be denoted with: $\mathbf{h} = \{\cdots \mathbf{h}_{-2}, \mathbf{h}_{-1}, \mathbf{h}_0, \mathbf{h}_1, \mathbf{h}_2, \cdots\}$. Here we restrict our attention to causal systems: $\mathbf{h}_k = 0,\ k < 0$; furthermore it is assumed that $\mathbf{u}(t) = 0,\ t < 0$. Hence, one can write that

$$\mathbf{y}(t) = \mathbf{h}_0\mathbf{u}(t) + \mathbf{h}_1\mathbf{u}(t-1) + \cdots + \mathbf{h}_k\mathbf{u}(t-k) + \cdots, \quad t \in \mathbb{Z}_+. \tag{18}$$

[6] Additional details and the data are available at https://morwiki.mpi-magdeburg.mpg.de/morwiki/index.php/Fluid_Flow_Linearized_Open_Cavity_Model.

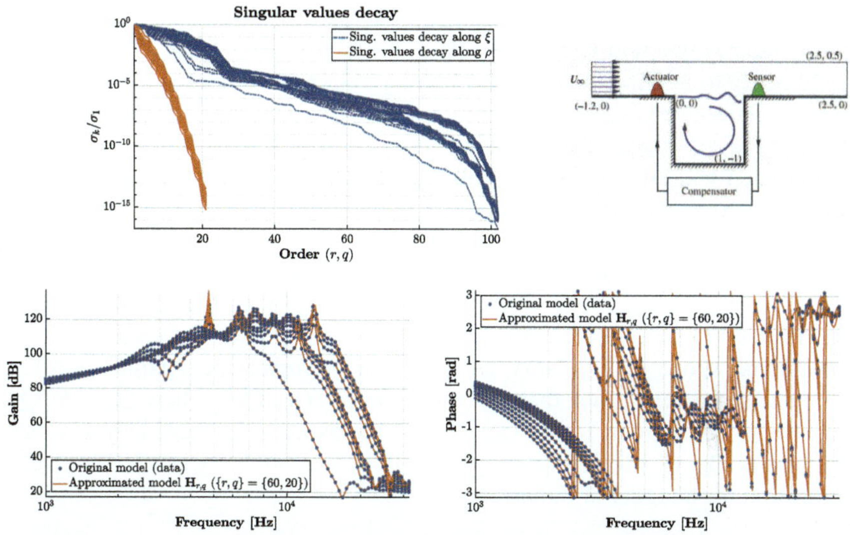

FIGURE 1 Top left: singular values drop of the one variable Loewner matrices (15). Top right: schematic view of the geometry (with illustration of the control structure used in Poussot-Vassal et al., 2018). Bottom frames: frequency response gain and phase of the original sampled data (blue dots) and resulting parametric model $\mathbf{H}_{60,20}$ for some parametric values (solid orange lines).

In the formulation above, $\mathbf{h}_j$ denotes the jth Markov parameter of the underlying system. In the time domain, the data are samples of input and output signals

$$\mathbf{u}_N = [\mathbf{u}_0, \ \cdots, \ \mathbf{u}_{N-1}], \ \ \mathbf{y}_N = [\mathbf{y}_0, \ \cdots, \ \mathbf{y}_{N-1}], \tag{19}$$

where, for simplicity, we have used the shortened expressions $\mathbf{u}_k := \mathbf{u}(k)$ and $\mathbf{y}_k := \mathbf{y}(k)$. The system identification problem consists in recovering a discrete-time linear time invariant system compatible with the data in (19). We seek a minimal realization $(\mathbf{E}, \mathbf{A}, \mathbf{B}, \mathbf{C}, \mathbf{D})$:

$$\mathcal{S}_D : \mathbf{E}\,\mathbf{x}(t+1) = \mathbf{A}\mathbf{x}(t) + \mathbf{B}\mathbf{u}(t), \ \ \mathbf{y}(t) = \mathbf{C}\mathbf{x}(t) + \mathbf{D}\mathbf{u}(t), \tag{20}$$

where $\mathbf{E}, \mathbf{A} \in \mathbb{R}^{n\times n}$, $\mathbf{B}, \mathbf{C}^T \in \mathbb{R}^{n\times 1}$, $\mathbf{D} \in \mathbb{R}$, with the transfer function

$$\mathbf{H}(z) = \mathbf{C}(z\mathbf{E} - \mathbf{A})^{-1}\mathbf{B} + \mathbf{D} = \frac{b_m z^m + \cdots + b_1 z + b_0}{z^n + \cdots + a_1 z + a_0}, \ m \leq n. \tag{21}$$

The Markov parameters in (18) can be explicitly written in terms of matrices from the realization in (20), as follows:

$$\mathbf{h}_0 = \mathbf{D}, \ \ \mathbf{h}_k = \mathbf{C}\mathbf{A}^{k-1}\mathbf{B}, \ \forall k \geq 1. \tag{22}$$

Moreover, another interpretation of Markov parameters is that they encode the behavior of the transfer function in $\mathbf{H}(z)$ in (21) at $z = \infty$. More precisely, the values $\mathbf{h}_k$'s represent the coefficients of the following Laurent series expansion of the transfer function $\mathbf{H}(z) = \mathbf{h}_0 + \mathbf{h}_1 z^{-1} + \mathbf{h}_2 z^{-2} + \cdots + \mathbf{h}_k z^{-k} + \cdots$.

In order to be able to accurately extract system invariants (poles, residues, Markov parameters, etc.) from input output data, there are certain conditions that need to be imposed to sequence of control inputs applied. For example, one of such conditions is the so-called *persistence of excitation*. However, as explained in Ionita and Antoulas (2012), this requirement of the input is not necessary when the initial conditions are zero. For any $k \geq 0$, we introduce Hankel matrices $\mathbf{U}_k \in \mathbb{R}^{M \times L}$, $\mathbf{Y}_k \in \mathbb{R}^{M \times L}$, as follows $(\mathbf{U}_k)_{i,j} = u_{k+i-1,k+j-1}$, $(\mathbf{Y}_k)_{i,j} = y_{k+i-1,k+j-1}$, for all $1 \leq i \leq M$, $1 \leq j \leq L$. Let also $\mathbf{Q}_0 = (\mathbf{I} - \mathbf{\Pi}_{\mathbf{U}_0})\mathbf{Y}_0$ and $\mathbf{Q}_1 = (\mathbf{I} - \mathbf{\Pi}_{\mathbf{U}_0})\mathbf{Y}_1$, where $\mathbf{\Pi}_{\mathbf{U}_0}$ be the orthogonal projection onto the column space of $\mathbf{U}_0$. Next, as explained in Ionita and Antoulas (2012), there exists matrix $\mathbf{Y}$, such that the matrix pencil $(\widehat{\mathbf{Q}}_0, \widehat{\mathbf{Q}}_1)$, where $\widehat{\mathbf{Q}}_0 = \mathbf{Y}^*\mathbf{Q}_0$, $\widehat{\mathbf{Q}}_1 = \mathbf{Y}^*\mathbf{Q}_1$, is regular (often $\widehat{\mathbf{Q}}_0$, $\widehat{\mathbf{Q}}_1$ can be taken as the leading $n \times n$ submatrices of $\mathbf{Q}_0$, $\mathbf{Q}_1$). The following result in Ionita and Antoulas (2012) gives a realization for a model of dimension n:

Theorem 1. *For zero initial conditions, the system has a minimal realization*

$$\widetilde{\mathbf{E}} = \widehat{\mathbf{Q}}_0, \quad \widetilde{\mathbf{A}} = \widehat{\mathbf{Q}}_1, \quad \widetilde{\mathbf{B}} = \mathbf{q}_0, \quad \widetilde{\mathbf{C}} = [\mathbf{h}_1, \cdots, \mathbf{h}_n], \quad \widetilde{\mathbf{D}} = \mathbf{h}_0,$$

where $\mathbf{q}_0$ is the first column of $\widehat{\mathbf{Q}}_0$ and the Markov parameters $\mathbf{h}_j$'s are obtained by solving the following linear system of equations

$$\begin{bmatrix} \mathbf{u}_0 & & & \\ \mathbf{u}_1 & \mathbf{u}_0 & & \\ \vdots & \ddots & \ddots & \\ \mathbf{u}_n & \cdots & \mathbf{u}_1 & \mathbf{u}_0 \end{bmatrix} \begin{bmatrix} \mathbf{h}_0 \\ \mathbf{h}_1 \\ \vdots \\ \mathbf{h}_n \end{bmatrix} = \begin{bmatrix} \mathbf{y}_0 \\ \mathbf{y}_1 \\ \vdots \\ \mathbf{y}_n \end{bmatrix}. \tag{23}$$

The Markov parameters can hence be computed for any input $\mathbf{u}$. The result in Theorem 1 can indeed be further specialized for the case of $\mathbf{u} = [1, 0, \cdots, 0]$. Hence, when the input is an impulse, the output is a finite sequence of Markov parameters, *i.e.*, $\mathbf{y} = [\mathbf{h}_0, \mathbf{h}_1, \cdots, \mathbf{h}_{N-1}]$. The realization in Theorem 1 is hence appropriately modified; let $\tilde{\mathcal{S}}_n$ be the new realization given by

$$\widetilde{\mathbf{E}} = \begin{bmatrix} \mathbf{h}_1 & \mathbf{h}_2 & \cdots & \mathbf{h}_n \\ \mathbf{h}_2 & \mathbf{h}_3 & \cdots & \mathbf{h}_{n+1} \\ \vdots & \vdots & \ddots & \vdots \\ \mathbf{h}_n & \mathbf{h}_{n+1} & \cdots & \mathbf{h}_{2n-1} \end{bmatrix}, \widetilde{\mathbf{A}} = \begin{bmatrix} \mathbf{h}_2 & \mathbf{h}_3 & \cdots & \mathbf{h}_{n+1} \\ \mathbf{h}_3 & \mathbf{h}_4 & \cdots & \mathbf{h}_{n+2} \\ \vdots & \vdots & \ddots & \vdots \\ \mathbf{h}_{n+1} & \mathbf{h}_{n+2} & \cdots & \mathbf{h}_{2n} \end{bmatrix}, \tag{24}$$

$$\widetilde{\mathbf{C}} = \left[\, \mathbf{h}_1, \ \mathbf{h}_2, \cdots, \ \mathbf{h}_n \,\right], \ \widetilde{\mathbf{B}} = \widetilde{\mathbf{C}}^T, \ \widetilde{\mathbf{D}} = \mathbf{h}_0.$$

As in Section 2.2, we could further reduce the dimension of the fitted model in (24) by means of projection (compressing the realization of order n to one of order r by means of orthogonal matrices computed using the **SVD** Antoulas, 2005, chap. 3). In this case, we enforce approximation, *i.e.* by fitting a model which approximately explains the data. Hence, let $\mathbf{Y} \in \mathbb{R}^{n\times r}$ (resp. $\mathbf{X} \in \mathbb{R}^{n\times r}$) be the matrix containing the first r left and respectively, right singular vectors of the Hankel matrix denoted in (24) by $\widetilde{\mathbf{E}}$. The reduced-order realization $\tilde{\mathcal{S}}_r$: $(\tilde{\mathbf{E}}_r, \tilde{\mathbf{A}}_r, \tilde{\mathbf{B}}_r, \tilde{\mathbf{C}}_r, \mathbf{0})$ is computed:

$$\tilde{\mathbf{E}}_r = \mathbf{Y}^T\tilde{\mathbf{E}}\mathbf{X}\,,\ \tilde{\mathbf{A}}_r = \mathbf{Y}^T\tilde{\mathbf{A}}\mathbf{X}\,,\ \tilde{\mathbf{B}}_r = \mathbf{Y}^T\tilde{\mathbf{B}}\,,\ \tilde{\mathbf{C}}_r = \tilde{\mathbf{C}}\mathbf{X}, \text{ and } \tilde{\mathbf{D}}_r = \mathbf{h}_0. \tag{25}$$

Example 4 (A structural mechanics model). As a numerical test case, we consider the model of a building (the Los Angeles University Hospital) from the NICONET benchmark examples collection (Niconet, 2002). The original model is a second-order linear system of dimension $n_0 = 24$. It can be written equivalently as a first-order linear system of dimension $n = 48$. We modify the original model by scaling the vector $\mathbf{B} \in \mathbb{R}^{48}$ with 10^4. Then, the original continuous-time LTI model of dimension $n = 48$ is discretized using a classical Backward Euler first order scheme. The simulation time horizon is $[0, 5]$s, while the time step is $\Delta t = 4 \cdot 10^{-3}$. The control input is chosen to be $u(t) = \frac{1}{10}\big(\cos(50t) + 2\cos(20t) + 3\cos(10t)\big)$. Hence, by means of this time-domain simulation, we collect $N = 2001$ measurements of the discretized input and output, *i.e.*, as in (19). These values are depicted in the left pane of Fig. 2. The Markov parameters are extracted by following the approach in Ionita and Antoulas (2012), and are depicted in the right pane of Fig. 2 (there, the magnitude of the error between the true Markov parameters and the estimated ones is shown in orange).

FIGURE 2 Samples of the input and output signals (left) and the true and recovered Markov parameters (right).

Next, form a 1000×1000 Hankel matrix as in (24). The decay of its singular values is displayed in the left pane of Fig. 3. Then, choose the truncation order $r = 20$, and construct a realization of order r as presented in (25). Finally, convert this discrete-time model back to the continuous time, and compare the frequency response of the original model of order n, with that of the reduced one of order r (on a range of 500 frequency points in the interval $[10^0, 10^2]$). The results (frequency responses and the approximation error) are presented in the right pane of Fig. 3. Indeed, the model is well approximated by means of the proposed method.

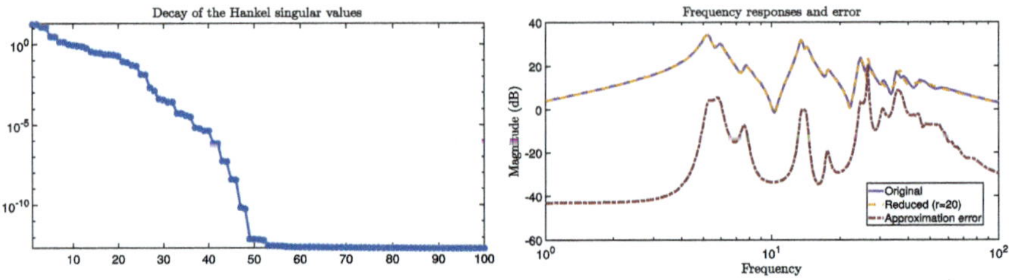

FIGURE 3 Decay of the Hankel singular values (left) and frequency responses computations: original, reduced and the approximation error (right).

Remark 5. As shown in this experiment, the estimated Markov parameters illustrated in Fig. 2 match the original ones quite well (up to $O(10^{-14})$ errors). This accurate matching is explained by the fact that the input and output were assumed to be measured without noise. However, in practical applications such measurements could be indeed perturbed by various exogenous sources. An exact quantification of how the noise affects the Markov parameter estimation will be studied in future works specifically for the method presented here (a generic analysis is readily available in Ljung, 1987, Chapter 8).

2.5 Extensions to nonlinear systems

Consider a nonlinear system described by the following equations $\mathcal{S}_N$

$$\dot{\mathbf{x}}(t) = \mathbf{f}(\mathbf{x}(t)) + \mathbf{g}(\mathbf{x}(t))\mathbf{u}(t), \quad \mathbf{y}(t) = \mathbf{C}\mathbf{x}(t), \tag{26}$$

where $t \geqslant 0$, $\mathbf{x}(0) = \mathbf{x}_0$ and the nonlinear functions $\mathbf{f}, \mathbf{g} : \mathbb{R}^n \to \mathbb{R}^n$ are assumed to be analytic in $\mathbf{x}(t)$. Here, we will focus on an extension of the Loewner framework to reducing bilinear systems. The motivation for this is that any smooth, nonlinear system with analytical nonlinearities can be approximated by a bilinear system. This is accomplished by means of Carleman linearization (Carleman, 1932; Rugh, 1981). Since this is based on Taylor expansion and truncation, the resulting bilinear system will approximate the original nonlinear system depending on the number of terms kept in the expansion. We proceed by writing the truncated *Taylor series* for the non-linear functions **f** and **g**, *i.e.*

$$\begin{cases} \mathbf{f}(\mathbf{x}) \approx \sum_{k=0}^{N} \mathbf{F}_k \mathbf{x}^{(N)} = \mathbf{F}_0 + \mathbf{F}_1\mathbf{x} + \mathbf{F}_2\mathbf{x}^{(2)} + \ldots + \mathbf{F}_N\mathbf{x}^{(N)}, \\ \mathbf{g}(\mathbf{x}) \approx \sum_{k=0}^{N-1} \mathbf{G}_k \mathbf{x}^{(k)} = \mathbf{G}_0 + \mathbf{G}_1\mathbf{x} + \ldots + \mathbf{G}_{N-1}\mathbf{x}^{(N-1)}, \end{cases} \tag{27}$$

where $\mathbf{F}_0, \mathbf{G}_0 \in \mathbb{R}^{n \times 1}$, $\mathbf{F}_j, \mathbf{G}_j \in \mathbb{R}^{n^j \times n^j}$, $j \geqslant 1$ and N is the truncation index. Here, the term $\mathbf{F}_0$ is usually chosen to be $\mathbf{0}$ in the case of non-existing forcing terms (if indeed is non-zero, it can be incorporated in the state vector by means of shifting). Additionally, $\mathbf{F}_1$, $\mathbf{G}_1$ denote the Jacobian matrices of **f** and **g**, respectively, and $\mathbf{F}_k$, $\mathbf{G}_k$ denote the matrices of higher derivatives. Moreover $\mathbf{x}^{(k)}$

denotes the Kronecker product of the state variable $\mathbf{x}$ with itself (k times). Then, introduce a new state variable

$$\mathbf{x}^{\otimes}(t) = \left[\mathbf{x}(t) \quad \mathbf{x}^{(2)}(t) \quad \ldots \quad \mathbf{x}^{(N)}(t)\right]^T \in \mathbb{R}^{n^{(N)}},$$

where $n^{(N)} = n + n^2 + ... + n^N = \frac{n^N - n}{n-1}$. This is obtained by concatenating all higher powers of vector $\mathbf{x}$ (up to N). In this way, by computing derivatives of $\mathbf{x}^{(k)}$, we obtain a bilinear system with the following realization

$$\dot{\mathbf{x}}^{\otimes}(t) \approx \mathbf{A}^{\otimes}\mathbf{x}^{\otimes}(t) + \mathbf{N}^{\otimes}\mathbf{x}^{\otimes}(t)\mathbf{u}(t) + \mathbf{B}^{\otimes}\mathbf{u}(t), \quad \mathbf{y} = \mathbf{C}^{\otimes}\mathbf{x}^{\otimes}(t), \tag{28}$$

where $\mathbf{x}^{\otimes}(0) = \mathbf{0}$ and the matrices $\mathbf{A}^{\otimes}, \mathbf{N}^{\otimes} \in \mathbb{R}^{n^{(N)} \times n^{(N)}}, \mathbf{B}^{\otimes}, \left(\mathbf{C}^{\otimes}\right)^T \in \mathbb{R}^{n^{(N)}}$ are as in (Gosea, 2017, Section 2.1.1). In what follows, we employ a more generic definition of *b*ilinear systems $\mathcal{S}_{\mathrm{B}} : (\mathbf{C}, \mathbf{E}, \mathbf{A}, \mathbf{N}, \mathbf{B})$, characterized by:

$$\mathbf{E}\dot{\mathbf{x}}(t) = \mathbf{A}\mathbf{x}(t) + \mathbf{N}\mathbf{x}(t)\mathbf{u}(t) + \mathbf{B}\mathbf{u}(t), \quad \mathbf{y}(t) = \mathbf{C}\mathbf{x}(t), \tag{29}$$

where $\mathbf{E}, \ \mathbf{A}, \ \mathbf{N} \in \mathbb{R}^{n \times n}$, $\mathbf{B} \in \mathbb{R}^{n \times m}$, $\mathbf{C} \in \mathbb{R}^{p \times n}$, and $\mathbf{x} \in \mathbb{R}^n$, $\mathbf{u}, \ \mathbf{y} \in \mathbb{R}$. The matrix $\mathbf{E}$ is assumed to be non-singular. Also, for simplicity of exposition, we will discuss only the **SISO** case. More details on bilinear system model order reduction can be found in Breiten and Damm (2010); Benner and Breiten (2012); Flagg and Gugercin (2015). Bilinear systems as in (29) are equivalent to an infinite collection of coupled linear time-varying systems:

$$\mathbf{E}\dot{\mathbf{x}}_1(t) = \mathbf{A}\mathbf{x}_1(t) + \mathbf{B}\mathbf{u}(t), \quad \mathbf{E}\dot{\mathbf{x}}_i(t) = \mathbf{A}\mathbf{x}_i(t) + \mathbf{N}\mathbf{x}_{i-1}(t)\mathbf{u}(t), \ i \geq 2. \tag{30}$$

The time-varying factor appears only in the matrices that scale the control input $\mathbf{u}(t)$ at each level $i \geq 2$. Based on (30), the solution of (29) is decomposed as $\mathbf{x}(t) = \sum_{i=1}^{\infty} \mathbf{x}_i(t)$. Furthermore, the input-output representation of the bilinear system $\mathcal{S}_{\mathrm{B}}$ can be expressed in terms of the *Volterra series representation* (Rugh, 1981; Flagg and Gugercin, 2015). Moreover, considering $\mathbf{x}_{i-1}(t)$ in the ith equation as a pseudo-input for $i = 1, 2, \ldots$, the frequency-domain behavior is described by a series of generalized transfer functions as given also in Rugh (1981); Flagg and Gugercin (2015); Antoulas et al. (2016a):

$$\mathbf{H}_i(s_1, s_2, \ldots, s_i) = \mathbf{C}\,\mathbf{\Phi}(s_1)\,\mathbf{N}\,\mathbf{\Phi}(s_2)\,\mathbf{N} \ \cdots \ \mathbf{N}\,\mathbf{\Phi}(s_i)\,\mathbf{B}, \tag{31}$$

where the resolvent of the pencil $(\mathbf{A}, \mathbf{E})$ is denoted by $\mathbf{\Phi}(\xi) = (\xi\mathbf{E} - \mathbf{A})^{-1}$. The characterization of bilinear systems by means of the rational functions in (31) suggests that reduction of such systems can be performed by means of the Loewner framework. In what follows, we will review some highlights of the procedure originally presented in Antoulas et al. (2016a). We use the concept of multi-tuples, composed of multiple interpolation points corresponding to evaluations of the transfer functions in (31). For simplicity, we will assume that one set of right multi-tuples $\boldsymbol{\lambda}$, and one set of left multi-tuples $\boldsymbol{\mu}$ with the same number of interpolation points (denoted with k), are given as

$$\begin{aligned} \boldsymbol{\lambda} &= \{\{\lambda_1\}, \{\lambda_2, \lambda_1\}, \ \ldots, \ \{\lambda_k, \ldots, \lambda_2, \lambda_1\}\}, \\ \boldsymbol{\mu} &= \{\{\mu_1\}, \{\mu_1, \mu_2\}, \ \ldots, \ \{\mu_1, \mu_2, \ldots, \mu_k\}\}. \end{aligned} \tag{32}$$

For the tuples in (32), we introduce the associated generalized controllability and observability matrices, denoted with $\mathcal{R} \in \mathbb{C}^{n\times k}$, and respectively with $\mathcal{O} \in \mathbb{C}^{k\times n}$, as in Antoulas et al. (2016a), (Antoulas, 2005, chap. 4).

Given the above notations, we introduce the following matrices, *i.e.*, the generalized Loewner matrix $\mathbb{L}$, and the generalized shifted Loewner matrix $\mathbb{M}$

$$\mathbb{L} = -\mathcal{O}\mathbf{E}\mathcal{R} \in \mathbb{C}^{k\times k}, \ \ \mathbb{M} = -\mathcal{O}\mathbf{A}\mathcal{R} \in \mathbb{C}^{k\times k}. \tag{33}$$

In addition, we define matrices $\mathbb{T} = \mathcal{O}\mathbf{N}\mathcal{R} \in \mathbb{C}^{k\times k}$, $\mathbb{V} = \mathcal{O}\mathbf{B} \in \mathbb{C}^{k}$, and $\mathbb{W} = \mathbf{C}\mathcal{R} \in \mathbb{C}^{1\times k}$. Note that $\mathbb{L}$ and $\mathbb{M}$ as defined above are indeed Loewner matrices, that is, they can be expressed as divided differences of appropriate transfer function values of the underlying bilinear system; the following equalities hold:

$$\mathbb{L}(j,i) = \frac{\mathbf{H}_{j+i-1}(\mu_1, \ldots, \mu_j, \lambda_{i-1}, \ldots, \lambda_1) - \mathbf{H}_{j+i-1}(\mu_1, \ldots, \mu_{j-1}, \lambda_i, \ldots, \lambda_1)}{\mu_j - \lambda_i}$$

$$\mathbb{M}(j,i) = \frac{\mu_j\mathbf{H}_{j+i-1}(\mu_1, \ldots, \mu_j, \lambda_{i-1}, \ldots, \lambda_1) - \lambda_i\mathbf{H}_{j+i-1}(\mu_1, \ldots, \mu_{j-1}, \lambda_i, \ldots, \lambda_1)}{\mu_j - \lambda_i},$$

while $\mathbf{V}(j,1) = \mathbf{H}_j(\mu_1, \ldots, \mu_{j-1}, \mu_j)$, $\mathbf{W}(1,i) = \mathbf{H}_i(\lambda_i, \lambda_{i-1}, \ldots, \lambda_1)$, and $\mathbb{T}(j,i) = \mathbf{H}_{j+i}(\mu_1, \ldots, \mu_{j-1}, \mu_j, \lambda_i, \lambda_{i-1}, \ldots, \lambda_1)$. This result shows that all quantities of the bilinear Loewner surrogate model can be indeed computed using only data, and the realization is written concisely as

$$\hat{\mathbf{E}} = -\mathbb{L}, \ \ \hat{\mathbf{A}} = -\mathbb{M}, \ \ \hat{\mathbf{N}} = \mathbb{T}, \ \ \hat{\mathbf{B}} = \mathbb{V}, \ \ \hat{\mathbf{C}} = \mathbb{W}. \tag{34}$$

It was shown in Antoulas et al. (2016a), that the bilinear model of dimension k in (34) matches a total of $2k + k^2$ transfer function values of the original bilinear system of dimension n. If necessary, the model given is (34) is further reduced similarly to the classical linear case, e.g., as in (10). This is done by projecting with special matrices using the singular value decay of the Loewner pencil involved. This provides a useful indicator for choosing the truncation order (Antoulas et al., 2016a).

Example 5 (Viscous (bi)linearized Burgers' equation model). We consider a discretized model of the viscous Burgers' equation (previously presented also in Antoulas et al. (2016a)). The original partial differential equation is given by

$$\frac{\partial v(x,t)}{\partial t} + v(x,t)\frac{\partial v(x,t)}{\partial x} = \frac{\partial}{\partial x}\left(\nu\frac{v(x,t)}{\partial x}\right), \quad (x,t) \in (0,1) \times (0,T), \tag{35}$$

with initial and boundary conditions $v(x,0) = f(x)$, $x \in [0,1]$, $v(0,t) = u(t)$, $v(1,t) = 0$, $t \geqslant 0$. The viscosity coefficient $\nu(x,t) = \nu$ is assumed to be constant and we consider zero initial conditions. Finally, we assume that the left boundary is subject to a control.

Start with a spatial discretization of Eq. (35), using an equidistant step size $h = \frac{1}{n+1}$ where n denotes the number of interior points of the interval (0, 1). By using first-order derivative approximations schemes, a nonlinear model is obtained (with quadratic-bilinear nonlinearities). Next, use the Carleman bilinearization technique to approximate this n^{th} order nonlinear system with a bilinear system of order $\mathcal{N} = n^2 + n$.

Denote with $\mathcal{S}_B$ the 4970th order initial bilinear system obtained by means of the Carleman bilinearization. The first step is to collect samples from generalized bilinear transfer functions up to order two; the 400 interpolations points are chosen logarithmically spaced in the interval $[10^{-3}, 10^{3}]\imath$. Next, we construct the bilinear Loewner matrices as presented in this section, and display the singular value decay in the left pane of Fig. 4. We construct a reduced-order model of order $r = 32$; the poles are depicted in the right pane of Fig. 4.

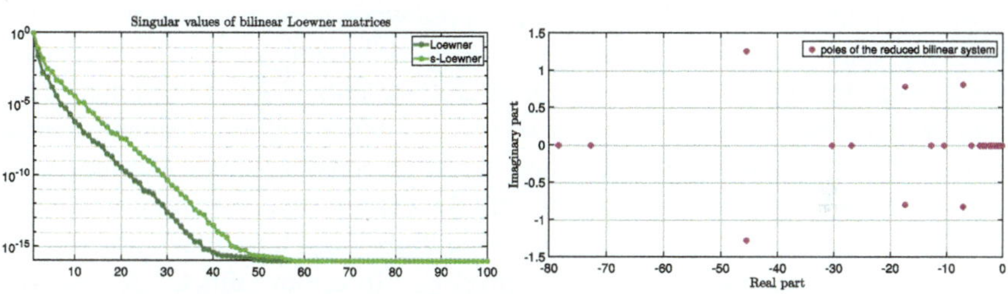

FIGURE 4 The first 100 sv's of the Loewner matrices (left) and the poles of the ROM (right).

Finally, perform time-domain simulation for a control input given by $u(t) = \frac{1}{5}(\cos(2\pi t) + \sin(20\pi t)e^{-t/5})$, on time span [0, 10]s. The observed outputs for both the original and of the reduced-order bilinear systems are displayed in the left pane of Fig. 5, while the error is depicted in the right pane of Fig. 5.

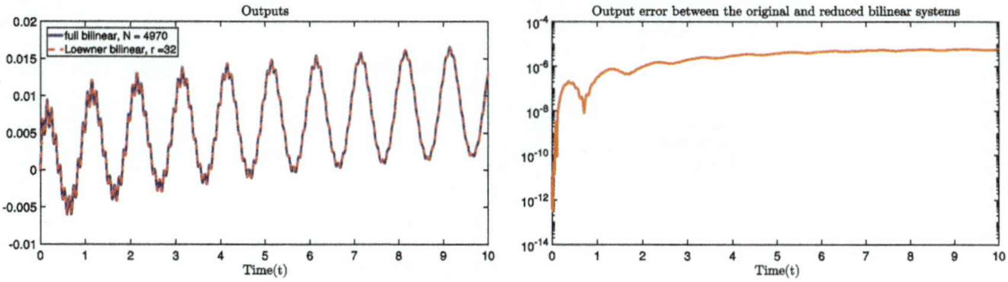

FIGURE 5 The observed outputs (left) and the approximation error in the time domain (right).

3 Model reduction examples (large-scale systems)

In this section, we illustrate how Loewner-based rational approximation and reduction features have been successfully applied to real-life industrial problems. First, two benchmarks sequentially involving a generic business jet aircraft model and measurements data obtained by Dassault-Aviation, a French aircraft supplier, are considered (Meyer et al., 2017; Poussot-Vassal et al., 2021c). Second, a benchmark involving a simplified open channel model constructed by Electricité De France, the French electricity supplier is involved (Dalmas et al., 2016). A gust oriented model described by a non-rational transfer function is considered (Section 3.1), ground vibration experimental data (Section 3.2), and a linear partial differential equation (Section 3.3).

3.1 Gust load oriented generic business jet aircraft model

The *gust load envelope* monitoring, is an important element to guarantee in aircraft structural integrity. One important certificate is to preserve and limit the worst case loads along the wings in response to vertical gust episodes. To this aim, we consider vertical gust disturbances $\mathbf{w}$, modeled through the so-called "1-cosine" profiles. The *gust load envelope* is the worst case load responses along the wing span in reaction to the set of many differently chosen time-domain vertical wind gust profiles affecting the aircraft structure. In the preliminary conception step, the aircraft is designed so that the wings support a given nominal load envelope, dictated by physical and industrial considerations. The larger the supported loads are, the larger the structural stiffeners and mass reinforcements should be, increasing its consumption during flight. Gust load alleviation (**GLA**) control is aimed at lowering the loads envelope. To achieve this **GLA** function, as illustrated in Fig. 6, model-based control design approaches are usually preferred. Following Poussot-Vassal et al. (2021c), we illustrate through a generic business jet aircraft model constructed by Dassault-Aviation, how the Loewner framework is a pivotal tool used in the industry to simplify the complexity of these dynamical models, prior control design and analysis.

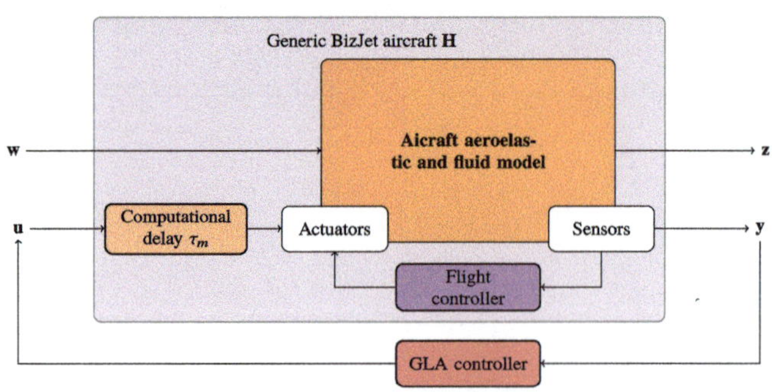

FIGURE 6 Closed-loop architecture of the **GLA** problem.

At each flight and mass configuration, a linear dynamical model is constructed. Generic aircraft models have the following continuous-time realization $\mathcal{S}$

$$\begin{aligned}\mathbf{E}\dot{\mathbf{x}}(t) &= \mathbf{A}_0\mathbf{x}(t) + \mathbf{A}_1\mathbf{x}(t-\tau_1) + \mathbf{A}_2\mathbf{x}(t-\tau_2) + \mathbf{B}_u\mathbf{u}(t) + \mathbf{B}_w\mathbf{w}(t)\,,\\ \mathbf{y}(t) &= \mathbf{C}_0\mathbf{x}(t) + \mathbf{C}_1\mathbf{x}(t-\tau_m)\end{aligned} \tag{36}$$

where $\mathbf{E}, \mathbf{A}_0, \mathbf{A}_1, \mathbf{A}_2 \in \mathbb{R}^{n\times n}$, $\mathbf{B}_u \in \mathbb{R}^{n\times n_u}$, $\mathbf{B}_w \in \mathbb{R}^{n\times n_w}$, $\mathbf{C}_0, \mathbf{C}_1 \in \mathbb{R}^{p\times n}$ and $\mathbf{x}(t) \in \mathbb{R}^n$, $\mathbf{u}(t) \in \mathbb{R}^{n_u}$, $\mathbf{w}(t) \in \mathbb{R}^{n_w}$ $(m = n_u + n_w)$ and $\mathbf{y}(t) \in \mathbb{R}^p$ are the internal variables, control input, exogenous gust input and output signals, respectively. In the considered case, $n_u = 3$, $n_w = 1$ $(m = 4)$, $p = 5$, and $n \approx 500$. The presence of internal delays is caused by the physical restitution of the gust impact over the fuselage at three different locations function of the aircraft velocity. Moreover, due to the model construction method (see *e.g.* Quero et al. (2021) or Poussot-Vassal et al. (2021c)), the $\mathbf{E}$ matrix may also be rank deficient. Here, due to the additional double derivative and delay structure added to accurately describe the gust disturbance effect along the fuselage, $\mathbf{rank}\,\mathbf{E} = n-6$. Following (36), the gust load model transfer associated function $\mathbf{H}$, from $[\mathbf{u}^T, \mathbf{w}^T]^T$ to $\mathbf{y}$ thus reads, $\mathbf{H}(s) = \left(\mathbf{C}_0 + \mathbf{C}_1 e^{-\tau_m s}\right)\left(s\mathbf{E} - \mathbf{A}_0 - \mathbf{A}_1 e^{\tau_1 s} - \mathbf{A}_2 e^{\tau_2 s}\right)^{-1}\mathbf{B} \in \mathbb{C}^{p\times m}$. We seek a simplified rational and polynomial model description to be used in place of $\mathbf{H}(s)$ for fast simulation, control design and (modal) analysis. The first step in the process consists in gridding the interpolation (support points) along the imaginary axis and collecting the associated response (with $\overline{n} = \underline{n} = n = 500$, $2n = N$, and $\omega_i \neq \omega_j$): $\{z_k\}_{k=1}^N = \{\imath\omega_i, -\imath\omega_i\}_{i=1}^{n/2} \cup \{\imath\omega_j, -\imath\omega_j\}_{j=1}^{n/2}$ and $\{\Phi_k\}_{k=1}^N = \{\Phi_i, -\overline{\Phi_i}\}_{i=1}^{n/2} \cup \{\Phi_j, -\overline{\Phi_j}\}_{j=1}^{n}$, where $\omega_i, \omega_j \in \mathbb{R}_+$ are the frequencies at which one evaluates each transfer $\mathbf{H}$. In our application, the ω_i, ω_j's are logarithmically spaced.

Remark 6 (About a Padé approximation). One option is to replace the delays with a Padé approximation. This preserves the gain but modifies the phase. While this is classically used in many applications, it is, to the authors experience, not the most accurate way to deal with internal and external delays. Indeed, Padé often results in significant error in the phase, which can be inappropriate for flexible structures. In addition, Padé will drastically increase the model internal dimension which in turn is not appropriate for model reduction. The accuracy / complexity ratio is not in favor of Padé (see also Fig. 7).

Fig. 7 illustrates the transfer function from the gust disturbance to a wing bending moment output, used to monitor the gust envelope. It compares the responses of the original irrational model $\mathbf{H}$ with its rational approximate model $\mathbf{H}_n$ constructed with Loewner matrices and its rational approximation obtained with Padé $\mathbf{H}_{\text{Padé}}$.

Fig. 7 emphasizes the good performance of the rational model obtained after reducing the complexity of the model. The phase is much well captured

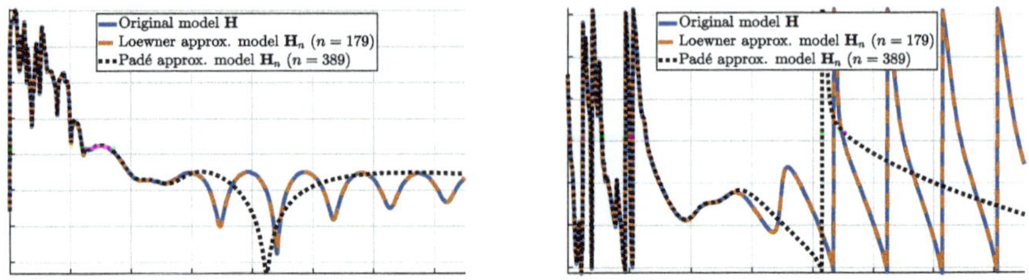

FIGURE 7 Left: frequency response gain. Right: frequency phase response. Comparison of the original model with the rational approximation obtained by Loewner and Padé.

by the Loewner approach than with Padé, while using less internal variables. In Poussot-Vassal et al. (2021c), this rational model is then used for **GLA** controller synthesis, leading to a load envelope reduction which is achievable thanks to the Loewner framework.

3.2 Ground vibration tests on business jet aircraft

We continue on the business aircraft benchmark provided by Dassault-Aviation, moving to the vibration problem, related to fatigue and comfort issues. Antivibration controllers are designed using model, targeting undesirable amplifications of the aerodynamical effects on the fuselage around some specified frequencies (Poussot-Vassal et al., 2013). After control design, Ground Vibration Tests (**GVT**) are performed to both validate the control performances and the original model. The benchmark considered here illustrates the generic business jet **GVT**, performed on a Falcon 7X at Istres, France, in 2015 (Meyer et al., 2017).

Dassault-Aviation engineers implemented the control law on the real business jet aircraft. Then, using shakers applied at some aircraft locations, the structure was excited, thus simulating aerodynamic disturbances. Hundreds of sensors were positioned on the aircraft and used for analysis.[7]

Fig. 8 (left) shows the singular frequency response of the (open-loop) data collected between a single-input and 100-outputs, compared with the frequency response of a rational models (with different complexity). The singular values drop is also illustrated on Fig. 8 (right).

In this industrial case, the Loewner framework is shown to be able to accurately recover the transfer function from raw data. It allows engineers to re-adjust the theoretical models with experiments, to detect some new phenomena and re-tune the control laws. This step contributes to the quest for a so-called *digital twin*.[8]

7 https://drive.google.com/file/d/1H2GqlYkiny_PZND2ekB6swSetoGcmFTK/view video illustrates the kinematic effect of the control law acting on the tail surface to reduce the vibrations.

8 Additional information may be found in Meyer et al. (2017) or in §2.4.7 of Poussot-Vassal (2019).

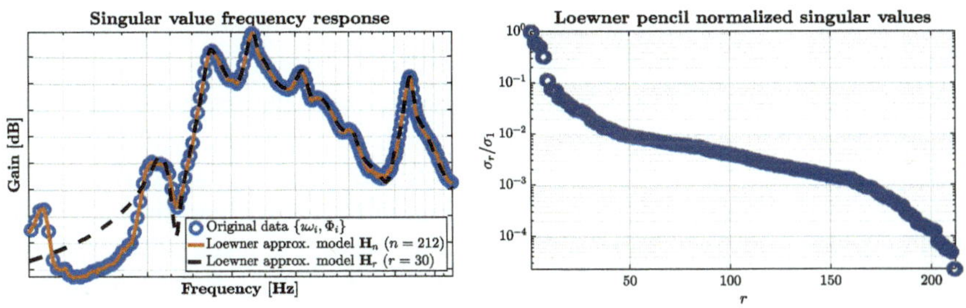

FIGURE 8 Left: singular value frequency response of the data (doted blue), minimal McMillan degree model (solid orange) and 30-th order model (dashed black). Right: Loewner singular values.

3.3 Hydroelectricity open-channel benchmark

Let consider a model representing the level h of an open-channel as a function of the inflow q_i and outflow q_o inputs. Such a model is used by hydro-electricity engineers from Electricité De France to monitor the level of a river in order to control the available energy. In France, in May 2021, the hydraulic energy represented about 10% of the total produced energy. These models in such benchmarks belong to the class of linear partial differential equations (**PDE**), coming from two nonlinear hyperbolic Saint-Venant equations, given as:

$$\frac{\partial S}{\partial t} + \frac{\partial Q}{\partial x} = 0 \text{ and } \frac{\partial Q}{\partial t} + \frac{\partial (Q^2/S)}{\partial x} + gS\frac{\partial H}{\partial x} = gS(I - J), \tag{37}$$

where $x \in [0, L]$ is the spatial variable, t the time variable, $H(x,t)$ the water depth, $S(x,t)$ the wetted area, $Q(x,t)$ the discharge, g the gravity acceleration and J the Manning-Strickler friction.[9] Under mild assumptions a linearization around an equilibrium point (Q_0, H_0), detailed in Dalmas et al. (2016), expresses the variation relations (q, h), between inflow (q_e, being q at $x = 0$), outflow (q_s, being q at $x = L$) and the water depth (h, at a given measurement point x) as follows, $h(x,s) = \mathbf{G}_e(x,s)q_e(s) - \mathbf{G}_s(x,s)q_s(s)$, where

$$\begin{aligned}\mathbf{G}_i(x,s) &= \frac{\lambda_1(s)e^{\lambda_2(s)L+\lambda_1(s)x} - \lambda_2(s)e^{\lambda_1(s)L+\lambda_2(s)x}}{B_0 s(e^{\lambda_1(s)L} - e^{\lambda_2(s)L})} \text{ and} \\ \mathbf{G}_o(x,s) &= \frac{\lambda_1(s)e^{\lambda_1(s)x} - \lambda_2(s)e^{\lambda_2(s)x}}{B_0 s(e^{\lambda_1(s)L} - e^{\lambda_2(s)L})}.\end{aligned} \tag{38}$$

For a frozen measurement point $x = x_m$, one has $h_{x_m}(s) = \mathbf{H}(s)\mathbf{u}(s) = \mathbf{G}_i(x_m,s)q_i(s) + \mathbf{G}_o(x_m,s)q_o(s)$, where $\mathbf{u}(s)$ contains the $q_i(s)$ and $q_o(s)$ and where $\mathbf{H}$ is now a one output two inputs complex-valued transfer function. Fig. 9 illustrates the approximation features and accurate reconstruction of the

[9] Numerical values of this model are provided at https://morwiki.mpi-magdeburg.mpg.de/morwiki/index.php/Hydro-Electric_Open_Channel.

open-channel phenomenon. To obtain this result, we consider complex conjugated points $\{z_k\}_{k=1}^{N} = \{\imath\omega_k, -\imath\omega_k\}_{k=1}^{N/2}$ (where $\overline{n} = \underline{n} = 300 = N/2$) sampled between 10^{-4} and $10^{1.5}$ in logarithmic space. Then, the responses $\mathbf{H}(s)$ and $\widetilde{\mathbf{H}}(s) = \mathbf{H}(s)s/(s+10^{-2})(s+10^{-3})$, are computed. Dealing with $\mathbf{H}$ remains standard with the framework presented so far. By approximating $\widetilde{\mathbf{H}}$ removes the integral action and enforces roll-off in high frequency, and thus allows to deal with limited energy functions. In this latter case, the resulting interpolated model should be post processed as $\widetilde{\mathbf{H}}_n \leftarrow \mathbf{H}_n(s+10^{-2})(s+10^{-3})/s$ to recover the original one.

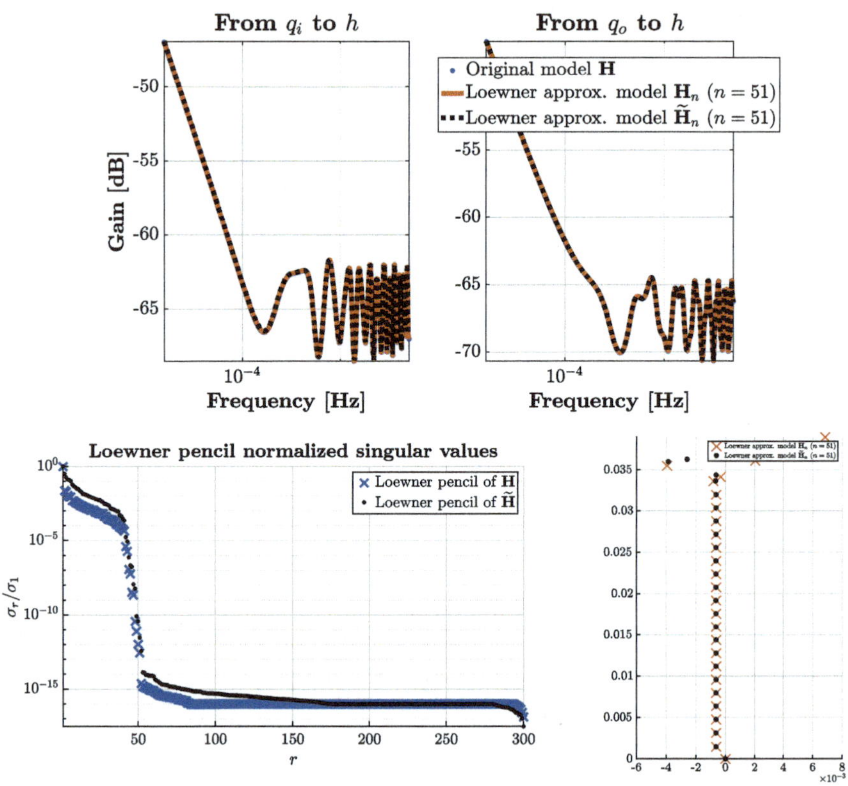

FIGURE 9 Top: frequency response comparison between the original irrational model and two approximated Loewner models. Bottom left: Singular values drop of the Loewner pencil for the two models. Bottom right: eigenvalues of the resulting minimal order rational approximation.

Both approaches lead to a perfect matching of the irrational transfer. Interestingly, working with $\widetilde{\mathbf{H}}$ instead of $\mathbf{H}$ leads to a model with all singularities on the left hand side plus the 0 one. Working with the shifted function $\widetilde{\mathbf{H}}$ illustrates how one can perform gray box identification by simply shifting the original data. Here, the integral action (physically known from open-channel models) is removed and added afterward.

4 Control in the Loewner framework

Let us now deviate from the original purpose of the Loewner framework, initially introduced to provide solutions to the identification, approximation and reduction problems through the lens of rational and polynomial function construction. Here instead, Loewner is used for feedback controller design. More specifically it is used as in some traditional loop shaping methods, to fit a reference controller. In the proposed setup, the reference controller is not computed by means of a model but in terms of input-output data. Two applications are illustrated. The first one involves experimental data and considers the design of a reference tracking controller applied on a pulsed fluidic actuator (**PFA**), see Section 4.2, (Poussot-Vassal et al., 2021a). The second considers a numerical benchmark representing the boundary control a wave equation, described by an infinite dimensional equation.

4.1 Data-driven control, virtual reference model and L-DDC rationale

In this section, the Loewner framework is used for synthesizing a controller directly from measured data, being then a data-driven control (**DDC**) framework.[10] **DDC** consists in recasting the control design problem as an identification one. The technique under consideration belongs to the so-called *reference model approaches* and more specifically relies on the definition of a so-called *ideal controller*, derived from a reference model. This framework is deployed in the frequency-domain, with the use of the Loewner framework as the identification tool in Kergus et al. (2017); Vuillemin et al. (2020), allowing to skip the controller complexity selection thanks to its rank properties (see Section 2). The Loewner data-driven control (**L-DDC**) is thus a combination of determining the ideal controller from frequency-domain data via a reference model and the use of the Loewner framework (Mayo and Antoulas, 2007) to construct a reduced order controller. Such an interpolatory-based data-driven control design solves problems faced by practitioners: *(i)* the controller design is directly obtained using open-loop raw data collected on the experimental setup, *(ii)* without any prior controller structure or order specification.

The **L-DDC** procedure boils down to two steps: first deriving the *ideal controller* definition and second the *controller identification* via interpolation in the Loewner framework. We recall the mains steps in the **SISO** case. Following Fig. 10, the objective is to find a controller with transfer function $\mathbf{K} : \mathbb{C}\backslash\mathbf{\Lambda_K} \to \mathbb{C}$ that minimizes the difference between the resulting closed-loop and a given user-defined reference model $\mathbf{M} : \mathbb{C}\backslash\mathbf{\Lambda_M} \to \mathbb{C}$. This is made possible through the definition of the ideal controller $\mathbf{K}^\star$, being the **LTI** controller that would have given the desired reference model behavior if inserted

[10] The reader may notice that **DDC** methods have a long history dating to the proportional, integral, derivative (**PID**) tuning method by Ziegler-Nichols in early 40's or the self tuning regulator by Åström in the 90's (see *e.g.* Ziegler and Nichols, 1942 for more details and references).

in the closed-loop. The latter is defined as $\mathbf{K}^\star = \mathbf{H}^{-1}\mathbf{M}(I-\mathbf{M})^{-1}$. In the data-driven case, this definition may be recast as a discrete set of equations (where $\{z_k\}_{k=1}^N \in \mathbb{C}, k=1,\ldots,N$)

$$\mathbf{K}^\star(z_k) = \mathbf{H}(z_k)^{-1}\mathbf{M}(z_k)(I-\mathbf{M}(z_k))^{-1}, \tag{39}$$

where $\mathbf{H}(z_k)$ is the evaluation of the considered model, if available. In an experimental context, one usually considers sampling $\mathbf{H}$ at $z_k = \imath\omega_k$ ($\omega_k \in \mathbb{R}_+$). Finding a controller $\mathbf{K}$ that fits $\mathbf{K}^\star(z_k)$ can be considered to be an identification problem. In the Loewner framework, the control design boils down to finding a transfer function $\mathbf{K}$ satisfying condition (39).

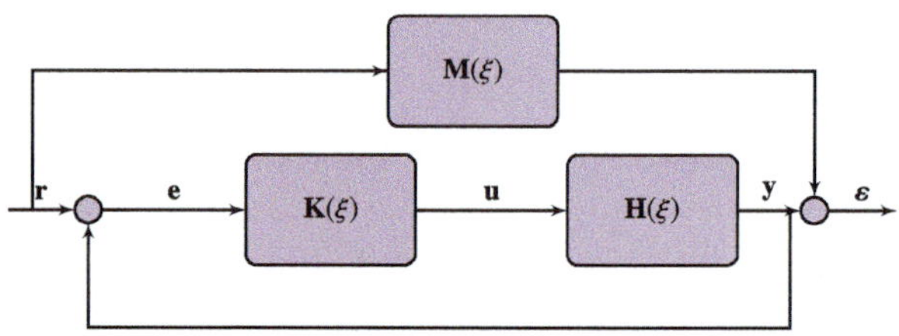

FIGURE 10 Data-driven control problem formulation.

4.2 Pulsed fluidic actuator control

The design of active closed-loop flow controllers constitutes an important field of research in fluid mechanics (Sipp and Schmid, 2016). In most contributions, both the sensor(s) and the actuator(s) are supposed to be lumped and ideal. To move towards experimental applications and real-life validations, it is essential to consider realistic set-ups and is the core contribution of Poussot-Vassal et al. (2021a), where the **L-DDC** is applied on a **PFA**. **PFA** are on/off actuators that blow air to modify the pressure in a flow setup. The control setup considered is schematized on Fig. 11.

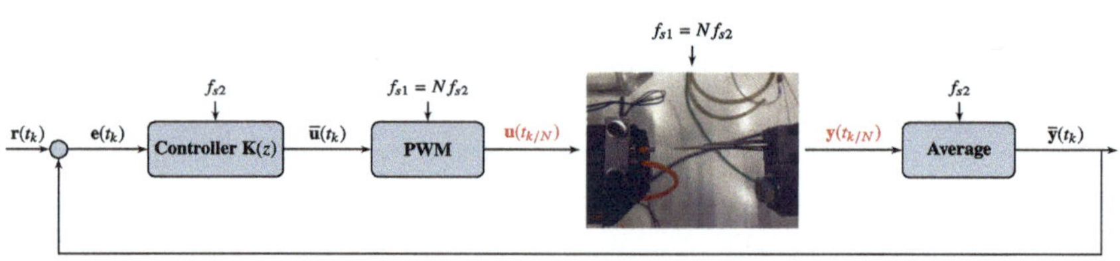

FIGURE 11 **PFA** control setup (details are in Poussot-Vassal et al., 2021a).

After exciting the **PFA** using a pseudo random binary sequence $\mathbf{u}(t_{k/n})$, output data $\mathbf{y}(t_{k/n})$ are collected. The corresponding frequency responses $\overline{\mathbf{u}}$ and $\overline{\mathbf{y}}$ are computed and transfer function values $\mathbf{H}(\imath\omega_k)$ are thus obtained. Applying (39) with $z_k = \imath\omega_k$ and the Loewner approach, it leads to a singular value decay indicating that a first or third order model is sufficient to recover the main dynamics (see Fig. 12). One important result is the ability of the **L-DDC** to construct, directly from raw open-loop data, a control law performing well on an experimental setup. Relevant in this context is that the **L-DDC** structure and complexity is almost automatically chosen by the Loewner framework, and no pole pre-assignment is required.

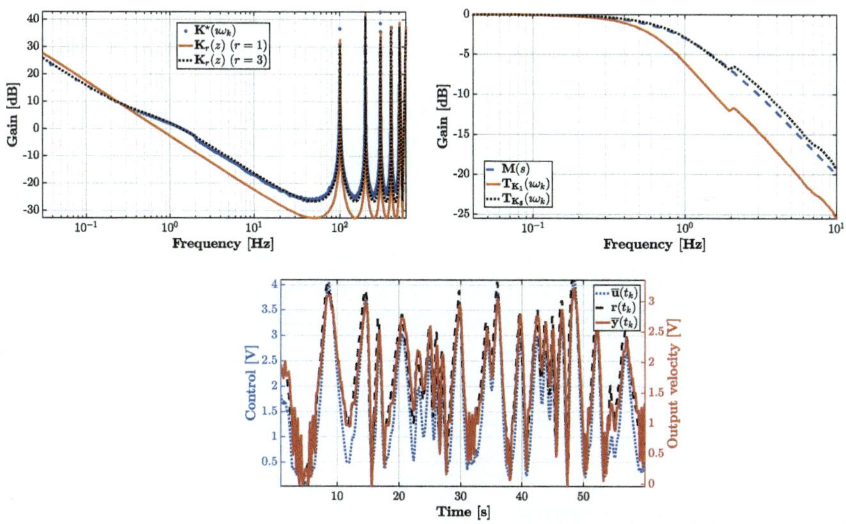

FIGURE 12 Top left: gain of the frequency responses of the ideal controller $\mathbf{K}^\star$ (blue dots) and of the estimated controller $\tilde{\mathbf{K}}_\mathbf{r}(s)$ of order $r = 1$ (solid orange) and $r = 3$ (dotted black). Top right: closed-loop response estimation using controller $\mathbf{K_1}$ and $\mathbf{K_3}$ of the averaged output $\overline{\mathbf{y}}(t_k)$ (solid orange and dotted black). Bottom: time-domain closed-loop response to a reference trajectory $\mathbf{r}(t_k)$ (dashed black), averaged control signal $\overline{\mathbf{u}}(t_k)$ (dotted blue) and averaged output (solid orange).

4.3 Transport phenomena benchmark

Let us consider the case of a one dimensional transport equation controlled at its left boundary through a second order actuator. This model is used in Poussot-Vassal et al. (2021b) and detailed in §2, Example 7 of Poussot-Vassal (2019). This phenomenon is represented by a linear **PDE** with constant coefficients interconnected with a second order linear **ODE** actuator, as described in (40).

$$\frac{\partial \tilde{y}(x,t)}{\partial x} + 2x\frac{\partial \tilde{y}(x,t)}{\partial t} = 0\,,\ \tilde{y}(x,0) = 0 \text{ and } \tilde{y}(0,t) = \frac{1}{\sqrt{t}}\tilde{u}_f(0,t) \tag{40}$$

where $\omega_0^2/(s^2 + m\omega_0 s + \omega_0^2)u(0,s) = u_f(0,s)$ and $x \in [0\ L]$ $(L = 3)$ is the spatial variable. Then, $\omega_0 = 3$ and $m = 0.5$ are the input actuator parameters. The scalar model input is the vertical force applied at the left boundary, *i.e.* at $x = 0$. We denote the input $\tilde{\mathbf{u}}(0,t)$ in the time-domain or $\mathbf{u}(0,s)$ in the complex one. Similarly, the output at location x is given as $\tilde{\mathbf{y}}(x,t)$ for the time-domain and $\mathbf{y}(x,s)$ in the complex one. By applying the Laplace transform, one obtains the transfer function from the input $\mathbf{u}(0,s)$ to the output $\mathbf{y}(x,s)$:

$$\mathbf{y}(x,s) = \frac{\sqrt{\pi}}{\sqrt{s}} e^{-x^2 s} \frac{\omega_0^2}{s^2 + m\omega_0 s + \omega_0^2} \mathbf{u}(0,s) = \mathbf{G}(x,s) u(0,s). \tag{41}$$

Let us now consider that one single sensor is available and is located at $x_m = 1.9592$ along the x-axis.[11] The transfer from $\mathbf{u}(0,s)$, denoted by $\mathbf{u}(s)$ to $\mathbf{y}(x_m,s)$, denoted by $\mathbf{y}(s)$ then reads $\mathbf{y}(s) = \mathbf{y}(x_m,s) = \mathbf{G}(s,x_m)\mathbf{u}(0,s) = \mathbf{H}(s)\mathbf{u}(s)$, where $\mathbf{H}(s)$ is now a **SISO** complex-valued irrational transfer function.

4.3.1 A model-driven approximation and control

By Loewner interpolation, the transfer function $\mathbf{H}$ can be approximated by a rational function $\mathbf{H}_r$ ($r = 33$). Then, standard feedback synthesis methods can be applied. In this example, the HINFSTRUCT function has been used (Apkarian and Noll, 2006). Details may be found in Poussot-Vassal et al. (2021b). Here the model-based $\mathcal{H}_\infty$-norm minimization oriented control design allows to construct a filtered proportional integral (**PI**) of the form $\mathbf{K}(s) = (k_p + k_i \frac{1}{s}) \frac{1}{s/a+1}$, where $k_p, k_i, a \in \mathbb{R}$.

4.3.2 Data-driven control

Let us now apply the **L-DDC** rationale, instead of a model-based control design. The reference model choice is a key factor for the **L-DDC** success, as for any other model reference control procedure. Indeed, the latter should not only represent a desirable closed-loop behavior, but also achievable dynamics of the considered system (*i.e.* the ideal controller should not internally destabilize the plant). A reference model is said to be achievable by the plant if the corresponding ideal controller internally stabilizes the plant. Here let us skip this point and focus on the equivalence of model vs. data-based design. Let the reference model $\mathbf{M}$ be the closed-loop rational function obtained by the previous approach interconnecting $\mathbf{H}_r$ with the obtained filtered **PI** control law obtained in the above section.

By computing the ideal controller through (39), we again compute the Loewner pencil, leading to a minimal realization with $n = 42$. Obviously, such a control order is prohibitive for classical control applications. The singular values decay indicates that an order $r = 2$ is enough to catch the main dynamics of the underlying controller. One obtains transfer function $\mathbf{K}_2(s) = (1082.7(s+0.1313))/(s(s+5656))$, being close to the values obtained by the

[11] In the rest of the chapter, x will be discretized with 50 points from 0 to $L = 3$, and x_m has been chosen to be located at $x(\lfloor 50 \times 2/3 \rfloor)$.

model-based approach.[12] The controller and resulting close-loop frequency response gains are shown in Fig. 13.

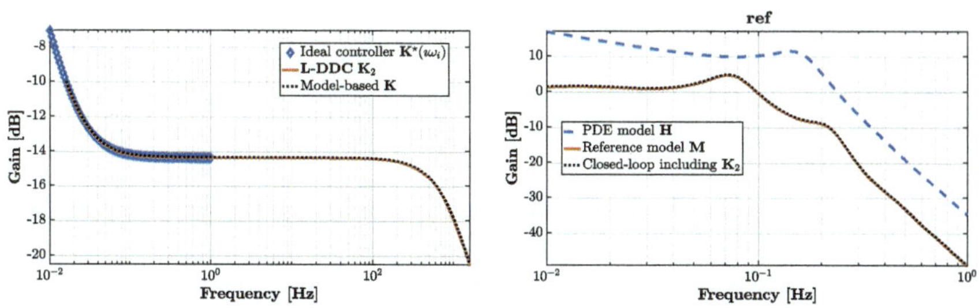

FIGURE 13 Left: frequency response of the controller (ideal, model-based and data-driven). Right: open-loop vs. closed-loop frequency responses.

Interestingly, with reference to Fig. 13, $\mathbf{K}_r$ perfectly recovers the model-based requested performances of $\mathbf{M}$ with a controller of rational order two (indeed, we expected to observe this result since we knew from the model-based approach presented in Section 4.3.1 that a rational control of this order leading to this performance is achievable).

This example demonstrates how the Loewner framework can be effectively used, either for model-based, or for data-driven control. Interestingly, by choosing the closed-loop performances $\mathbf{M}$ obtained with the model-based approach, the controller $\mathbf{K}_r$ exactly recovers the original properties, while skipping the model construction step and the order selection. This property reduces the model construction step and allows a quick design of the controller. However, this main advantage is balanced by the fact that in the model-based approach, the stability assessment is usually carried out using the approximate model, here $\mathbf{H}_r$. The latter being very accurate, the eigenvalues computation is traditionally enough for concluding stability. On the contrary, in the second data-driven approach, stability cannot be analyzed as easily. Still, Poussot-Vassal et al. (2021b) suggests an approach based on the combination of Loewner with optimal $\mathcal{H}_\infty$ projections.

Finally, we mention that all numerical experiments reported in this manuscript were performed on a laptop computer with the following specifications: 16 GB RAM and an Intel(R) Core(TM) i7-10510U CPU running at 1.80 GHz 2.30 GHz, while the software platform used was MATLAB® R2019a.

5 Summary and conclusions

In this work, we have provided an inventory of selected extensions and applications of the Loewner framework. The main philosophy of this approach is as follows: use the available data to construct a model or a controller; if needed, apply compression techniques to reduce the complexity of the model

---

[12] The model based approach yield to $1084.9(s+0.1313)/s(s+5667)$.

or of the controller. The Loewner framework was shown to be applicable for reducing large-scale dynamical systems from computational fluid dynamics (such as the linearized Navier-Stokes model with more than half a million degrees of freedom), to data-driven modeling in aeronautics applications, and to various benchmarks described by complicated dynamics (characterized by irrational transfer functions, having multiple delays, with many input or output ports, with nonlinear terms etc.). The key observation here is that one can accomplish all of these successful endeavors by having access only to compressed data (transfer function measurements, Markov parameters, etc.). Moreover, the Loewner data-driven control approach was shown to faithfully recover the performance attained by other classical model-based control approaches. Thus, one advantage is the data-driven characteristic, and another is the robustness of the approach. The Loewner framework is hence a valid alternative to intrusive methodologies, and can be successfully used when data are available.

References

Antoulas, A.C., 2005. Approximation of Large-Scale Dynamical Systems. SIAM, Philadelphia.

Antoulas, A.C., Anderson, B.D.O., 1986. On the scalar rational interpolation problem. IMA Journal of Mathematical Control and Information 3 (2–3), 61–88.

Antoulas, A.C., Beattie, C.A., Gugercin, S., 2020a. Interpolatory Methods for Model Reduction. Society for Industrial and Applied Mathematics, Philadelphia.

Antoulas, A.C., Gosea, I.V., Heinkenschloss, M., 2019. On the Loewner framework for model reduction of Burgers' equation. In: King, R. (Ed.), Active Flow and Combustion Control, Notes on Numerical Fluid Mechanics and Multidisciplinary Design. Springer, Cham, Switzerland, pp. 255–270.

Antoulas, A.C., Gosea, I.V., Heinkenschloss, M., 2020b. Reduction of systems with polynomial nonlinearities in the Loewner framework. In: Book of Abstracts of XXI Householder Symposium on Numerical Linear Algebra. Selva di Fasano, Italy, June 14–19.

Antoulas, A.C., Gosea, I.V., Ionita, A.C., 2016a. Model reduction of bilinear systems in the Loewner framework. SIAM Journal on Scientific Computing 38 (5), B889–B916.

Antoulas, A.C., Ionita, A.C., Lefteriu, S., 2012. On two-variable rational interpolation. In: Special Issue dedicated to Danny Sorensen's 65th birthday. Linear Algebra and Its Applications 436 (8), 2889–2915.

Antoulas, A.C., Lefteriu, S., Ionita, A.C., 2017. A tutorial introduction to the Loewner framework for model reduction. In: Model Reduction and Approximation. SIAM, pp. 335–376. Chapter 8.

Antoulas, A.C., Zhu, B., Zhang, Q., York, B., O'Malley, B.W., Dacso, C., 2018. A novel mathematical method for disclosing oscillations in gene transcription: a comparative study. PLoS ONE 13 (9), 1–20.

Apkarian, P., Noll, D., 2006. Nonsmooth $\mathcal{H}_\infty$ synthesis. IEEE Transactions on Automatic Control 51 (1), 71–86.

Barbagallo, A., Sipp, D., Schmid, P.J., 2008. Closed-loop control of an open cavity flow using reduced-order models. Journal of Fluid Mechanics 641, 1–50.

Benner, P., Breiten, T., 2012. Interpolation-based $\mathcal{H}_2$-model reduction of bilinear control systems. SIAM Journal on Matrix Analysis and Applications 33, 859–885.

Benner, P., Gugercin, S., Willcox, K., 2015. A survey of projection-based model reduction methods for parametric dynamical systems. SIAM Review 57 (4), 483–531.

Benner, P., Ohlberger, M., Cohen, A., Willcox, K., 2017. Model Reduction and Approximation. Society for Industrial and Applied Mathematics, Philadelphia, PA.

Berrut, J-P., Trefethen, L.N., 2006. Barycentric Lagrange interpolation. SIAM Review 46 (3), 501–517.

Breiten, T., Damm, T., 2010. Krylov subspace methods for model order reduction of bilinear control systems. Systems & Control Letters 59, 443–450.

Carleman, T., 1932. Application de la théories des équations intégrales linéaires aux systèmes d'équations différentielles non linéaires. Acta Mathematica 59, 63–87.

Dalmas, V., Robert, G., Poussot-Vassal, C., Pontes Duff, I., Seren, C., 2016. From infinite dimensional modelling to parametric reduced order approximation: application to open-channel flow for hydroelectricity. In: Proceedings of the 15th European Control Conference. Aalborg, Denmark, pp. 1982–1987.

Flagg, G., Gugercin, S., 2015. Multipoint Volterra series interpolation and $\mathcal{H}_2$ optimal model reduction of bilinear systems. SIAM Journal on Matrix Analysis and Applications 36 (2), 549–579.

Gallivan, K.A., Vanderope, A., Van Dooren, P., 2004. Model reduction of MIMO systems via tangential interpolation. SIAM Journal on Matrix Analysis and Applications 26 (2), 328–349.

Gosea, I.V., 2017. Model order reduction of linear and nonlinear systems in the Loewner framework. PhD thesis. Jacobs University Bremen.

Gosea, I.V., Antoulas, A.C., 2015. Model reduction of linear and nonlinear systems in the Loewner framework: a summary. In: 14th European Control Conference (ECC). Linz, Austria, July 15–17, pp. 345–349.

Gosea, I.V., Antoulas, A.C., 2016. Stability preserving post-processing methods applied in the Loewner framework. In: IEEE 20th Workshop on Signal and Power Integrity (SPI). Turin, Italy, May 8–11, pp. 1–4.

Gosea, I.V., Antoulas, A.C., 2018. Data-driven model order reduction of quadratic-bilinear systems. Numerical Linear Algebra with Applications 25 (6), e2200.

Gosea, I.V., Karachalios, D.S., Antoulas, A.C., 2021a. Learning reduced-order models of quadratic control systems from input-output data. In: Proceedings of the IEEE European Control Conference ECC, Virtual, pp. 1426–1431.

Gosea, I.V., Poussot-Vassal, C., Antoulas, A.C., 2021b. On Loewner data-driven control for infinite-dimensional systems. In: Proceedings of the IEEE European Control Conference, Virtual, pp. 93–99.

Gosea, I.V., Zhang, Q., Antoulas, A.C., 2020. Preserving the DAE structure in the Lowner model reduction and identification framework. Advances in Computational Mathematics 46 (3).

Gugercin, S., Antoulas, A.C., Beattie, C.A., 2008. $\mathcal{H}_2$ model reduction for large scale linear dynamical systems. SIAM Journal on Matrix Analysis and Applications 30 (2), 609–638.

Ionita, A.C., Antoulas, A.C., 2012. Matrix Pencils in Time and Frequency Domain System Identification. Developments in Control Theory Towards Glocal Control. The Institution of Engineering and Technology. Chapter 9.

Ionita, A.C., Antoulas, A.C., 2014. Data-driven parametrized model reduction in the Loewner framework. SIAM Journal on Scientific Computing 36 (3), A984–A1007.

Karachalios, D.S., Gosea, I.V., Antoulas, A.C., 2020a. The Loewner framework for system identification and reduction. In: Benner, P., Grivet-Talocia, S., Quarteroni, A., Rozza, G., Schilders, W.H.A., Silveira, L.M. (Eds.), Methods and Algorithms. In: Handbook on Model Reduction, vol. 1. De Gruyter, pp. 181–228.

Kergus, P., Poussot-Vassal, C., Demourant, F., Formentin, S., 2017. Frequency-domain data-driven control design in the Loewner framework. In: Proceedings of the 20th IFAC World Congress. Toulouse, France, pp. 2095–2100.

Lefteriu, S., Antoulas, A.C., Ionita, A.C., 2011. Parametric model reduction in the Loewner framework. In: 18th IFAC World Congress. IFAC Proceedings Volumes 44 (1), 12751–12756.

Ljung, L., 1987. System Identification: Theory for the User. Prentice Hall, Upper Saddle River, New Jersey. 2nd edition in 1999.

Mayo, A.J., Antoulas, A.C., 2007. A framework for the solution of the generalized realization problem. Linear Algebra and Its Applications 425 (2–3), 634–662.

Meyer, C., Broux, G., Prodigue, J., Cantinaud, O., Poussot-Vassal, C., 2017. Demonstration of innovative vibration control on a Falcon Business Jet. In: Proceedings of the International Forum on Aeroelasticity and Structural Dynamics. Como, Italy.

Nakatsukasa, Y., Sete, O., Trefethen, L.N., 2018. The AAA algorithm for rational approximation. SIAM Journal on Scientific Computing 40 (3), A1494–A1522.

Niconet, e.V., 2002. Benchmark examples for model reduction of linear time invariant dynamical systems.

Peherstorfer, B., Gugercin, S., Willcox, K., 2017. Data-driven reduced model construction with time-domain Loewner models. SIAM Journal on Scientific Computing 39 (5), A2152–A2178.

Poussot-Vassal, C., 2019. Large-scale dynamical model approximation and its applications. HDR, habilitation thesis. Onera, INP Toulouse, Toulouse, France.

Poussot-Vassal, C., Kergus, P., Kerhervé, F., Sipp, D., Cordier, L., 2021a. Interpolatory-based data-driven pulsed fluidic actuator control design and experimental validation. In: IEEE Transactions on Control Systems Technology.

Poussot-Vassal, C., Kergus, P., Vuillemin, P., 2021b. Interpolation-Based Irrational Model Control Design and Stability Analysis. Springer. In press. Preprint available arXiv:2012.01040.

Poussot-Vassal, C., Leclercq, C., Sipp, D., 2018. Structured linear fractional parametric controller $\mathcal{H}_\infty$ design and its applications. In: Proceedings of the European Control Conference. Limassol, Cyprus.

Poussot-Vassal, C., Loquen, T., Vuillemin, P., Cantinaud, O., Lacoste, J-P., 2013. Business jet large-scale model approximation and vibration control. In: Proceedings of the 11th IFAC ALCOSP. Caen, France, pp. 199–204.

Poussot-Vassal, C., Sipp, D., 2015. Parametric reduced order dynamical model construction of a fluid flow control problem. In: Proceedings of the 1st IFAC Workshop on Linear Parameter Varying Systems. Grenoble, France, pp. 133–138.

Poussot-Vassal, C., Vuillemin, P., Cantinaud, O., Sève, F., 2021c. Interpolatory methods for generic BizJet gust load alleviation function. SIAM Journal on Applied Dynamical Systems 20 (4), 2391–2411.

Quero, D., Vuillemin, P., Poussot-Vassal, C., 2021. A generalized eigenvalue solution to the flutter stability problem with true damping: the p-L method. Journal of Fluids and Structures 103, 103266.

Rugh, W.J., 1981. Nonlinear System Theory - the Volterra/Wiener Approach. The Johns Hopkins University Press.

Sipp, D., Schmid, P.J., 2016. Linear closed-loop control of fluid instabilities and noise-induced perturbations: a review of approaches and tools. Applied Mechanical Revue 68.

Vuillemin, P., Kergus, P., Poussot-Vassal, C., 2020. Hybrid Loewner data driven control. In: Proceedings of the IFAC World Congress. Berlin, Germany.

Ziegler, J.G., Nichols, N.B., 1942. Optimum settings for automatic controllers. Transactions of the American Society of Mechanical Engineers 64, 759–768.

Chapter 16

Machine learning and control theory

Alain Bensoussan[a,b], Yiqun Li[c], Dinh Phan Cao Nguyen[d,e], Minh-Binh Tran[d], Sheung Chi Phillip Yam[c], and Xiang Zhou[b,f,*]

[a]*International Center for Decision and Risk Analysis, Jindal School of Management, The University of Texas at Dallas, Richardson, TX, United States,* [b]*School of Data Science, City University of Hong Kong, Kowloon Tong, Hong Kong,* [c]*Department of Statistics, The Chinese University of Hong Kong, Shatin, Hong Kong,* [d]*Department of Mathematics, Southern Methodist University, Dallas, TX, United States,* [e]*Faculty of Information Technology, Nha Trang University, Nha Trang, Viet Nam,* [f]*Department of Mathematics, City University of Hong Kong, Kowloon Tong, Hong Kong*

**Corresponding author: e-mail address: xizhou@cityu.edu.hk*

Contents

Abstract

We survey in this chapter the connections between Machine Learning and Control Theory. Control Theory provide useful concepts and tools for Machine Learning. Conversely Machine Learning can be used to solve large control problems. In the first part of the pa-

Handbook of Numerical Analysis, Volume 23, ISSN 1570-8659, https://doi.org/10.1016/bs.hna.2021.12.016

per, we develop the connections between reinforcement learning and Markov Decision Processes, which are discrete time control problems. In the second part, we review the concept of supervised learning and the relation with static optimization. Deep learning which extends supervised learning, can be viewed as a control problem. In the third part, we present the links between stochastic gradient descent and mean field theory. Conversely, in the fourth and fifth parts, we review machine learning approaches to stochastic control problems, and focus on the deterministic case, to explain, more easily, the numerical algorithms.

Keywords
Control theory, Machine learning, Deep learning

MSC Codes
49J15, 49L20, 93E20, 93E99, 68T05, 68T99, 90C39, 90C15

1 Introduction

The Big Data phenomenon is at the origin of a new expansion of Artificial Intelligence. Machine learning (Jordan and Mitchell, 2015) is a way to implement AI, by providing the machine with the capability of learning and decision making, which characterize humans. The fact that humans use algorithms to help performing these two tasks is not new, by itself. As soon as computing possibilities appeared, algorithms have been developed. The ambition of AI came also early. However, during the last decades, the momentum has been spectacular, and Machine Learning has become the new Graal. Its introduction has revolutionized all kinds of fields in science, in engineering, in medicine, in management. Image processing, pattern recognition, text mining, speech recognition, automatic translation have benefited considerably from this development. An important breakthough occurred with the methodology of deep neural network (DNN).

Conceptually, since the objective is to improve the knowledge of environment and improve decision making, we are naturally dealing with optimization and statistics. This is clearly apparent in supervised learning. On the other hand, reinforcement learning and DNN add an additional variable, which is time or ordered like time. Control theory comes in as the framework of dynamic optimization.

Control theory, see for instance, Bensoussan (2018), is about how to design optimal actions for dynamical models, in continuous or discrete time. However, it is notoriously acknowledged that the numerical computation is the main barrier of putting these control theories to work in practice and many applications are unfortunately limited to the linear quadratic regulator. The curse of dimensionality as Bellman, the creator of Dynamic Programming, coined it has been haunting the numerical methods of control theory for quite a long time. It is therefore natural that the new possibilities of ML be considered to overcome the challenge of dimension. This explains why, in the past few years, we have been witnessing many exciting ideas and innovative results from the perspective of merging the above two research areas, with the efforts from different communities like applied and computational mathematics, optimal control, stochastic optimization as well as computer science. The two sides, researchers from machine learning and optimal control, start to explore the techniques, tools as well as problem formulations, from each other. We can roughly divide these works into two categories: *control theory for machine learning* and *machine learning for control theory*. Generally speaking, the former refers to the use of control theory as a mathematical tool to formulate and solve theoretical and practical problems in machine learning, such as optimal parameter tuning, training neural network; while the latter is how to use machine learning practice such as kernel method and DNN to numerically solve complex models in control theory which can become intractable by traditional methods (Han et al., 2018).

There are many evidences to support our argument of close connections between machine learning and control theory. We begin with reinforcement learning (RL), which became famous when AlphaGo Zero (Silver et al., 2017) was invented. Reinforcement learning (Sutton and Barto, 2018) is a subfield of machine learning that studies how to use past data to enhance the future manipulation of a dynamical system. The control communities target for the same problems as RL. However, the RL and control communities are practically disjoint due to the distinctive language and culture; see Recht (2019) for a recent effort to unify this gap.

In RL, one of the simplest strategies is to first estimate such models from the given data, which is called system identification in control community. This can be achieved by supervised learning (Chiuso and Pillonetto, 2019). Then in the second stage dynamical programming principle in control theory can be applied and to derive many popular RL algorithms such as Q-learning and Temporal Difference algorithms (Sutton and Barto, 2018).

As said above, Dynamic Programming is hard to implement numerically, for high dimensional dynamic systems. Machine learning and DNN can be helpful. For example, Han et al. (2018) proposed an efficient machine learning algorithm by using DNN to approximate the value function in the high dimensional Hamilton-Jacobi-Bellman equation, based on the equivalent stochastic control formulation of the PDE.

The bond that ties machine learning and control theory more closely in recent years gets critically strengthened from continuous perspective in various contexts (E, 2017; E et al., 2019b; Recht, 2019). For example, deep residual neural network (ResNet) (He et al., 2016) can be obtained by recasting it as dynamical systems with network layers considered as time discretization (Chang et al., 2018a,b; Chen et al., 2018; Haber and Ruthotto, 2017; Li and Hao, 2018; Li and Shi, 2017; Sonoda and Murata, 2017). Based on this point of view, machine learning algorithm for ResNet can be viewed as part of static and dynamic optimization for an ordinary differential equation controlled by network parameters (E et al., 2019a). This continuous model immediately triggered several new training methods based on well-known techniques in control theory: Li et al. (2017a) from the Pontryagin Maximum Principle and Chen et al. (2018) from the adjoint approach. This viewpoint of continuous modeling is also becoming more and more popular in optimization community for machine learning, particularly for the stochastic gradient descent (SGD), in which a stochastic differential equation (SDE) emerges as the continuous model (Li et al., 2019). The acceleration of the SGD is then regarded as an optimal control problem for the SDE to reach minimum point as early as possible (Li et al., 2017b). The contribution of control theory is certainly not restricted to the training algorithm. For RL, the trade-off between exploration and exploitation is a very serious and daunting practical problem. Wang et al. (2018) recently studied the analysis of this problem in theory through the lens of stochastic control. Similarly, the need to provide a solid mathematical framework to analyze deep neural networks is very pressing. Recent works have pointed out that new mathematical properties of deep neural network can be obtained by recasting deep learning as dynamical systems (cf. Chang et al., 2018a,b; Chen et al., 2018; Haber and Ruthotto, 2017; Li et al., 2017a; Li and Hao, 2018; Li and Shi, 2017; Lu et al., 2017; Sonoda and Murata, 2017).

Nowadays, it is difficult to ignore the intervene and synthesis between machine learning and control theory and the fusion of these two fields at certain boundaries is pushing forward tremendous research progress with accelerating momentum now. This paper is to give a brief introduction to and a short review of some selective works on the overlap of these communities. The interaction between the data-driven approach in machine learning and the model-based control theory is still at the very early age and there are certainly many challenges at the control-learning interface to advance the deeper development both in theory and in practice. We hope that the gap between the learning-centric views of ML and the model-centric views of control can diminish in the foreseen future on an arduous journey of understanding machine learning and artificial intelligence. As a result, a new territory may emerge (e.g. *actional intelligence* in Recht, 2019) from these joint efforts across the disciplines.

In the first part of the chapter, we discuss Markov Decision Processes (MDP), which normally provide mathematical frameworks for modeling decision making in stochastic environment where outcomes are partly random and partly under the control of a decision maker. MDPs can indeed be solved via Dynamic Programming and provide a very useful framework for Reinforcement Learning.

The second part of the chapter is devoted to Supervised Learning and Deep Learning, that concerns the approximation of a function given some preliminary observations. Supervised Learning is an optimization problem. Deep Learning can be recast into a control theory problem and can be solved using various strategies, including the Pontryagin Maximum Principle approach.

Recent mean field and stochastic control views for Stochastic Gradient Descent methods will be provided in the third part of this paper.

In the next section, we study a Stochastic Control Problem, in which the state is that of a controlled diffusion. We then propose a Machine Learning approach for this problem.

We finally focus on the deterministic case in Section 6 to simplify the theory. We provide some related theoretical results in companion with a few high-dimensional numerical illustrations to demonstrate the effectiveness of the algorithms.

2 Reinforcement learning

2.1 General concepts

In the language of Control Theory, we consider a dynamical system, which evolves in an uncertain environment. The evolution of this system is called a process, which can be characterized by its state. A controller decides a strategy of actions, called feedback, and there is at each time a cost or profit attached to the current state and the current action. In the language of RL, every time the action is made, the controller receives an award. The controller will try to choose the actions such that the sum of rewards is maximized. Since time is discrete, the control problem is called a Markov Decision Process (MDP) and can be solved by Dynamic Programming approach. The award is then a function of the state and the action.

2.2 Mathematical model without action

We suppose that the states belong to a space X. On X, there is a σ-algebra, denoted by $\mathcal{X}$. A transition probability is a (regular enough) function $\pi(x;\Gamma)$ on $(X,\mathcal{X})$. For any fixed x we define the probability of $\Gamma\in\mathcal{X}$ to be $\pi(x;\Gamma)$. If B is the space of bounded functions on X, equipped with the norm $\|f\| = \sup_x |f(x)|$, we associate to the transition probability a linear operator Φ from B to B as follows:

$$\Phi f(x) = \int_X f(\eta)\pi(x;\mathrm{d}\eta), \tag{2.1}$$

and clearly $||\Phi|| \leqslant 1$. A Markov chain $\{X_i\}_{i=1}^{\infty}$ on X with transition probability $\pi(x;\mathrm{d}\eta)$ is a stochastic process on X such that

$$\mathbb{E}[f(X_{n+1})|X_n = x] = \int_X f(\eta)\pi(x;\mathrm{d}\eta), \text{ for } n = 1,2,\cdots. \tag{2.2}$$

Assuming stationarity in (2.2), and choosing α to be a discount factor, we can express the function

$$u(x) = \mathbb{E}\left[\sum_{n=1}^{+\infty}\alpha^{n-1} f(X_n)\middle| X_1 = x\right]. \tag{2.3}$$

This is the sum of rewards, in the terminology of RL. There is no action to modify the trajectory. We just add the discounted rewards. We can give an explicit analytic expression (not probabilist) of the function $u(x)$. It is the unique solution of the analytic problem

$$u = f + \alpha \Phi u. \tag{2.4}$$

It then follows that

$$u = (I - \alpha \Phi)^{-1} f, \tag{2.5}$$

and using the generator Φ, we can also rewrite,

$$u = \sum_{n=1}^{\infty} \alpha^{n-1} \Phi^{n-1} f. \tag{2.6}$$

2.3 Approximation

Our main task now is to compute the function $u(x)$. Eqs. (2.4) and (2.6) are explicit and straightforward. However, the challenge is when the dimension d of X is large, these formulations are not of practical use. Here comes the other aspect of machine learning: how to approximate a function given by formulae (2.3) or (2.4). Since Supervised Learning does exactly that, approximate a function, we follow the ideas of SL. There are basically two methods, the *parametric* method and the *non parametric* method. In the parametric method, we look for an approximation of the form

$$u(x) \approx \sum_{i=1}^{I} \theta_i \varphi_i(x), \tag{2.7}$$

where $\varphi_i(x)$ are given functions, so that the family $\{\varphi_i(\cdot)\}_{i=1}^{\infty}$ forms a basis of the functional space to which $u(x)$ belongs, and θ_i are coefficients to be determined. We need to compute parameters minimizing the error

$$\left\| \sum_{i=1}^{I} \theta_i (\varphi_i - \alpha \Phi \varphi_i) - f \right\|^2, \tag{2.8}$$

where $\| \cdot \|$ is the sup-norm $\|G\| = \sup |G|$. To guarantee the existence and uniqueness of the parameters, we minimize the quadratic functional, with a quadratic regularization.

$$\gamma \sum_{i=1}^{I} \theta_i^2 + \left\| \sum_{i=1}^{I} \theta_i (\varphi_i - \alpha \Phi \varphi_i) - f \right\|^2. \tag{2.9}$$

In the non parametric method, used in supervised learning, we do not refer to a functional equation for $u(x)$. We assume that we can compute the value at a finite number of points. For a given x, $u(x)$ can be calculated by formula (2.3), by Monte Carlo simulation. We then find

$$u(x) \approx \frac{1}{N}\sum_{\nu=1}^{N}\sum_{n=1}^{+\infty}\alpha^{n-1}f(X_n^\nu), \tag{2.10}$$

where $X_1^\nu = x, \cdots, X_n^\nu = X_n(\omega^\nu), \cdots$, represents one trajectory indexed by ν of the Markov chain, corresponding to one sample point ω^ν. We choose M points $x^1, \cdots, x^M$ in $\mathbb{R}^d$, and then compute $u(x^1) = y^1, \cdots u(x^M) = y^M$ by using Monte-Carlo method and the approximation formula (2.10). The number M is chosen arbitrarily. If we assume that f is continuous and bounded, then $u(x)$ is also bounded and continuous. The goal is to extrapolate $u(x)$ from the knowledge of $y^1, \cdots, y^M$.

We now choose a subset $\mathcal{H}$ of $C(\mathbb{R}^d)$. This subset is called the hypothesis space. We select an element in $\mathcal{H}$ such that it is the closest possible to $y^1, \cdots, y^M$ at points $x^1, \cdots, x^M$. We assume naturally that $\mathcal{H}$ is a nice enough functional space. The theory of reproducing kernels allows us to define $\mathcal{H}$ as a Hilbert space, with a continuous injection in $C(\mathbb{R}^d)$. The function $u(x)$ can be defined as the solution of the minimization problem

$$\min_{u\in\mathcal{H}}\left\{\gamma\|u\|_{\mathcal{H}}^2 + \sum_{m=1}^{M}(u(x^m) - y^m)^2\right\}. \tag{2.11}$$

Remark 1. In RL, one claims that a significant difference with MDP is that the Markov Chain may not be known. The controller, however makes trials, which is similar to Monte Carlo, without referring to a selection of trajectories according to a given probability transition.

2.4 Mathematical model with action

The Markov chain has a probability transition depending on an auxiliary variable a called the action, $\pi(x, a; \mathrm{d}\eta)$. When the action is a function of the state, also called the feedback, $a(x)$, we then get $\pi(x, a(x); \mathrm{d}\eta)$. We now define the operator $\Phi^a f(x)$ or $\Phi^{a(x)} f(x)$ by

$$\Phi^a f(x) = \int_X f(\eta)\pi(x, a; \mathrm{d}\eta); \quad \Phi^{a(x)} f(x) = \int_X f(\eta)\pi(x, a(x); \mathrm{d}\eta). \tag{2.12}$$

We consider an award depending on the state and the action $f(x, a)$. For convenience, this award is supposed to be a cost and we assume $f(x, a) \geqslant 0$. We then set the aggregate cost to be

$$J_{a(\cdot)}(x) = \mathbb{E}\left[\sum_{n=1}^{+\infty} \alpha^{n-1} f(X_n, a(X_n)) \middle| X_1 = x\right], \tag{2.13}$$

in which X_n evolves as a function of the probability transition $\pi(x, a(x); \mathrm{d}\eta)$. We also define the value function

$$u(x) = \inf_{a(\cdot)} J_{a(\cdot)}(x), \tag{2.14}$$

which is the solution of the following Bellman equation

$$u(x) = \inf_a [f(x, a) + \alpha \Phi^a u(x)]. \tag{2.15}$$

It is also interesting to introduce the cost (Q-function) when the first action is arbitrary and the following actions are optimized, namely

$$Q(x, a) = f(x, a) + \alpha \Phi^a u(x). \tag{2.16}$$

Taking into account the fact that

$$u(x) = \inf_a Q(x, a), \tag{2.17}$$

we arrive at

$$Q(x, a) = f(x, a) + \alpha[\Phi^a(\inf_{a'} Q(\cdot, a'))](x). \tag{2.18}$$

There are two basic types of iteration to solve the above Bellman equation, the *value iteration* and the *policy iteration*. The *value iteration* is defined by

$$u_{k+1}(x) = \inf_a [f(x, a) + \alpha \Phi^a u_k(x)], \tag{2.19}$$

and $u_0(x) = 0$. When $f(x, a)$ is bounded, the solution of the Bellman equation is unique and the sequence $u_k(x)$ converges to the value function monotonically. When $f(x, a)$ is not bounded, the solution of the Bellman equation is not unique. The sequence $u_k(x)$ converges monotonically to the value function, which is the minimum solution. We can also interpret $u_k(x)$ as the value function for the control problem with k periods. To see this, we define

$$J^k_{a(\cdot)}(x) = \mathbb{E}\left[\sum_{n=1}^{k} \alpha^{n-1} f(X_n, a(X_n)) \middle| X_1 = x\right], \tag{2.20}$$

then

$$u_k(x) = \inf_{a(\cdot)} J^k_{a(\cdot)}(x). \tag{2.21}$$

On the other hand, the *policy iteration* technique starts with a given feedback control $a^k(x)$ and solves the linear (fixed point) problem similar to (2.4)

$$u^{k+1}(x) = f(x, a^k(x)) + \alpha \Phi^{a^k(x)} u^{k+1}(x). \tag{2.22}$$

With the $u^{k+1}(x)$, then $a^{k+1}(x)$ is defined by

$$\inf_a [f(x, a) + \alpha \Phi^a u^{k+1}(x)],$$

in which we start with a function $a^0(x)$, which minimizes

$$\inf_a f(x, a).$$

In both types of iterations, we have to solve the respective minimization problems

(i) value iteration: $\inf_a[f(x, a) + \alpha \Phi^a u_k(x)]$;
(ii) policy iteration: $\inf_a[f(x, a) + \alpha \Phi^a u^{k+1}(x)]$,

in which $u_k(x)$ and $u^{k+1}(x)$ are known respectively (obtained a-priori by some approximation method such as the use of reproducing kernals).

For both cases, the minimization problem can be resolved by a gradient decent technique. We set

$$Q^{k+1}(x, a) = f(x, a) + \alpha \Phi^a u^{k+1}(x)$$

which is an approximation of $Q(x, a)$.

Since we cannot find the infimum exactly, we use the following approximation for $a^{k+1}(x)$

$$a^{k+1}(x) = a^k(x) - \rho^k D_a Q^{k+1}(x, a^k(x)), \tag{2.23}$$

in which the coefficient ρ^k is chosen such that it solves the following scalar optimization problem

$$\inf_\rho Q^{k+1}(x, a^k(x) - \rho D_a Q^{k+1}(x, a^k(x))). \tag{2.24}$$

By definition,

$$u^{k+1}(x) = Q^{k+1}(x, a^k(x)),$$

We can then, instead of solving the linear problem for $u^{k+1}(x)$, use the approximation $Q^k(x, a^k(x))$. We then proceed as follows: knowing $a^k(x)$ and $Q^k(x, a)$, we define

$$\begin{cases} \bar{u}^{k+1}(x) := Q^k(x, a^k(x)), \\ Q^{k+1}(x, a) = f(x, a) + \alpha \Phi^a \bar{u}^{k+1}(x), \end{cases} \tag{2.25}$$

and $a^{k+1}(x)$ can then be obtained approximately through (2.23) and (2.24). In the above procedure, we start with $Q^0(x,a) = f(x,a)$ and $a^0(x)$ is chosen to be the minimizer of $f(x,a)$.

3 Control theory and deep learning

3.1 Supervised learning

Supervised learning concerns basically approximating an unknown function $F(x) : \mathbb{R}^d \to \mathbb{R}$, given some noisy observations $y^m = F(x^m) + \varepsilon^m$, in which x^m is known and the noise ε^m models the uncertainty (unknown). There are two methods that can be used to solve this approximation problem. In the first method, we consider a function $f(x;\theta)$, where $\theta \in \mathbb{R}^n$ for some n and we try to perform a minimization problem with the parameter θ

$$\min_{\theta} \left\{ \gamma |\theta|^2 + \sum_{m=1}^{M} (f(x^m;\theta) - y^m)^2 \right\}. \tag{3.1}$$

This is the well-known *parametric method*. In the simplest case of neural networks, the function $f(x;\theta)$ is defined as follows:

We first introduce $X = \chi(x)$, where $\chi : \mathbb{R}^d \to \mathbb{R}^n$. Of course χ can be identity. Now, choose σ to be a scalar function, called the *activation function* and W to be a matrix in $\mathcal{L}(\mathbb{R}^n;\mathbb{R}^d)$. Suppose that b is a vector in $\mathbb{R}^n$. The pair (W,b) represents the parameter θ. We now define the vector $\widetilde{X}$ by

$$\widetilde{X}_i = \sigma\left(\sum_{j=1}^{d} W_{ij} X_j + b_i\right), \tag{3.2}$$

and

$$f(x;\theta) = g(\widetilde{X}), \tag{3.3}$$

where $g : \mathbb{R}^n \to \mathbb{R}$ is the output function. The minimization in (3.1) is performed by a gradient method with iterative application of chain rules.

In the *non-parametric method*, particularly, the *kernel method*, one finds a functional space $\mathcal{H}$, to which the approximation of $F(x)$ belongs. One then chooses that approximation $f(x)$ by solving

$$\min_{f\in\mathcal{H}} \left\{ \gamma \|f\|_{\mathcal{H}}^2 + \sum_{m=1}^{M} (f(x^m) - y^m)^2 \right\}. \tag{3.4}$$

3.2 Deep learning

Deep learning is a generalization of supervised learning with a sequence of layers. We generalize (3.3) to K layers as follows with $X^k \in \mathbb{R}^n$, $\theta^k :=$

$(W^{(k+1)}, b^{(k+1)}) \in \mathcal{L}(\mathbb{R}^n; \mathbb{R}^n) \times \mathbb{R}^n$,

$$X^{k+1} = f_k(X^k; \theta^k),\ k = 0, \cdots K-1, \tag{3.5}$$

where

$$f_k(X^k; \theta^k) := \sigma(W^{(k+1)} X^k + b^{(k+1)})$$

and

$$X^0 = \chi(x), \tag{3.6}$$

with

$$f(x, \theta) = g(X^K). \tag{3.7}$$

The parameter θ is the collection of $\{\theta^0, \cdots \theta^{K-1}\}$. We have written the case of several layers of neural networks, but it is just an example.

3.3 Control theory approach

This approach has been introduced by Li et al. (2015, 2017b, 2019). The idea is to consider a continuous time extension of (3.5), (3.6), and (3.7). We write

$$\begin{cases} \dfrac{\mathrm{d}X}{\mathrm{d}t} = f(X_t, \theta_t, t), \\ X_0 = \chi(x). \end{cases} \tag{3.8}$$

The approximation of $F(x)$ is then $f(x, \theta) = g(X_T)$. The loss in this scenario is $(y - g(X_T))^2 = \Phi(X_T)$, recalling that y will be a known value. The idea is to consider θ_t as a control and we want to minimize an analog of (3.4), which is expressed as:

$$J(\theta) := \sum_{m=1}^{M} \Phi(X_T^m) + \int_0^T L(\theta_t)\mathrm{d}t, \tag{3.9}$$

where $L(\theta)$ is a regularization function, for instance, $\gamma \|\theta\|^2$. Using a Pontryagin Maximum Principle approach, we can write a necessary condition of optimality for the control $\widehat{\theta}_t$. Define $(\widehat{X}_t^m, \widehat{p}_t^m)$, we have the optimal state and optimal adjoint state solutions of

$$\begin{cases} \dfrac{\mathrm{d}\widehat{X_t}^m}{\mathrm{d}t} = f(\widehat{X}_t^m, \widehat{\theta}_t, t), \quad \widehat{X}_0^m = x^m, \\ -\dfrac{\mathrm{d}\widehat{p}_t^m}{\mathrm{d}t} = (D_x f)^*(\widehat{X}_t^m, \widehat{\theta}_t, t)\widehat{p}_t^m, \quad \widehat{p}_T^m = D_x \Phi(\widehat{X}_T^m), \end{cases} \tag{3.10}$$

and the optimality condition

$$\widehat{\theta}_t \text{ minimizes } H(\widehat{X}_t, \widehat{p}_t, \theta, t), \text{ a.e. } t,$$

with

$$H(\widehat{X}_t, \widehat{p}_t, \theta, t) = \sum_{m=1}^{M} \widehat{p}_t^m f(\widehat{X}_t^m, \theta, t) + L(\theta).$$

To solve (3.10), one can use the following approximation recursively: let $\widehat{\theta}_t^k$ be given, define $(\widehat{X}_t^{m,k}, \widehat{p}_t^{m,k})$ by

$$\begin{cases} \dfrac{\mathrm{d}\widehat{X}_t^{m,k}}{\mathrm{d}t} = f(\widehat{X}_t^{m,k}, \widehat{\theta}_t^k, t), \quad \widehat{X}_0^{m,k} = x^m, \\ -\dfrac{\mathrm{d}\widehat{p}_t^{m,k}}{\mathrm{d}t} = (D_x f)^*(\widehat{X}_t^{m,k}, \widehat{\theta}_t^k, t)\widehat{p}_t^{m,k}, \quad \widehat{p}_T^{m,k} = D_x\Phi(\widehat{X}_T^{m,k}). \end{cases} \tag{3.11}$$

We look for $\widehat{\theta}_t^{k+1}$ that minimizes

$$\sum_m \widehat{p}_t^{m,k} f(\widehat{X}_t^{m,k}, \theta, t) + L(\theta). \tag{3.12}$$

Note that the above approximation may fail to converge. We refer to E et al. (2019a,b); Han and E (2016); Sutton and Barto (2018) for recent improvements and techniques to deal with this issue.

4 Stochastic gradient descent and control theory

4.1 Comments

The gradient descent algorithm plays an essential role in the various parts of ML, as we have seen above. There has been a considerable amount of work in order to improve its efficiency. The stochastic version it, described below, offers another example of connection to control theory, stochastic control and even mean field control theory. We limit ourselves to some basic considerations. In particular we do not discuss the case of the use of SG for DNN, as in Mei et al. (2018). This is because the connection is of a different nature. It does not involve control theory, but introduces interesting PDE. In this section, we will define SG and relate the choice of the optimal descent parameters to an MDP problem. We will then give a continuous version and also connect with mean field control.

4.2 Stochastic gradient and MDP

We recall the definition of gradient descent. If $f(x)$ is a function on $\mathbb{R}^d$, for which we want to find a minimum x^*. The gradient descent algorithm is defined by the sequence

$$x_{k+1} = x_k - \eta_k Df(x_k), \tag{4.1}$$

where η_k is a positive constant, which can be chosen independent of k. This is simpler, but by all means not optimal. Suppose now that the function $f(x)$ is an expected value

$$f(x) = \mathbb{E}(f(x, Z)). \tag{4.2}$$

Applying the gradient method to this function leads to

$$x_{k+1} = x_k - \eta_k \mathbb{E}(D_x f(x_k, Z)).$$

In the SG descent method, one chooses a sequence of independent versions of Z, called Z_k and define

$$X_{k+1} = X_k - \eta_k D_x f(X_k, Z_k). \tag{4.3}$$

We clearly obtain a controlled Markov chain, in which η_k is the control. If we define the σ-algebra $\mathcal{F}^k = \sigma(Z_1, \cdots, Z_k)$, then η_k is adapted to $\mathcal{F}^{k-1}$, $k \geqslant 1$. $\mathcal{F}^0$ is the trivial σ-algebra. The process X_k is also adapted to $\mathcal{F}^{k-1}$. We have to define the pay off to optimize. Suppose we stop at K. We naturally want X_K as close as possible to x^*. One way to proceed would be to minimize $\mathbb{E} f(X_K)$. But this requires the computation of $f(x)$, for random values of the argument, which we want to avoid. In fact, if we insure that X_K is close to a constant, that constant will be necessarily x^*, provided η_k is larger than a fixed positive constant. So a good criterion will be to minimize

$$\mathbb{E}|X_K - EX_X|^2. \tag{4.4}$$

This is not a standard MDP, but a Mean Field type control problem in discrete time.

4.3 Continuous version

We first write (4.3) as follows:

$$X_{k+1} = X_k - \eta_k Df(X_k) + \eta_k \widetilde{Y}_k,$$

with

$$\widetilde{Y}_k = Df(X_k) - D_x f(X_k, Z_k).$$

Note that we have $\mathbb{E}[\widetilde{Y}_k | \mathcal{F}^{k-1}] = 0$ and

$$\mathbb{E}[\widetilde{Y}_k (\widetilde{Y}_k)^* | \mathcal{F}^{k-1}] = \Sigma(X_k),$$

with

$$\Sigma(x) = \mathbb{E}(D_x f(x, Z)(D_x f(x, Z))^*) - Df(x)(Df(x))^*. \tag{4.5}$$

We shall write

$$\Sigma(x) = \sigma(x)\sigma(x)^*. \tag{4.6}$$

If we write $\widetilde{Y}_k = \sigma(X_k)Y_k$, then the process Y_k satisfies

$$\mathbb{E}[Y_k|\mathcal{F}^{k-1}] = 0, \ \mathbb{E}[Y_k(Y_k)^*|\mathcal{F}^{k-1}] = 0. \tag{4.7}$$

We obtain the algorithm

$$X_{k+1} = X_k - \eta_k Df(X_k) + \eta_k \sigma(X_k)Y_k. \tag{4.8}$$

We can then follow Li et al. (2017b) to define a diffusion approximation of (4.8) as follows

$$dX = -u(t)Df(X(t))dt + u(t)\eta\sigma(X(t))dB(t) \tag{4.9}$$

where $B(t)$ is a brownian motion and $u(t)$ adapted to the filtration generated by the brownian motion, with values in [0, 1]. The number η is a scaling constant. We can then choose the control to minimize the payoff $\mathbb{E}f(X(T))$ which defines a stochastic control problem, or

$$\min_{u(\cdot)\geqslant u_0>0} \mathbb{E}\left(|X(T) - \mathbb{E}X(T)|^2\right), \tag{4.10}$$

which defines a mean field type control problem.

5 Machine learning approach of stochastic control problems

5.1 General theory

Let us now consider the following problem, in which the state equation is a controlled diffusion

$$\begin{cases} \mathrm{d}x(t) = g(x(t), a)\mathrm{d}t + \mathrm{d}w, \\ x(0) = x, \end{cases} \tag{5.1}$$

and the pay-off is given by

$$J_x(a(\cdot)) = \mathbb{E}\left[\int_0^{+\infty} \exp(-\alpha t)\, f(x(t), a(t))\mathrm{d}t\right]. \tag{5.2}$$

There are two approaches to resolve the above problem: *Dynamic Programming* and *Stochastic Pontryagin Maximum Principle*. The theory shows that the optimal control is described by a feedback. The value function is defined by

$$u(x) = \inf_{a(\cdot)} J_x(a(\cdot)). \tag{5.3}$$

In the above problem, there are 3 functions of interest, the value function $u(x)$, the optimal feedback $\widehat{a}(x)$ (if exists), the gradient of $u(x)$, $\lambda(x) = Du(x)$. Introducing $u(x)$ and its gradient independently may look superfluous. It turns out that the gradient has a very interesting interpretation, the shadow price in economics. Surprisingly, the gradient is solution of a self-contained vector equation. On the numerical side, approximating the gradient of $u(x)$ by the gradient of the approximation of $u(x)$ results in a source of errors. This justifies the interest in the system of equations for $\lambda(x)$. We may think of *parametric and non-parametric approximations* for these functions. We shall discuss a parametric approach for the optimal feedback, and a non-parametric approach for the value function and its gradient.

5.2 Parametric approach for the feedback

Let us now replace candidacy of $a(x)$ by a special function of the form $\widetilde{a}(x,\theta)$ with θ being a parameter in $\mathbb{R}^p$. The function $g(x,a)$ is then replaced by $g(x,\widetilde{a}(x,\theta))$, which can be renamed as $g(x,\theta)$ by abuse of notation. We obtain the control problem

$$\begin{cases} \mathrm{d}x(t) = g(x(t),\theta)\mathrm{d}t + \mathrm{d}w, \\ x(0) = x, \end{cases} \tag{5.4}$$

$$J_x(\theta(\cdot)) = \mathbb{E}\left[\int_0^{+\infty} \exp(-\alpha t)\, f(x(t),\theta(t))\mathrm{d}t\right], \tag{5.5}$$

where $f(x,\theta)$ abbreviates for $f(x,\widetilde{a}(x,\theta))$. The important simplification of this procedure is that $\theta(t)$ is regarded as deterministic.

We can write a necessary condition of optimality for the optimal new control $\widehat{\theta}(t)$. Define the optimal state $\widehat{x}(t)$ and the adjoint state $\widehat{p}(t)$ by:

$$\begin{cases} \mathrm{d}\widehat{x} = g(\widehat{x},\widehat{\theta})\mathrm{d}t + \mathrm{d}w, \quad \widehat{x}(0) = x, \\ -\dfrac{\mathrm{d}\widehat{p}}{\mathrm{d}t} + \alpha\widehat{p} = g_x^*(\widehat{x},\widehat{\theta})\widehat{p} + f_x(\widehat{x},\widehat{\theta}), \end{cases} \tag{5.6}$$

and $\widehat{\theta}(t)$ satisfies

$$\inf_\theta \mathbb{E}\,[\widehat{p}(t)\cdot g(\widehat{x}(t),\theta) + f(\widehat{x}(t),\theta)],\ t\text{-a.e.}. \tag{5.7}$$

To obtain $\widehat{\theta}(t)$, we can use an iterative approximation coupled with a gradient method

$$\begin{cases} \mathrm{d}\widehat{x}^k = g(\widehat{x}^k,\widehat{\theta}^k)\mathrm{d}t + \mathrm{d}w, \quad \widehat{x}(0) = x, \\ -\dfrac{d\widehat{p}^k}{dt} + \alpha\widehat{p}^k = g_x^*(\widehat{x}^k,\widehat{\theta}^k)\widehat{p}^k + f_x(\widehat{x}^k,\widehat{\theta}^k), \end{cases} \tag{5.8}$$

$$\widehat{\theta}^{k+1}(t) = \widehat{\theta}^k(t) - \rho^k(t)\mathbb{E}[g_\theta^*(\widehat{x}^k, \widehat{\theta}^k)\widehat{p}^k + f_\theta(\widehat{x}^k, \widehat{\theta}^k)], \ t\text{-a.e.},$$

where $\rho^k(t)$ minimizes in ρ the following function

$$\begin{aligned}&\mathbb{E}\,[\widehat{p}(t)\cdot g(\widehat{x}(t), \theta) + f(\hat{x}(t), \theta)],\\ &\text{in which } \theta = \widehat{\theta}^k(t) - \rho\mathbb{E}[g_\theta^*(\widehat{x}^k, \widehat{\theta}^k)\widehat{p}^k + f_\theta(\widehat{x}^k, \widehat{\theta}^k)].\end{aligned}$$

5.3 Non-parametric approach for the value function and its gradient

First, we notice that the value function $u(x)$, the gradient $\lambda(x) = Du(x)$ and the optimal feedback $\widehat{a}(x)$ are linked as follows[1]

$$\begin{cases}\alpha u(x) = \lambda(x)\cdot g(x, \widehat{a}(x)) + f(x, \widehat{a}(x)) + \dfrac{1}{2}\mathrm{tr}(D\lambda(x)),\\ \alpha\lambda(x) = D\lambda(x)g(x, \widehat{a}(x)) + D_x^* g(x, \widehat{a}(x))\lambda(x) + D_x f(x, \widehat{a}(x)) + \dfrac{1}{2}\Delta\lambda(x),\end{cases} \tag{5.9}$$

where

$$\widehat{a}(x) \text{ minimizes in } a \text{ of the function } \lambda(x)\cdot g(x,a) + f(x,a).$$

The above system has an interesting structure, in which there is coupling only for the last two equations. The first equation allows one to define the value function. Note that we have used the fact that $D\lambda(x) = (D\lambda(x))^*$.

We now define the following iteration: suppose that we know $(\widehat{a}^k(x), \lambda^k(x))$, we can find $\lambda^{k+1}(x)$ by solving the differential equation system:

$$\begin{aligned}&\alpha\lambda^{k+1}(x) - D\lambda^{k+1}(x)g(x, \widehat{a}^k(x)) - \frac{1}{2}\Delta\lambda^{k+1}(x)\\ &\quad = D_x^* g(x, \widehat{a}^k(x))\lambda^k(x) + D_x f(x, \widehat{a}^k(x))\end{aligned} \tag{5.10}$$

such that

$$\widehat{a}^{k+1}(x) \text{ minimizes in } a \text{ of the function } \lambda^{k+1}(x)\cdot g(x,a) + f(x,a).$$

The equations for the components of $\lambda^{k+1}(x)$ are completely decoupled, and can be solved in parallel. One possibility is to use simulation to define $\lambda^{k+1}(x)$ in a finite number of points and to use an extrapolation by a kernel method.

6 Focus on the deterministic case

In this section, we shall simplify by considering the case of a deterministic dynamics. Some theoretical and numerical results will be presented to illustrate the efficiency of the numerical algorithms.

---

1. $(5.9)_1$ is the HJB system and $(5.9)_2$ follows by differentiating $(5.9)_1$ with respect to x.

6.1 Problem and algorithm

We first define the relation between the three functions $u(x)$, $\lambda(x)$ and $\widehat{a}(x)$ (as the special case of (5.9)):

$$\begin{cases} \alpha u(x) = f(x, \widehat{a}(x)) + \lambda(x) \cdot g(x, \widehat{a}(x)), \\ \alpha \lambda(x) = D\lambda(x) g(x, \widehat{a}(x)) + D_x^* g(x, \widehat{a}(x))\lambda(x) + D_x f(x, \widehat{a}(x)), \end{cases} \tag{6.1}$$

where

$$\widehat{a}(x) \text{ minimizes in } a \text{ of the function } \lambda(x) \cdot g(x, a) + f(x, a).$$

We propose two iterations.

(i) The first one is: for given functions $(\widehat{a}^k(x), \lambda^k(x))$, we define $u^k(x)$ as

$$\alpha u^k(x) = f(x, \widehat{a}^k(x)) + \lambda^k(x) \cdot g(x, \widehat{a}^k(x)). \tag{6.2}$$

Now, we find $\lambda^{k+1}(x)$ by solving

$$\alpha \lambda^{k+1}(x) - D\lambda^{k+1}(x) g(x, \widehat{a}^k(x)) = D_x^* g(x, \widehat{a}^k(x))\lambda^k(x) + D_x f(x, \widehat{a}^k(x)). \tag{6.3}$$

We next resolve $\widehat{a}^{k+1}(x)$ by minimizing

$$\text{the function } \lambda^{k+1}(x) \cdot g(x, a) + f(x, a) \text{ in } a,$$

and $u^{k+1}(x)$ is constructed by

$$\alpha u^{k+1}(x) = f(x, \widehat{a}^{k+1}(x)) + \lambda^{k+1}(x) \cdot g(x, \widehat{a}^{k+1}(x)). \tag{6.4}$$

(ii) The second one is to describe the policy iteration as follows: given the functions $(\widehat{a}^k(x), u^k(x))$, we then set

$$\lambda^k(x) = Du^k(x). \tag{6.5}$$

We obtain $u^{k+1}(x)$ by solving

$$\alpha u^{k+1}(x) = f(x, \widehat{a}^k(x)) + Du^{k+1}(x) \cdot g(x, \widehat{a}^k(x)). \tag{6.6}$$

We now set

$$\lambda^{k+1}(x) = Du^{k+1}(x), \tag{6.7}$$

and the values of the function $\widehat{a}^{k+1}(x)$ can be obtained by minimizing

$$\text{the function } \lambda^{k+1}(x) \cdot g(x, a) + f(x, a) \text{ in } a. \tag{6.8}$$

Since $\widehat{a}^k(x)$ satisfies the necessary condition of optimality

$$D_a f(x,\widehat{a}^k(x)) + (D_a g)^*(x,\widehat{a}^k(x))\lambda^k(x) = 0, \tag{6.9}$$

we can use a gradient descent method

$$\widehat{a}^{k+1}(x) = \widehat{a}^k(x) - \theta^{k+1}[D_a f(x,\widehat{a}^k(x)) + (D_a g)^*(x,\widehat{a}^k(x))\lambda^{k+1}(x)]. \tag{6.10}$$

The suitable scalar θ^{k+1} can now be obtained by a one-dimensional optimization problem by setting

$$w^{k+1}(\theta)(x) = \widehat{a}^k(x) - \theta[D_a f(x,\widehat{a}^k(x)) + (D_a g)^*(x,\widehat{a}^k(x))\lambda^{k+1}(x)], \tag{6.11}$$

and

$$H^{k+1}(\theta)(x) = f(x, w^{k+1}(\theta)(x)) + \lambda^{k+1}(x)\cdot g(x, w^{k+1}(\theta)(x)), \tag{6.12}$$

where we can now find θ^{k+1} by minimizing the function $H^{k+1}(\theta)(x)$ in θ. As a result, θ^{k+1} depends on x and plugging back in (6.10) to obtain $\widehat{a}^{k+1}(x)$.

6.2 Splitting up method

As a part of both iterations (6.3) and (6.6) described above we have to solve a generic linear PDE

$$\alpha\lambda(x) - D\lambda(x)\cdot G(x) = F(x). \tag{6.13}$$

For which we propose a parallel splitting up method[2]: knowing $\lambda^j(x)$, and writing $x = (x_1, \cdots, x_d)$, we define $\lambda^{j+\frac{l}{d}}(x), l = 1, \cdots, d$, by

$$\alpha\lambda^{j+\frac{l}{d}}(x) - \frac{\partial\lambda^{j+\frac{l}{d}}(x)}{\partial x_l}G_l(x) = Z^{j+\frac{l}{d}}(x), \tag{6.14}$$

where $Z^{j+\frac{l}{d}}(x) = \sum_{h\neq l}\frac{\partial\lambda^j(x)}{\partial x_h}G_h(x) + F(x)$. And $\lambda^{j+1}(x)$ is defined by

$$\lambda^{j+1}(x) = \frac{1}{d}\sum_{l=1}^{d}\lambda^{j+\frac{l}{d}}(x). \tag{6.15}$$

Note that (6.14) is a one dimensional first order differential equation, which has an explicit solution

2 The parallel splitting up method not only reduces the original problems into a number of separable one dimensional linear problems, but also enables us to compute all these one dimensional linear problems by parallel computing, for which the calibration of the fractional steps are independent of each other (Lu et al., 1991).

$$\lambda^{j+\frac{l}{d}}(x) = -\int_{-\infty}^{x_l} \frac{Z^{j+\frac{l}{d}}(\xi_l, \bar{x}_l)}{G_l(\xi_l, \bar{x}_l)} \exp\left(\alpha \int_{\xi_l}^{x_l} \frac{d\eta_l}{G_l(\eta_l, \bar{x}_l)}\right) d\xi_l. \tag{6.16}$$

Here, we have used the notation $x = (x_l, \bar{x}_l)$ where $\bar{x}_l \in \mathbb{R}^{d-1}$.

7 Convergence results

7.1 Setting of the problem

We take

$$g(x, a) = A(x) + Ba, \tag{7.1}$$

such that

$$x \mapsto A(x) : \mathbb{R}^n \to \mathbb{R}^n, |A(x)| \leqslant \gamma|x|, \; B \in \mathcal{L}(\mathbb{R}^d; \mathbb{R}^n),$$

and

$$||DA(x_1) - DA(x_2)|| \leqslant \frac{b|x_1 - x_2|}{1 + |x_1| + |x_2|}.$$

The pay-off functional

$$f(x, a) = F(x) + \frac{1}{2}a^* N a, \tag{7.2}$$

$$x \mapsto F(x) : \mathbb{R}^n \to \mathbb{R}, \; |DF(x)| \leqslant M|x|,$$

$$N \in \mathcal{L}(\mathbb{R}^d; \mathbb{R}^d), \text{ symmetric and invertible.}$$

We then have

$$\widehat{a}(x) = -N^{-1} B^* \lambda(x), \tag{7.3}$$

and thus the second relation (6.1) becomes

$$\alpha\lambda(x) - DA^*(x)\lambda(x) - D\lambda(x)(A(x) - BN^{-1}B^*\lambda(x)) = DF(x). \tag{7.4}$$

7.2 Preliminaries

We will need conditions on α and b: α sufficiently large and b sufficiently small. We first assume that

$$\alpha - 2\gamma > 2\sqrt{M||BN^{-1}B^*||}. \tag{7.5}$$

We set

$$\beta = \frac{(\alpha - 2\gamma)^2}{4M||BN^{-1}B^*||} > 1. \tag{7.6}$$

We define

$$\varpi = \frac{\alpha - 2\gamma - \sqrt{(\alpha - 2\gamma)^2 - 4M||BN^{-1}B^*||}}{2||BN^{-1}B^*||}, \tag{7.7}$$

which is a solution of

$$\varpi^2||BN^{-1}B^*|| - (\alpha - 2\gamma)\varpi + M = 0. \tag{7.8}$$

We next need to solve for the equation

$$\nu^2||BN^{-1}B^*|| - (\alpha - 2\gamma)\nu + (M + b\varpi) = 0, \ \nu > \varpi. \tag{7.9}$$

We need that

$$b < \sqrt{\frac{||BN^{-1}B^*||}{M}} \frac{\beta - 1}{\sqrt{\beta} - \sqrt{\beta - 1}}. \tag{7.10}$$

We then define

$$\nu = \frac{\alpha - 2\gamma - \sqrt{(\alpha - 2\gamma)^2 - 4(M + b\varpi)||BN^{-1}B^*||}}{2||BN^{-1}B^*||}. \tag{7.11}$$

7.3 Main result

We can state

Theorem 2. *We assume (7.1), (7.2), (7.5), and (7.10). Then Eq. (7.4) has a unique solution such that*

$$|\lambda(x)| \leqslant \varpi|x|, \ ||D\lambda(x)|| \leqslant \nu. \tag{7.12}$$

Proof. We will use a contraction mapping argument. Let $\lambda(x)$ be a vector of functions satisfying (7.12). We shall define a function $\Gamma(x)$ as follows. We consider the differential equation

$$\begin{cases} \frac{dy}{ds} = A(y) - BN^{-1}B^*\lambda(y), \\ y(0) = x. \end{cases} \tag{7.13}$$

Since $A(x)$ and $\lambda(x)$ are uniformly Lipschitz, this equation has a unique solution. We then define $\Gamma(x)$ by the formula

$$\Gamma(x) = \int_0^{+\infty} \exp(-\alpha s)\left(DF(y(s)) + DA^*(y(s))\lambda(y(s))\right) ds. \tag{7.14}$$

This integral is well-defined. Indeed, from (7.13), the second assumption (7.1) and the first property (7.12), we can assert that

$$|y(s)| \leqslant |x| \exp(\gamma + ||BN^{-1}B^*||\varpi)s \tag{7.15}$$

and, from (7.14) we get

$$\begin{aligned}|\Gamma(x)| &\leqslant (M+\varpi\gamma)\int_0^{+\infty}\exp(-\alpha s)\,|y(s)|ds\\&\leqslant (M+\varpi\gamma)|x|\int_0^{+\infty}\exp(-(\alpha-\gamma-||BN^{-1}B^*||\varpi)s)ds\\&=\frac{(M+\varpi\gamma)|x|}{\alpha-\gamma-||BN^{-1}B^*||\varpi}=\varpi|x|,\end{aligned}\tag{7.16}$$

from the definition of ϖ of (7.7) and (7.8). In particular $\Gamma(x)$ satisfies the first property (7.12). We next differentiate in x the formula (7.14). We set $Y(s) = D_x y(s)$. From Eq. (7.13) we obtain

$$\begin{cases}\dfrac{dY(s)}{ds}=DA(y(s))Y(s)-BN^{-1}B^*D\lambda(y(s))Y(s),\\ Y(0)=I,\end{cases}\tag{7.17}$$

then,

$$\begin{aligned}D\Gamma(x)=\int_0^{+\infty}&\exp(-\alpha s)\Big(D^2F(y(s))+DA(y(s))D\lambda(y(s))\\&+D^2A(y(s)\lambda(y(s))\Big)Y(s)ds.\end{aligned}\tag{7.18}$$

So

$$||D\Gamma(x)||\leqslant(M+\gamma\nu+b\varpi)\int_0^{+\infty}\exp(-\alpha s)||Y(s)||ds,$$

and from (7.17) it follows that

$$\begin{aligned}||D\Gamma(x)|| &\leqslant (M+\gamma\nu+b\varpi)\int_0^{+\infty}\exp(-(\alpha-\gamma-||BN^{-1}B^*||\nu)s)\,ds\\&=\frac{M+\gamma\nu+b\varpi}{\alpha-\gamma-||BN^{-1}B^*||\nu}=\nu,\end{aligned}$$

and thus $\Gamma(x)$ satisfies the second condition (7.12).

We consider the Banach space of functions $\lambda(x):\mathbb{R}^n\to\mathbb{R}^n$, with the norm

$$||\lambda||=\sup_x\frac{|\lambda(x)|}{|x|},$$

and the closed subset

$$\mathcal{C}=\{\lambda(\cdot)|\ ||\lambda||\leqslant\varpi,\ ||D\lambda(x)||\leqslant\nu,\forall x\}.$$

We consider the map $\mathcal{T}$: $\lambda\to\Gamma$, defined by the formula (7.14). We want to show that it is a contraction from $\mathcal{C}$ to $\mathcal{C}$. We pick two functions λ_1, λ_2 in C, let

$y_1(s)$, $y_2(s)$ be defined by (7.13) with $\lambda = \lambda_1, \lambda_2$ respectively, and $\Gamma_1 = \mathcal{T}\lambda_1$, $\Gamma_2 = \mathcal{T}\lambda_2$. We have

$$\frac{d}{ds}(y_1 - y_2) = A(y_1) - A(y_2) - BN^{-1}B^*(\lambda_1(y_1) - \lambda_2(y_2)).$$

Noting that

$$\begin{aligned}|\lambda_1(y_1) - \lambda_2(y_2)| &\leqslant |\lambda_1(y_1) - \lambda_2(y_1)| + |\lambda_2(y_1) - \lambda_2(y_2)| \\ &\leqslant ||\lambda_1 - \lambda_2||\,|y_1| + \nu|y_1 - y_2|,\end{aligned}$$

we get, using the estimate (7.15),

$$\begin{cases}\dfrac{d}{ds}|y_1 - y_2| \leqslant (\gamma + \nu||BN^{-1}B^*||)|y_1 - y_2| \\ \qquad\qquad + ||BN^{-1}B^*||\,||\lambda_1 - \lambda_2||\,|x|\exp(\gamma + \varpi||BN^{-1}B^*||)s, \\ (y_1 - y_2)(0) = x,\end{cases}$$

therefore

$$\begin{aligned}|y_1(s) - y_2(s)| \leqslant\ & ||BN^{-1}B^*||\,||\lambda_1 - \lambda_2||\,|x|\exp((\gamma + \nu||BN^{-1}B^*||)s) \\ & \times \int_0^s \exp(-(\nu - \varpi)||BN^{-1}B^*||\tau)\,d\tau.\end{aligned}$$

Finally, we obtain that

$$\begin{aligned}|y_1(s) - y_2(s)| \leqslant\ & \frac{||\lambda_1 - \lambda_2||\,|x|}{\nu - \varpi}(\exp((\gamma + \nu||BN^{-1}B^*||)s) \\ & - \exp((\gamma + \varpi||BN^{-1}B^*||)s)).\end{aligned} \tag{7.19}$$

Next, from the definition of $\Gamma(x)$, we get

$$\begin{aligned}\Gamma_1(x) - \Gamma_2(x) = \int_0^{+\infty} & \exp(-\alpha s)[DF(y_1(s)) - DF(y_2(s)) \\ & + DA^*(y_1(s))\lambda_1(y_1(s)) - DA^*(y_2(s))\lambda_2(y_2(s))]ds.\end{aligned}$$

Writing

$$\begin{aligned}& DA^*(y_1(s))\lambda_1(y_1(s)) - DA^*(y_2(s))\lambda_2(y_2(s)) \\ &= \left(DA^*(y_1(s)) - DA^*(y_2(s))\right)\lambda_1(y_1(s)) \\ &\quad + DA^*(y_1(s))(\lambda_1(y_1(s)) - \lambda_2(y_2(s))).\end{aligned}$$

From the third line of assumption (7.1) we obtain

$$|\left(DA^*(y_1(s)) - DA^*(y_2(s))\right)\lambda_1(y_1(s))| \leqslant b\varpi|y_1(s) - y_2(s)|.$$

Moreover

$$|\lambda_1(y_1(s)) - \lambda_2(y_2(s))| \leqslant \nu|y_1(s) - y_2(s)| \\ + ||\lambda_1 - \lambda_2||\,|x|\exp((\gamma + \varpi||BN^{-1}B^*||)s).$$

Collecting results, we can write

$$|\Gamma_1(x) - \Gamma_2(x)| \leqslant (M + b\varpi + \gamma\nu)\int_0^{+\infty} \exp(-\alpha s)\,|y_1(s) - y_2(s)|ds \\ + \frac{\gamma||\lambda_1 - \lambda_2||\,|x|}{\alpha - \gamma - ||BN^{-1}B^*||\varpi}.$$

Making use of (7.19), it follows that

$$|\Gamma_1(x) - \Gamma_2(x)| \\ \leqslant \frac{\gamma||\lambda_1 - \lambda_2||\,|x|}{\alpha - \gamma - ||BN^{-1}B^*||\varpi} + (M + b\varpi + \gamma\nu) \\ \leqslant \frac{||\lambda_1 - \lambda_2||\,|x|}{\nu - \varpi}\left[\frac{1}{\alpha - \gamma - ||BN^{-1}B^*||\nu} - \frac{1}{\alpha - \gamma - ||BN^{-1}B^*||\varpi}\right].$$

Rearranging and using the definition of ν, see (7.9), we finally obtain that

$$||\Gamma_1 - \Gamma_2|| \leqslant \frac{\gamma + ||BN^{-1}B^*||\nu}{\alpha - \gamma - ||BN^{-1}B^*||\varpi}||\lambda_1 - \lambda_2||. \tag{7.20}$$

We need to check that

$$\frac{\gamma + ||BN^{-1}B^*||\nu}{\alpha - \gamma - ||BN^{-1}B^*||\varpi} < 1 \tag{7.21}$$

which is equivalent to

$$\alpha - 2\gamma - ||BN^{-1}B^*||(\varpi + \nu) > 0,$$

which is true, from the definition of ϖ and ν, see (7.7) and (7.11). If we call $\lambda(x)$ the unique fixed point of $\mathcal{T}$, it satisfies (7.12) and the equation

$$\lambda(x) = \int_0^{+\infty} \exp -\alpha s\left(DF(y(s)) + DA^*(y(s))\lambda(y(s))\right)ds. \tag{7.22}$$

It is standard to check that (7.12) and (7.22) is equivalent to (7.12) and (7.4). This concludes the proof of the Theorem. □

7.4 Algorithm

We can write the algorithm (6.3) which leads to

$$\alpha\lambda^{k+1}(x) - D\lambda^{k+1}(x)(A(x) - BN^{-1}B^*\lambda^k(x)) = DF(x) + DA^*(x)\lambda^k(x). \tag{7.23}$$

From the contraction property obtained in Theorem 2, we can obtain immediately

Corollary 3. *Under the assumptions of Theorem 2, if we start the iteration with λ_0 such that $|\lambda_0(x)| \leqslant \varpi |x|$ and $||D\lambda_0(x)|| \leqslant \nu$, we have*

$$||\lambda^k - \lambda|| \to 0, \tag{7.24}$$

where λ is the solution of (7.4).

7.5 Linear quadratic case

We take $A(x) = Ax$, $F(x) = \frac{1}{2}x^*Mx$, then Eq. (7.4) becomes

$$\alpha\lambda(x) = Mx + A^*\lambda(x) + D\lambda(x)(Ax - BN^{-1}B^*\lambda(x)), \tag{7.25}$$

and its solution is $\lambda(x) = Px$, with P the solution of the Riccati equation

$$\alpha P = M + A^*P + PA - PBN^{-1}B^*. \tag{7.26}$$

We have $\gamma = ||A||$ and $b = 0$. Assumption (7.5) becomes

$$\alpha > 2||A|| + 2\sqrt{M||BN^{-1}B^*||}. \tag{7.27}$$

We have

$$\varpi = \nu = \frac{\alpha - 2||A| - \sqrt{(\alpha - 2||A|)^2 - 4M||BN^{-1}B^*||}}{2||BN^{-1}B^*||}. \tag{7.28}$$

The iteration (7.23) becomes $\lambda^k(x) = P^k x$, with

$$P^{k+1}(\alpha I - A + BN^{-1}B^*P^k) = M + A^*P^k, \tag{7.29}$$

and if $||P^0|| \leqslant \varpi$, we obtain $||P^k - P|| \to 0$, as $k \to +\infty$.

8 Numerical results

We now present numerical tests for the algorithm. We consider the following values for m, n: $m = 10, n = 30$; the matrices M, N, A, B are chosen arbitrarily and their values are not displayed here.

In Fig. 1, we choose $\alpha = 1$ and we pick 4 samples of the initial guess $P^{(0)}$. These choices do not satisfy the two conditions (7.5) and (7.6). Using the results of our Python code, we display the difference between $\|P^{(5)} - P^{(6)}\|$, $\|P^{(10)} - P^{(11)}\|$, $\|P^{(15)} - P^{(16)}\|$, $\|P^{(20)} - P^{(21)}\|$, $\|P^{(25)} - P^{(26)}\|$. We can see that as the number of iterations k increases, the difference between $\|P^{(k)} - P^{(k+1)}\|$ does not become small. This shows that the algorithm does not converge.

Num	$P^{(0)}$	$\|P^{(5)} - P^{(6)}\|$	$\|P^{(10)} - P^{(11)}\|$	$\|P^{(15)} - P^{(16)}\|$	$\|P^{(20)} - P^{(21)}\|$	$\|P^{(25)} - P^{(26)}\|$	$\|P^{(30)} - P^{(31)}\|$
1		9.473760	4.068934	15.22311	0.653054	13.32068	2.236581
2		69.12447	245.5828	16.69521	9.010269	4.967819	0.713422
3		36.53236	39.42164	13.61281	56.88253	2.214869	0.707675
4		9.473760	4.068934	15.22311	0.653054	13.32068	2.236581

FIGURE 1 Solving for P: 4 tests where $\alpha = 1$.

In Fig. 2, we choose $\alpha = 1770.3688$, $0 < \|P^{(0)}\| < 6.8153$ for the 4 samples of the initial guess $P^{(0)}$. These choices satisfy the two conditions (7.5) and (7.6). We can see that as the number of iterations k increases, the difference between $\|P^{(k)} - P^{(k+1)}\|$ becomes very small. And after 15 iterations, these differences are essentially 0.

Num	$P^{(0)}$	$\|P^{(5)} - P^{(6)}\|$	$\|P^{(10)} - P^{(11)}\|$	$\|P^{(15)} - P^{(16)}\|$	$\|P^{(20)} - P^{(21)}\|$	$\|P^{(25)} - P^{(26)}\|$	$\|P^{(30)} - P^{(31)}\|$
1		4.19×10^{-8}	3.97×10^{-16}	0.0	0.0	0.0	0.0
2		4.24×10^{-8}	4.06×10^{-16}	0.0	0.0	0.0	0.0
3	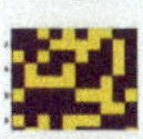	3.78×10^{-8}	3.61×10^{-16}	0.0	0.0	0.0	0.0
4		3.77×10^{-8}	3.55×10^{-16}	0.0	0.0	0.0	0.0

FIGURE 2 Solving for P: 4 tests where $\alpha = 1770.3688$.

In Fig. 3, we choose $\alpha = 225$. In this test, the condition (7.5) corresponds to $\alpha > 270$. Using the results of our Python code, we display the difference between $\|P^{(k)} - P^{(k+1)}\|$. We can see that the difference does not converge to 0 even after 10000 iterations. Therefore, the condition (7.5) is quite good.

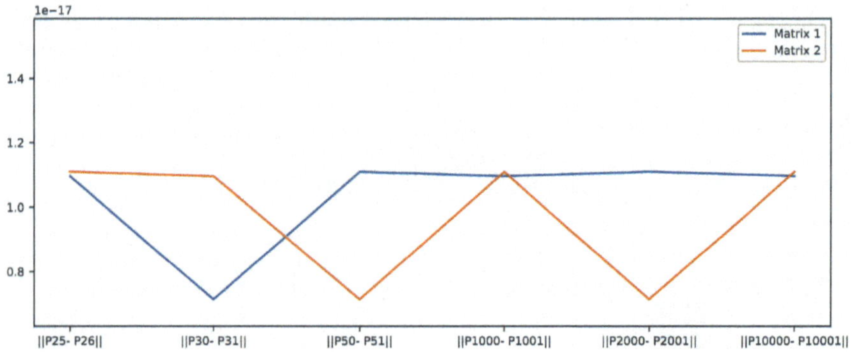

FIGURE 3 Solving for P: 4 tests where $\alpha = 225$.

In Fig. 4, we take the first choice of $P^{(0)}$ arbitrarily, and varies the values of α to be 1, 5, 20, 100, 200, 300. Using the results of our Python code, we plot the values of $\|P^{(5)} - P^{(6)}\|$, $\|P^{(10)} - P^{(11)}\|$, $\|P^{(15)} - P^{(16)}\|$, $\|P^{(20)} - P^{(21)}\|$, $\|P^{(25)} - P^{(26)}\|$ for each value of α as a curve. We can see that the algorithm starts to converge very fast if α is big: the curves for $\alpha = 100, 200, 300$ (red, purple and brown) are almost the 0-line.

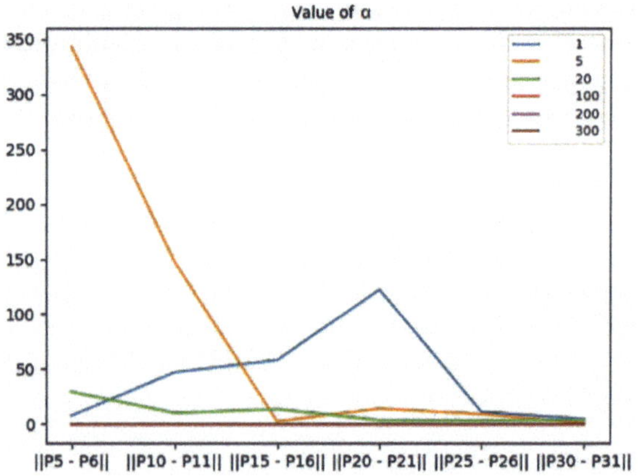

FIGURE 4 Solving for P: The convergence rate for different values of α.

Acknowledgment

Alain Bensoussan acknowledges the financial support from the National Science Foundation under grants DMS-1612880, DMS-1905449, and the Research Grant Council of Hong Kong Special Administrative Region under grant GRF 11303316. Minh-Binh Tran is partially supported by NSF Grant DMS-1854453, SMU URC Grant 2020, SMU DCII Research Cluster Grant, Dedman College Linking Fellowship, Alexander von Humboldt Fellowship. Dinh Phan Cao Nguyen and Minh-Binh Tran would like to thank Prof. T. Hagstrom and Prof. A.

Aceves for the computational resources. Phillip Yam acknowledges the financial supports from HKGRF-14300717 with the project title "*New kinds of Forward-backward Stochastic Systems with Applications*", HKGRF-14300319 with the project title "*Shape-constrained Inference: Testing for Monotonicity*", and Direct Grant for Research 2014/15 (Project No. 4053141) offered by CUHK. Xiang Zhou acknowledges the support of Hong Kong RGC GRF grants 11337216 and 11305318.

References

Bensoussan, A., 2018. Estimation and Control of Dynamical Systems. Interdisciplinary Applied Mathematics. Springer International Publishing.

Chang, B., Meng, L., Haber, E., Ruthotto, L., Begert, D., Holtham, E., 2018a. Reversible architectures for arbitrarily deep residual neural networks. In: The Thirty-Second AAAI Conference on Artificial Intelligence (AAAI-18), p. 2811.

Chang, Bo, Meng, Lili, Haber, Eldad, Tung, Frederick, Begert, David, 2018b. Multi-level residual networks from dynamical systems view.

Chen, Ricky T.Q., Rubanova, Yulia, Bettencourt, Jesse, Duvenaud, David, 2018. Neural ordinary differential equations. In: Advances in Neural Information Processing Systems, 2018-Decem, pp. 6571–6583.

Chiuso, A., Pillonetto, G., 2019. System identification: a machine learning perspective. Annual Review of Control, Robotics, and Autonomous Systems 2 (1), 281–304.

E, W., 2017. A proposal on machine learning via dynamical systems. Communications in Mathematics and Statistics 5 (1), 1–11.

E, W., Han, J., Li, Q., 2019a. A mean-field optimal control formulation of deep learning. Research in the Mathematical Sciences 6 (1), 1–41.

E, W., Ma, C., Wu, L., 2019b. Machine learning from a continuous viewpoint. arXiv:1912.12777.

Haber, E., Ruthotto, L., 2017. Stable architectures for deep neural networks. Inverse Problems 34 (1), 014004.

Han, J., E, W., 2016. Deep learning approximation for stochastic control problems. arXiv:1611.07422.

Han, Jiequn, Jentzen, Arnulf, E, Weinan, 2018. Solving high-dimensional partial differential equations using deep learning. Proceedings of the National Academy of Sciences 115 (34), 8505–8510.

He, Kaiming, Zhang, Xiangyu, Ren, Shaoqing, Sun, Jian, 2016. Deep residual learning for image recognition. In: The IEEE Conference on Computer Vision and Pattern Recognition (CVPR), pp. 770–778.

Jordan, M.I., Mitchell, T.M., 2015. Machine learning: trends, perspectives, and prospects. Science 349 (6245), 255–260.

Li, Q., Chen, L., Tai, C., E, W., 2017a. Maximum principle based algorithms for deep learning. Journal of Machine Learning Research 18 (1), 5998–6026.

Li, Q., Hao, S., 2018. An optimal control approach to deep learning and applications to discrete-weight neural networks. arXiv preprint. arXiv:1803.01299.

Li, Q., Tai, C., E, W., 2017b. Stochastic modified equations and adaptive stochastic gradient algorithms. In: Proceedings of the 34th International Conference on Machine Learning-Volume 70. JMLR.org, pp. 2101–2110.

Li, Q., Tai, C., E, W., 2019. Stochastic modified equations and dynamics of stochastic gradient algorithms I: mathematical foundations. Journal of Machine Learning Research 20 (40), 1–40.

Li, Qianxiao, Tai, Cheng, E, Weinan, 2015. Dynamics of stochastic gradient algorithms. arXiv preprint. arXiv:1511.06251.

Li, Z., Shi, Z., 2017. Deep residual learning and PDEs on manifold. arXiv preprint. arXiv:1708.05115.

Lu, T., Neittaanmaki, P., Tai, X.-C., 1991. A parallel splitting up method and its application to Navier-Stokes equations. Applied Mathematics Letters 4, 25–29.

Lu, Y., Zhong, A., Li, Q., Dong, B., 2017. Beyond finite layer neural networks: bridging deep architectures and numerical differential equations. arXiv preprint. arXiv:1710.10121.

Mei, S., Montanari, A., Nguyen, P.-M., 2018. A mean field view of the landscape of two-layer neural networks. Proceedings of the National Academy of Sciences 115 (33), E7665–E7671.

Recht, Benjamin, 2019. A tour of reinforcement learning: the view from continuous control. Annual Review of Control, Robotics, and Autonomous Systems 2 (1), 253–279.

Silver, David, Schrittwieser, Julian, Simonyan, Karen, Antonoglou, Ioannis, Huang, Aja, Guez, Arthur, Hubert, Thomas, Baker, Lucas, Lai, Matthew, Bolton, Adrian, Chen, Yutian, Lillicrap, Timothy, Hui, Fan, Sifre, Laurent, van den Driessche, George, Graepel, Thore, Hassabis, Demis, 2017. Mastering the game of go without human knowledge. Nature 550 (7676), 354.

Sonoda, S., Murata, N., 2017. Double continuum limit of deep neural networks. In: ICML Workshop Principled Approaches to Deep Learning.

Sutton, R.S., Barto, A.G., 2018. Reinforcement Learning: An Introduction, 2 edition. MIT Press.

Wang, Haoran, Zariphopoulou, Thaleia, Zhou, Xunyu, 2018. Exploration versus exploitation in reinforcement learning: a stochastic control approach. arXiv:1812.01552.

Index

C

G

H

I

J

K

L

O